Lecture Notes in Computer Science 1373

Edited by G. Goos, J. Hartmanis and J. van Leeuwen

T0189184

Springer
Berlin
Heidelberg
New York
Barcelona
Budapest
Hong Kong
London
Milan
Paris
Santa Clara
Singapore
Tokyo

Michel Morvan   Christoph Meinel
Daniel Krob (Eds.)

STACS 98

15th Annual Symposium
on Theoretical Aspects of Computer Science
Paris, France, February 25-27, 1998
Proceedings

Springer

Series Editors

Gerhard Goos, Karlsruhe University, Germany
Juris Hartmanis, Cornell University, NY, USA
Jan van Leeuwen, Utrecht University, The Netherlands

Volume Editors

Michel Morvan
Daniel Krob
LIAFA, Université Denis Diderot Paris 7
2, place Jussieu, F-75251 Paris Cedex 05, France
E-mail: {morvan/dk}@liafa.jussieu.fr

Christoph Meinel
Universität Trier, FB IV - Theoretische Informatik
D-54286 Trier, Germany
E-mail: meinel@uni-trier.de

Cataloging-in-Publication data applied for

Die Deutsche Bibliothek - CIP-Einheitsaufnahme

STACS <15, 1998, Paris>:
Proceedings / STACS 98 / 15th Annual Symposium on Theoretical Aspects of
Computer Science, Paris, France, February 25 - 27, 1998. Michel Morvan ...
(ed.). - Berlin ; Heidelberg ; New York ; Barcelona ; Budapest ; Hong Kong ;
London ; Milan ; Paris ; Santa Clara ; Singapore ; Tokyo : Springer, 1998
  (Lecture notes in computer science ; Vol. 1373)
  ISBN 3-540-64230-7

CR Subject Classification (1991): F, D.1, D.4, G.1-2, I.3.5, E.3, E.1

ISSN 0302-9743
ISBN 3-540-64230-7 Springer-Verlag Berlin Heidelberg New York

© Springer-Verlag Berlin Heidelberg 1998
Printed in Germany

Typesetting: Camera-ready by author
SPIN 10631918   06/3142 – 5 4 3 2 1 0   Printed on acid-free paper

Preface

STACS, the Symposium on Theoretical Aspects of Computer Science, is held annually, alternating between France and Germany. STACS is organized jointly by the Special Interest Group for Theoretical Computer Science of the Gesellschaft für Informatik (GI) in Germany and the Maison de l'Informatique et des Mathématiques Discrètes (MIMD) in France. It takes place at the end of February each year.

STACS '98 is the 15th in the series. It will be held in Paris from February 25th to 27th, 1998. Previous STACS symposia took place in Paris (1984), Saarbrücken (1985), Orsay (1986), Passau (1987), Bordeaux (1988), Paderborn (1989), Rouen (1990), Hamburg (1991), Cachan (1992), Würzburg (1993), Caen (1994), München (1995), Grenoble (1996), and Lübeck (1997).

STACS has become one of the most important annual meetings in Europe for the theoretical computer science community. It covers a wide range of topics in the area of foundations of computer science. For STACS '98, 155 submissions from all over the world and most European countries were received, almost all in electronic form, thanks to *Jochen Bern* in Trier, who made the electronic procedure possible and so efficient. The submitted papers address basic problems from many areas of computer science: algorithms and data structures, automata and formal languages, complexity, and logic. The sessions of the conference have been chosen to represent these main topics.

The members of the program committee are: *Jean Berstel* (Marne-la-Vallée), *Thomas Ehrhard* (Marseille), *Jean-Marc Fédou* (Nice), *Erich Grädel* (Aachen), *Monika Henzinger* (Palo Alto), *Ulrich Hertrampf* (Stuttgart), *Giuseppe Italiano* (Venice), *Juhani Karhumäki* (Turku), *Daniel Krob* (Paris), *Jack H. Lutz* (Ames), *Christoph Meinel* (Trier, co-chair), *Michel Morvan* (Paris, chair), *Friedrich Otto* (Kassel), *Jose Rolim* (Geneva), and *Angelika Steger* (München). Every submitted paper has been evaluated by at least four members of the program committee and a large majority of them by five, partly using the help of colleagues who served as subreferees.

The program committee met in Paris on October 31st and November 1st, 1997. Fifty-four papers were selected for presentation. (Two of them have been cancelled by their authors, so the number of contributed papers in this volume is 52.) Because of the constraints imposed by the format of the conference (two and a half days), a number of good papers could not be included in the program.

We would like to thank the program committee for its demanding work in the evaluation process. We also want to thank all the subreferees who assisted us in this process (listed separately).

We also thank the invited speakers (by order of appearance), *Richard Karp* (Seattle), *Serge Vaudenay* (Paris), and *Torben Hagerup* (Saarbrücken) for accepting our invitation and presenting us their point of view on their research area.

Thanks to all members of the *Laboratoire de l'Informatique Algorithmique:*

Fondements et Applications (LIAFA) who have helped us to organize the meeting: *Karell Bertet, Fabrice Derepas, Nicole Godard, Isabelle Guérin, Florent Hivert, Jean-Christophe Novelli, Ha Duong Phan,* and *Laurent Viennot.*

Thanks to *Vincent Pancol*, author of the photography for the STACS '98 poster.

Thanks also to the various sources who have supported STACS '98: *European Community, Ministère des Affaires Etrangères, Ministère de l'Education Nationale, de la Recherche et de la Technologie, Laboratoire de l'Informatique Algorithmique: Fondements et Applications* and other organizations.

Paris, January 1998

Michel Morvan
Christoph Meinel
Daniel Krob

List of Subreferees

Agarwala R.
Albers S.
Allender E.
Allouche J.-P.
Alt H.
Ambos-Spies K.
Arapis C.
Arora S.
Asarin E.
Avenhaus J.
Avermiddig A.
Babaoglu O.
Bacik R.
Baeza-Yates R.
Baker B.
Barbosa V.
Baudon O.
Baumeister H.
Beauquier D.
Becker B.
Bender M. A.
Bergen A.
Bermond J.-C.
Berry G.
Bertet K.
Bidoit M.
Bilardi G.
Birget J.-C.
Birkendorf A.
Bischof S.
Blum A.
Blum N.
Boasson L.
Bodlaender H.
Bollig B.
Bonacina M.
Bond J.
Bonuccelli M.
Boppana R.
Borchert B.
Boucheron S.

Bozga M.
Bretelle B.
Brodal G. S.
Buchholz F.
Buening K.
Buhrman H.
Buntrock G.
Burnikel Ch.
Buss J.
Cai L.
Canetti R.
Chang R.
Charatonik W.
Chaudhuri S.
Chlebus B.
Choffrut C.
Chor B.
Chretienne P.
Clementi A.
Clote P.
Comon H.
Compton K.
Contejean E.
Courcelle B.
Cramer R.
Creignou N.
Cremanns R.
Crescenzi P.
Crochemore M.
Cucker F.
Czumaj A.
Dahlhaus E.
Dam M. F.
Damaschke P.
Damm C.
Dassow J.
Dauchet M.
Daude H.
de Rougemont M.
Dehne J.
Denis F.

Deransart P.
Derepas F.
Devienne P.
Dietzfelbinger M.
Drewes F.
Dreyfus A. F. R.
Droste M.
Dulucq S.
Eberhard Z.
Engelfriet J.
Erlebach T.
Esik Z.
Etessami K.
Etheridge A.
Fagin R.
Falquet G.
Farach M.
Feautrier P.
Fellows M.
Felsner S.
Fenner S.
Fernau H.
Ferreira A.
Fischlin M.
Flajolet P.
Flammini M.
Fleischer R.
Focardi R.
Fortnow L.
Fournier J.-C.
Fraigniaud P.
Frank J.
Fribourg L.
Friedetzky T.
Gargano L.
Gatermann K.
Gaujal B.
Gavoille C.
Geffert V.
Gergov J.
Giakoumakis V.

Giammarresi D.

Goel A.

Goerdt A.

Goldberg A. V.

Goldman S.

Goldsmith J.

Goles E.

Gonthier G.

Gouyou-Beauchamps D.

Green F.

Gritzmann P.

Grumbach S.

Gruska J.

Guérin Lassous I.

Gustedt J.

Habib M.

Hagerup T.

Hanus M.

Harju T.

Hastad J.

Hegedus T.

Hella L.

Hemaspaandra L.

Hempel H.

Hennicker R.

Herlihy M.

Hermo M.

Hirvensalo M.

Hivert F.

Hofbauer D.

Hoffmann F.

Hofmeister T.

Holzer M.

Honkala J.

Howell R.

Hromkovic J.

Huebel E.

Ilie L.

Immerman N.

Impagliazzo R.

Irani S.

Istrate G.

Jacobs B.

Jacquemard F.

Jakoby A.

Jansen K.

Jantzen M.

Jenner B.

Jiang T.

Juedes D.

Jukna S.

Kanarek P.

Kann V.

Kari J.

Kärkkäinen J.

Karpinski M.

Kautz S.

Kemp R.

Kenyon C.

Kessler C.

Kivinen J.

Klasing R.

Klasner N.

Kleine Büning H.

Köbler J.

Köhler E.

Koiran P.

Krajicek J.

Kratsch D.

Krause M.

Kreowski H.

Kucera L.

Kühnle K.

Kunde M.

Kuske D.

Kutrib M.

Lafitte G.

Lange S.

Lautemann C.

Lazard E.

Lebris L.

Lefmann H.

Leonardi S.

Levy M.

Liotta G.

Litovsky I.

Liu Z.

Loebbing M.

Lorys K.

Louboutin S.

Louchard G.

Lozano A.

Luccio F.

Lucks S.

Luquet P.

Lynch J.

Möhring R. H.

Médina R.

Madlener K.

Maffioli F.

Mairesse J.

Maler O.

Malmstroem A.

Maraninchi F.

Marchiori E.

Margenstern M.

Marlin N.

Martin B.

Martinez C.

Mateescu A.

Mayordomo E.

Mayr E.

Mazoyer J.

McColm G.

Meer K.

Mentrasti P.

Meyer auf der Heide F.

Michaux C.

Michel P.

Mignosi F.

Miltersen P.B.

Mitas J.

Mitzenmacher M.

Mnuk M.

Muller J.-M.

Müller-Hannemann M.

Mundhenk M.

Munier A.

Muscholl A.

Néraud J.

Naish L.

Namyst R.

Nanni U.

Narbel P.

Narendran P.

Navarro G.
Niedermeier R.
Nielsen M.
Niemann G.
Nisan N.
Novelli J.-C.
Ogihara M.
Ohlebusch E.
Orponen P.
Ossona de Mendez P.
Osterloh A.
Panconesi A.
Pardubská D.
Parente M.
Pellegrini M.
Perennes S.
Petersen H.
Petit A.
Peyrat C.
Phan H. D.
Picouleau C.
Pin J.-E.
Pinzani R.
Plandowski W.
Pocchiola M.
Podelski A.
Prömel H. J.
Preneel B.
Pudlak P.
Radzick T.
Rampon J.-X.
Ramshaw L.
Rao S.
Raspaud A.
Rauzy A.
Ravi S. S.
Ravi R.
Razborov A.
Reed B.
Regan K.
Reinhardt K.
Reischuk R.
Rensink A.
Restivo A.
Robert P.

Robson M.
Roncato A.
Rosen E.
Rossmanith P.
Royer J.
Rutten J.
Sagot M.-F.
Salibra A.
Salomaa A.
Salomaa K.
Salvy B.
Santha M.
Sauerhoff M.
Schindelhauer C.
Schirra S.
Schlick C.
Schnorr C. P.
Scholz P.
Schöning U.
Schröder K.
Schuler R.
Schulz A. S.
Schwentick T.
Seese D.
Seibert S.
Seidl H.
Sénizergues G.
Settle A.
Shukla S. K.
Sieling D.
Sifakis J.
Silvestri R.
Sivakumar D.
Skordev G.
Skutella M.
Slobodová A.
Smid M.
Sopena E.
Sorenson J.
Sotteau D.
Spielmann M.
Sreedhar V.
Staiger L.
Steinby M.
Stephan F.

Stern J.
Stewart I.
Steyaert J.-M.
Straubing H.
Strauss M.
Sudan M.
Sutner K.
Syska M.
Tan S.
Tel G.
Terwijn S.
Therien D.
Thierauf T.
Thomas W.
Thorup M.
Thuillier H.
Torán J.
Torenvliet L.
Träff J. L.
Tran N.
Trevisan L.
Tyszkiewicz J.
Vaccaro U.
Vallée B.
Vallee B.
van Melkebeek D.
Vaudenay S.
Viennot L.
Vigna S.
Vinay V.
Voll U.
Vollmer J.
Vollmer H.
Wätjen D.
Wagner D.
Wagner K.
Wagner K. W.
Walukiewicz I.
Wang J.
Wanka R.
Wareham T.
Watier G.
Wegner R.
Wegner L.
Weil P.

Westermann M.

Wilke T.

Williamson D.

Worsch T.

Yamakami T.

Yen H.-C.

Yunes J.-B.

Yvinec M.

Zeitoun M.

Zola E.

Table of Contents

Automata and Formal Languages I

Complexity II

Algorithms and Data Structures II

Invited Talk

Algorithms and Data Structures III

Automata and Formal Languages II

Invited Talk

Algorithms and Data Structures IV

Logic II

Complexity III

Automata and Formal Languages III

Algorithms and Data Structures V

Complexity IV

Random Graphs, Random Walks, Differential Equations and the Probabilistic Analysis of Algorithms

Richard M. Karp

We present recent results of the author and his colleagues on the following three problems related to the probabilistic analysis of algorithms:

1 Multiway Partitioning of a Graph

In this problem a graph G with n vertices is given, and we wish to partition the vertex set into k sets of size $\frac{n}{k}$ so as to minimize the number of edges whose end points lie in different sets. In the spirit of earlier work on graph bisection we adopt a "planted partition" model, in which a good partition is planted in the graph but not revealed to the algorithm. This partition has k parts of equal size. Edge $\{u, v\}$ is present with probability p if u and v are in the same part of the planted partition, and with probability $r < p$ if u and v are in different parts of the planted partition. Our main result is that, if $p - r = \Omega(n^{-\frac{1}{2}+\epsilon})$, where ϵ is a positive constant then, with probability tending exponentially to 1, the planted partition will be the optimal one and a simple linear-time algorithm will find the planted partition. This is joint work with Anne Condon.

2 The threshold for k-orientability

A graph is called *k-orientable* if its edges can be oriented so that no more than k edges are oriented towards any vertex. We present a general conjecture as to the threshold for k-orientability in random graphs, and prove the conjecture for many particular values of k. This is joint work with Michael Saks. The problem was suggested by Juan Alemany and Jayram Thathachar in a setting where one is given a set of servers such as videodisks and a set of tasks such as showing a given film strip. Each task is assigned at random to two servers and can be executed by either one. All the tasks are to be executed, subject to the constraint that each server may execute only k tasks. The problem is represented by a graph in which each server corresponds to a vertex and each task to an edge.

3 Proofs of Unsatisfiability for Random 3-CNF Formulas

We present reasonably tight upper and lower bounds on the length of the shortest resolution proof of unsatisfiability for a random unsatisfiable 3-CNF formula with n variables and m clauses. The upper bound is based on an analysis using random graphs of a well-known algorithm for testing the satisfiability of a 2-CNF formula. This is joint work with Paul Beame, Toniann Pitassi and Michael Saks.

In the course of deriving these results we illustrate a number of techniques, including the use of differential equations, for analyzing random walks and other stochastic processes.

Distributed Online Frequency Assignment in Cellular Networks *

(Extended Abstract)

Jeannette Janssen[1], Danny Krizanc[2], Lata Narayanan[3], Sunil Shende[4]

[1] Department of Mathematics, Acadia University, Wolfville, Nova Scotia B0P 1X0, Canada, email: jeannette.janssen@acadiau.ca

[2] School of Computer Science, Carleton University, Ottawa, Ontario K1S 5B6, Canada, email: krizanc@scs.carleton.ca

[3] Department of Computer Science, Concordia University, Montreal, Quebec H3G 1M8, Canada, email: lata@cs.concordia.ca, FAX (514) 848-2830

[4] Department of Computer Science and Eng., University of Nebraska-Lincoln, Lincoln, NE 68588, USA. email: sunil@calypso.unl.edu, FAX (402) 472-7767

Abstract. In this paper, we develop a general framework for studying *distributed online frequency assignment* in cellular networks. The problem can be abstracted as a multicoloring problem on a node-weighted graph whose weights change over time. The graph, with nodes corresponding to network cells, is usually modeled as a subgraph of the triangular lattice and the instantaneous weight at a node models the number of calls requiring service at the corresponding network cell. In this setting, we present several distributed online algorithms for this problem and prove bounds on their competitive ratios. Specifically, we demonstrate a series of such algorithms that utilize information about increasingly larger neighborhoods around nodes, and thereby achieve progressively better competitive ratios. We also exhibit lower bounds on the competitive ratios of some natural classes of distributed online algorithms for the problem; in some cases, our bounds are shown to be optimal.

1 Introduction

Cellular data and communication networks are usually modeled as planar graphs with each node representing a base station (sometimes called a *cell*) in the network. At any given time, a certain number of active connections (or *calls*) are serviced by their nearest base station. In most cases (especially in FDMA technology), service involves the assignment of a frequency to each client call in a manner that minimizes or avoids radio interference between different calls in the network. A similar problem also arises in cellular networks employing CDMA technology. A common abstraction that captures most of the physical features of the network is the assumption that if two calls are assigned the same frequency, they would interfere with one another if and only if they originate in the same cell

---

* Research supported by NSERC, Canada.

or in physically adjacent cells. For this reason, we refer to the graphs that model these networks as *interference graphs*. However, cellular networks have a limited spectrum of radio frequencies available to handle calls and the efficient shared utilization of the bandwidth is critical to the smooth operation of the network. In this paper we study the *distributed online frequency assignment* problem which consists of designing a distributed online interference-free frequency allocation protocol for a network where the number of calls per cell changes over time.

The planar graphs most often used to model cellular networks are finite portions of the infinite triangular lattice. The reason for adopting this particular geometry stems from the fact that cells are uniformly distributed in the geographic area of the network, and an individual cell generally has six directional transceivers. Hence, the cell's calling area can be idealized as a regular hexagon. The triangular lattice representing the network is simply the planar dual of the resulting Voronoi diagram. We refer to a finite induced subgraph of the triangular lattice as a *hexagon graph*. Unless otherwise specified, in the rest of this paper, the interference graphs we consider are always *hexagon graphs*.

The static frequency assignment problem incorporating interference constraints can be abstracted as follows. Let $G = (V, E, w)$ denote an interference graph where each node $v \in V$ has an associated nonnegative integer weight, $w(v) \geq 0$. The graph G models a static snapshot of the network at some instant in time, with the nodes representing cells and the weights representing the number of calls that require service in the cell. Our problem is to properly *multicolor* the graph G, i.e. we are required to assign $w(v)$ *distinct* colors to each v such that for every edge, $(u, v) \in E$, the set of colors assigned to the endpoints u and v are *disjoint*. The *span* of a multicoloring is the total number of colors used in the coloring. In particular, we are interested in a proper multicoloring of G whose span is equal to the minimum number of colors required to multicolor G, denoted by $\chi(G)$. In the context of frequency assignment, a multicoloring as defined above, provides a useful abstraction of the essential interference constraints: each color represents a distinct frequency and it is assumed that two calls may use the same frequency if and only if they originate in distinct cells that are not neighbors. It is convenient to treat the color palette of available colors as the set of natural numbers; we further assume without loss of generality that any such palette can be suitably reordered or partitioned.

The complexity of the static version of the problem has received considerable recent attention. Let the weight of a maximal clique in G be defined as the sum of the weights of the nodes belonging to the clique; note that G being a subgraph of the triangular lattice, the only maximal cliques are isolated nodes, edges or triangles. It is easy to see that $\chi(G)$ must be greater than the weight of any maximal clique in the graph. It has been shown recently that the problem of multicoloring hexagon graphs optimally is NP-hard [9]. In terms of upper bounds, there is a vast literature on algorithms for frequency assignment on graphs (especially hexagon graphs) which claim to use few colors in practice, but have no proven bounds on their performance [1, 5, 7, 11, 13]. A well-known algorithm often referred to as *Fixed Allocation (FA)*, uses the fact that the underlying graph can be 3-colored. The algorithm uses three fixed sets of colors, one for each base color. A node that has base color 1 uses colors from the first set, and a node that base color 2 or 3 uses colors from the second or third sets respectively. It is easy to see that this algorithm is an approximation algorithm with performance ratio 3. Janssen *et. al.* [4] propose a different algorithm called *Fixed Preference Allocation (FPA)* that is guaranteed to use no more than $\frac{3}{2}$ times the minimum number of colors required. Approximation algorithms with performance ratio $\frac{4}{3}$ have recently been shown in [9] and [10].

In the online case, the interference graph to be multicolored changes over time. We model these changes as an ordered sequence of interference graphs, $\{G_t = (V, E, w_t) : t \geq 0\}$, where w_t represents the set of calls to be serviced at time t. At time instant t, an online algorithm must arrange to color the graph G_t before moving on to the graph G_{t+1} at the next time instant $t + 1$. It must perform this coloring with no knowledge of the later graphs in the sequence. Very little is known about the online version of the problem. There has been a lot of work on online graph *coloring* (see for example, [12, 8, 2]); however our interests are in the *multicoloring* problem. The results of [3] imply that for general k-colorable graphs, no online multicoloring algorithm can have a competitive ratio smaller than $k/2$, a bound which is met by an online version of FPA. They also show that every algorithm which is not allowed to recolor has a competitive ratio at least k on such graphs, a bound which is met by FA. However, the lower bounds involve constructions of non-planar graphs; there are no known lower bounds on the competitive ratio of algorithms on hexagon graphs. Furthermore, the online algorithms that have proven upper bounds on the competitive ratio use information about the changing state of the entire graph and are therefore not distributed.

In this paper we develop a reasonable operational model in which the problem of distributed online frequency assignment in cellular networks can be studied. In particular, this involves a number of considerations: a precise delineation of the various kinds of admissible online algorithms for the problem (Section 1.1), the framework and efficiency measures under which the performance of any such algorithm can be meaningfully evaluated (Section 1.2), and models in which lower bounds can be shown (Section 1.3). A brief summary of our results is given in Section 1.4. We present our upper bounds in Section 2 and lower bounds in Section 3. Conclusions and future directions for research are discussed in Section 4.

1.1 Admissible distributed online algorithms

Since most practical frequency assignment algorithms are required to be distributed in nature, it is convenient to describe the algorithm as though it were running simultaneously on servers, one server per active node in the network. In fact, the scope of our paper is limited to *distributed* and *deterministic* algorithms: each server running the algorithm is independently responsible for the color assignment at its resident node at any given time instant. Furthermore, this local assignment is computed deterministically based upon a limited amount of information. The time-indexed sequence of interference graphs is the online input to the algorithm - but presented in a manner that is completely distributed, i.e. as though each node v is presented synchronously with weight $w_t(v)$ at time $t \geq 0$ indicating the number of calls that need color assignment. Additionally, v may get requests to *drop* certain calls, which can identified by the colors assigned to them in the previous step. For example, if $w_t(v) < w_{t-1}(v)$, then at least $w_{t-1}(v) - w_t(v)$ calls to be dropped are specified as part of the input at time t. In response, v may gather some local information and use it to provide service at time t by allocating $w(v)$ distinct colors to its local calls without conflicting with assignments at neighboring nodes. Overall, between successive time steps only a constant amount of communication and computation is permissible.

Various subtle issues present themselves in this framework. For instance, what kind of information ought to be reasonably admitted? While it is difficult to answer this question in its full generality, we propose the following restrictions motivated by practical considerations typical in most cellular networks. We insist that no *global* knowledge of the current state of the network be available at any node or small set of nodes in the network. However, it is assumed that nodes may be permitted to gather some limited amount of information from their local neighborhood between successive time instants. In particular, for integers $k \geq 1$, we define the *k-locality* of a node v to be the induced subgraph consisting of those nodes in G whose graph distance from v is less than or equal to k. The maximum weight taken over all the maximal cliques in the k-locality of v, is denoted by $D_k(v)$. It is easy to see that the maximum of $D_k(v)$ over all nodes v in the graph, is a lower bound on $\chi(G)$. For some small constant $k \geq 0$ independent of the input weight sequence, we say that an algorithm is k-*local* if between successive time instants, the values of the weights at time t at every node in v's k-locality, and only those weights, are available at v to decide its allocation for time t. Furthermore, it can be assumed that at the very beginning (at time $t = 0$), some "hard-wired" or pre-computed information, independent of the input weight sequence, is available to each node for free. In general, this pre-computed information may be an arbitrary finite function of a node's label in a fixed labeling of the triangular lattice. The algorithms discussed below use only the fact that the nodes of the lattice can be partitioned into a small constant number of stable sets depending on their distance from a fixed origin. For example, the Fixed Allocation Local (FA-Local) and Fixed Preference Allocation Local (FPA-Local) algorithms discussed below depend only on the existence of a 3-coloring of the lattice. On the other hand, the remaining two algorithms additionally use a 2-coloring of every directional axis in the graph.

A second important issue concerns whether or not a node, when allocating colors for the next time step, can change the colors it has already assigned to its local calls on previous steps. Recall that in practice, this means changing the frequency previously assigned to an ongoing call. We say an algorithm is *non-recoloring* if once having assigned a color in response to a particular new call it never changes that assignment (i.e., recolors the call). The algorithm FA-Local is an example of a non-recoloring algorithm. Recent technical developments, however, allow for a limited rearrangement of frequencies. All the other algorithms discussed below are *recoloring* algorithms, i.e. a node may change the assigned color of a call. For the k-local algorithms we discuss, such changes occur only in response to changes in demand within a node's k-locality.

1.2 Performance measures

We adapt a standard yardstick for measuring the efficacy of online algorithms: that of *competitive ratios* [6]. Given an online algorithm P that processes a sequence of N interference graphs G_t, $t = 0, \ldots, N$, let $\mathcal{S}(P_t)$ denote the span of the multicoloring computed by P after step t, i.e. after graph G_t has been processed. Let $\mathcal{S}_N(P) = \max_t\{\mathcal{S}(P_t)\}$ and $\chi_N(G) = \max_t\{\chi(G_t)\}$. We say that P is a c-*competitive* algorithm if and only if there is a constant b independent of N such that for any sequence,

$$\mathcal{S}_N(P) \leq c \cdot \chi_N(G) + b.$$

In other words, a c-competitive algorithm uses at most c times as many colors (frequencies) overall as the optimal offline algorithm would. We note that all of the algorithms discussed in this paper (with the exception of FA-Local) in fact satisfy the stricter requirement

$$\mathcal{S}(P_t) \leq c \cdot \chi(G_t) + b$$

for all $t \geq 0$, i.e. they approximate the optimal span within a factor of c at *all* times while still processing the input sequence online. All of our lower bounds hold for the above definition of c-competitive (and therefore imply lower bounds on algorithms satisfying the stricter requirement).

1.3 Lower bound models

We provide lower bounds on the competitive ratio of algorithms in a number of models. Notice that if the online algorithm is provided at each step with a complete description of the weights at every node in G the problem reduces to that of solving a series of static frequency assignment or multi-coloring problems. To capture the distributed nature of the problem we must restrict the possible actions taken by nodes during the running of the algorithm, while still allowing every node to have knowledge of the current state of the entire network. By constraining the algorithms in very natural ways, we are able to provide lower bounds in models that capture the properties of most reasonable distributed algorithms including the algorithms discussed in this paper. In some cases the algorithms we provide are optimal for their class.

The first restriction we consider is that of the recoloring ability of the online algorithms. We show lower bounds for both recoloring and non-recoloring algorithms. In the recoloring case we add the constraint that recoloring can only occur in response to a change in demand within a node's immediate neighborhood. We say a recoloring algorithm has *recoloring distance* ℓ if a node recolors its calls during a time step only if a change of demand has occurred within its ℓ-locality.

We further make a distinction between models based upon the kind of information the nodes can use in making their assignments. In particular we consider a class of algorithms in which the pre-computed information is limited to a fixed 3-coloring of the lattice and for which nodes with similar localities act the same. This class includes the algorithms FA-Local and FPA-Local discussed below. More precisely, assume a fixed 3-coloring of the triangular lattice. Call two nodes *k-view-equivalent* if they are in the same color class and there is an isomorphism which maps one node's k-locality to the other's preserving the colors assigned to calls. An algorithm is said to be *k-view and color class determined* if on each step, k-view-equivalent nodes make precisely the same color assignments.

1.4 Our results

In this paper we present the first distributed online algorithms for frequency assignment on hexagon graphs with proven bounds on their competitive ratio along with lower bounds on the performance of online algorithms falling in naturally constrained classes.

All of our upper bounds are obtained by modifying known (global) approximation algorithms for the static frequency assignment problem. The required modifications have the effect of making the coloring decisions depend only on local information. The results indicate that the larger the locality taken into account, the better the competitive ratio the algorithm attains. The algorithm *FA-Local* is a straightforward modification of FA which performs no recoloring, is 0-local and has a competitive ratio of 3. *FPA-Local* (a modification of FPA) is a recoloring 1-local algorithm with competitive ratio bounded by $\frac{3}{2}$. Two further algorithms are presented which are modifications of the global static algorithm first described by Narayanan and Shende [10]. The first of these is a 2-local recoloring algorithm with a competitive ratio of $\frac{17}{12}$. By expanding the locality under consideration to radius 4, we give a 4-local recoloring algorithm which achieves a competitive ratio of $\frac{4}{3}$.

We present lower bounds for recoloring distance bounded algorithms and for view and color class determined recoloring and non-recoloring algorithms. For recoloring algorithms limited to recoloring distance k, we show a lower bound of $1 + \frac{1}{4(k+1)}$. (For the special case of $k = 0$ this can be improved to $\frac{9}{7}$.) This implies that any algorithm that depends only on information from a constant radius neighborhood around each node in making recoloring decisions for that node, can never achieve competitive ratio 1. Note that all the local recoloring algorithms described above have recoloring distance limited to the locality that they have knowledge of. In the setting of arbitrary non-recoloring algorithms we show a lower bound of 2 on the competitive ratio. We show that for any k and any $\epsilon > 0$, a k-view and color class determined non-recoloring algorithm must have competitive ratio at least $3 - \epsilon$. This implies that the algorithm FA-Local is optimal for this class. We also show that for any k, a k-view and color class determined recoloring algorithm (with any recoloring distance ℓ) must have competitive ratio at least 3/2, showing that FPA-local is optimal for this class. Our results imply that both the ability to recolor and the restriction to the k-view and color class determined algorithms have *provable* effects on the performance of an algorithm for multicoloring on hexagon graphs.

2   Online multicoloring of the triangular grid

In this section, we give online algorithms for frequency assignment on hexagon graphs. We describe k-local online algorithms for $k = 0, 1, 2$, and 4 and show upper bounds on the competitive ratio of these algorithms. We begin with a key technical definition. A *static k*-local distributed algorithm for multicoloring gets as input a graph G corresponding to a snapshot of the network at some time step; it has the further property that the color assignment at any node depends only on the weights in the k-locality of the node and some pre-computed information about the lattice. The following general lemma then aids the development of online algorithms:

Lemma 1. *Let A be a k-local static approximation algorithm for multicoloring with performance ratio α. Then A can be converted to a k-local α-competitive online algorithm for multicoloring.*

For the online case, each node runs the k-local static algorithm independently at every step. If the color spectrum obtained by node v is the same as the one used by it in the previous step, then v does not have to recolor. Otherwise, if some colors previously assigned to currently active calls do not appear in the newly computed set of colors, then the algorithm has to recolor those calls.

By Lemma 1, we only need to describe a correct static algorithm which then translates to a corresponding online algorithm. We note, however, that the conversion is only guaranteed if the algorithm has the ability to recolor. That *FA-Local* (the local version of FA) is a 0-local algorithm with performance ratio 3 is a folklore result; to color *new* calls at time t, red, blue, and green nodes simply use the smallest available colors from the sets 0, 1 and 2 mod 3 respectively. Note that FA-Local is a non-recoloring algorithm. In the sequel, we describe three static algorithms for multicoloring, and prove bounds on their performance ratio.

Next, we show that a 1-local version of FPA described below has performance ratio $\frac{3}{2}$.

THE FPA-LOCAL ALGORITHM

Local Information: The colors are divided into three palettes: the red colors are the colors 0 mod 3, the blue colors are 1 mod 3 and the green colors are 2 mod 3. Each node v knows its base color (red, blue or green), its weight and its neighbors' weights.

(1) Let $D_1(v)$ be the 1-local maximal clique weight as computed by v.

(2) v constructs a local spectrum of size $3\lceil D_1(v)/2\rceil$ equally split into red, blue and green colors as described above.

(3) If v's base color is

red: v uses the first $\lceil w(v)/2\rceil$ colors from its red spectrum and the last $\lfloor w(v)/2\rfloor$ colors from its blue spectrum.

blue: v uses the first $\lceil w(v)/2\rceil$ colors from its blue spectrum and the last $\lfloor w(v)/2\rfloor$ colors from its green spectrum.

green: v uses the first $\lceil w(v)/2\rceil$ colors from its green spectrum and the last $\lfloor w(v)/2\rfloor$ colors from its red spectrum.

It is straightforward to see that *FPA-Local* is a 1-local algorithm and uses at most $1.5\max_v D_1(v) \leq 1.5\chi(G)$ colors. We next argue that no two adjacent nodes assign common colors to their local demands. Without loss of generality, consider two such nodes, a red node v and a blue node u both of whom assign blue colors to their calls but from opposite sides of their local blue spectra. Since $w(v) + w(u) \leq \min\{D_1(v), D_1(u)\}$, it follows that the number of blue colors used between u and v, $(\lfloor w(v)/2\rfloor + \lceil w(u)/2\rceil)$, is at most $\lceil\min\{\frac{D_1(v)}{2}, \frac{D_1(u)}{2}\}\rceil$; the latter quantity is the minimum among the sizes of local blue spectra at u and v. This establishes the correctness of FPA-Local.

To obtain online algorithms with better performance, we adapt an offline algorithm described in [10] and derive a 2-local algorithm and a 4-local algorithm from it. Assume that each node knows its base color (red, green or blue) and its parity in a 2-coloring along each of the three directional axes it is an element of. Moreover, we allow each node to know the weights of all nodes in its 2-locality. For technical reasons, we arbitrarily impose a priority scheme over the nodes: red nodes have priority over blue nodes which in turn have priority over green nodes.

The algorithm uses colors that are organized into red, blue, green, purple, and yellow palettes. Each of the first four palettes are four times the size of the yellow one; thus, we can assume that the red palette has the colors $\{0, 1, 2, 3\}$ mod 17, the blue palette the colors $\{4, 5, 6, 7\}$ mod 17, and so on ending with the yellow palette which has the colors $\{16\}$ mod 17. Each node computes its local spectrum consisting of $17\lceil\frac{D_2(v)}{12}\rceil$ colors (recall that $D_2(v) \leq \chi(G)$ is the maximum weight of a clique in v's 2-locality).

Every node independently executes a five-phase algorithm to determine what colors it should use. Essentially, a node always uses colors from its own base color palette first. The remaining colors are assigned either by borrowing colors from other base color palettes, or by using colors from the purple or yellow palettes. The priority scheme and the 2-coloring along different axes mentioned above are used to ensure that there is no conflict between neighbors in the color assignment. Details of the algorithm can be found in the full paper.

A further modification allows us to convert the offline algorithm that formed the basis of *NSA-Local* to a 4-local algorithm (which we identify as *NSB-Local*) with a competitive ratio of $\frac{4}{3}$. It can be shown that the yellow colors become unnecessary if each node can access its 4-locality. The details of the color assignment and the proof of its correctness will be shown in the full version of the paper.

The discussion above and Lemma 1, lead to the following result:

Theorem 2. *There is a distributed online algorithm for multicoloring that is:*

1. *0-local, non-recoloring and 3-competitive.*
2. *1-local and $\frac{3}{2}$-competitive.*
3. *2-local and $\frac{17}{12}$-competitive.*
4. *4-local and $\frac{4}{3}$-competitive.*

3 Lower bounds

In this section, we show lower bounds on the competitive ratio of any online algorithm for multicoloring hexagon graphs. We consider first algorithms with recoloring distance k, and then algorithms that do not allow recoloring. We also show lower bounds on algorithms that are k-view and color class determined.

The following technical lemma is useful in proving Theorem 4 and is stated here without proof.

Lemma 3. *Let P be a path of length ℓ, with demand n on each of its $\ell + 1$ nodes. Then the minimal number of colors required to color P such that the end nodes have exactly α colors in common, is at least $2n + \frac{2\alpha}{\ell-1}$ when ℓ is odd, and at least $2n + \frac{2(n-\alpha)}{\ell}$ when ℓ is even.*

We show a lower bound on any online algorithm with recoloring distance $k > 0$.

Theorem 4. *Any online algorithm with recoloring distance $k > 0$ has competitive ratio at least $1 + \frac{1}{4(k+1)}$.*

Proof. Fix $k > 0$. We will exhibit a strategy for the adversary which forces the algorithm to use at least $2n + \frac{n}{2(k+1)}$ colors, while the offline algorithm needs only $2n$ colors. The graph used by the adversary is shown in Fig 1. Let u and v be two nodes at distance 3 of each other along one of the axes of the lattice. All nodes in the graph initially have weight 0. The adversary chooses an arbitrary even integer $n > 0$ and starts by raising the demand on u and v to n. The adversary continues to raise the demand to n on all nodes along two parallel axes of length $k-1$ which make an angle of $\frac{\pi}{3}$ with the axis uv, and which start at u and v, and end at u' and v' respectively. The algorithm may color and recolor as desired. Next, the adversary raises the demand to n on the two nodes, a and b.

The next moves of the adversary will only involve nodes at distance greater than k from u and v, so the colors on u and v are now fixed. Let α be the number of colors that u and v have in common. The strategy of the adversary now depends on α. If $\alpha \geq n/2$, then the adversary raises the demand to n on

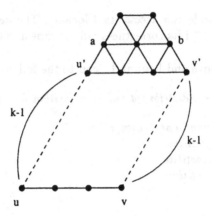

Fig. 1. Graph used by adversary to show a lower bound on any online algorithm with recoloring distance k.

the two nodes which lie on a path of length 3 between a and b and which have distance greater than k to both u and v. The nodes of positive weight now lie on a path of length $2k + 3$. By Lemma 3, the algorithm must now use at least $2n + \frac{n}{2k+2}$ colors. If $\alpha < n/2$, the adversary raises the demand of the common neighbor of a and b to n. By Lemma 3, the algorithm will have to use at least $2n + \frac{n}{2(k+1)}$ colors. The number of colors needed by the offline algorithm is $2n$. The above construction can be repeated as many times as desired, proving that there are arbitrarily long sequences on which the ratio bound of $1 + \frac{1}{4(k+1)}$ is achieved. $\qquad\qquad\square$

Owing to lack of space, we can only list here the other lower bound results we have. The proofs will appear in the full version.

Theorem 5. *(i) Any online algorithm with recoloring distance 0, has competitive ratio at least $1 + \frac{2}{7}$.*

(ii) Any non-recoloring online algorithm has competitive ratio at least 2.

(iii) $\forall k \geq 0, \forall \ell \geq 0$, any recoloring k-view and color class determined algorithm with recoloring distance ℓ has competitive ratio at least $\frac{3}{2}$.

(iv) For any constant $\epsilon > 0, k \geq 0$, any non-recoloring k-view and color class determined algorithm has competitive ratio at least $3 - \epsilon$.

The results in this section, together with Theorem 2 imply that FA-Local and FPA-Local are optimal for the class of non-recoloring and recoloring k-view and color class determined algorithms respectively.

4   Discussion

While we showed tight bounds on the competitive ratio of any k-view and color class determined algorithm (recoloring and non-recoloring) on hexagon graphs, we were unable to prove such tight results in the absence of restrictions on the algorithm. Even in the static case, the best known approximation algorithms for the problem have performance ratio $\frac{4}{3}$. While we derived a 4-local algorithm with the same competitive ratio, it would be interesting to know if the same ratio can be achieved by a 0-local or a 1-local algorithm.

Another natural requirement on online algorithms might be to restrict the *amount* of recoloring at any vertex to be proportional to the change in demand in the vertex's k-locality in every step. Among the recoloring algorithms we describe, *FPA-Local* meets this requirement, but the remaining two algorithms do not. Finally, in practice, the available frequency spectrum is a contiguous linear sub-interval of the radio spectrum, and frequency reuse is controlled by a sequence of non-negative integers, $c_0 \geq c_1 \ldots$, with $c_0 \geq 1$, called *distance reuse constraints*, where any two frequencies assigned to nodes that are distance i apart in the graph are required to differ by at least c_i. This paper assumes $c_0 = c_1 = 1$ and $c_i = 0$ for $i > 1$; tight bounds for the more general case would be interesting.

References

1. D. Dimitrijević and J. Vučetić. Design and performance analysis of algorithms for channel allocation in cellular networks. *IEEE Transactions on Vehicular Technology*, 42(4):526–534, 1993.
2. S. Irani. Coloring inductive graphs online. *Algorithmica*, 11:53–72, 1994.
3. J. Janssen and K. Kilakos. Optimal multicolouring algorithms with limited recolouring. Submitted for publication, April 1995.
4. J. Janssen, K. Kilakos, and O. Marcotte. Fixed preference frequency allocation for cellular telephone systems. Unpublished manuscript, April 1995.
5. T. Kahwa and N. Georganas. A hybrid channel assignment scheme in large-scale cellular-structured mobile communication systems. *IEEE Transactions on Communications*, 4:432–438, 1978.
6. A. Karlin, M. Manasse, L. Rudolph, and D. Sleator. Competitive snoopy caching. *Algorithmica*, 3(1):70–119, 1988.
7. S. Kim and S. L. Kim. A two-phase algorithm for frequency assignment in cellular mobile systems. *IEEE Transactions on Vehicular Technology*, 1994.
8. L. Lovasz, M. Saks, and W. Trotter. An online graph coloring algorithm with sublinear performance ratio. *Discrete Math*, 75:319–325, 1989.
9. C. McDiarmid and B. Reed. Channel assignment and weighted colouring. Submitted for publication, 1997.
10. L. Narayanan and S. Shende. Static frequency assignment in cellular networks. In *SIROCCO 97*, 1997.
11. P. Raymond. Performance analysis of cellular networks. *IEEE Transactions on Communications*, 39(12):1787–1793, 1991.
12. S. Vishwanathan. Randomized online graph coloring. *Journal of Algorithms*, 13:657–669, 1992.
13. W. Wang and C. Rushforth. An adaptive local-search algorithm for the channel-assignment problem. Technical Report, August 1995.

Floats, Integers, and Single Source Shortest Paths*

Mikkel Thorup

Department of Computer Science, University of Copenhagen, Universitetsparken 1,
2100 Kbh.˙Ø, Denmark. (mthorup@diku.dk, http://www.diku.dk/~mthorup)

Abstract. Floats are ugly, but to everyone but theoretical computer
scientists, they are the real thing. A linear time algorithm is presented
for the undirected single source shortest paths problem with positive
floating point weights.

1 Introduction

The technical goal of this paper is to present a linear time solution to the undirected single source shortest paths problem (USSSP) where the weights are positive floating points, or just floats. On a more philosophical level, the goal is to draw attention to the problem of making efficient algorithms for floats. Suppose, for example, we have an algorithm for the max-flow problem whose running time includes a factor $\log C$, where C is the maximal capacity. If we allow floating points, such an algorithm is not even polynomial, i.e. $\log C$ is the exponent of C, and the exponent is stored with $\log \log C$ bits.

Floating points are at least as used as integers, by everybody but theoretical computer scientists, who seem to prefer integers. To multiply two integers the faster way is often to convert them to double floats and send them to the floating point co-processor. Why? Because people simply don't care enough about integers to make an integer co-processor. Also, as theoretical computer scientists, we have to admit that floats do provide an elegant way of dealing with numbers in a large range. It is for good reasons that all other scientists and engineers have used them for centuries.

The rounding of floating point arithmetic is ugly in that, for example, addition is neither associative, nor commutative. Nevertheless, in this paper, we hope to indicate, that finding the structure of the rounding for a given problem, such as USSSP, can be an appealing challenge.

It should be noted that often floats do not cause any problems relative to integers. For problems like sorting, priority queues, and searching, there is no difference. The IEEE floating point standard is made such that interpreting the bit-string representation of floating points as representing integers, is order preserving. Hence we can feed floating points to an integer priority queue, if we just don't tell it that it is floats. Similarly, van Emde Boas' data structure

---

* Most of this work was done while the author visited the Max-Planck-Institut für Informatik.

[vBKZ77] works in time $O(\log \omega)$, where ω is the word length, no matter whether the words represent integers or floats.

Up to recently, all theoretical developments in the single source shortest paths problem (SSSP) were based in Dijkstra's algorithm [Dij59], where vertices are visited in increasing order of distance to the source using a priority queue. Since the priority queue doesn't care whether the input is integers or floats, all implementations of Dijkstra's algorithm work equally well for integers and floats. However, recently, the author [Tho97] presented a linear time non-Dijkstra algorithm for undirected SSSP with integer weights. It is crucial to this algorithm that the weights are sorted with respect to their exponents. For integers in words, the exponent is at most ω, and then the exponents are easily sorted in linear time. However, for floating points, sorting the exponents is as hard as integer sorting, and this we do not know how to do in linear time.

Here, we show how to move edges around in the graph, preserving the floating point distances, ending with a series of independent subgraphs for which the exponents can be sorted in linear time. Given the sorted exponents, it is not too difficult to modify the algorithm from [Tho97] for integer USSSP to solve the floating point USSSP in linear time. Some remarks on this will be made in the journal version of [Tho97]. Given the sorted exponents, such a floating point version of the algorithm from [Tho97] may work very well in practice because of the efficiency of floating point arithmetic. However, here we give a self contained presentation, showing something in principle stronger; namely, that USSSP with floating point weights can be solved in linear time using a linear time oracle for USSSP with integer weights. Finally, we will make a few remarks on floats and max-flow.

2 Preliminaries

Our algorithm runs on a RAM, which models what we program in imperative programming languages such as C. The memory is divided into addressable words of length ω. Addresses are themselves contained in words. Hence, in order to address the n different nodes of the graph we are working on, we assume $\omega \geq \log n$. Moreover, we have a constant number of registers, each with capacity for one word. The basic assembler instructions are: conditional jumps, direct and indirect addressing for loading and storing words in registers, and some computational instructions, such as comparisons, addition, and multiplication, for numbers in registers. The space complexity is the maximum memory address used, and the time complexity is the number of instructions performed. All weights are floating point numbers, each contained in $O(1)$ words.

The numbers may be either integers, represented the usual way, of floating point numbers. A *floating point*, or just a *float*, is a pair $x = (e, m)$, where e is an integer, and m is a bit string $b_1 \cdots b_p$. Then x represents the real number $2^e(1 + \sum_{i=1}^{p} b_i/2^i)$. Thus $e = \lfloor \log_2 x \rfloor$. We call e the *exponent*, denoted $\mathbf{expo}(x)$, and m the *mantissa*, denoted $\mathbf{mant}(x)$. Often we will identify a float with the real number it represents. Both e and m are assumed to fit in a constant number of words. The number $\wp = O(\omega)$ is fixed throughout a given computation and is referred to as the *precision*. We have previously mentioned that according to the IEEE floating point standard, we get the correct ordering of floats by perceiving them as integers. This is obtained by first having the sign-bit of m, then e, and finally m without the sign-bit.

We let $\oplus$ denote floating point addition. In this paper, for simplicity, when two floats x and y are added, they are always rounded *down* to nearest float (determined by the precision $\wp$). Thus $x \oplus y \leq x + y$. This gives us the following basic rule:

$$\textbf{expo}(x) > \textbf{expo}(y) + \wp \Rightarrow x \oplus y = y \oplus x = x \qquad (1)$$

If instead we were rounding either up or down to nearest float, the rule would only apply if $\textbf{expo}(x) > \textbf{expo}(y) + \wp + 1$, and all the calculations below, would have to be changed accordingly.

Let $G = (V, E)$, $|V| = n$, $|E| = m$, be an undirected connected graph with a distinguished source vertex s. Each edge $e \in E$ has a positive floating point weight $\ell(e)$ associated with it. The length of a path from s to some vertex v, is the floating point sum of the weights added up starting from s. More specifically, if the path is $P = (v_0, v_1, \ldots, v_l)$, $s = v_0$, $v_l = v$, then the length of P is

$$((((\ell(v_0, v_1) \oplus \ell(v_1, v_2)) \oplus \ell(v_2, v_3)) \cdots) \oplus \ell(v_{l-1}, v_l)$$

By $d(v)$ we denote the length of the shortest path from s to v. The floating point USSSP problem is that of finding $d(v)$ for all v. Different orders of adding the weights could give different answers, so it is natural to ask how big the errors can get. Let $d^*(v)$ denote the distance where the weights are summed using normal addition of the reals represented by the floats. Then $d^*(v)$ represents the ideal answer. Clearly $d^*(v) \geq d(v)$.

Observation 1 $\dfrac{d^*(v) - d(v)}{d(v)} \leq 2^{-\wp + \log_2 n}$

Proof: Set $e = \textbf{expo}(d(v))$. We are adding at most $n - 1$ numbers. Since the maximal exponent is $\leq e$, the maximal loss per number is $< 2^{e-\wp}$. Thus, the total error is $< (n-1)2^{e-\wp} < 2^{e-\wp+\log_2 n}$. ∎

In theory, since $\omega \geq \log_2 n$, we can always simulate $\log_2 n$ extra bit of precision without affecting the asymptotic running time. In practice, according to the IEEE standard format, with long floats, we have $\wp = 52$. Hence the relative error $\dfrac{d^*(v) - d(v)}{d(v)}$ is at most $2^{-52 + \log_2 n}$, which is normally OK. Throughout the paper, we will assume $p > \log_2 n$.

3 Sorting the weight exponents

First, by a *source component* of a graph with source s, we mean a maximal subgraph H with $H \setminus \{s\}$ connected. We will now apply some different transformations to G constructing a graph G' so that the d-values of the nodes are unchanged (though some vertices may be identified), but so that for each source component of G', the weight exponents vary by at most $n(\wp+1)$. Since the source components only intersect in s, we can solve the USSSP problem independently for each source component H. The small variation in the weight exponents within H implies that they can be sorted in linear time.

Our first step is to construct a minimum spanning tree T for G in deterministic linear time and space [FW94]. The algorithm from [FW94] assumes integer weights, but since we are only interested in the ordering, we can just perceive our floats as representing integers when running the algorithm from [FW94].

Let $d_T(v)$ denote the weight of the path from s to v in T. Let $f_T(v)$ denote the weight of the first edge on this path. Clearly $d_T(v) \geq d(v)$. Also, since T is a minimum spanning tree, $f_T(v) \leq d(v)$. Using these simple bounds on $d(v)$, the algorithm below identifies places where the graph can be simplified without affecting the distance to any vertex.

Algorithm A: Transforms G to a graph G' so that the d-values of the nodes are unchanged, though some vertices may be identified, and so that for each source component of G', the weight exponents vary by at most $n(\wp + 1)$.

A.1.As described in [FW94], compute a minimum spanning tree T of G.

A.2.Compute $d_T(\cdot)$.

A.3.For all $(v, w) \in E$,

A.3.1. If $v \neq s$, $\mathbf{expo}(\ell(v,w)) > \mathbf{expo}(d_T(v)) + \wp$ and $d_T(w) \geq d_T(v)$, replace (v, w) by an edge (s, w) of the same weight. If (v, w) was in T, it is replaced by (s, w) in T.

A.4.Compute $f_T(\cdot)$

A.5.For all $(v, w) \in T$ with v nearest s and $v \neq s$,

A.5.1. If $(v, w) \in T$ and $\mathbf{expo}(\ell(v,w)) < \mathbf{expo}(f_T(v)) - \wp$, $\ell(v,w) := 0$

A.6.Contract all components induced by edges of weight 0, removing any loops created.

As stated above, the algorithm may introduce multiple edges when adding an edge (s, w) in step A.3.1. However, really we will only have the lightest edge (s, w) in the graph. For each vertex w, we put an edge to s first in the incidence list of w. Then, when a new edge (s, w) is introduced, it is easily compared with the previous one, if any.

Algorithm A is analysed in the remainder of this section. In Lemmas 1–3, we will first prove that Algorithm A does not change any d-values, that T is maintained as a minimum spanning tree, that Algorithm A runs in linear time and space, and that for each edge (v, w), $v \neq s$ in the produced graph G',

(i) $\mathbf{expo}(\ell(v,w)) \leq \mathbf{expo}(d_T(v)) + \wp$
(ii) $\mathbf{expo}(\ell(v,w)) \geq \mathbf{expo}(f_T(v)) - \wp$.

Then in Lemmas 4–7, we will show that (i) and (ii) imply the desired bounded variance of $n(\wp + 1)$ on the weight exponents within each source component of G'. Finally, in Lemma 8, we will show that the bounded variance on the weight exponents of a source component imply that they can be sorted in linear time.

Lemma 1. *During the course of Algorithm A, all d-values remain unchanged, and all d_T- and f_T values remain unchanged after once computed. Finally T is maintained as a minimum spanning tree.*

Proof: We need to argue that steps A.3.1, A.5.1, and A.6 cannot violate the claim of the lemma. First, suppose (v, w) pass the test of step A.3.1. Then $\text{expo}(\ell(v, w)) > \text{expo}(d_T(v)) + \wp \geq \text{expo}(d(v)) + \wp$. Hence, by (1), $d(v) \oplus \ell(v, w) = \ell(v, w)$ and $d(v)_T \oplus \ell(v, w) = \ell(v, w)$. Also, $d_T(w) \oplus \ell(v, w) \geq d(w) \oplus \ell(v, w) \geq \ell(v, w) > d_T(v) \geq d(v)$, so no shortest path in G or T comes to v via (w, v). Thus, replacing (v, w) by (s, w) with $\ell(s, w) = \ell(v, w)$ in G and in T does not change any d- or d_T-value. In contrast, step A.3.1 may change f_T-values, and this is why the f_T values are not computed until step A.4.

Concerning the maintainance of T as a minimum spanning tree in step A.3.1, suppose first that $(v, w) \in T$. Since $d_T(v) \leq d_T(w)$, this implies that v is nearest to s; for otherwise, $d_T(v) = d_T(w) \oplus \ell(v, w) \geq \ell(v, w) > d_T(v)$. Thus, deleting (v, w) cuts T in two components with s in the same component as v. We know that (v, w) was a minimal edge crossing this cut, and since (s, w) crosses the same cut with the same weight, we get a new minimal spanning tree when replacing (v, w) by (s, w) in T. Now, if (v, w) was not in T, we know that $\ell(v, w)$ was at least as big as any weight on the path from v to w in T. Then $\ell(s, w) = \ell(v, w)$ is also bigger than all the weights on the path from s to w, for all edges on this path that are not on the path from v to w are on the path from s to v, and all these edges have weight $\leq d_T(v) < \ell(v, w) = \ell(s, w)$. Thus (s, w) is at least as heavy as any edge on the path from s to w, and hence we do not need to involve it in T. Thus, we conclude that step A.3.1 cannot violate the lemma.

Suppose (v, w) pass the test of step A.5.1 and gets its weight decreased to 0. Since $(v, w) \in T$, T remains a minimum spanning tree. Since $\text{expo}(\ell(v, w)) < \text{expo}(f_T(v)) - \wp$ and $d(v) \geq f_T(v)$, $d(v) \oplus \ell(v, w) = d(v)$, so $d(w) \leq d(v)$. Also, $d_T(v) \geq f_T(v)$, so $d_T(w) \leq d_T(v)$. However, $f_T(v) = f_T(w)$, so symmetrically, $d(v) \leq d(w)$ and $d_T(v) \leq d_T(w)$. Thus $d(v) = d(w)$ and $d_T(v) = d_T(w)$, and hence it does not change any d- or d_T-value to set $\ell(v, w) := 0$. Further, since the structure of T is unchanged and $v \neq s$, no f_T-value is changed by step A.5.1.

Concerning the contraction in step A.6, note that all 0 weight edges are in T and not incident to s. Hence, the contraction of 0 weight edges does not change any d-, d_T-, or f_T-values, and neither does the contraction of edges in T stop T from being a minimal spanning tree. ∎

Lemma 2. *Algorithm A runs in linear time and space.*

Proof: Step A.1 is known to take $O(m)$ time and space [FW94]. Steps A.2 and A.4 are done in time $O(n)$, each by a simple scan from starting in the root. By Lemma 1, we do not need to worry about updating d_T- and f_T-values when once computed, and hence it is straightforward to implement each of step A.3.1 and step A.5.1 in constant time. Hence step A.3 takes $O(m)$ time and step A.5 takes $O(n)$ time. For step A.6, we identify the components of the graph induced by 0 weight edges in time $O(m)$. Then we map the nodes of each component to a surviving representative of that component. Finally, we contract the components in $O(m)$ total time and space by renaming the end-points according to the map, removing all loops created. ∎

Lemma 3. *All edges $(v, w) \in G'$ with $v \neq s$ satisfy (i) and (ii).*

Proof: By Lemma 1, d_T- and f_T-values are not changed once computed. Hence, no edge in the resulting graph G' satisfy the condition of step A.3.1 or of step A.5.1.

Let $(v, w) \in G'$, $v \neq s$. Suppose for a contradiction that (v, w) violates (i), that is, $\text{expo}(\ell(v, w)) > \text{expo}(d_T(v)) + \wp$. If $d_T(w) \geq d_T(v)$, (v, w) should have passed the test of step A.3.1, contradicting that $(v, w) \in G'$. However, if $d_T(w) < d_T(v)$, we have $\text{expo}(\ell(v, w)) > \text{expo}(d_T(v)) + \wp > \text{expo}(d_T(w)) + \wp$, so then (w, v) should have passed the test of step A.3.1. Thus, we conclude that (i) is not violated.

Suppose instead that (v, w) violates (ii), that is, $\text{expo}(\ell(v, w)) < \text{expo}(f_T(v)) - \wp$. If $(v, w) \in T$ with v nearest to s, (v, w) should have passed the test of step A.5.1, contradicting $(v, w) \in G'$. If instead w is nearest to s and $w \neq s$, (w, v) should have passed the test since $f_T(v) = f_T(w)$. Finally, we cannot have $(w, v) \in T$ and $w = s$, for then $\text{expo}(\ell(v, w)) = \text{expo}(f_T(v))$, contradicting $\text{expo}(\ell(v, w)) < \text{expo}(f_T(v)) - \wp$.

Now, suppose $(v, w) \in G' \backslash T$. Since T is a minimum spanning tree, the weight of $\ell(v, w)$ is at least as big as the weight of any edge on the path from v to w in T. Consider the path P from v to w in T. If $s \in P$, let (s, u) be the first edge on the path from s to v in P. Then $f_T(v) = \ell(s, u)$, which together with $\text{expo}(\ell(v, w)) < \text{expo}(f_T(v)) - \wp < \text{expo}(f_T(v))$ contradicts that $\ell(v, w)$ is as big as any weight in P. Thus $s \notin P$.

Consider any edge $(x, y) \in P$ with x nearest s. Since $s \notin P$, $f_T(v) = f_T(x)$. Hence, $\text{expo}(\ell(x, y)) \leq \text{expo}(\ell(v, w)) < \text{expo}(f_T(v)) - \wp = \text{expo}(f_T(x)) - \wp$. Thus, (x, y) satisfies the condition of step A.5.1. Consequently, all edges in P should have been contracted in step A.6, and then (v, w) should have been removed as a loop. This contradicts $(v, w) \in G'$, and completes the proof of the lemma. ∎

We can now solve the USSSP problem for G, by solving it independently for each source component of G'. Thus, in the following, let H by any source component of G'. Then $H \backslash \{s\}$ is connected and (i) and (ii) are satisfied for all $(v, w) \in H$ with $v \neq s$. In Lemmas 4–7, we will prove that these properties suffice for a bounded variation of $n(\wp + 1)$ on the weight exponents within H.

Lemma 4. *If $v \in H$ has depth $d > 0$, $\text{expo}(d_T(v)) \leq \text{expo}(f_T(v)) + (d-1)(\wp+1)$.*

Proof: (induction on d) The statement is trivially true for $d = 1$. If $d > 1$ and u is the ancestor of v, by induction, $\text{expo}(d_T(u)) \leq \text{expo}(f_T(u)) + (d-2)(\wp+1) = \text{expo}(f_T(v)) + (d-2)(\wp+1)$. Applying (i), $\text{expo}(\ell(u, v)) \leq \text{expo}(d_T(u)) + \wp \leq \text{expo}(f_T(v)) + (d-1)(\wp+1) - 1$. Since $d_T(v) = d_T(u) \oplus \ell(u, v)$, $\text{expo}(d_T(v)) \leq \max\{d_T(u), \ell(u, v)\} + 1 \leq \text{expo}(f_T(v)) + (d-1)(\wp+1)$. ∎

Lemma 5. *If $(v, w) \in H$ where $v \neq s$, $\text{expo}(\ell(v, w)) \leq \text{expo}(f_T(v)) + \text{depth}_T(v)(\wp+1) - 1$.*

Proof: By (i) and Lemma 4, $\text{expo}(\ell(v, w)) \leq \text{expo}(d_T(v)) + \wp \leq \text{expo}(f_T(v)) + (\text{depth}_T(v) - 1)(\wp+1) + \wp$. ∎

Lemma 6. *If (s, u) and (v, w) are in H, $\mathrm{expo}(\ell(v, w)) \leq \mathrm{expo}(\ell(s, u)) + (|V(H)| - 1)(\wp + 1) - 1$.*

Proof: Let $T_1, \ldots, T_i$ be distinct trees in $T \setminus \{s\}$ such that $u \in T_1$, and for $i > 1$, there is an edge (x, y) from T_{i-1} to T_i. By induction on i, we will argue that if $(a, b) \in H$ and $a \in T_i$,

$$\mathrm{expo}(\ell(a, b)) \leq \mathrm{expo}(\ell(s, u)) + (\sum_{j=1}^{i} |V(T_j)|(\wp + 1)) - i. \qquad (2)$$

Suppose $i = 1$. Since $depth_T(a) \leq |V(T_i)|$, (2) follows directly from Lemma 5.

Suppose $i > 1$. Let (x, y) be the edge entering T_i from T_{i-1}. By induction,

$$\mathrm{expo}(\ell(x, y)) \leq \mathrm{expo}(\ell(s, u)) + (\sum_{j=1}^{i-1} |V(T_j)|(\wp + 1)) - (i - 1).$$

Since T is a minimum spanning tree $\ell(x, y) \geq f_T(y) = f_T(a)$. Thus, by Lemma 5,

$$\begin{aligned}
\mathrm{expo}(\ell(a, b)) &\leq \mathrm{expo}(f_T(a)) + depth_T(a)(\wp + 1) - 1 \\
&\leq \mathrm{expo}(\ell(x, y)) + |V(T_i)|(\wp + 1) - 1 \\
&\leq \mathrm{expo}(\ell(s, u)) + (\sum_{j=1}^{i} |V(T_j)|(\wp + 1)) - i,
\end{aligned}$$

completing the proof of (2). The trees $T_1, \ldots, T_i$ can be chosen for any $(a, b) \in H$, and since $\bigcup_j T_j \subseteq H \setminus \{s\}$, the lemma follows. $\blacksquare$

Lemma 7. *The minimum and maximum weight exponent of an edge in H differ by at most $|V(H)|(\wp + 1)$.*

Proof: Let (s, u) be the minimum weight edge in H leaving s. Consider any edge (v, w). By (ii), $\mathrm{expo}(\ell(v, w)) \geq \mathrm{expo}(f_T(v)) - \wp \geq \mathrm{expo}(\ell(s, u)) - \wp$. At the same time, by Lemma 6, $\mathrm{expo}(\ell(v, w)) \leq \mathrm{expo}(\ell(s, u)) + (|V(H)| - 1)(\wp + 1) - 1$. Hence, exponents of weights in H can vary by at most $|V(H)|(\wp + 1) - 1$. $\blacksquare$

Lemma 8. *We can sort the weight exponents of the edges in H in linear time and space.*

Proof: First subtracting the minimum weight exponent from all other weight exponents, the maximum weight exponent becomes $n(\wp + 1)$, which is represented by $\log_2 n + \log_2(\wp + 1)$ bits. If $\wp \leq n - 1$, each weight exponent is viewed as two $\leq \log_2 n$ bit characters, and we radix sort in two rounds, each round taking linear time and space.

In the extreme case where $\wp \geq n$, since $\wp = O(\omega)$, we have $\log_2 n + \log_2 \wp = O(\omega / (\log n \log \log n))$. Then we can apply the linear time and space packed sorting from [AH97]. $\blacksquare$

Combining Lemma 1, 2, 7, and 8, we conclude,

Theorem 9. *Given a floating point weight graph G with source vertex s, in linear time and space, we can transform G into a graph G' with the same d-values, though possibly with some vertices identified, such that for each source component H, the minimum and maximum weight exponent of an edge in H differ by at most $|V(H)|(\wp + 1)$. Further, we can sort the weight exponents of each source component H in linear time and space.*

Concerning a concrete implementation of our algorithm, it should be noted that it would suffice with the minimal spanning tree T just being minimal with respect to the exponents of the weights. Dealing just with the exponents rather than the whole floats may be easier. For example, for 64-bit floats in the IEEE floating point standard, the exponent an 11-bit integer, or really, the exponent is represented as an unsigned 11-bit integer from which one has to subtract 1023 to get the exponent. For the ordering of the exponents, however, it suffices just to mask out the 11 exponent bits. Also, in practice, one would certainly prefer the randomized linear time algorithm of Karger, Klein, and Tarjan [KKT95] to the deterministic one of Fredman and Willard [FW94].

4 Using an integer oracle

Having applied Theorem 9, we complete solving the USSSP problem for G, by solving it in $O(|E(H)|)$ time and space for each of the source components H of G'. In our solution, we will apply the linear time integer USSSP oracle from [Tho97]. First we will estimate the exponents ± 1. Second we will make exact calculations.

Distances with a bit too much precision

We will calculate the distances from s but sometimes using some extra precision. Since we always round down, extra precision means larger values. For each vertex v, we denote the obtained distance by $D(v)$. Referring to Observation 1, we get $d(v) \leq D(v) \leq d^*(v)$. Since $\wp \geq \log_2 n \geq \log_2 |V(H)|$, we further get $D(v) \leq 2d(v)$. Consequently, $\texttt{expo}(d(v)) \leq \texttt{expo}(D(v)) \leq \texttt{expo}(d(v)) + 1$.

Our first step is to subtract the exponent e of the smallest weight from all weight exponents (corresponding to division by 2^e). By Theorem 9, all weight exponents are now in the interval $[0, |V(H)|(\wp + 1)) \subseteq [0, 2|V(H)|\wp)$.

We are going to proceed in rounds for $i = 1, \ldots, 2|V(H)|$. In round i, we are going to find the distances from s along paths where the maximal exponent of an edge weight is $< i\wp$. We assume that this has already been done over paths where the maximal exponent is $< (i - 1)\wp$.

Consider the set $S = \{v \mid \texttt{expo}(D(v)) < (i - 1)\wp\}$. Edges with weight exponents $\geq (i - 1)\wp$ are not going to give any better distances for vertices $v \in S$, so the distances to vertices in S may now be output. Next we *contract* S as follows. For all edges $(v, w) \in E$, $v \in S \setminus \{s\}$, $w \notin S$, we replace (v, w) by a *distance edge* (s, w) whose weight is $D(v) \oplus \ell(v, w)$. Afterwards $S \setminus \{s\}$ is removed. Clearly

(v, w) is only once replaced by a distance edge (s, w). Multiple distance edges (s, w) are assumed removed as described immediately below Algorithm A.

Now contract all edges (v, w) with $\texttt{expo}(\ell(v, w)) < (i-2)\wp$. This corresponds to reducing their weight to 0. We claim that this does not not change the floating point distance to any vertex. By symmetry, it suffices to show that it does not change the distance to w. Suppose the shortest path to w goes through v. Since v was not contracted in S, $\texttt{expo}(D(v)) \geq (i-1)\wp$, but this immediately implies $D(v) \oplus \ell(v, w) = D(v)$

Here in round i, concerning the original edges, we restrict our attention to the surviving ones with weight exponent $< i\wp$. Since we have contracted all original edges with weight exponent $< (i-2)\wp$, this implies that edges will be considered in their original form for at most 2 rounds. Concerning distance edges, we restrict our attention to the surviving ones where the maximal edge weight on the corresponding path has exponent $< i\wp$. The weight of these distance edges is then $< (|V(H)| - 1)2^{i\wp} < 2^{(i+1)\wp}$, which means that their exponents are $< (i+1)\wp$. At the same time, we have contracted every distance edge whose exponent is $< (i-1)\wp$. Hence distance edges are considered for at most 2 rounds. In conclusion, each edge will be considered for a total of at most 4 times.

Before converting our floats to integers, we subtract $(i-3)\wp$ from every exponent, corresponding to dividing all weights by $2^{(i-3)\wp}$. Since the smallest weight exponent was $\geq (i-2)\wp$, it is now $\wp$, so all weights are now integers. All distances computed in round i, are based on original edges whose weight exponents are now $< i\wp - (i-3)\wp = 3\wp$, hence of weight $< 2^{3\wp}$. Thus, the maximal distance computed in round i is $< (|V(H)| - 1)2^{3\wp} \leq 4\wp$ Thus, we call our integer USSSP, where each integer is represented by $4\wp = O(\omega)$ bits. This takes time linear in the number of edges considered, and since each edge is considered in at most 4 rounds, the total running time over all rounds is linear in the total number of edges. The distances produced are essentially correct, except that in the calculations, we vary between $\wp$ and $4\wp$ bits of precision, where we should really only have had a precision of $\wp$ in every single addition. As pointed out above, this means that the exponents of the calculated distances are either correct, or at most one too large.

Exact results

In order to get the exact floating point distances $d(v)$, as defined in Section 2, we are going to proceed in rounds as above, but working with one exponent e at the time. We will go through the exponents e in increasing order. Recall that the exponents of the edge weights were sorted in Section 3. Inserting exponents of computed distances in the ordering is straightforward; for if $d(w) = d(v) \oplus \ell(v, w)$, $\texttt{expo}(d(w)) \in \{\texttt{expo}(d(v)), \texttt{expo}(d(v)) + 1, \texttt{expo}(\ell(v, w)), \texttt{expo}(\ell(v, w)) + 1\}$. For each vertex v, we are going to start with our estimated distance $D(v) \geq d(v)$ computed in the previous subsection. We are going to decrease $D(v)$ to $d(v)$ over several rounds. Recall that when we start $\texttt{expo}(d(v)) \leq \texttt{expo}(D(v)) \leq \texttt{expo}(d(v)) + 1$.

The goal of round e is to get $D(v) = d(v)$ for all v with $\mathbf{expo}(d(v)) \leq e$. Inductively, we assume this has already been achieved for all v with $\mathbf{expo}(d(v)) \leq e - 1$. Let S be the set of vertices v with $\mathbf{expo}(d(v)) \leq e - 1$. These vertices are contracted, as in the previous section, that is, each outgoing edge (v, w), $v \in S \setminus \{s\}$, $w \notin S$, is replaced by a distance edge (s, w) with $\ell(s, w) = d(v) \oplus \ell(v, w)$. Moreover, we set $D(w) = \min\{D(w), D(v) \oplus \ell(v, w)\}$.

Construct the graph G_e with the distance edges (s, v) with $\mathbf{expo}(D(v)) = e$, and all edges (v, w) with $\{\mathbf{expo}(D(v)), \mathbf{expo}(D(w))\} \subseteq \{e, e + 1\}$. The latter includes all edges (v, w) with $\mathbf{expo}(d(v)) = \mathbf{expo}(d(w)) = e$. Subtract $e - \wp$ from all the exponents and convert to integers, rounding down, i.e. $x \mapsto \lfloor x/2^{e-\wp} \rfloor$. Call the integer USSSP algorithm, convert back to floats, and add $e - \wp$ to the exponents.

To see that the above is correct for any w with $\mathbf{expo}(d(w)) = e$, consider a shortest path $v_0 \cdots v_l$ from $s = v_0$ to $w = v_l$. Let v_i be the last vertex with $\mathbf{expo}(d(v_i)) < e$. By induction, $D(v_i) = d(v_i)$, and also, we contracted $v_i \in S$ correctly, setting $D(v_{i+1}) = D(v_i) \oplus \ell(v_i, v_{i+1}) = d(v_i) \oplus \ell(v_i, v_{i+1}) = d(v_{i+1})$. For $j = i + 1, \ldots, l - 1$, we have $\mathbf{expo}(d(v_j)) = \mathbf{expo}(d(v_{j+1})) = e$, so (v_j, v_{j+1}) is an edge in G_e. Moreover, for each of the desired floating point additions $D(v_j) \oplus \ell(v_j, v_{j+1})$, the first term and the sum both have exponent e. This implies that the integer additions performed by the integer USSSP algorithm simulate the desired floating point additions with exactly the right precision $\wp$. Thus, we end up with $D(w) = d(w)$, as desired.

Clearly, each edge is considered at most twice in its original form, and at most once as distance edge. Thus, the total time spent above is linear in the total number of edges.

Theorem 10. *There is a linear time Turing reduction from the floating point USSSP problem to the integer USSSP problem.*

Combining with the linear time algorithm for integer USSSP from [Tho97], we get

Corollary 11. *There is a linear time algorithm for USSSP problem with floating point weights.*

5   A remark on max-flow

Above it was shown how we can deal with floating points in connection with USSSP. In the introduction, we made some remarks concerning max-flow. The problem is that many max-flow algorithms include a factor $O(\log C)$ in their time bound, where C is the maximal capacity in the network. However, if the capacities are floating points, $\log C = \mathbf{expo}(C)$, and $\mathbf{expo}(C)$ may be exponentially large. We will address the problem briefly by showing that by at most doubling the error, we can restrict our attention to a network with maximal capacity C' where $\log C' = O(\log n + \wp)$.

Let f^* denote the (unknown) maximal flow value. First we find an s-t path with maximal minimal capacity D, that is, D is the maximal flow that can be pushed along a single path. Then $D \leq f^* \leq mD$. We can therefore reduce all capacities bigger than mD to mD without affecting the maximal flow. Concerning the small capacities, we can exploit that we are anyway going to accept an error in the order of $D/2^\wp$. At most doubling this error, we can take all weights and round down to the nearest multiple of $D/(m2^\wp)$. More precisely, we set $e = \mathbf{expo}(D) - \lfloor \log_2 m \rfloor - \wp$, subtract e from all exponents and round down to nearest integer. If C' is now the maximal capacity, $\log_2 C' \leq 2\lceil \log_2 m \rceil + \wp = O(\log n + \wp)$. Then we solve the max-flow problem with the reduced weights, and add e to the exponent of the resulting flow value.

Note above that we convert a floating point max-flow problem to one with bounded integer capacities. In practice, one may want to keep working with floats in order to benefit from an efficient floating point co-processor. However, working directly with floats in integer max-flow algorithms may, for example, cause infinite loops due to rounding problems. Such problems are addressed in [AM97].

Acknowledgement

I would particularly like to thank referee 4 from STACS'98 for some very thorough and useful comments.

References

[AH97]   S. ALBERS AND T. HAGERUP, Improved parallel integer sorting without concurrent writing, *Information and Control* **136** (1997) 25–51.

[AM97]   E. ALTHAUS AND K. MEHLHORN, Maximum Network Flow with Floating Point Arithmetic, Max-Plank-Institut für Informatik, Technical Report MPI-I-97-1-022, 1997.

[Dij59]   E.W. DIJKSTRA, A note on two problems in connection with graphs, *Numer. Math.* **1** (1959), 269–271.

[FW94]   M.L. FREDMAN AND D.E. WILLARD, Trans-dichotomous algorithms for minimum spanning trees and shortest paths, *J. Comp. Syst. Sc.* **48** (1994) 533–551.

[KKT95]   D.R. KARGER, P.N. KLEIN, AND R.E. TARJAN, A Randomized Linear-Time Algorithm to Find Minimum Spanning Trees *J. ACM*, 42:321–328, 1995.

[Tho97]   M. THORUP. Undirected Single Source Shortest Paths in Linear Time. In *Proceedings of the 38th IEEE Symposium on Foundations of Computer Science (FOCS)*, pages 12–21, 1997.

[vBKZ77] P. VAN EMDE BOAS, R. KAAS, AND E. ZIJLSTRA, Design and implementation of an efficient priority queue, *Math. Syst. Th.* **10** (1977), 99–127.

A Synthesis on Partition Refinement: A Useful Routine for Strings, Graphs, Boolean Matrices and Automata

Michel HABIB*   Christophe PAUL*   Laurent VIENNOT†

Abstract

Partition refinement techniques are used in many algorithms. This tool allows efficient computation of equivalence relations and is somehow dual to union-find algorithms. The goal of this paper is to propose a single routine to quickly implement all these already known algorithms and to solve a large class of potentially new problems. Our framework yields to a unique scheme for correctness proofs and complexity analysis. Various examples are presented to show the different ways of using this routine.

1 Introduction

A *partition* of a finite set E is a collection of disjoint subsets of E called *classes* whose union is E. *Refining* a partition consists in splitting its classes into smaller classes. Partition refinement techniques have been studied in four main papers [7, 15, 13, 6]. Hopcroft [7] may be the very first designer of such a technique. He used it in order to minimize the number of states of a deterministic

*LIRMM, Montpellier, France; email: (habib,paul)@lirmm.fr
†LIAFA, Paris, France; email: lavie@liafa.jussieu.fr

finite automaton. Spinrad [15] investigated the graph partitioning field with application in substitution decomposition and transitive orientation. Paige and Tarjan [13] used partition refinement techniques in three different applications: strings lexicographic sort, doubly lexicographic ordering of a boolean matrix and relational coarsest partition (the authors of [13] point out that this last problem is very close to deterministic finite automaton minimization). Habib, McConnell, Paul and Viennot [6] proposed new efficient algorithms based on partition refinement for the recognition of various classes of graphs and boolean matrices that have the consecutive one's property.

It turns out that partition refinement is used in many different area of computer science dealing with graphs, strings, boolean matrices or automata. Indeed, many articles rely on partition refinement subroutines even if they do not develop them. The goal of this paper is to propose a single routine to quickly implement all these algorithms and to solve a large class of problems. A sample of examples is presented to show different ways of using this routine: computing twins of a graph, lexicographic string sorting, consecutive ones test of boolean matrix and minimization of a deterministic automata. Just small basic procedures have to be adapted for each applications. This compilation allows similar correctness proofs and a general scheme for complexity issues.

Section 2 presents the partition refinement paradigm and the main routine in details. A classification of the different applications is proposed in Section 3. The last sections develop involved examples that make a powerful use of the partition refinement technique: automata minimization and consecutive ones test. Detailed proofs are only given for Hopcroft's Algorithm for automaton minimization. By the way, we show how this proof that has the reputation of being difficult becomes quite simple. The reader interested in detail proofs of the other algorithms is invited to consult the references given in this paper.

2 Partition Refinement

All the algorithms we are going to propose are based on the following routine that iteratively refines a partition of a set E according to a subset S of E called *pivot set*: each class $\mathcal{X}$ is replaced by $\mathcal{X} \cap S, \mathcal{X} - S$. Partitions will be implemented by sorted lists. Therefore our partitions are implicitly ordered. A partition Q is *compatible* with a partition P if every class of Q is included in a class of P and if the ordering in P respects the ordering in Q (i.e. if in P the class $\mathcal{X}$ is before the class $\mathcal{Y}$ then any class $\mathcal{X}' \subseteq \mathcal{X}$ of Q is before any class $\mathcal{Y}' \subseteq \mathcal{Y}$). Refining a partition produces a compatible partition. We say by extension that an ordering $x_1, \ldots, x_n$ of the elements of E is *compatible* with a partition L if the partition $(\{x_1\}, \ldots, \{x_n\})$ is compatible with L. Given a subset $S \subseteq E$, a partition L of E is said to be S-*stable* when no class properly overlap S. After the refinement step of L by S, L is S-stable: each class $\mathcal{X} \in L$ verifies either $\mathcal{X} \subseteq S$ or $\mathcal{X} \cap S = \emptyset$.

Algorithm 1: Procedure *PartitionRefinement(L)*

Input: a partition L of a set E in classes; L is ordered from left to right

Output: a refined partition $L = (\mathcal{X}_1, \ldots, \mathcal{X}_h)$ of E

begin

    pivots $= \emptyset$ is an empty stack of pivots (each pivot is associated to a subset of E via the procedure *PivotSet*)

    while *the LaunchPartition procedure does not break the loop* **do**

        LaunchPartition(L)

        while pivots $\neq \emptyset$ **do**

            pick a pivot p in *pivots*

            $S = PivotSet(P)$

            refinement step by S

                let M be the set of classes properly overlapping S (i.e. intersecting S and not included in S)

                let N be the set of classes included in S

                for each *class $\mathcal{X} \in M$* **do**

                    let $\mathcal{Y}$ be the members of $\mathcal{X}$ that are in S

                    remove $\mathcal{Y}$ from $\mathcal{X}$

                    if *InsertRight($\mathcal{X}, \mathcal{Y}, p, M, N$)* **then**

                        insert $\mathcal{Y}$ immediately on the right of $\mathcal{X}$

                    else insert $\mathcal{Y}$ immediately on the left of $\mathcal{X}$

                    AddPivot($\mathcal{X}, \mathcal{Y}$)

end

Algorithm 1 shows an implementation of the partition refinement routine that allows to simulate many algorithms. Depending on the use of the routine, four basic procedures, *PivotSet, LaunchPartition, InsertRight* and *AddPivot* have to be implemented. *PivotSet* should compute a pivot set (i.e. a subset of E) from some small information called *pivot*. In simple applications, the pivot is a pointer on some subset of E given in the input data and *PivotSet* returns this set. In others, the pivot set must be computed when needed. A stack *pivots* stores the pivots that may be used to refine the partition. There are two ways of adding pivots to the stack: with the procedure *AddPivot*, whenever a class is splited, and with the procedure *LaunchPartition*, whenever the stack is empty. The management of the pivots is critical for the complexity issues. The last procedure concerns the order of the classes in the partition. Whenever a class $\mathcal{X}$ is splited by a pivot set S, $\mathcal{X}$ is replaced either by $\mathcal{X} \setminus S, \mathcal{X} \cap S$ or $\mathcal{X} \cap S, \mathcal{X} \setminus S$ in L according to the result of the procedure *InsertRight*.

Partition Refinement Correctness Proof

All the algorithms of this paper can be proved with this invariant of the inner while loop of Algorithm 1. Properties A and B are defined for each application:

- **A:** *some property implying the existence of a solution compatible with the partition*
- **B:** *some property obtained when the partition is enough refined*

Invariant: *The partition L always verifies A and if L does not verify B   then some pivot in pivots will strictly refine L.*

To prove the correctness of the algorithms, we just have to prove that any refinement step maintains Property A and that enough pivots are added so that B is verified when there are no more pivots. After the inner while loop A and B are verified. The correctness of *LaunchPartition* is a key point of the proofs.

Complexity Issues

The refinement step can be performed in $O(|S|)$ using the following data structure. All the elements in E are stored in a doubly linked list. Each class consists in an interval of this list and is made of two pointers to its first element and to its last element. All the classes are stored in a doubly linked list L. The integers bounds of the intervals can also be maintained to allow constant computation of the cardinality of the classes and their relative positions in L. Each element keeps a pointer to its class. This data structure is illustrated by the above figure.

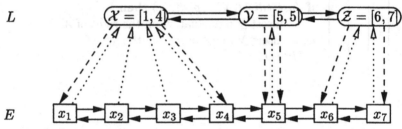

During the refinement step, each element in the pivot set S is simply removed from the list and inserted at the end of its new class. This allows to preserve the initial ordering of the vertices inside the classes when S is sorted according to this ordering. Notice that the classes are in fact totally ordered subsets of E. The set M of classes intersecting S is computed while scanning S. Afterwards M is traversed and the empty classes are un-splited, removed from M, and added to N (to unsplit a class, simply give it the bounds values of the associated new class which is deleted from L).

A bound on the overall execution time of the *PartitionRefinement* procedure is thus easily obtained from a bound on the sum of the sizes of the pivot sets and on the overall time spent in the procedures *PivotSet*, *LaunchPartition*, *InsertRight* and *AddPivot*. *InsertRight* will always consist in a test computable in constant time.

Bounding the complexity in time and space of the partition refinement routine resides in bounding the number of pivots added in the stack. All the pivot sets are either distinct parts of the input data or are computed from the elements of one of the two subclasses newly created by splitting a class. In the first case, pivots are known in advance and each one is used once yielding a linear time and space algorithm. In the second case, the Hopcroft "process the smaller half" strategy allows to get an $O(m \log n)$ time (where m is the size of the input) algorithm. Breaking the pivot set computation in two steps with *AddPivot* and *PivotSet* allows to bound the stack size to get a linear space algorithm. This discussion inspires a classification of the partition refinement applications presented in the paragraph "Pivot Rule" of the next section.

3 Classifications of Partition Refinement Applications

There are mainly three criteria to classify the partition refinement applications. The first one concerns the type of problem that the application solves and will be used as a main classification along the paper. The second classification concerns the way the pivots are chosen and the third one concerns the use of the procedure *LaunchPartition*.

Ordered and Unordered Partition Refinement

There are mainly two classes of problems that can be solved with partition refinement depending on the nature of the solution: an unordered partition (the classes of some equivalence relation have to be computed) or an ordering of the elements (sorting for example). In the first case, the order of the partition is not important with regard to the problem. In the second case the ordering is obtained by computing an ordered partition where all classes are singletons.

Unordered Partition Refinement:

Unordered refinement partition algorithms allow to compute congruence relations when the information "x, y do not belong to the same class" is easy to compute. In our framework, a pivot separates x and y. Dually, when the information "x, y belong to the same class" is easier to compute, a natural paradigm is to use union-find techniques [16].

Unordered partition refinement is used in twins computation, automaton minimization [7, 2], modular decomposition [15, 11, 5] and coarsest partition computation [13]. All these problems can be described as the computation of congruence classes of some congruence (or equivalence relation). This relation is the Nerode equivalence in automaton minimization; the coarsest partition problem is very similar and a similar relation can be defined. Two vertices of a graph are twins if they have same neighborhood, "being twins" is clearly an equivalence relation and its computation is given as an example of unordered partition refinement in Algorithm 2.

Algorithm 2: Computing twins

Input: the adjacency lists of a graph with vertex set V

Output: a partition $L = (\mathcal{X}_1, \ldots, \mathcal{X}_h)$ of the vertices in classes of twins

begin

    let L be the one element list (V)

    run *PartitionRefinement*(L) with the following definitions (the pivots are vertices)

end

Procedure *PivotSet(p)*
L return the adjacency list of the vertex *p*

Procedure *LaunchPartition(L)*
| add all the vertices in *V* to *pivots* at the first call to *LaunchPartition* and
L exit and return *L* at the second call

Procedure *InsertRight(X, Y, p, M, N)*
L return true

Procedure *AddPivot(X, Y)*
L do nothing

- **A:** *if the vertices u and v are twins, then they are in the same class*
- **B:** *L is stable for every neighborhood N(v) of a vertex v*

Invariant: *The partition L always verifies A and if L does not verify B then some pivot in pivots will strictly refine L.*

The correctness of Algorithm 2 is due to the above invariant. If two vertices are twins (they have the same neighbors), then no vertex can separate then. That proves property *A*. If property *B* is not true then there exists a vertex *w* that splits some class X of *L* that must still be in the stack *pivots*. Algorithm 2 is linear since each vertex is used once as pivot and the refinement step by a pivot *v* takes $O(|N(v)|)$ time where $N(v)$ is the adjacency list of *v*.

Ordered Partition Refinement: In that case, the partition is generally refined until all classes are singletons yielding a total ordering of the elements. Partition refinement can be seen here as a sort where the elements of a class are considered equal with respect to computation done up to that point.

Ordered partition refinement may also compute a congruence relation in addition. This is the case when the final classes are not necessarily singletons. The final classes are then the classes of elements that can be ordered independently and the final ordering associated to the partition is a solution.

Ordered partition refinement is used in transitive orientation [11], Lex-BFS (lexicographic breadth first search) [14], consecutive ones property testing [8, 6], string sorting [13] and sorting in general. We give a simple algorithm for lexicographic string sorting (see Algorithm 3) as example.

- **A:** *a lexicographic sort of the string is compatible with L*
- **B:** *two different strings cannot appear in the same class*

Invariant: *The partition L always verifies A and if L does not verify B then some pivot in pivots will strictly refine L.*

Algorithm 3: Lexicographic string sorting

Input: *n* strings $x_1, \ldots, x_n$
Output: a partition $L = (X_1, \ldots, X_h)$ of the strings
begin
| let *L* be the one element list $(\{x_1, \ldots, x_n\})$
| run *PartitionRefinement(L)* with the following definitions (the pivots are
| sets of strings)
end

Procedure *PivotSet(p)*
⌊ return p

Procedure *LaunchPartition(L)*
│ if this is the second call to *LaunchPartition* then exit and return L
│ precompute the non empty sets $S_{i,a}$ of string having a fixed letter a at a
│ fixed position i (create a couple $(-i, a)$ for each letter of any string and
│ radix sorting them)
│ add all these sets to *pivots* so that sets corresponding to largest ordered
⌊ pairs $(-i, a)$ will be picked first

Procedure *InsertRight($\mathcal{X}, \mathcal{Y}, p, M, N$)*
⌊ return true

Procedure *AddPivot($\mathcal{X}, \mathcal{Y}$)*
⌊ do nothing

This algorithm does not terminate with singletons but with classes of equal strings. The radix sort procedures takes $O(n + k + m)$ time where n is the number of strings, k is the size of the alphabet and m is the sum of the string sizes. The sum of the pivot set cardinalities is m and the partition refinement procedure thus takes $O(n+m)$ time. It is shown in [13] how to get a $O(n+k+m')$ algorithm by carefully computing the pivot sets during the partition refinement process where m' is the sum of the distinguishing prefix sizes ($m = m'$ in the worst case). Similarly a simulation of quick-sort with partition refinement is left to the reader.

Pivot Rule

As we have seen before, the partition refinement applications can use different strategy to choose the pivots (i.e. to add them in time in the stack *pivots*). There are two basic cases: the pivots are known in advance or they are computed during the algorithm depending on the current partition. The first case is simplier and as we previously mentioned leads generally to a linear time algorithm (*e.g.* twins computation, lexicographic string sorting, Lex-BFS).

The second case corresponds to more involved problems like automaton minimization, coarsest partition computation, consecutive ones test, modular decomposition and transitive orientation. For these problems, the Hopcroft's "process the smaller half" strategy allows to get very simple $O(m \log n)$ algorithms (m is the size of the data and n the number of elements). It is surprising to see that all these problems except automaton minimization and coarsest partition computation can be solved in linear time by finding a clever pivot choice [8, 6, 12] (Finding the pivots may be extremely complex, such as in linear transitive orientation for example [12]). This suggests that it should be possible to minimize a deterministic finite automaton in linear time.

The order in which the pivots are picked from the stack can be important or not. Sometimes only one pivot must be chosen among a set of possible ones. A rule has to be included in the procedure *LaunchPartition* to break this tie. This is the subject of the next classification.

Tie-break Rule

Some further differences come from the existence of a tie-break rule. In some algorithms the process is launched just once. It means that when the stack of pivot is empty and the algorithm ends (the inner while loop of Algorithm 1 is executed only once). This is the case in twins computation, lexicographic string sorting, automaton minimization and modular decomposition. The other algorithms have to be launched again until the resulting partition is a set of singletons. In that case, a tie-break rule has to be designed. This is the case for Lex-BFS, transitive orientation and the consecutive ones test. The way of breaking the tie is often a key part of the algorithm. Based on Lex-BFS, some interval graph recognition algorithms have been proposed by computing several successive Lex-BFS with more elaborate tie-break rules [4].

Considerations about the Representation of the Input

An interesting property of partition refinement is that refining a partition by a subset S or its complement $E - S$ is equivalent (the procedure *InsertRight* may have to be tuned slightly differently from one case to the other). This allows some degrees of freedom with the input data structure.

Algorithm 4: Clique Lex-BFS

Input: a graph $G = (V, \mathcal{E})$ given by its maximal cliques $C_1, \ldots, C_k$

Output: a lex-BFS ordering $L = C'_1, \ldots, C'_k$ of the cliques

begin

    let L be the one element list $(\{C_1, \ldots, C_k\})$

    precompute the set of cliques containing each vertex; and set $i = |V|$

    run *PartitionRefinement*(L) with the following definitions (the pivots are vertices and are all set as "unused" at the beginning)

end

Procedure *PivotSet(p)*

    number p by i and set i to $i - 1$

    return the list of cliques containing the vertex p

Procedure *LaunchPartition(L)*

    if *there is only singleton classes in* L **then** exit and return L

    let C be a clique in the rightmost class $\mathcal{X}$ with cliques containing unnumbered vertices

    if $\mathcal{X}$ is not a singleton then replace $\mathcal{X}$ by $\mathcal{X} \setminus \{C\}, \{C\}$ in L

    add all the unused vertices in C to *pivots* and set them as "used"

Procedure *InsertRight($\mathcal{X}, \mathcal{Y}, p, M, N$)*

    return true

Procedure *AddPivot($\mathcal{X}, \mathcal{Y}$)*

    do nothing

In graph algorithms, E is often the vertex set and S is the neighborhood of a vertex p. In that case, it is possible to work on a graph using its complementary as a data structure that represents it. In other words, we can run the partition refining routine on the complement of a graph without computing it, using only the edges of the graph itself. This nice property was used in [11] to recognize permutation graphs which are the comparability graphs such that the complementary graph is also a comparability graph and in [6] to recognize co-chordal and co-interval graphs. This idea is further developed in [5] where it is proposed to represent a graph by giving for each vertex either the list of its neighbors or the list of its non neighbors. Partition refinement allows to compute with such a representation as easily as with a classical one. Algorithm 4 is an adaptation of an existing partition refinement algorithm (namely Lex-BFS) when its input is given in an other form.

Lex-BFS [14] is a partition refinement algorithm that computes a special ordering of the vertices of a graph called a Lex-BFS ordering. When the input is chordal, this ordering allows to compute in linear time the maximal cliques of the input (a clique is a set of vertices inducing a complete graph). Moreover, this ordering induces an ordering of the maximal cliques called Clique Lex-BFS ordering. Algorithm 4 allows to compute a Clique Lex-BFS when the maximal cliques are directly given as input. This algorithm is useful for the consecutive ones test algorithm presented in [6]. Note that we cannot compute a classical representation of the graph in order to run the classical Lex-BFS on it for complexity reasons since its size may be significantly greater.

4 An Unordered Partition Refinement Problem

Hopcroft's Algorithm for Deterministic Finite Automata Minimization Revisited

We now introduce a partition refinement version of Hopcroft's Algorithm [7]. The hard part of automata minimization is to compute the classes of the Nerode equivalence. Two states q and q' of an automata are Nerode equivalent if and only if $q.w = q'w$ for every word w ($q.w$ denotes the state reached by reading w when the automata is in state q).

Algorithm 5 shows how to simulate the Hopcroft algorithm for computing these classes with a call to our partition refinement procedure. A partition of the states set of the automata is refined according to pivot sets of the form $a^{-1}\mathcal{X}$ where a is a letter, $\mathcal{X}$ is a class of the current partition and $a^{-1}S$ denotes the set of the states q such that $q.a \in S$ (for any letter a and any state subset S).

Algorithm 5 runs in time $O(nk \log n)$ where n is the number of states and k the number of letters since each set $a^{-1}q$ is traversed at most $\log n$ times.

The correctness of Algorithm 5 comes from the fact that the partition L always verifies the following invariant:

  - **A:** *if two states p and q are Nerode equivalent then they are in the same class*
  - **B:** *L is stable for $a^{-1}\mathcal{X}$ for every class $\mathcal{X} \in L$ and every letter $a \in A$*

Invariant: *The partition L always verifies A and if L does not verify B then some pivot in pivots will strictly refine L.*

The proof of Hopcroft's Algorithm has the reputation of being quite intricate. We now give the complete proof of Algorithm 5 to show how our formalism makes it simple.

Algorithm 5: Deterministic Finite Automata Minimization

Input: a complete accessible deterministic automata (Q, i, T)

Output: a partition $L = (\mathcal{X}_1, \ldots, \mathcal{X}_h)$ of the states into Nerode classes

begin

> let L be the one element list (Q)
> precompute for each state q and each letter a the set $a^{-1}q$ of the states p such that there exists a transition $q = p.a$
> run *PartitionRefinement*(L) with the following definitions (the pivots are couples (class, letter))

end

Procedure *PivotSet(p)*

> p is the couple $(\mathcal{X}, a)$
> compute $a^{-1}\mathcal{X}$ by merging the sets $a^{-1}q$ for $q \in \mathcal{X}$ and return this set

Procedure *LaunchPartition(L)*

> if this is the second call to *LaunchPartition*, exit and return L
> remove T from Q and insert it as a new class in L
> *AddPivot*$(Q - T, T)$

Procedure *InsertRight*$(\mathcal{X}, \mathcal{Y}, p, M, N)$

> return true

Procedure *AddPivot*$(\mathcal{X}, \mathcal{Y})$

> let $\mathcal{Z}$ be the smallest class between $\mathcal{X}$ and $\mathcal{Y}$
> **for each** $a \in A$ **do**
>> **if** $(\mathcal{X}, a)$ *is already in* pivots*(i.e* $(\mathcal{X}, a)$ *has been added to* pivots *before* $(\mathcal{Y}, a)$ *was removed from it)* **then** add $(\mathcal{Y}, a)$ to *pivots*
>> **else** add $(\mathcal{Z}, a)$ to *pivots*

Proof: We say that a word w *separates* two states q and q' when $q.w \in T$ and $q'.w \notin T$ or when $q.w \notin T$ and $q'w \in T$. Two states are Nerode equivalent when no word separates them. The empty word is denoted ε.

Let us first show the conservation of Property A. Non terminal and terminal states are not Nerode Equivalent since ε separates them. A is thus verified after the first call to *LaunchPartition*. Suppose that A is true before an iteration of the inner while loop, we show that it is still after the refinement step by a pivot set $a^{-1}\mathcal{X}$ where a is some letter and $\mathcal{X}$ some class of the partition. Suppose by contradiction that two Nerode equivalent states q and q' are splited apart in two different classes. This means that $q.a$ and $q'.a$ appear in different classes. This contradicts the induction hypothesis.

We now show that if the property B is false then some pivot in *pivots* will strictly refine L. Suppose that L is not stable for $a^{-1}\mathcal{Z}$ for some letter a and some class $\mathcal{Z}$. There must exist two states q and q' in the same class such that $q.a \in \mathcal{Z}$ and $q'.a \notin \mathcal{Z}$. Consider the first time $q.a$ and $q'.a$ have been splited apart in two different classes. The smallest one contained either $q.a$ or $q'.a$ and has been added to *pivots* with the letter a. This implies that either the class of $q.a$ or the class of $q'.a$ appears in *pivots* with the letter a. This ordered pair will produce a pivot set containing either q or q' and pivoting on it will strictly refine the partition. Notice that whenever a class $\mathcal{Z}$ is splited in two classes $\mathcal{X}$ and $\mathcal{Y}$, if $(\mathcal{Z}, a)$ was in *pivots* then $(\mathcal{X}, a)$ and $(\mathcal{Y}, a)$ are in *pivots* after the call to *AddPivot*.

Let us finally prove that the conservation of the invariant implies that the final partition is made of the Nerode classes. We just have to prove that non Nerode equivalent states cannot be in the same class of the final partition. Suppose by contradiction that two states q and q' are in the same class but there exists a word w separating them. Let u be the longest prefix of w such that $q.u$ and $q'.u$ are in the same class of the final partition. We have $u \neq w$ since the final partition is a refinement of $(T, Q - T)$ implying that $q.w$ and $q'.w$ cannot be in the same class. Let a be the letter following u in w. Let $\mathcal{X}$ be the class of $q.ua$. The property B is then false since the partition is not stable for $a^{-1}\mathcal{X}$: contradiction since *pivots* is empty at the end of the algorithm. $\qquad\square$

5 An Ordered Partition Refinement Problem

Consecutive Ones Property Testing

A boolean matrix has the consecutive ones property if its columns can be permuted such that in each row the one entries occur consecutively. Such a permutation will be called a *consecutive permutation*. The problem consists in testing wether a given boolean matrix has the consecutive ones property and to compute a consecutive permutation if there exists one.

The first linear time algorithm for this problem [3] used the PQ-trees, a complicated data structure. A simpler algorithm was also presented in [8]. In [6] a new linear algorithm avoiding PQ-trees is proposed. A consecutive ones test can be used for interval graph recognition but it is a more general problem. A graph G is an interval graph iff its maximal cliques can be ordered such that the maximal cliques containing a given vertex appears consecutively. Thus if we represent an interval graph by its incidence vertex-clique matrix M, such an ordering of the maximal clique is exactly a consecutive permutation of the columns of M. Since Lex-BFS can be adapted for a clique representation of the graph (see algorithm 4), this correspondence allows the same adaptation for a vertex-clique matrix representation.

Algorithm 6 shows how to solve the consecutive ones problem thanks to a call to the partition refinement routine. Here, a partition of the columns is refined by pivoting on the rows where the pivot set associated to a row p is the set of columns containing a one entry at row p. The structure of the algorithm is similar to the algorithm proposed in [6] but here, the strategy for choosing the pivots is inspired from the Hopcroft Algorithm rule and is much more simple.

The result is an extremely simple $O(n + n' + m \log n)$ algorithm for testing the consecutive ones property (m is the number of one entries, n is the number of columns, and n' the number of rows).

The input is given by the coordinates of the m one entries of the matrix (the other entries are zeros). An efficient algorithm for consecutive ones test must have a tight bound on m rather than nn'. All zeros columns and rows can easily be treated separately. We say that a column C contains a row r if the corresponding entry of the matrix M is one ($M(r, C) = 1$). This allows to consider the columns as sets of rows. We can also associate to each row r the set $C(r)$ of columns containing them. These sets are easily computed in linear time by radix sorting the coordinates of the one entries. Those which are not empty are given as input to Algorithm 6. The input matrix verifies the consecutive ones property if the computed permutation of the column is consecutive (this can be tested in linear time by scanning each set $C(r)$).

Algorithm 6: Consecutive ones testing

Input: a boolean matrix M with no all zeros column and no all zero row

Output: a consecutive 1's permutation of the columns if there exists one

begin

> compute a Lex-BFS ordering $C_1 < \cdots < C_k$ of the columns of M
> let L be the one element list $(\{C_1, \ldots, C_k\})$
> bucket sort all the sets $C(r)$ of columns containing a given row r according to the Lex-BFS ordering of the columns
> run $PartitionRefinement(L)$ with the following definitions (the pivots are rows and are all set as "unused" at the beginning)

end

Procedure $PivotSet(p)$

> **if** *all the columns in $C(p)$ are in the same class of L* **then**
>> set p as "unused" and return $\emptyset$
> **else** return $C(p)$

Procedure $LaunchPartition(L)$

> **if** *there is only singleton classes in L* **then** exit and return L
> let $\mathcal{X}$ be a non singleton class in L and let C be the smallest column in $\mathcal{X}$ according to the Lex-BFS ordering of the columns
> replace $\mathcal{X}$ by $\mathcal{X} \setminus \{C\}, \{C\}$ in L
> add the "unused" rows in C to *pivots* and set them as "used"

Procedure $InsertRight(\mathcal{X}, \mathcal{Y}, p, M, N)$

> let $\mathcal{Z}$ be a class distinct from $\mathcal{X}$ in $M \cup N$
> return ($\mathcal{Y}$ is somewhere on the left of $\mathcal{Z}$)

Procedure $AddPivot(\mathcal{X}, \mathcal{Y})$

> let $\mathcal{Z}$ be the smallest class between $\mathcal{X}$ and $\mathcal{Y}$
> add the "unused" rows in the union of the columns in $\mathcal{Z}$ to *pivots* and set them as "used"

- **A**: *if M has the consecutive ones property, then there exist a consecutive permutation compatible with the ordered partition L*
- **B**: *the set of "used" rows is the same for every column of a given class $\mathcal{X}$ of L*

Invariant: *The partition L always verifies A and if L does not verify B then some pivot in* pivots *will strictly refine L.*

We now assume that M has the consecutive ones property. Let us remark that Property A implies that for any row p, the set S of columns containing p as one entry appears in consecutive classes of L. To give an idea of the proof, we just mention of properties proved in [6] that shows the correctness of the procedure *LaunchPartition*. First, there exist a consecutive permutation that ends with the last column numbered by a Lex-BFS ordering σ of the columns of M [9]. Then the first call to *LaunchPartition* preserve the invariant. When B is true, the authors of [6] proves that for any class $\mathcal{X}$ of L, σ induces a Lex-BFS ordering of the sub-matrices induced by the columns of $\mathcal{X}$. In that case, the refinement process can be launch again by the procedure *LaunchPartition*.

The complexity of Algorithm 6 is $O(m \log n)$. The reader should note that this complexity can be improved to linear time. In the presented algorithm the bottleneck is the choice of the pivot. It is shown in [6], that a clever choice of the pivot can be done using a special tree structure. By applying the Hopcroft's "process the smaller half" strategy we have obtained an extremely simple algorithm for testing the consecutive ones property.

6 Conclusions

Paige and Tarjan [13] conclude their introduction by "Although these applications are very different, the similarities among them are compelling and suggest further investigation of the underlying principles." We hope that this paper provides a step in this direction.

Acknowledgement: The authors wish to thank J.E. Pin for drawing their attention to the ties between automata minimization and partition refinement.

References

[1] A.V. Aho, J.E. Hopcroft, and J.D. Ullman. *Design and analysis of computer algortihms.* Addison-Wesley, 1974.

[2] J. Amilhastre. Représentation par automates de l'ensemble des solutions d'un problème de satisfaction de contraintes. Technical Report 94056, LIRMM, 1994.

[3] K.S. Booth and G.S. Lueker. Testing for the consecutive ones property, interval graphs and graph planarity using pq-tree algorithms. *J. Comput. and Syst. Sciences*, 13:335–379, 1976.

[4] D.G Corneil, S. Olariu, and L. Stewart. The ultimate interval graph recognition algorithm. Extended abstract, 1997.

[5] E. Dahlhaus, J. Gustedt, and R.M. McConnell. Efficient and practical modular decomposition. In *Proc. of SODA*, 1997.

[6] M. Habib, R. McConnell, C. Paul, and L. Viennot. Lex-bfs and partition refinement, with applications to transitive orientation, interval graph recognition and consecutive ones testing. *Theoritical Computer Science*, 1997. to appear.

[7] J.E. Hopcropft. A nlogn algorithm for minizing states in a finite automaton. *Theory of Machine and Computations*, pages 189–196, 1971.

[8] W.L. Hsu. A simple test for the consecutive ones property. In *LNCS 650*, pages 459–468, 1992.

[9] N. Korte and R. Möhring. An incremental linear-time algorithm for recognizing interval graphs. *SIAM J. of Comp.*, 18:68–81, 1989.

[10] A. Lubiw. Doubly lexical orderings of matrices. *SIAM Journ. of Algebraic Disc. Meth.*, 17:854–879, 1987.

[11] R.M. McConnell and J.P. Spinrad. Linear-time modular decomposition and efficient transitive orientation of comparability graphs. In *Proc. of SODA*, pages 536–545, 1994.

[12] R.M. McConnell and J.P. Spinrad. Linear-time modular decomposition and efficient transitive orientation of undirected graphs. In *Proc. of SODA*, 1997.

[13] R. Paige and R.E. Tarjan. Three partition refinement algorithms. *SIAM Journ. of Comput.*, 16(6):973–989, 1987.

[14] Donald J. Rose, R. Endre Tarjan, and George S. Leuker. Algorithmic aspects of vertex elimination on graphs. *SIAM J. of Comp.*, 5(2):266–283, June 1976.

[15] J.P. Spinrad. Graph partitioning. preprint, 1986.

[16] R.E. Tarjan. Efficiency of a good but not linear set union algorithm. *J. of ACM*, 22:215–225, 1975.

Simplifying the Modal
Mu-Calculus Alternation Hierarchy

J. C. Bradfield

Laboratory for Foundations of Computer Science,
Department of Computer Science, University of Edinburgh,
Edinburgh, United Kingdom, EH9 3JZ (email: jcb@dcs.ed.ac.uk)

Abstract: In [Bra96], the strictness of the modal mu-calculus alternation hierarchy was shown by transferring a hierarchy from arithmetic; the latter was a corollary of a deep and highly technical analysis of [Lub93]. In this paper, we show that the alternation hierarchy in arithmetic can be established by entirely elementary means; further, simple examples of strict alternation depth n formulae can be constructed, which in turn give very simple examples to separate the modal hierarchy. In addition, the winning strategy formulae of parity games are shown to be such examples.

1 Introduction

The modal mu-calculus, or Hennessy–Milner logic with fixpoints, is a popular logic for expressing temporal properties of systems. It was first studied by Kozen in [Koz83], and since then there has been much work on both theoretical and practical aspects of the logic. The feature of the logic that gives it both its simplicity and its power is that it is possible to have mutually dependent minimal and maximal fixpoint operators. This makes it simple, as the fixpoints are the only non-first-order operators, and powerful, as by such nesting one can express complex properties such as 'infinitely often' and fairness. A measure of the complexity of a formula is the *alternation depth*, that is, the number of alternating blocks of minimal/maximal fixpoints. Formulae of alternation depth higher than 2 are notoriously hard to understand, and in practice one rarely produces them—not least because they are so hard to understand. For some years, it was not even known whether formulae of high alternation depth were necessary, that is, whether the *alternation hierarchy* was indeed a strict hierarchy of expressive power—a problem with several interesting ramifications, as well as its intrinsic interest. In 1996 the strictness of the hierarchy was established by the present author [Bra96], and independently by Lenzi [Len96]. The proof technique in [Bra96] relied on the existence of a similar fixpoint alternation hierarchy in arithmetic with fixpoints (mu-arithmetic), which follows from a very deep and technical recursion-theoretic study of mu-arithmetic by Robert Lubarsky [Lub93]. Thus the proof was not self-contained; furthermore, it was apparently not feasible to exhibit examples of strict alternation depth n formulae, as the strict mu-arithmetic formulae of [Lub93] are not constructible with any reasonable amounts of paper, ink and patience—there is only a high level description of the complex coding required.

The purpose of this paper is to mend these drawbacks of [Bra96]. Our first result is an elementary proof of the alternation hierarchy in mu-arithmetic, thus removing the reliance on [Lub93]. Furthermore, the proof constructs very simple examples of strict level n formulae of mu-arithmetic; and by using a simplified version of the techniques of [Bra96], we are able to construct even simpler examples of strict level n modal mu-calculus formulae. These examples are of just the form that one expects, if one is a modal mu-calculus hacker. In addition, we can also show that the formulae defining the existence of a winning strategy in a parity game are examples of strict formulae—indeed, a referee has observed that this can be shown already from [Bra96].

2 Preliminaries

2.1 Modal Mu-Calculus

We assume some familiarity with the modal mu-calculus, so in this section we give brief definitions to establish notations and conventions. Expository material on the modal mu-calculus may be found in [Bra91,Sti91].

The modal mu-calculus, with respect to some countable set $\mathscr{L}$ of *labels*, has formulae Φ defined inductively thus: variables Z and the boolean constants tt, ff are formulae; if Φ_1 and Φ_2 are formulae, so are $\Phi_1 \vee \Phi_2$ and $\Phi_1 \wedge \Phi_2$; if Φ is a formula and l a label, then $[l]\Phi$ and $\langle l \rangle \Phi$ are formulae; and if Φ is a formula and Z a variable, then $\mu Z.\,\Phi$ and $\nu Z.\,\Phi$ are formulae.

Note that we adopt the convention that the scope of the binding operators μ and ν extends as far as possible. For consistency, we also apply this convention to the $\forall$ and $\exists$ of first-order logic, writing $\forall x.\,(\exists y.\,P) \vee Q$ rather than the logicians' traditional $\forall x\,[\exists y\,[P] \vee Q]$.

Observe that negation is not in the language, but any closed mu-formula can be negated by using the usual De Morgan dualities—μ and ν are dual by $\neg \mu Z.\,\Phi(Z) = \nu Z.\,\neg\Phi(\neg Z)$. Where necessary, we shall assume that free variables can be negated just by adjusting the valuation. We shall use $\Rightarrow$ etc. freely, though we must ensure that bound variables only occur positively.

We use the symbol $\wp$ to mean 'μ or ν as appropriate'.

Given a labelled transition system $\mathscr{T} = (\mathscr{S}, \mathscr{L}, \longrightarrow)$, where $\mathscr{S}$ is a set of states, $\mathscr{L}$ a set of labels, and $\longrightarrow \subseteq \mathscr{S} \times \mathscr{L} \times \mathscr{S}$ is the transition relation (we write $s \xrightarrow{l} s'$), and given also a valuation $\mathscr{V}$ assigning subsets of $\mathscr{S}$ to variables, the denotation $\|\Phi\|^{\mathscr{T}}_{\mathscr{V}} \subseteq \mathscr{S}$ of a mu-calculus formula Φ is defined in the obvious way for the variables and booleans, for the modalities by

$$\|[l]\Phi\|^{\mathscr{T}}_{\mathscr{V}} = \{\, s \mid \forall s'.\, s \xrightarrow{l} s' \Rightarrow s' \in \|\Phi\|^{\mathscr{T}}_{\mathscr{V}} \,\}$$

$$\|\langle l \rangle \Phi\|^{\mathscr{T}}_{\mathscr{V}} = \{\, s \mid \exists s'.\, s \xrightarrow{l} s' \wedge s' \in \|\Phi\|^{\mathscr{T}}_{\mathscr{V}} \,\},$$

and for the fixpoints by

$$\|\mu Z. \Phi\|_{\mathcal{V}}^{\mathcal{I}} = \bigcap \{ S \subseteq \mathcal{S} \mid \|\Phi\|_{\mathcal{V}[Z:=S]}^{\mathcal{I}} \subseteq S \}$$
$$\|\nu Z. \Phi\|_{\mathcal{V}}^{\mathcal{I}} = \bigcup \{ S \subseteq \mathcal{S} \mid S \subseteq \|\Phi\|_{\mathcal{V}[Z:=S]}^{\mathcal{I}} \} .$$

It is often useful to think of μZ and νZ as meaning respectively finite and infinite looping from Z back to μZ (νZ) as one 'follows a path of the system through the formula'. Examples of properties expressible by the mu-calculus are 'always (on a-paths) P', as $\nu Z. P \wedge [a]Z$, 'eventually (on a-paths) P', as $\mu Z. P \vee \langle a \rangle Z$, and 'there is an $\{a, b\}$-path along which b happens infinitely often', as $\nu Y. \mu Z. \langle b \rangle Y \vee \langle a \rangle Z$. (For the latter, we can loop around Y for ever, but each internal loop round Z must terminate.)

There are several notions of alternation. The naive notion is simply to count syntactic alternations of μ and ν, resulting in the following definition: A formula Φ is said to be in the classes $\Sigma_0^{S\mu}$ and $\Pi_0^{S\mu}$ iff it contains no fixpoint operators ('S' for 'simple' or 'syntactic'). The class $\Sigma_{n+1}^{S\mu}$ is the least class containing $\Sigma_n^{S\mu} \cup \Pi_n^{S\mu}$ and closed under the following operations: (i) application of the boolean and modal combinators; (ii) the formation of $\mu Z. \Phi$, where $\Phi \in \Sigma_{n+1}^{S\mu}$. Dually, to form the class $\Pi_{n+1}^{S\mu}$, take $\Sigma_n^{S\mu} \cup \Pi_n^{S\mu}$, and close under (i) boolean and modal combinators, (ii) $\nu Z. \Phi$, for $\Phi \in \Pi_{n+1}^{S\mu}$. Thus the examples above are in $\Pi_1^{S\mu}$, $\Sigma_1^{S\mu}$, and $\Pi_2^{S\mu}$ (but not $\Sigma_2^{S\mu}$) respectively. We shall say a formula is *strict* $\Sigma_n^{S\mu}$ if it is in $\Sigma_n^{S\mu} - \Pi_n^{S\mu}$.

For the modal mu-calculus, it is usual to define stronger notions of alternation [EmL86,Niw86], which capture the true interdependency of alternating fixpoints, rather than just their syntactic position. In [Bra96], the analysis is carried out for the strongest notion, that of [Niw86], giving the classes called $\Sigma_n^{N\mu}$ in [Bra96], as well as for the simple notion. In this paper, we shall not worry about the distinction, as the arguments apply whichever notion is used. Hence we shall just write Σ_n^{μ}.

2.2 The Arithmetic Mu-Calculus

In [Lub93] Robert Lubarsky studies the logic given by adding fixpoint constructors to first-order arithmetic. Precisely, the logic ('mu-arithmetic' for short) has as basic symbols the following: function symbols f, g, h; predicate symbols P, Q, R; first-order variables x, y, z; set variables X, Y, Z; and the symbols $\vee, \wedge, \exists, \forall, \mu, \nu, \neg, \in$. As with the modal mu-calculus, $\neg$ can be pushed inwards to apply only to atomic formulae, by De Morgan duality.

The language has expressions of three kinds, individual terms, set terms, and formulae. The individual terms comprise the usual terms of first-order logic. The set terms comprise set variables and expressions $\mu(x, X). \phi$ and $\nu(x, X). \phi$, where X occurs positively in ϕ. Here μ binds both an individual variable and a set variable; henceforth we shall write just $\mu X. \phi$, and assume that the individual variable is the lower-case of the set variable. The formulae are built by the usual first-order construction, together with the rule that if τ is an individual term and Ξ is a set term, then $\tau \in \Xi$ is a formula.

This language is interpreted over the structure $\mathbb{N}$ of first-order arithmetic with all recursive functions and predicates—in particular, let $\langle -, - \rangle$, $(-)_0$ and $(-)_1$ be standard pairing and unpairing functions. The semantics of the first-order connectives is as usual; $\tau \in \Xi$ is interpreted naturally; and the set term $\mu X. \phi(x, X)$ is interpreted as the least fixpoint of the functional $\mathbf{X} \mapsto \{ m \in \mathbb{N} \mid \phi(m, \mathbf{X}) \}$ (where $\mathbf{X} \subseteq \mathbb{N}$).

The simplest examples of mu-arithmetic just use least fixpoints to represent an inductive definition. For example, $\mu X. x = 0 \vee (x > 1 \wedge (x - 2) \in X)$ is the set of even numbers. Of course, the even numbers are also the complement of the odd numbers: the odd numbers are defined by $\mu X. x = 1 \vee (x > 1 \wedge (x - 2) \in X)$, so by negating we can express the even numbers as a maximal fixpoint $\nu X. x \neq 1 \wedge (x > 1 \Rightarrow (x - 2) \in X)$. To produce natural examples involving alternating fixpoints is rather difficult, since even one induction is already very powerful, and most natural mathematical objects are simple.

One can define the syntactic alternation classes for arithmetic just as for the modal mu-calculus: First-order formulae are Σ_0^μ and Π_0^μ, as are set variables. The Σ_{n+1}^μ formulae and set terms are formed from the $\Sigma_n^\mu \cup \Pi_n^\mu$ formulae and set terms by closing under (i) the first-order connectives and (ii) forming $\mu X. \phi$ for $\phi \in \Sigma_{n+1}^\mu$.

A crucial lemma is the following:

Lemma 1 [Lub93,Bra96] *A Σ_n^μ formula of mu-arithmetic can be put into a normal form of the following shape:*

$$\tau_n \in \mu X_n . \tau_{n-1} \in \nu X_{n-1} . \tau_{n-2} \in \mu X_{n-2} . \ldots \tau_1 \in \mu X_1 . \phi$$

where ϕ is first-order—that is, a string of alternating fixpoint quantifiers, and a first-order body. □

(See [Bra96] for detailed definitions and proof.)

The analysis of [Lub93] provides the following

Theorem 2 [Lub93] *The hierarchy of the sets of integers definable by Σ_n^μ formulae of the arithmetic mu-calculus is a strict hierarchy.* □

2.3 Summary of [Bra96]

The results of this paper require the results and proof techniques of [Bra96], so we now give a summary of these, skipping the fine details.

We define a *recursively presented transition system (r.p.t.s.)* to be a labelled transition system $(\mathcal{S}, \mathcal{L}, \longrightarrow)$ such that $\mathcal{S}$ is (recursively codable as) a recursive set of integers, $\mathcal{L}$ likewise, and $\longrightarrow$ is recursive. Henceforth we consider only recursively presented transition systems, with recursive valuations for the free variables. We have the following theorem, which is proved by a trivial translation of the semantics of the modal mu-calculus into mu-arithmetic:

Theorem 3 [Bra96] *For a modal mu-calculus formula $\Phi \in \Sigma_n^\mu$, the denotation $\|\Phi\|$ in any r.p.t.s. is a Σ_n^μ definable set of integers.* □

Theorem 4 [Bra96] *Let $\phi(z)$ be a Σ_n^μ formula of mu-arithmetic. There is a r.p.t.s. $\mathcal{T}$ with recursive valuation $\mathcal{V}$ and a Σ_n^μ formula Φ of the modal mu-calculus such that $\phi((s)_0)$ iff $s \in \|\Phi\|_{\mathcal{V}}^{\mathcal{T}}$. (Thus if ϕ is not Σ_{n-1}^μ-definable, neither is $\|\Phi\|$.)* □

This theorem is not inherently difficult; it is established by coding the evaluation of mu-arithmetic formulae into a r.p.t.s. and a modal mu-calculus formula, in such a way that arithmetic computation is handled by the transitions of the system, and the fixpoints of ϕ are handled by the fixpoints of Φ. The proof is then a fairly straightforward induction. In this paper, we shall see a simplified version of this technique.

These two theorems establish the modal alternation hierarchy: use Theorem 4 to code an arithmetic strict Σ_n^μ set of integers by a strict Σ_n^μ modal mu-formula Φ on a r.p.t.s. $\mathcal{T}$; by Theorem 3, no Σ_{n-1}^μ modal formula can have the same denotation in $\mathcal{T}$, and so no Σ_{n-1}^μ modal formula is logically equivalent to Φ.

3 A Simple Proof of the Alternation Hierarchy in Mu-Arithmetic

The first result of this paper is to observe that the alternation hierarchy theorem in mu-arithmetic can be proved simply along the lines of the proof of the strictness of the Kleene arithmetic hierarchy. The technique is to show that the truth of Σ_n^μ formulae can itself be expressed by a Σ_n^μ formula, and to use a diagonalization argument to show that this formula cannot be equivalent to any Π_n^μ formula.

Firstly, take a suitable Gödel numbering of mu-arithmetic. We consider only formulae without free set variables; wlog, we may assume that all encoded formulae are in normal form, and are normalized so that the free individual variables are $z_0, \ldots, z_k$, the first-order quantifiers bind $z_{k+1}, \ldots$, and for a formula of alternation depth n, the fixpoint variables are $X_n, \ldots, X_1$, with associated individual variables $x_n, \ldots, x_1$. We use sans-serif type to indicate that the variable is being seen as part of an encoded object-level formula; normal italic type indicates a meta-level variable. We use corner quotes to denote the Gödel coding. We also need coded *assignments* which map an encoded variable to a value: we write $[v/z]$ for the assignment that maps z (strictly, the code $\ulcorner z \urcorner$) to the integer v, and $a[v/z]$ for the updating of a by $[v/z]$. We use double quotes to indicate the appropriate meta-language formalization of the informal statement inside the quotes.

Now suppose that $\mathrm{Sat}_n(x, y)$ is a formula of mu-arithmetic expressing the truth of Σ_n^μ formulae, so that if ϕ is a formula and a an assignment of values $\vec{v}$ to the free variables $\vec{z}$ of ϕ, then $\mathrm{Sat}_n(\ulcorner\phi\urcorner, a)$ is true just in case $\phi(\vec{v}/\vec{z})$ is true. We have the

Lemma 5 $\mathrm{Sat}_n(z_0, [z_0/z_0])$ *is not equivalent to any Π_n^μ formula.*

Proof. The proof is exactly as for the arithmetical hierarchy. Suppose the contrary, i.e. that $\neg\mathrm{Sat}_n(z_0, [z_0/z_0])$ is equivalent to some Σ_n^μ formula $\theta(z_0)$. Then we have $\theta(\ulcorner\theta\urcorner)$ iff $\neg\mathrm{Sat}_n(\ulcorner\theta\urcorner, [\ulcorner\theta\urcorner/z_0])$ iff $\neg\theta(\ulcorner\theta\urcorner)$, which is a contradiction. □

It remains to show that Sat_n exists and is indeed a Σ_n^μ formula.

Theorem 6 Sat_n *is a Σ_n^μ formula of mu-arithmetic, for $n > 0$.*

Proof. We start by constructing Sat_0, truth in first-order arithmetic, both as a Σ_1^μ formula and as a Π_1^μ formula. $\text{Sat}_0(x, y)$ is defined as:

$$\langle x, y\rangle \in \mu(w, W). \text{``}(w)_0 = \ulcorner P(\tau)\urcorner \text{ and } \text{pred}(\ulcorner P\urcorner, \text{eval}(\ulcorner \tau\urcorner, (w)_1))\text{''}$$

$$\vee \text{``}(w)_0 = \ulcorner \phi_1 \wedge \phi_2 \urcorner \text{ and } ((\langle \ulcorner \phi_1 \urcorner, (w)_1\rangle \in W \wedge \langle \ulcorner \phi_2 \urcorner, (w)_1\rangle \in W)\text{''}$$

$$\vee \text{``}(w)_0 = \ulcorner \phi_1 \vee \phi_2 \urcorner \text{ and } ((\langle \ulcorner \phi_1 \urcorner, (w)_1\rangle \in W \vee \langle \ulcorner \phi_2 \urcorner, (w)_1\rangle \in W)\text{''}$$

$$\vee \text{``}(w)_0 = \ulcorner \exists z_i. \phi_1 \urcorner \text{ and } \exists v. \langle \ulcorner \phi_1 \urcorner, (w)_1[v/z_i]\rangle \in W\text{''}$$

$$\vee \text{``}(w)_0 = \ulcorner \forall z_i. \phi_1 \urcorner \text{ and } \forall v. \langle \ulcorner \phi_1 \urcorner, (w)_1[v/z_i]\rangle \in W\text{''}$$

where $\text{eval}(t, a)$ is the recursive function which evaluates a coded term $t = \ulcorner \tau\urcorner$ under the variable assignment a, and $\text{pred}(p, x)$ is the computable predicate which is true if the value x satisfies the predicate coded by $p = \ulcorner P\urcorner$.

We have here skipped the details of the coding, which are standard. For example, if we look in more detail at the clause for $\forall$, it actually says:

$$f((w)_0) = \ulcorner \forall \urcorner \wedge \forall v. \langle g((w)_0), h((w)_1, v, g'((w)_0))\rangle \in W$$

where f extracts the top-level connective of a coded formula, g extracts the body of a $\forall$ formula and g' extracts the bound variable, and $h(a, v, z)$ takes the variable assignment a and updates the variable whose code is z by the value v. The fact that these functions f, g, h are recursive is obvious, and since we allow ourselves all recursive functions as primitives, that is sufficient; but explicit definitions in standard arithmetic may be found in references such as [Kay91].

It is clear that this fixpoint formula simply encodes directly the definition of truth in arithmetic. The formula is Σ_1^μ, but since the encoded recursive function terminates on all arguments—it is just a definition by induction on the structure of formulae—it does not matter whether we use a minimal or maximal fixpoint to achieve the recursion. Thus we may also obtain Sat_0 as a Π_1^μ formula.

In order to encode within mu-arithmetic the evaluation of formulae with fixpoints, it is necessary to have the same fixpoint structure in the Sat formula as in the formula it's evaluating. Recall that we assume pair-normal form, and suppose that we wish to evaluate Σ_n^μ formulae where n is odd, that is, formulae of the form

$$\tau_n \in \mu X_n. \tau_{n-1} \in \nu X_{n-1} \ldots \tau_2 \in \nu X_2. \tau_1 \in \mu X_1. \phi \qquad (*)$$

where ϕ is first-order. The interpretation of the pure first-order part of ϕ may be done with the Σ_1^μ version of Sat_0—but ϕ may also now contain formulae $\tau \in X_i$. We cannot code as integers the sets referred to by the X_i, so they must be represented by set variables in the meta-language. We use the meta-level variable X_i to represent the object variable X_i, and extend the body of Sat_0 by the clauses (for each $1 \leq i \geq n$)

$$\vee \text{``}(w)_0 = \ulcorner \tau \in X_i \urcorner \text{ and } \text{eval}(\ulcorner \tau\urcorner, (w)_1)) \in X_i\text{''}.$$

Let Sat_0' denote the adjusted Sat_0.

With these adjustments, we have that $(*)$ is true with free variable assigment a just in case

$$\text{eval}(\ulcorner \tau_n \urcorner, a) \in \mu X_n.$$

$$\text{eval}(\ulcorner \tau_{n-1} \urcorner, a[x_n/\mathsf{x}_n]) \in \nu X_{n-1}. \ldots$$

$$\text{eval}(\ulcorner \tau_1 \urcorner, a[x_n, \ldots, x_2/\mathsf{x}_n, \ldots, \mathsf{x}_2]) \in \mu X_1.$$

$$\text{Sat}_0'(\ulcorner \phi \urcorner, a[x_n, \ldots, x_1/\mathsf{x}_n, \ldots, \mathsf{x}_1])$$

Now we just parametrize on $(*)$: let $f_1(x, y)$ be the function that given x encoding a Σ_n^μ formula $(*)$ and an assignment y, computes $\text{eval}(\ulcorner \tau_n \urcorner, y)$, and so on, and let $g(x)$ extract the body of $(*)$. Then we have $\text{Sat}_n(x, y)$ in the form

$$f_n(x, y) \in \mu X_n. \, f_{n-1}(x, y[x_n/\mathsf{x}_n]) \in \nu X_{n-1}. \ldots$$

$$f_1(x, y, [x_n, \ldots, x_2/\mathsf{x}_n, \ldots, \mathsf{x}_2]) \in \mu X_1. \text{Sat}_0'(g(x), y[x_n, \ldots, x_1/\mathsf{x}_n, \ldots, \mathsf{x}_1])$$

which is Σ_n^μ as required. If n is even, we use the Π_1^μ version of Sat_0 instead.

The fact that Sat_n does indeed code truth is easily shown: show by induction on i that each meta-level fixpoint set X_i coincides with the object-level set X_i. The base case follows from the correctness of Sat_0', and the induction step is easy.

It may be noted that we have also skipped details of what the functions f_1 etc. should do if given ill-formed arguments. Any convenient trick may be used; the details are unimportant. $\qquad \square$

4 The Simple Examples in the Modal Mu-Calculus

To construct examples of strict Σ_n^μ formulae in the modal mu-calculus, it would suffice to apply the general construction of Theorem 4 to Sat_n. However, Sat_n contains a large number of function symbols, and the translation would contain many labels. By specializing the general construction, we can eliminate most of these labels, and obtain very simple examples. The following presentation is self-contained, but terse; for a longer explanation of the technique, see [Bra96].

We aim to construct a transition system $\mathcal{T}$ and a Σ_n^μ modal mu-calculus formula Φ such that the set of states satisfying Φ is defined by the strict Σ_n^μ arithmetic formula Sat_n.

The transition system $\mathcal{T}$ should be viewed as a machine for evaluating arithmetic expressions in the same way that Sat_n does: the computation happening in the body of Sat_0' will be dealt with by the definition of the transitions of $\mathcal{T}$, and the arithmetic fixpoints are translated into modal fixpoints in Φ.

The states of $\mathcal{T}$ encode several pieces of information. Namely, a state s contains: a formula ψ_s of the form $(*)$, and a variable assignment a_s, and a pointer p_s into ψ_s which keeps track of where we are in the evaluation. We use the notation of $(*)$ to refer to parts of ψ_s.

The labels of $\mathscr{T}$ are used to distinguish various steps of computation; we shall start with enough labels to make the construction clear, and then argue the number down a little.

The transitions of $\mathscr{T}$ from a state s are defined thus:

- If p_s points at τ_i (or after $\mu X_{i+1}.$, which we consider to be the same), then $s \xrightarrow{x_i} s'$ where $\psi_{s'} = \psi_s$, and $a_{s'} = a_s[\text{eval}(\tau_i, a_s)/x_i]$ and $p_{s'}$ points after μX_i. That is, the term τ_i is evaluated in the current assignment, x_i is set to its value, and we start evaluating the inner fixpoint.

Otherwise, p_s points at a subformula of ϕ. The transition from s mimics the appropriate clause of Sat'_0. The ψ component is not altered by any transition, and the a component is unchanged unless otherwise stated.

- If p_s points at $P(\tau)$, then $s \xrightarrow{a} s_a$ ('a' for atom), where s_a is a special state with no structure, only if $P(\tau)$ is true with variable assignment a_s; otherwise there are no transitions from s.
- If p_s points at $\phi_1 \wedge \phi_2$, then $s \xrightarrow{c} s_k$ ('c' for conjunction) for $k = 1, 2$, where p_{s_k} points at ϕ_k.
- If p_s points at $\forall z_i . \phi_1$, then $s \xrightarrow{c} s_k$ (universal quantification is treated as conjunction) for $k \in \mathbb{N}$, where p_{s_k} points at ϕ_1, and $a_{s_k} = a_s[k/z_i]$.
- If p_s points at $\phi_1 \vee \phi_2$, then $s \xrightarrow{d} s_k$ ('d' for disjunction) for $k = 1, 2$, where p_{s_k} points at ϕ_k.
- If p_s points at $\exists z_i . \phi_1$, then $s \xrightarrow{d} s_k$ (existential quantification is treated as disjunction) for $k \in \mathbb{N}$, where p_{s_k} points at ϕ_1, and $a_{s_k} = a_s[k/z_i]$.
- If p_s points at $\tau \in X_i$, then $s \xrightarrow{x_i} s'$, where $p_{s'}$ points after $\mu X_i.$, and $a_{s'} = a_s[\text{eval}(\tau, a_s)/x_i]$. That is, the term τ is evaluated, copied to the input variable x_i of the fixpoint X_i, and evaluation of the fixpoint started.

It is clear that $\mathscr{T}$ is a recursively presented transition system.

Now consider the following *modal* mu-calculus formula:

$$\text{MuSat}_n \stackrel{\text{def}}{=} \langle x_n \rangle \mu X_n . \langle x_{n-1} \rangle \nu X_{n-1} . \ldots . \langle x_1 \rangle \mu X_1 . \mu W.$$

$$\langle a \rangle \text{tt} \vee ((\langle c \rangle \text{tt} \wedge [c]W) \vee \langle d \rangle W$$

$$\vee \langle x_1 \rangle X_1 \vee \ldots \vee \langle x_n \rangle X_n$$

By the construction of $\mathscr{T}$, we have:

Theorem 7 $s \models \text{MuSat}_n$ *just in case* p_s *points at* ψ_s, *and* $\text{Sat}_n(\ulcorner \psi_s \urcorner, a_s)$. *Hence* MuSat_n *is a strict* Σ_n^μ *modal formula.*

Proof. The proof is a special case of the proof of Theorem 4, the details of which are in [Bra96]. $\quad\square$

MuSat_n is already quite a simple formula, but it is interesting to try to simplify it further, which we shall do in stages.

Firstly, is it necessary to have the double occurrence of $\langle x_i \rangle$, or can we remove the guards from the fixpoint formulae? Yes, we can: consider the formula

$$\text{MuSat}'_n \stackrel{\text{def}}{=} \mu X'_n . \nu X'_{n-1} . \ldots . \mu X'_1 . \mu W. \Psi$$

where Ψ is formed from the body of MuSat_n by priming the X_is. The relation between MuSat_n and MuSat'_n is that $X'_n = \ldots = X'_1 = X_1$) (note that in MuSat_n, we have $X_1 \supseteq X_2 \cup \ldots \cup X_n$), and conversely $X_i = \langle x_i \rangle X'_i$ for $i = n, \ldots, 2$. The denotation of MuSat'_n is still a strict Σ^μ_n set, since the denotation of MuSat_n is a projection of it.

Next, the occurrence of $\langle c \rangle \text{tt}$ is irritating. Its purpose is to assert that p_s is indeed pointing at an $\land$-subterm of ψ_s—of course, $[c]W$ is true at any state with no c-transitions from it. However, we can render it unnecessary by modifying $\mathcal{T}$: if s is any state *other than* an $\land$-subterm state, then add a transition $s \xrightarrow{c} s$. Since W is a least fixpoint variable, if W is true at a state with a c-loop, it is true by virtue of some other disjunct than $[c]W$, and it is not true if it was not true before the loop was added.

We can also eliminate the requirement for a separate a-transition, by modifying the modification: for all those states s with an a-transition, remove the c-loop added in the previous paragraph; now $[c]W$ is true at those states, so we can discard the $\langle a \rangle \text{tt}$ clause.

Finally, we note that $W = X'_1$, and they are adjacent least fixpoints, so we can amalgamate them; further, the job of the d transition can as well be done by x_1, since they work on disjoint sets of states.

Hence we arrive at the following very simple example of a strict Σ^μ_n modal formula (replacing X' by X again):

$$\text{MuSat}''_n \stackrel{\text{def}}{=} \mu X_n . \nu X_{n-1} . \ldots \mu X_1 . [c]X_1 \lor \langle x_1 \rangle X_1 \lor \ldots \lor \langle x_n \rangle X_n$$

5 Relation to Parity Games

In the originally submitted version of this paper, we showed that if one chooses to look at models with no action labels, but with atomic propositions, the above formula appears in a form that is the same as the formula describing the existence of a winning strategy in a parity game of rank n, and hence that formula is strict Σ^μ_n.

One of the referees has pointed out that the strictness of the winning strategy formula can be shown directly from [Bra96] and the game interpretation of modal mu-calculus [EmJ91], without requiring the explicit use of the arithmetic formula Sat_n. As this is an elegant proof, we outline it here, and then comment on the similarities to MuSat.

A parity game of rank n [EmJ91] is played on a directed graph with the following properties: every vertex belongs either to Player or Opponent, and every vertex has an index between 1 and n. If the current vertex belongs to player A, then A moves by choosing a successor vertex. (In [EmJ91], the graph is bipartite so that Player and Opponent alternate, but this is not essential.) In a given play, Player wins if either Opponent gets stuck, or if the greatest index occurring infinitely often is even. ('greatest' is sometimes replaced by 'least', e.g. in [Wal96].) For simplicity, assume henceforth that n is odd.

Given such a graph, let P be true at Player vertices, O be true at Opponent vertices, and R_i true at vertices of index i. It is easy to show [EmJ91,Wal96] that the modal mu-calculus formula

$$\text{Par}_n \stackrel{\text{def}}{=} \mu X_n . \nu X_{n-1} \ldots . \mu X_1 .$$

$$\left(P \Rightarrow \langle \rangle \bigwedge_{1 \leq i \leq n} (R_i \Rightarrow X_i)\right) \wedge \left(O \Rightarrow [] \bigwedge_{1 \leq i \leq n} (R_i \Rightarrow X_i)\right)$$

defines exactly the set of vertices from which Player has a winning strategy.

Now take a strict Σ_n^μ formula $\phi(z)$ of mu-arithmetic, and construct the r.p.t.s. $\mathscr{T}$ and modal formula Φ of Theorem 4. Given a Σ_n^μ modal formula, one can easily, and recursively, construct a parity game G of rank n, whose vertices are pairs (s, Ψ) of states of $\mathscr{T}$ and subformulae of Φ, such that Player wins from (s, Ψ) iff $s \vDash^{\mathscr{T}} \Psi$. Hence $(s, \Phi) \vDash^G \text{Par}_n$ iff $\phi(s_0)$. Therefore $\|\text{Par}_n\|^G$ is an arithmetic Σ_n^μ-hard set, and so by Theorem 3 we conclude the

Theorem 8 Par_n *is a strict* Σ_n^μ *modal formula.* $\qquad\square$

The alternative approach for showing the strictness of Par_n is to work from the transition system $\mathscr{T}$ of Theorem 7, and replace the action labels by atomic propositions, so that P is true at disjunctive states, O at conjunctive states, and R_i at X_i states. With a little manipulation along the lines of the construction of MuSat_n'' from MuSat_n, one obtains exactly the formula Par_n as the modal encoding of Sat_n. Thus we use Sat_n explicitly, and use a specialization of the proof of Theorem 4. The referee's suggestion avoids this work, and so produces examples without requiring the previous results of this paper.

6 Conclusion

In this paper, we have made our previous proof of the modal mu-calculus alternation hierarchy entirely self-contained, and provided simple examples of formulae which separate the hierarchy, including the winning strategy formulae for parity games. It is natural to wonder whether it is possible to avoid the use of arithmetic with fixpoints altogether, and somehow carry out the entire diagonalization process within the framework of modal mu-calculus, or perhaps modal mu-calculus and parity automata. So far, such a proof has not presented itself, but we hope that it may.

Another avenue for investigation is the relationship of these techniques with Lenzi's proof [Len96] of the alternation hierarchy in the mu-calculus of deterministic trees. Lenzi's technique is entirely different, and uses topological analyses of certain structures derived from formulae and their models. It also gives examples of strict formulae; these examples are not of the form we have here, but rather look more like the formulae of [Niw86], where an alternation hierarchy for a restricted calculus on trees was established. The exact connexions between this approach and ours, and between it and automata-theoretic approaches, is far from clear.

7 Acknowledgments

I especially thank Alex Simpson, who suggested to me that there must be a simple proof of the mu-arithmetic hierarchy along these lines. Thanks also to Igor Walukiewicz, who pointed me at the parity game formulae. In addition to the referee who provided the improved proof for parity game formulae, other STACS referees provided helpful suggestions; I am grateful to them.

References

[Bra91] J. C. Bradfield, *Verifying Temporal Properties of Systems* (Birkhäuser, Boston, 1991).

[Bra96] J. C. Bradfield, The modal mu-calculus alternation hierarchy is strict, in: U. Montanari and V. Sassone, eds., *Proc. CONCUR '96*, LNCS **1119** (Springer, Berlin, 1996) 233–246.

[EmJ91] E. A. Emerson and C. S. Jutla, Tree automata, mu-calculus and determinacy, in: *Proc. FOCS 91*. (1991)

[EmL86] E. A. Emerson and C.-L. Lei, Efficient model checking in fragments of the propositional mu-calculus, in: *Proc. 1st LICS* (IEEE, Los Alamitos, CA, 1986) 267–278.

[Kay91] R. Kaye, *Models of Peano Arithmetic*. (Oxford University Press, Oxford, 1991).

[Koz83] D. Kozen, Results on the propositional mu-calculus, *Theoret. Comput. Sci.* **27** (1983) 333–354.

[Len96] G. Lenzi, A hierarchy theorem for the mu-calculus, in: F. Meyer auf der Heide and B. Monien, eds., *Proc. ICALP '96*, LNCS **1099** (Springer, Berlin, 1996) 87–109.

[Lub93] R. S. Lubarsky, μ-definable sets of integers, *J. Symbolic Logic* **58** (1993) 291–313.

[Mos84] A. W. Mostowski, Regular expressions for infinite trees and a standard form of automata, in: A. Skowron, ed., *Fifth Symp. on Computation Theory*, LNCS **208** (Springer, Berlin, 1984) 157–168.

[Niw86] D. Niwiński, On fixed point clones, in: L. Kott, ed., *Proc. 13th ICALP*, LNCS **226** (Springer, Berlin, 1986) 464–473.

[Sti91] C. P. Stirling, Modal and temporal logics, in: S. Abramsky, D. Gabbay and T. Maibaum, eds., *Handbook of Logic in Computer Science*, Vol. 2 (Oxford University Press, 1991) 477–563.

[Wal96] I. Walukiewicz, Monadic second order logic on tree-like structures, in: C. Puech and Rüdiger Reischuk, eds., *Proc. STACS '96*, LNCS **1046** (Springer, Berlin, 1996) 401–414.

On Disguised Double Horn Functions and Extensions

Thomas Eiter[1], Toshihide Ibaraki[2] and Kazuhisa Makino[3]

[1] Institut für Informatik, Universität Gießen, Arndtstraße 2, D-35392 Gießen, Germany. (eiter@informatik.uni-giessen.de)
[2] Department of Applied Mathematics and Physics, Graduate School of Engineering, Kyoto University, Kyoto 606, Japan. (ibaraki@kuamp.kyoto-u.ac.jp)
[3] Department of Systems and Human Science, Graduate School of Engineering Science, Osaka University, Toyonaka, Osaka, 560, Japan. (makino@sys.es.osaka-u.ac.jp)

Abstract. As a natural restriction of disguised Horn functions (i.e., Boolean functions which become Horn after a renaming (change of polarity) of some of the variables), we consider the class $\mathcal{C}_{DH}^R$ of disguised double Horn functions, i.e., the functions which and whose complement are both disguised Horn. We investigate the syntactical properties of this class and relationship to other classes of Boolean functions. Moreover, we address the extension problem of partially defined Boolean functions (pdBfs) in $\mathcal{C}_{DH}^R$, where a pdBf is a function defined on a subset (rather than the full set) of Boolean vectors. We show that the class $\mathcal{C}_{DH}^R$ coincides with the class $\mathcal{C}_{1\text{-}DL}$ of 1-decision lists, and with the intersections of several well-known classes of Boolean functions. Furthermore, polynomial time algorithms for the recognition of a function in $\mathcal{C}_{DH}^R$ from Horn formulas and other classes of formulas are provided, while the problem is intractable in general. We also present an algorithm for the extension problem which, properly implemented, runs in linear time.

1 Introduction

The class of Horn functions is perhaps the most well-studied class of Boolean functions (Bfs) in artificial intelligence and logic programming. Horn functions are at the heart of knowledge-based systems, cf. [2, 16, 7]. This motivates increasing research on Horn functions, e.g., minimum representations, their learning and identification [2], constructing Horn approximations [18], and computing Horn extensions [20]. As a special subclass of Horn functions, the class of double Horn functions $\mathcal{C}_{DH}$ was introduced in [11, 12], where a Horn function f is double Horn if its complement $\overline{f}$ is also Horn. For a Bf f, denote by $T(f)$ (resp., $F(f)$) the set of true (resp., false) vectors of f. From a logical perspective, the double Horn functions are those functions f such that both $T(f)$ and $F(f)$, respectively, can be described by implications

$$x_{i_1} \wedge x_{i_2} \wedge \cdots \wedge x_{i_k} \rightarrow x_{i_0}, \tag{1}$$

where either the antecedents or the consequence may be empty. This is attractive in applications, because both $T(f)$ and $F(f)$ can be treated by Horn techniques. Furthermore, *abduction*, one of the basic operations of expert systems, can be done in polynomial time for double Horn functions [13], whereas abduction for general Horn functions is NP-hard [23].

Changing polarity of some variables in a given Boolean formula (i.e., interchanging literals x_i and $\overline{x}_i$) may benefit to computational advantages. A classical example is a *disguised* (or *renamable*) Horn formula [5, 19]. In this case, for example, after a proper renaming, checking the satisfiability of a disguised Horn formula can be done in polynomial time. In this sense, extending classes of formulas and functions by renamings is an important issue, in order to enlarge the applicability of algorithms and methods.

In this paper, based on the above two observations, we introduce the class $\mathcal{C}_{DH}^R$, which is the renaming closure of double Horn functions. In other words, a Bf f is disguised double Horn if it becomes a double Horn function by an appropriate renaming. Surprisingly, we show that $\mathcal{C}_{DH}^R$ coincides with the class $\mathcal{C}_{1\text{-}DL}$ of 1-decision lists. Here a 1-decision list f is a sequential decision process that determines $f(v)$ for each $v \in \{0,1\}^n$ by "if $l_1(v) = 1$ then $f(v) = b_1$; elseif $l_2(v) = 1$ then $f(v) = b_2$; else ...," where $l_1, l_2, \ldots$ are literals, and $b_1, b_2, \ldots$ are either 0 or 1 (see Section 2 for its formal definition). Decision lists play important roles in learning theory as well as in practical learning systems.

A partially defined Boolean function (pdBf) naturally generalizes a Boolean function, by allowing that the function values on some input vectors are unknown. Such pdBfs have many applications. One is with the representation of incomplete information about cause-effect relationships [8, 6]. E.g., the effect of a number of facts (a patient is male, is a smoker etc.) on a specific disease (e.g., cancer) can be modeled as a Bf $f(x_1, x_2, \ldots, x_n)$, where the arguments x_i represent presence of the facts, and the value of f tells whether the disease is present or not. However, in general the results of all combinations of the n facts on the disease will hardly be known, and thus a description of the relationship as a fully defined Bf is not possible, but may be properly modeled as a pdBf.

The pdBfs have also applications in machine learning. A typical problem ([3]) therein is that, given n Boolean valued attributes, find a hypothesis (i.e., a Bf f from a fixed class of Bfs $\mathcal{C}$) that accurately approximates the actual correlation (a Bf g from $\mathcal{C}$) after seeing a reasonably small number of examples. In our terms, a learning algorithm gradually refines a pdBf, until finally a total Bf is output. In this context, it is important to know if the pdBf given by the considered examples implicitly defines a Bf f from $\mathcal{C}$; if so, the algorithm can stop and output f.

More formally, a pdBf is described by a pair (T, F) of sets T and F of true and false vectors $v \in \{0,1\}^n$, respectively, where $T \cap F = \emptyset$. The above examples suggest the following problem: Given a pdBf (T, F), determine whether some f from a particular class of Bfs $\mathcal{C}$ exists, such that $T \subseteq T(f)$ and $F \subseteq F(f)$. Any such f is called an *extension* f in $\mathcal{C}$, and finding such an f is known as the

The main contributions of this paper can be summarized as follows.

- We give a precise syntactical characterization of a Bf $f \in C_{DH}^{R}$. We show that C_{DH}^{R} coincides with $C_{1\text{-}DL}$, and that C_{DH}^{R} is the intersection of other well-known classes of Bfs.
- We present polynomial time algorithms for important computational problems on C_{DH}^{R}, or prove the intractability. In particular, the extension problem for C_{DH}^{R} is proved to be solvable in linear time. We also address recognizing a function in C_{DH}^{R} from a given formula φ; this is proved tractable for a Horn formula φ and other classes of formulas, but intractable for an arbitrary formula φ.

Due to space limitations, we only sketch proofs of some results. Proofs for all results are given in the extended report [11], which contains more results.

2 Preliminaries

A *partially defined Boolean function* (pdBf) is a mapping $g : T \cup F \rightarrow \{0,1\}$ defined by $g(v) = 1$ if $v \in T$; 0 if $v \in F$, where $T \subseteq \{0,1\}^n$ denotes a set of true vectors (or positive examples) and $F \subseteq \{0,1\}^n$ denotes a set of false vectors (or negative examples) such that $T \cap F = \emptyset$. For simplicity, a pdBf is denoted by (T, F). It can be seen as a representation for all Boolean functions (Bfs) $f : \{0,1\}^n \rightarrow \{0,1\}$ such that $T \subseteq T(f) = \{v \mid f(v) = 1\}$ and $F \subseteq F(f) = \{v \mid f(v) = 0\}$; any such f is called an *extension* of (T, F).

The length of a formula φ, denoted by $|\varphi|$, is the number of symbols in φ; by default, n is the number of Boolean variables in φ. A term t is a conjunction $\bigwedge_{i \in P(t)} x_i \wedge \bigwedge_{i \in N(t)} \overline{x}_i$ of Boolean literals such that $P(t) \cap N(t) = \emptyset$; t is *positive* if $N(t) = \emptyset$ and *Horn* if $|N(t)| \leq 1$. For example, a term $t = x_1 \overline{x}_2 x_3 x_4$ has $P(t) = \{1, 3, 4\}$ and $N(t) = \{2\}$, and is Horn but not positive. A *disjunctive normal form* (DNF) $\varphi = \bigvee_i t_i$ is *positive*, if all t_i are positive, and *Horn* if all t_i are Horn. By default, m is the number of terms in φ.

A Bf f is *positive* (resp., *Horn*), if f is represented by some positive (resp., Horn) DNF; by $C_{\leq}$ (resp., C_{HORN}) we denote the class of these functions. It is known that f is Horn if and only if $F(f) = Cl_{\wedge}(F(f))$ holds, where $Cl_{\wedge}(S)$ denotes the closure of a set $S \subseteq \{0,1\}^n$ under intersection $\wedge$ of vectors. E.g., $Cl_{\wedge}(\{(0101), (1001), (1101)\}) = \{(0101), (1001), (1101), (0001)\}$.

A Bf f is called *double Horn* if $T(f) = Cl_{\wedge}(T(f))$ and $F(f) = Cl_{\wedge}(F(f))$. The class of these functions is denoted by C_{DH}. Note that f is double Horn if and only if f and $\overline{f}$ are Horn. E.g.,

$$f = \overline{x}_1 \vee x_2 x_3 \overline{x}_4 \vee x_2 x_3 x_5 x_6 \overline{x}_7$$

is double Horn, because

$$\overline{f} = x_1 (\overline{x}_2 \vee \overline{x}_3 \vee x_4)(\overline{x}_2 \vee \overline{x}_3 \vee \overline{x}_5 \vee \overline{x}_6 \vee x_7)$$

$$= x_1 \overline{x}_2 \vee x_1 \overline{x}_3 \vee x_1 x_4 \overline{x}_5 \vee x_1 x_4 \overline{x}_6 \vee x_1 x_4 x_7$$

is Horn.

Let $r = (r_1, r_2, \ldots, r_n) \in \{0,1\}^n$. Then we define $ON(r) = \{i \mid r_i = 1\}$ and $OFF(r) = \{i \mid r_i = 0\}$. Furthermore, for any vector $a = (a_1, a_2, \ldots, a_n) \in \{0,1\}^n$, denote by $a \oplus r$ the vector $b = (b_1, b_2, \ldots, b_n)$ such that $b_i = a_i$ if $r_i = 0$, and $b_i = 1 - a_i$, otherwise. Let r be called *renaming*. The *renaming of a Bf f by r*, denoted f^r, is the Bf defined by $a \in T(f^r) \iff a \oplus r \in T(f)$. For any class of Bfs $\mathcal{C}$, we denote by $\mathcal{C}^R$ the closure of $\mathcal{C}$ under renamings. The renaming of a formula φ by r, denoted φ^r, is the formula resulting from φ by replacing each literal involving a variable x_i with $r_i = 1$ by its opposite. For example, let $a = (0,1,1,0)$ and $f = x_1\overline{x}_2 \vee x_2 x_3 \vee \overline{x}_1\overline{x}_3\overline{x}_4$. Then, for a renaming $r = (1,1,0,0)$, $a \oplus r = (1,0,1,0)$ and $f^r = \overline{x}_1 x_2 \vee \overline{x}_2 x_3 \vee x_1\overline{x}_3\overline{x}_4$.

A 1-*decision list* L is a finite sequence of pairs $(l_1, b_1), (l_2, b_2), \ldots, (l_d, b_d)$ and (l_{d+1}, b_{d+1}), where for each $i \in \{1, 2, \ldots, d\}$, l_i is a literal, $b_i \in \{0,1\}$, $l_{d+1} = \top$ (i.e., truth), and $b_{d+1} \in \{0,1\}$. A 1-decision list L defines a function f by $f(v) = b_i$ for every $v \in \{0,1\}^n$, where i is the smallest index such that $l_i(v) = 1$. We call a Bf f 1-*decision list* if f can be defined by some 1-decision list; by $\mathcal{C}_{1\text{-}DL}$ we denote the class of these functions.

The following computational problems have been extensively studied for many classes of Bfs $\mathcal{C}$:

(Recognition) : Given a formula φ, does the function represented by φ belong to $\mathcal{C}$?

(Extension) : Given a pdBf (T, F), does there exist an extension f of (T, F) such that $f \in \mathcal{C}$?

Variants of the latter problem concern the uniqueness of an extension and generation of all extensions. In computational learning theory, recognition and extension problem are called *representation* and *consistency* problem, respectively [3, 1]. Furthermore, a pdBf having the unique extension is called *teaching set* [15].

For any assignment $A = (x_{i_1} \leftarrow a_1, \ldots, x_{i_k} \leftarrow a_k)$ to some variables x_{i_j}, where $a_i \in \{0,1\}$, the function obtained from f by fixing the x_{i_j} as specified by A is $f_A = f_{(x_{i_1} \leftarrow a_1, \ldots, x_{i_k} \leftarrow a_k)}$; we use the same notation for formulas φ. Implication $\varphi \Rightarrow \psi$ is also denoted by $\varphi \leq \psi$, and similar for functions ($f_1 \leq f_2$).

3 Renamable Double Horn Functions

It appears that renamable double Horn functions have a surprisingly regular syntactical characterization.

Define the class $\mathcal{F}_{LR\text{-}1}$ of *linear read-once formulas* by the following recursive form:

(1) $\top$ (empty conjunction, truth), $\perp$ (empty disjunction, falsity) $\in \mathcal{F}_{LR\text{-}1}$, and

(2) if $\varphi \in \mathcal{F}_{LR\text{-}1}$ and x_i is a variable not occurring in φ, then $x_i \vee \varphi$, $\overline{x}_i \vee \varphi$, $x_i \wedge \varphi$, $\overline{x}_i \wedge \varphi \in \mathcal{F}_{LR\text{-}1}$.

A Bf f is called *linear read-once*, if it can be represented by a linear read-once formula; by $C_{LR\text{-}1}$ we denote the class of all linear read-once functions. E.g., $x_1 x_2(\overline{x}_4 \vee x_3 \vee x_5 \overline{x}_6)$ is linear read-once, while $x_2 x_3 \vee x_4 \vee \overline{x}_1 \overline{x}_5$ is not. Note that two read-once formulas are equivalent if and only if they can be transformed through associativity and commutativity into each other [17]. Hence, the latter formula does not represent a linear read-once function. It can be easily seen that

$$C_{LR\text{-}1} = C_{1\text{-}DL}$$

holds.

We can show the somewhat unexpected result that the classes C_{DH}^R and $C_{LR\text{-}1}$ coincide (and hence $C_{DH}^R = C_{LR\text{-}1} = C_{1\text{-}DL}$), which gives a precise syntactical characterization of the semantically defined class C_{DH}^R.

The proof of this result is based on the following lemma. Let $V = \{1, 2, \ldots, n\}$ and $\pi : V \to V$ be any permutation of V. Then, Γ_π is the set of Horn terms $\Gamma_\pi = \{x_{\pi(1)} \cdots x_{\pi(i)} \overline{x}_{\pi(i+1)} \mid 0 \le i < n\} \cup \{x_{\pi(1)} \cdots x_{\pi(n)}\}$; e.g., for $V = \{1, 2\}$ and $\pi(1) = 2, \pi(2) = 1$, we have $\Gamma_\pi = \{\overline{x}_2, x_2 \overline{x}_1, x_2 x_1\}$.

Lemma 1. *Let f be a function on variables V. Then, $f \in C_{DH}$ if and only if f can be represented by a DNF $\varphi = \bigvee_{t \in S} t$ for some permutation π of V and $S \subseteq \Gamma_\pi$.*

Proof. (Sketch) The if-direction is immediate, since φ is obviously Horn and so is $\varphi' = \bigvee_{t \in \Gamma_\pi \setminus S} t$, which represents the complementary function.

The only-if direction is proved by induction on the number n of variables. It is easily checked that the assertion holds for $n = 1$. Assume that it holds for $n = k$, and consider the case $n = k + 1$. We consider two cases (i) and (ii) for the value of f on $\mathbf{0} = (0, \ldots, 0)$.

(i) $f(\mathbf{0}) = 1$. Since f is Horn, there exists a Horn term $t \le f$ such that $t(\mathbf{0}) = 1$. The fact $t(\mathbf{0}) = 1$ implies either $t = \top$ or $t = \overline{x}_j$ for some j. Since $\top \ge \overline{x}_j$ holds for all $j \in V$, there is an index $j \in V$ such that $\overline{x}_j \le f$. Now f can be represented by

$$f = \overline{x}_j \vee f_{(x_j \leftarrow 1)} x_j. \tag{2}$$

As easily seen, $f_{(x_j \leftarrow 1)}$ is double Horn. Hence by the induction hypothesis, there exist a permutation π of $V \setminus \{j\}$ and a subset $S \subseteq \Gamma_\pi$ such that $f_{(x_j \leftarrow 1)} = \bigvee_{t \in S} t$. Define the permutation π' of V by $\pi'(i) = \pi(i) + 1, i \ne j$ and $\pi'(j) = 1$. Let

$$S' = \{\overline{x}_j\} \cup \{x_j t \mid t \in S\}.$$

Then we can see $S' \subseteq \Gamma_{\pi'}$, and by (2), we have $f = \bigvee_{t \in S'} t$, which proves the statement for $k + 1$.

(ii) $f(\mathbf{0}) = 0$. Then there exists an index $j \in V$ such that $\overline{x}_j \le \overline{f}$, and

$$f = \overline{\overline{f}} = \overline{\overline{x}_j \vee \overline{f}_{(x_j \leftarrow 1)} x_j} = f_{(x_j \leftarrow 1)} x_j. \tag{3}$$

Then, by a similar argument, f can be represented by $f = \bigvee_{t \in S} t$ for some π on V and $S \subseteq \Gamma_\pi$. This completes the induction. $\qquad\square$

By algebraic transformations of the formula φ, it can be rewritten to a linear read-once formula of the form

$$
\psi = \begin{cases}
x_{11}x_{12}\ldots x_{1n_1}(\bar{x}_{21} \vee \bar{x}_{22} \vee \ldots \vee \bar{x}_{2n_2} \\
\qquad \vee (x_{31}x_{32}\ldots x_{3n_3}(\ldots(\bar{x}_{d1} \vee \bar{x}_{d2} \vee \ldots \vee \bar{x}_{dn_d})))) & \text{if } d \text{ is even} \\
x_{11}x_{12}\ldots x_{1n_1}(\bar{x}_{21} \vee \bar{x}_{22} \vee \ldots \vee \bar{x}_{2n_2} \\
\qquad \vee (x_{31}x_{32}\ldots x_{3n_3}(\ldots(x_{d1}x_{d2}\ldots x_{dn_d})))) & \text{if } d \text{ is odd,}
\end{cases} \tag{4}
$$

where $d \geq 0$, $n_1 \geq 0$, $n_i \geq 1$ for $i = 2, 3, \ldots, d$, and variables $x_{11}, x_{12}, \ldots, x_{dn_d}$ are all different.

Since any linear read-once formula can be transformed to such a formula by changing the polarities of variables, we obtain the next result.

Theorem 2. $\mathcal{C}_{DH}^R = \mathcal{C}_{LR\text{-}1} = \mathcal{C}_{1\text{-}DL}.$ $\qquad\qquad\qquad\qquad\qquad\qquad\qquad\qquad\qquad\square$

Thus, there exists an interesting relationship between read-once formulas and disguised Horn functions. By means of this relationship, we are able to precisely characterize the prime DNFs of functions in $\mathcal{C}_{DH}^R$. This is an immediate consequence of the next theorem.

Theorem 3. *Every $f \in \mathcal{C}_{DH}^R$ has a renaming r such that f^r is positive and represented by the unique prime DNF*

$$
\varphi = \bigvee_{i=1}^m t_1 \cdots t_i x_{\ell_i} \tag{5}
$$

where t_i and x_{ℓ_i}, $i = 1, 2, \ldots, m$, are pairwise disjoint positive terms and t_i for $i = 1, 2, \ldots, m$ are possibly empty. In particular, (5) implies $\varphi = \bot$ if $m = 0$. Conversely, every such φ of (5) represents an $f \in \mathcal{C}_{DH}^R$. $\qquad\qquad\square$

It appears that $\mathcal{C}_{DH}^R$ has interesting relationships to well-known classes of Bfs. Denote by $\mathcal{C}_{TH}$ the class of threshold functions and by $\mathcal{C}_{2M}$ the class of 2-monotonic functions. A function f on variables $x_1, \ldots, x_n$ is *threshold* if there are weights w_i, $i = 1, 2, \ldots, n$, and a threshold w_0 such that $f(x_1, \ldots, x_n) = 1$ if $\sum_{i=1}^n w_i x_i \geq w_0$; otherwise 0. A function is *2-monotonic* if for each assignment A of size at most 2, either $f_A \leq f_{\bar{A}}$ or $f_A \geq f_{\bar{A}}$ holds, where $\bar{A}$ denotes the opposite assignment to A [21]. The 2-monotonicity and related concepts have been studied under various names in the fields such as threshold logic, hypergraph theory and game theory. The 2-monotonicity can be seen as an algebraic generalization of the thresholdness. Thus $\mathcal{C}_{TH} \subset \mathcal{C}_{2M}$ holds, where $\subset$ denotes proper inclusion. Note that $\mathcal{C}_{TH}^R = \mathcal{C}_{TH}$ and $\mathcal{C}_{2M}^R = \mathcal{C}_{2M}$. We have the following result.

Theorem 4. $\mathcal{C}_{DH}^R = \mathcal{C}_{LR\text{-}1} = \mathcal{C}_{1\text{-}DL} = \mathcal{C}_{TH} \cap \mathcal{C}_{R\text{-}1} = \mathcal{C}_{2M} \cap \mathcal{C}_{R\text{-}1}.$

Proof. (Sketch) Since $\mathcal{C}_{TH} \subseteq \mathcal{C}_{2M}$ and $\mathcal{C}_{1\text{-}DL} \subseteq \mathcal{C}_{TH}$ are known [21, 4], it remains by the results from above to show that $\mathcal{C}_{2M} \cap \mathcal{C}_{R\text{-}1} \subseteq \mathcal{C}_{DH}^R$.

Recall that a function g on $x_1, x_2, \ldots, x_n$ is called *regular* if $g(x) \geq g(y)$ holds for all $x, y \in \{0,1\}^n$ with $\sum_{j \leq k} x_j \geq \sum_{j < k} y_j$, for $k = 1, 2, \ldots, n$. It is known [21] that a regular function is positive and 2-monotonic, and furthermore, any

2-monotonic function becomes regular after permuting and renaming arguments. It is also known that the class of regular function is closed under any assignment A. From these, it is sufficient to show that each function f that is regular and read-once is in C_{DH}^R.

We claim that a function f, which is regular and read-once, can be written either as

$$\text{(i)} \quad f = x_{i_1} \vee x_{i_2} \vee \ldots \vee x_{i_k} \vee f' \qquad \text{or} \qquad \text{(ii)} \quad f = x_{i_1} x_{i_2} \ldots x_{i_k} f',$$

where f' is a regular read-once function not depending on any x_{i_j}, $1 \leq j \leq k$. An easy induction using Theorem 2 gives then the desired result and completes the proof.

Since f is read-once, it can be decomposed according to one of the following two cases:

Case 1: $f = f_1 \vee f_2 \vee \ldots \vee f_k$, where the f_i depend on disjoint sets of variables B_i and no f_i can be decomposed similarly. We show that $|B_i| \geq 2$ holds for at most one i, which means that f has form (i). For this, assume on the contrary that, w.l.o.g., $|B_1|, |B_2| \geq 2$. By considering an assignment A that kills all $f_3, f_4, \ldots, f_k$, it follows that function $g = f_1 \vee f_2$ is regular. Observe that any prime implicant of g is a prime implicant of f_1 or f_2, and that each of them has length ≥ 2 (since f is read-once and by the assumption on the decomposition). Let ℓ be the smallest index in $B_1 \cup B_2$ i.e., $\ell \leq k$ for all $k \in B_1 \cup B_2$, and assume w.l.o.g. that $\ell \in B_1$. Let t be any prime implicant of f_2 and a satisfy $ON(a) = P(t)$. Let $b = a + e^{(\ell)} - e^{(h)}$, where $h \in ON(a)$ and $e^{(k)}$, for any k, is the unit vector with 1 at position k. Note that $l < h$ and $l \in OFF(a)$ by definition. Then $g(b) = 0$ holds. Indeed, $ON(b) \not\supseteq P(t_2)$ for every prime implicant t_2 of f_2, since $ON(b) \cap B_2 \subset P(t)$, and also $ON(b) \not\supseteq P(t_1)$ for every prime implicant t_1 of f_1, since $|ON(b) \cap B_1| = 1$. Consequently, vectors a and b with $\sum_{j \leq k} b_j \geq \sum_{j \leq k} a_j$ for all $k = 1, 2, \ldots, n$ satisfy $g(a) = 1$ and $g(b) = 0$. Thus g is not regular, which is a contradiction. This proves our claim.

Case 2: $f = f_1 f_2 \ldots f_k$, where the f_i depend on disjoint sets of variables B_i and no f_i can be decomposed similarly. Then, the dual function f^d has the form in case 1. Since the dual of a regular function is also regular [21], it follows that f^d has the form (i), which implies that f has form (ii). $\square$

Thus, every renamable double Horn function is read-once and threshold (hence 2-monotonic). C_{DH}^R can be characterized by intersections of those important classes of Bfs.

3.1 Recognition of a renamable double Horn function

Renamable double Horn functions can be recognized in polynomial time from a Horn DNF. The basis of such an algorithm is the following lemma:

Lemma 5. A Bf f is in C_{DH}^R if and only if either (ia) $\bar{x}_j \leq f$, (ib) $\bar{x}_j \leq \bar{f}$, (ic) $x_i \leq f$ or (id) $x_i \leq \bar{f}$ holds for some j, and (ii) $f_{(x_i \leftarrow 1)} \in C_{DH}$ (resp.,

Given a φ, the algorithm picks an index j such that one of (ia)–(id) holds, and then recursively proceeds with $\varphi_{(x_j \leftarrow a)}$ as in (ii). It is polynomial for a variety of formulas.

Theorem 6. *Let $\mathcal{F}$ be a class of formulas closed under assignments, such that checking equivalence of φ to $\top$ and $\bot$, respectively, can be done in $O(t(n, |\varphi|))$ time for any $\varphi \in \mathcal{F}$.[4] Then, deciding whether a given $\varphi \in \mathcal{F}$ represents an $f \in C_{DH}^R$ can be done in $O(n^2 t(n, |\varphi|))$ time.*   □

Since testing $\varphi \equiv \top$ and $\varphi \equiv \bot$ for a Horn DNF φ is possible in $O(|\varphi|)$ time [9], we have

Corollary 7. *Deciding whether a given Horn DNF φ represents an $f \in C_{DH}^R$ can be done in $O(n^2 |\varphi|)$ time.*   □

Theorem 6 has yet another interesting corollary.

Corollary 8. *Deciding if an arbitrary positive formula φ represents an $f \in C_{LR-1}$ can be done in polynomial time.*   □

In fact, deciding if a positive formula φ represents an $f \in C_{R-1}$, where C_{R-1} denotes the class of read-once functions, is co-NP-complete [10, 17]. C_{LR-1} appears to be a maximal subclass of C_{R-1} w.r.t. an inductive (i.e., context-free) bound on disjunctions and conjunctions in a read-once formula such that deciding $f \in C_{R-1}$ from a positive formula φ is polynomial.

However, in general, the recognition problem is intractable.

Theorem 9. *Given any DNF φ, deciding if it represents a renamable double Horn function is co-NP-complete.*

Proof. (Sketch) Recognition problems for C_{R-1} and C_{2M} are in co-NP [1], and since co-NP is closed under conjunction, membership in co-NP follows. Hardness for co-NP is easily shown.   □

3.2 Finding a renamable double Horn extension

The extension problem for C_{DH}^R has already idependently been studied to prove the PAC-learnability. It is known [22] that this problem can be solved in polynomial time. In this subsection, we obtain that the results in [22] can be further improved, i.e., the extension problem for C_{DH}^R can be solved in linear time. This can be regarded as a positive result, since the extension problem for the renaming closures of classes that contain C_{DH}^R is mostly intractable, e.g., for C_{HORN}^R, $C_{\leq}^R$, $C_{R-1}^R = C_{R-1}$, $C_{2M}^R = C_{2M}$ [8, 6].

We describe here an algorithm RDH-EXTENSION, which uses Lemma 5 for a recursive extension test. Informally, it examines the vectors of T and F, respectively, to see whether a decomposition of form $L \wedge \varphi$ or $L \vee \varphi$ is possible,

---

[4] As usual, $t(n, |\varphi|)$ is monotonic in both arguments.

where L is a literal on a variable x_i; if so, then it discards from T and F the vectors which are covered or excluded by this decomposition, and recursively looks for an extension at the projection of (T, F) to the remaining variables. Cascaded decompositions $L_1 \wedge (L_2 \wedge L_3 \wedge (\cdots))$ etc are handled simultaneously.

Algorithm RDH-EXTENSION
Input: A pdBf (T, F), where $T, F \subseteq \{0,1\}^n$.
Output: A linear read-once formula ψ for f if there is an extension $f \in \mathcal{C}_{DH}^R$ of (T, F); otherwise, "No".

    Step 1. Call RDH-AUX for $T, F, \{1, 2, \ldots, n\}$, and return the obtained result;
               Halt. $\qquad\square$

Procedure RDH-AUX
Input: $T, F \subseteq \{0,1\}^n$ and a set $I \subseteq \{1, 2, \ldots, n\}$ of indices.
Output: A linear read-once formula ψ on variables x_i, $i \in I$ if there is an extension $f \in \mathcal{C}_{DH}^R$ of $(T[I], F[I])$, where $T[I]$ and $F[I]$ are the projections of T and F to I, respectively; otherwise, "No".

    Step 1. if $T[I] = \emptyset$ then return $\psi := \bot$ (exit) (* no true vectors, $\bot$ is an extension *)
               else if $F[I] = \emptyset$ then $\psi := \top$ (exit) fi. (* no false vectors, $\top$ is an extension *)
    Step 2. $I^+ := \cap_{v \in T[I]} ON(v)$ and $I^- := \cap_{v \in T[I]} OFF(v)$;
               (* try $x_i(\cdots)$, $\overline{x}_j(\cdots)$, $i \in I^+$, $j \in I^-$ *)
               if $I^+ \cup I^- = \emptyset$ then go to Step 3 (* no extension $x_i(\cdots)$, $\overline{x}_i(\cdots)$ possible *)
               else (* go into recursion *)
                    $F' := F \setminus \{w \in F \mid OFF(w) \cap I^+ \neq \emptyset \text{ or } ON(w) \cap I^- \neq \emptyset\}$;
                    $T' := T$; $I' := I \setminus (I^+ \cup I^-)$;
                    Call RDH-AUX(T', F', I') and let ψ' be the answer;
                    if the answer is ψ' ($\neq \top$) then return $\psi := \psi' \wedge \bigwedge_{i \in I^+} x_i \wedge \bigwedge_{j \in I^-} \overline{x}_j$ (exit)
                         else if the answer is $\top$ then return $\psi := \bigwedge_{i \in I^+} x_i \wedge \bigwedge_{j \in I^-} \overline{x}_j$ (exit)
                         else return "No" (exit) (* decomposition failed, no extension exists *) fi
        fi;
    Step 3. $J^+ := \cap_{w \in F[I]} ON(w)$ and $J^- := \cap_{w \in F[I]} OFF(w)$;
               (* try $x_i \vee \cdots$, $\overline{x}_j \vee \cdots$, $i \in J^+$, $j \in J^-$ *)
               if $J^+ \cup J^- = \emptyset$ then return "No" (exit) (* no decomposition possible, no extension exists *)
               else (* go into recursion *)
                      $T' := T \setminus \{v \in T \mid OFF(v) \cap J^+ \neq \emptyset \text{ or } ON(v) \cap J^- \neq \emptyset\}$;
                    $F' := F$; $I' := I \setminus (J^+ \cup J^-)$;
                    Call RDH-AUX(T', F', I') and let ψ' be the answer;
                    if the answer is ψ' ($\neq \bot$) then return $\psi := \psi' \vee \bigvee_{i \in J^+} \overline{x}_i \vee \bigvee_{j \in J^-} x_j$ (exit)
                         else if the answer is $\bot$ then return $\psi := \bigvee_{i \in J^+} \overline{x}_i \vee \bigvee_{j \in J^-} x_j$ (exit)
                         else return "No" (exit) (* decomposition failed, no extension exists *) fi
        fi; $\qquad\square$

Theorem 10. *Given a pdBf* (T, F), *where* $T, F \subseteq \{0,1\}^n$, *algorithm RDH-EXTENSION correctly finds an extension* $f \in \mathcal{C}_{DH}^R$ *in* $O(n^2(|T| + |F|))$ *time.* $\quad\square$

It is possible to speed up the above algorithm by using proper data structures so that it runs in time $O(n(|T| + |F|))$, i.e., linear time (see [11] for details). In particular, the data structures assure that the same bit of the input is looked up only few times.

3.3 Generating all renamable double Horn extensions

We also have an algorithm for enumerating *all* extensions $f \in \mathcal{C}_{DH}^R$ of a pdBf with a polynomial delay, i.e., the time between subsequent outputs and before as well as after the last output is bounded by a polynomial. A similar algorithm for $\mathcal{C}^R_{HORN}$ does not exist (unless P = NP) [6]; even for $\mathcal{C}_{HORN}$, no such algorithm is known.

Theorem 11. *There is a polynomial delay algorithm for enumerating the (unique) prime DNFs for all renamable double Horn extensions of a pdBf (T, F).* □

Our algorithm is a backtracking procedure similar to RDH-EXTENSION, but far more complicated. The reason is that multiple output of the same extension must be avoided. Indeed, syntactically different formulas ϕ and ψ may represent the same function. Roughly, the algorithm recursively outputs all extensions with a common prefix γ in their linear read-once formulas.

Theorem 11 has important corollaries. Applying the algorithm on (T, F), where $T = F = \emptyset$ for an n, yields the class $\mathcal{C}^R_{DH}(n)$ of all Bfs of n variables in $\mathcal{C}^R_{DH}$. Hence,

Corollary 12. *There is a polynomial delay algorithm for enumerating the (unique) prime DNFs of all $f \in \mathcal{C}^R_{DH}(n)$.* □

Another corollary is that checking if a pdBf (T, F) implicitly defines precisely one $f \in \mathcal{C}^R_{DH}$ is polynomial.

Corollary 13. *Deciding whether a pdBf (T, F) has the unique extension $f \in \mathcal{C}^R_{DH}$ can be done in polynomial time.* □

4 Related Work and Conclusion

Other notions of two-face Horn functions have been considered. For example, bidual Horn functions [11], which require that both f and its dual f^d are Horn, and submodular Horn functions [14], which require that f is Horn and co-Horn, i.e., $f(\overline{x}_1, \overline{x}_2, \ldots, \overline{x}_n)$ is Horn.

It appears that the renaming closure of bidual Horn functions, $\mathcal{C}^R_{BH}$, is a rich class. It contains the class of positive functions $\mathcal{C}_\leq$, as each positive function f is Horn and the dual f^d of f is also positive; hence, by the above results, the relationship

$$\mathcal{C}^R_{DH} \subseteq \mathcal{C}^R_\leq \subseteq \mathcal{C}^R_{BH}$$

follows, and both inclusions are in fact proper. The recognition problem for $\mathcal{C}^R_{BH}$ is polynomial from a Horn DNF [11] and co-NP-complete from an arbitrary DNF, while the extension problem for $\mathcal{C}^R_{BH}$ is co-NP-complete. The nonobvious membership part of the latter result uses a polynomial time algorithm for finding a bidual Horn extension.

Our results show that the renamable (disguised) double Horn functions possess appealing mathematical and computational properties. They have a stringent syntactical characterization, and they coincide with completely differently defined classes of Bfs as well as with intersections of well-known classes of

Boolean functions. This provides new insights into the syntactic and semantical properties of these classes of Boolean functions. Moreover, the class C_{DH}^R has fast polynomial time algorithms for some problems (e.g., the pdBf-extension problem) which are intractable for more general classes.

It would be interesting to explore other problems for which the class C_{DH}^R allows fast algorithms, and where similar improvements can be obtained for more general classes C.

Acknowledgements. The authors thank Martin Anthony for pointing out the equivalence of $C_{LR\text{-}1}$ and $C_{1\text{-}DL}$. We greatly appreciate the comments given by the anonymous reviewers. Especially, one reviewer pointed out the membership in co-NP for the problem in Theorem 9.

References

1. H. Aizenstein, T. Hegedűs, L. Hellerstein and L. Pitt, Complexity theoretic hardness results for query learning, to appear in *Journal of Complexity*.
2. D. Angluin, M. Frazier, and L. Pitt, Learning conjunctions of Horn clauses, *Machine Learning*, 9:147-164, 1992.
3. M. Anthony and N. Biggs, *Computational Learning Theory*, Cambridge University Press, 1992.
4. M. Anthony, G. Brightwell and J. Shawe-Taylor, On specifying Boolean functions by labelled examples, *Discr. Appl. Math.*, 61:1-25, 1995.
5. B. Aspvall, Recognizing disguised NR(1) instance of the satisfiability problem, *J. Algorithms*, 1:97-103, 1980.
6. E. Boros, T. Ibaraki and K. Makino, Error-free and best-fit extensions of partially defined Boolean functions, RUTCOR RRR 14-95, Rutgers University, 1995. *Information and Computation*, to appear.
7. S. Ceri, G. Gottlob, L. Tanca, *Logic Programming and Databases*, Springer, 1990.
8. Y. Crama, P. L. Hammer and T. Ibaraki, Cause-effect relationships and partially defined Boolean functions, *Annals of Operations Research*, 16:299-326, 1988.
9. D. W. Dowling and J.H. Gallier, Linear-time algorithms for testing the satisfyability of propositional Horn formulae, *J. Logic Programming*, 3:267-284, 1984.
10. T. Eiter, Generating Boolean μ-expressions, *Acta Informatica*, 32:171-187, 1995.
11. T. Eiter, T. Ibaraki, and K. Makino, Multi-Face Horn Functions, CD-TR 96/95, CD Lab for Expert Systems, TU Vienna, Austria, iii + 97 pages, 1996.
12. T. Eiter, T. Ibaraki, and K. Makino, Double Horn functions, RUTCOR Research Report RRR 18-97, Rutgers University 1997; to appear in *Information and Computation*.
13. T. Eiter, T. Ibaraki, and K. Makino, Two-face Horn extensions, to appear in Proceedings of *ISAAC'97*, Springer LNCS.
14. O. Ekin, P.L. Hammer and U. N. Peled, Horn functions and submodular Boolean functions, *Theoretical Computer Science*, 175(2):257-270, 1997.
15. S. A. Goldman, On the complexity of teaching, *J. Computer and System Sciences*, 50:20-31, 1995.
16. M. Golumbic, P.L. Hammer, P. Hansen, and T. Ibaraki (eds), Horn Logic, search and satisfiability, *Annals of Mathematics and Artificial Intelligence* 1, 1990.
17. H. Hunt III and R. Stearns, The complexity of very simple Boolean formulas with applications, *SIAM J. Computing*, 19:44-70, 1990.
18. H. A. Kautz, M. J. Kearns, and B. Selman, Horn approximations of empirical data, *Artificial Intelligence*, 74:129-145, 1995.
19. H. Lewis, Renaming a set of clauses as a Horn set, *JACM*, 25:134-135, 1978.
20. K. Makino, K. Hatanaka and T. Ibaraki, Horn extensions of a partially defined Boolean function, RUTCOR RRR 27-95, Rutgers University, 1995.
21. S. Muroga, *Threshold Logic and Its Applications*, Wiley-Interscience, 1971.
22. R. L. Rivest, Learning decision lists, *Machine Learning*, 2:229-246, 1996.
23. B. Selman and H. J. Levesque, Support set selection for abductive and default reasoning, *Artificial Intelligence*, 82:259-272, 1996.

The Complexity of Propositional Linear Temporal Logics in Simple Cases (Extended Abstract)

S. Demri[1] and Ph. Schnoebelen[2]

[1] Leibniz-IMAG, Univ. Grenoble & CNRS UMR 5522,
46, av. Félix Viallet, 38031 Grenoble Cedex, France
email: demri@imag.fr

[2] Lab. Specification and Verification, ENS de Cachan & CNRS URA 2236,
61, av. Pdt. Wilson, 94235 Cachan Cedex, France
email: phs@lsv.ens-cachan.fr

Abstract. It is well-known that model-checking and satisfiability for PLTL are **PSPACE**-complete. By contrast, very little is known about whether there exist some interesting fragments of PLTL with a lower worst-case complexity. Such results would help understand why PLTL model-checkers are successfully used in practice.

In this paper we investigate this issue and consider model-checking and satisfiability for all fragments of PLTL one obtains when restrictions are put on (1) the temporal connectives allowed, (2) the number of atomic propositions, and (3) the temporal height.

1 Introduction

Background. PLTL is the standard linear-time propositional temporal logic used in the specification and automated verification of reactive systems [MP92,Eme90]. It is well-known that model-checking and satisfiability for PLTL are **PSPACE**-complete [SC85,HR83]. However, many research groups were not deterred from implementing PLTL model-checkers. They often comment about the **PSPACE** complexity by emphasizing that, in practice, PLTL specifications are not very complex, have a low temporal height (number of nested temporal connectives) and are mainly boolean combinations of simple eventuality, safety, responsiveness, fairness, ... properties. Actually a systematic theoretical study deserves to be made in order to understand whether some natural classes of PLTL formulas have lower complexity. We know of no such systematic study of this kind in the literature. This is all the more surprising because PLTL is extensively used in the specification and automated verification of reactive systems.

Our objectives. In this paper, our goal is to revisit the complexity questions from [SC85] when there is a bound on the number of propositions and/or on the temporal height of formulas. The first aim is to obtain a better understanding of where does the complexity come from. For instance [SC85] notes that satisfiability for L(F), the fragment of PLTL using only the F (sometimes) operator, is **NP**-hard because already satisfiability is **NP**-hard for the propositional calculus. This is not very enlightening. It does not say anything about how much added complexity is brought by introducing F into the propositional calculus. For the propositional calculus, and even for many modal logics [Hal95], satisfiability becomes linear-time when at most n propositions can be used. What about L(F) ?

Another aim is to see whether there is a formal way of stating that "practical applications only use *simple* PLTL formulas". For example, in practical applications the temporal height often turns out to be at most 3 (when fairness is involved) even when the specification is quite large and combines a large number of temporal constraints. Could such a bounded height be used to argue about reduced complexity ?

Our contribution. Our contribution is twofold. On one hand we provide a number of polynomial-time reductions allowing us to answer all the questions we put forward (only a few remaining ones are solved with ad-hoc methods). As a matter of fact, we show that (1) when the number of propositions is fixed, satisfiability can be transformed in polynomial-time into model-checking, (2) n propositional variables can be encoded into only one if F (sometimes) and X (next) are allowed, (3) into only two if U (until) is allowed. When arbitrarily many propositions are allowed, (4) temporal height can be reduced to 2 if F is allowed, and (5) model-checking for logics with X can be transformed into model-checking without X. Besides, when the formula φ has temporal height at most 1, (6) knowing whether $S \models \varphi$ only depends on a $O(|\varphi|)$ number of places in S.

On the other hand, we give new proofs showing that satisfiability and model-checking for L(U) and L(F, X) are **PSPACE**-hard. These proofs are "simpler" than the construction in [SC85] (e.g. we directly transform any QBF problem into a linear-sized L(U) formula of temporal height 2) and they directly apply to restricted fragments of PLTL. Also, these proofs transform QBF into a model-checking problem, so that they are new "master reductions" for PLTL, interesting in their own right.

Related work. It is common to find papers considering *extensions* of earlier temporal logics. The search for *fragments* with lower complexity is less common. [EES90] investigates (very restricted) fragments of CTL (a branching-time logic) where satisfiability is polynomial-time. [Hal95] investigates, in a systematic way, the complexity of satisfiability (*not model-checking*) for various multimodal logics when the modal height or the number of atomic propositions is restricted. We found that PLTL behaves differently. As far as PLTL is concerned, some complexity results for some particular restricted fragments of PLTL can be found in [EL87,CL93,Spa93,DFR97] but these are not a systematic study sharing our objectives. [Har85] has a simple proof, based on a general reduction from tiling problems into modal logics, that *satisfiability* for L(F, X) is **PSPACE**-hard. In fact, his proof shows that **PSPACE**-hardness is already obtained with temporal height 2 but bounding temporal height is not a concern in this paper.

Plan of the paper. Section 2 recalls various definitions we need throughout the paper. Sections 3 and 4 study the complexity of PLTL fragments when the number of atomic propositions is bounded. Polynomial-time transformations from QBF into model-checking problems can be found in Section 5. Section 6 studies the complexity of PLTL fragments when the temporal height is bounded. Section 7 contains concluding remarks and provides a table summarizing the complete picture we have established about complexity for PLTL fragments.

2   Basic definitions and results

Regarding complexity, we assume that the reader understands what is meant by classes such as **P**, **NP** and **PSPACE**, see e.g. [Joh90]. As usual, given two problems $\mathcal{P}_1$ and $\mathcal{P}_2$, we write $\mathcal{P}_1 \leq_p \mathcal{P}_2$ when there exists a polynomial-time transformation ("many-one reduction") from $\mathcal{P}_1$ into $\mathcal{P}_2$.

Regarding temporal logic, we follow notations and definitions from [Eme90]. Some of them are recalled below.

Syntax. PLTL is a propositional linear-time temporal logic based on a countably infinite set $\mathcal{P} = \{A_1, A_2, \ldots, P_1, P_2, \ldots\}$ of propositional variables, the classical connectives ¬ and ∧ (negation and conjunction), and the temporal operators X (next), U (until), F (sometimes). The set $\{\varphi, \ldots\}$ of formulas is defined in the standard way. We use the connectives ∨, ⇒, ⇔ and G (always) as abbreviations with their standard meaning. We write $\mathcal{P}(\varphi)$ (resp. $sub(\varphi)$) for the set of propositions occurring in (resp. the set of subformulas of) φ. The temporal height of formula φ, written $th(\varphi)$, is the maximum number of nested temporal operators (among X, U, F) in φ. We write $|\varphi|$ to denote the *length* (or *size*) of φ, assuming a reasonably succinct encoding. Following the usual notations (see e.g. [SC85,Eme90]), we let $\mathsf{L}(H_1, H_2, \ldots)$ denote the fragment of PLTL for which only the temporal operators $H_1, H_2, \ldots$ are allowed. For instance $\mathsf{L}(\mathsf{U})$ is "PLTL without X". We write $\mathsf{L}_n^k(H, \ldots)$ to denote the fragment of $\mathsf{L}(H, \ldots)$ where at most n propositions are used, and at most temporal height k is allowed. We write nothing for n and/or k, or we use ω, when no bound is required: $\mathsf{L}(H, \ldots) = \mathsf{L}_\omega^\omega(H, \ldots)$.

Semantics. A *linear-time structure* (also called a *model*) is a pair (S, ϵ) of an ω-sequence $S = s_0, s_1, \ldots$ of *states*, with a mapping $\epsilon : \{s_0, s_1, \ldots\} \to 2^{\mathcal{P}}$ labeling each state s_i with the set of propositions that hold in s_i. We often only write S for a structure, and we often use the fact that a structure S can be viewed as an infinite string of subsets of $\mathcal{P}$. Let S be a structure, $i \in \mathbb{N}$ and a PLTL formula φ, the satisfiability relation $\models$ is inductively defined as follows (we omit the usual conditions for the propositional connectives):

- $S, i \models A \overset{\text{def}}{\Leftrightarrow} A \in \epsilon(s_i)$ (when $A \in \mathcal{P}$) ;
- $S, i \models \mathsf{X}\varphi \overset{\text{def}}{\Leftrightarrow} S, i+1 \models \varphi$;
- $S, i \models \mathsf{F}\varphi \overset{\text{def}}{\Leftrightarrow}$ for some $j \geq i, S, j \models \varphi$;
- $S, i \models \varphi\mathsf{U}\psi \overset{\text{def}}{\Leftrightarrow}$ there is a $j \geq i$ s.t. $S, j \models \psi$ and for all $i \leq j' < j, S, j' \models \varphi$.

φ is *satisfiable* iff $S, 0 \models \varphi$ (also written $S \models \varphi$ or $S, s_0 \models \varphi$) for some S. The *satisfiability problem* for a fragment $L(\ldots)$, written $SAT(L(\ldots))$, is the set of all satisfiable formulas in $L(\ldots)$.

Two models are *equivalent modulo stuttering*, written $S \approx S'$, if they display the same sequence of subsets of $\mathcal{P}$ when repeated (consecutive) elements are seen as one element only. Lamport [Lam83] argued that one should not distinguish between stutter-equivalent models and he advocated prohibiting X in high-level specifications. Indeed $S \approx S'$ iff S and S' satisfy the same $L(U)$ formulas.

Model-checking. A *Kripke structure* $T = (N, R, \epsilon)$ is a triple such that N is a non-empty set of *states*, $R \subseteq N \times N$ is a total *next-state relation*, and $\epsilon : N \to 2^{\mathcal{P}}$ labels each state s with the set of propositions that hold in s. A *path* (or an *execution*) in T is an ω-sequence $S = s_0, s_1, \ldots$ of states of N such that $s_i R s_{i+1}$ for all $i \in \mathbb{N}$. (A path in T is a linear-time structure and a linear-time structure is an infinite Kripke structure.) We follow [Eme90,SC85] and write $T, s \models \varphi$ when there *exists* a path S starting from s s.t. $S \models \varphi$. This existential formulation is what we need for our complexity study. It is the dual of the more common "all paths from s satisfy φ" used in verification. All complexity results can be translated, modulo duality, between the two formulations. The *model-checking problem* for a fragment $L(\ldots)$, written $MC(L(\ldots))$, is the set of all $\langle T, s, \varphi \rangle$ s.t. $T, s \models \varphi$ where T is finite and φ is in $L(\ldots)$.

As far as computational complexity is concerned we make a substantial use of the already known upper bounds: $SAT(L(F))$ and $MC(L(F))$ are **NP**-complete. $SAT(L(F,X))$, $MC(L(F,X))$, $SAT(L(U))$ and $MC(L(U))$ are **PSPACE**-complete. As a consequence, most of our proofs establish lower bounds.

3 Bounding the number of atomic propositions

In this section we evaluate the complexity of satisfiability and model-checking when the number of propositions is bounded. As a consequence, we show that there exist instances of a (linear temporal) logic for which satisfiability is **NP**-complete (resp. **PSPACE**-complete) but whose restriction to the formulas with at most n atomic propositions for some fixed $n \geq 2$ (resp. with exactly one proposition) is still **NP**-complete (resp. is in **P**). This is in contrast with the results obtained with the standard modal logics S5, KD45 (resp. the modal logic S4) in [Hal95].

We start by observing that, when the number of propositions is bounded, satisfiability can be polynomial-time reduced to model-checking.

Proposition 1. *For any* $n \in \mathbb{N}$ *and* $\overline{H} \subseteq \{F, X, U\}$, $SAT(L_n^\omega(\overline{H})) \leq_p MC(L_n^\omega(\overline{H}))$.

Proof. Take $\varphi \in L_n^\omega(\overline{H})$ such that $\mathcal{P}(\varphi) \subseteq \{A_1, \ldots, A_n\}$. Let $T = (N, R, \epsilon)$ be the Kripke structure such that, $N \overset{\text{def}}{=} 2^{\{A_1, \ldots, A_n\}}$ is the set of all 2^n valuations, $R \overset{\text{def}}{=} N \times N$ relates any two states and for all $s \in N$, s is its own valuation: $\epsilon(s) \overset{\text{def}}{=} s$. One can see that φ is satisfiable iff for some $s \in N$, $T, s \models \varphi$. The reduction is in polynomial-time since n and then $|T|$ are constants. Then the polynomial-time transformation can be easily defined.

Proposition 1 is used extensively in the rest of the paper. It only holds when n is bounded and should not be confused with the reductions from model-checking into satisfiability one can find in the literature (e.g. [SC85,Eme90]).

We show that n propositional variables can be encoded into one if F and X are allowed and into only two if U is allowed.

Proposition 2. *For* $\mathsf{H}_1, \ldots$ *a set of temporal operators,* (1) $MC(\mathsf{L}_w(\mathsf{H}_1, \ldots)) \leq_p MC(\mathsf{L}_2(\mathsf{U}, \mathsf{H}_1, \ldots))$, *and* (2) $MC(\mathsf{L}_w(\mathsf{H}_1, \ldots)) \leq_p MC(\mathsf{L}_1(\mathsf{F}, \mathsf{X}, \mathsf{H}_1, \ldots))$.

Proof. We show (1) here. (2) can be found in [DS97]. To a Kripke structure $T = (N, R, \epsilon)$ on $\mathcal{P} = \{P_1, \ldots, P_n\}$ we associate a Kripke structure $D_n(T) \stackrel{\text{def}}{=} (N', R', \epsilon')$ over $\mathcal{P}' = \{A, B\}$ given by

$$N' \stackrel{\text{def}}{=} \{\langle s, i \rangle \mid s \in N, 1 \leq i \leq 2n + 2\}$$

$$\langle s, i \rangle R' \langle s', i' \rangle \stackrel{\text{def}}{\Leftrightarrow} \begin{cases} s = s' \text{ and } i' = i + 1, \text{ or} \\ sRs' \text{ and } i = 2n + 2 \text{ and } i' = 1, \end{cases}$$

$$\epsilon'(\langle s, 1 \rangle) \stackrel{\text{def}}{=} \{A, B\}, \qquad \epsilon'(\langle s, 2j + 1 \rangle) \stackrel{\text{def}}{=} \{A\},$$

$$\epsilon'(\langle s, 2 \rangle) \stackrel{\text{def}}{=} \{\}, \qquad \epsilon'(\langle s, 2j + 2 \rangle) \stackrel{\text{def}}{=} \begin{cases} \{B\} \text{ if } P_j \in \epsilon(s), \\ \{\} \text{ otherwise.} \end{cases}$$

where $j = 1, \ldots, n$. Fig. 1 displays an example.

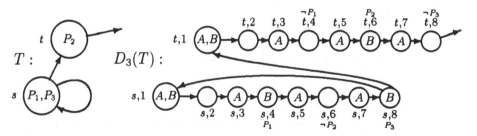

Fig. 1. T and $D_3(T)$

Here alternations between A and $\neg A$ mark slots in $D_n(T)$. B in the i-th encodes that P_i holds. Define $At_D \stackrel{\text{def}}{=} A \wedge B$, $Alt_n^0 \stackrel{\text{def}}{=} At_D$ and $Alt_n^{k+1} \stackrel{\text{def}}{=} A \wedge \neg B \wedge (A \wedge \neg B) \mathsf{U} \left(\neg A \wedge \left(\neg A \mathsf{U} Alt_n^k \right) \right)$ for $k \in \{1, \ldots, n - 1\}$. Clearly, At_D is satisfied in $D_n(T)$ at all $\langle s, j \rangle$ with $j = 1$ and only there. Alt_n^k expresses the fact that there remains k "A-$\neg A$" alternations before the next state satisfying At_D. We now translate formulas over T into formulas over $D_n(T)$ via the following inductive definition:

$$D_n(P_i) \stackrel{\text{def}}{=} A \mathsf{U} \left(\neg At_D \wedge \neg At_D \mathsf{U} \left(Alt_n^{n+1-i} \wedge A \mathsf{U} B \right) \right);$$

D_n is homomorphic for the boolean connectives;
$$D_n(\varphi \mathsf{U} \varphi') \overset{\text{def}}{=} (At_D \Rightarrow D_n(\varphi)) \mathsf{U} (At_D \wedge D_n(\varphi'))$$

We have for $s \in N$, $T, s \models \varphi$ iff $D_n(T), \langle s, 1 \rangle \models D_n(\varphi)$.

Proposition 3. *1. $SAT(\mathsf{L}(\ldots)) \leq_p SAT(\mathsf{L}_2(\mathsf{U}, \ldots))$ and 2. $SAT(\mathsf{L}(\ldots)) \leq_p SAT(\mathsf{L}_1(\mathsf{F}, \mathsf{X} \ldots))$.*

Proof. By way of example let us show 1. (see [DS97] for the full details). Let ψ'_n be the formula

$$\psi'_n \overset{\text{def}}{=} At_D \wedge \mathsf{G}\big(\neg A \Rightarrow (B \Rightarrow B\mathsf{U}A) \wedge (\neg B \Rightarrow \neg B\mathsf{U}A)\big)$$
$$\wedge \mathsf{G}\Big[At_D \Rightarrow At_D \mathsf{U}(\neg A \wedge \neg B \wedge ((\neg A \wedge \neg B)\mathsf{U}Alt^n_n))\Big]$$

One can show that for any model S over $\mathcal{P} = \{P_1, \ldots, P_n\}$, $D_n(S) \models \psi'_n$ and for any S' over $\{A, B\}$, if $S' \models \psi'_n$ then there exists a (unique) S such that $S' \approx D_n(S)$. Then φ, an $\mathsf{L}_n(\ldots)$ formula, is satisfiable iff $\psi'_n \wedge D_n(\varphi)$, an $\mathsf{L}_2(\mathsf{U}, \ldots)$ formula, is satisfiable.

In the full version [DS97], we also show that $MC(\mathsf{L}_2(\mathsf{F}))$ and $SAT(\mathsf{L}_2(\mathsf{F}))$ are **NP**-hard using

Proposition 4. $SAT(\mathsf{L}(\mathsf{F})) \leq_p SAT(\mathsf{L}_2(\mathsf{F}))$.

We also provide in [DS97], a polynomial-time transformation from boolean SAT into $MC(\mathsf{L}^\omega_2(\mathsf{F}))$.

4 One proposition and U is in P

In this section, we give a linear-time algorithm for $\mathsf{L}^\omega_1(\mathsf{U})$ that relies on linear-sized Büchi automata. Recall that the standard method for PLTL satisfiability and model-checking is to compute, for a given PLTL formula φ, a Büchi automaton [1] $\mathcal{A}_\varphi$ recognizing exactly the models of φ and then checking whether a Kripke structure T satisfies φ by computing a synchronous product of T and $\mathcal{A}_{\neg\varphi}$ and checking whether the resulting system (a larger Büchi automaton) recognizes an empty language or not. This method was first presented in [WVS83], where a first algorithm for computing $\mathcal{A}_\varphi$ was given. **PSPACE**-completeness comes from the fact that $\mathcal{A}_\varphi$ can have exponential size. Indeed, once we have $\mathcal{A}_\varphi$ the rest is easy:

Lemma 5. *It is possible, given a Büchi automaton $\mathcal{A}$ recognizing the models of formula φ, and a Kripke structure T, to say in time $O(|T| \cdot |\mathcal{A}|)$ whether there is a computation in T which satisfies φ.*

---

[1] or a Muller automaton, or an alternating Büchi automaton, or ...

We consider a single proposition: $\mathcal{P} = \{A\}$. Any linear model is equivalent, modulo stuttering, to one of the following:

$$S_1^n \stackrel{\text{def}}{=} (A.\neg A)^n.A^\omega \quad S_2^n \stackrel{\text{def}}{=} \neg A.(A.\neg A)^n A^\omega \quad S_3^n \stackrel{\text{def}}{=} (A.\neg A)^\omega$$
$$S_4^n \stackrel{\text{def}}{=} (\neg A.A)^n.\neg A^\omega \quad S_5^n \stackrel{\text{def}}{=} A.(\neg A.A)^n \neg A^\omega \quad S_6^n \stackrel{\text{def}}{=} (\neg A.A)^\omega$$

For $1 \leq i \leq 6$, the size of a Büchi automaton recognizing S_i^n (modulo stuttering) is in $O(n)$.

Lemma 6. *For any $i = 1, \ldots, 6$, $\varphi \in L_1^\omega(U, X)$ and $n \geq th(\varphi)$, we have $S_i^{n+1} \models \varphi$ iff $S_i^n \models \varphi$.*

Proof. By structural induction on φ and using the fact that the first suffix of a S_i^n is a $S_j^{n'}$ with $n - 1 \leq n' \leq n$, e.g. $(S_1^n)'$ is S_2^{n-1} (for $n > 0$) and $(S_2^n)'$ is S_1^n.

Lemma 7. *1. There exists an algorithm which, given $T, s_0, n \in \mathbb{N}$ and $1 \leq i \leq 6$, checks whether, starting from s_0, T has a path $S \approx S_i^n$ in time $O(n. |T|)$.*
2. There exists an algorithm which, given $T, s_0, n \in \mathbb{N}$ and $1 \leq i \leq 6$, checks whether there is a $m \geq n$ s.t., starting from s_0, T has a path $S \approx S_i^m$ in time $O(n. |T|)$.

Proof. Given $1 \leq i \leq 6$ and $n \in \mathbb{N}$, it is easy to build a Büchi automaton with size $O(n)$ recognizing all models S s.t. $S \approx S_i^n$ (resp. s.t. $S \approx S_i^m$ for some $m \geq n$). Then Lemma 5 concludes the proof.

Theorem 8. *Model-checking for $L_1^\omega(U)$ is in $\mathbf{P}$.*

Proof. We consider a Kripke structure $T = (N, R, \epsilon)$ and some state $s_0 \in N$. If there is a path S from s_0 satisfying $\varphi \in L_1^\omega(U)$ then $S \approx S_i^n$ for some $n \in \mathbb{N}$ and some $i = 1, \ldots, 6$ and $S_i^n \models \varphi$. Conversely, if $S_i^n \models \varphi$ and there is a path $S \approx S_i^n$ starting from s_0, then $T, s_0 \models \varphi$.

It is possible to check whether T contains such a path in polynomial-time: We consider all S_i^k for $k < th(\varphi)$. When $S_i^k \models \varphi$, seen in time $O(k. |\varphi|)$, we check in time $O(k. |T|)$, whether, from s_0, T admits a path $S \approx S_i^k$. We also consider all S_i^k for $k = th(\varphi)$. When $S_i^k \models \varphi$ we know that $S_i^{k+m} \models \varphi$ for all m (Lemma 6) so that it is correct to check whether there is a m s.t. T admits a path $S \approx S_i^{k+m}$. Thanks to Lemma 7, this can be done in polynomial-time. Because $k \leq |\varphi|$, the complete algorithm only needs $O(|T| .|\varphi|^2)$.

With Proposition 1, we get $SAT(L_1^\omega(U))$ is in $\mathbf{P}$. In order to be exhaustive one can show that $SAT(L_1^\omega(X))$ and $MC(L_1^\omega(X))$ are $\mathbf{NP}$-complete [DS97]. Moreover since there are only a finite number of essentially distinct formulas in a given $L_n^k(U, X)$ (for any fixed $k, n < \omega$), $SAT(L_n^k(X))$ and $MC(L_n^k(X))$ can be proved in $\mathbf{P}$ (see [DS97]). This concludes the study of all fragments with a bounded number of propositions. In the remaining of the paper, this bound is removed.

5 From QBF into $MC(\mathsf{L}(\mathsf{U}))$

In this section, we offer a polynomial-time transformation from validity of Quantified Boolean Formulas (QBF) into model-checking for $\mathsf{L}(\mathsf{U})$ that involves rather simple constructions of models and formulas. This reduction can be adapted to various fragments and, apart from the fact that it offers a simple means to get **PSPACE**-hardness, we obtain a new master reduction from a well-known logical problem. As a side-effect, we establish that $MC(\mathsf{L}_\omega^2(\mathsf{U}))$ is **PSPACE**-hard, which is not subsumed by any reduction from the literature. In the full version we show reduction from QBF into model-checking for $\mathsf{L}(\mathsf{F}, \mathsf{X})$ and give a construction showing that $MC(\mathsf{L}_2^\omega(\mathsf{F}))$ is **NP**-hard.

Consider an instance of QBF. It has the form $P \equiv Q_1 x_1 \ldots Q_n x_n \overbrace{\wedge_{i=1}^m \vee_{j=1}^{k_i} l_{i,j}}^{P_0}$ where every Q_r $(1 \le r \le n)$ is a universal, $\forall$, or existential, $\exists$, quantifier. P_0 is a propositional formula without any quantifier. Here we consider w.l.o.g. that P_0 is a conjunction of clauses, i.e. every $l_{i,j}$ is a propositional variable $x_{r(i,j)}$ or the negation $\neg x_{r(i,j)}$ of a propositional variable from $X = \{x_1, \ldots, x_n\}$. The question is to decide whether P is valid or not. Recall that

Lemma 9. *P is valid iff there exists a non-empty set $\mathcal{V} \subseteq \{\top, \bot\}^X$ of valuations (truth-value assignments) s.t.*
correctness: *$\forall v \in \mathcal{V}, v \models P_0$, and*
closure: *for all $v \in \mathcal{V}$, for all r s.t. $Q_r = \forall$, there is a $v' \in \mathcal{V}$ s.t. $v'[x_r] = not(v[x_r])$ and for all $r' < r$, $v'[x_{r'}] = v[x_{r'}]$.*

To P we associate the Kripke structure T_P as given in Figure 2, using labels from $\mathcal{P} = \{A_0, A_1, \ldots, x_1^T, \ldots, L_1^1, \ldots\}$.

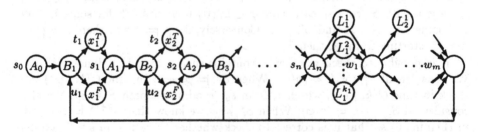

Fig. 2. The structure T_P associated to $P \equiv Q_1 x_1 \ldots Q_n x_n \bigwedge_{i=1}^m \bigvee_{j=1}^{k_i} l_{i,j}$

Assume S is an infinite path starting from s_0. Between s_0 and s_n, it picks a boolean valuation for all variables in X, then reaches w_m and goes back to some B_r-labeled state $(1 \le r \le n)$ where (possibly distinct) valuations for $x_r, x_{r+1}, \ldots, x_n$ are picked.

In S, at any position lying between a s_n and the next w_m, we have a notion of *current valuation* which associates $\top$ or $\bot$ to any x_r depending on *the latest*

u_r or t_r node we visited. To S we associate the set $\mathcal{V}(S)$ of all valuations that are current at positions where S visits s_n (there are infinitely many such positions).

Now consider some r with $Q_r = \forall$ and assume that whenever S visits s_{r-1} then it visits both t_r and u_r before any further visit to s_{r-1}. In $\mathsf{L}(\mathsf{U})$, this can be written $S \models \psi_r$ with ψ_r given by

$$\psi_r \stackrel{\text{def}}{=} \mathsf{G}\Big(A_{r-1} \Rightarrow (\neg B_{r-1}\mathsf{U}x_r^T) \wedge (\neg B_{r-1}\mathsf{U}x_r^F)\Big).$$

Let $\psi_{\text{clo}} \stackrel{\text{def}}{=} \bigwedge\{\psi_r \mid Q_r = \forall\}$: if S satisfies ψ_{clo} then $\mathcal{V}(S)$ is closed in the sense of Lemma 9.

Now, whenever S visits a L_i^j-state, we say it agrees with the current valuation v if $v \models l_{i,j}$. This too can be written in $\mathsf{L}(\mathsf{U})$, using the fact that the current valuation for x_r cannot be changed without first visiting the B_r-state. For $i = 1, \ldots, m$, for $j = 1, \ldots, k_i$, let

$$\psi_{i,j} \stackrel{\text{def}}{=} \begin{cases} \mathsf{G}[x_r^F \Rightarrow \mathsf{G}\neg L_i^j \vee \neg L_i^j \mathsf{U}B_r] & \text{if } l_{i,j} = x_r, \\ \mathsf{G}[x_r^T \Rightarrow \mathsf{G}\neg L_i^j \vee \neg L_i^j \mathsf{U}B_r] & \text{if } l_{i,j} = \neg x_r. \end{cases}$$

Lemma 10. *Let $\varphi_P \stackrel{\text{def}}{=} \psi_{\text{clo}} \wedge \left(\bigwedge_{i=1}^{m} \bigwedge_{j=1}^{k_i} \psi_{i,j}\right)$. Then $T_P, s_0 \models \varphi_P$ iff P is valid.*

Proof. If $S \models \varphi_P$ then $\mathcal{V}(S)$ is closed and correct for P so that P is valid. Conversely, if P is valid, there exists a validating $\mathcal{V}$ (Lemma 9). We can build an infinite path S starting from s_0 s.t. $\mathcal{V}(S) = \mathcal{V}$ and $S \models \varphi_P$: from a lexicographical enumeration of $\mathcal{V}$, S is easily constructed so that $S \models \psi_{\text{clo}}$. Then, to ensure $S \models \varphi_P$, between any visit to s_n and to the next w_m we choose to visit L_i^j-states validated by the current valuation v, which is possible because $v \models P_0$.

Now, because $|T_P|$ and $|\varphi_P|$ are in $O(|P|)$, and because $th(\varphi_P) \leq 2$ (and using Proposition 2), we get

Corollary 11. QBF $\leq_p MC(\mathsf{L}_\omega^2(\mathsf{U})) \leq_p MC(\mathsf{L}_2^\omega(\mathsf{U}))$.

Corollary 12. *$MC(\mathsf{L}_\omega^2(\mathsf{U}))$ and $MC(\mathsf{L}_2^\omega(\mathsf{U}))$ are **PSPACE**-hard.*

6 Bounding the temporal height

In this section we investigate the complexity of satisfiability and model-checking when the temporal height is bounded. From Section 5, we already know that $MC(\mathsf{L}_\omega^2(\mathsf{U}))$ is **PSPACE**-hard.

Elimination of X We show how problems for $\mathsf{L}(\mathsf{X}, \ldots)$ can be transformed into problems for $\mathsf{L}(\ldots)$. Say a formula φ has *inner-nexts* if the only occurrences of X are in subformulas of the form $\mathsf{XX}\ldots\mathsf{X}A$ (where A is a propositional variable). Assume φ has inner-nexts, with at most k nested X and that T is a Kripke structure. It is possible to partially unfold T into a Kripke structure T^k where a state $\overline{s}$ (in T^k) codes for a state s_0 in T with the k next states $s_1, \ldots, s_k$ already chosen. We can now replace all $X^i A$ in φ by new propositions A^i and label T^k so that $A^i \in \epsilon(\overline{s})$ iff $A \in \epsilon(s_i)$. Finally, $T, s \models \varphi$ iff $T^k, \overline{s} \models \varphi^k$ for some $\overline{s}$ starting with s. Because the size of T^k is in $O(|T|^k)$ and $|\varphi^k|$ is in $O(|\varphi|)$, we have

Proposition 13. [DS97] $MC(\mathsf{L}^k_\omega(\mathsf{X}, \ldots)) \leq_p MC(\mathsf{L}^k_\omega(\ldots))$ *for any fixed* $k \geq 0$.

As a corollary, $MC(\mathsf{L}^k_\omega(\mathsf{X}))$ is in **P** and $MC(\mathsf{L}^k_\omega(\mathsf{F}, \mathsf{X}))$ is in **NP** for any fixed $k \geq 0$. $MC(\mathsf{L}^1_\omega(\mathsf{F}))$ is **NP**-hard as can be seen from the proof for $\mathsf{L}^\omega_\omega(\mathsf{F})$ in [SC85]. Hence for $k \geq 1$, $MC(\mathsf{L}^k_\omega(\mathsf{F}, \mathsf{X}))$ is **NP**-complete. Elimination of X can also be performed for satisfiability. If φ is satisfiable, then φ^k is. Now if φ^k is satisfiable, it is perhaps satisfiable in a model that is not a S^k for some S. But we can express the fact that a given model is a S^k with an $\mathsf{L}^2_\omega(\mathsf{F}, \mathsf{X})$ formula. Then φ is satisfiable iff $\varphi^k \wedge \mathsf{G}(\bigwedge_{j=1}^n \bigwedge_{i=1}^k A^i_j \Leftrightarrow \mathsf{X}A^{i-1}_j)$ is. Actually, by using systematically the standard renaming techniques (the operator G propagates the constraints of renaming), we can show that $SAT(\mathsf{L}(\overline{\mathsf{H}})) \leq_p SAT(\mathsf{L}^2_\omega(\overline{\mathsf{H}}))$ for $\overline{\mathsf{H}} \in \{\{\mathsf{F}\}, \{\mathsf{F}, \mathsf{X}\}, \{\mathsf{U}\}, \{\mathsf{U}, \mathsf{X}\}\}$. As a corollary, $SAT(\mathsf{L}^2_\omega(\mathsf{F}, \mathsf{X}))$ and $SAT(\mathsf{L}^2_\omega(\mathsf{U}))$ are **PSPACE**-hard. It is worth observing that [Spa93,DFR97] have another proof that $SAT(\mathsf{L}^2_\omega(\mathsf{F}, \mathsf{X}))$ is **PSPACE**-hard.

Temporal height less or equal to 1: upper bounds in **NP** Below temporal height 2, the upper bounds can be improved. For any $\varphi \in \mathsf{L}^1_\omega(\mathsf{U}, \mathsf{X})$, one can show that φ is satisfiable iff φ is satisfied in a model $S = s_0, s_1, \ldots$ such that for any $i, j \geq 1+|\varphi|$, for $A \in \mathcal{P}(\varphi)$, $A \in \epsilon(s_i)$ iff $A \in \epsilon(s_j)$. As a corollary, for $\overline{\mathsf{H}} \in \{\{\mathsf{F}\}, \{\mathsf{F}, \mathsf{X}\}, \{\mathsf{U}\}, \{\mathsf{U}, \mathsf{X}\}\}$, $SAT(\mathsf{L}^1_\omega(\overline{\mathsf{H}}))$ is in **NP**. Since those fragments contain the propositional calculus, **NP**-hardness is immediate. Now let us turn to model-checking when the temporal height is at most 1. We already know that $MC(\mathsf{L}^1_\omega(\mathsf{F}))$ is **NP**-hard [SC85]. We can also show that $MC(\mathsf{L}^1_\omega(\mathsf{U}, \mathsf{X}))$ is in **NP** [DS97]. As a corollary, for $\overline{\mathsf{H}} \in \{\{\mathsf{F}\}, \{\mathsf{F}, \mathsf{X}\}, \{\mathsf{U}\}, \{\mathsf{U}, \mathsf{X}\}\}$, $MC(\mathsf{L}^1_\omega(\overline{\mathsf{H}}))$ is **NP**-complete.

7 Concluding remarks

In the paper we have investigated the complexity of satisfiability and model-checking for all fragments of PLTL obtained (1) by bounding the number of atomic propositions, (2) the temporal height, and (3) restricting the temporal operators one allows.

Our results take advantage of a few general techniques that might be reused to tackle similar problems for other temporal logics. Most of the time, these techniques are used to strengthen earlier hardness results so that they also apply to specific fragments. In some cases we develop specific arguments showing that the complexity really decreases under the identified threshold values.

The table at the end of this section contains the results of the full paper and some general conclusions can be read. In most cases no reduction in complexity occurs when two propositions are allowed, or when temporal height two is allowed. Moreover in most cases, for equal fragments, satisfiability and model-checking belong to the same complexity class. See the table for some exceptions.

$n, k < \omega$		Model-Checking	Satisfiability
$L(\ldots)$	$L_n^k(\ldots)$	P	P
	$L_\omega^0(\ldots)$	P	NP-complete
$L(F)$	$L(F)$	NP-complete [SC85]	NP-complete [NO80]
	$L_\omega^1(F)$	NP-complete	NP-complete
	$L_2^\omega(F)$	NP-complete	NP-complete
	$L_1^\omega(F)$	P	P
$L(U)$	$L(U)$	PSPACE-complete [SC85]	PSPACE-complete [SC85,HR83]
	$L_\omega^2(U)$	PSPACE-complete	PSPACE-complete
	$L_\omega^1(U)$	NP-complete	NP-complete
	$L_2^\omega(U)$	PSPACE-complete	PSPACE-complete
	$L_1^\omega(U)$	P	P
$L(X)$	$L(X)$	NP-complete	NP-complete
	$L_\omega^k(X)$	P	NP-complete
	$L_1^\omega(X)$	NP-complete	NP-complete
$L(F,X)$	$L(F,X)$	PSPACE-complete [SC85]	PSPACE-complete [SC85,HR83]
	$L_\omega^{2+k}(F,X)$	NP-complete	PSPACE-complete [Har85,Spa93]
	$L_\omega^1(F,X)$	NP-complete	NP-complete
	$L_1^\omega(F,X)$	PSPACE-complete	PSPACE-complete
$L(U,X)$	$L(U,X)$	PSPACE-complete [SC85]	PSPACE-complete [SC85,HR83]
	$L_\omega^2(U,X)$	PSPACE-complete	PSPACE-complete [Har85,Spa93]
	$L_\omega^1(U,X)$	NP-complete	NP-complete
	$L_1^\omega(U,X)$	PSPACE-complete	PSPACE-complete

The frequent preservation of lower bounds when fragments are taken into account does not explain the alleged simplicity of "simple practical applications". The fact that $L_1^\omega(U)$ (only one proposition) is in **P** can be used inside a PLTL verifier as an efficient method *for a few special cases*, but it is too restricted for practical applications. The fact that $MC(L_\omega^k(F,X))$ is in **NP** has more potential explanatory power, but **NP**-hardness is still intractable. The fact that $L_n^k(U,X)$ is in **P** really means that the complexity depends at least on the number of propositions or the temporal height. This dependence must be scrutinized in more details.

Understanding and taming the complexity of linear temporal logics remains an important issue and the present work can be seen as some steps in this direction. The ground is open for further investigations. We think future work could investigate

- different, finer definitions of fragments (witness [EES90]) that can be inspired by practical examples, or that aim at defeating one of our hardness proofs, e.g. forbidding the renaming technique we use in section 6,
- other problems that are important for verification: module checking, semantic entailment, ... ,
- other complexity measures: e.g. average complexity, or separated complexity measure for models and formulas,
- restrictions on the models rather than the formulas.

Acknowledgments. We thank B. Bérard, E. A. Emerson, F. Laroussinie, M. Reynolds, A. Zanardo, and the anonymous referees for their suggestions.

References

[CL93]   C.-C. Chen and I.-P. Lin. The computational complexity of satisfiability of temporal Horn formulas in propositional linear-time temporal logic. *Information Processing Letters*, 45(3):131–136, March 1993.

[DFR97]   C. Dixon, M. Fisher, and M. Reynolds. Execution and proof in a Horn-clause temporal logic. In *Proc. 2nd Int. Conf. on Temporal Logic (ICTL'97), Manchester, UK, July 1997*, 1997. to appear.

[DS97]   S. Demri and Ph. Schnoebelen. The complexity of propositional linear temporal logics in simple cases. Research Report LSV-97-11, Lab. Specification and Verification, ENS de Cachan, Cachan, France, December 1997. Available at http://www.lsv.ens-cachan.fr/~phs.

[EES90]   E. A. Emerson, M. Evangelist, and J. Srinivasan. On the limits of efficient temporal decidability. In *Proc. 5th IEEE Symp. Logic in Computer Science (LICS'90), Philadelphia, PA, USA, June 1990*, pages 464–475, 1990.

[EL87]   E. A. Emerson and C. Lei. Modalities for model checking: Branching time logic strikes back. *Science of Computer Programming*, 8:275–306, 1987.

[Eme90]   E. A. Emerson. Temporal and modal logic. In J. van Leeuwen, editor, *Handbook of Theoretical Computer Science, vol. B*, chapter 16, pages 995–1072. Elsevier Science Publishers, 1990.

[Hal95]   J. Y. Halpern. The effect of bounding the number of primitive propositions and the depth of nesting on the complexity of modal logic. *Artificial Intelligence*, 75(2):361–372, 1995.

[Har85]   D. Harel. Recurring dominos: Making the highly undecidable highly understandable. *Annals of Discrete Mathematics*, 24:51–72, 1985.

[HR83]   J. Y. Halpern and J. H. Reif. The propositional dynamic logic of deterministic, well-structured programs. *Theor. Comp. Sci.*, 27(1–2):127–165, 1983.

[Joh90]   D. S. Johnson. A catalog of complexity classes. In J. van Leeuwen, editor, *Handbook of Theoretical Computer Science, vol. A*, chapter 2, pages 67–161. Elsevier Science Publishers, 1990.

[Lam83]   L. Lamport. What good is temporal logic ? In R. E. A. Mason, editor, *Information Processing'83. Proc. IFIP 9th World Computer Congress, Sep. 1983, Paris, France*, pages 657–668. North-Holland, 1983.

[MP92]   Z. Manna and A. Pnueli. *The Temporal Logic of Reactive and Concurrent Systems: Specification*. Springer-Verlag, 1992.

[NO80]   A. Nakamura and H. Ono. On the size of refutation Kripke models for some linear modal and tense logics. *Studia Logica*, 39(4):325–333, 1980.

[SC85]   A. P. Sistla and E. M. Clarke. The complexity of propositional linear temporal logics. *Journal of the ACM*, 32(3):733–749, July 1985.

[Spa93]   E. Spaan. *Complexity of Modal Logics*. PhD thesis, ILLC, Amsterdam University, NL, March 1993.

[WVS83]   P. Wolper, M. Y. Vardi, and A. P. Sistla. Reasoning about infinite computation paths (extended abstract). In *Proc. 24th IEEE Symp. Found. of Computer Science (FOCS'83), Tucson, USA, Nov. 1983*, pages 185–194, 1983.

Searching Constant Width Mazes Captures the AC^0 Hierarchy

David A. Mix Barrington[1] Chi-Jen Lu[1] Peter Bro Miltersen[2]* Sven Skyum[2]*

[1] Computer Science Department, University of Massachusetts.
[2] BRICS, Basic Research in Computer Science, Centre of the Danish National Research Foundation, Department of Computer Science, University of Aarhus.

Abstract. We show that searching a width k maze is complete for Π_k, i.e., for the k'th level of the AC^0 hierarchy. Equivalently, st-connectivity for width k grid graphs is complete for Π_k. As an application, we show that there is a data structure solving dynamic st-connectivity for constant width grid graphs with time bound $O(\log \log n)$ per operation on a random access machine. The dynamic algorithm is derived from the parallel one in an indirect way using algebraic tools.

1 Introduction

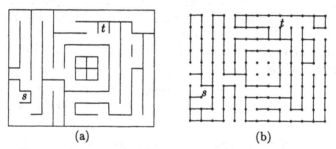

Fig. 1. (a) A maze and (b) the corresponding grid graph

Blum and Kozen [4] considered the problem of searching a *maze*. A maze is an object as depicted in Figure 1(a). Formally, we will define a maze of width m and length n as follows: Let $S_{n,m} = \{1, \ldots, n\} \times \{1, \ldots, m\}$. We call an element s of $S_{n,m}$ a *square* and identify s with the unit square with center s in the plane. A maze of width m and length n is a set M of line segments (*walls*) of length exactly 1, each separating two squares of $S_{n,m}$. Figure 1(a) depicts a maze of width 10 and length 13 (we consider the longer line segments as consisting of several atomic walls of length 1). A path in the maze between two squares s and t is a path inside the rectangle $[0, n] \times [0, m]$ connecting the centers of s and t and not intersecting any of the walls in M. The reader is invited to verify that there is a path between s and t in Figure 1(a). Blum and Kozen gave bounds on the power of systems of automata capable of searching a maze, i.e. capable of deciding whether a path between two given squares in the maze exists. In complexity theoretic terms, one of their results was that searching a maze is in deterministic logspace.

* Supported by the ESPRIT Long Term Research Programme of the EU under project number 20244 (ALCOM-IT).

In this paper we consider the complexity of searching a *constant width* maze, i.e., rather than letting both n and m be parameters, we fix m to a constant $k \geq 1$. Let MAZE_k be the problem which takes as input (a Boolean encoding of) a maze of width k, two squares s and t, and decides if there is a path from s to t.

We relate the complexity of MAZE_k in a strong way to the levels of the AC^0 hierarchy. Recall the following definitions: Non-uniform AC^0 is the class of languages recognizable by families of AND/OR/NOT-circuits of constant depth, polynomial size, and unbounded fan-in. Inside AC^0 we find the following hierarchy: Non-uniform Σ_k is the class of languages recognizable by circuits with k alternating levels of unbounded fan-in AND and OR gates, with the output an OR-gate and a "zeroth level" of input gates and their negations. Non-uniform Π_k is defined analogously, but with the output gate being an AND-gate. Following [1], we define a uniform version of the hiearchy as follows: Uniform Π_k (Σ_k) is the class of languages accepted by alternating Turing machines running in logarithmic time and making exactly k alternations, the first being universal (existential).

An appropriate class of reductions to use for the non-uniform classes in the AC^0 hierarchy is the class of (non-uniform) p-projections [15]. Similarly, an appropriate class of reductions to use for the uniform classes is the class of DLOGTIME-uniform projections (for a precise definition, see Section 2).

Our main result is:

Theorem 1. *For $k \geq 3$, MAZE_k is complete for non-uniform Π_k under p-projections and complete for uniform Π_k under DLOGTIME-uniform projections.*

This seems to be the first example of natural graph problems, complete for the levels of the AC^0 hierarchy.

There is a close correspondence between mazes and *grid graphs*, as defined by Itai *et al.* [11]. An $n \times k$ grid graph is an undirected graph G with vertex set $V_{n,k} = \{1, \ldots, n\} \times \{1, \ldots, k\}$ and with the property that if $\{(a,b), (c,d)\}$ is an edge in G, we have $|a - c| + |b - d| = 1$. The length of the grid graph is n and the width is k. A grid graph is shown in Figure 1(b). The st-connectivity problem USTCON_k for width k grid graphs is the following: Given a grid graph, and two vertices s and t, decide if s and t are connected in G. There is a trivial isomorphism between MAZE_k and USTCON_k: To get from a maze problem to a grid graph problem, simply make a vertex for each square of the maze, and put an edge between two vertices if and only if there is *not* a wall between the corresponding squares. We shall use the grid graph formulation in the main part of the paper.

A third setting for these problems is the following variant of bounded-width branching programs. An $n \times k$ *switching network* is an *undirected* labelled graph whose vertices form a rectangular array with k rows and n columns and whose edges are restricted to be between vertices in adjacent columns. (Switching networks are also called "contact schemes" — see the survey of Razborov [13] for further background.) Each edge is labelled by an input variable, its negation, or the value 1, and the network accepts a given input string iff there is a path from a fixed vertex s to another fixed vertex t such that the label of each edge on the path evaluates to 1 on the input. It is not hard to show that the grid graph problem is closely related to *planar* switching networks as follows: USTCON$_k$ is complete, under p-projections, for the class of languages decidable by families of width-k, polynomial-size planar switching networks. This is because an $n \times k$ planar switching network can be simulated by a $kn \times k$ grid graph, and an $n \times k$ grid graph can be simulated by a $kn \times k$ planar switching network. We omit the details of these simulations in this version of the paper.

In our second result, we consider the following *dynamic* graph problem: Maintain, on a random access machine with word size $O(\log n)$, a data structure representing an $n \times k$ grid graph under insertions and deletions of edges and connectivity queries, i.e. queries asking whether there is a path between two vertices, given as input. For *non-constant* width $m \leq n$, Eppstein *et al* provide a solution to this problem with a time bound $O(\log n)$ per operation [6]. We show:

Theorem 2. *For any constant k, there is a solution to the dynamic connectivity problem for width k grid graphs with time complexity $O(\log \log n)$ per operation. On the other hand, no solution to the dynamic connectivity problem for width 2 grid graphs has time complexity $o(\log \log n / \log \log \log n)$.*

We derive the dynamic algorithm from the parallel one in an indirect way: We note that the *existence* of the parallel algorithm implies that a certain monoid, G_k, associated with the width-k problem is aperiodic by results of Barrington and Therien [2]. Combining this with a result of Thomas [16], we in fact show that G_k has dot-depth exactly k, providing a natural example of such a monoid which may be of independent interest. We then use results on dynamic word problems by Frandsen, Miltersen and Skyum [7] to derive the dynamic algorithm. Unfortunately, the algorithm obtained is rather impractical; the constant in the big-O is $2^{2^{O(k)}}$. The lower bound we get as a corollary to work of Beame and Fich [3], again by looking at the problem from an algebraic point of view.

2 The completeness result

An instance of the USTCON$_k$ problem consists of a grid graph, a vertex s and a vertex t. We represent the graph by a number of Boolean *edge indicator variables*, one for each edge position in the grid. We pack them in two binary relations, $E_h \subseteq \{1, \ldots, n-1\} \times \{1, \ldots, k\}$ representing the horizontal edges; $E_h(i, j)$ is true if and only if there is an edge between (i, j) and $(i+1, j)$, and $E_v \subseteq \{1, \ldots, n\} \times \{1, \ldots, k-1\}$; $E_v(i, j)$ is true if and only if there is an edge between (i, j) and $(i, j+1)$. The vertices s and t are represented by two indicator variables for each vertex v, one which is true iff $v = s$ and one which is true iff $v = t$. With the encoding made clear, we can now show the non-uniform part of Theorem 1.

Lemma 3. USTCON$_k$ *is Π_k, for $k \geq 3$. The constructed circuit is positive (monotone) in the edge variables.*

Proof. We first show that for all $k \geq 1$, the statement "there is a path from vertex s to vertex t in G" for *fixed boundary* vertices s and t can be computed by a positive Π_k circuit; that is, we do not let s and t be part of the input and we assume them to be on the boundary of the grid. Then, we generalize, first to the case of non-boundary vertices, and then to s and t being given as input.

Base, $k = 1$: There is a path between s and t if and only if, for all a, if a is an edge position between s and t, a is an edge. This is a Π_1 statement in the edge indicator variables, as desired.

Now suppose $k > 1$. Given a grid graph G on $V_{n,k}$, we define its *dual* G^* as follows: G^* has a vertex s^* for each square s of the grid and a vertex ∞ representing the region outside the grid. We put an edge between two vertices u^* and v^* of G^* if and only if the edge position separating u and v in G is *not* an edge. Thus, for every edge position e of G there is an edge position e^* of G^* and exactly one of G or G^* has an edge at that position. G and G^* can be simultaneously embedded in the plane. Note that $G^* - \{\infty\}$ is a grid graph on $V_{n-1,k-1}$.

Now, for any given vertices s and t of G, there is a path between s and t in G if and only if there is *not* a simple cycle in G^* so that if the cycle is drawn in the plane, s is on the outside of the cycle and t is on the inside of the cycle. Since s and t are border vertices, such a cycle must go through the vertex ∞. Let C be some cycle going through ∞ and let e_1^* and e_2^* be the two edges adjacent to ∞ on the cycle. C separates s and t if and only the edge positions e_1 and e_2 separate s and t in the following sense: If one tracks the border clockwise from s back to itself, one of e_1 and e_2 is found before hitting t and the other is found after.

Thus, there is a path from s to t if and only if for all border edge positions e_1 and e_2, such that e_1 and e_2 separate s and t, there is *not* a path in $G^* - \{\infty\}$ from u^* to v^*, where u is the square of G adjacent to e_1 and v is the square of G adjacent to e_2.

The statement "e_1 and e_2 separate s and t" is independent of the input. Since $G^* - \{\infty\}$ is a grid graph of width $k - 1$ there is a positive Π_{k-1} circuit deciding whether a path between two fixed border vertices exists. Note that the inputs of this circuit are edge indicators for G^*, i.e. negations of edge indicators for G. Thus, using DeMorgan's law, checking whether *no* path between two fixed border vertices exists can be done by a positive Σ_k circuit in the primal edge indicator variables. We conclude that the validity of the entire statement can be checked by a positive Π_k circuit, as desired.

Now consider the more general problem, where s and t are not on the boundary, but still fixed. Assume without loss of generality that s is to the left of t or immediately above t. Split the graph into 3 parts — the part left of s, the part between s and t, and the part right of t. Compute the transitive closure for each component, restricted to the vertical border vertices. By the above, this can be done by $O(k^2)$ Π_k circuits, i.e. a constant number. The end result is now a monotone Boolean function of the computed information. Since the amount of information is constant, we can compute this function with a positive NC^0

circuit. Since Π_k is closed under positive finite Boolean combinations, the entire thing is Π_k.

Finally, consider the USTCON$_k$ problem with s and t being part of the input. Recall that they are given by two indicator variables for each vertex. For each value of s and t we can construct a gate $E_{s,t}$ which evaluates to 1 if and only if the inputs are s and t; this gate is just an AND of two indicator variables. For each possible value of (s,t), construct the Π_k circuit $C_{s,t}$ solving the problem for this value. Now, we adjust $C_{s,t}$ so that it outputs 1, if s or t do not match the actual input. We do this by giving each of the OR gates of the second layer from the top of $C_{s,t}$ one additional input, namely the negation of $E_{s,t}$. The end result is the AND of all these adjusted $C_{s,t}$ circuits. There is no penalty in depth if $k \geq 3$. Note that the final circuit is no longer positive, but the only negative literals are these $E_{s,t}$'s.

Lemma 4. *For $k \geq 1$, every problem in non-uniform Π_k reduces to USTCON$_k$ by a non-uniform p-projection.*

Proof. We will show the following stronger statement: For every $k \geq 1$, every problem in non-uniform Π_k reduces to USTCON$_k$ and every problem in non-uniform Σ_k reduces to USTCON$_{k+1}$ by p-projections. Furthermore, the value of the node s in the reduction is the bottommost left corner of the grid and the value of the node t in the reduction is the bottommost right corner of the grid.

Given a Π_k circuit of size s, we can construct a Π_k formula of size $s^{O(1)}$ computing the same function, so we can assume without loss of generality that we are given a function which can be computed by a Π_k formula. By the definition of p-projection [15], an alternative formulation of the statement is then this:

Given a Π_k formula C, we can construct a polynomial sized, width k grid graph $G(C)$ where some of the edges are labelled with input variables or their negations, the bottommost left corner of the grid is labelled s and the bottommost right corner of the grid is labelled t, so that, given an input vector $\mathbf{x}$, if we remove the edges labelled with variables assigned 0, there is a path form s to t in $G(C)$ if and only if $C(\mathbf{x})$ evaluates to true. Similarly, given a Σ_k circuit, we can construct a width $k+1$ grid graph with corresponding properties. We will construct this mapping G by recursion in k.

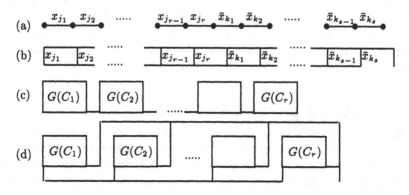

Fig. 2. The reduction

First suppose a Π_1 formula C is given. We can write C as $\bigwedge_{i=1}^{r} x_{j_i} \wedge \bigwedge_{i=1}^{s} \bar{x}_{k_i}$, where $x_{j_i}, i = 1 \ldots r$ and $x_{j_i}, i = 1 \ldots s$ are input variables. The corresponding width 1 grid graph $G(C)$ is shown in figure 2(a). Similarly, if a Σ_1 formula C is given, we write C as $\bigvee_{i=1}^{r} x_{j_i} \vee \bigvee_{i=1}^{s} \bar{x}_{k_i}$ and let $G(C)$ be the width 2 grid graph of Figure 2(b).

Now, let $k > 1$ and assume we have both the Σ_j and Π_j constructions for all $j < k$. Let a Π_k formula C be given. We can write it as $\bigwedge_{i=1}^{r} C_i$, where the C_i's are Σ_{k-1} formulae. Construct the width k graphs $G(C_i)$ corresponding to the C_i's and let $G(C)$ be the graph of Figure 2(c). Note that this graph also has width k, as desired. Finally, let a Σ_k formula C be given. Write it as $\bigvee_{i=1}^{r} C_i$, where the C_i's are Π_{k-1} formulae. Construct the width $k-1$ graphs $G(C_i)$ corresponding to the C_i's and let $G(C)$ be the width $k+1$ graph of Figure 2(d).

The correctness of the construction is easily checked.

Lemma 3 and Lemma 4 together gives us the non-uniform part of Theorem 1. We now make precise the content of the uniform part.

As in Barrington, Immerman, and Straubing [1], we define a *log-time* Turing machine to have a read-only input tape of length n, a constant number of read-write work tapes of total length $O(\log n)$, and a read-write input address tape of length $\log n$. On a given time step the machine has access to the bit of the input tape denoted by the contents of the address tape (or to the fact that there is no such bit, if the address tape holds too large a number). An *alternating* log-time machine has universal and existential states with the usual semantics. Furthermore, the alternating machine queries its input only once in a computation, in its last step. We now define uniform Π_k as the class of languages accepted by alternating log-time Turing machines, making exactly k alternations, the first being universal. It is shown in [1] that this hierarchy is in fact a uniform version of the AC^0 circuit hierarchy, where specific questions about the circuit family can be answered by a log-time Turing machine.

Recall that a family of p-projections can be viewed syntactically as a family of maps $\sigma_n : \{y_1, y_2, \ldots, y_{m(n)}\} \rightarrow \{0, 1, x_1, x_2, \ldots, x_n, \bar{x}_1, \bar{x}_2, \ldots, \bar{x}_n\}$, where $m(n)$ is bounded by a polynomial in n. A DLOGTIME-uniform projection is a family of projections σ_n, so that there is a log-time Turing machine which on input $\langle i, 1^n \rangle$ outputs the binary encoding of the values of $m(n)$ and $\sigma_n(y_i)$ on a specified work tape. It is easy to see that DLOGTIME-uniform projections are closed under composition and that uniform Π_k and Σ_k are closed under DLOGTIME-uniform projections. To obtain the uniform part of Theorem 1, we essentially now have to redo Lemma 3 and Lemma 4, checking that they are valid in the uniform setting. Due to lack of space, this is omitted from this version of the paper.

3 The width-k grid graph monoid

In this section we consider an algebraic interpretation of our results. We explore its consequences in Section 4, but we also consider it interesting in its own right.

We define the width k grid graph monoid G_k. It will be a submonoid of the following monoid M_k: The ground set of M_k is the set of equivalence relations on $V_{2,k} = \{1,2\} \times \{1,\ldots,k\}$, or, equivalently, the set of transitively closed undirected graphs with vertex set $V_{2,k}$. Let G and H be members of M_k, viewed as graphs. We now define the composition of G and H. Let $U = \{1,\frac{3}{2},2\} \times \{1,\ldots,k\}$. Let R be the graph on U obtained by embedding G in U by the embedding $(1,y) \to (1,y),(2,y) \to (\frac{3}{2},y)$ and embedding H in U by the embedding $(1,y) \to (\frac{3}{2},y),(2,y) \to (2,y)$. Let R^* be the transitive closure of R. The composition $G \circ H$ is the restriction of R^* to $V_{2,k} = \{1,2\} \times \{1,\ldots,k\}$. G_k is now defined to be the submonoid of M_k generated by the transitive closures of the set of grid graphs on $V_{2,k}$.

A very intuitive way of viewing G_k is as follows. An element of G_k is a collection of plane blobs inside the $[1,2] \times [1,k]$ rectangle in the plane, where a blob is identified with the set of grid points it contains. In Figure 3, (a) are (b) are two such elements. To multiply two elements, we concatenate them and scale down the resulting picture by a factor of two on the x-axis. In Figure 3, (a) and (b) are concatenated to form (c) and then scaled down to (d). Finally, since two blobs are equivalent if they contain the same elements, we can make a nicer picture (e) which is equivalent to (d).

Fig. 3. Elements in G_k and their product

Note that every member of G_k can be described as a word $a_1 \circ a_2 \circ \ldots a_{f(k)}$ where the a_i's are closures of $2 \times k$ grid graphs for some function f (a trivial upper bound on $f(k)$ is $|G_k|$). Another way of viewing this: Every member of G_k can be described by the transitive closure of an $f(k) \times k$ grid graph, restricted to the vertical border vertices.

The size of G_k is $c_{2k} = \frac{1}{2k+1}\binom{4k}{2k}$, i.e. the $2k$'th Catalan number. We omit the proof of this in this version of the paper. Our earlier analysis of grid graphs can now be interpreted in terms of the algebraic structure of G_k. To use the vocabulary of formal language theory, we may think of a grid graph problem as a string where the individual letters are elements of G_k. The following result tells us something about the language of strings representing graphs with a particular connectivity property. It is star-free (meaning that it can be formed from one-letter languages by concatenation and boolean operations including complementation), and has dot-depth k (meaning that the optimal depth of nesting of concatenation operations is k). See, for example, [12] for further background on algebraic automata theory.

Lemma 5. G_k is an aperiodic monoid with dot-depth exactly k.

Proof. (sketch) We first show that the dot-depth is at most k. In our proof of Theorem 1, we essentially showed that connectivity in a width-k grid graph could be expressed by a logical formula with k quantifiers in prenex normal form, and atomic predicates that either referenced individual edges (properties of individual "letters" in the input string) or compared two column numbers (positions of letters in the string). By a theorem of Thomas [16], any language so describable has a syntactic monoid that is aperiodic with dot-depth k. But this syntactic monoid is G_k itself, since G_k was designed to exactly capture this connectivity information.

If the dot-depth of G_k were less than k, we could derive a contradiction as follows. Consider any circuit C of depth k and size s. By our construction in Lemma 4, we can construct a word over G_k, of size polynomial in s, whose product determines the value of C. But if G_k has dot-depth $k-1$ or less, this product can be evaluated by a circuit of depth $k-1$ and size polynomial in s, using a construction of Barrington and Thérien [2]. Since C was arbitrary, we have collapsed two distinct levels of the AC^0 hierarchy, contradicting a theorem of Sipser [14].

In Section 4, we only use the aperiodicity of G_k, not its dot-depth. Still, G_k gives us an example of a natural (or at least easily visualizable) monoid which we know is aperiodic with dot depth exactly k — such examples are rare.

4 The dynamic grid graph connectivity problem

We consider the following *dynamic* graph problem: Maintain, on a random access machine with word size $O(\log n)$, a data structure representing an $n \times k$ grid graph under insertions and deletions of edges and connectivity queries, i.e. queries asking whether there is a path between two given vertices.

We are now ready to show Theorem 2 from the introduction. For the upper bound, we shall use a result of Frandsen, Miltersen, and Skyum [7]. Let S be a finite monoid. The dynamic range query problem for S is the problem of maintaining, on a random access machine with word size $O(\log n)$, a sequence $(a_1, a_2, \ldots, a_n) \in S^n$ under a change(i, b)-operation which changes a_i to $b \in S$, and an operation query(i, j) which returns $a_i \circ a_{i+1} \circ \cdots \circ a_{j-1} \circ a_j$. Frandsen, Miltersen, and Skyum show that if S is aperiodic, there is a solution to the dynamic range query problem for S with time bound $O(\log \log n)$ per operation. The constant in the big-O is $2^{O(|S|)}$.

We show that dynamic reachability reduces to dynamic range queries over G_k with a constant overhead. By Lemma 5, we have shown the upper bound of Theorem 2. Suppose we are to maintain a graph on $\{1, .., n\} \times \{1, ..., k\}$. We maintain the product $a_1 \circ a_2 \circ ... \circ a_{n-1}$ where a_i is the element of G_k correponding to the subgraph in $\{i, i+1\} \times \{1, ..., k\}$. A change in the graph corresponds to a single change of an a_i. If we are to answer if there is a path from (x_1, y_1) to (x_2, y_2) we do the following (assuming $x_1 < x_2$): We query the subproducts $a = a_1 \circ ... \circ a_{x_1-1}, b = a_{x_1} \circ ... \circ a_{x_2}$ and $c = a_{x_2+1} \circ ... \circ a_n$. Whether or not there is a path between (x_1, y_1) and (x_2, y_2) is completely determined by (a, b, c, y_1, y_2), and since these are all in a constant range, we can hardwire the answer for each possible value of the tuple into the algorithm.

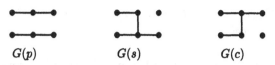

$$G(p) \qquad\qquad G(s) \qquad\qquad G(c)$$

Fig. 4. Gadgets for reducing the dynamic prefix problem for L to dynamic connectivity

We now show the lower bound part of Theorem 2. We shall use a result by Beame and Fich [3]. A regular language L is called *indecisive* if and only if for all strings x, there exist z and z' such that $xz \in L$ and $xz' \notin L$. Given a language L over an alphabet Σ, the dynamic prefix problem for L is the problem of maintaining a string $x = x_1 \ldots x_n \in \Sigma^n$ under a change(i, a) operation which changes x_i to a and a prefix(j) operation which answers the question "Is $x_1 x_2 \ldots x_j \in L$?". Beame and Fich show that if L is indecisive, then, in any implementation of the dynamic prefix problem for L on a RAM with word size $O(\log n)$, if the change operation takes time at most $2^{(\log n)^{1-\Omega(1)}}$ in the worst case, then the query operation takes time at least $\Omega(\log \log n / \log \log \log n)$ in the worst case.

Now let L be the regular language $(c + s + p)^* sp^* + p^*$, i.e. the language over $\{c, s, p\}$, where $x \in L$ if and only if the last letter of x which is not a p is an s. This language is indecisive, so Beame and Fich's result apply. Now assume that the dynamic connectivity problem for width 2 grid graphs can be solved with time bound $o(\log n \log n / \log \log \log n)$ per operation. We will show that the dynamic prefix problem for L can also be solved with time bound $o(\log n \log n / \log \log \log n)$, a contradiction.

Given an instance $a_1 a_2 \cdots a_n$ of the prefix problem to maintain, we maintain a grid graph G, of width 2 and length $2n$, defined as follows. Consider G to be divided into n blocks of length 2. The i'th block of G is $G(a_i)$, where G is the mapping defined in Figure 4. Clearly, a change of a symbol in the maintained string corresponds to a constant number of insert and delete operations in the dynamic graph. Now, in order to determine if $a_1 a_2 \ldots a_i \in L$, we remove all edges in the $i + 1$'st block of G using the delete operation of the dynamic connectivity operation. Then we ask if there is a path from the bottom left vertex of G to the bottom right vertex of the i'th block of G. This is the case if and only if $a_1 a_2 \ldots a_i \in L$. After getting the right answer, we restore the data structure by reinserting the edges of block i. This completes the reduction, and the proof of Theorem 2.

5 Generalization to directed graphs

In a directed grid graph, each edge present has one or both of the two possible orientations, and we consider finding directed paths from s to t. (Such graphs correspond to planar nondeterministic branching programs or planar "switching-and-rectifier networks" [13], except that a directed grid graph need not have horizontal arrows in only one direction. It follows from the analysis here, of course, that this additional ability is of no use in the case of constant width.)

All of our theorems about constant-width grid graphs hold for directed grid graphs. Of course, the lower bounds are trivial extensions, but we must revisit the upper bounds:

Lemma 6. *STCON$_k$ is in uniform Π_k, with the constructed circuit being positive in the edge variables.*

Proof. (sketch) Given a directed grid graph G, we will define its dual G^* as follows. The possible edge positions of G^* are exactly as in the undirected case. Now let e^* be an edge position of G^*, we define which orientations of e^* are present in G^* as a function of the orientations of e present in G: An orientation of e^* is present if and only if the orientation, turned 90° degrees clockwise, is *not* an orientation of e.

Now it is easy to see that there is a directed path from s to t if and only if there is *not* a directed cycle in G^*, going clockwise around t, with s on the outside. The rest of the proof proceeds exactly as in the proof of Lemma 3 — the only operations needed on the column numbers are comparisons.

As before, we can define a monoid whose elements are now directed $2 \times k$ grid graphs, and show that this monoid is aperiodic with dot-depth exactly k. As a corollary, we also get the same upper bound on the directed dynamic grid graph connectivity as on undirected dynamic grid graph connectivity. (Interestingly, for general graphs, the directed version of the dynamic problem seems to be much harder than the undirected version).

6 Discussion and open problems

- As mentioned in the introduction, Blum and Kozen showed that the general problem, where the width is not fixed, is in L. By carrying out our construction in Section 4 for a width of $\log n$, we get that even this restricted version of the general problem is hard for NC^1. An obvious open question is to determine the complexity of the general problem precisely: Is it complete for NC^1, or for L, or does it have intermediate complexity?
- We can also consider the general version of the problem for directed grid graphs, which is still hard for NC^1 but which we only know to be in NL. Our notion of duality gives a simple positive reduction of this problem to its complement, suggesting (but of course not proving, as NL is in fact closed under complement, by a more complicated reduction) that it is not NL-complete. There are potentially interesting restrictions of this problem as well, where we prohibit edges in one or two of the four directions.
- A similar gap occurs in our understanding of the dynamic grid graph connectivity problem, if the width of the graph is non-constant. If the width is a free parameter m, with the restriction $2 \leq m \leq n$, the following is known: Eppstein *et al* [6] construct a data structure with a time bound of $O(\log n)$ per operation and Eppstein [5] shows a lower bound of $\Omega(\log m/\log\log m)$. This lower bound is improved by Husfeldt and Rauhe [8] to $\Omega(m)$, provided $m \leq \log n/\log\log n$. Our upper bound is $2^{2^{O(m)}} \log\log n$ and our lower bound is $\Omega(\log\log n/\log\log\log n)$, provided $m \geq 2$. Combining everything, we get a lower bound of $\Omega(\min(m, \log n/\log\log n) + \log\log n/\log\log\log n)$ and an upper bound of $O(\min(\log n, 2^{2^{O(m)}} \log\log n))$ with a big gap to close.

Acknowledgements

We would like to thank Paul Beame and Arny Rosenberg for help with historical references, Howard Straubing and Denis Therien for very helpful discussions about automata theory and monoids, and Sairam Subramanian for very helpful discussions about dynamic graph problems. This work was greatly facilitated by the March 1997 Dagstuhl workshop in Boolean Function Complexity, attended by the first and third authors.

References

1. D. A. M. Barrington, N. Immerman and H. Straubing. On uniformity within NC^1 *Journal of Computer and System Sciences*, 41(3):274–306.
2. D. A. M. Mix Barrington and D. Thérien. Finite monoids and the fine structure of NC^1. *Journal of the ACM*, 35(4):941–952, October 1988.
3. P. Beame and F. Fich. On searching sorted lists: A near-optimal lower bound. Manuscript, 1997.
4. M. Blum and D. Kozen. On the power of the compass (or why mazes are easier to search than graphs). In *19th Annual Symposium on the Foundations of Computer Science*, pages 132–142, October 1978.
5. D. Eppstein. Dynamic connectivity in digital images. Technical Report 96-13, Univ. of California, Irvine, Department of Information and Computer Science, 1996.
6. D. Eppstein, G. Italiano, R. Tamassia, R. E. Tarjan, J. Westbrook, and M. Yung. Maintenance of a minimum spanning forest in a dynamic planar graph. *Journal of Algorithms*, 13:33–54, 1992.
7. G. S. Frandsen, P. B. Miltersen, and S. Skyum. Dynamic word problems. *Journal of the ACM* 44:257–271, 1997.
8. T. Husfeldt and T. Rauhe. Hardness results for dynamic problems by extensions of Fredman annd Saks' chronogram method.. Manuscript, 1997.
9. N. Immerman. Languages that capture complexity classes. *SIAM Journal on Computing*, 16(4):760–778, 1987.
10. N. Immerman and S. Landau. The complexity of iterated multiplication. *Information and Computation*, 116(1):103–116, January 1995.
11. A. Itai, C. H. Papadimitriou, and J. L. Szwarcfiter. Hamilton paths in grid graphs. *SIAM Journal on Computing*, 11(4):676–686, 1982.
12. J. E. Pin. *Varieties of Formal Languages*. New York: Plenum Press, 1986.
13. A. A. Razborov. Lower Bounds for deterministic and nondeterministic branching programs. In L. Budach, ed., *Fundamentals of Computation Theory, 8th International Conference: FCT '91*. Lecture Notes in Computer Science 529, 47–60. Berlin, Springer Verlag, 1991.
14. M. Sipser. Borel sets and circuit complexity. In *Proceedings, 15th ACM Symposium on the Theory of Computing*, 1983, 61–69.
15. S. Skyum and L. G. Valiant. A complexity theory based on Boolean algebra. *Journal of the ACM*, 32(2):484–502, April 1985.
16. W. Thomas. Classifying regular events in symbolic logic. *J. Comput. System Sci.* 25, 1982, 360–376.

Nearly Optimal Language Compression Using Extractors

Lance Fortnow* and Sophie Laplante**

University of Chicago

Topics: Computational and structural complexity, Kolmogorov complexity.

Abstract. We show two sets of results applying the theory of extractors to resource-bounded Kolmogorov complexity:

- Most strings in easy sets have nearly optimal polynomial-time CD complexity. This extends work of Sipser [Sip83] and Buhrman and Fortnow [BF97].
- We use extractors to extract the randomness of strings. In particular we show how to get from an arbitrary string, an incompressible string which encodes almost as much polynomial-time CND complexity as the original string.

1 Introduction

The Kolmogorov complexity of a string x, denoted $C(x)$, is the length of the shortest program which prints out that string. Researchers have used Kolmogorov complexity in extremely powerful ways to help prove results in many areas of theoretical computer science (see the textbook by Li and Vitányi [LV97]). One of the most important tools is the following lemma that allows us to measure sizes of sets with Kolmogorov complexity.

Lemma 1. *Let A be a recursive set. For all n and for all $x \in A \cap \Sigma^n$,*

$$C(x) \le \log |A \cap \Sigma^n| + O(\log n).$$

A language L is called *compressible* if there is a recursive function which is one-to-one on L and whose inverse is also recursive. The function must have the additional property that it is length-decreasing except for possibly finitely many

* University of Chicago, Department of Computer Science, 1100 E. 58th St., Chicago, IL 60637. Email: fortnow@cs.uchicago.edu. URL: http://www.cs.uchicago.edu/~fortnow. Supported in part by NSF grant CCR 92-53582, the Fulbright scholar program and the Dutch Foundation for Scientific Research (NWO). Some of this research was done while on leave at CWI in Amsterdam.
** Current address: Laboratoire de Recherche en Informatique, Bât. 490, Université Paris Sud, 91405 Orsay Cedex, France. Email: laplante@lri.fr. Supported in part by NSF grant CCR 92-53582.

instances of L. The question becomes interesting when we consider polynomial-time language compression, and in this paper, we achieve bounds close to that of Lemma 1.

Sipser [Sip83] is the first to have considered extending Lemma 1 to resource-bounded Kolmogorov complexity. He gets a time-bounded result similar to Lemma 1 with a few differences. Instead of A recursive, we have A computable in polynomial-time as expected. Instead of using polynomial-time $C(x)$ complexity, Sipser uses polynomial-time $CD(x)$ complexity where we look at the smallest program that *recognizes* x when x is given as its input, instead of *printing* it as in $C(x)$ complexity. Fortnow and Kummer [FK96] give strong evidence that CD complexity is necessary here.

Sipser requires one more ingredient—an additional polynomially long string r, depending only on the length of x, that can be used by the program recognizing x. Even with all of these restrictions, Sipser uses the resource-bounded version of Lemma 1 to give the first proof that BPP lies in the polynomial-time hierarchy.

Is this extra string r needed? Buhrman and Fortnow show how to eliminate the r at the cost of doubling the bound on the Kolmogorov complexity. They give several applications, relating CD complexity to important questions in complexity related to enumerating, computing and isolating satisfying assignments such as an oracle relative to which $EXP = NEXP$ but where accepting paths in a $NEXP$ machine cannot always be found in exponential time.

In this paper we continue to improve bounds for language compression, and eliminate the extra string r with bounds that nearly match the tight bounds given in Lemma 1. We unfortunately do have the caveat that our bounds only work for most strings in A instead of all of them. To prove our result we use recent combinatorial work on extractors. Our results extend to nondeterministic polynomial-time as well.

Extractors are a collection of bipartite graphs with many more nodes on the left than the right and of relatively small degree. Extractors have the property that for any large enough collection S of nodes of the left, the process of randomly choosing a neighbor of a node in S generates a close to uniform distribution of the nodes on the right. We give more details in Section 2.1. We also recommend the excellent survey on extractors by Nisan [Nis96].

Extractors get their name from their ability to "extract" randomness from a distribution on strings. One would expect some connections with Kolmogorov complexity, one of the best known measures of "randomness". We show how to use extractors to achieve this connection in time-bounded Kolmogorov complexity.

How hard is it given a string x to find a string y such that y is random and y has roughly the same length as x? Note that y "extracts" out the randomness from x. In traditional Kolmogorov complexity one can describe y by x and only $\log n$ additional bits—the size of the smallest program for x. For polynomial-time complexity this attack appears not to work. However, we can use extractors to extract the randomness. We show that given a string x of high CND^p complexity we can find a string y that captures most of the randomness of x using only a small additional number of bits.

2 Preliminaries

We start by giving some basic definitions of Kolmogorov complexity and to introduce the notion of an extractor graph. We follow the presentation in Nisan's survey paper [Nis96].

2.1 Extractors

An extractor can be thought of as a bipartite graph, whose first color class is larger than the second color class. By convention we think of the first color class as being on the left, and the second on the right. The vertices on the left side are all the strings of length n, so the first color class can be equated with the set $[N]$, where $N = 2^n$. Likewise, the vertices on the right side of the graph are labeled by strings of length $m \leq n$, so we let $M = 2^m$ and $[M]$ is equated with the vertices in the second color class.

Distributions We will be choosing a node on the left side of the graph at random according to a distribution X. The result of choosing a neighbor uniformly at random in the graph will produce a distribution Y on vertices on the right.

The *min-entropy* of a distribution X over $[N]$ can be thought of as a measure of the randomness present in a string x chosen according to X. The min-entropy of X is defined to be $\min\{-\log_2 X(x)|x \in [N]\}$.

A distribution Y is said to be ε−close to Z if both distributions are over the same space $[M]$, and that for any $S \subseteq [M], |Y(S) - Z(S)| \leq \varepsilon$.

Definition of extractors A bipartite graph G with (independent) vertex sets $[N]$ and $[M]$, $N = 2^n, M = 2^m$, and for which the degree of all the vertices in the first color class is bounded by $D = 2^d$ is an $(n, k, d, m, \varepsilon)$ *extractor* if given any distribution X on the N vertices whose min-entropy is at least k, the result of choosing an x according to this distribution and a random neighbor y of x in the graph is ε-close to the uniform distribution over $[M]$. In our setting, the distribution X will be the uniform distribution over a subset $A \subseteq [N]$, so k will be $\log |A|$.

$\Gamma(x)$ denotes the set of neighbors of x in G when x is a vertex on the left side of the graph. The number of edges originating at some vertex x on the left side of the graph is called the *outdegree* of x, whereas the number $w(y)$ of edges adjacent with a vertex y on the right side of the graph is called the *indegree* (or weight) of y. $G(x, r)$ represents the rth neighbor of x in the graph, where multiple edges are allowed. When y is a vertex on the right side of the graph, $\Gamma_A^{-1}(y)$ is the subset of preimages of y which lie in A. The notation extends to sets in the natural way.

Best known explicit constructions The results we state are subject to improvement if better explicit extractor constructions are found. We have stated our results in general terms so that new results on extractors will be immediately applicable.

The current best known explicit constructions for extractors are due to Ta-Shma and Zuckerman [Ta-96,Zuc96]. The Ta-Shma construction is more useful for our purposes.

Theorem 2. (Ta-Shma) *There is an explicit construction which for every* $n, \varepsilon = \varepsilon(n)$ *and every* $m = m(n) \leq n$ *yields an extractor with parameters* $(n, m, \log^{O(1)}(n/\varepsilon), m, \varepsilon)$.

It is useful to compare this construction to the current lower bound on extractors, due to Nisan and Zuckerman [NZ93].

Theorem 3. (Nisan-Zuckerman) *There is a constant* c *such that for all* $n, m, k \leq n - 1, \varepsilon < 1/2$, *if there is an extractor whose parameters are* $(n, k, d, m, \varepsilon)$, *then it must be the case that* $d \geq \min\{m, c \log(n/\varepsilon)\}$.

This lower bound also gives a good indication as to the limits of the techniques described in this paper.

For more information on extractors we recommend the survey paper by Nisan [Nis96].

2.2 Kolmogorov and CD Complexity

Several variants on Kolmogorov complexity are used in this paper. We give the definitions below, following the notation put forth in the textbook of Li and Vitányi [LV97].

The *time-bounded Kolmogorov complexity* of a string x relative to a string y, written $C^t(x|y)$, is the length of the shortest program p which on input y, prints out the string x in time bounded by $t(|x| + |y|)$. When y is the empty string, we write $C^t(x)$.

The *time-bounded Kolmogorov distinguishing complexity* of a string x relative to a string y, written $CD^t(x|y)$, is the length of the shortest program p which on input y, z, outputs 1 when $z = x$ and 0 on all other strings. The time taken on all inputs y, z must be bounded by $t(|p| + |y| + |z|)$. When y is the empty string, we write $CD^t(x)$.

The *nondeterministic time-bounded Kolmogorov distinguishing complexity* of a string x relative to a string y, written $CND^t(x|y)$, is the length of the shortest program p which has the following behavior. If $z = x$, then there is a witness w of length bounded by $t(|x|)$ such that p accepts on input y, z, w. If $z \neq x$, then no such witness can cause p to accept. The time taken on all inputs y, z, w must be bounded by $t(|p| + |y| + |z| + |w|)$. When y is the empty string, we write $CND^t(x)$.

For more information on Kolmogorov complexity we recommend the comprehensive book by Li and Vitányi [LV97].

3 Complexity Bounds on Easy Sets

Our theorems improve upon the results of Sipser [Sip83] and Buhrman and Fortnow [BF97] in the sense that our bounds are stronger and the strings we obtain have high mutual information. The price we pay for these improvements is that our bounds apply only to "most" strings, not to all strings as in the work of Sipser, and Buhrman and Fortnow [Sip83,BF97].

Theorem 4 (Sipser). *For every set $A \in \mathcal{P}$, there is a polynomial p and a constant c such that for every n and for most r of length $p(n)$, and for every $x \in A \cap \Sigma^n$, $CD^p(x|r) \leq \log(|A \cap \Sigma^n|) + c \log n$.*

Is the random string r necessary to prove Theorem 4? Buhrman and Fortnow show how to eliminate r at the cost of doubling the complexity.

Theorem 5 (Buhrman-Fortnow). *For any set $A \in \mathcal{P}$, there is a polynomial p and a constant c such that for all strings $x \in A \cap \Sigma^n$, $CD^p(x) \leq 2 \log(|A \cap \Sigma^n|) + c \log n$.*

We extend the work of Buhrman and Fortnow [BF97] by getting nearly the bound of Sipser [Sip83] without the random string used by Sipser. However, our result only works for most strings in A.

Using the extractor construction of Ta-Shma [Ta-96] we get the following theorem.

Theorem 6. *For any set $A \in \mathcal{P}$, $\varepsilon = \varepsilon(n)$, there is a polynomial p such that for all n and for all but a 2ε fraction of the $x \in A \cap \Sigma^n$, $CD^p(x) \leq \log |A \cap \Sigma^n| + \log^{O(1)}(n/\varepsilon)$.*

We give a few extensions of this result. In one extension, we consider sets in $\mathcal{NP}$ and find short strings whose CD complexity is close to the nondeterministic CD complexity of the original string. We also give a randomized version of these theorems, stating that the shorter string can be chosen at random and the probability of getting a short string which encodes as much information as the original string is bounded away from $1/2$.

Buhrman and Fortnow also bound the nondeterministic CD complexity of strings in sets in time-bounded nondeterministic classes. We state their result here and use it in the proof of one of our theorems.

Theorem 7. (Buhrman-Fortnow) *For any set $A \in \text{NTIME}[t(n)]$, there is a polynomial p and a constant c such that for all strings $x \in A \cap \Sigma^n$, $CND^{p \cdot t}(x) \leq 2 \log(|A \cap \Sigma^n|) + c \log n$.*

3.1 Extracting CD complexity for sets in $\mathcal{P}$

Theorem 6 follows immediately from the following result, using the explicit extractor construction of Ta-Shma. For this theorem we assume that there is an explicit extractor construction whose parameters are $(n, k, d, m, \varepsilon)$.

Theorem 8. *Fix a set A in $\mathcal{P}$, a polynomial $q(n)$ and $\varepsilon = \varepsilon(n)$. Then there is a polynomial $p(n)$ such that for all n and for all but a 2ε fraction of the $x \in A \cap \Sigma^n$, there is a y such that*

1. $|y| = m$
2. $CP(y|x) \le \log D + O(1)$
3. $CD^p(x) \le C(y) + 3 \log D + 2 \log(K/M) + c \log(n + \log(DK/M)) + O(1),$

where the extractor's parameters are determined by n and $k = \log|A \cap \Sigma^n|$, and c is an absolute constant from Theorem 5.

For the remainder of this section, we fix n and we let $S = |A \cap \Sigma^n|$. In our setting, we will think of the set $A \cap \Sigma^n$ as defining a distribution of min-entropy $k = \log S$. The string x represents an element of $A \cap \Sigma^n$ and y is one of its neighbors in the graph G. Hence, y has length m; computing y from x requires knowing only a short ("random") string of length $\log D$; and as we will see, y together with some short additional distinguishing information will suffice to distinguish the string x (in the sense of CD complexity).

In order to get a short description for x, we need to find a string y in its range which has small indegree (counting only those edges originating in A.)

Lemma 9. *Consider the restriction of the extractor to the set of edges originating in $A \cap \Sigma^n$. Let w_0 be an indegree threshold, $\frac{DS}{M} < w_0 \le DS$, and Y be a subset of vertices on the right hand side of the extractor graph. If $\forall y \in Y, w(y) > w_0$, then $|Y| \le \varepsilon \left(\frac{w_0}{DS} - \frac{1}{M} \right)^{-1}$.*

Proof. This is an easy counting argument. Let Y be the set of vertices whose indegree (in the restricted graph) exceeds w_0. Because the graph is an extractor, it must be the case that $\frac{w(Y)}{DS} - \frac{|Y|}{M} \le \varepsilon$. Since $w(Y) \ge w_0|Y|$, we get by simple rearrangement that $|Y| \le \varepsilon \left(\frac{w_0}{DS} - \frac{1}{M} \right)^{-1}$ as claimed.

Lemma 10. *If Y is a set on the right side of the graph containing at most $(1 - \varepsilon)S$ vertices, then $|\{x \in A \cap \Sigma^n : \Gamma(x) \subseteq Y\}| \le \left(\varepsilon + \frac{|Y|}{M} \right) S$.*

Proof. Once again this is a simple counting argument. The fact that the graph is an extractor tells us that $w(Y) \le \left(\varepsilon + \frac{|Y|}{M} \right) DS$. Each x for which $\Gamma(x) \subseteq Y$ contributes D edges to the indegree of Y, therefore the total number of such x cannot exceed $\left(\varepsilon + \frac{|Y|}{M} \right) S$, as claimed.

To conclude, we give the proof of Theorem 8.

Proof. Let A be a set in $\mathcal{P}$ and ε, n be given as in the statement of the theorem. By Lemma 9, applied with $w_0 = 2DS/M$ and Lemma 10 with Y as in the hypothesis of Lemma 9, the size of the subset $B \subseteq A \cap \Sigma^n$ such that $\forall x \in B$, $\forall y \in \Gamma(x)$, y has indegree at least w_0 can have size at most $2\varepsilon S$. Therefore for all but $2\varepsilon S$ of the x in $A \cap \Sigma^n$, there is a y in its range whose indegree is at most $2DS/M$. For each such x, let r_x be the label of one of the edges in G which connects x to such a y.

We need to verify 3 properties for each of these pairs x, y.

1. $|y| = m$: This is by choice of the extractor G.
2. $C(y|x) \leq \log D + O(1)$: $y = G(x, r_x)$ for some $r_x \in \Sigma^d$, so the algorithm to print y will contain an encoding of r_x, and on input x computes $G(x, r_x)$ and outputs the result.
3. $CD(x) \leq C(y) + 3\log D + 2\log(S/M) + c\log n + O(1)$: The program to recognize x will contain an encoding for an r_x and y for which $G(x, r_x) = y$ and the indegree of y is at most $2DS/M$. It must also contain a distinguishing program p which recognizes x, r_x among the $2DS/M$ edges originating in A that are adjacent to y. The length of p is bounded by $2\log(2DS/M) + c\log(n + \log(2DS/M))$, as given by Theorem 5. The total length of the program is $C(y) + \log(D) + 2\log(2DS/M) + c\log(n + \log(2DS/M)) + O(1)$.

3.2 Extracting CND complexity for sets in $\mathcal{NP}$

Using a slight variant of the proof of Theorem 8, we can get the following result about CND complexity. It is stronger in that it applies to sets in $\mathcal{NP}$, and although the bound on the complexity of x is for CND instead of CD complexity, the trade-off is that the bound is smaller by a term of $\log D$ and it involves only $CD(y)$ instead of $C(y)$.

Theorem 11. *Fix a set A in $\mathcal{NP}$, a polynomial $q(n)$, and $\varepsilon = \varepsilon(n)$. Then there is a polynomial $p(n)$ such that for every n and for all but a 2ε fraction of the $x \in A \cap \Sigma^n$, there is a y such that*

1. $|y| = \log|A \cap \Sigma^n|$
2. $C^p(y|x) \leq \log D + O(1)$
3. $CND^p(x) \leq CD^q(y) + 2\log D + c\log(n + \log D) + O(1)$.

The proof is essentially the same as that of Theorem 8. To simpify the notation we make the assumption that the extractor used achieves $k = m$, as does Ta-Shma's construction. To obtain property 3, we need only guess y, and verify our guess using a distinguishing program for y whose length is bounded by $CD^q(y)$. Likewise, we can simply guess r and omit its encoding, and use the distinguishing program p to verify our guess for r.

3.3 Randomly extracting CD complexity

Another trade-off we obtain to save a $\log D$ term is to choose a counterpart y to a string x in a set in $\mathcal{P}$ at random. We will only require that for most x, at least half of the edges from x map to a "good" y. Although this comes at the cost of only applying to "most" strings x, this improves upon the result of Sipser [Sip83] by reducing the length of the random string from $n^{O(1)}$ to $\log^{O(1)}(n/\varepsilon)$. The proof is similar to that of Theorem 8; it requires only a slight modification to the counting argument.

Theorem 12. *Fix a set A in $\mathcal{P}$, a polynomial $q(n)$, and $\varepsilon = \varepsilon(n)$. Then there is a polynomial $p(n)$ such that for every n and for all but a 4ε fraction of the $x \in A \cap \Sigma^n$, and at least half of the strings r of length D, there is a y such that:*

1. $|y| = \log |A \cap \Sigma^n|$
2. $C^p(y|x, r) \leq O(1)$
3. $CD^p(x|r) \leq C^q(y) + 2 \log D + 2 \log(n + \log D) + O(1).$

4 Extracting random strings

In the previous section, we used the fact that the strings examined were in a small set of bounded complexity, and we showed the existence of strings for which the mutual information was roughly the CND complexity of the original string. Here we use extractor techniques to a achieve a slightly different goal. We obtain an *incompressible* string whose length is close to the CD complexity of x and which can be computed from x using only $\log(n/\varepsilon)$ bits.

In the case of unbounded Kolmogorov complexity, it is easy to see that the following proposition is true.

Proposition 13. [LV97, Ex. 2.1.5, p. 102] *For any string x of length n, there is a y such that:*

1. $|y| = C(x)$
2. $C(y|x) \leq \log(n)$
3. $C(y) > |y| - O(1).$

Namely, y is a minimal-length program for x, and can be obtained from x by dovetailing, given the value of $C(x)$. In the time-bounded setting however, this argument fails, since dovetailing would take too much time. Our use of extractors is far afield from the above approach, yet it yields results surprisingly close to Proposition 13. (Non-explicit extractors actually allow us to give an alternate proof of Proposition 13, although this is more an artifact than a useful new proof.)

Theorem 14. *Fix a polynomial $q(n)$ and $\varepsilon = \varepsilon(n)$, and let $c > \log\left(\frac{1}{1-\varepsilon}\right)$ for large enough n. Then there exists a polynomial $p(n)$ such that for any string x, there is a string y such that:*

1. $|y| = m$
2. $C^p(y|x) \leq \log D + O(1)$
3. $C^q(y) > |y| - O(1),$

provided there is an explicit extractor with parameters $(n, k, d, m, \varepsilon)$ where $k = \frac{CND^{p \cdot D}(x) - O(\log n)}{2}.$

Proof. (Sketch) We consider a family of extractors with parameters $n, k, m(k)$. For fixed n, m, we let $A_{n,m} = \{x | \Gamma(x) \subseteq C[q(n), m - c]\}$, where $C[t, l] = \{z | C^t(z) \leq l\}$.

The fact that G is an extractor prohibits the set $A_{n,m}$ from being large, as we see now. If $|A_{n,m}| > 2^k$, then by the properties of the extractor graph, the weight on $C[q(n), m - c]$ induced by $A_{n,m}$ must be close to uniform, namely:

$$\frac{w(C[q(n), m - c])}{D|A_{n,m}|} \leq \frac{C[q(n), m - c]}{2^m} + \varepsilon.$$

Using the fact that $w(C[q(n), m - c]) = D|A_{n,m}|$ by definition of $A_{n,m}$ and that $|C[q(n), m - c]| \leq 2^{m-c}$, we get $1 \leq 2^{-c} + \varepsilon$. However, we have chosen $c > \log\left(\frac{1}{1-\varepsilon}\right)$ precisely to eliminate this possibility. Hence we must conclude that $|A_{n,m}| \leq 2^k$.

Now we may apply Theorem 7 to conclude that all $x \in A_{n,m}$ must have small CND complexity. First notice that verifying membership in $A_{n,m}$ is in NTIME$[D \cdot p]$ for some polynomial p, since it suffices to guess, for each neighbor y of x in $G_{n,m}$ a program of length $m - c$ which prints out y. Hence, for every $x \in A_{n,m}$, $CND^{D \cdot p}(x) \leq 2\log(|A_{n,m}|) + 2\log n + O(1)$.

Now consider x with respect to the extractor $G_{n,k,m(k)}$, where $k = \frac{1}{2}(CND^{D \cdot p}(x) - 2\log n - O(1) - 1)$ and m is maximal for this k. By the observation above, it must be the case that $x \notin A_{n,m}$. Therefore there must be a y not in $C[q(n), m - c]$ to which x is mapped under $G_{n,k,m}$. It is easy to verify that y satisfies the properties claimed in the statement of the theorem.

5   Extensions

We can state the results above in a yet more general form. Instead of requiring that the reference set A be in $\mathcal{P}$, all results carry over to the setting where the instance complexity of all instances in A is small, adding the appropriate (small) term to the bounds we obtain. We refer the reader to the paper of Orponen, Ko, Schöning, and Watanabe [OKSW94] or to section 7.4 of the textbook by Li and Vitányi [LV97] for more information on instance complexity. Note that this observation applies to the results of Buhrman and Fortnow [BF97] as well. Of particular interest are the sets in the class $IC[\log, \text{poly}]$, which is known to sit properly between the nonuniform classes $\mathcal{P}/\log$ and $\mathcal{P}/\text{poly}$.

6   Acknowledgments

We would like to thank Stuart Kurtz, Amber Settle, Harry Buhrman and David Zuckerman for several helpful discussions.

References

[BF97] H. Buhrman and L. Fortnow. Resource-bounded kolmogorov complexity revisited. In *Proceedings of the 14th Symposium on Theoretical Aspects of Computer Science*, volume 1200 of *Lecture Notes in Computer Science*, pages 105–116. Springer, Berlin, 1997.

[FK96] L. Fortnow and M. Kummer. Resource-bounded instance complexity. *Theoretical Computer Science A*, 161:123–140, 1996.

[OKSW94] P. Orponen, K. Ko, U. Schöning, and O. Watanabe. Instance Complexity *Journal of the ACM*, 41(1):96–121, 1994.

[LV97] M. Li and P. Vitányi. *An Introduction to Kolmogorov Complexity and Its Applications*. Graduate Texts in Computer Science. Springer, New York, second edition, 1997.

[Nis96] N. Nisan. Extracting randomness: How and why (a survey). In *Proceedings of the 11th IEEE Conference on Computational Complexity*, pages 44–58. IEEE, New York, 1996.

[NZ93] N. Nisan and D. Zuckerman. More deterministic simulation in logspace. In *Proceedings of the 25th ACM Symposium on the Theory of Computing*, pages 235–244. ACM, New York, 1993.

[Sip83] M. Sipser. A complexity theoretic approach to randomness. In *Proceedings of the 15th ACM Symposium on the Theory of Computing*, pages 330–335. ACM, New York, 1983.

[Ta-96] A. Ta-Shma. On extracting randomness from weak random sources (extended abstract). In *Proceedings of the 28th ACM Symposium on the Theory of Computing*, pages 276–285. ACM, New York, 1996.

[Zuc96] D. Zuckerman. Randomness-optimal sampling, extractors, and constructive leader election. In *Proceedings of the 28th ACM Symposium on the Theory of Computing*, pages 286–295. ACM, New York, 1996.

Random Sparse Bit Strings at the Threshold of Adjacency

Joel H. Spencer[1] and Katherine St. John[2]

[1] Courant Institute, New York University, 251 Mercer Street, New York, NY 10012
[2] Department of Mathematics, Santa Clara University, Santa Clara, CA 95053-0290

1 Abstract

We give a complete characterization for the limit probabilities of first order sentences over sparse random bit strings at the threshold of adjacency. For strings of length n, we let the probability that a bit is "on" be $\frac{c}{\sqrt{n}}$, for a real positive number c. For every first order sentence ϕ, we show that the limit probability function:

$$f_\phi(c) = \lim_{n \to \infty} \Pr[U_{n,\frac{c}{\sqrt{n}}} \text{ has the property } \phi]$$

(where $U_{n,\frac{c}{\sqrt{n}}}$ is the random bit string of length n) is infinitely differentiable. Our methodology for showing this is in itself interesting. We begin with finite models, go to the infinite (via the almost sure theories) and then characterize $f_\phi(c)$ as an infinite sum of products of polynomials and exponentials. We further show that if a sentence ϕ has limiting probability 1 for some c, then ϕ has limiting probability identically 1 for every c. This gives the surprising result that the almost sure theories are identical for every c.

2 Introduction

Expressibility is a central question in computer science. Over classes of ordered finite structures, membership in a complexity classes is often equivalent to the expressibility of the desired set in a given logic. For example, Immerman [6] showed that the expressibility in transitive closure logic is equivalent to NLOGSPACE, and Fagin [4] proved Σ_1^1 captures NPTIME. The characterizations of logics and the limit probabilities of their sentences over ordered structures could shed light on issues in complexity theory.

We focus on the class of ordered structures with a single unary predicate– that is, bit strings. Besides being a natural class to consider, logic, over bit strings, offers a useful tool to characterize the languages accepted by finite state automata. If we let the alphabet of our automata be $\{0, 1\}$, then the words in the language are bit strings. First order logic captures exactly the plus free regular languages, while monadic second order logic expresses the regular languages (see [7] and chapter 5 of [3]). We discuss the behavior of first order logic over random sparse bit strings and raise some interesting open problems about monadic second order logic in Section 5.

If we allow all bit strings to occur with equal probability, then for every first order sentence, ϕ,

$$\lim_{n \to \infty} \frac{\text{\# of models of size } n \text{ with property } \phi}{\text{total \# of models of size } n}$$

converges. We can focus on bit strings where a small number of bits are on by allowing a bit to be "on" with probability $p(n)$ that depends on n, the length of the string. The random unary predicate $U_{n,p}$ is a probability space over predicates U on $[n] = \{1, \ldots, n\}$ with the probabilities determined by $\Pr[U(x)] = p(n)$, for $1 \leq x \leq n$, and the events $U(x)$ are mutually independent over $1 \leq x \leq n$. $U_{n,p}$ is also called the *random bit string*. To see the correspondence, write each structure as a sequence of 0's and 1's, with the ith element in the sequence a 1 if and only if $U(i)$ holds in the structure. For example, if $n = 5$ and the unary predicate holds only on the least element, we write: [10000].

In [11], Shelah and Spencer showed that for every such sentence ϕ and for $p(n) \ll n^{-1}$ or $n^{-1/k} \ll p(n) \ll n^{-1/(k+1)}$, there is convergence for the limit probability. That is, there exists a constant a_ϕ such that

$$\lim_{n \to \infty} \Pr[U_{n,p} \models \phi] = a_\phi \tag{1}$$

Dolan [2] showed that for $p(n) \ll n^{-1}$ and $n^{-1} \ll p(n) \ll n^{-1/2}$ for every ϕ, $a_\phi = 0$ or 1 in Equation 1. This stronger convergence is called a Zero-One Law for $U_{n,p}$. Dolan also showed that the Zero-One Law does not hold for $n^{-1/k} \ll p(n) \ll n^{-1/(k+1)}$, $k > 1$.

In [14], we examine the random sparse bit strings with probability $p(n) = c/n$ and give a finer analysis than convergence. For this choice of p, we have the limit probabilities of ϕ are either

$$\sum_{i=1}^{i=m} e^{-c} \frac{c^{t_i}}{t_i!} \quad \text{or} \quad 1 - \sum_{i=1}^{i=m} e^{-c} \frac{c^{t_i}}{t_i!}$$

for some (possibly empty) sequence of positive integers $t_1, \ldots, t_m$. We achieve a simpler characterization for $p = c/n$ due to the simpler underlying structures. Other interesting structures that have also been examined in this fashion are random graphs (without order) with edge probability $p(n) = c/n$ and $p(n) = \ln n/n + c/n$ (see the work of Lynch, Spencer, and Thoma: [9], [10], and [13]).

For each real constant c, let S_c be the almost sure theory of the linear ordering with $p(n) = \frac{c}{\sqrt{n}}$. That is,

$$S_c = \{\phi \mid \lim_{n \to \infty} \Pr[U_{n, \frac{c}{\sqrt{n}}} \models \phi] = 1\}$$

Let T_1 be the almost sure theory for $n^{-1} \ll p(n) \ll n^{-1/2}$ and T_2 be the almost sure theory for $n^{-1/2} \ll p(n) \ll n^{-1/3}$. By [2], we have that T_1 is a complete theory. We characterize the theories at the threshold of adjacency, namely the

S_c's. These, in some sense, lie between T_1 and T_2. For each first order formula, ϕ, we define the function:

$$f_\phi(c) = \lim_{n \to \infty} \Pr[U_{n, \frac{c}{\sqrt{n}}} \models \phi]$$

where c ranges over the real, positive numbers. We show that $f_\phi(c)$ is infinitely differentiable. Moreover, we show:

Theorem 1 *For every first-order sentence ϕ, $f_\phi(c)$ is*

$$\sum_m e^{-tc^2} \frac{c^m}{m!} g(t, m)$$

where for every fixed t, $g(t, m)$ is the sum of the product of polynomials and exponentials in m.

In fact, the above theorem holds for any $t > 2^r$ where r is the quantifier rank of the sentence ϕ. The function $g(t, m)$ counts the number of equivalence classes of models (with respect to the maximum number, m, of pairs of 1's of distance at most t from one another) that satisfy ϕ.

We further show that if a sentence ϕ has limiting probability 1 for some c, then ϕ has limiting probability identically 1 for every c. This gives the surprising result that the almost sure theories are identical for every c.

Theorem 2 *Let $S = \bigcap_c S_c$ be the intersection of all the almost sure theories. Then, for every real, positive c, $S_c = S$.*

To prove these theorems, we look first at the countable models of the almost sure theories (for more on this, see [12]). Let $\mathcal{U} \models S_c$ be such a model. Each of these models satisfy a set of basic axioms Δ (defined in Section 3). Let $\mathbf{a} = (a_0, \ldots, a_{m-1})$ be a finite sequence of non-negative integers, representing the distance pairs of 1's occur apart. We show, using Ehrenfeucht-Fraisse games, that for every first order sentence, ϕ,

$$\Delta \cup \{\sigma_{\mathbf{a}, t}\} \models \phi \quad \text{or} \quad \Delta \cup \{\sigma_{\mathbf{a}, t}\} \models \neg\phi$$

where $t > 2^r$ for r the quantifier rank of ϕ, and $\sigma_{\mathbf{a}, t}$ is the first-order sentence that states the 1's that occur within t of one another are exactly those specified in the finite sequence $\mathbf{a}$ in exactly the order in $\mathbf{a}$. For example, if $\mathbf{a} = (2, 0)$, then $\sigma_{\mathbf{a}, 3}$ would say that the first pair of 1's, that occur within three of one another, occur with exactly 2 0's in between, and the next (and only other) pair of 1's that occurs within three are adjacent. We show that

$$q_{\mathbf{a}, t} = \Pr[U_{n, \frac{c}{\sqrt{n}}} \models \sigma_{\mathbf{a}, t}] = e^{-tc^2} \frac{c^{2m}}{m!}$$

To show our results, we "transfer" to another language. Let $\mathcal{L}$ be the language of random bit strings, and for each positive integer t, let $\mathcal{L}_t$ be the language with equality, linear ordering, and a t-valued function, d. We will define a function $\mathcal{F}_t$

that takes the countable models of the almost sure theories, in the language $\mathcal{L}$, to structures in the language, $\mathcal{L}'_t$. Roughly, $\mathcal{F}_t$ takes each model $\mathcal{U}$ to an ordered set of positive integers that characterize the pairs of 1's that occur within t of one another. If this set is finite, then it is the (unique) sequence of positive integers $\mathbf{a}$ such that $\mathcal{U} \models \sigma_{\mathbf{a},t}$. For example, if $\mathcal{U} \models \sigma_{(2,0),3}$, then $\mathcal{F}_t(\mathcal{U})$ is the two element structure $[0, 1]$ such that $d(0) = 2$ and $d(1) = 0$. Just as we wrote the unary structures as bit strings, we can write the models of $\mathcal{L}'_t$ as ordered sequences of $\{0, \ldots, t-1\}$. So, for our example, we would write $[2, 0]$. We show:

Theorem 3 *Fix a real, positive constant, c. Let $\mathcal{U}_1, \mathcal{U}_2$ be models of an almost sure theory S_c. If $\mathcal{F}_t(\mathcal{U}_1)$ and $\mathcal{F}_t(\mathcal{U}_2)$ agree on $\mathcal{L}'_t$-sentences of quantifier rank at most t (that is, $\mathcal{F}_t(\mathcal{U}_1) \equiv_t \mathcal{F}_t(\mathcal{U}_2)$), then $\mathcal{U}_1$ and $\mathcal{U}_2$ agree on all $\mathcal{L}$-sentences of quantifer rank at most t (that is, $\mathcal{U}_1 \equiv_t \mathcal{U}_2$).*

These facts give the desired form for $f_\phi(c)$ in Theorem 1 and are used to show Theorem 2.

3 Examples

To give some intuition about what these models and theories look like, we begin with an informal discussion of the almost sure theories T_1 and T_2. For T_1, we have $\frac{1}{n} \ll p(n) \ll \frac{1}{\sqrt{n}}$, and almost surely isolated 1's occur. To see this, let A_i be the event that $U(i)$ holds, X_i be the random indicator variable, and $X = \sum_i X_i$, the total number of 1's that occur (i.e. the total number of elements for which the unary predicate holds). Then, $E(X_i) = p(n)$, and by linearity of expectation,

$$E(X) = \sum_i E(X_i) = np(n).$$

We have $E(X) \to \infty$ as $n \to \infty$. Since all the events are independent, $\mathrm{Var}[X] \le E[X]$. By the Second Moment Method (see [1], chapter 4 for details),

$$\Pr[X = 0] \le \frac{\mathrm{Var}[X]}{E[X]^2} \le \frac{E[X]}{E[X]^2} = \frac{1}{E[X]} \to 0$$

Thus, $\Pr[X > 0] \to 1$. So, almost surely, arbitrarily many 1's occur. We can write this in first order logic as a schema of sentences α_r, each of which states "there is at least r 1's":

$$\alpha_r : (\exists x_1 \ldots x_r)(x_1 < x_2 < \cdots < x_r \wedge U(x_1) \wedge \cdots \wedge U(x_r))$$

Each of these 1's occurs arbitrarily far apart. To see this, let B_i be the event that i and $i+1$ are 1's, let Y_i be its random indicator variable, and $Y = \sum_i Y_i$. Then, $E(Y_i) = \Pr[B_i] = p^2$ and $E(Y) \sim np^2 \to 0$. So, almost surely, 1's occur, but no 1's occur adjacent in the order. If, for each $r > 0$, we let $C_{i,r}$ be the event that i and $i+r$ are 1's and $C_r = \sum_i C_{i,r}$, we can show, by similar argument, that $C_r \to 0$. This works for any fixed r, so, the 1's that do occur are isolated from one another by arbitrarily many 0's. So, almost surely, the schema of sentences β_r that state that "between every pair of 1's there is r 0's" hold:

$$\beta_r : (\forall x_1, x_2)[(U(x_1) \wedge U(x_2) \wedge x_1 < x_2) \rightarrow$$
$$(\exists y_1, \ldots, y_r)(\neg U(y_1) \wedge \ldots \wedge \neg U(y_r) \wedge x_1 < y_1 < \cdots < y_r < x_2)]$$

Thus, for every r, $\alpha_r, \beta_r \in T_1$, the almost sure theory.

The almost sure theory also contains sentences about the ordering. Since every $U_{n,p}$ is linearly ordered with a minimal and maximal element, the first-order sentences that state these properties are in T_1, T_2, and each S_c. Let Γ_l be the order axioms for the linear theory, that is, the sentences:

$$(\forall xyz)[(x \leq y \wedge y \leq z) \rightarrow x \leq z]$$
$$(\forall xy)[(x \leq y \wedge y \leq x) \rightarrow x = y]$$
$$(\forall x)(x \leq x)$$
$$(\forall xy)(x \leq y \vee y \leq x)$$

The following sentences guarantee that there is a minimal element and a maximal element:

$$\mu_1 : (\exists x \forall y)(x \leq y)$$
$$\mu_2 : (\exists x \forall y)(x \geq y)$$

There is also a minimal and maximal 1, which can be stated as:

$$\mu_1' : (\exists x)(\forall y)[(U(x) \wedge U(y)) \rightarrow (x \leq y)]$$
$$\mu_2' : (\exists x)(\forall y)[(U(x) \wedge U(y)) \rightarrow (x \geq y)]$$

Further, every element, except the maximal element, has a unique successor under the ordering, and every element, except the minimal element, has a unique predecessor. This can be expressed in the first-order language as:

$$\eta_1 : (\forall x)[(\forall y)(x \geq y) \vee (\exists y \forall z)((x \leq y \wedge \neg(x = z)) \rightarrow y \leq z)]$$
$$\eta_2 : (\forall x)[(\forall y)(x \leq y) \vee (\exists y \forall z)((x \geq y \wedge \neg(x = z)) \rightarrow y \geq z)]$$

As $n \rightarrow \infty$, the number of elements also goes to infinity. To capture this, we add for each positive r the axiom:

$$\delta_r : (\exists x_1 \ldots x_r)(x_1 < x_2 < \cdots < x_r)$$

For $n \geq r$, $U_{n,p} \models \delta_r$. Thus, for every $U_{n,p}$,

$$U_{n,p} \models \Gamma_l \wedge \mu_1 \wedge \mu_2 \wedge \mu_1' \wedge \mu_2' \wedge \eta_1 \wedge \eta_2 \wedge \delta_1 \wedge \delta_2 \wedge \ldots \wedge \delta_n$$

We also have that for every j, there exists an n such that

$$U_{n,p} \models \alpha_l \wedge \ldots \wedge \alpha_j \wedge \beta_1 \wedge \ldots \wedge \beta_j$$

Let $\Delta = \{\Gamma_l, \mu_1, \mu_2, \mu_1', \mu_2', \eta_1, \eta_2, \bigwedge_r \alpha_r, \bigwedge_r \delta_r\}$. Then $\Delta \subset T_1, T_2$ and for each $c > 0$, $\Delta \subset S_c$. The set of sentences, $\Sigma \cup \{\bigwedge_r \beta_r\}$, axiomatizes T_1. This follows from an Ehrenfeucht-Fraisse game argument (see [12] for more details).

For countable models of T_1, we cannot have a single infinite chain, since all the 1's must be isolated. So, we must have infinitely many chains, ordered like the integers (called **Z**-chains) that contain a single 1 with an infinite increasing chain of 0's at the beginning and an infinite decreasing chain of 0's at the end. Between these can be any number of **Z**-chains that contain no 1's. Call any **Z**-chain that contains a 1 **distinguished**. For any distinguished **Z**-chain, except the maximal distinguished chain, almost surely, there's a least distinguished **Z**-chains above it (this follows from the discreteness of the finite models). In other words, every distinguished **Z**-chain, except the maximal 1, has a distinguished successor **Z**-chain. This rules out a "dense" ordering of the distinguished **Z**-chains and leads to a "discreteness" of 1's, similar to the discreteness of elements we encountered above. It says nothing about **Z**-chains without 1's– those could have any countable order type they desire. So, the simplest model is pictured in Figure 1.

$$[00\cdots)(\cdots00100\cdots)\underbrace{(\cdots00100\cdots)}_{\text{"a Z-chain"}}\cdots\quad\cdots(\cdots00100\cdots)(\cdots00100\cdots)(\cdots00]$$

Fig. 1. A model of T_1

When $\frac{1}{\sqrt{n}} \ll p(n) \ll \frac{1}{\sqrt[3]{n}}$, almost surely isolated 1's occur, as well as more complicated occurrences of 1's. The more complicated occurrences, which we will refer to as level 2 occurrences, are $11, 101, 1001, \ldots, 10^r1 \ldots$, where "$10^r1$" is an interval $[i, i+r+1]$ with $U(i)$, $U(i+r+1)$, and for each $1 \le j \le i+r$, $\neg U(i+j)$. Using the notation from above, note $E(Y) = np^2 \to \infty$ and $\Pr[Y > 0] \to 1$. By similar argument, we can also show that three 1's cannot occur "close" together. Again, the distinguished **Z**-chains (i.e. those that contain at least 1 1) in a model of T_2 cannot be dense. The argument above can be extended to give that for every $r, s > 0$, almost surely for any occurrence of 10^r1, except the maximal 1, there exists a least occurrence of 10^s1 above it. In between any two level 2 occurrences, we have arbitrarily many isolated 1's. These cannot be densely ordered since almost surely every 1 has a successor. So, these sequences cannot be densely ordered either. That leaves only the **Z**-chains without 1's. Since we have no way to say things about them, they can have any countable order type they wish. Further, every finite sequence of level 2 occurrences must occur. Since $T_2 \models \Delta$, any model of T_2 begins with an ascending chain of 0's. In fact, each model will begin with a model of T_1, followed by a level 2 occurrence. Which level 2 occurrence occurs first is not fixed. [12] gives more details about the countable models of the almost sure theories T_1 and T_2.

When $p = \frac{c}{\sqrt{n}}$ and using the notation from above, the expected number of 1's in $U_{n,\frac{c}{\sqrt{n}}}$ is

$$E(X) = n \cdot \frac{c}{\sqrt{n}} = c\sqrt{n} \to \infty.$$

The expected number of pairs of 1's in $U_{n,\frac{c}{\sqrt{n}}}$ is

$$E(Y) = n \cdot p^2 = n \cdot \frac{c^2}{n} = c^2.$$

In any countable model of the almost sure theory, S_c, we have infinitely many isolated 1's, and we also have a non-zero probability of pairs of 1's occurring close together. Figure 2 shows a possible model of S_c. Note that the model in Figure 2 has two level 2 occurrences, namely, of length 2 and of length 0.

$$
\begin{array}{cc}
\vdots & \vdots \\
(\cdots 00100 \cdots) & (\cdots 00100 \cdots) \\
(\cdots 00100 \cdots) & (\cdots 00100 \cdots) \\
\end{array}
$$

$$[000 \cdots)(\cdots 010 \cdots) \cdots (\cdots 00100100 \cdots) (\cdots 001100 \cdots) \cdots (\cdots 010 \cdots)(\cdots 000]$$

$$
\begin{array}{cc}
(\cdots 00100 \cdots) & (\cdots 00100 \cdots) \\
(\cdots 00100 \cdots) & (\cdots 00100 \cdots) \\
\vdots & \vdots \\
\end{array}
$$

Fig. 2. A model of S_c

Let $\mathbf{a} = (a_0, \ldots, a_{m-1})$ be a finite sequence of non-negative integers. Let $m = |\mathbf{a}|$ be the length of the sequence and $M = \max\{a_0, \ldots, a_{m-1}\}$ be the maximum value of the sequence $\mathbf{a}$. Then, for each $t > M$, we can define a first order sentence $\sigma_{\mathbf{a},t}$ that says the only pairs of 1's that occur within t of one another are exactly the distance prescribed by $\mathbf{a}$ and in that order.

For example, if $\mathbf{a} = (2, 0)$, then $\mathbf{a}$ represents the level 2 occurrences "1001" and "11", occurring in that order. $\sigma_{\mathbf{a},3}$ is the sentence:

$$(\exists x_1 x_2 x_3 x_4)[U(x_1) \wedge U(x_2) \wedge U(x_3) \wedge U(x_4) \wedge x_1 < x_2 < x_3 < x_4$$
$$\wedge (\exists y_1 y_2)(x_1 < y_1 < y_2 < x_2 \wedge (\forall z)(x_1 < z < x_2 \rightarrow (y_1 = z \vee y_2 = z)))$$
$$\wedge (\forall z)(x_3 < z \rightarrow (x_4 = z \vee x_4 < z))$$
$$\wedge (\forall w_1 w_2)(U(w_1) \wedge U(w_2) \wedge w_1 < w_2$$
$$\wedge (\neg(w_1 = x_1) \wedge \neg(w_1 = x_3)) \rightarrow (\exists y_1 y_2 y_3 y_4)(x_1 < y_1 < y_2 < y_3 < y_4 < x_2))]$$

which states that the only 1's occuring within three of one another are "1001" and "11," in that order.

Definition 1 *Fix t and c and let $\mathcal{U}$ be a model of S_c (that is, $\mathcal{U} \models S_c$). Let*

$$D = \{(x_i, a_i) \mid \mathcal{U} \models U(x_i) \wedge U(x_i + a_i + 1)$$
$$\wedge \neg U(x_i + 1) \wedge \neg U(x_i + 2) \wedge \ldots \neg U(x_i + a_i) \text{ for } a_i < t \}$$

and let $A = \{x \mid \exists a, (x, a) \in D\}$. We define the function $\mathcal{F}_t$ from models of S_c to models of $\mathcal{L}'_t$ as follows: Define $\mathcal{F}_t(\mathcal{U})$ as the structure $< A, \leq', d >$ where $\leq'$ is the order induced from $\mathcal{U}$ (that is, $x \leq' y$ in $\mathcal{F}_t(\mathcal{U})$ iff $x \leq y$ in $\mathcal{U}$) and d is the t-valued function with the set D as its graph (that is, $d(x) = a$ iff $(d, a) \in D$).

For example, if $\mathcal{U} \models \sigma_{(2,0),3}$, then $\mathcal{F}_3(\mathcal{U})$ can be written as $[2, 0]$.

4   The Results

We use several facts and theorems from [12]. First, we can view $\mathcal{U}$ as a sequence of models of T_1, separated by pairs of 1's occurring within t of one another. That is:

Theorem 4 *([12]) Fix a positive integer t and a real positive constant c. Let $\mathcal{U} \models S_c$. Let x_i be the element at which the pair of length $a_i < t$ begins in $\mathcal{U}$, x_m be element at which the last pair of length $< t$ occurs, and $\max$ be the maximal element in $\mathcal{U}$. Then,*

$$< [0, x_0 - 1], \leq, U >, < [x_0 + a_0, x_1 - 1], \leq, U >, \ldots, < [x_m + a_m, \max], \leq, U >$$

are models of the almost sure theory T_1.

We also have that the almost sure theory T_1 is complete, which gives:

Theorem 5 *([12]) Fix a positive integer t. Let $\mathcal{U}_1$ and $\mathcal{U}_2$ be models of T_1. Then, $\mathcal{U}_1 \equiv_t \mathcal{U}_2$.*

We can now prove the transfer theorem, Theorem 3. Recall Theorem 3 states that if $\mathcal{U}_1, \mathcal{U}_2 \models S_c$ and $\mathcal{F}_t(\mathcal{U}_1) \equiv_t \mathcal{F}_t(\mathcal{U}_2)$, then $\mathcal{U}_1 \equiv_t \mathcal{U}_2$. That is, we can "transfer" the winning strategy from the finite structures of $\mathcal{L}'_t$ to the corresponding infinite models of $\mathcal{L}$.

Proof of Theorem 3: Assume $\mathcal{U}_1, \mathcal{U}_2 \models S_c$. By assumption, $\mathcal{F}_t(\mathcal{U}_1) \equiv_t \mathcal{F}_t(\mathcal{U}_2)$. Thus, we have a winning strategy for the t-move EF game played on $\mathcal{F}_t(\mathcal{U}_1)$ and $\mathcal{F}_t(\mathcal{U}_2)$.

We need to show that this winning strategy can be used to give Duplicator a winning strategy for the t-move game on $\mathcal{U}_1$ and $\mathcal{U}_2$. We show this by induction on q, the number of moves remaining in the EF game on $\mathcal{U}_1$ and $\mathcal{U}_2$. We play with two sets of pebbles, one for the actual game on $\mathcal{U}_1$ and $\mathcal{U}_2$, and a "shadow" set for the game on $\mathcal{F}_t(\mathcal{U}_1)$ and $\mathcal{F}_t(\mathcal{U}_2)$.

Without loss of generality, assume Spoiler plays on the element x in $\mathcal{U}_1$. Let i be the index of the closest pair of 1's within distance t of one another to x (if x is infinitely far from the pair above and below it, let i be the index of the one below it). Place the Spoiler's shadow pebble on i in $\mathcal{F}_t(\mathcal{U}_1)$. By hypothesis, we have a winning strategy for Duplicator for any game on $\mathcal{F}_t(\mathcal{U}_1)$ and $\mathcal{F}_t(\mathcal{U}_2)$. Let j be the move corresponding to i and place Duplicator's shadow pebble on j. Returning to the actual game, roughly Duplicator plays on the same relative distance from the pair b_j as Spoiler did from a_i.

In more detail, we need to take in account distance (up to 2^q, where q is the number of moves remaining) from the endpoints, the other placed pebbles, and the pairs of 1's of distance less than t. Here, we appeal to Theorem 4. From it, we have that every interval $\mathcal{M}_1 = < [x_i + a_i, x_{i+1} - 1], \leq, U_1 >$ and $\mathcal{M}_2 = < [y_j + b_j, y_{j+1} - 1], \leq, U_2 >$ are models of T_1, where x_i is the element that begins the pair a_i, y_j is the element that begins the pair b_j, U_1 is the restriction of the unary predicate of U_1, and U_2 is the restriction of the unary predicate of U_2. By Theorem 5, Duplicator has a winning strategy for any q-move EF game played on $\mathcal{M}_1$ and $\mathcal{M}_2$. So, for any move of Spoiler in such a interval, Duplicator has a winning move. This gives $U_1 \equiv_t U_2$. ⊣

Using the Janson Inequalities

The Janson Inequalities (see [1], chapter 8 for more details) says that events that are "mostly" independent sometimes have probability "nearly equal" to the truly independent case. We use these inequalities to give the limiting probability that a sequence of pairs $\mathbf{a}$ are the only occuring of length up to t.

Lemma 1 *Let c be a real positive constant, $\mathbf{a} = (a_0, \ldots, a_{m-1})$, a finite sequence of positive integers and t a positive integer such that $t > 2^{\max\{a_1, \ldots, a_m, m\}}$. Then,*

$$q_{\mathbf{a},t} = \lim_{n \to \infty} Pr[U_{n, \frac{c}{\sqrt{n}}} \models \sigma_{\mathbf{a},t}] = e^{-tc^2 \frac{c^{2|\mathbf{a}|}}{|\mathbf{a}|!}}$$

The above lemma gives that the likelihood of a sequence occurring depends solely on its length and becomes less likely as the sequence length increases. As a corollary, we have:

Lemma 2 *Let A be an ordered, countable set from $\{0, 1, \ldots, t-1\}$. Then the probability, as $n \to \infty$, that $U_{n,c/\sqrt{n}}$ contains the pairs listed in A is zero.*

With this in mind, we focus on the limit probabilities of finite sequences, since those are the only sequences that make positive contributions to the limit probability.

Definition 2 *For each first-order sentence ϕ and $t > qr(\phi)$, let*

$$M(\phi, t) = \{\mathbf{a} \mid S \cup \{\sigma_{\mathbf{a},t}\} \models \phi\} \text{ and } f_{\phi,t}(c) = \sum_{\mathbf{a} \in M(\phi,t)} q_{\mathbf{a},t}.$$

Recall that $f_\phi(c) = \lim_{n \to \infty} Pr[U_{n, \frac{c}{\sqrt{n}}} \models \phi]$. We show that for every $s, t > qr(\phi)$, $f_{\phi,t} = f_{\phi,s}$. That is, the value of $f_{\phi,t}$ is fixed for sufficiently large t.

Lemma 3 *For a finite sequence $\mathbf{a} = (a_0, \ldots, a_{m-1})$ with maximum value M and for every $s > t > M$, let $\theta = \sigma_{\mathbf{a},M}$, then $f_{\theta,s}(c) = q_{\mathbf{a},t}$.*

Thus, excluding pairs up to M, the limit probability of $\mathbf{a}$ has the same limit probability for any $t > M$. It follows immediately:

Corollary 1 *For every first order sentence ϕ, and for every $s > t > qr(\phi)$, we have: $f_{\phi,s}(c) = f_{\phi,t}(c)$.*

Proofs of the Theorems

Proof of Theorem 1 :

Let ϕ be a first-order sentence. By Lemma 1, for each $\mathbf{a}$, t:

$$q_{\mathbf{a},t} = \lim_{n \to \infty} \Pr[U_{n,\frac{c}{\sqrt{n}}} \models \sigma_{\mathbf{a},t}] = e^{-tc^2}\frac{c^{2m}}{m!}$$

where $m = |\mathbf{a}|$. Then, for fixed t,

$$\lim_{m_0 \to \infty} \sum_{m=0}^{m_0} \sum_{|\mathbf{a}|=m, a_i < t} q_{\mathbf{a},t} = \sum_{m=0}^{\infty} t^m e^{-tc^2}\frac{c^{2m}}{m!}$$
$$= e^{-tc^2}e^{tc^2} = 1$$

So, for any positive $\epsilon > 0$, there exists $m_0 > t$ such that

$$\left| 1 - \sum_{m=0}^{m_0} \sum_{|\mathbf{a}|=m, a_i < t} q_{\mathbf{a},t} \right| < \epsilon$$

Let $\beta_{s_0} = \neg \bigvee_{m \le m_0} \bigvee \sigma_{\mathbf{a},t} \mid |\mathbf{a}| = m, a_i < t$. We have

$$\lim_{n \to \infty} \Pr[U_{n,\frac{c}{\sqrt{n}}} \models \beta_{s_0}] = \left| 1 - \sum_{m=0}^{m_0} \sum_{|\mathbf{a}|=m, a_i < t} q_{\mathbf{a},t} \right| < \epsilon$$

Claim 1 ϕ *has limiting probability* $\displaystyle\sum_{\mathbf{a} \in M(\phi,t)} q_{\mathbf{a},t}$ *where* $t > qr(\phi)$.

Using the claim,

$$f_\phi(c) = \lim_{n \to \infty} \Pr[U_{n,\frac{c}{\sqrt{n}}} \models \bigvee_{\mathbf{a} \in M(\phi,t)} \sigma_{\mathbf{a},t}]$$
$$= \sum_{\mathbf{a} \in M(\phi,t)} \lim_{n \to \infty} \Pr[U_{n,\frac{c}{\sqrt{n}}} \models \sigma_{\mathbf{a},t}]$$
$$= \sum_{\mathbf{a} \in M(\phi)} e^{-c}\frac{c^{2m}}{m!}, \text{ where } m = |\mathbf{a}|$$

⊣

Proof of Theorem 2: By definition, $S \subseteq \bigcap_c S_c$. To show $\bigcap_c S_c \subseteq S$, assume $\phi \in S_c$ for some c. Then, $f_\phi(c) = 1$. So, for $t > 2^{qr(\phi)}$:

$$1 = f_\phi(c) = e^{-tc^2} \sum_{m=0}^{\infty} \left(\sum_{|\mathbf{a}|=m, \mathbf{a} \in M(\phi,t)} \frac{c^{2m}}{m!} \right)$$

This happens if and only if

$$\sum_{m=0}^{\infty} \sum_{|\mathbf{a}|=m, \mathbf{a} \in M(\phi,t)} \frac{c^{2m}}{m!} = e^{tc^2}$$

This occurs, if and only if, for each m, the number of $\mathbf{a} \in M(\phi)$ of length m is t^m. But, for each m, this is the total number possible in $M(\phi,t)$, so, we must have that every $\mathbf{a} \in M(\phi)$. Thus, for every possible finite sequence $\mathbf{a}$, $S \cup \sigma_{\mathbf{a},t} \models \phi$. So, $f_\phi(c)$ is constantly 1, and $\phi \in S_c$ for every c. Therefore, $S = \bigcap_c S_c$. ⊣

5 Future Work

The work of [11] and [12] characterize the almost sure theories and their countable models for $p(n) \ll n^{-1}$ and $n^{-1/k} \ll p(n) \ll n^{-1/(k+1)}$ for $k \geq 1$. In [14] and this paper, we fill some of the "gaps" between these theories by characterizing the almost sure theories of $U_{n,\frac{c}{n}}$ and $U_{n,\frac{c}{\sqrt{n}}}$ and giving the form of the function $f_\phi(c)$ for each first order sentence ϕ. Monadic second order logic is more expressive than first order logic over bit strings. For example, "evenness" can be expressed in monadic second order logic but not in first order logic. Is there a characterization for the limit probabilities of monadic second order logic over random sparse bit strings with $p = c/n$?

Let $\mathcal{L}_P$ be the language with the basic operations of addition (viewed as a ternary relation), ordering, and the unary predicate. What happens to the limit probabilities of sentences over this extended language? In related work, Lynch [8] gave sufficient conditions on the unary predicates to be indistinguishable under sentences of quantifier rank less than k, for a fixed k over the natural numbers. Grädel [5] linked subclasses of Presburger arithmetic (the first order theory of the model $< N, +, \leq>$) to the polynomial time hierarchy and related the truth value of sentences of quantifier rank k to the truth on an initial segment of N whose length is dependent on k. These techniques might be useful on this question and related 1's in random sequences.

References

1. N. Alon, J. Spencer, & P. Erdős. *The Probabilistic Method*. Wiley and Sons, 1992.
2. P. Dolan. A zero-one law for a random subset. *Random Structures & Algorithms*, 2:317–326, 1991.
3. H. Ebbinghaus & J. Flum. *Finite Model Theory*. Springer, Berlin, 1995.
4. R. Fagin. Generalized first-order spectra and polynomial-time recognizable sets. In *Complexity of Computation, SIAM-AMS Proceedings*, pages 43–73, 1974.
5. E. Graedel. Subclasses of Presburger arithmetic and the polynomial-time hierarchy. *Theoretical Computer Science*, 56:289–301, 1988.
6. N. Immerman. Expressibility as a complexity measure: results and directions. In *Second Structure in Complexity Conference*, pages 194–202. Springer, 1987.
7. H. Lewis & C. Papadimitriou. *Elements of the theory of computation*. Prentice-Hall, 1981.
8. J. Lynch. On sets of relations definable by addition. *J. Sym. Logic*, 47:659–668.
9. J. Lynch. Probabilities of sentences about very sparse random graphs. *Random Structures & Algorithms*, 3:33–54, 1992.
10. S. Shelah & J. Spencer. Can you feel the double jump. *Random Structures & Algorithms*, 5(1):191–204, 1994.
11. S. Shelah & J. Spencer. Random sparse unary predicates. *Random Structures & Algorithms*, 5(3):375–394, 1994.
12. J. Spencer & K. St. John. Random unary predicates: Almost sure theories and countable models. Submitted for publication, 1997.
13. J. Spencer & L. Thoma. On the limit values of probabilities for the first order properties of graphs. Technical Report 97-35, DIMACS, 1997.
14. K. St. John. Limit probabilities for random sparse bit strings. *Electronic Journal of Combinatorics*, 4(1):R23, 1997.

Lower Bounds for Randomized Read-k-Times Branching Programs

(Extended Abstract)

Martin Sauerhoff[*]

Fachbereich Informatik, Universität Dortmund, 44221 Dortmund, Germany
e-Mail: sauerhoff@cs.uni-dortmund.de

Abstract. Randomized branching programs are a probabilistic model of computation defined in analogy to the well-known probabilistic Turing machines. In this paper, we contribute to the complexity theory of randomized read-k-times branching programs.

We first consider the case $k = 1$ and present a function which has nondeterministic read-once branching programs of polynomial size, but for which every randomized read-once branching program with two-sided error at most $27/128$ is exponentially large. The same function also exhibits an exponential gap between the randomized read-once branching program sizes for different constant worst-case errors, which shows that there is no "probability amplification" technique for read-once branching programs which allows to decrease the error to an arbitrarily small constant by iterating probabilistic computations.

Our second result is a lower bound for randomized read-k-times branching programs with two-sided error, where $k > 1$ is allowed. The bound is exponential for $k < c \log n$, c an appropriate constant. Randomized read-k-times branching programs are thus one of the most general types of branching programs for which an exponential lower bound result could be established.

1 Introduction

Branching programs are a theoretically and practically interesting data structure for the representation of Boolean functions. In complexity theory, among other problems, lower bounds for the size of branching programs for explicitly defined functions and the relations of the various branching program models are investigated.

A *branching program (BP)* on the variable set $\{x_1, \ldots, x_n\}$ is a directed acyclic graph with one source and two sinks, the latter labelled by the constants 0 and 1. Each non-sink node is labelled by a variable x_i and has exactly two outgoing edges labelled by 0 or 1. This graph represents a Boolean function $f \colon \{0,1\}^n \to \{0,1\}$ in the following way. To compute $f(a)$ for some input $a \in \{0,1\}^n$, start at the source node. For a non-sink node labelled by x_i, check the value of this variable and follow the edge which is labelled by this value. Iterate this until a sink node is reached. The value of f on input a is the value of the reached sink. The *size* of a branching program G is the number of its non-sink nodes and is denoted by $|G|$.

* This work has been supported by DFG grants We 1066/7-3 and We 1066/8-1.

Note that an edge of a branching program can be regarded as an assignment of a variable, and each path corresponds to a sequence of assignments of variables.

A *read-k-times branching program* is a branching program with the restriction that on each path from the source to a sink each variable is allowed to be tested at most k times. This model is sometimes termed *syntactic* read-k-times BP, in contrast to the "non-syntactic" variant with the restriction that only on each *consistent* path from the source to a sink each variable is allowed to be tested at most k times (a path is called consistent if all assignments of variables on it are consistent). In this paper, we only consider syntactic read-k-times BPs. For syntactic read-k-times BPs, exponential lower bounds have been independently proved by Okolnishnikova [11] for $k \leq c \log n / \log \log n$, $c < 1$ arbitrarily chosen, and by Borodin, Razborov and Smolensky [4] even for nondeterministic syntactic read-k-times BPs and $k \leq c \log n$, for an appropriate constant c.

The theory of read-once branching programs (i. e., the case $k = 1$) is especially well understood. This is the branching program model for which the first exponential lower bounds have been proved (see [13] for an overview). OBDDs (ordered binary decision diagrams) introduced by Bryant [5] are restricted read-once branching programs which have turned out to be extremely useful in practice. An *OBDD* is a read-once branching program with an additional ordering of variables. On each path from the source to the sinks, the variables have to appear in the prescribed order.

In this paper we are concerned with randomized branching programs, i. e., branching programs with additional "coin-tossing nodes." We will give a formal definition of this model in the next section. In the context of Turing machines, randomized models have been studied since the introductory work of Gill [7]. But to clarify the relations of the respective complexity classes among each other and to the polynomial hierarchy belongs to the famous open problems in complexity theory. In spite of this, these questions could be solved for some restricted computation models, most important perhaps communication protocols (see [3], [12]). By the analysis of these restricted models we hope to be able to improve our tools for proving lower bounds and thus also to gain deeper insights into the structure of the more general models.

It is therefore natural to ask what can be done for randomized variants of restricted branching programs. Ablayev and Karpinski [2] have made the first step by presenting a function for which randomized OBDDs are exponentially smaller than their deterministic counterpart. Lower bounds for randomized OBDDs with two-sided error have been proved by Ablayev [1] and the author [14] using tools from communication complexity theory.

Our present work goes a step further and attacks the lower bound problem for the more general model of randomized read-once and even randomized read-k-times branching programs.

We first deal with the question whether nondeterminism is more powerful than randomness or vice versa for read-once branching programs. We show that both concepts are incomparable if the error allowed for the randomized BPs is not too large, which partially answers a question raised in [8].

As for Turing machines, we can decrease the error of randomized OBDDs and randomized (general) branching programs below an arbitrarily small constant by iteration. We will prove that there is no such "probability amplification" technique for read-once branching programs.

Finally, we prove the first exponential lower bound for randomized read-k-times branching programs, where $k < c \log n$, c an appropriate constant.

We now give an overview on the rest of the paper. In Section 2, we formally define randomized branching programs. In Section 3, we present the proof technique which will be used for our main results. Section 4 deals with the question of "nondeterminism versus randomness" for read-once branching programs, and Section 5 contains the lower bound result for randomized read-k-times branching programs.

2 Definitions and Basic Facts

In this section, we give the definitions for randomized branching programs and nondeterministic branching programs.

Definition 1. A *randomized branching program* G syntactically is a branching program with two disjoint sets of variables $x_1, \ldots, x_n$ and $z_1, \ldots, z_r$. We will call the latter "stochastic" variables. By the usual semantics for deterministic branching programs (defined in the introduction), G represents a function g on $n + r$ variables.

Now we introduce an additional probabilistic semantics for G. We say that G as a randomized branching program represents a function $f \colon \{0, 1\}^n \to \{0, 1\}$ with

- *one-sided error* at most ε, $0 \leq \varepsilon < 1$, if for all $x \in \{0, 1\}^n$ it holds that
  $\Pr\{g(x, z) = 0\} = 1$, if $f(x) = 0$;
  $\Pr\{g(x, z) = 1\} \geq 1 - \varepsilon$, if $f(x) = 1$;
- *two-sided error* at most ε, $0 \leq \varepsilon < 1/2$, if for all $x \in \{0, 1\}^n$ it holds that
  $\Pr\{g(x, z) = f(x)\} \geq 1 - \varepsilon$.

In these expressions, z is an assignment to the stochastic variables which is chosen according to the uniform distribution from $\{0, 1\}^r$.

A *randomized read-k-times BP* is a randomized branching program with the restriction that on each path from the source to a sink, each variable x_i and each variable z_i is tested at most k times. For a *randomized OBDD*, an ordering on the variables $x_1, \ldots, x_n$ and $z_1, \ldots, z_r$ is given.

Definition 2. A *nondeterministic branching program* G has the same syntax as described for randomized branching programs in the previous definition. Again, let g be the function on $n + r$ variables computed by G as a deterministic branching program. Then we say that G as a nondeterministic branching program computes a function $f \colon \{0, 1\}^n \to \{0, 1\}$ if for all $x \in \{0, 1\}^n$

$$\Pr\{g(x, z) = 0\} = 1, \quad \text{if } f(x) = 0;$$
$$\Pr\{g(x, z) = 1\} > 0, \quad \text{if } f(x) = 1;$$

where z is an assignment to the stochastic variables chosen according to the uniform distribution from $\{0, 1\}^r$.

This is equivalent to other definitions nondeterministic branching programs, e. g., that of Meinel [10]. Definitions for nondeterministic read-k-times BPs and nondeterministic OBDDs are derived from this definition in the same way as done for the randomized types above.

In analogy to the well-known complexity classes for Turing machines, let $\text{RP}_\varepsilon\text{-BP}k$ be the class of sequences of functions computable by polynomial size randomized read-k-times branching programs with one-sided error at most ε, $\varepsilon < 1$. Let $\text{BPP}_\varepsilon\text{-BP}k$ be the class of sequences of functions computable by polynomial size randomized read-k-times branching programs with two-sided error at most ε, $\varepsilon < 1/2$. Furthermore, let

$$\text{RP-BP}k := \bigcup_{\varepsilon \in [0,1)} \text{RP}_\varepsilon\text{-BP}k, \quad \text{and} \quad \text{BPP-BP}k := \bigcup_{\varepsilon \in [0,\frac{1}{2})} \text{BPP}_\varepsilon\text{-BP}k.$$

Analogous classes can be defined for randomized OBDDs. Finally, for each of the considered complexity classes $\mathcal{C}$ let co-$\mathcal{C}$ be the class of sequences of functions (f_n) for which $(\neg f_n) \in \mathcal{C}$.

As for Turing machines, it holds that $\text{RP-BP}k \subseteq \text{NP-BP}k$. We can also adapt the well-known technique of iterating probabilistic computations to improve the error probability of randomized branching programs. The following can be proved by using standard construction techniques for branching programs (see [14] for more details and a similar theorem for OBDDs).

Lemma 3 (Probability amplification).
Let G be a randomized read-k-times BP for a function $f : \{0,1\}^n \to \{0,1\}$ with two-sided error at most $\varepsilon \in [0, \frac{1}{2})$. Let $0 \le \varepsilon' \le \varepsilon$. Then a randomized read-(mk)-times BP G' for f with two-sided error less than ε' can be constructed which has size $|G'| = O(m^2|G|)$, with $m = O\left(\log((\varepsilon')^{-1})\left(\frac{1}{2} - \varepsilon\right)^{-2}\right)$.

Note that this "probability amplification technique" only allows to decrease the error by increasing the number of tests of variables.

3 A Lower Bound Technique for Randomized Read-k-Times BPs

In this section, we describe a proof technique that allows us to establish exponential lower bounds on the size of randomized read-k-times branching programs. We extend a technique of Borodin, Razborov and Smolensky [4] which yields lower bounds on the size of nondeterministic read-k-times BPs. The main idea is to relate the number of nodes of a branching program to the number of certain subsets of the input set. As Borodin, Razborov and Smolensky, we describe the technique for a generalized variant of branching programs, called s-way branching programs, which use s-valued variables instead of Boolean ones. For the whole section, let S be a finite set and $s := |S|$,

Definition 4. An s-way branching program on the variable set $\{x_1, \ldots, x_n\}$ is a directed acyclic multigraph which has one source and two sinks, the latter labelled by the constants 0 and 1. Each non-sink node is labelled by a variable x_i and has exactly s outgoing edges labelled by "1" to "s." The semantics of an s-way BP is an obvious generalization of the semantics of 2-way BPs.

We omit explicit definitions of read-k-times s-ways BPs and randomized s-way BPs, since these are analogous to the 2-way case. Also note that Lemma 3 holds in an analogous form for s-way BPs.

Next we introduce the notion of *rectangles* which is central to our proof technique. This definition is from [4].

Definition 5 ((k,p)-Rectangle). Let X be a set of variables, $n := |X|$. Let k be an integer and $1 \le p \le n$. Let sets $X_1, \ldots, X_{kp} \subseteq X$ be given with

(1) $X_1 \cup \ldots \cup X_{kp} = X$ and $|X_i| \le \lceil n/p \rceil$, for $i = 1, \ldots, kp$;

(2) each variable from X appears in at most k of the sets X_i.

Let $R \subseteq S^n$ be given. If there are functions $f_i \colon S^n \to \{0,1\}$ depending only on the variables from X_i such that for the characteristic function $f_R \colon S^n \to \{0,1\}$ of R (with $f_R(x) = 1$ iff $x \in R$) it holds that $f_R = f_1 \wedge f_2 \wedge \ldots \wedge f_{kp}$, then we call R a (k,p)-rectangle in S^n (with respect to the sets $X_1, \ldots, X_{kp}$).

Notation: We will regard rectangles as sets or as characteristic functions, depending on what is more convenient, and we will use the same name for the set as well as for its characteristic function.

By letting $k = 1$ and $p = 2$ we obtain the rectangles considered in communication complexity theory, which we will call 2-*dimensional rectangles* in the following. In this special case, we have a balanced partition (X_1, X_2) of the variable set X, i.e., it holds that $\|X_1\| - |X_2\| \le 1$ and $X_1 \cup X_2 = X$, $X_1 \cap X_2 = \emptyset$. A rectangle R with respect to this partition can be written as $R = A \times B$, where $A \subseteq 2^{X_1}$ and $B \subseteq 2^{X_2}$.

Our goal is to map the "complicated" structure of a branching program to a combinatorial representation based on rectangles which is "simpler" to analyze. The desired type of representation is described by the definition below.

Definition 6. Let X be a set of variables, $n := |X|$, and let k be an integer and $1 \le p \le n$. A function $\varphi \colon S^n \to \{0,1\}$ is called a *step function (with respect to X, k and p)*, if there is a partition of S^n into (k,p)-rectangles $R_1, \ldots, R_m$ (where the respective sets $X_1, \ldots, X_{kp} \subseteq X$ from Def. 5 may be different for all R_i) and constants $c_1, \ldots, c_m \in \{0,1\}$ such that $\varphi(x) = c_i$ for all $x \in R_i$, $i = 1, \ldots, m$. For a step function φ we call the least number m such that there are rectangles as described above *the number of rectangles used by φ*.

Let $f \colon S^n \to \{0,1\}$ be defined on the variable set X and let φ be a step function as described above. Define

$$\varepsilon := \frac{1}{|S^n|} \cdot \sum_{i=1}^m |\{x \in R_i \mid f(x) \ne c_i\}|.$$

Then we say that φ *approximates f with (total) error ε.*

The next lemma links the number of nodes in an arbitrary randomized branching program to the number of rectangles used by an appropriate step function.

Lemma 7. *Let $f \colon S^n \to \{0,1\}$ be defined on the variable set X, $|X| = n$. Let G be a randomized read-k-times s-way BP for f with two-sided error at most ε. Let $p \in \{1, \ldots, n\}$. Then there is a step function which approximates f with total error at most ε and which uses at most $(s|G|)^{kp-1}$ rectangles.*

Sketch of Proof. The first step of the proof is to turn G into a *deterministic* branching program. By a simple counting argument (due to Yao [16]) one can prove that there is a deterministic read-k-times s-way BP G' representing a function $f' \colon S^n \to \{0,1\}$ such that

$$\Pr\{f'(x) = f(x)\} \ge 1 - \varepsilon, \tag{1}$$

where the input x is chosen according to the uniform distribution from S^n.

The second step of the proof operates on the deterministic BP G'. We map paths in G' to (k, p)-rectangles in S^n as described in the proof of Theorem 1 in the paper of Borodin, Razborov and Smolensky [4]. Borodin, Razborov and Smolensky consider *nondeterministic* BPs and use a "non-standard" BP-model where tests of variables occur at the edges and not at the nodes. The adaption to our setting is mainly technical work. We only sketch the mapping of paths to rectangles, using ideas of Okolnishnikova [11] which simplify the presentation.

Each path from the source to a sink in the BP G' is partitioned into kp segments, say by "intermediate" nodes $v_1, \ldots, v_{kp+1}$, where v_1 is the source of G' and v_{kp+1} is the sink reached by the path. Let X_i be set of variables tested on the segment between v_i and v_{i+1}, $i = 1, \ldots, kp$. It can be shown that there is a partition of the path such that the set of all inputs of S^n for which the computation path runs through the intermediate nodes $v_1, \ldots, v_{kp+1}$ is a (k, p)-rectangle with respect to $X_1, \ldots, X_{kp}$. Moreover, the collection of rectangles obtained for all paths forms a partition of S^n, since the path for any input runs through exactly one sequence of intermediate nodes.

The function f' computed by G' is constant within each rectangle of the partition constructed above. Hence, f' is a step function. The upper bound on the number of rectangles used by f' follows by estimating the number of sequences of intermediate nodes, and the claim on the error bound follows from (1). □

In order to prove large lower bounds on the size of randomized BPs, we have to choose functions which are "hard" to approximate by step functions. One type of such functions is described in the hypothesis of the following theorem, which summarizes our proof technique.

Theorem 8. *Let $f : S^n \to \{0, 1\}$ be defined on the variable set X, $|X| = n$. Let k be an integer and $1 \leq p \leq n$. Assume that there is a constant ϱ such that*

$$\frac{|f^{-1}(1)|}{|S^n|} \geq \varrho > 0. \tag{H1}$$

Furthermore, assume that for every (k, p)-rectangle R in S^n (with respect to sets $X_1, \ldots, X_{kp} \subseteq X$ as in Def. 5, which may depend on R) it holds that

$$\frac{|R \cap f^{-1}(0)|}{|S^n|} \geq \alpha \cdot \frac{|R \cap f^{-1}(1)|}{|S^n|} - \delta(n), \tag{H2}$$

where $\alpha \geq 1$ is a constant and δ is a real-valued function. Then it holds for every randomized read-k-times s-way BP G for f with two-sided error at most ε that

$$|G| \geq \frac{1}{s} \left(\frac{\alpha(\varrho - \varepsilon)}{\delta(n)} \right)^{1/(kp-1)}.$$

Note that in the applications of this theorem, $\delta(n)$ will be exponentially small in n.

Sketch of Proof. Let φ be a step function which approximates f with total error at most ε. Chose an arbitrary partition of S^n into (k, p)-rectangles such that φ is constant within each rectangle. For $c \in \{0, 1\}$ let $R_1^c, \ldots, R_{r_c}^c$ be the rectangles for which φ computes the result c. It holds that

$$|f^{-1}(1)| = \sum_{i=1}^{r_0} |R_i^0 \cap f^{-1}(1)| + \sum_{i=1}^{r_1} |R_i^1 \cap f^{-1}(1)|, \tag{2}$$

since the R_i^0, R_i^1 are a partition of S^n. Since φ approximates f with total error at most ε, we have

$$\varepsilon \geq \frac{1}{|S^n|} \cdot \left(\sum_{i=1}^{r_0} |R_i^0 \cap f^{-1}(1)| + \sum_{i=1}^{r_1} |R_i^1 \cap f^{-1}(0)| \right). \tag{3}$$

Using equation (2) and the hypothesis (H2), we can derive a lower bound on r_1 from inequality (3). We additionally estimate the fraction of 1-inputs of f by (H1). The final result of these calcuations (which we omit here) is the bound

$$r_1 \geq \frac{\alpha(\varrho - \varepsilon)}{\delta(n)}. \tag{4}$$

The claimed lower bound on $|G|$ follows from this by Lemma 7. □

The above theorem says that such functions are "hard" for randomized read-k-times BPs for which (i) the fraction of 1-inputs is "not too small" (especially, it does not tend to zero with increasing input size); and (ii) each (k, p)-rectangle which covers a "large" fraction of 1-inputs also contains a "large" fraction of 0-inputs (rectangles cannot consist solely of 1-inputs unless they are "very small").

4 Nondeterminism versus Randomness for Read-Once BPs

In this section, we show that nondeterminism and randomness are incomparable for read-once branching programs if the error allowed for the randomized BPs is not too large. The complete proofs of the theorems within this section can be found in [15].

We start by exhibiting a function which has only nondeterministic read-once BPs of exponential size, whereas its complement has randomized read-once BPs with small one-sided error of polynomial size.

We consider the function PERM defined on an $n \times n$-matrix $X = (x_{ij})_{1 \leq i, j \leq n}$ of Boolean variables. Let $\text{PERM}(X) = 1$ if and only if X is a permutation matrix, i.e., if each row and each column contains exactly one entry equal to 1.

Theorem 9.

(1) PERM $\in \text{coRP}_{\varepsilon(n)}$-OBDD *for all* $\varepsilon(n) \in [0, 1)$ *with* $\varepsilon(n)^{-1} = O(\text{poly}(n))$, *but*
(2) PERM $\notin$ NP-BP1.

For the proof of this theorem, we have to refer to [15]. It follows that RP-BP1 $\neq$ coRP-BP1 and also BPP-BP1 $\not\subseteq$ NP-BP1. It is easy to improve this result to show that BPP-BP1 $\not\subseteq$ (NP-BP1 $\cup$ coNP-BP1). The function 2PERM: $\{0, 1\}^{2n^2} \to \{0, 1\}$, defined on two Boolean $n \times n$-matrices X and Y by $\text{2PERM}(X, Y) := \text{PERM}(X) \wedge \neg\text{PERM}(Y)$ is contained in the class BPP-BP1 but neither in NP-BP1 nor in coNP-BP1 (this immediately follows from Theorem 9).

It is much harder to show that nondeterminism can be more powerful than randomness for read-once BPs, since we have to consider a function which is "easy" enough to be computable by nondeterministic read-once BPs, but for which nevertheless the proof technique of Section 3 for lower bounds on randomized BPs works. We claim that the following function has these properties.

Definition 10. Define MS: $\{0,1\}^{n^2} \to \{0,1\}$ ("Mod-Sum") on the $n \times n$-matrix $X = (x_{i,j})_{1 \le i,j \le n}$ of Boolean variables by $\mathrm{MS}(X) := \mathrm{RT}(X) \wedge \mathrm{RT}(X^T)$, where $\mathrm{RT}: \{0,1\}^{n^2} \to \{0,1\}$ ("Row-Test") is defined by

$$\mathrm{RT}(X) := [s(X) \equiv 0 \bmod 2], \quad s(X) := \sum_{i=1}^{n} [x_{i,1} + \cdots + x_{i,n} \equiv 0 \bmod 3].$$

(By the expression $[P]$, P a predicate, we denote the Boolean function which is equal to 1 iff P is true.)

Theorem 11.

(1) MS $\in$ coRP$_{1/2}$-BP1;

(2) MS $\notin$ BPP$_\varepsilon$-BP1, *for* $\varepsilon < \frac{27}{128} < 0.211$.

Sketch of Proof. Part (1): We can easily construct a randomized read-once BP with error at most $1/2$ for $\neg$ MS as follows. We use two OBDDs computing $\neg \mathrm{RT}(X)$ and $\neg \mathrm{RT}(X^T)$ as submodules. These modules are connected by a single node labelled by a stochastic variable, which chooses whether $\neg \mathrm{RT}(X)$ or $\neg \mathrm{RT}(X^T)$ is correctly evaluated.

Part (2): This is the hard part of the proof. We are going to apply the technique of Section 3. We only consider 2-dimensional rectangles and 2-way branching programs here. In order to be able to apply Theorem 8, we have to show that the function MS fulfills the two hypotheses (H1) and (H2).

We estimate $|\mathrm{MS}^{-1}(1)|$ by counting the assignments of a 4×4-submatrix of X which can be completed to 1-inputs. We obtain that $|\mathrm{MS}^{-1}(1)| \ge \varrho \cdot 2^{n^2}$, where $\varrho := 27/128 > 0.21$. Hence, hypothesis (H1) is fulfilled.

We comment on the more interesting part of the proof that hypothesis (H2) is also fulfilled. We claim that for an arbitrary 2-dimensional rectangle R in $\{0,1\}^{n^2}$ it holds that

$$2^{-n^2} \cdot |\mathrm{MS}^{-1}(0) \cap R| \ge \alpha \cdot 2^{-n^2} \cdot |\mathrm{MS}^{-1}(1) \cap R| - \delta(n), \tag{5}$$

where we choose $\alpha := 1$ and $\delta(n) := (\sqrt{14}/4)^{n/4}$.

For the proof of this fact, let an arbitrary 2-dimensional rectangle R be given. R is associated with a balanced partition (X_1, X_2) of the input matrix X. The function MS is a conjunction of a "row-wise" and a "column-wise" test of the matrix X. Our strategy is to prove that for any choice of the partition (X_1, X_2), at least one of these tests is "difficult."

It is easy to show that either "many" rows or "many" columns are "split" by the given partition, i. e., some variables lie on either side of the partition. W. l. o. g. let this happen for the rows. In this case, we expect the evaluation of $\mathrm{RT}(X)$ to be difficult.

As the next step, we restrict ourselves to a class of subfunctions of RT obtained by assigning constant values to a fixed subset of the variables in X. This class contains functions $\mathrm{RTC}_{c_0,c_1,\ldots,c_m}: \{0,1\}^{4m} \to \{0,1\}$, $m := n/4$, defined on vectors of variables $x = ((x_1^1, x_1^2), \ldots, (x_m^1, x_m^2))$ and $y = ((y_1^1, y_1^2), \ldots, (y_m^1, y_m^2))$ as follows:

$$\mathrm{RTC}_{c_0,c_1,\ldots,c_m}(x,y) := [s(x,y) \equiv c_0 \bmod 2],$$

$$s(x,y) := \sum_{i=1}^{m} [x_i^1 + x_i^2 + y_i^1 + y_i^2 + c_i \equiv 0 \bmod 3],$$

where $c_0 \in \{0,1\}$ and $c_1,\ldots,c_m \in \mathbf{Z}_3$ are arbitrary constants.

The main work of the proof is to show that already these subfunctions are "difficult" for an arbitrary choice of the constants. As the well-known inner product function from communication complexity theory (see, e. g., [9]), the function RTC has the property that the numbers of 0- and 1-inputs in any 2-dimensional rectangle are "nearly balanced." More precisely, for any 2-dimensional rectangle R' in $\{0,1\}^{4m}$ it holds that

$$2^{-m^2} \cdot (|\operatorname{RTC}^{-1}(1) \cap R'| - |\operatorname{RTC}^{-1}(0) \cap R'|) \leq (\sqrt{14}/4)^m. \qquad (6)$$

We claim that fact (6) concerning the subfunctions RTC of RT can be used to derive (5) for the complete function MS (proof omitted).

Finally, we apply Theorem 8 with parameters ϱ, α and δ as defined above and $k := 1$, $p := 2$ (since we consider 2-dimensional rectangles). We obtain that

$$|G| \geq \frac{1}{2} \left(\frac{\varrho - \varepsilon}{(\sqrt{14}/4)^{n/4}} \right) = 2^{cn - O(\log n)}, \quad c := (1/4) \cdot \log_2(4/\sqrt{14}) \approx 0.024,$$

for $\varepsilon < \varrho = 27/128$. $\qquad \Box$

Corollary 12. *For $\varepsilon < \frac{27}{128}$ it holds that*

(1) NP-BP1 $\not\subseteq$ BPP$_\varepsilon$-BP1;

(2) RP$_\varepsilon$-BP1 $\subsetneq$ RP$_{1/2}$-BP1 $\subseteq$ NP-BP1.

The second part of this corollary shows that an increase in the number of tests of variables as described in Lemma 3 is indeed unavoidable if we want to decrease the error below a small constant for read-once BPs. This is contrary to the situation for OBDDs or general branching programs, where arbitrary small constant errors may be obtained.

5 A Lower Bound for Randomized Read-k-Times BPs

Now we are going to present our second main result, an exponential lower bound on the size of randomized read-k-times BPs. The complete proofs of this section can be found in the paper [14].

We consider the function SIP: $\mathbf{Z}_3^n \times \mathbf{Z}_3^n \to \{0,1\}$ ("Sylvester inner product"), $n = 2^d$, with

$$\operatorname{SIP}(x,y) = 1 \quad :\Leftrightarrow \quad x^T A y \equiv 0,$$

where $A = (a_{i,j})_{1 \leq i,j \leq 2^d}$ is the Sylvester matrix of dimension $2^d \times 2^d$, i. e.,

$$a_{i+1,j+1} := (-1)^{<\operatorname{bin}(i),\operatorname{bin}(j)>},$$

for $0 \leq i,j \leq 2^d - 1$, where $\operatorname{bin}(i)$ is the binary representation of i and $< \cdot , \cdot >$ the inner product in $\mathbf{Z}_2^d$.

Borodin, Razborov and Smolensky [4] have proved that this function has no polynomial size nondeterministic read-k-times BP for $k \leq c \log n$ for appropriate c. We can show that this function also has no polynomial size randomized read-k-times BPs with two-sided error.

Theorem 13. *Let G be a randomized 3-way read-k-times BP for* SIP *with two-sided error at most ε, ε < 1/9. Then*

$$|G| = \exp\left(\Omega\left(\frac{n}{k^3 \cdot 4^k}\right)\right).$$

Sketch of Proof. We can only give some comments highlighting the main ideas of the proof of this theorem here. We require the technique of Section 3 in its general form for (k, p)-rectangles and 3-way branching programs. We choose $p := 4k$. The proof follows the same pattern as the proof of Theorem 11. Again, we have to show that hypotheses (H1) and (H2) of Theorem 8 are fulfilled. It is easy to see that for hypothesis (H1); it holds that

$$3^{-2n} \cdot |\text{SIP}^{-1}(1)| = 1/3 - o(1). \tag{7}$$

As in the previous section, the main work is to establish that hypothesis (H2) holds. For the function SIP, we can show that for an arbitrary (k, p)-rectangle R it holds that

$$3^{-2n} \cdot |R \cap \text{SIP}^{-1}(0)| \geq \alpha \cdot 3^{-2n} \cdot |R \cap \text{SIP}^{-1}(1)| - \delta(n), \tag{8}$$

with $\alpha := 2$ and $\delta(n) := \Omega\left(n/(k \cdot 4^k)\right)$.

Besides some combinatorial facts from the original paper of Borodin, Razborov and Smolensky [4], the core of the proof of the above claim consists of a generalization of a lemma attributed to Lindsey (see, e. g., [6]). In its familiar form, this lemma states that in every submatrix of a Hadamard matrix which is "not too small" the number of 1's and (-1)'s is "nearly balanced." (A Hadamard matrix is an orthogonal matrix with entries equal to -1 or 1.)

The generalized form of this lemma (proved in [14]) says that if B is a $t \times u$-matrix over the field $\mathbf{Z}_3$ with "large" rank, then the $3^t \times 3^u$-matrix $M = (m_{x,y})$ defined by $m_{x,y} := x^T B y$, $x \in \mathbf{Z}_3^t$, $y \in \mathbf{Z}_3^u$, has the property that in every submatrix of M which is "not too small" the number of 0's, 1's and (-1)'s is "nearly balanced."

Now consider the inner product in $\mathbf{Z}_3$ defined by $x^T A y$, $x, y \in \mathbf{Z}_3^n$, where A is the matrix of SIP. The generalized form of Lindsey's Lemma can be applied to show that in each (k, p)-rectangle, the fractions of inputs for which this inner product yields the result 0, 1 or -1, resp., are equal up to an exponentially small term. Hence, the number of 0-inputs for SIP in each (k, p)-rectangle is "essentially" at least twice as large as the number of 1-inputs, which proves (8). □

By Lemma 3, the above lower bound also holds for arbitrary $ε < 1/2$. Finally, we mention that it is also possible to derive a similar exponential lower bound for a Boolean variant of SIP where each value from $\mathbf{Z}_3$ is encoded by two Boolean variables. The precise definition of this function together with the proof of the lower bound can again be found in [14].

Conclusion and Open Problems

We have introduced a technique for proving lower bounds on the size of randomized read-k-times branching programs, which has turned out to be powerful enough to prove an exponential lower bound for arbitrary $k < c \log n$.

We have also shown that $\text{BPP}_ε\text{-BP1}$ is incomparable to NP-BP1 if the error $ε$ is not too large. This partially solves the open problem raised in [8] to separate the classes

BPP-BP1 and NP-BP1. We even have obtained an exponential gap between the randomized read-once BP sizes for different constant worst-case errors.

Nevertheless, some interesting problems concerning randomized read-once BPs still remain open, e. g. :

(1) Find a function f with $f \in$ NP-BP1, but $f \notin$ BPP$_{1/2-\varepsilon}$-BP1 for arbitrarily small $\varepsilon > 0$, showing that NP-BP1 $\not\subseteq$ BPP-BP1.

(3) Show that for arbitrary ε and ε' with $0 \leq \varepsilon < \varepsilon' < 1$ holds that RP$_\varepsilon$-BP1 $\subsetneq$ RP$_{\varepsilon'}$-BP1.

Acknowledgement

I would like to thank Ingo Wegener and Martin Dietzfelbinger for helpful discussions on the subject of this paper.

References

1. F. Ablayev. Randomization and nondeterminism are incomparable for polynomial ordered binary decision diagrams. In *Proc. of ICALP*, LNCS 1256, 195–202. Springer, 1997.
2. F. Ablayev and M. Karpinski. On the power of randomized branching programs. In *Proc. of ICALP*, LNCS 1099, 348 – 356. Springer, 1996.
3. L. Babai, P. Frankl, and J. Simon. Complexity classes in communication complexity theory. In *Proc. of the 27th IEEE Symp. on Foundations of Computer Science*, 337 – 347, 1986.
4. A. Borodin, A. A. Razborov, and R. Smolensky. On lower bounds for read-k-times-branching programs. *Computational Complexity*, 3:1–18, 1993.
5. R. E. Bryant. Graph-based algorithms for Boolean function manipulation. *IEEE Trans. Computers*, C-35(8):677–691, Aug. 1986.
6. B. Chor and O. Goldreich. Unbiased bits from sources of weak randomness and probabilistic communication complexity. *SIAM J. Comput.*, 17(2):230 – 261, Apr. 1988.
7. J. Gill. *Probabilistic Turing Machines and Complexity of Computations*. Ph. D. dissertation, U. C. Berkeley, 1972.
8. S. Jukna, A. Razborov, P. Savický, and I. Wegener. On P versus NP ∩ coNP for decision diagrams and read-once branching programs. In *Proc. of the 22th Int. Symp. on Mathematical Foundations of Computer Science (MFCS)*, LNCS 1295, 319–326. Springer, 1997. Submitted to *Computational Complexity*.
9. E. Kushilevitz and N. Nisan. *Communication Complexity*. Cambridge University Press, 1997.
10. C. Meinel. The power of polynomial size Ω-branching programs. In *Proc. of the 5th Ann. ACM Symp. on Theoretical Aspects of Computer Science*, LNCS 294, 81–90. Springer, 1988.
11. E. A. Okolnishnikova. On lower bounds for branching programs. *Siberian Advances in Mathematics*, 3(1):152 – 166, 1993.
12. C. H. Papadimitriou and M. Sipser. Communication complexity. In *Proc. of the 14th Ann. ACM Symp. on Theory of Computing*, 196 – 200, 1982.
13. A. A. Razborov. Lower bounds for deterministic and nondeterministic branching programs. In *Proc. of Fundamentals of Computation Theory*, LNCS 529, 47–60. Springer, 1991.
14. M. Sauerhoff. A lower bound for randomized read-k-times branching programs. Technical Report TR97-019, Electronic Colloquium on Computational Complexity, 1997. Available via WWW from http://www.eccc.uni-trier.de/.
15. M. Sauerhoff. On non-determinism versus randomness for read-once branching programs. Technical Report TR97-030, Electronic Colloquium on Computational Complexity, 1997. Available via WWW from http://www.eccc.uni-trier.de/.
16. A. C. Yao. Lower bounds by probabilistic arguments. In *Proc. of the 24th IEEE Symp. on Foundations of Computer Science*, 420 – 428, 1983.

Inducing an Order on Cellular Automata by a Grouping Operation *

Jacques Mazoyer and Ivan Rapaport

LIP-École Normale Supérieure de Lyon
46 Allée d'Italie, 69364 Lyon Cedex 07, France
email: {mazoyer,irapapor}@lip.ens-lyon.fr

Abstract. A grouped instance of a cellular automaton (CA) is another one obtained by grouping several states into blocks and by letting interact neighbor blocks. Based on this operation a preorder $\leq$ on the set of one dimensional CA is introduced. It is shown that (CA,$\leq$) admits a global minimum and that on the bottom of (CA,$\leq$) very natural equivalence classes are located. These classes remind us the first two well-known Wolfram ones because they capture global (or dynamical) properties as nilpotency or periodicity. Non trivial properties as the undecidability of $\leq$ and the existence of bounded infinite chains are also proved. Finally, it is shown that (CA,$\leq$) admits no maximum. This result allows us to conclude that, in a "grouping sense", there is no universal CA.

1 Preliminaries

A one dimensional cellular automaton with unitary radius, or simply a CA, is an infinite array of finite state machines called cells and indexed by $\mathbb{Z}$. These identical cells evolve synchronously at discrete time steps following a local rule by which the state of a cell is determined as a function of its own state together with the states of its two neighbors. These devices are capable to simulate any Turing machine ([Smi71]) and, despite their simplicity, they may exhibit very complex behavior. The goal of this paper is to introduce an order on the set of CA by considering them as algebraic objects.

A first approach is to say that a CA A is a subautomaton of a CA B if the transition table of A is contained (after a suitable relabeling of the states) on the transition table of B. This notion takes into account only the finite-state-machine nature of the CA cells but not their spatial dimension.

On the other hand, the evolution of a CA from a particular initial configuration is usually represented by a space-time diagram in $\mathbb{Z}^2$. When observing it, the human eye typically changes the scale in order to remove irrelevant microscopic details and to discover a macroscopic behavior. We call to the CA that generate these scaled space-time diagrams grouped instances (or powers) of the original one. They are obtained by grouping several states into blocks and by considering as transitions the interactions of neighbor blocks.

---

* This work was partially supported by Program ECOS-97.

Previous grouping operation (together with the subautomaton notion) allows us to introduce a preorder $\leq$ on the set of CA. In fact, it suffices to note $A \leq B$ when a grouped instance of A is a subautomaton of a grouped instance of B. This preorder induces a canonical equivalence relation. More precisely, we say that two CA A and B are equivalent if $A \leq B$ and $B \leq A$. Notice that, as we expected it, at least all the grouped instances of a CA are equivalent.

We start Section 2 by showing that for $(CA, \leq)$ there exists a global minimum. Then it is shown that on the bottom of $(CA, \leq)$ very natural equivalence classes are located. These classes remind us the first two well-known Wolfram ones ([Wol84]) because they capture global (or dynamical) properties as nilpotency or periodicity. As a corollary, the undecidability of $\leq$ is stated.

In Section 3 a non trivial property concerning $(CA, \leq)$ is proved: the existence of two incomparable infinite chains with a common upper bound. Finally in Section 4 a natural question is answered. In fact, it is shown that $(CA, \leq)$ has no maximum. In other words, for any CA, the subsystems of all its grouped instances will never cover all the CA classes. This result allows us to conclude that, in a "grouping sense", there is no universal CA.

Formally, a CA is a couple (Q, δ) where Q is a finite set of states and $\delta :$ $Q^3 \to Q$ is a transition function. A configuration of a CA (Q, δ) is a bi-infinite sequence $\mathcal{C} \in Q^{\mathbb{Z}}$, and its global transition function $G_\delta : Q^{\mathbb{Z}} \to Q^{\mathbb{Z}}$ is such that $(G_\delta(\mathcal{C}))_i = \delta(\mathcal{C}_{i-1}, \mathcal{C}_i, \mathcal{C}_{i+1})$.

Given a state $s \in Q$, we denote its corresponding homogeneous configuration by $\bar{s} = (\cdots, s, s, s, \cdots) \in Q^{\mathbb{Z}}$. We denote $IN^* = IN - \{0\}$ and, to each $n \in IN^*$, we associate the set of states $S_n = \{0, \cdots, n-1\}$.

We say that (Q_1, δ_1) is a subautomaton of (Q_2, δ_2), and we note $(Q_1, \delta_1) \subseteq (Q_2, \delta_2)$, if there exists an injection $\varphi : Q_1 \to Q_2$ such that for all $x, y, z \in Q_1$: $\varphi(\delta_1(x, y, z)) = \delta_2(\varphi(x), \varphi(y), \varphi(z))$. When the function φ is a bijection we say that (Q_1, δ_1) and (Q_2, δ_2) are isomorph, and we note $(Q_1, \delta_1) \cong (Q_2, \delta_2)$.

For any CA (Q, δ) the evolution of a finite block of states looks like a light-cone (see Figure 1). This basic fact inspires the notion of the n-block evolution function $\delta^n : Q^{2n+1} \to Q$, which is recursively defined for all $n \in IN^*$ as follows:

$$\delta^1(w_{-1}, w_0, w_1) = \delta(w_{-1}, w_0, w_1)$$
$$\delta^n(w_{-n} \cdots, w_0, \cdots, w_n) = \delta^{n-1}(\delta(w_{-n}, w_{-n+1}, w_{-n+2}) \cdots \delta(w_{n-2}, w_{n-1}, w_n))$$

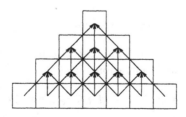

Fig. 1. Dependences diagram representing a block of states evolution as a light-cone.

By grouping several states into a block and by letting interact triplets of blocks as schematically appears in Figure 2, we generate CA with (exponentially) more states.

Formally, the n-grouped instance of a CA (Q, δ) is the CA $(Q, \delta)^n = (Q^n, \delta_g^n)$, where $\vec{q} \in Q^n$ is denoted $(q_1, \cdots, q_n)$ and for all $\vec{x}, \vec{y}, \vec{z} \in Q^n$:

$$(\delta_g^n(\vec{x}, \vec{y}, \vec{z}))_i = \delta^n(x_i, \cdots, x_n, y_1, \cdots, y_i, \cdots, y_n, z_1, \cdots, z_i)$$

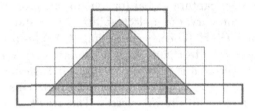

Fig. 2. Blocks interaction.

Let $\leq$ be the binary relation that associates two CA when a grouped instance of the first is a subautomaton of a grouped instance of the second. More precisely,

$$(Q_1, \delta_1) \leq (Q_2, \delta_2) \iff \exists n, m \in IN^* : (Q_1, \delta_1)^n \subseteq (Q_2, \delta_2)^m$$

By some algebraic manipulations it can be proved the following proposition:

Proposition 1. *The relation $\leq$ is a preorder on CA.*

Remark. The preorder $\leq$ defined on CA induces:

- An equivalence relation $\sim$ on CA, with $(Q_1, \delta_1) \sim (Q_2, \delta_2)$ if and only if $(Q_1, \delta_1) \leq (Q_2, \delta_2)$ and $(Q_2, \delta_2) \leq (Q_1, \delta_1)$.
- A strict preorder $<$ on CA, with $(Q_1, \delta_1) < (Q_2, \delta_2)$ if and only if $(Q_1, \delta_1) \leq (Q_2, \delta_2)$ and $(Q_1, \delta_1) \not\sim (Q_2, \delta_2)$.
- The canonical order on $(CA/\sim)$ compatible with $\leq$.

As for previous proposition, it is not difficult to prove that all the grouped instances of a cellular automaton are equivalent. More precisely,

Proposition 2. *For any CA (Q, δ) and for all $i, j \in IN^* : (Q, \delta)^i \sim (Q, \delta)^j$.*

In order to obtain a local maximum for a finite family of CA, let us consider the following "superposition" operation:

Definition 3. Let $\{B\}$ be such that $\{B\} \not\subseteq Q$ for every CA (Q, δ). Let (Q_1, δ_1), (Q_2, δ_2) be two CA. Then $(Q, \delta) = (Q_1, \delta_1) \otimes (Q_2, \delta_2)$ if $Q = (Q_1 \cup \{B\}) \times (Q_2 \cup \{B\})$ and for all $\vec{x} = (x_1, x_2), \vec{y} = (y_1, y_2), \vec{z} = (z_1, z_2) \in Q$:

$$\delta(\overrightarrow{x}, \overrightarrow{y}, \overrightarrow{z}) = \begin{cases} (\delta_1(x_1, y_1, z_1), B) & \text{if } (\overrightarrow{x}, \overrightarrow{y}, \overrightarrow{z}) \in (Q_1 \times \{B\})^3 \\ (B, \delta_2(x_2, y_2, z_2)) & \text{if } (\overrightarrow{x}, \overrightarrow{y}, \overrightarrow{z}) \in (\{B\} \times Q_2)^3 \\ (B, B) & \text{otherwise} \end{cases}$$

Proposition 4. *For every two CA (Q_1, δ_1) and (Q_2, δ_2) the CA $(Q, \delta) = (Q_1, \delta_1) \otimes (Q_2, \delta_2)$ satisfies $(Q_1, \delta_1) \leq (Q, \delta)$ and $(Q_2, \delta_2) \leq (Q, \delta)$.*

Proof. It suffices to consider the injections $\varphi_1 : Q_1 \to Q$ with $\varphi(x_1) = (x_1, B)$ and $\varphi_2 : Q_2 \to Q$ with $\varphi(x_2) = (B, x_2)$. $\qquad\qquad\square$

2 On the Bottom of (CA,$\leq$)

In this section we study the equivalence classes represented by the simplest CA. We show that for trivial transition functions (constant, identity, shift) very natural classes which capture global (or dynamical) properties as periodicity or nilpotency are generated. The undecidability of $\leq$ is concluded. In order to localize previous classes on the bottom of (CA,$\leq$) we must start by showing that (CA,$\leq$) effectively has a bottom. In fact, the next proposition says that there exists a global minimum consisting on all the isomorph CA having only one state:

Proposition 5. *The canonical order on $(CA/\sim)$ compatible with $\leq$ admits a global minimum which corresponds to the class of CA having a single state.*

Proof. Let $(\{s\}, \delta_s)$ be such that $\delta_s(s, s, s) = s$. Let (Q, δ) be a CA. By the finiteness of Q there exist $\tilde{q} \in Q$ and $P \in I\!N^*$ with $1 \leq P \leq |Q|$ such that $\delta^P(\tilde{q}, \cdots, \tilde{q}) = \tilde{q}$, and therefore $(\{s\}, \delta_s) \subseteq (Q, \delta)^P$. Finally notice that if $|Q| > 1$ then $(\{s\}, \delta_s) < (Q, \delta)$ because any grouped instance of a singleton CA is also a singleton CA. $\qquad\qquad\square$

For a CA (Q, δ) its limit set $\Omega(Q, \delta)$ is defined as the set of all the configurations that can occur after arbitrarily many computation steps. More precisely, if we define $\Omega^0 = Q^{\mathbb{Z}}$ and $\Omega^i = G_\delta(\Omega^{i-1})$ for $i \geq 1$, then $\Omega(Q, \delta) = \bigcap_{i=1}^{\infty} \Omega^i$. We say that a CA belongs to the class NIL, and we call it nilpotent, if its limit set is a singleton. In other words, $NIL = \{(Q, \delta) : (|Q| > 1) \wedge (|\Omega(Q, \delta)| = 1)\}$.

Obviously, when the limit set is a singleton it corresponds to an homogeneous configuration. In [CPY89] it is proved that when nilpotency holds then this configuration is reached from any other one in a finite and fixed number of steps. More precisely,

$$NIL = \{(Q, \delta) : (|Q| > 1) \wedge (\exists s_0 \in Q, n \in I\!N^*)(\forall \mathcal{C} \in Q^{\mathbb{Z}})(G_\delta^n(\mathcal{C}) = \bar{s}_0)\}$$

We introduce now the simplest nilpotent CA: those reaching the homogeneous configuration in one step. Let therefore $n > 1$. The CA $(S_n, 0_n)$ is such that for all $x, y, z \in S_n$: $0_n(x, y, z) = 0$. The fact that $(S_2, 0_2)^n \cong (S_{2^n}, 0_{2^n})$ for all $n \in I\!N^*$ allows us to conclude the following lemma:

Lemma 6. For all $n > 1$ it holds: $(S_2, 0_2) \sim (S_n, 0_n)$.

Lemma 7. If $(Q, \delta) \leq (S_2, 0_2)$ then $(Q, \delta) \in NIL$ or $|Q| = 1$.

Proof. If $(Q, \delta) \leq (S_2, 0_2)$ then $\exists i, j \in IN^* : (Q, \delta)^i \subseteq (S_2, 0_2)^j$. It follows:

$$\exists i, j \in IN^* : (Q, \delta)^i \subseteq (S_{2j}, 0_{2j})$$
$$\Longrightarrow \exists i \in IN^*, \vec{s} \in Q^i : \forall \vec{c}_1, \vec{c}_2, \vec{c}_3 \in Q^i \quad \delta_g^i(\vec{c}_1, \vec{c}_2, \vec{c}_3) = \vec{s}$$
$$\Longrightarrow \exists i \in IN^*, s_0 \in Q : \forall \vec{c} \in Q^{2i+1} \quad \delta^i(\vec{c}) = s_0$$
$$\Longrightarrow (Q, \delta) \in NIL \vee |Q| = 1 \qquad \qquad \square$$

Proposition 8. Let (Q, δ) be a CA. It holds:
$$(Q, \delta) \sim (S_2, 0_2) \Longleftrightarrow (Q, \delta) \in NIL$$

Proof. If $(Q, \delta) \in NIL$ then, by definition, $\exists n \in IN^*$ and $s_0 \in Q$ such that $\forall \vec{c} \in Q^{2n+1} \; \delta^n(\vec{c}) = s_0$. It follows that $(Q, \delta)^n \cong (S_{|Q|^n}, 0_{|Q|^n})$ with $|Q|^n > 1$ and therefore, by Lemma 6, $(Q, \delta) \sim (S_2, 0_2)$. The other implication corresponds to Lemma 7 and to the fact that if $(S_2, 0_2) \leq (Q, \delta)$ then $|Q| > 1$. $\qquad \square$

Proposition 9. $(Q, \delta) < (S_2, 0_2)$ if and only if (Q, δ) is isomorph to the minimum.

Proof. Let us suppose that $(Q, \delta) < (S_2, 0_2)$ and $|Q| > 1$. By Lemma 7, $(Q, \delta) \in NIL$, and therefore $(Q, \delta) \sim (S_2, 0_2)$. $\Rightarrow \Leftarrow$ $\qquad \square$

Corollary 10. Given a CA (Q, δ), it is undecidable whether $(Q, \delta) \leq (S_2, 0_2)$.

Proof. By the fact that the nilpotency problem is undecidable ([Kar92]). $\qquad \square$

Some other global properties concerning cyclic behavior are considered. First we say that a CA belongs to the class PER, and we call it periodic, if every configuration belongs to a cycle. More precisely,

$$PER = \{(Q, \delta) : (|Q| > 1) \wedge (\forall C \in Q^{\mathbb{Z}}, \exists n \in IN^* : G_\delta^n(C) = C)\}$$

On the other hand we introduce the R_{SHIFT} and L_{SHIFT} classes. In this case, for every configuration there exists an $n \in IN^*$ for which the configuration reappears n cells shifted after n time steps. In other words,

$$R_{SHIFT} = \{(Q, \delta) : (|Q| > 1) \wedge (\forall C \in Q^{\mathbb{Z}}, \exists n \in IN^* : ((G_\delta)^n(C))_i = C_{i-n})\}$$
$$L_{SHIFT} = \{(Q, \delta) : (|Q| > 1) \wedge (\forall C \in Q^{\mathbb{Z}}, \exists n \in IN^* : ((G_\delta)^n(C))_i = C_{i+n})\}$$

As in the nilpotency case, for these classes the length of the cycles does not depend on the considered configurations. This result is stated in the next lemma:

Lemma 11. It holds the following:

$$PER = \{(Q, \delta) : (|Q| > 1) \wedge (\exists n \in IN^*, \forall C \in Q^{\mathbb{Z}} : G_\delta^n(C) = C)\}$$
$$R_{SHIFT} = \{(Q, \delta) : (|Q| > 1) \wedge (\exists n \in IN^*, \forall C \in Q^{\mathbb{Z}} : ((G_\delta)^n(C))_i = C_{i-n})\}$$
$$L_{SHIFT} = \{(Q, \delta) : (|Q| > 1) \wedge (\exists n \in IN^*, \forall C \in Q^{\mathbb{Z}} : ((G_\delta)^n(C))_i = C_{i+n})\}$$

Proof. Let $(Q, \delta) \in PER$. Let us consider any configuration C^* in which all the words over Q appear (it suffices to construct it as a suitable concatenation). Denoting the period of the cycle to which C^* belongs by n^*, it follows that

$$\forall \vec{c} = (c_{-n^*}, \cdots, c_0, \cdots, c_{n^*}) \in Q^{2n^*+1} : \delta^{n^*}(\vec{c}) = c_0$$

and therefore any other configuration $C \in Q^{\mathbb{Z}}$ is n^*-periodic. For the R_{SHIFT} and the L_{SHIFT} classes the proof is exactly the same. $\square$

We introduce now the simplest periodic and shift-like CA: those having unitary length cycles. More precisely, let $n > 1$ and let (S_n, I_n), (S_n, σ_n) (S_n, σ_n^{-1}) be the CA such that, for all $x, y, z \in S_n$:

$$I_n(x, y, z) = y \quad \sigma_n(x, y, z) = x \quad \sigma_n^{-1}(x, y, z) = z$$

It follows the same as for the nilpotency case. In fact, the proofs of the next two propositions are completely equivalent to those of Proposition 8 and Proposition 9:

Proposition 12. *Let (Q, δ) be a CA. It holds:*

$$(Q, \delta) \sim (S_2, I_2) \iff (Q, \delta) \in PER$$
$$(Q, \delta) \sim (S_2, \sigma_2) \iff (Q, \delta) \in R_{SHIFT}$$
$$(Q, \delta) \sim (S_2, \sigma_2^{-1}) \iff (Q, \delta) \in L_{SHIFT}$$

Proposition 13. *Let $(S_2, \rho_2) \in \{(S_2, I_2), (S_2, \sigma_2), (S_2, \sigma_2^{-1})\}$. Then $(Q, \delta) < (S_2, \rho_2)$ if and only if (Q, δ) is isomorph to the minimum.*

Corollary 14. *The canonical order on $(CA/\sim)$ compatible with $\leq$ is partial. Moreover, the classes R_{SHIFT}, L_{SHIFT}, NIL, and PER are two by two incomparables.*

Proof. Consider any non (spatialy) periodic configuration and notice that its behavior could never be simultaneously of two types. $\square$

3 Two Incomparable Infinite Chains with a Common Upper Bound

A non trivial property concerning $(CA, \leq)$ is proved in this section: the existence of two incomparable infinite chains with a common upper bound. The two chains are introduced in the next definition:

Definition 15. Let the two CA families $\{(S_n, \eta_n)\}_{n>1}$ and $\{(S_n, \mu_n)\}_{n>1}$ be such that:

$$\eta_n(x, y, z) = \begin{cases} x & \text{if } x = y = z \\ 0 & \text{otherwise} \end{cases} \text{ and } \mu_n(x, y, z) = \min\{x, y, z\}$$

Before proving that previous families are incomparable and infinite chains, notice that there exists a pair of points belonging to different chains which are comparable. In fact, the initial points (S_2, η_2) and (S_2, μ_2) are isomorph. On the other hand, they are located above the NIL class as it is showed in the next proposition:

Proposition 16. For all $(Q_{NIL}, \delta_{NIL}) \in NIL$ it holds:

$$(Q_{NIL}, \delta_{NIL}) < (S_2, \eta_2) \text{ and } (Q_{NIL}, \delta_{NIL}) < (S_2, \mu_2)$$

Proof. Let $(Q_{NIL}, \delta_{NIL}) \in NIL$. First $(S_2, \eta_2) \not\leq (Q_{NIL}, \delta_{NIL})$ because (S_2, η_2) is not nilpotent. On the other hand, notice that $(S_2, 0_2) \subseteq (S_2, \eta_2)^2$ because it suffices to consider $\varphi : S_2 \to (S_2)^2$ such that $\varphi(x) = (0x)$. $\qquad\square$

Lemma 17. Let $n > 1$. For all $i \in IN^*$ it holds:

$$n = |\{\overrightarrow{x} \in (S_n)^i : (\eta_n)_g^i(\overrightarrow{x}, \overrightarrow{x}, \overrightarrow{x}) = \overrightarrow{x}\}|$$
$$= |\{\overrightarrow{x} \in (S_n)^i : (\mu_n)_g^i(\overrightarrow{x}, \overrightarrow{x}, \overrightarrow{x}) = \overrightarrow{x}\}|$$

Proof. Notice that $(\eta_n)_g^i(\overrightarrow{x}, \overrightarrow{x}, \overrightarrow{x}) = \overrightarrow{x} \iff (\mu_n)_g^i(\overrightarrow{x}, \overrightarrow{x}, \overrightarrow{x}) = \overrightarrow{x} \iff \exists x \in S_n : \overrightarrow{x} = (x \cdots x)$. $\qquad\square$

Proposition 18. For all $n > 1$ it holds:

$$(S_n, \eta_n) < (S_{n+1}, \eta_{n+1}) \text{ and } (S_n, \mu_n) < (S_{n+1}, \mu_{n+1})$$

Proof. First $(S_n, \eta_n) \leq (S_{n+1}, \eta_{n+1})$ because $\eta_{n+1}|_{S_n} = \eta_n$. Let us suppose that there exist $i, j \in IN^*$ such that $(S_{n+1}, \eta_{n+1})^i \subseteq (S_n, \eta_n)^j$. Let $\varphi : (S_{n+1})^i \to (S_n)^j$ be the suitable injection. It follows that, if $\overrightarrow{x} \in (S_{n+1})^i$ is such that $(\eta_{n+1})_g^i(\overrightarrow{x}, \overrightarrow{x}, \overrightarrow{x}) = \overrightarrow{x}$ then $(\eta_n)_g^j(\varphi(\overrightarrow{x}), \varphi(\overrightarrow{x}), \varphi(\overrightarrow{x})) = \varphi(\overrightarrow{x})$ and we contradict Lemma 17. For $(S_n, \mu_n) < (S_{n+1}, \mu_{n+1})$ the argument is exactly the same. $\qquad\square$

Lemma 19. Let $i, n \in IN^*$ with $n > 1$. For all $\overrightarrow{a} = (a_1 \cdots a_i) \in (S_n)^i$ there exist $p, q \in S_n$ such that:

$$(\eta_n)_g^i(\overrightarrow{a}, \overrightarrow{a}, \overrightarrow{a}) = (p \cdots p) \text{ and } (\mu_n)_g^i(\overrightarrow{a}, \overrightarrow{a}, \overrightarrow{a}) = (q \cdots q)$$

Proof. It suffices to notice that for all $\overrightarrow{x} = (x_{-i} \cdots x_0 \cdots x_i) \in (S_n)^{2i+1}$ it holds:

$$(\eta_n)^i(x_{-i} \cdots x_0 \cdots x_i) \neq 0 \iff x_{-i} = \cdots = x_0 = \cdots = x_i \neq 0$$
$$(\mu_n)^i(x_{-i} \cdots x_0 \cdots x_i) = \min\{x_{-i}, \cdots, x_0, \cdots, x_i\} \qquad\square$$

Proposition 20. For all $n > 2, m > 2$: $(S_m, \mu_m) \not\leq (S_n, \eta_n)$ and $(S_n, \eta_n) \not\leq (S_m, \mu_m)$.

Proof. Let us suppose that there exist $i, j \in IN^*$ such that $(S_m, \mu_m)^i \subseteq (S_n, \eta_n)^j$ and let us denote $\varphi : (S_m)^i \to (S_n)^j$ the suitable injection. It follows that:

$$\forall x \in S_m, \exists \varphi_x \in S_n \text{ such that } \varphi(x \cdots x) = (\varphi_x \cdots \varphi_x)$$

In fact,

$$\begin{aligned}
\varphi(x \cdots x) &= \varphi((\mu_m)^i_g(x \cdots x, x \cdots x, x \cdots x)) \\
&= (\eta_n)^j_g(\varphi(x \cdots x), \varphi(x \cdots x), \varphi(x \cdots x)) \\
&= (\varphi_x \cdots \varphi_x) \text{ (by Lemma 19)}
\end{aligned}$$

Let $x \in S_m$ be such that $0 \leq x < m - 1$ and $\varphi(x \cdots x) \neq (0 \cdots 0)$. It follows:

$$\begin{aligned}
\varphi(x \cdots x) &= \varphi((\mu_m)^i_g(m - 1 \cdots m - 1, x \cdots x, m - 1 \cdots m - 1)) \\
&= (\eta_n)^j_g(\varphi_{m-1} \cdots \varphi_{m-1}, \varphi_x \cdots \varphi_x, \varphi_{m-1} \cdots \varphi_{m-1}) \\
&= (0 \cdots 0) \Rightarrow\Leftarrow
\end{aligned}$$

For $(S_n, \eta_n) \not\leq (S_m, \mu_m)$ the argument is similar. $\qquad\square$

Now in order to obtain an upper bound for the two infinite chains we are going to compose a pair of CA. One of them is related to the classical *firing squad* problem introduced in [Moo64] and which consists to design a CA capable to synchronize "as soon as possible" an array of cells of arbitrary size. In Lemma 21 a result that appears in [MR92] concerning a slightly modified version of the original problem known as *two-ends firing squad* is formally stated:

Lemma 21. [MR92] *There exists a CA (Q_{FS}, δ_{FS}) such that $\{G_l, G_r, q_0\} \subseteq Q_{FS}$ and which satisfies for all $n \in IN^*$ the following:*

- $(\delta_{FS})^{n+2}_g(G_l q_0 \cdots q_0 G_r, G_l q_0 \cdots q_0 G_r, G_l q_0 \cdots q_0 G_r) = (G_l q_0 \cdots q_0 G_r)$
- *For all substring $\overrightarrow{x}$ of the triple concatenation $(G_l q_0 \cdots q_0 G_r G_l q_0 \cdots q_0 G_r G_l q_0 \cdots q_0 G_r) \in (Q_{FS})^{3(n+2)}$ such that $|\overrightarrow{x}| = 2k + 1$ with $k < (n + 2)$, it holds:*

$$(\delta_{FS})^k(\overrightarrow{x}) \neq q_0$$

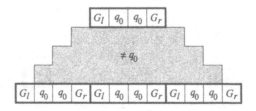

Fig. 3. Two-ends firing squad.

In Figure 3 it is represented the case $n + 2 = 4$. The CA to be composed with (Q_{FS}, δ_{FS}) is introduced in the next definition. Its cells simply transmit the signals (or states) coming from its left (resp. right) neighbor to its right (resp. left) neighbor keeping only its own information. More precisely,

Definition 22. Let Q be an arbitrary set of states. We define the CA $(S_Q^{signal}, \delta_Q^{signal})$ such that $S_Q^{signal} = Q^3$ and for all $x, y, z \in S_Q^{signal}$:

$$\delta_Q^{signal}(x, y, z) = (x_l y_c z_r)$$

where all the states of S_Q^{signal} are denoted by $s = (s_l s_c s_r)$.

The composed CA is almost the "superposition" of the two previously introduced ones (with the set of signals $Q = \{0, 1\}$): the exception is done at the last step of the firing squad period. More precisely, the only 1 signals not destroyed (transformed into 0) are those arriving simultaneously to a cell. Formally:

Definition 23. Let $(Q, \mathcal{D})$ be the CA such that $Q = Q_{FS} \times S_{\{0,1\}}^{signal}$ and, if we denote $\mathcal{D}((x_1, x_2), (y_1, y_2), (z_1, z_2))$ by $\mathcal{D}_{x,y,z}$, it follows:

$$\mathcal{D}_{x,y,z} = \begin{cases} (\delta_{FS}(x_1, y_1, z_1), 000) & \text{if } \delta_{FS}(x_1, y_1, z_1) = q_0 \\ & \text{and } \delta_{\{0,1\}}^{signal}(x_2, y_2, z_2) \neq (111) \\ (\delta_{FS}(x_1, y_1, z_1), \delta_{\{0,1\}}^{signal}(x_2, y_2, z_2)) & \text{otherwise} \end{cases}$$

Proposition 24. *For all $n > 1$: $(S_n, \eta_n) < (Q, \mathcal{D})$ and $(S_n, \mu_n) < (Q, \mathcal{D})$.*

Proof. If $n \in I\!N^*$ then $(S_n, \eta_n) \subseteq (Q, \mathcal{D})^{n+1}$ and $(S_n, \mu_n) \subseteq (Q, \mathcal{D})^{n+1}$ by the injections $\varphi_\eta, \varphi_\mu : S_n \to Q^{n+1}$ explicited in the following (see Figure 4):

$$\varphi_\eta(0) = \varphi_\mu(0) = ((G_l, 000), (q_0, 000), \cdots, (q_0, 000), (G_r, 000))$$
$$\varphi_\eta(x) = ((G_l, 000), \underbrace{(q_0, 000), \cdots, (q_0, 000), (q_0, 111)}_{x \text{ with } 0 < x \leq n-1}, (q_0, 000), \cdots, (G_r, 000))$$
$$\varphi_\mu(x) = ((G_l, 000), \underbrace{(q_0, 111), \cdots, (q_0, 111)}_{x \text{ with } 0 < x \leq n-1}, (q_0, 000), \cdots, (G_r, 000)) \qquad \square$$

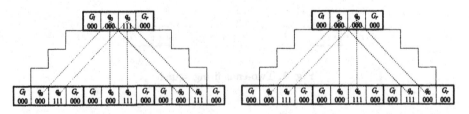

Fig. 4. "Simulating" by $(Q, \mathcal{D})^4$ the transitions of (S_3, η_3) for $(2, 2, 2)$ and $(2, 2, 1)$.

4 An Unbounded Infinite Chain

In this section we prove that $(CA, \leq)$ has no maximum. In other words, for any CA, the subsystems of all its grouped instances will never cover all the CA classes. This result allows us to conclude that, in a "grouping sense", there is no universal CA. The proof is implicitly based on the existence of an unbounded infinite chain obtained after "processing" the next one:

Definition 25. $\{(S_n, \Delta_n)\}_{n>1}$ is the family of CA such that, for each $n > 1$:

$$\Delta_n(x, y, z) = \begin{cases} x & \text{if } x = z \\ y & \text{if } x \neq z \end{cases}$$

The bottom of previous family is located above the NIL class. More precisely,

Proposition 26. $(S_2, 0_2) < (S_2, \eta_2) < (S_2, \Delta_2)$.

Proof. First it suffices to check that $(S_2, \eta_2) \subseteq (S_2, \Delta_2)^2$ by the injection $\varphi : S_2 \to (S_2)^2$ such that $\varphi(x) = (0x)$. On the other hand, let us suppose that there exist $i, j \in IN^*$ such that $(S_2, \Delta_2)^i \subseteq (S_2, \eta_2)^j$. It is easy to prove that then $(S_2, \Delta_2)^{2i} \subseteq (S_2, \eta_2)^{2j}$. By Lemma 17, $|\{\overrightarrow{x} \in (S_2)^{2j} : (\eta_2)_g^{2j}(\overrightarrow{x}, \overrightarrow{x}, \overrightarrow{x}) = \overrightarrow{x}\}| = 2$. However, $|\{\overrightarrow{x} \in (S_2)^{2i} : (\Delta_2)_g^{2i}(\overrightarrow{x}, \overrightarrow{x}, \overrightarrow{x}) = \overrightarrow{x}\}| \geq 4$, because for all $x, y \in S_2 : (\Delta_2)_g^{2i}(xy \cdots xy, xy \cdots xy, xy \cdots xy) = (xy \cdots xy)$. $\square$

The following is the key result of the present section. It says, in a general way, that the Δ_n's transition functions are too complicated to be simulated by grouped instances of a CA. This impossibility is based on the fact that, for any grouped instance of a CA, any cell belonging to a block has no access during the interaction process to the whole information contained on the neighbor blocks. Formally:

Proposition 27. *For every CA (Q, δ) and for every $n \in IN^*$ such that $n > |Q|$ it holds:*

$$\forall i \in IN^* : (S_n, \Delta_n) \not\subseteq (Q, \delta)^i$$

Proof. Suppose that there exist (Q, δ) and $n, i \in IN^*$ such that $n > |Q|$ and $(S_n, \Delta_n) \subseteq (Q, \delta)^i$. Then, by definition, there exists an injection $\varphi : S_n \to Q^i$ such that:

$$\forall x, y, z \in S_n : \varphi(\Delta_n(x, y, z)) = \delta_g^i(\varphi(x), \varphi(y), \varphi(z))$$

Let i_0 be the smallest index of Q^i for which there exist at least two elements of $\varphi(S_n)$ having different values. Formally:

$$i_0 = \min\{k \in IN^* : \exists (x_1, \cdots, x_i), (y_1, \cdots, y_i) \in \varphi(S_n) \text{ such that } x_k \neq y_k\}$$

Notice that $i_0 \in \{1, \cdots i\}$ is well defined because $|S_n| > 1$. It follows:

$$\forall \overrightarrow{x}, \overrightarrow{z} \in \varphi(S_n) : \overrightarrow{x} \neq \overrightarrow{z} \Rightarrow x_{i_0} \neq z_{i_0}$$

In fact, suppose that there exist $\overrightarrow{x}, \overrightarrow{z} \in \varphi(S_n)$ with $\overrightarrow{x} \neq \overrightarrow{z}$ such that $x_{i_0} = z_{i_0}$. By construction of i_0 there always exists $\overrightarrow{y} \in \varphi(S_n)$ such that $x_{i_0} \neq y_{i_0}$. On the other hand:

$$
\begin{aligned}
x_{i_0} = (\delta_g^i(\overrightarrow{x}, \overrightarrow{y}, \overrightarrow{x}))_{i_0} &= \delta^{2i+1}(x_{i_0}, \cdots, x_i, y_1 \cdots, y_i, x_1, \cdots, x_{i_0}) \\
&= \delta^{2i+1}(x_{i_0}, \cdots, x_i, y_1 \cdots, y_i, z_1, \cdots, z_{i_0}) \\
&= (\delta_g^i(\overrightarrow{x}, \overrightarrow{y}, \overrightarrow{z}))_{i_0} = y_{i_0} \quad \Rightarrow\Leftarrow
\end{aligned}
$$

Finally it follows that $\beta : S_n \to Q$ with $\beta(x) = (\varphi(x))_{i_0}$ is an injection and therefore $n \leq |Q|. \Rightarrow\Leftarrow$ □

In order to conclude that $(CA, \leq)$ admits no maximum we are going to show that every CA is contained in all the grouped instances of a suitable composition of itself with the CA that transmits signals introduced in Definition 22. More precisely,

Lemma 28. *For any CA* (Q, δ) *there exists a normalized version* $(Q, \delta)^* = (Q^*, \delta^*)$ *satisfying, for all* $i \in IN^*$: $(Q, \delta) \subseteq (Q^*, \delta^*)^i$.

Proof. Let us denote B a state not belonging to any CA. Let (Q, δ) be a CA. We define $Q^* = S_{\{Q \cup \{B\}\}}^{signal}$ and for all $x, y, z \in Q^*$:

$$
\delta^*(x, y, z) = \begin{cases} (\delta(x_l, y_c, z_r), \delta(x_l, y_c, z_r), \delta(x_l, y_c, z_r)) & \text{if } x_l, y_c, z_r \in Q \\ \delta_{\{Q \cup \{B\}\}}^{signal}(x, y, z) & \text{otherwise} \end{cases}
$$

For any $i \in IN^*$, the injection $\varphi : Q \to (Q^*)^i$ that allows us to conclude the property is the following: $\varphi(x) = ((xxx), (BBB), \cdots (BBB))$ (see Figure 5). □

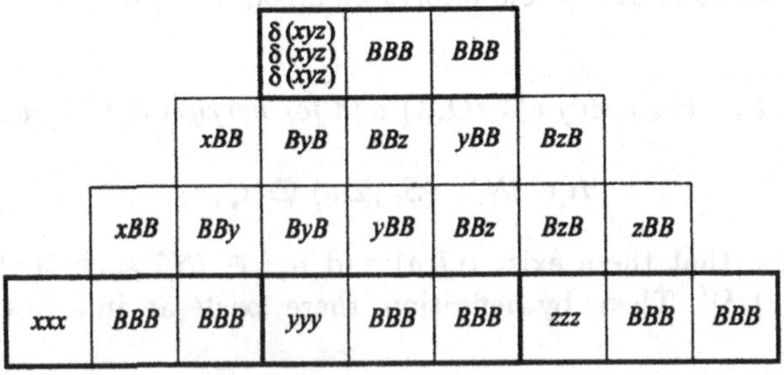

				$\delta(xyz)$ $\delta(xyz)$ $\delta(xyz)$	BBB	BBB			
			xBB	ByB	BBz	yBB	BzB		
		xBB	BBy	ByB	yBB	BBz	BzB	zBB	
	xxx	BBB	BBB	yyy	BBB	BBB	zzz	BBB	BBB

Fig. 5. Embedding (Q, δ) into $(Q^*, \delta^*)^i$ for $i = 3$.

Proposition 29. *For every CA* (Q, δ) *and for every* $n > |Q|$: $(S_n^*, \Delta_n^*) \not\leq (Q, \delta)$.

Proof. Suppose that there exist $(Q, \delta), n > |Q|$ and $i, j \in IN^*$ such that $(S_n^*, \Delta_n^*)^i \subseteq (Q, \delta)^j$. Then, by Lemma 28, $(S_n, \Delta_n) \subseteq (Q, \delta)^j$, which contradicts Proposition 27. □

Corollary 30. $(CA, \leq)$ *has no maximum.*

Proposition 31. *It holds, for all* $n \in IN^* : (S_n^*, \Delta_n^*) < (S_{(n+1)^3+1}^*, \Delta_{(n+1)^3+1}^*)$.

Proof. First $(S_n^*, \Delta_n^*) \leq (S_{(n+1)^3+1}^*, \Delta_{(n+1)^3+1}^*)$ because $S_n^* \subseteq S_{(n+1)^3+1}^*$ and $\Delta_{(n+1)^3+1}^* |_{S_n^*} = \Delta_n^*$. Let us suppose that $(S_{(n+1)^3+1}^*, \Delta_{(n+1)^3+1}^*) \leq (S_n^*, \Delta_n^*)$. Considering that $|S_n^*| = (n+1)^3$ we contradict Proposition 29. □

Proposition 32. *For every CA* (Q, δ) *there exists another one* $(\tilde{Q}, \tilde{\delta})$ *such that* $(Q, \delta) < (\tilde{Q}, \tilde{\delta})$.

Proof. It suffices to consider $(\tilde{Q}, \tilde{\delta}) = (Q, \delta) \otimes (S_{|Q|+1}^*, \Delta_{|Q|+1}^*)$. By Proposition 4 $(Q, \delta) \leq (\tilde{Q}, \tilde{\delta})$. On the other hand, if $(\tilde{Q}, \tilde{\delta}) \leq (Q, \delta)$ then $(S_{|Q|+1}^*, \Delta_{|Q|+1}^*) \leq (Q, \delta)$. $\Rightarrow\Leftarrow$ □

5 Further Research

- A deeper understanding of the $(CA, \leq)$ structure could help us to develop a complexity notion on CA. For instance, it is interesting to notice that the only algorithmically non-trivial CA which appeared in this work (a modified version of the one that solves the firing squad problem) admits an infinite chain separating it from the minimum.

- The concept of intrinsic or self-referenced simulation on CA was introduced in [AC87]. It simply means that we "simulate directly a CA without passing through Turing machines". In [Mar94] appears an intrinsic universal CA working on quasi linear time but restricted to totalistic transitions. Our relation $\leq$ may be interpreted as a particular kind of simulation working on real time. A clarifying study about the different notions of simulations (and the time they take) should be done in the future.

References

[AC87] Albert J., Culik II K.: A simple universal cellular automaton and its one-way and totalisting version. Complex Systems 1 (1987) 1-16.

[CPY89] Culik II K., Pachl J., Yu S.: On the limit sets of cellular automata. SIAM J. Computing 18 (1989) 831-842.

[Kar92] Kari J.: The nilpotency problem of one-dimensional cellular automata. SIAM J. Computing 21 (1992) 571-586.

[Mar94] Martin B.: A universal cellular automaton in quasi-linear time and its s-m-n form. Theoretical Computer Science 123(2) (1994) 199-237.

[Moo64] Moore E.F.: Sequential machines, selected papers. Addison Wesley Reading Mass. (1964) 213-214.

[MR92] Mazoyer J., Reimen N.: A linear speed-up theorem for cellular automata. Theoretical Computer Science 101 (1992) 59-98.

[Smi71] Smith III A.R.: Simple computation-universal cellular spaces. Journal ACM 18 (1971) 339-353.

[Wol84] Wolfram S.: Universality and complexity in cellular automata. Physica D 10 (1984) 1-35.

Attractors of D-dimensional Linear Cellular Automata

Giovanni Manzini[1,2], Luciano Margara[3,4]

[1] Dipartimento di Scienze e Tecnologie Avanzate, Università di Torino, Via Cavour 84, 15100 Alessandria, Italy. manzini@mfn.al.unipmn.it
[2] Istituto di Matematica Computazionale, Via S. Maria, 46, 56126 Pisa, Italy.
[3] Dipartimento di Scienze dell'Informazione, Università di Bologna, Mura Anteo Zamboni 7, 40127 Bologna, Italy. margara@cs.unibo.it
[4] International Computer Science Institute (ICSI), Berkeley CA.

Abstract. In this paper we study the asymptotic behavior of D-dimensional linear cellular automata over the ring $\mathbf{Z}_m$ ($D \geq 1$, $m \geq 2$). In the first part of the paper we consider non-surjective cellular automata. We prove that, after a transient phase of length at most $\lfloor \log_2 m \rfloor$, the evolution of a linear non-surjective cellular automata F takes place completely within a subspace Y_F. This result suggests that we can get valuable information on the long term behavior of F by studying its properties when restricted to Y_F. We prove that such study is possible by showing that the system (Y_F, F) is topologically conjugated to a linear cellular automata F^* defined over a different ring $\mathbf{Z}_{m^*}$. In the second part of the paper, we study the attractor sets of linear cellular automata. Recently, Kurka [8] has shown that CA can be partitioned into five disjoint classes according to the structure of their attractors. We present a procedure for deciding the membership in Kurka's classes for any linear cellular automata. Our procedure requires only gcd computations involving the coefficients of the local rule associated to the cellular automata.

1 Introduction

Cellular Automata (CA) are dynamical systems consisting of a D-dimensional lattice of variables which can take a finite number of discrete values. The *global state* of the CA, specified by the values of all the variables at a given time, evolves in synchronous discrete time steps according to a given *local rule* which acts on the value of each single variable. For an introduction to the CA theory and an extensive and up-to-date bibliography see [5].

In this paper we restrict our attention to the class of linear CA (CA based on a linear local rule defined over the ring $\mathbf{Z}_m$). Despite of their simplicity that makes it possible a detailed algebraic analysis, linear CA exhibit many of the complex features of general CA. Recently, many important properties of linear CA have been completely characterized (see Fig. 1). These properties have been introduced for the study of discrete time dynamical systems (see for example [4]). Among other things, they provide valuable information on the long

Property	Characterization	Reference
Surjectivity	$\gcd(m, \lambda_1, \ldots, \lambda_s) = 1$	[7]
Injectivity	$(\forall p \in \mathcal{P})\, (\exists! \lambda_i) : p \nmid \lambda_i$	[7]
Ergodicity	$\gcd(m, \lambda_2, \ldots, \lambda_s) = 1$	[11]
Transitivity	$\gcd(m, \lambda_2, \ldots, \lambda_s) = 1$	[3]
Sensitivity	$(\exists p \in \mathcal{P}) : p \nmid \gcd(\lambda_2, \ldots, \lambda_s)$	[10]
Pos. Expansivity	$\gcd(m, a_1, \ldots, a_r) = \gcd(m, a_{-1}, \ldots, a_{-r}) = 1$	[10]
Equicontinuity	$(\forall p \in \mathcal{P})\, p \mid \gcd(\lambda_2, \ldots, \lambda_s)$	[10]
Strong Trans.	$(\forall p \in \mathcal{P})\, (\exists \lambda_i, \lambda_j) : p \nmid \lambda_i \wedge p \nmid \lambda_j$	[10]
Expansivity	$\gcd(m, a_{-r}, \ldots, a_{-1}, a_1, \ldots, a_r) = 1$	[9]
Regularity	$\gcd(m, a_{-r}, \ldots, a_r) = 1$	[2]

Fig. 1. Characterization of set theoretic and topological properties of linear CA over $\mathbf{Z}_m$ in terms of the coefficients λ_i's (for D-dimensional CA) or a_i's (for 1-dimensional CA). $\mathcal{P}$ denotes the set of prime factors of m.

term behavior of a complex system. The results mentioned above show that it is often possible to make a detailed analysis of the *global* dynamical behavior of a linear CA by analyzing the coefficients of its *local* rule.

In the first part of this paper we study the asymptotic behavior of non-surjective linear CA. By looking at Fig. 1 we notice that a non-surjective CA cannot be neither ergodic, nor (strongly) transitive, nor (positively) expansive, nor regular. This makes most of the above listed results inapplicable to the study of the long term behavior of non-surjective linear CA over $\mathbf{Z}_m$. However, we prove (Theorem 5) that for any non-surjective linear CA F, there exists a subspace Y_F such that, for any configuration x, $F^k(x) \in Y_F$ for all $k \geq \lfloor \log_2 m \rfloor$. That is, after a transient phase of length at most $\lfloor \log_2 m \rfloor$, the evolution of the system takes place completely within the subspace Y_F. This result indicates that, in order to study the asymptotic behavior of non-surjective linear CA, one should analyze the behavior of the map F over the subspace Y_F. We show how to carry out this analysis by proving (Theorem 6) that the behavior of F over Y_F is identical to the behavior of a linear surjective map F^* defined over a configuration space isomorphic to Y_F. We also give an explicit formula for the coefficients of the local rule associated to F^*. The knowledge of these coefficients makes it possible to easily recognize if the map F restricted to Y_F satisfies any of the properties reported in Fig. 1.

In the second part of this paper, we make a further step in the analysis of the long term behavior of linear CA by studying the structure of their attractors. Informally, an attractor for a dynamical system (X, F) is a subset $Z \subseteq X$ of configurations such that the *forward trajectory* under iterations of F of any configuration which is *sufficiently* close to Z gets closer and closer to Z. Attractors of CA have been studied for example by Hurley [6], Kurka [8], and Blanchard *et al.* [1]. In particular, Kurka partitions the set of CA into five disjoint classes, labeled A_1–A_5, according to the structure of their attractors. We prove that for linear (surjective and non-surjective) CA it is possible to determine the membership in these classes by looking at the coefficients of the associated local rule. In particular, we prove (Theorem 10) that a linear surjective CA belongs to A_5 if it is transitive, otherwise it belongs to A_1. For non-surjective linear CA we use the results established in the first part of the paper. We prove (Theorem 12) that any attractor for a non-surjective map F is also an attractor for F restricted to the invariant subspace Y_F. As a consequence, the attractors for F can be determined by looking at the behavior of the associated surjective map F^*. If F^* is transitive then, F belongs to class A_4, otherwise F belongs to A_1 (Theorem 13). Since transitivity of F and F^* can be easily checked with gcd computations involving only the coefficients of the associated local rules, these results provide an effective procedure for deciding the membership of linear CA in Kurka's classes.

2   Basic definitions

For $m \geq 2$, let $\mathbf{Z}_m = \{0, 1, \ldots, m-1\}$ denote the ring of integers modulo m. We consider the *space of configurations*

$$\mathcal{C}_m^D = \left\{ c \mid c \colon \mathbf{Z}^D \to \mathbf{Z}_m \right\},$$

which consists of all functions from $\mathbf{Z}^D$ into $\mathbf{Z}_m$. Each element of $\mathcal{C}_m^D$ can be visualized as an infinite D-dimensional lattice in which each cell contains an element of $\mathbf{Z}_m$. A special configuration is the *null* configuration $\mathbf{0}$ which has the property that $\mathbf{0}(\mathbf{v}) = 0$ for all $\mathbf{v} \in \mathbf{Z}^D$.

Let $s \geq 1$. A *neighborhood frame* of size s is an ordered set of distinct vectors $\mathbf{u}_1, \mathbf{u}_2, \ldots, \mathbf{u}_s \in \mathbf{Z}^D$. Given $f \colon \mathbf{Z}_m^s \to \mathbf{Z}_m$, a D-dimensional CA based on the *local rule* f is the pair $(\mathcal{C}_m^D, F)$, where $F \colon \mathcal{C}_m^D \to \mathcal{C}_m^D$, is the *global transition map* defined as follows. For every $c \in \mathcal{C}_m^D$ the configuration $F(c)$ is such that for every $\mathbf{v} \in \mathbf{Z}^D$

$$[F(c)](\mathbf{v}) = f\left(c(\mathbf{v} + \mathbf{u}_1), \ldots, c(\mathbf{v} + \mathbf{u}_s)\right). \tag{1}$$

In other words, the content of cell $\mathbf{v}$ in the configuration $F(c)$ is a function of the content of cells $\mathbf{v} + \mathbf{u}_1, \ldots, \mathbf{v} + \mathbf{u}_s$ in the configuration c. Note that the local rule f and the neighborhood frame completely determine F. In this paper we consider mainly *linear* CA, that is, CA which have a local rule of the form

$$f(x_1, \ldots, x_s) = \sum_{i=1}^{s} \lambda_i x_i \bmod m, \tag{2}$$

with $\lambda_1, \ldots, \lambda_s \in \mathbf{Z}_m$ Note that for a linear D-dimensional CA (1) becomes

$$[F(c)](\mathbf{v}) = \sum_{i=1}^{s} \lambda_i c(\mathbf{v} + \mathbf{u}_i) \bmod m. \tag{3}$$

Throughout the paper, $F(c)$ will denote the result of the application of the map F to the configuration c, and $c(\mathbf{v})$ will denote the value assumed by c in $\mathbf{v}$. For $n \geq 0$, we recursively define $F^n(c)$ by $F^n(c) = F(F^{n-1}(c))$, where $F^0(c) = c$.

Given a CA $(\mathcal{C}_m^D, F)$ we say that $X \subseteq \mathcal{C}_m^D$ is an *invariant* subspace iff $F(X) \subseteq X$. If X is invariant, we say that (X, F) is a *subsystem* of $(\mathcal{C}_m^D, F)$.

The configuration space $\mathcal{C}_m^D$ is usually endowed with the topology which has as a basis of open and closed sets the D-dimensional *cylinders*. A D-dimensional cylinder $\langle (\mathbf{v}_1, a_1), \ldots, (\mathbf{v}_l, a_l) \rangle$ is the subset of $\mathcal{C}_m^D$ defined by

$$\langle (\mathbf{v}_1, a_1), \ldots, (\mathbf{v}_l, a_l) \rangle = \left\{ c \in \mathcal{C}_m^D \colon c(\mathbf{v}_i) = a_i, \ i = 1, \ldots, l \right\}.$$

One can easily verify that this topology coincides with the product topology induced on $\mathcal{C}_m^D$ by the discrete topology on $\mathbf{Z}_m$. With this topology, the space of configurations is compact and totally disconnected and every CA is a (uniformly) continuous map. A topological property which is of particular interest for us is *transitivity* since, as we will see, it is related to the structure of the attractors of a CA.

Definition 1. A dynamical system (X, F) is topologically transitive if and only if for all non-empty open subsets $U, V \subseteq X$ there exists a natural number n such that $F^n(U) \cap V \neq \emptyset$.

Intuitively, a transitive map F has points which eventually move under iteration of F from one arbitrarily small neighborhood to any other. As a consequence, the dynamical system cannot be decomposed into two disjoint open sets which are invariant under the map. The following result, proven in [3], shows that for linear CA transitivity can be recognized by looking at the coefficients of the local rule.

Theorem 2. *Let F denote the global transition map of a linear D-dimensional CA over $\mathbf{Z}_m$ defined by*

$$[F(c)](\mathbf{v}) = \sum_{i=1}^{s} \lambda_i c(\mathbf{v} + \mathbf{u}_i) \bmod m.$$

Assume $\mathbf{u}_1 = \mathbf{0}$, that is, λ_1 is the coefficient associated to the null displacement. The map F is topologically transitive iff $\gcd(m, \lambda_2, \ldots, \lambda_s) = 1$. $\square$

An important tool for the study of dynamical systems is the concept of topological conjugation. We say that two dynamical systems (X, F) and (X', F') are *topologically conjugated* if there exists a bijective function $\theta \colon X \to X'$ such that $\theta(F(x)) = F'(\theta(x))$ and both θ and θ^{-1} are continuous (that is, θ is a homeomorphism between X and X').

Informally, an attractor for a dynamical system (X, F) is a subset $Z \subseteq X$ of configurations such that the *forward trajectory* under iterations of F of any configuration $x \in X$ that is *sufficiently* close to Z gets closer and closer to Z.

Definition 3. Let (X, F) be a dynamical system. A nonempty subset $Z \subseteq X$ is an attractor for F iff there exists an open set $U \subseteq X$ such that

$$F(\overline{U}) \subseteq U \quad \text{and} \quad Z = \bigcap_{j \geq 0} F^j(U).$$

A nonempty set $Y \subseteq X$ is a *quasi-attractor* for F if it is a countable intersection of attractors but is itself not an attractor. An attractor (quasi-attractor) is *minimal* iff it does not contain any other attractor (quasi-attractor) as a proper subset. Let (X, F) be a dynamical system. We define the ω-limit of a set $Y \subseteq X$ according to F by

$$\omega_F(Y) = \bigcap_{n \geq 0} \overline{\bigcup_{m \geq n} F^m(Y)}.$$

In a zero-dimensional space ($\mathcal{C}_m^D$ is a zero-dimensional space) a set Z is an attractor for F iff there exists an invariant closed and open set (clopen from now) C with $Z = \omega_F(C)$. It is straightforward to verify that, if the dynamical systems (X, F) and (X', F') are topologically conjugated via the homeomorphism θ, then Z is an attractor for (X, F) iff $\theta(Z)$ is an attractor for (X', F').

The following result is basically due to Hurley [6] and has been refined by Kurka [8]. It provides an attractor based classification of CA.

Theorem 4. *Every CA ($\mathcal{C}_m^D, F$) satisfies exactly one of the following properties.*
(A_1) *F has a pair of disjoint attractors.*
(A_2) *F has a unique minimal quasi-attractor.*
(A_3) *F has a unique minimal attractor different from $\omega_F(X)$.*
(A_4) *F has a unique attractor $\omega_F(X)$ different from X.*
(A_5) *F has a unique attractor X.* □

3 Asymptotic behavior of non-surjective linear CA

In this section we study the asymptotic behavior of non-surjective linear CA. As a first result, we show that after a transient phase the evolution of a non-surjective CA takes place completely within an invariant subspace which can be exactly characterized.

Theorem 5. *Let F denote the global transition map of a linear D-dimensional CA over $\mathbf{Z}_m$ defined by*

$$[F(c)](\mathbf{v}) = \sum_{i=1}^{s} \lambda_i c(\mathbf{v} + \mathbf{u}_i) \bmod m. \tag{4}$$

If F is non-surjective there exists an invariant subspace $Y_F \subset \mathcal{C}_m^D$ such that for any $c \in \mathcal{C}_m^D$ and $k \geq \lfloor \log_2 m \rfloor$, we have $F^k(c) \in Y_F$.

Proof. Let $d = \gcd(m, \lambda_1, \ldots, \lambda_s)$. For the characterization of surjective linear CA (see Fig. 1) we know that $d > 1$. Let $m = p_1^{k_1} p_2^{k_2} \cdots p_h^{k_h}$. Without loss of generality we can assume that $d = p_1^{v_1} p_2^{v_2} \cdots p_l^{v_l}$ with $1 \leq v_i \leq k_i$ and $l \leq h$. Let

$$q = p_1^{k_1} \cdots p_l^{k_l}, \tag{5}$$

and define

$$Y_F = \{c \in \mathcal{C}_m^D \mid c(\mathbf{v}) \equiv 0 \pmod{q}, \forall \mathbf{v} \in \mathbf{Z}^D\}. \tag{6}$$

We now show that for $k \geq \max_{1 \leq i \leq l} k_i$ we have $F^k(c) \in Y_F$ for any $c \in \mathcal{C}_m^D$. By (4), we know that $[F(c)](\mathbf{v})$ is a multiple of $d = \gcd(m, \lambda_1, \ldots, \lambda_s)$. Hence, $[F(c)](\mathbf{v})$ is a multiple of $p_1 p_2 \cdots p_l$. Similarly, $[F^k(c)](\mathbf{v})$ is a multiple of $(p_1 p_2 \cdots p_l)^k$. Thus, $[F^k(c)](\mathbf{v}) \equiv 0 \pmod{q}$ and $F^k(\mathcal{C}_m^D) \subseteq Y_F$. With a similar argument one can easily see that $c \in Y_F$ implies $F(c) \in Y_F$. Hence Y_F is an invariant subspace as claimed. $\qquad\square$

Our next theorem shows that the study of the subsystem (Y_F, F) can be easily done since it is topologically conjugated to a surjective linear CA for which we are able to compute the coefficients of the associated local rule.

Theorem 6. *Let $(\mathcal{C}_m^D, F)$ denote a non-surjective linear CA, and let Y_F be defined as in the proof of Theorem 5. The subsystem (Y_F, F) is topologically conjugated to a surjective linear CA $(\mathcal{C}_{m^*}^D, F^*)$, with $m^* < m$.*

Proof. Let d, q be defined as in the proof of Theorem 5, and let $m^* = m/q$. Assume first $m^* > 1$. Since $\gcd(q, m^*) = 1$, the ring $\mathbf{Z}_m$ is isomorphic to the direct product $\mathbf{Z}_q \otimes \mathbf{Z}_{m^*}$. That is, each element $x \in \mathbf{Z}_m$ can be replaced by the pair $(x \bmod q, x \bmod m^*)$ with sums and products done componentwise. Replacing $\mathbf{Z}_m$ with $\mathbf{Z}_q \otimes \mathbf{Z}_{m^*}$, the set Y_F defined by (6) can be characterized as

$$Y_F = \{c \in \mathcal{C}_m^D \mid c(\mathbf{v}) = (0, t_{\mathbf{v}}), \ \forall \mathbf{v} \in \mathbf{Z}^D\}, \qquad (7)$$

where $t_{\mathbf{v}}$ denotes a generic element of $\mathbf{Z}_{m^*}$. It is straightforward to verify that (Y_F, F) is topologically conjugated to $(\mathcal{C}_{m^*}^D, F^*)$, where F^* denotes the linear map with the same neighborhood frame as F and coefficients $\lambda_1', \ldots, \lambda_s' \in \mathbf{Z}_{m^*}$ defined by $\lambda_i' = (\lambda_i \bmod m^*)$, $i = 1, \ldots, s$.

To complete the proof it remains to be shown that F^* is surjective. We use the characterization of surjective linear CA given in Fig. 1. Assume by contradiction that there exists a prime factor p_j $(l < j \leq h)$ of m^* such that

$$p_j | \lambda_i', \qquad i = 1, 2, \ldots, s. \qquad (8)$$

By construction, for $i = 1, \ldots, s$, we have $\lambda_i' = \lambda_i - v_i m^*$. Since $p_j | m^*$, we get

$$p_j | \lambda_i' \implies p_j | (\lambda_i - v_i m^*) \implies p_j | \lambda_i.$$

Hence, (8) implies $p_j | \gcd(m, \lambda_1, \ldots, \lambda_s)$ which is impossible since, by construction m^* contains only the prime factors of m which are not in $\gcd(m, \lambda_1, \ldots, \lambda_s)$.

Finally, if $m^* = 1$ it is straightforward to verify that $Y_F = \{0\}$. Hence, (Y_F, F) is topologically conjugated to the trivial CA $(\mathcal{C}_1^D, I)$ where I denotes the identity map. $\qquad\square$

To complete the picture, we show that there cannot be an invariant subspace smaller than Y_F for which Theorem 5 holds.

Corollary 7. *Let $(\mathcal{C}_m^D, F)$ denote a non-surjective linear CA, and let Y_F be defined as in the proof of Theorem 5. For any $k \geq \lfloor \log_2 m \rfloor$ we have $F^k(\mathcal{C}_m^D) = Y_F$.*

Proof. By Theorem 5 we already know that $F^k(\mathcal{C}_m^D) \subseteq Y_F$. To prove the other inclusion, notice that, since (Y_F, F) is topologically conjugated to the surjective CA $(\mathcal{C}_{m^*}^D, F^*)$, the map F must be surjective over Y_F. Hence, $F^k(\mathcal{C}_m^D) \supseteq F^k(Y_F) \supseteq Y_F$ as claimed. $\qquad\square$

4 Attractor structure of linear CA

In this section we study the attractor structure of linear CA. First, we consider the simpler case of surjective linear CA. Then, we use the results of Section 3 to analyze the attractors of non-surjective CA.

Lemma 8. *Let F denote the global transition map of a linear D-dimensional CA over $\mathbf{Z}_m$ defined by*

$$[F(c)](\mathbf{v}) = \sum_{i=1}^{s} \lambda_i c(\mathbf{v} + \mathbf{u}_i) \bmod m. \tag{9}$$

Assume $\mathbf{u}_1 = \mathbf{0}$, that is, λ_1 is the coefficient associated to the null displacement. If

$$\gcd(m, \lambda_1, \ldots, \lambda_s) = 1 \qquad and \qquad \gcd(m, \lambda_2, \ldots, \lambda_s) = d > 1, \tag{10}$$

then for every $c \in \mathcal{C}_m^D$ and $\mathbf{v} \in \mathbf{Z}^D$ we have

$$\gcd([F(c)](\mathbf{v}), d) = \gcd(c(\mathbf{v}), d). \tag{11}$$

Proof. Let $c \in \mathcal{C}_m^D$. From (9) we get $[F(c)](\mathbf{v}) = \lambda_1 c(\mathbf{v}) + \sum_{i=1}^{s} \lambda_i c(\mathbf{v} + \mathbf{u}_i) \bmod m$. By (10) we know that d divides every λ_i for $i > 2$, whereas $\gcd(\lambda_1, d) = 1$. Hence, for the properties of the gcd, we have

$$\gcd([F(c)](\mathbf{v}), d) = \gcd(\lambda_1 c(\mathbf{v}) + dH, d) = \gcd(\lambda_1 c(\mathbf{v}), d) = \gcd(c(\mathbf{v}), d).$$

$\qquad\square$

Corollary 9. *Let F denote the global transition map of a linear D-dimensional CA over $\mathbf{Z}_m$ defined by (9) and assume the hypotheses of Lemma 8 hold. For any divisor i of d consider the set*

$$C_i = \{c \in \mathcal{C}_m^D \mid \gcd(c(\mathbf{0}), d) = i\}. \tag{12}$$

Then C_i is a clopen set and $F(C_i) = C_i$.

Proof. The set C_i is clearly clopen since it is the finite union of cylinders. Moreover, by Lemma 8 we know that $F(C_i) \subseteq C_i$. To prove the opposite inclusion, we note that, by the characterization reported in Fig 1, the hypothesis $\gcd(m, \lambda_1, \ldots, \lambda_s) = 1$ implies that F is surjective. Thus, for each $x \in C_i$ there exists $y \in \mathcal{C}_m^D$ such that $x = F(y)$. By Lemma 8 we must have $y \in C_i$. Hence, $C_i \subseteq F(C_i)$ as claimed. $\qquad\square$

We are now able to determine the attractor structure for any surjective linear CA.

Theorem 10. *Let F denote the global transition map of a linear surjective D-dimensional CA over $\mathbf{Z}_m$. We have*

$$F \text{ transitive} \implies F \in A_5, \qquad F \text{ not transitive} \implies F \in A_1.$$

Proof. The first implication follows from a result by Kurka [8] which states that every transitive CA belongs to A_5. For the second implication assume F is not transitive. Since F is surjective, by the characterization of Fig. 1 and Theorem 2 we have that F satisfies the hypotheses of Lemma 8. Let $d = \gcd(m, \lambda_2, \ldots, \lambda_s)$. For any divisor i of d let C_i be defined according to (12). By Corollary 9 we have

$$F(\overline{C_i}) \subseteq C_i \qquad \text{and} \qquad C_i = \bigcap_{j \geq 0} F^j(C_i).$$

By Definition 3 we have that each C_i is an attractor and since they are clearly disjoint F belongs to A_1. □

We now study the attractor structure of non-surjective linear CA. The following simple lemma shows that every attractor for F is a subset of the invariant subspace Y_F introduced in Section 3.

Lemma 11. *Let $(\mathcal{C}_m^D, F)$ denote a non-surjective linear CA, and let Y_F be defined by (6). If Z is an attractor for F then $Z \subseteq Y_F$.*

Proof. Let $k = \lfloor \log_2 m \rfloor$. By Definition 3 we have that Z is a subset of $F^k(U)$ for a suitable $U \subseteq \mathcal{C}_m^D$. By Corollary 7 we have $Z \subseteq F^k(U) \subseteq F^k(\mathcal{C}_m^D) = Y_F$. □

In order to study the attractors of $(\mathcal{C}_m^D, F)$ we extend Lemma 11 by proving a much stronger result. We prove (Theorem 12) that Z is an attractor for $(\mathcal{C}_m^D, F)$ iff it is an attractor for (Y_F, F). We use the notation introduced in Section 3. Let q, m^* be defined as in the proofs of Theorems 5 and 6. We have already observed that, since $\gcd(q, m^*) = 1$, the ring $\mathbf{Z}_m$ is isomorphic to $\mathbf{Z}_q \otimes \mathbf{Z}_{m^*}$, and each element $x \in \mathbf{Z}_m$ can be replaced by the pair $(x \bmod q, x \bmod m^*)$ with sums and products done componentwise. Using this notation the set Y_F can be characterized as in (7). We define the map $\Pi : \mathcal{C}_m^D \to Y_F$ such that for any $c \in \mathcal{C}_m^D$

$$[\Pi(c)](\mathbf{v}) = (0, c(\mathbf{v}) \bmod m^*) \qquad \forall \mathbf{v} \in \mathbf{Z}^D.$$

It is straightforward to verify that the map Π satisfies the following properties:

P1. Π restricted to Y_F is the identity map;
P2. $\Pi \circ F = F \circ \Pi$;
P3. if V is a cylinder of $\mathcal{C}_m^D$ then $\Pi(V)$ is a cylinder of the subspace Y_F;
P4. if V' is a cylinder of Y_F then $\Pi^{-1}(V')$ is a finite union of cylinders of $\mathcal{C}_m^D$.

We are now ready to prove the main result of this subsection.

Theorem 12. *The set Z is an attractor for $(\mathcal{C}_m^D, F)$ if and only if it is an attractor for (Y_F, F).*

Proof. Assume Z is an attractor for $(\mathcal{C}_m^D, F)$. For the characterization of Section 2 we know that there exists an invariant clopen set $C \subseteq \mathcal{C}_m^D$ such that

$$Z = \bigcap_{n>0} \overline{\bigcup_{k\geq n} F^k(C)}.$$

We prove that Z is an attractor also for (Y_F, F) be showing that $\Pi(C)$ is an invariant clopen subset of Y_F, and that

$$Z = \bigcap_{n>0} \overline{\bigcup_{k\geq n} F^k(\Pi(C))}^{Y_F}, \tag{13}$$

where we use $\overline{W}^{Y_F}$ to denote the closure of the set W with respect to the topology of Y_F. Note, however, that since Y_F is a closed subset of $\mathcal{C}_m^D$, we have $\overline{W}^{Y_F} = \overline{W}$ for any $W \subseteq Y_F$.

Since C is a clopen subset of $\mathcal{C}_m^D$ it is the finite union of cylinders and by property P3 $\Pi(C)$ it is a clopen subset of Y_F. Moreover, since C is invariant, using property P2 we have

$$F(\Pi(C)) = \Pi(F(C)) \subseteq \Pi(C)$$

which shows that $\Pi(C)$ is invariant as well.

Let $n^* = \lfloor \log_2 m \rfloor$. By Theorem 5 we know that, for $k \geq n^*$, $F^k(C) \subseteq Y_F$. Hence, by properties P1–P2 we get

$$Z = \bigcap_{n\geq n^*} \overline{\bigcup_{k\geq n} \Pi(F^k(C))} = \bigcap_{n\geq n^*} \overline{\bigcup_{k\geq n} F^k(\Pi(C))} = \bigcap_{n\geq n^*} \overline{\bigcup_{k\geq n} F^k(\Pi(C))}^{Y_F},$$

which proves (13). Hence Z is an attractor for (Y_F, F) as claimed.

Assume now Z' is an attractor for (Y_F, F). Then, there exists a clopen invariant set $C' \subseteq Y_F$ such that

$$Z' = \bigcap_{n>0} \overline{\bigcup_{k\geq n} F^k(C')}^{Y_F}.$$

Reasoning as above, using properties P1–P4, one can easily see that $\Pi^{-1}(C')$ is an invariant clopen subset of $\mathcal{C}_m^D$ such that $Z' = \omega_F(\Pi^{-1}(C'))$. This proves that Z' is an attractor for $(\mathcal{C}_m^D, F)$ and the theorem follows. $\square$

Theorem 13. *Let $(\mathcal{C}_m^D, F)$ denote a non-surjective linear CA, and let $(\mathcal{C}_{m^*}^D, F^*)$ denote the surjective linear CA topologically conjugated to (Y_F, F) defined in Theorem 6. We have*

$$F^* \text{ transitive} \implies F \in \mathcal{A}_4, \qquad F^* \text{ not transitive} \implies F \in \mathcal{A}_1.$$

Proof. By Theorem 12 we know that $(\mathcal{C}_m^D, F)$ and (Y_F, F) have the same attractors. Since (Y_F, F) is topologically conjugated to $(\mathcal{C}_{m^*}^D, F^*)$, it has the same number of attractors as $(\mathcal{C}_{m^*}^D, F^*)$. The thesis follows by Theorem 10. $\square$

Attractor Class	Characterization
$A_{1.1}$	$\gcd(m, \lambda_1, \ldots, \lambda_s) > 1, \quad \gcd(m^*, \lambda_2', \ldots, \lambda_s') > 1$
$A_{1.2}$	$\gcd(m, \lambda_1, \ldots, \lambda_s) = 1, \quad \gcd(m, \lambda_2, \ldots, \lambda_s) > 1$
A_2	empty
A_3	empty
A_4	$\gcd(m, \lambda_1, \ldots, \lambda_s) > 1, \quad \gcd(m^*, \lambda_2', \ldots, \lambda_s') = 1$
A_5	$\gcd(m, \lambda_2, \ldots, \lambda_s) = 1$

Fig. 2. Characterization of Kurka's classes for linear CA. λ_i represent the coefficients of the local rule associated to F while λ_i' represent the coefficients of the local rule associated to F^*.

4.1 Kurka's classification for linear CA

Using the results of the previous sections we can now easily classify every linear CA according to Kurka's scheme. However, we notice that Kurka's class A_1 contains both surjective and non-surjective CA. Note that for surjective CA the union of all attractors is the whole space, whereas this is never true for non-surjective CA. For symmetry with the other classes, we split Kurka's class A_1 into two subclasses: $A_{1.1}$ and $A_{1.2}$. Class $A_{1.1}$ contains those CA which have at least two disjoint attractors and for which the union of all the attractors is a proper subset of the entire space of configuration. Class $A_{1.2}$ contains those CA which have at least two disjoint attractors for which the union of all the attractors is equal to the entire space of configuration. As an immediate corollary to Theorems 10 and 13 we have the following results.

Corollary 14. *Let $(\mathcal{C}_m^D, F)$ denote a linear CA and, if F is not surjective, let $(\mathcal{C}_{m^*}^D, F^*)$ denote the linear CA defined in Theorem 6. We have*

$$F \in A_{1.1} \Leftrightarrow F \text{ is not surjective and } F^* \text{ is not transitive}$$
$$F \in A_{1.2} \Leftrightarrow F \text{ is surjective but not transitive}$$
$$F \in A_4 \Leftrightarrow F \text{ is not surjective and } F^* \text{ is transitive}$$
$$F \in A_5 \Leftrightarrow F \text{ is transitive}$$

$\square$

Corollary 15. *Classes A_2 and A_3 do not contain linear CA.* $\square$

Combining Corollary 14 with the results summarized in Fig. 1 we get a characterization of Kurka's classes for linear CA based on the coefficients $\lambda_1, \ldots, \lambda_s$ of F and $\lambda_1', \ldots, \lambda_s'$ of F^*. This characterization is shown in Fig. 2

5 Conclusions

This paper continues a stream of previous works (see for example [2, 3, 7, 9, 10, 11]) where criteria are provided for analyzing the topological behavior of linear

CA over the entire space of configurations. In the case of non-surjective CA the above mentioned criteria cannot be applied and the CA under consideration can only be labeled as non-surjective. Here, we overcome this problem by proving the following results: 1) all the forward trajectories of any D-dimensional linear CA over $\mathbf{Z}_m$ move into an invariant subspace Y_F after at most $\lfloor \log_2(m) \rfloor$ time steps; 2) the dynamics of any linear CA on Y_F is topologically conjugated to the dynamics of another (explicitly given) surjective linear CA defined over a smaller alphabet.

These results allow us to apply the topological characterizations summarized in Fig. 1 also to non-surjective CA. As a byproduct, we have been able to completely classify linear CA according to the attractor based Kurka's classification scheme.

References

1. F. Blanchard, P. Kurka, and A. Maass. Topological and measure-theoretic properties of one-dimensional cellular automata. *Physica D*, 103:86–99, 1997.
2. G. Cattaneo, E. Formenti, G. Manzini, and L. Margara. Ergodicity and regularity for linear cellular automata over Z_m. *Theoretical Computer Science*. To appear.
3. G. Cattaneo, E. Formenti, G. Manzini, and L. Margara. On ergodic linear cellular automata over Z_m. In *14th Annual Symposium on Theoretical Aspects of Computer Science (STACS '97)*, pages 427–438. LNCS n. 1200, Springer Verlag, 1997.
4. R. L. Devaney. *An Introduction to Chaotic Dynamical Systems*. Addison-Wesley, Reading, MA, USA, second edition, 1989.
5. M. Garzon. *Models of Massive Parallelism*. EATCS Texts in Theoretical Computer Science. Springer Verlag, 1995.
6. M. Hurley. Attractors in cellular automata. *Ergodic Theory and Dynamical Systems*, 10:131–140, 1990.
7. M. Ito, N. Osato, and M. Nasu. Linear cellular automata over Z_m. *Journal of Computer and System Sciences*, 27:125–140, 1983.
8. P. Kurka. Languages, equicontinuity and attractors in cellular automata. *Ergodic theory and dynamical systems*, 17:417–433, 1997.
9. G. Manzini and L. Margara. Invertible linear cellular automata over Z_m: Algorithmic and dynamical aspects. *Journal of Computer and System Sciences*. To appear.
10. G. Manzini and L. Margara. A complete and efficiently computable topological classification of D-dimensional linear cellular automata over Z_m. In *24th International Colloquium on Automata Languages and Programming (ICALP '97)*. LNCS n. 1256, Springer Verlag, 1997.
11. T. Sato. Ergodicity of linear cellular automata over Z_m. *Information Processing Letters*, 61(3):169–172, 1997.

Optimal Simulations Between Unary Automata*

Carlo Mereghetti and Giovanni Pighizzini

Dipartimento di Scienze dell'Informazione – Università degli Studi di Milano
via Comelico 39 – 20135 Milano, Italy – {mereghc,pighizzi}@dsi.unimi.it

Abstract. We consider the problem of computing the costs — in terms of states — of optimal simulations between different kinds of finite automata recognizing *unary* languages. Our main result is a tight simulation of unary n–state two–way nondeterministic automata by $O(e^{\sqrt{n \ln n}})$–state one–way deterministic automata. In addition, we show that, given a unary n–state two–way nondeterministic automaton, one can construct an equivalent $O(n^2)$–state two–way nondeterministic automaton performing both input head reversals and nondeterministic choices at the endmarkers only. Further results on simulating unary alternating finite automata are pointed out. Our results give answers to some questions left open in the literature.

1 Introduction

Finite automata [RS59] are one of the simplest computational models. Many of their properties have been widely investigated. Notwithstanding their simplicity, some important questions concerning finite automata are still open. In particular, not all the costs of optimal simulations between different kinds of automata are known. For instance, we recall the problem, proposed in 1978 by W. Sakoda and M. Sipser [SS78], of proving the existence of a polynomial — with respect to the number of states — simulation of two–way nondeterministic finite automata (2nfa, for short) by two–way deterministic finite automata (2dfa). As pointed out by P. Berman and A. Lingas in [BL77], such a problem is not only interesting *per se*, but is tightly related to the DLOGSPACE = NLOGSPACE question.

On the other hand, the costs of other simulations have been precisely stated. Besides the optimal simulation of n–state one–way nondeterministic finite automata (1nfa) by 2^n–state one–way deterministic finite automata (1dfa), we can mention the optimal simulation of 2dfa's by 1dfa's ($O(2^{n \log n})$ states), of 2nfa's by 1dfa's ($O(2^{n^2})$ states), and of one–way alternating finite automata (1afa) by 1dfa's ($O(2^{2^n})$ states). (For lack of space, we do not provide credits for all these results. The reader can find precise references in [Chr86].)

It is important to stress that the optimality of such simulations has been established by witness languages built over alphabets of two or more symbols. As a matter of fact, the situation turns out to be quite different whenever we

---
* This work is partially supported by Ministero dell'Università e della Ricerca Scientifica e Tecnologica (MURST).

restrict to *unary* automata, i.e., automata with a single letter input alphabet. For example, we know that $O(e^{\sqrt{n \ln n}})$ states suffice in order to simulate a unary n–state 1nfa or 2dfa by a 1dfa. Furthermore, a unary n–state 1nfa can be simulated by a 2dfa having $O(n^2)$ many states. All these results and their optimality have been proved in 1986 by M. Chrobak [Chr86].

In this paper, we further deepen the study of optimal simulations between different kinds of unary automata. The first part of the paper is devoted to study 2nfa's. We closely analyze the structure of their computation paths. In particular, using number theoretical arguments, we show that, for sufficiently large inputs, it is possible to consider only computation paths in which states are repeated in a very regular way. This enables us to state our main result concerning the optimal simulation of unary 2nfa's via 1dfa's: each unary n–state 2nfa can be simulated by a 1dfa with $O(e^{\sqrt{n \ln n}})$ states. Note that the complexity is the same as the above mentioned optimal simulations of 1nfa's and 2dfa's by 1dfa's. Thus, the simultaneous elimination of both two–way motion and nondeterminism has the same cost as the elimination of either of them. As another consequence of our analysis of 2nfa's, we are able to prove that each n–state unary 2nfa can be simulated by a 2nfa with $O(n^2)$ states which reverses the input head direction and makes nondeterministic decisions *only* when the input head visits the left or the right end of the input. This result can be regarded as a further contribution to a better understanding of the 2nfa model. In particular, it could be useful in the investigation of the Sakoda–Sipser open question recalled above, at least for unary inputs.

In the second part of the paper, we study the relationships between unary 1afa's and dfa's, nfa's. This problem was proposed in [Chr86], and partly solved in [Bir93], where it is proved that 2^n states are necessary for 2nfa's to simulate n–state 1afa's. We point out an optimal simulation of n–state unary 2nfa's by $O(\sqrt{n \ln n})$–state 1afa's.

By combining our results with those in the literature, we are able to draw an almost complete description of the costs of optimal simulations between the different types of unary automata here considered. We do not list all the resulting optimal bounds, which can be better understood by observing Figure 1.

2 Preliminary definitions and results

In this section, we recall definitions and results used throughout the paper. Given an alphabet Σ, Σ^* denotes the set of strings on Σ, with the empty string. Given a string $x \in \Sigma^*$, $|x|$ denotes its length. A language $L \subseteq \Sigma^*$ is said to be *unary* (or *tally*) whenever L is built over a single letter alphabet (usually, $\Sigma = \{1\}$). We assume that the reader is familiar with one–way/two–way, deterministic/nondeterministic finite automata (1dfa, 2dfa, 1nfa, 2nfa, for short). For a detailed exposition see, e.g., [HU79]. Moreover, we also assume familiarity with one–way/two–way alternating finite automata (1afa, 2afa, for short), as presented in [CKS81,FJY90]. We stipulate that input strings for two–way

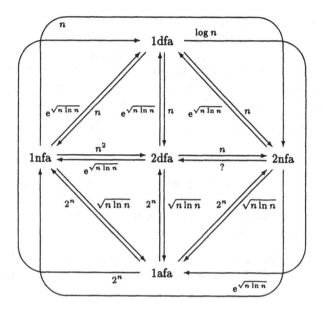

Fig. 1. Costs of the optimal simulations between different kinds of unary automata. An arc labeled $f(n)$ from a vertex x to a vertex y means that a unary n–state automaton of type x can be simulated by a $O(f(n))$–state automaton of type y.

devices are enclosed between a left and a right endmarker symbol. The left end-marker is contained in the 0th input square. A finite automaton is said to be *unary* whenever the input alphabet contains exactly one symbol. Our automata are assumed to be complete, i.e., their transition functions are total.

We need few notions of graph theory. A directed graph (digraph) G is *strongly connected* if there exists a directed path between any two vertices. A *strongly connected component* of G is a sub–graph of G which is strongly connected, and not contained in any other strongly connected sub–graph. Following [Ber85], we call *elementary* any cycle in which no vertex is encountered more than once (except, of course, the initial vertex which is also the terminal vertex). Our estimations on simulation costs rely on some number theoretical topics that are now to be quickly presented. First, some notation. We let $\mathbf{Z}^+$ ($\mathbf{Z}^-$) the set of positive (negative) integers, and $\mathbf{Z}$ the set of whole integers. Natural numbers are denoted with $\mathbf{N}$. For any $z \in \mathbf{Z}$, $|z|$ denotes its absolute value. For any $z \in \mathbf{Z}^+$, $\ln z$ denotes the natural logarithm of z, while $\log z$ is the logarithm of z taken to the base 2. The *greatest common divisor* of $a_1, \ldots, a_s \in \mathbf{Z}$ writes as $\gcd(a_1, \ldots, a_s)$. Their *least common multiple* is denoted by $\mathrm{lcm}(a_1, \ldots, a_s)$. Both gcd and lcm are, actually, meant to be taken on $|a_1|, \ldots, |a_s|$.

In the simulation of unary automata, a crucial role is played by the function $F(n) = \max\{\mathrm{lcm}(x_1, \ldots, x_s) \mid x_1, \ldots, x_s \in \mathbf{Z}^+ \text{ and } x_1 + \ldots + x_s = n\}$. Evaluating the growth of $F(n)$ is known as Landau's problem, and is addressed in [Chr86]. For our purposes, it is enough to know that

Lemma 1. $F(n) = O(e^{\sqrt{n \ln n}})$.

The other arithmetical tool we shall make use of is the theory of linear Diophantine equations. We consider equations in the form $a_1 x_1 + \ldots + a_s x_s = z$, where $a_1, \ldots, a_s, z \in \mathbf{Z}$, and $x_1, \ldots, x_s$ are variables. It is well-known that, whenever $x_1, \ldots, x_s$ range over $\mathbf{Z}$, the set of integers in the form $a_1 x_1 + \ldots + a_s x_s$ coincides with the set of all integral multiples of $\gcd(a_1, \ldots, a_s)$. Instead, whenever $x_1, \ldots, x_s$ range over $\mathbf{N}$, one can prove the following

Lemma 2. Let $a_1, \ldots, a_s \in \mathbf{Z}$, with $|a_i| \leq n$, and let

$$X = \{a_1 x_1 + \ldots + a_s x_s \mid x_1, \ldots, x_s \geq 0\}.$$

Define $X^+ = X \cap \{z \in \mathbf{Z}^+ \mid z > n^2\}$, and $X^- = X \cap \{z \in \mathbf{Z}^- \mid z < -n^2\}$. Then, $X^+ \neq \emptyset$ ($X^- \neq \emptyset$) if and only if there exists at least one positive (negative) a_i. Moreover, if $X^+ \neq \emptyset$ ($X^- \neq \emptyset$), then it coincides with the set of positive (negative) multiples of $\gcd(a_1, \ldots, a_s)$ greater than n^2 (smaller than $-n^2$).

The proof of Lemma 2 easily follows from results presented in [Chr86, Sec. 3], where the reader can find more on linear Diophantine equations. Moreover, as a routine exercise, one can show that

Lemma 3. Let $a_1, \ldots, a_s \in \mathbf{Z}^+$, with $a_i \leq n$, and let $z \geq 0$. If the equation $a_1 x_1 + a_2 x_2 + \ldots + a_s x_s = z$ has a solution over the natural numbers, then it has a solution over natural numbers satisfying $a_2 x_2 + \ldots + a_s x_s \leq 2n^2$.

3 Computation paths of 2nfa's and simulation by 1dfa's

In this section, we focus on unary 2nfa's. By discovering some regularities in the structure of their computation paths, we show that any unary language L accepted by a 2nfa with n states, forms a definitively periodic set of period $p = O(e^{\sqrt{n \ln n}})$. More precisely, we find an integer $\overline{k} = O(n^2)$ such that, for any $k \geq \overline{k}$, $1^k \in L$ if and only if $1^{k+p} \in L$. This immediately leads to a $O(e^{\sqrt{n \ln n}})$ simulation of unary 2nfa's via 1dfa's. The optimality of such a simulation is a direct consequence of a result in [Chr86]. The results here obtained will be useful also in the next section, where we show that by just squaring the number of the states, it is possible to consider 2nfa's with a very particular structure.

To state our results, we need some tools to examine computations over unary inputs. Specifically, we refer to three lemmas that V. Geffert proved in [Gef91] for two-way nondeterministic Turing machines accepting unary languages in sublogarithmic space. We are now going to reformulate these lemmas within the realm of unary 2nfa's. Their validity in this new form comes straightforwardly from the originals.

From now on, we always refer to a unary 2nfa A with n states.

The first lemma says that each computation path of A beginning and ending at the same input square and visiting neither of the endmarkers, can be replaced by an equivalent computation path not moving the input head more than n^2 positions to the right (or left).

Lemma 4. (U–Turn) *Let $m > 2n^2$, and suppose there exists a computation path of A on input 1^m where (a) the first state is q_1 with the input head at a position i, (b) the last state is q_2 with the input head at the same position, and (c) the input head never moves to the left (right) of i, neither it visits the right (left) endmarker. Then, there also exists a computation path on input 1^m satisfying (a), (b) and (c), where the input head never moves farther than n^2 positions to the right (left) of i.*

The second lemma states that we can freely "shift" a computation path that never visits the endmarkers to any position of the input tape, provided both the initial and final input head positions lie sufficiently far from the endmarkers.

Lemma 5. (Position Independence) *Let $m > 2n^2$, and suppose there exists a computation path of A on 1^m beginning in the state q_1 with the input head at a position i, ending in the state q_2 with the input head at the position $i + \ell$, $\ell \geq 0$, and visiting neither of the endmarkers. Then, there also exists a computation path on input 1^m beginning in q_1 with the input head at a position j, and ending in the state q_2 with the input head at the position $j + \ell$, for each $n^2 < j \leq j + \ell \leq m - n^2$. A similar condition can be formulated for moving to the left.*

The next lemma defines the structure of the computation paths that traverse the entire input. We give a new proof of this result for unary 2nfa's, relying on some properties of linear Diophantine equations. This approach enables us to reduce the $O(n^4)$ upper bound on s_1 and s_2 given in [Gef91] to $O(n^2)$. We point out that the same improvement has been obtained independently by Geffert, using a different argument [Gef97]. This improvement will be useful in the quadratic simulation stated in Theorem 13.

In the following, by *loop of length ℓ*, we mean a computation path beginning in a state p with the input head at a position i, and ending in the same state with the input head at the position $i + \ell$.

Lemma 6. (Dominant Loop) *Let $m > 2n^2$, and suppose the input 1^m is traversed from left to right by a computation path Π of A beginning in the state q_1, ending in the state q_2, and such that the endmarkers are visited only in q_1 and q_2. Then, 1^m can be traversed by an equivalent computation path such that A, having traversed the left endmarker and s_1 positions, gets into a loop of length ℓ, which starts in a state p and is repeated λ times, then traverses the remaining s_2 input squares, and finally gets the right endmarker, for some s_1, s_2, ℓ satisfying $1 \leq \ell \leq n$, $s_1, s_2 \geq n^2$, and $s_1 + s_2 \leq 5n^2$. (Note that $m = s_1 + \lambda \ell + s_2$.) The same result holds for computation paths traversing the input from right to left.*

Proof. Define the digraph $G = (Q, E)$, where $(q', q'') \in E$ if and only if there is a computation path which passes through q' with the input head at a position $0 \leq i \leq m$, ends in q'' with the input head at the position $i + 1$, and never visits either the endmarkers or the $(i + 1)$th input square in the intermediate steps. (Notice that, by Lemma 5, for sufficiently large inputs, the graph G does not depend on the input length.)

Let C be a path of length x in G, from q' to q''. By repeatedly eliminating from C elementary circuits of G, we finally obtain a circuit–free path from q' to q'' of length $x_0 \leq n$. Let $\{a_1, \ldots, a_t\}$ be the set of the lengths of those elementary circuits of G eliminated from C in the previous process. Notice that $a_i \leq n$, for $1 \leq i \leq t$. Then, $x = x_0 + a_1 x_1 + \ldots + a_t x_t$, where, for $1 \leq i \leq t$, $x_i > 0$ is the number of times elementary circuits of length a_i has been eliminated from C.

By Lemma 3, there exist $x_1', \ldots, x_t' \geq 0$ such that $x = x_0 + a_1 x_1' + a_2 x_2' + \ldots + a_t x_t'$, and $a_2 x_2' + \ldots + a_t x_t' \leq 2n^2$. Hence, we can devise a new path C' in G of length x, from q' to q'', containing a circuit of length a_1 which is repeated x_1' times, and where the length of the sub–path of C' outside such a circuit is $x_0 + a_2 x_2' + \ldots + a_t x_t' \leq 2n^2 + n$. Using Lemma 5, it is easy to see that we can associate with C' a computation path in A from the state q' to the state q'', which moves the input head from a certain position j to the position $j + x$, repeats x_1' consecutive times a loop of length a_1, and traverses no more than $2n^2 + n$ input tape positions outside such a loop, provided that $n^2 < j < j + x \leq m - n^2$.

Now, consider the given computation path Π. For $0 \leq i \leq m+1$, let p_i be the state of A when the input head visits the ith input square for the first time along Π. Thus, $p_0 = q_1$ and $p_{m+1} = q_2$. Notice that $(p_i, p_{i+1}) \in E$, for $0 \leq i \leq m$.

Following the above reasoning, we can associate with the path in G which traverses vertices $p_{n^2+1}, p_{n^2+2}, \ldots, p_{m-n^2}$, a computation path Π' in A beginning in the state p_{n^2+1} with the input head on the $(n^2 + 1)$th square, ending in the state p_{m-n^2} with the input head on the $(m - n^2)$th square, where a loop of length $\ell(=a_1)$ from a state p is repeated $\lambda(=x_1')$ times, and where the input positions traversed outside this loop are at most $2n^2 + n$.

In conclusion, we can replace in the original path Π the part of the computation from state p_{n^2+1}, with the input head on the $(n^2 + 1)$th square, to state p_{m-n^2}, with the input head on the $(m - n^2)$th square, with Π'. We get a computation path of A that starts from q_1 with the input head at the left endmarker, traverses the left endmarker and $s_1 \geq n^2$ input squares, gets λ times into a loop of length $1 \leq \ell \leq n$, traverses $s_2 \geq n^2$ input squares, and finally reaches the right endmarker in the state q_2. It is easy to see that $s_1 + s_2 = 2n^2 + x_0 + a_2 x_2' + \ldots + a_t x_t' \leq 5n^2$. $\qquad \square$

Now, we closely study loops in computation paths. Our goal is to show that we can focus only on some loops or, more precisely, only on some lengths. By *weighted digraph*, we mean a digraph where edges have weights from $\{-1, 0, 1\}$. The weight of a path is the sum of weights of the edges the path consists of.

Lemma 7. *Let $G = (V, E)$ be a strongly connected weighted digraph with $|V| = n$, and let $\{a_1, \ldots, a_s\}$ be the set of weights of the s elementary circuits in G. For each $v \in V$, define X_v as the set of weights of all circuits containing v. If there exists at least one $a_i > 0$ $(a_i < 0)$ then the set X_v and the set of positive (negative) multiples of $\gcd(a_1, \ldots, a_s)$ agree on the elements greater than $3n^2$ (less than $-3n^2$, respectively).*

Proof. Since each circuit is the sum of elementary circuits, $x \in X_v$ implies $x = a_1 x_1 + \ldots + a_s x_s$, for some $x_1, \ldots, x_s \geq 0$. Then, x is an integral multiple of $\gcd(a_1, \ldots, a_s)$.

Conversely, if x is a sufficiently large multiple of $\gcd(a_1, \ldots, a_s)$, then, by Lemma 2, $x = a_1 x_1 + \ldots a_s x_s$, for some $x_1, \ldots, x_s \geq 0$. This fact does not imply the existence of one circuit of weight x in the graph. For instance, if $s = 3$, $x_1 = 1$, $x_2 = 0$, $x_3 = 1$, and the circuits of weight a_1 and a_3 are connected only via the circuit of weight a_2, then x_1, x_2, x_3 define two disjoint circuits but cannot define a circuit in G.

In order to solve this problem, we shall prove the existence of a circuit of weight x which visits all vertices of G. To this aim, we first observe that it is easy to construct a circuit C_0, visiting all vertices of G, and such that its weight x_0 satisfies $|x_0| < 2n^2$. Notice that $x_0 = a_1 x_1 + \ldots + a_s x_s$, for some $x_1, \ldots, x_s \geq 0$, and hence $x_0 = k_0 \gcd(a_1, \ldots, a_s)$, for some $k_0 \in \mathbf{Z}$. Given $x = k \gcd(a_1, \ldots, a_s)$, for a $k \in \mathbf{Z}$, with $|x| > 3n^2$, let $z = x - x_0 = (k - k_0) \gcd(a_1, \ldots, a_s)$. Then $|z| = |x - x_0| \geq |x| - |x_0| > n^2$. By Lemma 2, and by observing that, for $1 \leq i \leq s$, $|a_i| \leq n$, this implies that $z = a_1 z_1 + \ldots + a_s z_s$, for some $z_1, \ldots, z_s \geq 0$. Finally, by padding the circuit C_0 with a suitable amount of elementary circuits corresponding to coefficients $z_1, \ldots, z_s$, we obtain a circuit of weight x in G. □

Let us now consider the weighted digraph $\mathcal{A}$ consisting of the digraph representing our automaton A, after removing transitions on the endmarkers, where with each edge is associated a weight representing the movement of the input head in the corresponding transition. Namely, the weight $-1, 1, 0$ is associated with each transition where the input head is moved one position left, right, or kept stationary, respectively. Notice that any circuit of weight ℓ in $\mathcal{A}$ represents a loop of length ℓ in the automaton A. By applying Lemma 7 to such a weighted digraph, we can easily obtain the following:

Lemma 8. *Let $p \in Q$, and let g be the greatest common divisor of the weights of the circuits in the strongly connected component of $\mathcal{A}$ containing p. Let X_p^+ the set of integers $x \geq 3n^2$ such that the automaton A, starting in the state p, with the input head at the ith input square can reach the $(i+x)$th input square in the same state p, without visiting the endmarkers, provided that $i + x < m - n^2$. If $X_p^+ \neq \emptyset$, then it coincides with the set of all integral multiples of g greater than $3n^2$. A symmetrical result holds for the set X_p^- defined in the obvious way.*

The following result is crucial in order to obtain our simulations:

Theorem 9. *There exists a set $\{\ell_1, \ldots, \ell_s\} \subseteq \{1, \ldots, n\}$ with $\ell_1 + \ldots + \ell_s \leq n$, such that, for any $m > 8n^2$, if the input 1^m can be traversed from left to right by a computation path of A beginning in a state q_1, ending in a state q_2, and where the endmarkers are visited only in q_1 and q_2, then there exists an index $i \in \{1, \ldots, s\}$ such that, for any $\mu > \frac{8n^2 - m}{\ell_i}$, there is a computation path from q_1 to q_2 which traverses from left to right the input $1^{m + \mu \ell_i}$.*

Proof. Let $\{\ell_1, \ldots, \ell_s\}$ be the set of the greatest common divisors of the weights of circuits of all the strongly connected components of the digraph $\mathcal{A}$. Thus, ℓ_i is bounded by the number of states of the ith strongly connected component, and, since strongly connected components are disjoint, $\ell_1 + \ell_2 + \ldots + \ell_s \leq n$.

Now, consider $s_1, s_2, \ell, \lambda, p$ as in Lemma 6. Let $\ell_i \in \{\ell_1, \ldots, \ell_s\}$ be the greatest common divisor of the weights of the circuits in the strongly connected component of $\mathcal{A}$ containing p. By Lemma 8, the set of integers $x > 3n^2$ for which there is a path from p to itself that visits neither of the endmarkers and which moves the input head from the ith to the $(i + x)$th square, coincides with the multiples of ℓ_i greater than $3n^2$. This leads us to conclude that for integers $\eta > \frac{3n^2}{\ell_i}$, there exists a path traversing the whole input $1^{s_1 + \eta \ell_i + s_2}$ from q_1 to q_2.

Now, given $\mu > \frac{8n^2 - m}{\ell_i}$ and $\eta = \mu + \frac{\lambda \ell}{\ell_i}$, we have $\eta > \frac{8n^2 - s_1 - \lambda \ell - s_2 + \lambda \ell}{\ell_i} \geq \frac{3n^2}{\ell_i}$, and $s_1 + \eta \ell_i + s_2 = m + \mu \ell_i$. Then, there exists a path traversing from left to right the whole input $1^{m + \mu \ell_i}$ from state q_1 to state q_2. $\square$

In the following theorem, we suitably adopt the well–known $n \to n + n!$ pumping technique. In the light of Theorem 9, we are able to state a bound on the length of the pumped part which enables us to show that unary languages accepted by n–state 2nfa's form definitively periodic sets of period $O(e^{\sqrt{n \ln n}})$:

Theorem 10. *Let L be a unary language accepted by a n–state 2nfa. Then, there exists a constant ℓ bounded by $O(e^{\sqrt{n \ln n}})$ such that, for any integer $k > 8n^2$,*

$$1^k \in L \text{ if and only if } 1^{k+\ell} \in L.$$

Proof. Let $\{\ell_1, \ldots, \ell_s\}$ as in Theorem 9, and let $\ell = \text{lcm}(\ell_1, \ldots, \ell_k)$. By Lemma 1, ℓ is bounded above by $O(e^{\sqrt{n \ln n}})$.

Suppose $1^{k+\ell} \in L$, and let $\mathcal{C}$ be an accepting computation path of A on $1^{k+\ell}$. Along $\mathcal{C}$, consider $r_0, r_1, \ldots r_p$, the sequence of all states in which the input head scans either of the endmarkers. For $1 \leq j \leq p$, we have the following possibilities:

- In both r_{j-1} and r_j, the input head scans the left (right) endmarker. By Lemma 4, we can suppose that this can be done by visiting no more than n^2 input squares to the right (left). Thus, the part of the computation from r_{j-1} to r_j can be simulated even on the input 1^k.
- In r_{j-1} the input head scans the left (right) endmarker, while in r_j it scans the right (left) endmarker. By Theorem 9, for some ℓ_i, we can replace the computation path from r_{j-1} to r_j on input $1^{k+\ell}$, with a computation path from r_{j-1} to r_j on input $1^{k+\ell+\mu \ell_i}$, provided that μ is greater than $\frac{8n^2 - (k+\ell)}{\ell_i}$. Choose $\mu = \mu_i = -\frac{\ell}{\ell_i}$. Since $k > 8n^2$, μ_i satisfies the previous limitation. Thus, we obtain a computation path which, on input 1^k, traverses across the entire input from left to right (right to left).

From these observations, it is easy to conclude that $\mathcal{A}$ accepts even the input 1^k. The proof of the converse is similar. $\square$

As an immediate consequence, we are able to state our main result:

Corollary 11. *Each unary n–state 2nfa can be simulated by a $O(e^{\sqrt{n \ln n}})$–state 1dfa.*

It is possible to show that this simulation cannot be improved. Actually, a stronger result can be stated, which proves the optimality of all $O(e^{\sqrt{n \ln n}})$ bounds in Figure 1:

Theorem 12. [Chr86, Th. 6.1] *For each integer n, there exists a unary 2dfa with n states such that each equivalent 1nfa requires $\Omega(e^{\sqrt{n \ln n}})$ states.*

The trivial simulations of cost n in Figure 1 are easily seen to be tight. For the optimal quadratic simulation of unary 1nfa's by 2dfa's see [Chr86, Sec. 6].

4 Bringing reversals and nondeterminism at the endmarkers

In [GI79,GI82], E. Gurari and O. Ibarra prove that each unary 2nfa can be transformed into an equivalent 2nfa performing input head reversals only at the endmarkers. Now, using the analysis in Section 3, we are able to strengthen this result. We show that not only reversals, but also nondeterminism can be brought at the endmarkers, by just squaring the number of states.

Theorem 13. *Each unary n–state 2nfa A can be simulated by a $O(n^2)$–state 2nfa A', performing input head reversals and nondeterministic choices only when the input head scans either of the endmarkers.*

Proof. First observe that, by adding one more state, we can assume that A accepts with the input head parked on the left endmarker. Now, we informally describe how A' performs its computation on input 1^m.

In a first scan, A' deterministically checks whether $m \leq 8n^2$ and, in this case, it immediately accepts if and only if 1^m is accepted by A. We can assume that at the end of this phase, which requires $O(n^2)$ states, the input head is positioned again on the left endmarker. At this point, if $m > 8n^2$, the second part of the simulation starts from the initial state of A.

We briefly explain what happens when A' simulates the behavior of A from a state q_1, with the input head scanning the left endmarker (a similar behavior can be described starting from the right endmarker). A' nondeterministically chooses one of the following operations: (a) simulation of a "U–Turn" of A; (b) simulation of a left–to–right traversal of the input.

Since $m > 8n^2$, by Lemma 4 the simulation (a) does not depend on the input length. Hence, each U–Turn from the left endmarker can be "embedded" in the transition function of A'. Namely, it can be replaced by one stationary move. This part of the simulation does not require new states.

Now, we describe how to perform simulation (b). (By $a \bmod b$, we denote the remainder of the integer division of a by b.) Let $\{\ell_1, \ldots, \ell_s\}$ be the set introduced in Theorem 9, with respect to the automaton A. We observe that given an index

$i \in \{1, \ldots, s\}$, two integers m', m'' with $8n^2 < m' < m''$, $m' \bmod \ell_i = m'' \bmod \ell_i$, and two states q_1, q_2 of A, there exists a computation path from q_1 to q_2 scanning the whole input $1^{m'}$ if and only if there exists a computation path from q_1 to q_2 scanning the whole input $1^{m''}$. In fact, $m' = h'\ell_i + k$, $m'' = h''\ell_i + k$, for some $0 < h' \leq h''$ and $0 \leq k < \ell_i$. Thus $m' = m'' + \mu\ell_i$, where $\mu = h' - h'' = \frac{m'-k-m''+k}{\ell_i} > \frac{8n^2-m''}{\ell_i}$. Hence, by Theorem 9, from a path from q_1 to q_2 scanning the whole input $1^{m''}$, we can obtain a path from q_1 to q_2 scanning the whole input $1^{m'}$. The proof of the converse is similar.

As a consequence of this observation, it turns out that, given the state q_1 of A and the input length m, the set of states reachable by A after traversing from left to right the input depends only on the starting state q_1 and on the set of pairs $\{(\ell_i, m \bmod \ell_i) \mid i = 1, \ldots, s\}$. Hence, the simulation of the left–to–right traversal is performed as follows:

- A' nondeterministically selects an index $i \in \{1, \ldots, s\}$;
- A' scans the input tape, counting the input length m modulo ℓ_i;
- when the input head reaches the right endmarker, A' nondeterministically selects one of the states associated with the starting state q_1 and the pair $(\ell_i, m \bmod \ell_i)$.

Note that A' makes nondeterministic choices only at the endmarkers. During the traversal, A' should remember the starting state q_1, the index $i \in \{1, \ldots, s\}$ of the chosen loop, and an integer $0 \leq k < \ell_i$, used to count modulo ℓ_i. Hence, for each starting state q_1, $\ell_1 + \ldots + \ell_s \leq n$ states suffice. Thus, this part of the simulation uses $O(n^2)$ states. □

5 1afa's versus other models

In this section, we analyze the cost of simulating unary 1afa's by means of other models of automata, and the costs of converse reductions. We begin by considering the simulations between 1afa's and 1dfa's:

Theorem 14. *The class of unary languages accepted by n–state 1afa's coincides with the class of unary languages accepted by 2^n–state 1dfa's.*

Proof. See [Bir93, Thm. 2.1], and use the fact that $x = x^R$, for $x \in \{1\}^*$.

As an immediate consequence, we get the following:

Corollary 15. *Let L be a unary language, for which the minimum 1dfa has n states. Then, every 1afa accepting L requires at least $\lceil \log n \rceil$ states.*

The simulation of 1afa's via 1dfa's, stated in Theorem 14, is optimal. Actually, a stronger result holds:

Theorem 16. [Bir93, Fact 3.2] *For each integer n, there exists a unary 1afa with n states such that each equivalent 2nfa requires $\Omega(2^n)$ states.*

Hence, as already shown in [Bir93], 1afa's and 2dfa's are not polynomially equivalent, as conjectured in [Chr86]. However, we can show that the converse simulation, namely 2dfa's (and, more generally, 2nfa's) via 1afa's, is sublinear:

Theorem 17. *Each unary n–state 2nfa A can be simulated by a $O(\sqrt{n \ln n})$-state 1afa.*

Proof. By Corollary 11, we get from A an equivalent 1dfa with $O(e^{\sqrt{n \ln n}})$ states. By Theorem 14, there exists even an equivalent 1afa with $O(\sqrt{n \ln n})$ states. $\square$

The simulation in Theorem 17 is optimal, even in case of 1nfa's and 2dfa's:

Theorem 18. *For each integer n, there exists a unary 1nfa and a unary 2dfa with n states such that each equivalent 1afa requires $\Omega(\sqrt{n \ln n})$ states.*

Proof. By [Chr86, Th. 4.5], for each n there is a unary n–state 1nfa A such that each 1dfa recognizing $L(A)$ requires $\Omega(e^{\sqrt{n \ln n}})$ states. By Corollary 15, this implies that each 1afa accepting $L(A)$ has at least $\Omega(\sqrt{n \ln n})$ states. Furthermore, the language $L(A)$ is accepted even by a unary n–state 2dfa [Chr86, Th. 5.2]. $\square$

Acknowledgments. The authors wish to thank anonymous referees for helpful comments and remarks. In particular, a referee pointed out contribution [Bir93]. Moreover, authors kindly acknowledge Viliam Geffert for stimulating discussions.

References

[Ber85] C. Berge. *Graphs and Hypergraphs.* North–Holland, 1985.

[Bir93] J.-C. Birget. State-complexity of finite-state devices, state compressibility and incompressibility. *Mathematical Systems Theory,* 26:237–269, 1993.

[BL77] P. Berman and A. Lingas. On the complexity of regular languages in terms of finite automata. Technical Report 304, Polish Academy of Sciences, 1977.

[Chr86] M. Chrobak. Finite automata and unary languages. *Theoretical Computer Science,* 47:149–158, 1986.

[CKS81] A. Chandra, D. Kozen, and L. Stockmeyer. Alternation. *Journal of the ACM,* 28:114–133, 1981.

[FJY90] A. Fellah, H. Jürgensen, and S. Yu. Constructions for alternating finite automata. *International J. Computer Math.,* 35:117–132, 1990.

[Gef91] V. Geffert. Nondeterministic computations in sublogarithmic space and space constructibility. *SIAM J. Computing,* 20:484–498, 1991.

[Gef97] V. Geffert, 1997. Private communication.

[GI79] E. Gurari and O. Ibarra. Simple counter machines and number–theoretic problems. *Journal of Computer and System Sciences,* 19:145–162, 1979.

[GI82] E. Gurari and O. Ibarra. Two–way counter machines and Diophantine equations. *Journal of the ACM,* 29:863–873, 1982.

[HU79] J. Hopcroft and J. Ullman. *Introduction to automata theory, languages, and computation.* Addison–Wesley, Reading, MA, 1979.

[RS59] M.O. Rabin and D. Scott. Finite automata and their decision problems. *IBM J. Res. Develop,* 3:114–125, 1959.

[SS78] W. Sakoda and M. Sipser. Nondeterminism and the size of two–way finite automata. In *Proc. 10th ACM Symposium on Theory of Computing,* pages 275–286, 1978.

Shuffle of ω-Words: Algebraic Aspects *

(Extended Abstract)

Alexandru Mateescu

Turku Centre for Computer Science,
Lemminkäisenkatu 14 A, 20520 Turku, Finland
and Department of Mathematics, University of Bucharest, Romania,
email: mateescu@sara.utu.fi

Abstract. We introduce and investigate some sets of ω-trajectories that have the following properties: each of them defines an associative and commutative operation of shuffle of ω-words and, moreover, each of them satisfies a certain condition of fairness. The interrelations between these sets are studied as well as with other well-known classes of ω-words, like infinite Sturmian words, periodic and ultimately periodic ω-words.

1 Preliminaries

Usually, the operation of parallel composition of words or languages is modelled by the shuffle operation or restrictions of this operation, such as literal shuffle, insertion or shuffle on trajectories.

This paper continues our investigations on the operation of shuffle on trajectories of ω-words and ω-languages, see [7] and [6]. We introduce and investigate several sets of ω-trajectories such that each of these sets satisfies two important features: the shuffle operation associated to the set is associative and it fulfils a certain condition of fairness. Both conditions are important: associativity ensures that the set of ω-words has a structure of semiring, see [4], where the product is defined by the shuffle on this set of ω-trajectories, whereas the fairness condition ensures good properties for practical use of this parallel composition operation in parallel computations.

The intuition behind this approach is that an ω-word represents an infinite sequential process. The letters are the atomic actions of the process. A fairness condition states that the atomic actions of one process are performed (during the parallel composition of these processes) with not too much "delay" with respect to the atomic actions of the other process. The fairness conditions considered are natural. Moreover, the shuffle operations associated to each of them are associative and commutative operations. Also, the interrelations between these fairness conditions are studied. All these results are summarized in section 8 as the main result of this paper.

* The work reported here has been supported by the Project 137358 of the Academy of Finland

The shuffle-like operations considered below are defined using the notion of the ω-trajectory. A first approach of these shuffle-like operations was considered in [8].

An ω-trajectory defines the general strategy to switch from one ω-word to another ω-word. Roughly speaking, an ω-trajectory is a line in plane, starting in the origin and continuing parallel with the axis Ox or Oy. The line can change its direction only in points with nonnegative integer coordinates. An ω-trajectory defines how to move from an ω-word to another ω-word when carrying out the shuffle operation. Each set T of ω-trajectories defines in a natural way a shuffle operation over T. Given a set T of ω-trajectories the operation of shuffle over T is not necessarily an associative operation. However, for each set T there exists a smallest set of trajectories $\overline{T}$ such that $\overline{T}$ contains T and, moreover, shuffle over $\overline{T}$ is an associative operations. Such a set $\overline{T}$ is referred to as the associative closure of T.

The shuffle associative closure operation provides alternative definitions for the sets of periodic and ultimately periodic ω-words, see Theorem 3.2.

The set of nonnegative integers is denoted by ω. If A is a set, then the set of all subsets of A is denoted by $\mathcal{P}(A)$.

Let Σ be an alphabet, i.e., a finite nonempty set of elements called *letters*. The free monoid generated by Σ is denoted by Σ^*. Elements in Σ^* are referred to as *words*. The *empty word* is denoted by λ.

If $w \in \Sigma^*$, then $|w|$ denotes the length of w. Note that $|\lambda| = 0$. If $a \in \Sigma$ and $w \in \Sigma^*$, then $|w|_a$ denotes the number of occurrences of the symbol a in the word w. The *mirror* of a word $w = a_1 a_2 \ldots a_n$, where a_i are letters, $1 \leq i \leq n$, is $mi(w) = a_n \ldots a_2 a_1$ and $mi(\lambda) = \lambda$. A word w is a *palindrome* iff $mi(w) = w$.

Let Σ be an alphabet. An ω-*word* over Σ is a function $f : \omega \longrightarrow \Sigma$. Usually, the ω-word defined by f is denoted as the infinite sequence $f(0)f(1)f(2)\ldots$. An ω-word w is *ultimately periodic* iff $w = \alpha vvvvv \ldots$, where α is a (finite) word, possibly empty, and v is a nonempty word. In this case w is denoted as αv^ω. The set of all ultimately periodic ω-words over Σ is denoted by $UltPer(\Sigma)$ (or $UltPer$, if Σ is understood from context). An ω-word w is referred to as *periodic* iff $w = vvv \ldots$ for some nonempty word $v \in \Sigma^*$. In this case w is denoted as v^ω. The set of all periodic ω-words over Σ is denoted by $Per(\Sigma)$ (or Per, if Σ is understood from context). Moreover, the set of all periodic ω-words over Σ that have a palindrome as their period is denoted by $PalPer(\Sigma)$ (or $PalPer$, if Σ is understood from context).

The set of all ω-words over Σ is denoted by Σ^ω. An ω-*language* is a subset L of Σ^ω, i.e., $L \subseteq \Sigma^\omega$. Let w be an ω-word. The set of all (finite) prefixes of w is denoted by $Pref(w)$. The reader is referred to [9],[10] and [11] for general results on ω-words.

The *shuffle* operation, denoted by $\sqcup\!\sqcup$, is defined recursively by:

$$(ax \sqcup\!\sqcup by) = a(x \sqcup\!\sqcup by) \cup b(ax \sqcup\!\sqcup y) \qquad \text{and} \qquad x \sqcup\!\sqcup \lambda = \lambda \sqcup\!\sqcup x = \{x\},$$

where $x, y \in \Sigma^*$ and $a, b \in \Sigma$.

The above operation is extended in a natural way to languages.

2 Shuffle on ω-trajectories

In this section we introduce the notions of the ω-trajectory and shuffle on ω-trajectories. The shuffle of two ω-words has a natural geometrical interpretation related to lattice points in the plane (points with nonnegative integer coordinates) and with a certain "walk" in the plane defined by each ω-trajectory.

Firstly, we define the shuffle of (finite) words on (finite) trajectories.

Let $V = \{r, u\}$ be the set of *versors* in the plane: r stands for the *right* direction, whereas, u stands for the *up* direction.

Definition 1. A *trajectory* is an element t, $t \in V^*$.

Let Σ be an alphabet and let t be a (finite) trajectory, let d be a versor, $d \in V$, let α, β be two (finite) words over Σ.

Definition 2. The shuffle of α with β on the trajectory dt, denoted $\alpha \sqcup\!\sqcup_{dt} \beta$, is recursively defined as follows:

if $\alpha = ax$ and $\beta = by$, where $a, b \in \Sigma$ and $x, y \in \Sigma^*$, then:

$$ax \sqcup\!\sqcup_{dt} by = \begin{cases} a(x \sqcup\!\sqcup_t by), & \text{if } d = r, \\ b(ax \sqcup\!\sqcup_t y), & \text{if } d = u. \end{cases}$$

if $\alpha = ax$ and $\beta = \lambda$, where $a \in \Sigma$ and $x \in \Sigma^*$, then:

$$ax \sqcup\!\sqcup_{dt} \lambda = \begin{cases} a(x \sqcup\!\sqcup_t \lambda), & \text{if } d = r, \\ \emptyset, & \text{if } d = u. \end{cases}$$

if $\alpha = \lambda$ and $\beta = by$, where $b \in \Sigma$ and $y \in \Sigma^*$, then:

$$\lambda \sqcup\!\sqcup_{dt} by = \begin{cases} \emptyset, & \text{if } d = r, \\ b(\lambda \sqcup\!\sqcup_t y), & \text{if } d = u. \end{cases}$$

Finally,

$$\lambda \sqcup\!\sqcup_t \lambda = \begin{cases} \lambda, & \text{if } t = \lambda, \\ \emptyset, & \text{otherwise.} \end{cases}$$

Comment. Note that if $|\alpha| \neq |t|_r$ or $|\beta| \neq |t|_u$, then $\alpha \sqcup\!\sqcup_t \beta = \emptyset$.

Now we define the operation of shuffle of ω-words on ω-trajectories.

Definition 3. An ω-trajectory is an ω-word t over V, i.e., $t \in V^\omega$.

Let Σ be an alphabet and let α, β be ω-words over Σ. Let t be an ω-trajectory, $t \in V^\omega$.

Definition 4. The shuffle of α with β on the ω-trajectory t is defined as the limit of the sequence $(\alpha' \sqcup\!\sqcup_{t'} \beta')_{t' \in Pref(t)}$, where $\alpha' \in Pref(\alpha)$, $\beta' \in Pref(\beta)$ such that $|\alpha'| = |t'|_r$ and $|\beta'| = |t'|_u$.

One can easily verify that the sequence $(\alpha' \sqcup_{t'} \beta')_{t' \in Pref(t)}$ has always a limit.

Observe that there is an important distinction between the finite case, i.e., the shuffle on trajectories, and the infinite case, i.e., the shuffle on ω-trajectories: sometimes the result of shuffling of two words α and β on a trajectory t can be empty whereas the shuffle of two ω-words over an ω-trajectory is always nonempty and consists of only one ω-word.

Let Σ be an alphabet and let L_1, L_2 be ω-languages over Σ, i.e., $L_1, L_2 \subseteq \Sigma^\omega$. If T is a set of ω-trajectories, the *shuffle of L_1 with L_2 on the set T of ω-trajectories*, denoted $L_1 \sqcup_T L_2$, is:

$$L_1 \sqcup_T L_2 = \bigcup_{\alpha \in L_1, \beta \in L_2, t \in T} \alpha \sqcup_t \beta.$$

Example 1. Let α and β be the ω-words $\alpha = a_0 a_1 a_2 a_3 \ldots$, $\beta = b_0 b_1 b_2 b_3 \ldots$ and assume that $t = r^2 u^2 r^3 u r^2 u r u \ldots$. The shuffle of α with β on the trajectory t is:

$$\alpha \sqcup_t \beta = \{a_0 a_1 b_0 b_1 a_2 a_3 a_4 b_2 a_5 a_6 b_3 a_7 b_4 \ldots\}.$$

The result has the following geometrical interpretation (see Figure 1): the trajectory t defines a line starting in the origin and continuing one unit right or up, depending of the definition of t. In our case, first there are two units right, then two units up, then three units right, etc. Assign α on the Ox axis and β on the Oy axis of the plane. The result can be read following the line defined by the trajectory t, that is, if being in a lattice point of the trajectory, (the corner of a unit square) and if the trajectory is going right, then one should pick up the corresponding letter from α, otherwise, if the trajectory is going up, then one should add to the result the corresponding letter from β. Hence, the trajectory t defines a line in the plane, on which one has "to walk" starting from the origin O. In each lattice point one has to follow one of the versors r or u, according to the definition of t.

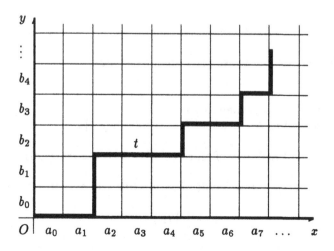

Figure 1

Assume now that $t' = ur^5u^4r^3 \ldots$ is another trajectory. Observe that:

$$\alpha \shuffle_{t'} \beta = \{b_0a_0a_1a_2a_3a_4b_1b_2b_3b_4a_5a_6a_7 \ldots\}.$$

Consider the set of trajectories, $T = \{t, t'\}$. The shuffle of α with β on the set T of trajectories is:

$$\alpha \shuffle_T \beta = \{a_0a_1b_0b_1a_2a_3a_4b_2a_5a_6b_3a_7b_4 \ldots, b_0a_0a_1a_2a_3a_4b_1b_2b_3b_4a_5a_6a_7 \ldots\}.$$

3 Associativity and commutativity

The results in this section deal with associativity and commutativity. After a few general remarks, we restrict the attention to the set V_+^ω of ω-trajectories t such that both r and u occur infinitely often in t. (It will become apparent below why this restriction is important.) It turns out that associativity can be viewed as stability under four particular operations, referred to as $\Diamond$-operations.

Definition 5. A set T of ω-trajectories is *associative* iff the operation $\shuffle_T$ is associative, i.e.,

$$(\alpha \shuffle_T \beta) \shuffle_T \gamma = \alpha \shuffle_T (\beta \shuffle_T \gamma),$$

for all $\alpha, \beta, \gamma \in \Sigma^\omega$.

A set T of ω-trajectories is *commutative* iff the operation $\shuffle_T$ is commutative, i.e.,

$$\alpha \shuffle_T \beta = \beta \shuffle_T \alpha,$$

for all $\alpha, \beta \in \Sigma^\omega$.

The following sets of ω-trajectories are associative:

1. $T = \{r, u\}^\omega$.
2. $T = \{t \in V^\omega \mid |t|_r < \infty\}$.
3. $T = \{\alpha_0\beta_0\alpha_1\beta_1 \ldots \mid \alpha_i, \beta_i$ are of even length, and $\alpha_i \in r^*, \beta_i \in u^*, i \geq 0\}$.

Nonassociative sets of ω-trajectories are for instance:

1. $T = \{(ru)^\omega\}$.
2. $T = \{t \in V^\omega \mid t$ is a Sturmian ω-word $\}$.
3. $T = \{w_0w_1w_2 \ldots \mid w_i \in L\}$, where $L = \{r^nu^n \mid n \geq 0\}$.

Notation. Let $\mathcal{A}$ be the family of all associative sets of ω-trajectories.

Proposition 6. *If $(T_i)_{i \in I}$ is a family of associative sets of ω-trajectories, then*

$$T' = \bigcap_{i \in I} T_i,$$

is an associative set of ω-trajectories.

Definition 7. Let T be an arbitrary set of ω-trajectories. The *associative closure* of T, denoted $\overline{T}$, is

$$\overline{T} = \bigcap_{T \subseteq T', T' \in \mathcal{A}} T'.$$

Observe that for all $T, T \subseteq \{r, u\}^\omega$, $\overline{T}$ is an associative set of ω-trajectories and $\overline{T}$ is the smallest associative set of ω-trajectories that contains T.

Notation. Let V_+^ω be the set of all ω-trajectories $t \in V^\omega$ such that t contains infinitely many occurrences both of r and of u.

Now we give another characterization of an associative set of ω-trajectories from V_+^ω. This is useful in finding an alternative definition of the associative closure of a set of ω-trajectories and also to prove some other properties related to associativity.

However, this characterization is valid only for sets of ω-trajectories from V_+^ω and not for the general case, i.e., not for sets of ω-trajectories from V^ω.

Definition 8. Let W be the alphabet $W = \{x, y, z\}$ and consider the following four morphisms, ρ_i, $1 \le i \le 4$, where $\rho_i : W \longrightarrow V_+^\omega$, $1 \le i \le 4$, and

$$\rho_1(x) = \lambda, \quad \rho_1(y) = r, \quad \rho_1(z) = u, \quad \rho_2(x) = r, \quad \rho_2(y) = u, \quad \rho_2(z) = u,$$

$$\rho_3(x) = r, \quad \rho_3(y) = u, \quad \rho_3(z) = \lambda, \quad \rho_4(x) = r, \quad \rho_4(y) = r, \quad \rho_4(z) = u.$$

Next, we consider four operations on the set of ω-trajectories, V_+^ω.

Definition 9. Let $\Diamond_i$, $1 \le i \le 4$, be the following operations on V_+^ω.

$$\Diamond_i : V_+^\omega \times V_+^\omega \longrightarrow V_+^\omega, \qquad 1 \le i \le 4,$$

Let t, t' be in V_+^ω. By definition:

$$\Diamond_1(t, t') = \rho_1((x^\omega \sqcup_t y^\omega) \sqcup_{t'} z^\omega), \qquad \Diamond_2(t, t') = \rho_2((x^\omega \sqcup_t y^\omega) \sqcup_{t'} z^\omega),$$

$$\Diamond_3(t', t) = \rho_3(x^\omega \sqcup_{t'} (y^\omega \sqcup_t z^\omega)), \qquad \Diamond_4(t', t) = \rho_4(x^\omega \sqcup_{t'} (y^\omega \sqcup_t z^\omega)).$$

The following theorem was firstly proved in [6].

Theorem 10. *Let T be a set of ω-trajectories, $T \subseteq V_+^\omega$. The following assertions are equivalent:*

(i) T is an associative set of ω-trajectories.
(ii) T is closed at $\Diamond$-operations, i.e., if $t_1, t_2 \in T$, then $\Diamond_i(t_1, t_2) \in T$, $1 \le i \le 4$.

Remark. We restricted our attention only to the set V_+^ω and not to the general case V^ω. However, if T contains a trajectory t that is not in V_+^ω, then $\Diamond_1(t, t)$ is not necessarily in V^ω. For instance $t = ru r^\omega$, then $\Diamond_1(t, t) = ur \notin V^\omega$. Thus the operation $\Diamond_1$ is not well defined. A similar phenomenon happens with the operation $\Diamond_3$.

Proposition 11. *Let $T \subseteq V_+^\omega$ be a set of ω-trajectories.*

(i) If each $t \in T$ is a periodic ω-word, then the associative closure of T, $\overline{T}$, has the same property, i.e., each ω-trajectory in $\overline{T}$ is periodic.

(ii) if additionally, each $t \in T$ has a palindrome as its period, then the associative closure of T, $\overline{T}$, has the same property.

(iii) If T is a set of ultimately periodic ω-trajectories, then the associative closure of T, $\overline{T}$, has the same property, i.e., each ω-trajectory in $\overline{T}$ is ultimately periodic.

The above proposition yields:

Corollary 12. *The following sets of ω-trajectories are associative:*

(i) the set of all periodic ω-trajectories from V_+^ω.

(ii) the set of all periodic ω-trajectories from V_+^ω that have as their period a palindrome.

(iii) the set of all ultimately periodic ω-trajectories from V_+^ω.

The following theorem gives characterizations of the sets of periodic and ultimately periodic ω-words in terms of the associative closure operation, see (i) and (ii). Also, this theorem shows that the associative closure of the Sturmian words strictly includes the set of all periodic ω-words, see (iii). We start by a definition:

Definition 13. The Fibonacci word f is defined as the limit of the sequence of words $(f_n)_{n \geq 0}$, where: $f_0 = r$, $f_1 = u$, $f_{n+1} = f_n f_{n-1}$, $n > 0$.

Theorem 14. *Let $V = \{r, u\}$ be the set of versors.*

(i) the associative closure of the ω-word $(ru)^\omega$ equals the set of all periodic ω-trajectories from V_+^ω, i.e., $\overline{\{(ru)^\omega\}} = Per(V)$.

(ii) the associative closure of the ω-word $r(ru)^\omega$ equals the set of all ultimately periodic ω-trajectories from V_+^ω, i.e., $\overline{\{r(ru)^\omega\}} = UltPer(V)$.

(iii) the associative closure of the Fibonacci word f strictly contains the set all periodic ω-trajectories from V_+^ω, i.e., $\overline{\{f\}} \supset Per(V)$.

4   Bounded increase of ω-trajectories

Definition 15. An ω-trajectory t has a bounded increase iff there exist two constants c_1 and c_2 such that:

B1 $|pref_p(t)|_u - |pref_q(t)|_u \leq c_1(|pref_p(t)|_r - |pref_q(t)|_r)$,
whenever $|pref_p(t)|_r - |pref_q(t)|_r > 0$, and

B2 $|pref_p(t)|_r - |pref_q(t)|_r \leq c_2(|pref_p(t)|_u - |pref_q(t)|_u)$,
whenever $|pref_p(t)|_u - |pref_q(t)|_u > 0$.

Notation. $BInc$ denotes the set of all ω-trajectories with bounded increase.

Proposition 16. *The trajectory $t = r^{i_1} u^{j_1} r^{i_2} u^{j_2} ... r^{i_n} u^{j_n} ... \in V_+^\omega$ has a bounded increase iff the sequences $\{i_n\}_{n \geq 1}$ and $\{j_n\}_{n \geq 1}$ are bounded, i.e., there are k_1 and k_2 such that $i_n \leq k_1$ and $j_n \leq k_2$, for all $n \geq 1$.*

Comment. Note that an ω-trajectory t is in $BInc$ iff there are two constants $k_1 > 0$ and $k_2 > 0$ such that, during the parallel composition of two processes on t, after at most k_1 occurrences of actions from the first process, it occurs at least one occurrence of an action from the second process and, after at most k_2 occurrences of actions from the second process, it occurs at least one occurrence of an action from the first process.

Theorem 17. *The set of all ω-trajectories with bounded increase, i.e., $BInc$, is an associative and commutative set of ω-trajectories.*

5 Asymptotic linear ω-trajectories

Definition 18. An ω-trajectory t is linear asymptotic iff the following limit does exist and belongs to $(0, +\infty)$

$$l = \lim \frac{|pref_n(t)|_u}{|pref_n(t)|_r}.$$

Notation. The set of all asymptotic linear ω-trajectories is denoted by $ALin$.
Comment. Note that an ω-trajectory t is in $ALin$ iff performing the parallel composition of two processes on t, in the prefix of length n of the resulting sequence, the value of the number of occurrences of actions from the first process divided by the number of occurrences of actions from the second process, is a sequence of real numbers, say $(x_n)_{n>0}$, such that $(x_n)_{n>0}$ has a finite limit l, $l > 0$, when $n \longrightarrow +\infty$. Intuitively, this means that the ω-trajectory is at infinity stable, in such a way that it keeps some "balance" in performing actions from the first/second process. This "balance" is defined as being the limit l.

Theorem 19. *The set of all asymptotic linear ω-trajectories, i.e., $ALin$, is an associative and commutative set of ω-trajectories.*

Next proposition shows that the sets $BInc$ and $ALin$ are incomparable with respect to the inclusion.

Proposition 20. $BInc - ALin \neq \emptyset$ and $ALin - BInc \neq \emptyset$.

6 Quasi linear ω-trajectories

Definition 21. We say that an ω-trajectory t is quasi linear iff there exist two constants, $c_1, c_2 > 0$ such that

$$||pref_n(t)|_u - c_1|pref_n(t)|_r| \leq c_2.$$

Notation. The set of all quasi linear ω-trajectories is denoted by $QLin$.
Comment. An ω-trajectory t is in $QLin$ iff the graph of t is bounded by two parallel lines.

Proposition 22. *The set of all quasi linear ω-trajectories, i.e., QLin, is included in both classes BInc and ALin.*

Remark. . From the above proposition, it follows that the class $QLin$ is included in the intersection of classes $BInc$ and $ALin$, i.e., $QLin \subseteq BInc \cap ALin$.

Proposition 23. $(BInc \cap ALin) - QLin \neq \emptyset$.

Theorem 24. *The set of all quasi linear ω-trajectories, QLin, is an associative and commutative set of ω-trajectories.*

Proposition 25. *UltPer is included in QLin and the inclusion is strict.*

Next theorem gives a characterization of those ω-trajectories that are in $QLin$. The theorem is based on an extension of the notion of a Sturmian word, see [1], [2] or [3].

Definition 26. The ω-trajectory t is quasi-Sturmian iff there is $c > 0$ such that:

$$||v|_u - |w|_u| \leq c, \forall v, w \in Sub(t), |v| = |w|$$

The set of all quasi-Sturmian ω-trajectories is denoted by QSturm.

Let $t \in V_+^\omega$. Consider the following function:

$$\pi_t : \omega \longrightarrow \omega, \ \pi_t(n) = |pref_n(t)|_u$$

Remark. The ω-trajectory t is quasi-Sturmian iff:

$$|\pi_t(n+p) - \pi_t(n) - (\pi_t(m+p) - \pi_t(m))| \leq c, \forall n, m, p \in \omega$$

Proposition 27. *If $t \in V_+^\omega$, $t = r^{i_1}u^{j_1}r^{i_2}u^{j_2}...r^{i_n}u^{j_n}...$ is a quasi-Sturmian ω-trajectory, then the sequence $\{j_n\}_n$ is bounded.*

Next theorem gives an alternative definition of quasi-Sturmian words.

Theorem 28. *QSturm = QLin.*

Corollary 29. *The set of all Sturmian ω-words is included in QLin and the inclusion is strict.*

Remark. Note that, as a consequence that there are noncountable many Sturmian ω-words, it follows that $QLin$ is a noncountable set, too.

7 Prefix bounded ω-trajectories

Definition 30. An ω-trajectory t is referred as prefix bounded iff there exist $a_1, a_2 > 0$ such that:
    P1 $|pref_n(t)|_u \leq a_1 |pref_n(t)|_r$, whenever $|pref_n(t)|_r > 0$ and
    P2 $|pref_n(t)|_r \leq a_2 |pref_n(t)|_u$, whenever $|pref_n(t)|_u > 0$.

Notation. The set of all prefix bounded ω-trajectories is denoted by $PrefB$.
Comment. An ω-trajectory is in $PrefB$ iff performing the parallel composition of two processes on t there are two constants $a_1 > 0$ and $a_2 > 0$ such that, in the prefix of length n of the resulting sequence, the value of the number of occurrences of actions from the first process is at most a_2-times the number of occurrences of actions from the second process and symmetrically.

Theorem 31. *The set of prefix bounded ω-trajectories, $PrefB$, is an associative and commutative set of ω-trajectories. Moreover, $PrefB$ strictly includes the sets $BInc$ and $ALin$.*

8 Main Result

Combining our results from the previous sections, we obtain the following:

Theorem 32. *All sets of ω-trajectories depicted in Figure 2 are associative, commutative and fair (with respect to a certain type of fairness). The arrows from Figure 2 denote strict inclusions.*

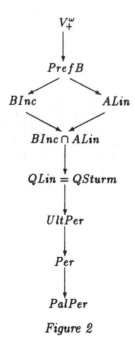

Figure 2

Notation. Let *Sturm* be the set of all infinite Sturmian words over the alphabet $V = \{r, u\}$. Denote by *ASturm* the associative closure of the set *Sturm*, i.e., $ASturm = \overline{Sturm}$.

Remark. From Corollary 29 it follows that $Sturm \subseteq QLin$. Note that $QLin$ is an associative set of ω-trajectories, see Theorem 24. Therefore, $ASturm \subseteq QLin$.

However, it is an *open problem* whether this inclusion is strict or not.

9 Conclusion

Ten sets of ω-trajectories, all of them associative, commutative and fulfilling a certain fairness condition were introduced and their interrelations were established. All these sets lead to a natural structure of a (commutative) semiring on the set $\mathcal{P}(\Sigma^\omega)$, the set of all ω-languages over an alphabet Σ. Moreover, the multiplicative law of these semirings obey a certain fairness condition.

References

[1]   J. Berstel, *Recent Results on Sturmian Words*, in Developments in Language Theory, eds. J. Dassow, G. Rozenberg and A. Salomaa, World Scientific, 1996, 13-24.

[2]   O. Carton, *Mots Infinis, ω-Semigroupes et Topologie*, Universite Paris 7, 1993.

[3]   C. Choffrout and J. Karhumäki, *Combinatorics of Words* in Handbook of Formal Languages, eds. G. Rozenberg and A. Salomaa, Springer, 1997, 329-438.

[4]   J. S. Golan, *The Theory of Semirings with Applications in Mathematics and Theoretical Computer Science*, Longman Scientific and Technical, Harlow, Essex, 1992.

[5]   M. Lothaire, *Combinatorics on Words*, Addison Wesley, vol. 17 Encyclopedia of Mathematics and its Applications, 1983.

[6]   A. Mateescu and G.D. Mateescu, "Associative shuffle of infinite words", TUCS Technical Report, 104, 1997.

[7]   A. Mateescu, G. Rozenberg and A. Salomaa, "Shuffle on Trajectories: Syntactic Constraints", Technical Report 96-18, University of Leiden, 1996.

[8]   D. Park, "Concurrency and automata on infinite sequences", in *Theoretical Computer Science*, ed. P. Deussen, LNCS 104, Springer-Verlag, 1981, 167-183.

[9]   D. Perrin and J. E. Pin, *Mots Infinis*, Report LITP 93.40, 1993.

[10] L. Staiger, "ω-Languages", in *Handbook of Formal Languages*, ed. G. Rozenberg and A. Salomaa, Springer, 1997, Vol. 3, 339-388.

[11] W. Thomas, "Automata on Infinite Objects", in *Handbook of Theoretical Computer Science*, Volume B, ed. J. van Leeuwen, Elsevier, 1990, 135-191.

A Generalization of Resource-Bounded Measure, With an Application (Extended Abstract)

Harry Buhrman[1], Dieter van Melkebeek[2], Kenneth W. Regan[3], D. Sivakumar[4], and Martin Strauss[5]

[1] CWI, Kruislaan 413, 1098SJ Amsterdam, The Netherlands. E-mail: buhrman@cwi.nl. Partly supported by the Dutch foundation for scientific research (NWO), by SION project 612-34-002, and by the European Union through NeuroCOLT ESPRIT Working Group Nr. 8556 and HC&M grant nr. ERB4050PL93-0516.

[2] Univ. of Chicago, Department of Computer Science, 1100 E. 58th St., Chicago, IL 60637 USA. E-mail: dieter@cs.uchicago.edu. Party supported by the European Union through Marie Curie Research Training Grant ERB-4001-GT-96-0783 at CWI and by NSF Grant CCR 92-53582.

[3] Computer Science, University at Buffalo, 226 Bell Hall, Buffalo, NY 14260-2000 USA. E-mail: regan@cs.buffalo.edu. Partly supported by NSF Grant CCR-9409104.

[4] Department of Computer Science, University of Houston, Houston, TX 77204-3475, USA. E-mail: siva@cs.uh.edu. Partly supported at Buffalo by NSF Grant CCR-9409104.

[5] AT&T Labs, Room C216, 180 Park Ave, Florham Park, NJ 07932-0971 USA. E-mail: mstrauss@research.att.com. Research performed at Rutgers University and Iowa State University, supported by NSF grants CCR-9204874 and CCR-9157382.

Abstract. We introduce *resource-bounded betting games*, and propose a generalization of Lutz's resource-bounded measure in which the choice of next string to bet on is fully adaptive. Lutz's martingales are equivalent to betting games constrained to bet on strings in lexicographic order. We show that if strong pseudo-random number generators exist, then betting games are equivalent to martingales, for measure on E and EXP. However, we construct betting games that succeed on certain classes whose Lutz measures are important open problems: the class of polynomial-time Turing-complete languages in EXP, and its superclass of polynomial-time Turing-autoreducible languages. If an EXP-martingale succeeds on either of these classes, or if betting games have the "finite union property" possessed by Lutz's measure, one obtains the non-relativizable consequence BPP $\neq$ EXP. We also show that if EXP $\neq$ MA, then the polynomial-time truth-table-autoreducible languages have Lutz measure zero, whereas if EXP = BPP, they have measure one.

1 Introduction

Lutz's theory of measure on complexity classes is now usually defined in terms of resource-bounded martingales. A martingale can be regarded as a gambling game played on unseen languages A. Let $s_1, s_2, s_3, \ldots$ be the standard lexicographic ordering of strings. The gambler G starts with capital $C_0 = \$1$ and places a bet $B_1 \in [0, C_0]$ on either "$s_1 \in A$" or "$s_1 \notin A$." Given a fixed particular language A, whether $s_1 \in A$ is used to resolve the bet. If the bet won, then the new capital C_1 equals $C_0 + B_1$; if the bet lost, then $C_1 = C_0 - B_1$. The gambler then places a bet $B_2 \in [0, C_1]$ on (or against) membership of the string s_2, then on s_3, and so forth. The gambler *succeeds* if G's capital C_i grows toward $+\infty$. The class $\mathcal{C}$ of languages A

on which G succeeds (and any subclass) is said to have *measure zero*. One also says G *covers* C. Lutz and others (see [13]) have developed a rich and extensive theory around this measure-zero notion, and have shown interesting connections to many other important problems in complexity theory.

We propose the generalization obtained by lifting the requirement that G must bet on strings in lexicographic order. That is, G may begin by choosing any string x_1 on which to place its first bet, and after the oracle tells the result, may choose any other string x_2 for its second bet, and so forth. Note that the sequences $x_1, x_2, x_3, \ldots$ (as well as $B_1, B_2, B_3, \ldots$) may be radically different for different oracle languages A—in complexity-theory parlance, G's queries are *adaptive*. The lone restriction is that G may not query (or bet on) the same string twice. We call G a *betting game*.

Our betting games remedy a possible lack in the martingale theory, one best explained in the context of languages that are "random" for classes $\mathcal{D}$ such as E or EXP. A language L is $\mathcal{D}$-*random* if L cannot be covered by a $\mathcal{D}$-martingale. Based on one's intuition about random 0-1 sequences, the language $L' = \{ flip(x) : x \in L \}$ should likewise be $\mathcal{D}$-random, where $flip(x)$ changes every 0 in x to a 1 and vice-versa. However, this closure property is not known for E-random or EXP-random languages, because of the way martingales are tied to the fixed lex ordering of Σ^*. Betting games can adapt to easy permutations of Σ^* such as that induced by $flip$. Similarly, a class C that is *small* in the sense of being covered by a $(\mathcal{D}$-$)$ betting game remains small if the languages $L \in C$ are so permuted. In the r.e./recursive theory of random languages, our generalization is similar to "Kolmogorov-Loveland place-selection rules" (see [11]). We make this theory work for complexity classes via a novel definition of "running in time $t(n)$" for an infinite process.

We also provide a useful new angle on Lutz's theory, in which a major open question is whether the class of EXP-complete sets—under polynomial-time Turing reductions—has EXP-measure zero. If so (in fact if this set does not have measure one), then by results of Allender and Strauss [1], BPP $\neq$ EXP. Since there are oracles A relative to which BPP^A = EXP^A [10], this kind of absolute separation would be a major breakthrough. We show that the EXP-complete sets *can* be covered by an EXP betting game—in fact, by an E-betting game. The one technical lack in our theory as a notion of measure is also interesting here: If the "finite unions" property holds for betting games (viz. C_1 small $\wedge$ C_2 small $\implies$ $C_1 \cup C_2$ small), then EXP $\neq$ BPP. Likewise, if Lutz's martingales do enjoy the permutation-invariance of betting games, then BPP $\neq$ EXP. Finally, we show that if a pseudorandom number generator (PRG) of security $2^{n^{\Omega(1)}}$ exists, then for every EXP-betting game G one can find an EXP-martingale that succeeds on all sets covered by G. PRGs of higher security $2^{\Omega(n)}$ likewise imply the equivalence of E-betting games and E-measure. Ambos-Spies and Lempp [4] proved that the EXP-complete sets have E-measure zero under a different hypothesis, namely P = PSPACE.

Measure theory and betting games help us to dig even deeper into questions about PRGs and complexity-class separations. Our pivot is the notion of an *autoreducible* set, whose importance in complexity theory was argued by Buhrman, Fortnow, and Torenvliet [7]. A language L is $\leq^p_T$-*autoreducible* if there is a polynomial-time oracle TM Q such that for all inputs x, Q^L correctly decides whether $x \in L$ without ever submitting x itself as a query to L. If Q is non-adaptive (i.e., computes a polynomial-time truth-table reduction), we say L is $\leq^p_{tt}$-*autoreducible*. We show that the class of

$\leq^p_T$-autoreducible sets is covered by an E-betting game. Since every EXP-complete set is $\leq^p_T$-autoreducible [7], this implies results given above. For the subclass of $\leq^p_{tt}$-autoreducible sets, we get a fairly tight picture:

- If MA $\neq$ EXP, then the $\leq^p_{tt}$-autoreducible sets have E-measure zero.
- If EXP = BPP, then the $\leq^p_{tt}$-autoreducible sets have E-measure one.

Here MA is the "Merlin-Arthur" class of Babai [5,9], which contains BPP and NP.

In sum, the whole theory of resource-bounded measure has progressed far enough to wind the issues of (pseudo-)randomness and stochasticity within exponential time very tightly. We turn the wheels a few more notches, and seek greater understanding of complexity classes in the places where the boundary between "measure one" and "measure zero" seems tightest. A full version of this paper with the proofs omitted here is being submitted concurrently to the ECCC Technical Reports Series.

2 Martingales

A *martingale* is abstractly defined as a function d from $\{0,1\}^*$ into the nonnegative reals that satisfies $d(w) = (d(w0) + d(w1))/2$ for all $w \in \{0,1\}^*$. The interpretation in Lutz's theory is that a string $w \in \{0,1\}^*$ stands for an initial segment of a language over an arbitrary alphabet Σ as follows: Let $s_1, s_2, s_3, \ldots$ be the standard lexicographic ordering of Σ^*. Then for any language $A \subseteq \Sigma^*$, write $w \sqsubseteq A$ if for all i, $1 \leq i \leq |w|$, $s_i \in A$ iff the ith bit of w is a 1. We also regard w as a function with domain $\{s_1, \ldots, s_{|w|}\}$ and range $\{0,1\}$, writing $w(s_i)$ for the ith bit of w. A martingale d *succeeds on* a language A if the sequence of values $d(w)$ for $w \sqsubseteq A$ is unbounded. Let $S^\infty[d]$ stand for the (possibly empty, often uncountable) class of languages on which d succeeds.

Definition 1 (cf. [12,14]). Let Δ be a complexity class of functions. A class $\mathcal{C}$ of languages *has Δ-measure zero*, written $\mu_\Delta(\mathcal{C}) = 0$, if there is a martingale d computable in Δ such that $\mathcal{C} \subseteq S^\infty[d]$. One also says that d *covers* $\mathcal{C}$.

For example, P has E-measure zero. Indeed, for any fixed $c > 0$, DTIME$[2^{cn}]$ has E-measure zero, and DTIME$[2^{n^c}]$ has EXP-measure zero [12].

Lutz defined complexity bounds in terms of the length of the argument w to d, which we denote by N. However, we prefer to work in terms of the largest length n of a string in the domain of w. For $N > 0$, n equals $\lfloor \log N \rfloor$; all we care about is that $n = \Theta(\log N)$ and $N = 2^{\Theta(n)}$. Because complexity bounds on languages we want to analyze will naturally be stated in terms of n, we prefer to use n for martingale complexity bounds. The following correspondence is helpful:

$$\text{Lutz's "} p \text{"} \quad \sim \quad N^{O(1)} = 2^{O(n)} \quad \sim \quad \text{measure on E}$$
$$\text{Lutz's "} p_2 \text{"} \quad \sim \quad 2^{(\log N)^{O(1)}} = 2^{n^{O(1)}} \quad \sim \quad \text{measure on EXP}$$

Our convention lets us simply write "μ_E" for E-measure (regarding Δ as E for functions), similarly "μ_{EXP}" for EXP-measure, and generally μ_Δ for any Δ that names both a language and function class. Abusing notation similarly, we define:

Definition 2 (after [12]). A class C has Δ-measure one, written $\mu_\Delta(C) = 1$, if $\mu_\Delta(\Delta \setminus C) = 0$.

The following lemma has appeared in various forms [14, 8]. It essentially says that we can assume a martingale grows almost monotonically (sure winnings) and not too fast (slow winnings).

Lemma 3. Let d be a time-$t(n)$ computable martingale, with $d(\lambda) = 1$. Then we can compute in time $O(2^n t(n))$ a martingale d' with $S^\infty[d] \subseteq S^\infty[d']$ such that

$$(\forall w)(\forall u) : d'(wu) > d'(w) - 2, \quad \text{and} \tag{1}$$

$$(\forall w) : d'(w) < 2(|w| + 1) . \tag{2}$$

3   Betting Games

To capture intuitions that have been expressed not only for Lutz measure but also in many earlier papers on random sequences, we formalize a betting game as an *infinite* process, rather than as a Turing machine that has *finite* computations on string inputs.

Definition 4. A *betting game* G is an oracle Turing machine that maintains a "capital tape" and a "bet tape," in addition to its standard query tape and worktapes, and works in *stages* $i = 1, 2, 3 \ldots$ as follows: Beginning each stage i, the capital tape holds a nonnegative rational number C_{i-1} — initially $C_0 = 1$. G computes a query string x_i to bet on, a *bet amount* B_i, $0 \le B_i \le C_{i-1}$, and a *bet sign* $b_i \in \{-1, +1\}$. The computation is *legal* so long as x_i does not belong to the set $\{x_1, \ldots, x_{i-1}\}$ of strings queried in earlier stages. G ends stage i by entering a special query state. For a given oracle language A, if $x_i \in A$ and $b_i = +1$, or if $x_i \notin A$ and $b_i = -1$, then the new capital is given by $C_i := C_{i-1} + B_i$, else by $C_i := C_{i-1} - B_i$. The query and bet tapes are blanked, and G proceeds to stage $i + 1$.

Since we require that G spend the time to write each bet out in full, it does not matter whether we suppose that the new capital is computed by G itself or updated instantly by the oracle. Note that every oracle set A determines a unique infinite computation of G, which we denote by G^A. This includes a unique infinite sequence $x_1, x_2, \ldots$ of query strings, and a unique sequence $C_0, C_1, C_2, \ldots$ telling how the gambler fares against A.

Definition 5. A betting machine G *runs in time* $t(n)$ if for all oracles A, every query of length n made by G^A is made in the first $t(n)$ steps of the computation.

A similar definition can be made for space usage, taking into account standard issues such as whether the query tape counts against the space bound, or whether the query itself is preserved in read-only mode for further computation by the machine.

Definition 6. A betting game G *succeeds* on a language A, written $A \in S^\infty[G]$, if the sequence of values C_i in the computation G^A is unbounded. If $A \in S^\infty[G]$, then we also say G *covers* A.

Our main motivating example where one may wish not to bet in lexicographic order, or according to any fixed ordering of strings, is deferred to Sect. 6.

We now want to argue that the more liberal requirement of being covered by a time $t(n)$ betting game, still defines a smallness concept for subclasses of DTIME[$t(n)$] in the intuitive sense Lutz established for his measure-zero notion. The following result is a good beginning.

Theorem 7. *For every time-$t(n)$ betting game G, we can construct a language in DTIME[$t(n)$] that is not covered by G.*

In particular, the class E cannot be covered by an E-betting game, nor EXP by an EXP-betting game. Put another way, the "measure conservation axiom" [12] of Lutz's measure carries over to betting games.

To really satisfy the intuition of "small," however, it should hold that the union of two small classes is small (moreover, "easy" countable unions of small classes should be small, as in [12]). Our lack of meeting this "finite union axiom" will later be excused insofar as it has the non-relativizing consequence BPP $\neq$ EXP. Theorem 7 is still good enough for the "measure-like" results in this paper.

To begin comparing betting games and martingales, we note first that the latter can be considered a direct special case of betting games. Say a betting game G is *lex-limited* if for all oracles A, the sequence $x_1, x_2, x_3 \ldots$ of queries made by G^A is in lex order. (It need not equal the lex enumeration $s_1, s_2, s_3, \ldots$ of Σ^*.)

Theorem 8. *Let $\mathcal{T}(n)$ be a collection of time bounds that is closed under multiplication by 2^n, such as $2^{O(n)}$ or $2^{n^{O(1)}}$. Then a class $\mathcal{C}$ has time-$\mathcal{T}(n)$ measure zero iff $\mathcal{C}$ is covered by a time-$\mathcal{T}(n)$ lex-limited betting game.*

Hence in particular for measure on E and EXP, martingales are equivalent to betting games constrained to bet in lex order.

A general betting game embodies a martingale in a different sense given by the following definition:

Definition 9. Let G be a betting games, and $i \geq 0$ an integer.

(a) A *play* α of length i is a sequence of i oracle answers. Note that α determines the first i-many stages of G, together with the query and bet for the next stage.
(b) $c_G(\alpha)$ is the capital C_i that G has at the end of the play α.

Note that the function c_G is a martingale over plays α. The following carryover of Lemma 3 is important.

Lemma 10 ("Slow-But-Sure" Lemma for betting games). *Let G be a betting game that runs in time $t(n)$. Then we can construct a betting game G' running in time $O(t(n))$ that always makes the same queries in the same order as G, such that $S^\infty[G] \subseteq S^\infty[G']$ and:*

$$(\forall \alpha)(\forall \beta) : c_{G'}(\alpha\beta) > c_{G'}(\alpha) - 2, \quad \text{and} \tag{3}$$

$$(\forall \alpha) : c_{G'}(\alpha) < 2|\alpha| + 2 . \tag{4}$$

4 From Betting Games to Martingales

This section associates to every betting game G a martingale d_G such that $S^\infty[G] \subseteq S^\infty[d_G]$, and begins examining the complexity of d_G. Before defining d_G, however, we discuss some tricky subtleties of betting games and their computations.

Given a finite initial segment w of an oracle language A, one can define the partial computation G^w of the betting game up to the stage i at which it first makes a query x_i that is not in the domain of w. Define $d(w)$ to be the capital C_{i-1} that G had entering this stage. It is tempting to think that d is a martingale and that d succeeds on all A for which G succeeds—but neither statement is true in general.

To see this, suppose x_i itself is the lexicographically least string not in the domain of w. That is, x_i is indexed by the bit b of wb, and $w1 \sqsubseteq A$ iff $x_i \in A$. It is possible that G^A makes a small (or even zero) bet on x_i, *and then goes back to make more bets in the domain of w, winning lots of money on them*. The definitions of both $d(w0)$ and $d(w1)$ will then reflect these added winnings, and both values will be greater than $d(w)$. For example, suppose G^A first puts a zero bet on $x_i = s_j$, then bets all of its money on $x_{i+1} = s_{j-1}$ not being in A, and then proceeds with $x_{i+2} = s_{j+1}$. If $w(s_{j-1}) = 0$, then $d(w0) = d(w1) = 2d(w)$.

Put another way, a finite initial segment w may carry much more "winnings potential" than the above definition of $d(w)$ reflects. To capture all of it, one needs to consider potential plays of the betting game outside the domain of w. Happily, one can bound the length of the considered plays via the running time function t of G. Let n be the maximum length of a string indexed by w; i.e., $n = \lfloor \log_2(|w|) \rfloor$. Then after $t(n)$ steps, G cannot query any more strings in the domain of w, so w's potential is exhausted. We will hence define $d_G(w)$ as an *average* value of those plays that can happen, given the query answers fixed by w. We use the following definitions and notation:

Definition 11. For any $t(n)$ time-bounded betting game G and string $w \in \Sigma^*$:

(a) A play α is *t-maximal* if G completes the first $|\alpha|$ stages, but *not* the query and bet of the next stage, within t steps.

(b) A play α is *G-consistent with* w, written $\alpha \sim_G w$, if for all stages j such that the queried string x_j is in the domain of w, $\alpha_j = w(x_j)$. That is, α is a play that could possibly happen given the information in w. Also let $m(\alpha, w)$ stand for the number of such stages j whose query is answered by w.

(c) Finally, put $d_G(\lambda) = 1$, and for nonempty w, with $n = \lfloor \log_2(|w|) \rfloor$ as above, let

$$d_G(w) = \sum_{\alpha\ t(n)-maximal,\, \alpha \sim_G w} c_G(\alpha)\, 2^{m(\alpha,w)-|\alpha|} . \tag{5}$$

The weight $2^{m(\alpha,w)-|\alpha|}$ in (5) has the following meaning: Suppose we extend the simulation of G^w by flipping a coin for every query outside the domain of w, for exactly i stages. Then the number of coin-flips in the resulting play α of length i is $i - m(\alpha, w)$, so $2^{m(\alpha,w)-i}$ is its probability. Thus $d_G(w)$ returns the suitably-weighted average of $t(n)$-step computations of G with w fixed. The interested reader may verify that this is the same as averaging $d(wv)$ over all v of length $2^{t(n)}$ (or any fixed longer length), where d is the non-martingale defined above.

Lemma 12. *The function $d_G(w)$ is a martingale.*

To ensure that d_G succeeds on all languages covered by G, however, we must first arrange that G satisfies the sure-winnings condition (3) of Lemma 10.

Lemma 13. *If G satisfies (3), then $S^{\infty}[G] \subseteq S^{\infty}[d_G]$.*

Now we turn our attention to the complexity of d_G. If G is a time-$t(n)$ betting game, it is clear that d_G can be computed deterministically in $O(t(n))$ *space*, because we need only cycle through all α of length $t(n)$, and all the items in (5) are computable in space $O(t(n))$. In particular, every E-betting game can be simulated by an ESPACE-martingale, and every EXP-betting game by an EXPSPACE-martingale. However, we show in the next section that one can *estimate* $d_G(w)$ well without having to cycle through all the α, using a pseudo-random generator to "sample" only a very small fraction of them.

5 Sampling Results

First we determine the accuracy to which we need to estimate the values $d(w)$ of a hard-to-compute martingale. Recall $N = 2^n$.

Lemma 14. *Let d be a martingale and $[\epsilon(i)]$ a sequence whose sum K converges. Suppose we can compute in time $t(n)$ the partial sum $\sum_{i=0}^{N} \epsilon(i)$ and a function $g(w)$ such that $|g(w) - d(w)| \le \epsilon(N)$ for all w of length N. Then there is a martingale d' computable in time $O(2^n t(n))$ such that for all w, $|d'(w) - d(w)| \le 2K$.*

Next, we will specify the function f_G that we will sample in order to estimate d_G.

Let G be a $t(n)$ time-bounded betting game. Consider a prefix w and let n denote the largest length of a string in the domain of w. With any string ρ of length $t(n)$, we can associate a unique "play of the game" G defined by using w to answer queries in the domain of w, and the successive bits of ρ to answer queries outside it. We can stop this play after $t(n)$ steps — so the stopped play is a $t(n)$-maximal α — and we define $f_G(w, \rho)$ to be the capital $c_G(\alpha)$. Note that we can compute $f_G(w, \rho)$ in linear time. The proportion of strings ρ of length $t(n)$ that map to the same play α is exactly the weight $2^{m(\alpha,w)-|\alpha|}$ in (5) for $d_G(w)$. Letting E stand for mathematical expectation, we have:

$$d_G(w) = E_{|\rho|=t(n)}[f_G(w, \rho)].$$

To estimate this mean, we apply concepts from pseudo-random generators. Alternatively, we can assume P = NP and apply Stockmeyer's method of approximate counting using alternation [16]—this yields results similar to Theorem 18 given in our full paper that improve those of [4], where P = PSPACE is assumed.

Definition 15 ([15]). (a) A *pseudo-random generator* (PRG) is a function D that, for each n, maps Σ^n into $\Sigma^{r(n)}$ where $r(n) > n$. The function r is called the *stretching* of D. We say that D is computable in a class $\mathcal{C}$ if every bit of $D(y)$ is computable in $\mathcal{C}$, given y and the index of the bit in binary.

(b) The *security* $S_D(n)$ of D at length n is the largest integer s such that for any circuit C of size at most s with $r(n)$ inputs

$$|\Pr_x[C(x) = 1] - \Pr_y[C(D(y)) = 1]| \le \frac{1}{s} \ ,$$

where x is uniformly distributed over $\Sigma^{r(n)}$ and y over Σ^n.

For our purposes, we will need a PRG computable in E that stretches seeds super-polynomially and has super-polynomial security at infinitely many lengths. Combining the results of Babai et al. [6] and of Nisan and Wigderson [15] with some padding yields:

Theorem 16. *If* MA $\ne$ EXP, *there is a PRG D computable in E with stretching $n^{\theta(\log n)}$ such that for any integer k, $S_D(n) \ge n^k$ for infinitely many n.*

We will also use PRGs with exponential security that are computable in exponential time.

Now we can apply PRGs to provide the accuracy and time bounds needed to get the desired martingale from Lemma 14.

Theorem 17. *Let D be a PRG computable in time $\delta(n)$ and with stretching $r(n)$. Let $f : \Sigma^* \times \Sigma^* \to (-\infty, \infty)$ be a linear-time computable function, and $s, R, m : \mathbb{N} \to \mathbb{N}$ be constructible functions such that $s(N) \ge N$ and the following relations hold for any integer $N \ge 0$, $w \in \Sigma^N$, and $\rho \in \Sigma^{s(N)}$:*

$$|f(w, \rho)| \le R(N)$$
$$r(m(N)) \ge s(N)$$
$$S_D(m(N)) \ge (s(N) + R(N))^6 \ . \tag{6}$$

Then we can approximate

$$h(w) = E_{|\rho|=s(N)}[f(w, \rho)] \tag{7}$$

to within N^{-2} in time $O(2^{m(N)} \cdot (s(N) + R(N))^4 \cdot \delta(m(N)))$.

Now, we would like to apply Theorem 17 to approximate efficiently $h = d_G$ given by (5) to within N^{-2}, by setting $f = f_G$ and $s(N) = t(\log N)$. The problem is that a given betting game G running in time $t(n)$ may only guarantee an upper bound of $R(N) = 2^{t(\log N)}$ on $|f(w, \rho)|$. Since S_D can be at most exponential, condition (6) would force $m(N)$ to be $\Omega(t(\log N))$, and Theorem 17 would only yield an approximation computable in time $2^{O(t(\log N))}$. However, we can assume wlog. that G satisfies the slow-winnings condition (4) of Lemma 10, in which case an upper bound of $R(N) \in O(N)$ holds. Then the term $s(N)$ in the right-hand side of (6) dominates, provided $t(n) \in 2^{\Omega(n)}$.

Taking everything together, we obtain the following result about transforming E- and EXP-betting games into equivalent E- respectively EXP-martingales:

Theorem 18. *If there is a PRG computable in E with security $2^{\Omega(n)}$, then for every E-betting game G, there exists an E-martingale d such that $S^\infty[G] \subseteq S^\infty[d]$. If there is a PRG computable in EXP with security $2^{n^{\Omega(1)}}$, then for every EXP-betting game G, there exists an EXP-martingale d such that $S^\infty[G] \subseteq S^\infty[d]$.*

6 Autoreducible Sets

An oracle Turing machine M is said to *autoreduce* a language A if $L(M^A) = A$, and for all strings x, M^A on input x does not query x. That is, one can learn the membership of x by querying strings other than x itself. If M runs in polynomial time, then A is P-*autoreducible*—we also write $\leq_T^p$-autoreducible. If M is also non-adaptive, then A is $\leq_{tt}^p$-*autoreducible*.

Autoreducible sets were brought to the polynomial-time context by Ambos-Spies [3]. Their importance was further argued by Buhrman, Fortnow, and Torenvliet [7], who showed that all $\leq_T^p$-complete sets for EXP are $\leq_T^p$-autoreducible (while some complete sets for other classes are not). Here we demonstrate that autoreducible sets are important for testing the boundaries of Lutz's measure theory. As stated in the Introduction, if the $\leq_T^p$-autoreducible sets in EXP (or sufficiently the $\leq_T^p$-complete sets for EXP) are covered by an EXP-martingale, then EXP $\neq$ BPP, a non-relativizing consequence. However, it is easy to cover them by an E-betting game. Indeed, the betting game uses its adaptive freedom only to "look ahead" at the membership of lexicographically greater strings, betting nothing on them.

Theorem 19. *There is an E-betting game that covers all $\leq_T^p$-autoreducible sets.*

Proof. One can effectively enumerate oracle TMs $M_1, M_2, \ldots$ that never query their input, with each M_i running in time $n^i + i$. Our betting game G regards its capital as composed of infinitely many "shares" c_i, one for each M_i. Initially, $c_i = 1/2^i$. Letting $\langle \cdot, \cdot \rangle$ be a standard pairing function, inductively define $n_0 = 0$ and $n_{\langle i,j \rangle + 1} = (n_{\langle i,j \rangle})^i + i$.

During a stage $s = \langle i, j \rangle$, G simulates M_i on input $0^{n_s - 1}$. Whenever M_i makes a query of length less than n_{s-1}, G looks up the answer from its table of past queries. Whenever M_i makes a query of length n_{s-1} or more, G places a bet of zero on that string and makes the same query. Then G bets all of the share c_i on $0^{n_s - 1}$ according to the answer of the simulation of M_i. Finally, G "cleans up" by putting zero bets on all strings with length in $[n_{s-1}, n_s)$ that were not queries in the previous steps.

If M_i autoreduces A, then share c_i doubles in value at each stage $\langle i, j \rangle$, and makes the total capital grow to infinity. And G runs in time $2^{O(n)}$—indeed, only the "cleanup" phase needs this much time. $\qquad\qquad\qquad\qquad\qquad\qquad\qquad\qquad\qquad\qquad\qquad\quad\Box$

Corollary 20. *Each of the following statements implies BPP $\neq$ EXP, as do the statements obtained on replacing "E" by "EXP."*

1. *The class of $\leq_T^p$-autoreducible sets has E-measure zero.*
2. *The class of $\leq_T^p$-complete sets for EXP has E-measure zero.*
3. *E-betting games and E-martingales are equivalent.*
4. *E-betting games have the finite union property.*

Since there is an oracle A giving EXP^A = BPP^A [10], this shows that relativizable techniques cannot establish the equivalence of E-martingales and E-betting games, nor of EXP-martingales and EXP-betting games. They cannot refute it either, since there are oracles relative to which strong PRGs exist—all "random" oracles, in fact.

It is tempting to think that the *non*-adaptively P-autoreducible sets should have E-measure zero, or at least EXP-measure zero, insofar as betting games are the adaptive cousins of martingales. However, it is not just adaptiveness but also the freedom to bet *out of the fixed lexicographic order* that adds power to betting games. If one carries out the proof of Theorem 19 to cover the class of $\leq_{tt}^p$-autoreducible sets, using an enumeration $[M_i]$ of $\leq_{tt}^p$-autoreductions, one obtains a *non-adaptive* E-betting game that (independent of its oracle) bets on all strings in order given by a single permutation of Σ^*. The permutation itself is E-computable. It might seem that an E-martingale should be able to "un-twist" the permutation and succeed on all these sets. However, our next results, which strengthen the above corollary, close the same "non-relativizing" door on proving this with current techniques.

Theorem 21. *For any k, the $\leq_{tt}^p$-complete sets for Δ_k^p are $\leq_{tt}^p$-autoreducible.*

Corollary 22. *Each of the following statements implies BPP $\neq$ EXP, as do the statements obtained on replacing "E" by "EXP."*

1. *The class of $\leq_{tt}^p$-autoreducible sets has E-measure zero.*
2. *The class of $\leq_{tt}^p$-complete sets for EXP has E-measure zero.*
3. *Non-adaptive E-betting games and E-martingales are equivalent.*
4. *If two classes can be covered by non-adaptive E-betting games, then their union can be covered by an E-betting game.*

This puts the spotlight on the question: Under what hypotheses can we show that the $\leq_{tt}^p$-autoreducible sets have E-measure zero? Our final results show that the hypothesis MA $\neq$ EXP suffices. This assumption is only known to yield PRGs of super-polynomial security (at infinitely many lengths) rather than exponential security (at almost all lengths). On the other hand, there exist oracles relative to which exponentially strong PRGs exist, but EXP = MA.

Theorem 23. *If MA $\neq$ EXP, then the class of languages A autoreducible by polynomial-time OTMs that make their queries in lex order has E-measure zero.*

Corollary 24. *If MA $\neq$ EXP, then the $\leq_{tt}^p$-autoreducible sets have E-measure zero.*

7 Conclusions

The initial impetus for this work was a simple question about measure: is the pseudo-randomness of a characteristic sequence invariant under simple permutations such as that induced by *flip* in the Introduction? Our "betting games" preserve Lutz's original idea of "betting" as a means of "predicting" membership in a language, without being tied to a fixed order of which instances one tries to predict, or to a fixed order of how one goes about gathering information on the language. We have shown some senses in which betting games are robust and well-behaved. We also contend that some current defects in the theory of betting games, notably the lack of a finite-unions theorem pending the status of pseudo-random generators, trade off with lacks in the resource-bounded measure theory, such as being tied to the lexicographic ordering of strings.

The research problems left in this paper that are most open to attack are to tighten even further the connections among PRGs, separation of classes within

EXP, and resource-bounded measure. Does EXP $\neq$ MA suffice to make the $\leq_T^p$-autoreducible sets have E-measure zero? Does that suffice to simulate every betting game by a martingale of equivalent complexity?

Another challenge is to determine how well these ideas work for measures on classes below E. Here even straightforward attempts to carry over Lutz's definitions run into difficulties, as described in [14] and [1, 2]. Perhaps the results in these papers can be re-cast in terms of betting games in ways that release new insights.

Acknowledgments The authors specially thank Klaus Ambos-Spies, Ron Book (pace), and Jack Lutz for organizing a special Schloss Dagstuhl workshop in July 1996, where preliminary versions of results and ideas in this paper were presented and extensively discussed. We also thank the STACS'98 referees for helpful comments.

References

1. Allender, E., Strauss, M.: Measure on small complexity classes, with applications for BPP. DIMACS TR 94-18, Rutgers University and DIMACS, April 1994.
2. Allender, E., Strauss, M.: Measure on P: Robustness of the notion. In *Proc. 20th International Symposium on Mathematical Foundations of Computer Science*, volume 969 of *Lect. Notes in Comp. Sci.*, pages 129–138. Springer Verlag, 1995.
3. Ambos-Spies, K.: P-mitotic sets. In E. Börger, G. Hasenjäger, and D. Roding, editors, *Logic and Machines, Lecture Notes in Computer Science 177*, pages 1–23. Springer-Verlag, 1984.
4. Ambos-Spies, K., Lempp, S.: Presentation at a Schloss Dagstuhl workshop on "Algorithmic Information Theory and Randomness," July 1996.
5. Babai, L.: Trading group theory for randomness. In *Proc. 17th Annual ACM Symposium on the Theory of Computing*, pages 421–429, 1985.
6. Babai, L., Fortnow, L., Nisan, N., Wigderson, A.: BPP has subexponential time simulations unless EXPTIME has publishable proofs. *Computational Complexity*, 3, 1993.
7. Buhrman, H., Fortnow, L., Torenvliet, L.: Using autoreducibility to separate complexity classes. In *36th Annual Symposium on Foundations of Computer Science*, pages 520–527, Milwaukee, Wisconsin, 23–25 October 1995. IEEE.
8. Buhrman, H., Longpré, L.: Compressibility and resource bounded measure. In *13th Annual Symposium on Theoretical Aspects of Computer Science*, volume 1046 of *lncs*, pages 13–24, Grenoble, France, 22–24 February 1996. Springer.
9. Babai, L., Moran, S.: Arthur-Merlin games: A randomized proof system, and a hierarchy of complexity classes. *J. Comp. Sys. Sci.*, 36:254–276, 1988.
10. Heller, F.: On relativized exponential and probabilistic complexity classes. *Inform. and Control*, 71:231–243, 1986.
11. Loveland, D. W.: A variant of the Kolmogorov concept of complexity. *Inform. and Control*, 15:510–526, 1969.
12. Lutz, J.: Almost everywhere high nonuniform complexity. *J. Comp. Sys. Sci.*, 44:220–258, 1992.
13. Lutz, J.: The quantitative structure of exponential time. In L. Hemaspaandra and A. Selman, eds., *Complexity Theory Retrospective II*. Springer Verlag, 1997.
14. Mayordomo, E.: *Contributions to the Study of Resource-Bounded Measure*. PhD thesis, Universidad Politécnica de Catalunya, Barcelona, April 1994.
15. Nisan, N., Wigderson, A.: Hardness versus randomness. *J. Comp. Sys. Sci.*, 49:149–167, 1994.
16. Stockmeyer, L.: The complexity of approximate counting. In *Proc. 15th Annual ACM Symposium on the Theory of Computing*, pages 118–126, Baltimore, USA, April 1983. ACM Press.

The Complexity of Modular Graph Automorphism

V. Arvind[1], R. Beigel[2*], and A. Lozano[3**]

[1] Institute of Mathematical Sciences, Chennai 600113, India
arvind@imsc.ernet.in
[2] Department of EE&CS, Lehigh University, Bethlehem PA 18015-3084, U.S.A.
beigel@eecs.lehigh.edu
[3] Universitat Politècnica de Catalunya, 08034 Barcelona, Catalonia, E.U.
antoni@lsi.upc.es

Abstract. Motivated by the question of the relative complexities of the Graph Isomorphism and the Graph Automorphism problems, we define and study the modular graph automorphism problems. These are the decision problems Mod_kGA which consist, for each $k > 1$, of deciding whether the number of automorphisms of a graph is divisible by k. The Mod_kGA problems all turn out to be intermediate in difficulty between Graph Automorphism and Graph Isomorphism.

We define an appropriate search version of Mod_kGA and design an algorithm that polynomial-time reduces the Mod_kGA search problem to the decision problem. Combining this algorithm with an IP protocol, we obtain a randomized polynomial-time checker for Mod_kGA, for all $k > 1$.

1 Introduction

The Graph Isomorphism problem (GI) consists of determining whether two graphs are isomorphic. It is well known that GI is in NP, but despite decades of study by mathematicians and computer scientists, it is not known whether GI is in P or whether GI is NP-complete. Many researchers conjecture that GI's complexity lies somewhere between P and NP-complete. Related to GI are several other decision problems (some graph-theoretic and others group-theoretic in nature) that are similarly not known to be in P or NP-complete. One such problem which is closely related to GI is Graph Automorphism (GA): Deciding whether a graph has a nontrivial automorphism. Regarding the relative complexities of GA and GI, it is known that GA is polynomial-time many-one reducible to GI. On the other hand, GI is not known to be even polynomial-time Turing reducible

* Supported in part by NSF grants CCR–8958528 and CCR–9415410, and NASA grant NAG 52895. Work done while on sabbatical from Yale University and affiliated with the Human–Computer Interaction Laboratory.
** Supported in part by the E.U. through ESPRIT project (no. LTR 20244, ALCOM-IT) and by DGICYT (Koala project, with no. PB95-0787), and CICYT (no. TIC97-1475-CE). Work completed while visiting the first author at the Institute of Mathematical Sciences.

to GA (see [9] for these and related results). However, in [11] it is shown that GI is polynomial-time reducible to the problem of computing the number of automorphisms of a graph.

The notion of program checking was introduced by Blum and Kannan [4] as an algorithmic alternative to program verification. Since then the design of efficient checkers for various computational problems has rapidly grown into a discipline of algorithm design [4, 5]. One of the first program checkers in [4] was a randomized polynomial-time checker for GI. It is an outstanding open question in the area if NP-complete problems have efficient program checkers. This can be construed as another evidence that GI is not NP-complete. Later, in [10] it was shown that GA has a *nonadaptive* checker. In other words, the checker can make all its queries to the program in parallel, hence enabling it to be fast in parallel (in NC, to be precise). It is an open question whether GI has also a nonadaptive checker, and the bottleneck here is that search does not reduce to decision for GI using parallel queries.

Thus, a natural next step in investigating the relationship between GI and GA is to consider exactly how much we need to know about the number of automorphisms of a graph in order to solve the Graph Isomorphism problem. This motivates us to define and study modular graph automorphism problems. Let $Aut(G)$ denote the automorphism group of the graph G.

Definition 1. For any k, let $\text{Mod}_k\text{GA} = \{G : |Aut(G)| \equiv 0 \pmod{k}\}$.

We show in Theorems 3 and 4 that for any $k > 1$, GA $\leq^p_m \text{Mod}_k\text{GA} \leq^p_m$ GI; thus the Mod_kGA problems are intermediate in difficulty between GA and GI. It is an open question whether any of the Mod_kGA problems is polynomial-time equivalent to GA or GI. We conjecture that Mod_kGA is *not* polynomial-time equivalent to GA or GI, for any $k > 1$. An evidence that some of the Mod_kGA problems could be actually harder than GA is our observation that Tournament Isomorphism (graph isomorphism for tournament graphs) is many-one reducible to Mod_2GA. This follows from the fact that the automorphism group of any tournament is of odd size [9], which in turn implies that two tournaments are isomorphic iff the automorphism group of their disjoint union contains an order two permutation (which must switch the two graphs).

The layout of the paper is as follows. Section 2 contains the preliminaries. In Section 3, we prove that the Mod_kGA problems are located between GA and GI. In Section 4, we show that search is polynomial-time Turing equivalent to decision for Mod_kGA, and in Section 5 we use this result in combination with an IP protocol for $\overline{\text{Mod}_p\text{GA}}$ to obtain an efficient program checker for Mod_kGA. Notice that although both GA and GI have program checkers ([10] and [4] resp.) and Mod_kGA is intermediate in complexity, it does not necessarily imply that Mod_kGA has a program checker [4].

In this extended abstract we omit several proofs due to lack of space[4].

---

[4] A fuller version is available at http://www.imsc.ernet.in/~arvind/modga.ps.gz

2 Preliminaries

In this paper by a graph we mean a finite directed graph[5] (see for example [8] or any other standard text on graph theory for basic definitions). For a graph G, let $V(G)$ denote its vertex set and $E(G)$ denote its edge set. A permutation π on the vertex set $V(G)$ of a graph G is an automorphism of G if $(u, v) \in E(G) \iff (\pi(u), \pi(v)) \in E(G)$. The set of automorphisms $Aut(G)$, of a graph G, is a subgroup of the permutation group on $V(G)$. The identity automorphism of any graph will be denoted by id.

Let X be a list of vertices in $V(G)$ for a given graph G. By $G_{[X]}$ we mean the graph G with distinct labels attached to the vertices in X. Given two lists of vertices $X, Y \subseteq V(G)$, the graphs $G_{[X]}$ and $G_{[Y]}$ have the same labels in vertices occupying the same positions in X and Y. It is not hard to see that in $G_{[X]}$ vertices of X are pointwise fixed in any automorphism[6]. Thus $Aut(G_{[X]})$ is isomorphic to the subgroup of $Aut(G)$ which pointwise fixes the vertices in X. Furthermore, given an automorphism of $Aut(G_{[X]})$ the corresponding automorphism of $Aut(G)$ can be easily constructed.

Definition 2. Let $G_1, \ldots, G_n$ be n graphs.

- Let P_n be a directed simple path of n new vertices $v_1, v_2, \ldots, v_n$, where each vertex v_i is labeled with a single label l. The graph $\text{Path}(G_1, \ldots, G_n)$ is obtained by taking one copy of each of the graphs $G_1, \ldots, G_n$ and, for $1 \le i \le n$, attaching all the vertices of G_i to v_i.
- Let C_n denote the directed simple cycle on n new vertices $v_1, v_2, \ldots, v_n$, with each vertex v_i, $1 \le i \le n$, labeled with a single label l. The graph $\text{Cycle}(G_1, \ldots, G_n)$ is obtained by taking one copy of each of the graphs $G_1, \ldots, G_n$ and, for $1 \le i \le n$, attaching all the vertices of G_i to v_i.

In both $\text{Path}(G_1, \ldots, G_n)$ and $\text{Cycle}(G_1, \ldots, G_n)$, since the new vertices $v_1, v_2, \ldots, v_n$ are labeled with l, any automorphism of these graphs must map the set $\{v_1, v_2, \ldots, v_n\}$ onto itself. Consequently, any automorphism of $\text{Path}(G_1, \ldots, G_n)$ $(\text{Cycle}(G_1, \ldots, G_n))$ when restricted to $\{v_1, v_2, \ldots, v_n\}$ is an automorphism of P_n (C_n) This means that an automorphism of $\text{Path}(G_1, \ldots, G_n)$ cannot permute the copies of $G_1, \ldots, G_n$, while an automorphism of $\text{Cycle}(G_1, \ldots, G_n)$ can permute them but only along the cycle C_n.

The reducibilities discussed in this paper are the standard polynomial-time Turing and many-one reducibilities. Formal definitions of these and other standard notions in complexity theory can be found in [2, 1]. The relative complexity of decision and search for NP problems is well studied [2, 1]. For instance, it is known that search and decision are polynomial-time Turing equivalent for all

---

[5] In this paper we consider the problems GI, GA, and Mod_kGA on directed graphs. However, all results of this paper hold for these problems on undirected graphs as well.

[6] Each label can be implemented with a graph gadget like a long path such that the overall size of the graph is still polynomially bounded. See, e.g. [9].

NP-complete problems. In particular, we recall that for GI, search is polynomial-time Turing reducible to decision [12] whereas for GA a stronger result holds: search is *nonadaptively* polynomial-time reducible to decision [10]. Finally, we recall from [6, 10] that GI is both a d-cylinder and a c-cylinder[7] [3].

3 Locating the Mod_kGA Problems

We show in this section that Mod_kGA is located between GA and GI, for all $k > 1$.

Theorem 3. *For all $k > 1$, GA $\leq_m^p \text{Mod}_k\text{GA}$.*

Proof. Given a graph G, we define for every i, j with $1 \leq i < j \leq n$, the graph $H_{i,j} = \text{Cycle}(G_{[i]}, G_{[j]}, \ldots, G_{[j]})$ which contains one copy of $G_{[i]}$ and $k - 1$ copies of $G_{[j]}$. Further, let H be obtained by applying the Path operator to all the graphs $H_{i,j}$ with $1 \leq i < j \leq n$. We claim that G has a nontrivial automorphism if and only if H is in Mod_kGA.

Suppose that G has a nontrivial automorphism φ. There exist two vertices i and j such that $\varphi(i) = j$. Notice that $H_{i,j}$ has the following nontrivial automorphism α that cyclically permutes the k graphs in $\text{Cycle}(G_{[i]}, G_{[j]}, \ldots, G_{[j]})$ as follows. The automorphism α maps the first graph $G_{[i]}$ to $G_{[j]}$ by φ. It maps each of the first $k - 2$ copies of $G_{[j]}$ to the next copy of $G_{[j]}$ by the identity automorphism. Finally, α maps the last copy of $G_{[j]}$ back to $G_{[i]}$ by the automorphism φ^{-1}.

The order of α is k since the vertices in $H_{i,j}$ are moved in a cyclic way through the different k subgraphs. In fact, the permutation α is a product of a bunch of k-cycles. Thus $H_{i,j} \in \text{Mod}_k\text{GA}$. Since $|Aut(H)| = \prod_{1 \leq i < j \leq n} |Aut(H_{i,j})|$, it follows that $H \in \text{Mod}_k\text{GA}$.

For the converse, assume that $H \in \text{Mod}_k\text{GA}$. Then, H has a nontrivial automorphism, say, α. Notice that α must induce an automorphism β in one of its subgraphs $H_{i,j}$. Since $H_{i,j} = \text{Cycle}(G_{[i]}, G_{[j]}, \ldots, G_{[j]})$, there are two possibilities: either β induces a nontrivial automorphism of $G_{[i]}$ or $G_{[j]}$, or else β maps the copy of $G_{[i]}$ to some copy of $G_{[j]}$. In either case, it is clear that we get a nontrivial automorphism of G. □

Mathon [11] has shown that $|Aut(G)|$ is polynomial-time computable with GI as oracle. From this it easily follows that $\text{Mod}_k\text{GA} \leq_T^p \text{GI}$. In the next theorem, we strengthen this to a $\leq_m^p$-reduction using some permutation group theory.

Theorem 4. *For all $k > 1$, $\text{Mod}_k\text{GA} \leq_m^p \text{GI}$.*

We need a couple of definitions and group-theoretic lemmas before we prove Theorem 4. Let A be a subgroup of S_n and let $[n]$ denote the set $\{1, 2, \ldots, n\}$. A

---

[7] Elsewhere in the literature, e.g. [9], these properties are called OR and AND functions respectively.

subset $X \subseteq [n]$ is *A-invariant* if $g(X) = X$ for all $g \in A$. If $X \subseteq [n]$ is *A*-invariant then consider the action of A *restricted* to X. This gives rise to a subgroup of the symmetric group S_X, which we denote by A^X. An obvious useful property is that $|A^X| \leq |A|$, for all *A*-invariant sets X. The proofs of the following two group-theoretic lemmas are omitted.

Lemma 5. *Let A be a subgroup of S_n s.t. $|A| = m$. Then there exists an A-invariant subset $X \subseteq [n]$ with $|X| \leq m \log m$, such that A is isomorphic to A^X.*

Lemma 6. *Let A be a finite group. Let $X = \{a_1, a_2, \ldots, a_t\}$ and $Y = \{b_1, b_2, \ldots, b_t\}$ be two subsets of A such that $\langle X \rangle \cap \langle Y \rangle =$ (id) and $a_i b_j = a_j b_i$, for $1 \leq i, j \leq t$. Then $|\langle X \rangle|$ divides the order of the group $\langle \{a_i b_i : 1 \leq i \leq t\} \rangle$.*

Proof of Theorem 4. First, we argue that it suffices to show that $\text{Mod}_{p^l} \text{GA} \leq_m^p$ GI for all prime p and $l > 0$. To see this, let $\prod_{1 \leq j \leq r} p_j^{l_j}$ be the prime factorization of k. Clearly, a graph $G \in \text{Mod}_k \text{GA}$ iff $G \in \bigcap_{1 \leq j \leq r} \text{Mod}_{p_j^{l_j}} \text{GA}$. Thus, if $\text{Mod}_{p_j^{l_j}} \text{GA} \leq_m^p$ GI for $1 \leq j \leq r$, it follows that $\text{Mod}_k \text{GA} \leq_m^p$ GI, since GI is a c-cylinder.

We first prove a useful group-theoretic claim. Let G be a graph on n vertices and f be a partial permutation on $[n]$ (i.e. f is defined on a subset of the domain $[n]$ and can be extended to a permutation in S_n). Then we call f a *partial automorphism* of G if f can be extended to an automorphism of G.

Claim. *Let p be a fixed prime and $l > 0$. A graph G on n vertices is in $\text{Mod}_{p^l} \text{GA}$ if and only if there exist a set $X \subseteq [n]$ with $|X| \leq p^l(\log p^l)$ and a subgroup $K = \{a_1, a_2, \ldots, a_{p^l}\}$ of S_X such that each $a_i \in K$ is a partial automorphism of G.*

Proof. Let $G \in \text{Mod}_{p^l} \text{GA}$ be an n vertex graph. Since p^l divides $|Aut(G)|$, by Sylow's theorem $Aut(G)$ has a subgroup A of size p^l. By Lemma 5 there is an *A*-invariant set $X \subseteq [n]$ with $|X| \leq p^l(\log p^l)$, such that A^X is isomorphic to A. Let $A^X = \{a_1, a_2, \ldots, a_{p^l}\}$. Furthermore, it also follows that A^X is a subgroup of S_X where each $a_i \in A^X$ is a partial automorphism of G. Conversely, suppose there is $X \subseteq [n]$ with $|X| \leq p^l(\log p^l)$ and a subgroup $K = \{a_1, a_2, \ldots, a_{p^l}\}$ of S_X where each $a_i \in K$ is a partial automorphism of G. Then for each i with $1 \leq i \leq p^l$, there is a $b_i \in S_{[n]-X}$ such that $a_i b_i \in Aut(G)$. It is not hard to see that Lemma 6 can be applied to the elements $\{a_i\}_{1 \leq i \leq p^l}$ and $\{b_i\}_{1 \leq i \leq p^l}$. Consequently, $|\langle \{a_i b_i : 1 \leq i \leq p^l\} \rangle|$ is divisible by p^l. Since $\langle \{a_i b_i : 1 \leq i \leq p^l\} \rangle$ is a subgroup of $Aut(G)$, it follows that p^l divides $|Aut(G)|$. $\square$

Now, note that the language $B = \{(G, f) : f$ is a partial automorphism of the graph $G\}$ is $\leq_m^p$-reducible to GI (for details see [9]). We will give a truth-table reduction from $\text{Mod}_{p^l} \text{GA}$ to B, where the truth-table is a disjunction of conjunctions. Since the language B is $\leq_m^p$-reducible to GI and since GI is both a c-cylinder and a d-cylinder, it follows that $\text{Mod}_{p^l} \text{GA}$ is $\leq_m^p$-reducible to GI. We describe below the said reduction of $\text{Mod}_{p^l} \text{GA}$ to B as a logical expression, which is easily seen to describe a disjunction-of-conjunctions kind of truth-table reduction:

$$G \in \mathrm{Mod}_{p^l}\mathrm{GA} \iff (\exists\, X \subseteq [n] : |X| \le p^l \log p^l)$$
$$(\exists \text{ subgroup } K < S_X : |K| = p^l)(\forall a \in K)[(G, a) \in B] \qquad \Box$$

4 Computing Solutions for $\mathrm{Mod}_k\mathrm{GA}$ Instances

The goal of this section is to design a polynomial-time algorithm that reduces the search problem for $\mathrm{Mod}_k\mathrm{GA}$ to the decision problem. Consider $\mathrm{Mod}_k\mathrm{GA}$ for an arbitrary $k > 1$. Notice that if the prime factorization of k is $\prod_{1 \le i \le m} p_i^{e_i}$, then the natural NP witness of the membership of a graph G in $\mathrm{Mod}_k\mathrm{GA}$ is a collection of m subgroups $\{A_1, A_2, \ldots, A_m\}$ of $Aut(G)$ where A_i is of order $p_i^{e_i}$, and is listed as a set of permutations. We consider such a witness as a solution for G for the $\mathrm{Mod}_k\mathrm{GA}$ search problem and we design a polynomial-time algorithm that computes this witness for any given instance of $\mathrm{Mod}_k\mathrm{GA}$ with oracle access to the $\mathrm{Mod}_k\mathrm{GA}$ decision problem. The next theorem is the main result of this section.

Theorem 7. *There is a polynomial-time algorithm $\mathcal{A}_k$ with $\mathrm{Mod}_p\mathrm{GA}$ as oracle such that given a graph $G \in \mathrm{Mod}_{p^k}\mathrm{GA}$ as input the algorithm $\mathcal{A}_k$ lists out the elements of a p^k-subgroup of $Aut(G)$.*

The proof of Theorem 7, which is a lengthy proof by induction on k and which exploits the structure of p-groups, is omitted from this extended abstract. In order to give a flavor of the sort of ideas involved in the proof of Theorem 7 we discuss in detail the base case, $k = 1$, in the following lemma.

Lemma 8. *There is a polynomial-time algorithm $\mathcal{A}_1$ with $\mathrm{Mod}_p\mathrm{GA}$ as oracle such that given a graph $G \in \mathrm{Mod}_p\mathrm{GA}$ as input the algorithm $\mathcal{A}_1$ outputs a cyclic group of order p contained in $Aut(G)$.*

Proof. For any list of vertices $X = \{i_1, \ldots, i_m\}$, let $r(X)$ be a right shift of X, this is $r(X) = \{i_m, i_1, \ldots, i_{m-1}\}$. Consider the following algorithm, which computes an order-p automorphism of an input graph $G \in \mathrm{Mod}_p\mathrm{GA}$.

Algorithm $\mathcal{A}_1$

Input G;
if $G \notin \mathrm{Mod}_p\mathrm{GA}$ **then stop**;
$X = \emptyset$;
for $i := 1$ **to** $|V(G)|$ **do**
    if $G_{[X \cup \{i\}]} \in \mathrm{Mod}_p\mathrm{GA}$ **then** $X := X \cup \{i\}$;
$S := V(G) - X$; $\mathcal{C} := \emptyset$; $G' := G$; $G'' := G$;
for each p-cycle $C \subseteq S - \bigcup_{D \in \mathcal{C}} D$ **do**
    if $Cycle(G'_{[C]}, G''_{[r(C)]}, \ldots, G''_{[r(C)]}) \in \mathrm{Mod}_p\mathrm{GA}$ **then**

(* There are $p-1$ copies of $G''_{[r(C)]}$ in the above Cycle definition *)
$$G' := G'_{[C]}; \ G'' := G''_{[r(C)]}; \ \mathcal{C} := \mathcal{C} \cup \{C\};$$
Output the order-p automorphism consisting of
p-cycles $\mathcal{C}$ and fixed-point set X

We now prove the correctness of the above algorithm. Notice that the first for-loop takes $G \in \text{Mod}_p\text{GA}$ as input and computes the graph $G_{[X]} \in \text{Mod}_p\text{GA}$ with X as its set of fixed points (such that no more points can be fixed preserving membership in Mod_pGA). It remains to show that there is an order-p automorphism with $\mathcal{C}$ as its collection of p-cycles and X as its fixed-point set. We will prove this as an invariant of the second for-loop in the algorithm. Clearly, before the loop is entered, there is an order-p automorphism of $G_{[X]}$ with $\mathcal{C} = \emptyset$ as subset of its p-cycle set. Suppose this property holds at the beginning of some iteration of the for-loop. Suppose in the next iteration a new p-cycle C gets included in $\mathcal{C}$. We have to show that there is an order-p automorphism of G with X as fixed-point set and such that $\mathcal{C} \cup \{C\}$ is contained in its p-cycle set. Consider $Cycle(G'_{[C]}, G''_{[r(C)]}, \ldots, G''_{[r(C)]})$, which is in Mod_pGA. Notice that the corresponding order-p automorphism ψ of $Cycle(G'_{[C]}, G''_{[r(C)]}, \ldots, G''_{[r(C)]})$ cannot map the copy of $G'_{[C]}$ to itself since $G'_{[C]}$ cannot have order-p automorphisms (because it forces G to have order-p automorphisms with $X \cup C$ as fixed points). Thus ψ must map the p graphs in $Cycle(G'_{[C]}, G''_{[r(C)]}, \ldots, G''_{[r(C)]})$ by a p-cyclic rotation. In particular, it implies that ψ maps $G'_{[C]}$ to some copy of $G''_{[r(C)]}$. Hence, ψ restricted to the nodes of G yields an automorphism ϕ of G with X as fixed-point set and such that $\mathcal{C} \cup \{C\}$ is contained in the p-cycle set of ϕ. This completes the proof. $\square$

Notice that an immediate consequence of Theorem 7 is that search is polynomial-time Turing reducible to decision for Mod_kGA. Another consequence of Theorem 7 is that $\text{Mod}_{p^k}\text{GA}$ is polynomial-time Turing reducible to Mod_pGA, for any prime p.

5 A Program Checker for Mod_kGA

The goal of this section is to show that for each $k > 1$ the decision problem Mod_kGA has a program checker in the sense of [4]. We first recall the definition of program checkers.

Definition 9. [4] A program checker C_A for a decision problem A is a (probabilistic) algorithm that for any program P (supposedly for A) that halts on all instances, for any instance x_0 of A, and for any positive integer k (the security parameter) presented in unary:

1. If P is a correct program, that is, if $P(x) = A(x)$ for all instances x, then with probability $\geq 1 - 2^{-k}$, $C_A(x_0, P, k)$=Correct.
2. If $P(x_0) \neq A(x_0)$ then with probability $\geq 1 - 2^{-k}$, $C_A(x_0, P, k)$=Incorrect.

The probability is computed over the sequences of coin flips that C_A could have tossed. Also C_A is allowed to make queries to the program P on some instances.

Before we proceed we also need the definition of IP protocols which was first introduced in [7].

Definition 10. An interactive proof system consists of a prover-verifier pair $P \leftrightarrow V$. The verifier V is a probabilistic polynomial time machine and the prover P is, in general, a machine of unlimited computational power which shares the input tape and a communication tape with V.

$P \leftrightarrow V$ is an interactive (i.e. IP) protocol for a language L, if for every $x \in \Sigma^*$:

$$x \in L \to \text{Prob}[P \text{ makes } V \text{ accept }] > 3/4,$$

$$x \notin L \to \forall \text{ provers } P' : \text{Prob}[P' \text{ makes } V \text{ accept }] < 1/4,$$

The design of our checker for Mod_kGA is based on the following theorem [4].

Theorem 11. [4] *If a decision problem A and its complement have both interactive proof systems, in each of which the honest prover can be simulated in polynomial time with queries to A, then A has a polynomial-time program checker.*

Notice that Theorem 7 already gives an IP protocol for Mod_kGA with the prover polynomial-time Turing reducible to Mod_kGA. Thus, it suffices to design an IP protocol for $\overline{\text{Mod}_k\text{GA}}$ with requisite properties. The overall IP protocol for $\overline{\text{Mod}_k\text{GA}}$ is based on the following IP protocol for $\overline{\text{Mod}_p\text{GA}}$, for prime p.

Lemma 12. *For any prime p, there is an IP protocol for $\overline{\text{Mod}_p\text{GA}}$ in which the honest prover is polynomial-time Turing reducible to Mod_pGA.*

Proof. We rewrite the definition of $\overline{\text{Mod}_p\text{GA}}$ as follows: $\overline{\text{Mod}_p\text{GA}} = \{G : G \text{ has no automorphism with a } p\text{-cycle}\}$. Given an input graph G, the aim is to design an IP protocol which accepts G with high probability if G has *no automorphism* with a p-cycle, and which rejects G with high probability otherwise. Notice that since the prime p is a constant, the total number of p-cycles in S_n is bounded by qn^p, where q is a constant. We will build the desired IP protocol from an IP protocol for the following related language $L = \{(G, C) : |V(G)| = n, C \in S_n \text{ is a } p\text{-cycle and } G \text{ has no automorphism with } C \text{ as one of its cycles }\}$.

2-round IP Protocol for L

Input (G, C);
Let $Y := [n] - \{i : i \in C\}$;
1. Verifier:
    Pick a permutation $\psi \in S_Y$ uniformly at random;
    Pick a random bit $b \in \{0, 1\}$;
    if $b = 0$ **then**

        send $G' = \psi(G)$ **to Prover**
        else send $G' = \psi \circ C(G)$ **to Prover**;

2. Prover:
    if there exists permutation $\pi \in S_Y$ such that $\pi(G) = G'$ **then**
        send back a bit $c = 0$
        else send back a bit $c = 1$;
if $c = b$ **then** the Verifier **accepts else** the Verifier **rejects**

We first show that if the prover is honest then the protocol accepts an input $(G, C) \in L$ with probability 1. Suppose b took the value 0 and the graph $\psi(G) = G'$ was sent to the prover. Then clearly, the prover will find a permutation, namely ψ, such that $\psi(G) = G'$ and send back $c = 0$ leading to the acceptance of the input. Next, suppose b took the value 1. In that case we claim that there does not exist any permutation $\pi \in S_Y$ such that $\pi(G) = G'$. Suppose there exists such a π. Then, since $\pi(G) = \psi \circ C(G)$, it follows that $(\pi)^{-1}\psi \circ C$ is in $Aut(G)$, which contradicts the assumption that $(G, C) \in L$. In this case the prover will send back $c = 1$ and the verifier will again accept.

Now, to prove the soundness of the protocol, we must show that for an input $(G, C) \notin L$, the verifier will reject the input with probability at least $1/2$, for any prover. We first need the following claim (proof omitted). In the sequel we use X to denote the set $\{i : i \in C\}$ and Y to denote $[n] - X$.

Claim A. *If G has an automorphism τ with C as one of its cycles then the random graphs $\psi(G)$ and $\psi \circ C(G)$ are identically distributed, where ψ is picked uniformly at random from S_Y.*

It follows from Claim A that if $(G, C) \notin L$ the prover cannot distinguish between whether G' came from the case $b = 0$ or from $b = 1$. In fact, whether $b = 0$ or $b = 1$ the prover will find a $\pi \in S_Y$ such that $\pi(G) = G'$. Therefore, the bit c that is sent back by any (even cheating) prover can agree with b with probability at most $1/2$. Consequently, the verifier will reject an input $(G, C) \notin L$ with probability at least $1/2$. The error probability can be made exponentially small (say 2^{-n}) in the above protocol by repeating the protocol [8] (in parallel or sequentially). We now describe the IP protocol for $\overline{\mathrm{Mod}_p\mathrm{GA}}$.

IP Protocol for $\overline{\mathrm{Mod}_p\mathrm{GA}}$

Input G; (* G has n nodes *)
for each p-cycle $C \in S_n$ **do**
        if the IP protocol for L rejects (G, C) **then**
            Reject (and stop);
Accept

It is easy to see that this IP protocol accepts $G \in \overline{\mathrm{Mod}_p\mathrm{GA}}$ with probability 1 and rejects $G \in \mathrm{Mod}_p\mathrm{GA}$ with probability at least $(1 - 2^{-n})^{qn^p}$ which is larger than $1/2$. The following claim completes the proof of the lemma.

---

[8] With some modifications we can easily get constant round IP protocols.

Claim B. *There is an honest prover that is polynomial-time Turing reducible to* $\mathrm{Mod}_p\mathrm{GA}$ *for the above IP protocol for* $\overline{\mathrm{Mod}_p\mathrm{GA}}$.

Proof of Claim B. First we observe that in bounding the complexity of the honest prover we are concerned about inputs $G \in \overline{\mathrm{Mod}_p\mathrm{GA}}$. More precisely, we must show that there is a polynomial-time algorithm with $\mathrm{Mod}_p\mathrm{GA}$ as oracle that can simulate the honest prover correctly for inputs $G \in \overline{\mathrm{Mod}_p\mathrm{GA}}$. Notice that the honest prover of the overall IP protocol must actually simulate the honest prover of the IP protocol for L for each input in the set $\{(G, C) : C$ is a p-cycle in $S_n\}$, where $G \in \overline{\mathrm{Mod}_p\mathrm{GA}}$. The honest prover in the protocol for L is supposed to try and compute a permutation $\pi \in S_Y$ such that $\pi(G) = G'$. We have already argued in the correctness proof that for $G \in \overline{\mathrm{Mod}_p\mathrm{GA}}$ such a permutation π exists if and only if the outcome of b is 0 and $G' = \psi(G)$ for the random permutation $\psi \in S_Y$. The honest prover constructs the graph $G'' = Cycle(G_{[X]}, G'_{[X]}, \ldots, G'_{[X]})$, with $p-1$ copies of $G'_{[X]}$. Using algorithm $\mathcal{A}_1$ of Lemma 8 the honest prover computes an automorphism of G'' of order p if it exists. Notice that if there is a permutation $\pi \in S_Y$ such that $\pi(G) = G'$ then there is a permutation π' such that $\pi'(G_{[X]}) = G'_{[X]}$. Hence we can find an order-p automorphism of G'' which cyclically permutes the p graphs in G'', by mapping the copy of $G_{[X]}$ to the first copy of $G'_{[X]}$ by π', and each of the first $p-2$ copies of $G'_{[X]}$ are mapped to the next copy of $G'_{[X]}$ by the identity permutation, and finally, the last copy of $G'_{[X]}$ is mapped back to $G_{[X]}$ by π'^{-1}. It is easy to see that this is an automorphism of G'' of order p. Conversely, suppose that G'' has an order-p automorphism τ computed by the honest prover. Since $G \notin \mathrm{Mod}_p\mathrm{GA}$ and $G' \notin \mathrm{Mod}_p\mathrm{GA}$, the p graphs defining G'' must be rotated in some p-cyclic order by the automorphism τ. It follows that the copy of $G_{[X]}$ is mapped by τ to some copy of $G'_{[X]}$. Let π' be the projection of τ to these two copies. We have $\pi'(G_{[X]}) = G'_{[X]}$. From π' we can easily recover a permutation $\pi \in S_Y$ such that $\pi(G) = G'$. Thus the honest prover finds an order-p automorphism τ of G'' iff there exists $\pi \in S_Y$ such that $\pi(G) = G'$, and moreover, from such a τ the corresponding π is easily computed. Hence, the honest prover is polynomial-time Turing reducible to $\mathrm{Mod}_p\mathrm{GA}$. $\square$

Theorem 13. *For each $k > 1$, $\mathrm{Mod}_k\mathrm{GA}$ has a program checker.*

Proof. Let $\prod_{1 \leq i \leq m} p_i^{e_i}$ be the prime factorization of k. Because the class of checkable sets is obviously closed under join and under Turing equivalence [4], by Lemma 12, it suffices to show that $\mathrm{Mod}_k\mathrm{GA} \equiv_T^p \mathrm{Mod}_{p_1}\mathrm{GA} \oplus \cdots \oplus \mathrm{Mod}_{p_m}\mathrm{GA}$. Observe that for any graph G, $G \in \mathrm{Mod}_k\mathrm{GA} \iff (\forall i \leq m)[G \in \mathrm{Mod}_{p_i^{e_i}}\mathrm{GA}]$. By Theorem 7, $\mathrm{Mod}_{p_i^{e_i}}\mathrm{GA} \equiv_T^p \mathrm{Mod}_{p_i}\mathrm{GA}$ for each i. Therefore, $\mathrm{Mod}_k\mathrm{GA} \leq_T^p \mathrm{Mod}_{p_1}\mathrm{GA} \oplus \cdots \oplus \mathrm{Mod}_{p_m}\mathrm{GA}$. It is easy to prove that $\mathrm{Mod}_{p_i}\mathrm{GA} \leq_T^p \mathrm{Mod}_k\mathrm{GA}$ for each i. Therefore, $\mathrm{Mod}_{p_1}\mathrm{GA} \oplus \cdots \oplus \mathrm{Mod}_{p_m}\mathrm{GA} \leq_T^p \mathrm{Mod}_k\mathrm{GA}$ as well. $\square$

6 Concluding Remarks

In this paper we define modular graph automorphism problems (Mod_kGA) and locate them between GA and GI. We also design an efficient program checker for Mod_kGA based on an algorithm that reduces search to decision for Mod_kGA and an IP protocol for $\overline{\text{Mod}_k\text{GA}}$. The bottleneck in making our checker nonadaptive is essentially the following: can search be reduced to decision via parallel queries for Mod_pGA, for prime p?

Indeed, our initial motivation in studying the Mod_kGA problems was to understand the difference between GI and GA by introducing problems of intermediate difficulty. In this context, a challenging question is whether search reduces to decision via parallel queries for GI (hence yielding nonadaptive checkers for GI). We believe that as a first step this question must be answered for Mod_pGA.

References

1. J. Balcázar, J. Díaz, and J. Gabarró. *Structural Complexity II*. Springer–Verlag, 1990.
2. J. Balcázar, J. Díaz, and J. Gabarró. *Structural Complexity I*. Springer–Verlag, second edition, 1995.
3. R. Beigel, M. Kummer, and F. Stephan. Approximable sets, *Information and Computation* 120(2):304–314, 1995.
4. M. Blum and S. Kannan. Designing programs that check their work, *Journal of the ACM*, 43:269–291, 1995.
5. M. Blum, M. M. Luby, and R. Rubinfeld. Self-testing/correcting with applications to numerical problems, *J. Comput. Syst. Sci.* 47:73–83, 1993.
6. R. Chang. On the structure of NP computations under boolean operators. Ph. D. Thesis, Cornell University, 1991.
7. S. Goldwasser, S. Micali, and C. Rackoff. The knowledge complexity of interactive proof systems. *SIAM Journal on Computing* 18:186–208, 1989.
8. F. Harary. *Graph Theory*, Addison Wesley, Reading, 1969.
9. J. Köbler, U. Schöning, and J. Torán. *The graph isomorphism problem: its structural complexity*, Birkhäuser, Boston, 1993.
10. A. Lozano and J. Torán. On the nonuniform complexity of the graph isomorphism problem. In *Proceedings of the 7th Structure in Complexity Theory Conference*, pp. 118–129, 1992.
11. R. Mathon. A note on the graph isomorphism counting problem. *Information Processing Letters*, 8:131–132, 1979.
12. C. P. Schnorr. On self-transformable combinatorial problems. *Math. Programming Study*, 14:95–103, 1982.

Unary Quantifiers, Transitive Closure, and Relations of Large Degree*

Leonid Libkin[1]   Limsoon Wong[2]

[1] Bell Laboratories/Lucent Technologies, 600 Mountain Avenue, Murray Hill, NJ
07974, USA, Email: libkin@research.bell-labs.com
[2] BioInformatics Center & Institute of Systems Science, Singapore 119597, Email:
limsoon@iss.nus.sg

Abstract. This paper studies expressivity bounds for extensions of first-
order logic with counting and unary quantifiers in the presence of relations of large degree. There are several motivations for this work. First, it
is known that first-order logic with counting quantifiers captures uniform
TC^0 over *ordered* structures. Thus, proving expressivity bounds for first-
order with counting can be seen as an attempt to show $TC^0 \subsetneq DLOG$
using techniques of descriptive complexity. Second, the presence of auxiliary built-in relations (e.g., order, successor) is known to make a big
impact on expressivity results in finite-model theory and database theory. Our goal is to extend techniques from "pure" setting to that of
auxiliary relations.

Until now, all known results on the limitations of expressive power
of the counting and unary-quantifier extensions of first-order logic dealt
with auxiliary relations of "small" degree. For example, it is known that
these logics fail to express some DLOG-queries in the presence of a successor relation. Our main result is that these extensions cannot define the
deterministic transitive closure (a DLOG-complete problem) in the presence of auxiliary relations of "large" degree, in particular, those which are
"almost linear orders." They are obtained from linear orders by replacing
them by "very thin" preorders on arbitrarily small number of elements.
We show that the technique of the proof (in a precise sense) cannot be
extended to provide the proof of separation of TC^0 from DLOG. We also
discuss a general impact of having built-in (pre)orders, and give some
expressivity statements in the pure setting that would imply separation
results for the ordered case.

1 Introduction

The development of Descriptive Complexity suggests a very close connection
between proving lower bounds in complexity theory and proving inexpressibility
results in logic. The latter are of the form "a property P cannot be expressed
in logic $\mathcal{L}$ over the class of finite models." Developing tools for proving such

---

* Part of this work was done when the first author was visiting Institute of Systems
Science.

expressivity bounds is one of the central problems in Finite-Model Theory. In this paper we show how tools based on locality of logics can be applied to the complexity class TC^0 and, more generally, how they allow us to derive new expressivity bounds in the presence of complex auxiliary relations.

The class TC^0 is an important complexity class: problems such as integer multiplication and division, and sorting belong to TC^0; this class has also been studied in connection with neural nets, cf. [19]. Despite serious efforts and a number of proved lower bounds (see [2] for a survey), it is still not known if $TC^0 \subsetneq NP$, and the results of [20] suggest that traditional approaches to lower bounds are unlikely to succeed in proving this separation.

A starting point for our study is a result of [4] stating that:

$$FO(C) + < = \text{uniform } TC^0.$$

Here, as usual, TC^0 is the class of problems solvable by polynomial-size, constant-depth threshold circuits, and uniform means DLOGTIME-uniform, see [4] for more details. From now on, whenever we write TC^0, we mean the uniform class.

By $FO(C)$ we mean the extension of first-order logic with counting quantifiers $\exists i$, where $\exists i x. \varphi(x)$ means that φ has at least i satisfiers. For example, $\exists i, j((j + j = i) \land \exists! i x. \varphi(x))$ (where $\exists! i$ is a shorthand for "exists exactly i") states that the number of satisfiers of φ is even — this is known not to be expressible in first-order logic alone. By $FO(C) + <$ we mean $FO(C)$ in the presence of a built-in order relation. Note that if we are interested in $FO(C) + <$ sentences, then it does not matter which linear order is used. However, it is known that the mere presence of an order relation increases expressiveness (cf. [3]).

Thus, the problem of separation of uniform TC^0 from classes such as DLOG, NLOG, P, etc, is reduced to proving that some problems in these classes are not expressible in $FO(C) + <$. However, it appears that the presence of an order relation is a major obstacle to proving such expressivity bounds. The first partial result was given in [10], using counting games of [15]:

Fact 1 *There exist a problem complete for* DLOG *under first-order reductions that cannot be defined by* $FO(C)$ *in the presence of a successor relation.*

The result of [10] also shows that *dtc*, deterministic transitive closure, is not in $FO(C) + succ$, while $FO + dtc + succ$ captures the class DLOG. This was extended in [16] as follows.

Fact 2 ([16]) *Deterministic transitive closure cannot be defined by* $FO(C)$ *in the presence of auxiliary relations, whose degrees are bounded by a fixed constant* k.

If we talk about directed graphs, by *degrees* we mean in- and out-degrees of nodes. (A more general definition can be given for arbitrary relational structures, cf. [7].) In the successor relation, every node has in- and out-degree either 0 or 1. In contrast to these two results, in a linear order on an n-element set, all n different (in- and out-) degrees from 0 to $n - 1$ are realized. Thus, in order to move closer to proving expressivity bounds in the presence of an order relation, one has to at least be able to lift the results from constant degrees to those that depend on the size of the input.

A result in this direction was proved in [16], using a definition on *moderate degree* from [11]. A class C of graphs (more generally, relational structures) is of moderate degree, if $degmax_C(n)$, the maximal in- or out-degree of an n-element graph from C, is at most $\log^{o(1)} n$. That is, for some function $\delta(n)$ such that $\lim_{n \to \infty} \delta(n) = 0$, we have $degmax_C(n) \leq \log^{\delta(n)} n$.

Fact 3 ([16]) *Deterministic transitive closure cannot be defined by* $\mathrm{FO}(\mathbf{C})$ *in the presence of auxiliary relations of moderate degree.* □

In [11], auxiliary relations of moderate degree were shown to be of no help for expressing connectivity of graphs in monadic Σ_1^1. This was extended to degrees $n^{o(1)}$ [22] and to a linear order [21]. So one may wonder if a similar program can be carried out for $\mathrm{FO}(\mathbf{C})$.

There is a significant difference between Facts 1, 2 and 3, and the desired separation for the ordered case: in those Facts, we only deal with auxiliary relations of *small* degrees – these are either constant, or very small compared to the size of the input structure. In contrast, a linear order realizes as many degrees as there are elements in the input. Hence, one needs techniques to lift the results for $\mathrm{FO}(\mathbf{C})$ from relations of *small* degrees to relations of *large* degrees, i.e. those comparable with the size of the input.

Organization After introducing the notation in Section 2, and the technical machinery based on local properties of logics in Section 3, we describe, in Section 4, a general approach to proving expressivity bounds for local logics in the presence of auxiliary relations. We then define the class of "almost linear orders" (shown in figure 1) and use the general technique to show that deterministic transitive closure (and thus other DLOG-complete problems) are not expressible in $\mathrm{FO}(\mathbf{C})$ in the presence of those relations. In Section 5, we show that, in a precise sense, this is the best partial result that can be obtained using locality techniques. In Section 6, we analyze expressivity of $\mathrm{FO}(\mathbf{C})$ in the pure case (without auxiliary relations) vs. built-in orders or preorders. We also describe problems whose inexpressibility in $\mathrm{FO}(\mathbf{C})$ (note the absence of an order relation!) would imply $\mathrm{TC}^0 \subsetneq \mathrm{DLOG}(\mathrm{NLOG})$. Complete proofs can be found in the full version, which also contains a more detailed comparison with known results, and shows some applications in database theory.

2 Notations

A relational signature σ is a set of relation symbols $\{R_1, ..., R_l\}$, with an associated arity function. In what follows, $p_i(> 0)$ denotes the arity of R_i. We write σ_n for σ extended with n new constant symbols. We use σ_{gr} for the signature of graphs (that is, one binary predicate E). A σ-structure is $\mathcal{A} = \langle A, R_1^{\mathcal{A}}, \ldots, R_l^{\mathcal{A}} \rangle$, where A is a *finite* set, and $R_i^{\mathcal{A}} \subseteq A^{p_i}$ interprets R_i. If $\mathcal{A}$ is understood, we will omit the superscript. The class of finite σ-structures is denoted by STRUCT$[\sigma]$. Isomorphism is denoted by $\cong$. The carrier of $\mathcal{A}$ is always denoted by A.

We deal with three logics: FO, FO(C) and FO($\mathbf{Q}_u$), the last one being first-order logic with unary quantifiers. First-order formulae are built-up from atomic formulae by using Boolean connectives and quantifiers $\exists$ and $\forall$. First-order logic with counting, FO(C), is defined as a two sorted logic, with second sort being the sort of natural numbers. That is, a structure $\mathcal{A}$ is of the form $\mathcal{A} = \langle\{1,\ldots,n\}, \{v_1,\ldots,v_n\}, <, \text{BIT}, 1, \max, R_1^{\mathcal{A}}, \ldots, R_l^{\mathcal{A}}\rangle$. Here the relations $R_i^{\mathcal{A}}$ are defined on the domain $\{v_1,\ldots,v_n\}$, while on the numerical domain $\{1,\ldots,n\}$ one has $1, \max, <$ and the BIT predicate available (BIT(i,j) iff the ith bit in the binary representation of j is one). It also has counting quantifiers $\exists ix.\varphi(x)$, meaning that φ has at least i satisfiers; here i refers to the numerical domain and x to the domain $\{v_1,\ldots,v_n\}$. These quantifiers bind x but not i.

Let σ_k^{unary} be a signature of k unary symbols, and let $\mathcal{K}$ be a class of σ_k^{unary}-structures which is closed under isomorphisms. Then $\mathcal{K}$ gives rise to a generalized quantifier $Q_{\mathcal{K}}$, and FO($Q_{\mathcal{K}}$) extends the set of formulae of FO with the following additional rule: if $\psi_1(x_1, \vec{y}_1), \ldots, \psi_k(x_k, \vec{y}_k)$ are formulae, then $Q_{\mathcal{K}} x_1 \ldots x_k.(\psi_1(x_1, \vec{y}_1), \ldots, \psi_k(x_k, \vec{y}_k))$ is a formula. Here $Q_{\mathcal{K}}$ binds x_i in the ith formula, for each $i = 1, \ldots, k$. The semantics is defined as follows: $\mathcal{A} \models Q_{\mathcal{K}} x_1 \ldots x_k.(\psi_1(x_1, \vec{a}_1), \ldots, \psi_k(x_k, \vec{a}_k))$ iff $(A, \psi_1[\mathcal{A}, \vec{a}_1], \ldots, \psi_k[\mathcal{A}, \vec{a}_k]) \in \mathcal{K}$, where $\psi_i[\mathcal{A}, \vec{a}_i] = \{a \in A \mid \mathcal{A} \models \psi_i(a, \vec{a}_i)\}$. In this definition, $\vec{a}_i$ is a tuple of parameters that gives the interpretation for those free variables of $\psi_i(x_i, \vec{y}_i)$ which are not equal to x_i. Examples of unary quantifiers include the usual $\exists$ and $\forall$, as well as Rescher (bigger cardinality) and Härtig (equicardinality) quantifiers. We use the notation FO($\mathbf{Q}_u$) for FO extended with *all* unary quantifiers.

Every FO(C) sentence can be expressed in FO($\mathbf{Q}_u$), while there exist properties definable in FO($\mathbf{Q}_u$) but not in FO(C).

While the results in this paper refer to these three logics, they can also be extended to abstract logics in the sense of [8], which are regular (e.g., closed under first-order operations and substitutions).

With each formula $\psi(x_1, \ldots, x_m)$ in the logical language whose symbols are in σ, we associate a *query* (semantic mapping) that maps a σ-structure $\mathcal{A}$ into a m-ary relation $\psi[\mathcal{A}] = \langle A, \{(a_1, \ldots, a_m) \in A^m \mid \mathcal{A} \models \psi(a_1, \ldots, a_m)\}\rangle$.

Given a relational signature σ and a class $\mathcal{R}$ of σ'-structures, where σ' is another relational signature, disjoint from σ, we say that a query q, producing an m-ary relation, is definable on σ-structures in the presence of $\mathcal{R}$-structures if there exists a $\sigma \cup \sigma'$-formula $\varphi(\vec{x})$ such that, for any σ-structure $\mathcal{A}$ with carrier A and for any structure $\mathcal{A}' \in \mathcal{R}$ on A, we have:

$$q(\mathcal{A}) = \{\vec{a} \in A^m \mid (\mathcal{A}, \mathcal{A}') \models \varphi(\vec{a})\}$$

where $(\mathcal{A}, \mathcal{A}')$ is the $\sigma \cup \sigma'$ structure obtained by putting $\mathcal{A}$ and $\mathcal{A}'$ together. We most often encounter the situation where $\mathcal{R}$ is the class of preorders (with special properties), or linear orders. Note that, according to this definition, a query definable in the presence of $\mathcal{R}$-structures is independent of a particular $\mathcal{R}$ structure being used. We use the notation $\mathcal{L} + \mathcal{R}$ for the class of queries definable in $\mathcal{L}$ the presence of relations from $\mathcal{R}$.

We shall use $\mathcal{O}_k$ for the class of preorders in which no equivalence class has more than k elements; these can be viewed as being very close to linear orders for small k. We also call them preorders of width k. In particular, $\mathcal{O}_1$ is the class of linear orders. We also write $\mathcal{L}+ <$ instead of $\mathcal{L} + \mathcal{O}_1$ for the class of queries definable in $\mathcal{L}$ in the presence of built-in order relation.

3 Local queries over finite models

A number of notions of locality have been introduced in finite-model theory in order to prove inexpressibility results, cf. [9, 12, 11, 7, 16]. Here we describe one of these notions, which will serve as a main technical tool.

Given a structure $\mathcal{A}$, its *Gaifman graph* [9, 12] $\mathcal{G}(\mathcal{A})$ is defined as $\langle A, E \rangle$ where (a, b) is in E iff there is a tuple $\vec{t} \in R_i^{\mathcal{A}}$ for some i such that both a and b are in $\vec{t}$. For example, if $\mathcal{A}$ is a graph itself, then $\mathcal{G}(\mathcal{A})$ is its reflexive-symmetric closure. The distance $d(a, b)$ is defined as the length of the shortest path from a to b in $\mathcal{G}(\mathcal{A})$; we assume $d(a, a) = 0$. Given $a \in A$, its r-*sphere* $S_r^{\mathcal{A}}(a)$ is $\{b \in A \mid d(a, b) \le r\}$. For a tuple $\vec{t}$, define $S_r^{\mathcal{A}}(\vec{t})$ as $\bigcup_{a \in \vec{t}} S_r^{\mathcal{A}}(a)$.

For $\vec{t} = (t_1, \ldots, t_n)$, its r-*neighborhood* $N_r^{\mathcal{A}}(\vec{t})$ is defined as a σ_n structure

$$\langle S_r^{\mathcal{A}}(\vec{t}), R_1^{\mathcal{A}} \cap S_r^{\mathcal{A}}(\vec{t})^{p_1}, \ldots, R_k^{\mathcal{A}} \cap S_r^{\mathcal{A}}(\vec{t})^{p_k}, t_1, \ldots, t_n \rangle$$

That is, the carrier of $N_r^{\mathcal{A}}(\vec{t})$ is $S_r^{\mathcal{A}}(\vec{t})$, the interpretation of the σ-relations is obtained by restricting them from $\mathcal{A}$ to the carrier, and the n extra constants are the elements of $\vec{t}$. If $\mathcal{A}$ is understood, we write $S_r(\vec{t})$ and $N_r(\vec{t})$.

We use the notation $\vec{a} \approx_r^{\mathcal{A}} \vec{b}$, or $\vec{a} \approx_r \vec{b}$ if $\mathcal{A}$ is understood, if $N_r^{\mathcal{A}}(\vec{a})$ and $N_r^{\mathcal{A}}(\vec{b})$ are isomorphic. Note that an isomorphism between these maps ith component of $\vec{a}$ onto ith component of $\vec{b}$.

A formula $\psi(x_1, \ldots, x_m)$ in a logic $\mathcal{L}$ is called *local* [7, 16] if there exists $r > 0$ such that, for every $\mathcal{A} \in \text{STRUCT}[\sigma]$ and for every two m-ary vectors $\vec{a}, \vec{b}$ of elements of A, $N_r(\vec{a}) \cong N_r(\vec{b})$ implies $\mathcal{A} \models \psi(\vec{a})$ iff $\mathcal{A} \models \psi(\vec{b})$. The minimum r for which this holds is called the *locality rank* of ψ, and is denoted by $\text{lr}(\psi)$. Based on results of [13, 17], the following was shown in [16]:

Fact 4 *Every* FO(C) *formula without free second-sort variables is local, and every* FO(Q_u) *formula is local.* $\qquad\square$

4 Expressivity bounds for FO(C) and FO(Q_u) in the presence of relations of large degree

We start by giving a general technique for proving expressivity bounds for local logics. Then we apply it to FO(C) to prove our main result that DLOG-complete problems (in particular, deterministic transitive closure) cannot be expressed in it in the presence of relations that are very close to linear orderings. In particular, it will follow that DLOG $\not\subseteq$ FO(C) + $\mathcal{O}_k$ for any $k > 1$.

Proving expressivity bounds in local logics Let q be a query that takes structures from STRUCT$[\sigma]$ as inputs and returns m-ary relations (e.g, transitive closure takes graphs from STRUCT$[\sigma_{gr}]$ as inputs and returns graphs). Let $\mathcal{R}$ be a class of relations, and $\mathcal{L}$ a logic. Suppose we want to prove that $q \notin \mathcal{L} + \mathcal{R}$. For that purpose, we introduce two conditions.

Def$_{\mathcal{L}[\sigma]}[\mathcal{R}, \mathcal{C}]$ Assume $\mathcal{C} \subseteq$ STRUCT$[\sigma]$. Then there exists a number n and an $\mathcal{L}$ formula φ in the vocabulary σ such that $\varphi[A] \in \mathcal{R}$ for every $A \in \mathcal{C}$ with $|A| > n$.

Sep$_{\mathcal{L}[\sigma]}[q, \mathcal{C}]$ For any two numbers $r, n > 0$, there exists $A \in \mathcal{C}$ with $|A| > n$ and two m-ary vectors $\vec{a}, \vec{b}$ of elements of A such that $\vec{a} \approx_r^A \vec{b}$, $\vec{a} \in q(A)$ and $\vec{b} \notin q(A)$.

That is, **Def$_{\mathcal{L}[\sigma]}[\mathcal{R}, \mathcal{C}]$** says that relations from $\mathcal{R}$ are definable by σ-formulae of $\mathcal{L}$ on large enough structures from $\mathcal{C}$, and **Sep$_{\mathcal{L}[\sigma]}[q, \mathcal{C}]$** says that q separates similarly looking (in a local neighborhood) tuples on arbitrarily large structures from $\mathcal{C}$.

Theorem 1. *Assume that $\mathcal{L}$ is FO, or FO(C), or FO($\mathbf{Q}_u$). Suppose for a given query q on σ-structures, one can find $\mathcal{C} \subseteq$ STRUCT$[\sigma]$ such that both* **Def$_{\mathcal{L}[\sigma]}[\mathcal{R}, \mathcal{C}]$** *and* **Sep$_{\mathcal{L}[\sigma]}[q, \mathcal{C}]$** *hold. Then $q \notin \mathcal{L} + \mathcal{R}$.*

Proof: Assume that q is definable in $\mathcal{L} + \mathcal{R}$ by a formula ψ in the vocabulary that includes σ and a symbol R for the relation from $\mathcal{R}$. Let ψ' be obtained from ψ by replacing each occurrence of $R(\cdots)$ by $\varphi(\cdots)$, where φ is given by **Def$_{\mathcal{L}[\sigma]}[\mathcal{R}, \mathcal{C}]$**. Then, for every $A \in \mathcal{C}$ with $|A| > n$, we have $\psi'[A] = q(A)$. Note that ψ' is a $\mathcal{L}$-formula in the vocabulary σ. By Fact 4, ψ' is local. Let $r = \mathrm{lr}(\psi')$. By **Sep$_{\mathcal{L}[\sigma]}[q, \mathcal{C}]$**, we find a structure $A \in \mathcal{C}$ such that, for two m-vectors, $\vec{a}$ and $\vec{b}$, one has $\vec{a} \approx_r \vec{b}$, $\vec{a} \in q(A)$ and $\vec{b} \notin q(A)$. Then $A \models \neg(\psi'(\vec{a}) \leftrightarrow \psi'(\vec{b}))$, which contradicts locality. $\square$

Note that this theorem can be straightforwardly extended to the case of several built-in relations of possibly different arities, by considering $\vec{\mathcal{R}}$ instead of $\mathcal{R}$, where $\vec{\mathcal{R}}$ is a tuple of classes of auxiliary relations. Then **Def$_{\mathcal{L}[\sigma]}[\vec{\mathcal{R}}, \mathcal{C}]$** says that relations from each component of $\vec{\mathcal{R}}$ can be defined by a σ-formula of $\mathcal{L}$ on sufficiently large structures from $\mathcal{C}$.

Theorem 1 can also be extended to any local logic that is closed under first-order operations and allows a notion of substitution in a way that was used in the proof. All naturally occurring extensions of FO that are known to be local have these properties.

Lower bounds for (deterministic) transitive closure: FO(C) and FO($\mathbf{Q}_u$) with "thin" preorders Deterministic transitive closure of a graph is obtained by closing its deterministic paths, that is, if $G = \langle V, E \rangle$ is a directed graph, then $dtc(G) = \langle V, E' \rangle$ where $(a, b) \in E'$ iff either $(a, b) \in E$ or there exists a path $(a, a_1), (a_1, a_2), \ldots, (a_{n-1}, a_n), (a_n, b) \in E$ such that a and each a_i, $i = 1, \ldots, n$ have outdegree 1. We shall use tc to denote the transitive closure of a graph.

According to [14], FO + dtc+ $<$ captures DLOG and FO + tc+ $<$ captures NLOG.

We next define the class of relations that we view as "almost linear orders." Let $g : \mathbb{N} \to \mathbb{R}$ be a nondecreasing function. Then $\mathcal{P}_g$ is the class of binary relations (A, R) such that there is a partition $A = B \cup C$ with the following properties: (1) $| B | \geq n - g(n)$; (2) R restricted to B is a linear order; (3) R restricted to C is a relation from $\mathcal{O}_2$, that is, a preorder where every equivalence class has at most two elements; and (4) For any $b \in B$ and $c \in C$, $(b, c) \in R$.

See Figure 1 for a preorder from $\mathcal{P}_g$. Actually, we show the associated successor relation in the Figure. A relation from $\mathcal{P}_g$ is really the transitive closure of the one shown in Figure 1. Intuitively, if g is very small, then this is the least possible "damage" that can be done to a linear ordering. In the result below, g can indeed be taken to be very small, for example, it could be $\log\log\ldots\log n$.

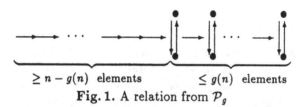

$\geq n - g(n)$ elements $\leq g(n)$ elements

Fig. 1. A relation from $\mathcal{P}_g$

Theorem 2. *Let $g : \mathbb{N} \to \mathbb{R}$ be a nondecreasing function that is not bounded by a constant. Then (deterministic) transitive closure is not definable in FO(C) or FO(Q_u) in the presence of relations from $\mathcal{P}_g$.* □

Corollary 3. *(Deterministic) transitive closure is not definable in FO(C) or FO(Q_u) in the presence of relations from $\mathcal{O}_k$ for any $k > 1$. In particular, DLOG $\not\subseteq$ FO(C) + $\mathcal{O}_k$.* □

This can be compared with the results of [6] where it was shown that first-order with fixpoint and counting fails to express some polynomial-time problems even in the presence of relations from $\mathcal{O}_4$ (of course first-order with fixpoint captures polynomial time in the presence of an order relation, cf. [9]).

To prove Theorem 2, we need the following:

Proposition 4. *Let q be (deterministic) transitive closure, and $\mathcal{L}$ be FO(C) or FO(Q_u). Assume that $g : \mathbb{N} \to \mathbb{R}$ is a nondecreasing function that is not bounded by a constant. Then there exists a class $\mathcal{C}$ of graphs such that both $\mathbf{Def}_{\mathcal{L}[\sigma_{gr}]}(\mathcal{P}_g, \mathcal{C})$ and $\mathbf{Sep}_{\mathcal{L}[\sigma_{gr}]}(q, \mathcal{C})$ hold.*

Bushy trees In what follows, trees are directed graphs with edges oriented from the root to the leaves. A tree is called *bushy* if, for any two non-leaf nodes $x \neq y$, $out\text{-}deg(x) \neq out\text{-}deg(y)$. A *k-bushy* tree is a bushy tree in which every path from the root to a leaf has the same length k. A *canonical k*-bushy tree is obtained as follows. We start with the root of outdegree 2. Its first child has 3 children, the second child has 4 children. This completes level 2, and we now have 7 elements at level 3. They will have 5, 6, 7, 8, 9, 10 and 11 children, respectively. This gives us 56 nodes at level 4, which will have 12(=11+1), 13, ..., 67(=11+56) children, resp. We continue until we fully filled all k levels. See the picture in Figure 2. We use B_k to denote the canonical k-bushy tree.

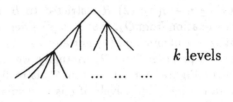

k levels

Fig. 2. Canonical k-bushy tree

Proof sketch of Proposition 4: We start by defining a family of graphs $G_{d,k}^0$, $d, k \in \mathbb{N}_+$, $d > k + 1$. Let s_k be the total number of nodes in the canonical k-bushy tree. The root of $G_{d,k}^0$ has $s_k + 1$ children. Two of them are roots of two copies of a canonical k-bushy tree, denoted here by B_k^1 and B_k^2. To other $s_k - 1$ nodes at the second level, we give $s_k + 2, s_k + 3, \ldots, s_k + (s_k - 1) = 2s_k - 1$ children respectively. Now, to those nodes at the second level that do not belong to the two canonical k-bushy trees, we give $2s_k, 2s_k + 1, \ldots$ children, as before, increasing the number by one. We continue this process until we fully fill the $k + 1$st level. After that, we look at the node at the level k with most children, say M of them, and start giving nodes at the $k+1$st level $M+1, M+2, M+3, \ldots$ children. We stop the process when we completely fill the dth level.

This is the graph $G_{d,k}^0$. Note that every two non-leaf nodes $x \neq y$ have different outdegrees, unless one of them is in B_k^1 and the other is in B_k^2. We define $G_{d,k}$ by adding graph edges that form a linear ordering on the leaves. When we speak of "leaf nodes" of $G_{d,k}$, we actually mean the leaf nodes of $G_{d,k}^0$.

Let B_1° and B_2° be the sets of non-leaf nodes in B_k^1 and B_k^2. Then, for any two distinct nodes $x, y \notin B_1^\circ \cup B_2^\circ$, it is the case that $(in\text{-}deg(x), out\text{-}deg(x)) \neq (in\text{-}deg(y), out\text{-}deg(y))$. Next, define two binary relations on the set of nodes: $x \prec_0 y$ iff $in\text{-}deg(x) < in\text{-}deg(y)$ or $in\text{-}deg(x) = in\text{-}deg(y)$ and $out\text{-}deg(x) < out\text{-}deg(y)$. Let B° be $B_1^\circ \cup B_2^\circ$. Then we let $x \prec y$ iff either $x \notin B^\circ, y \in B^\circ$, or $x, y \in B^\circ$ and $x \prec_0 y$, or $x, y \notin B^\circ$ and $x \prec_0 y$. This binary relation $\prec$ is definable in FO(C) and FO(Q_u).

For a given k, let d_k be the smallest number $d > k + 1$ such that $2s_k < g(n)$ for all $n \geq N_{d,k}$, where $N_{d,k}$ is the total number of nodes in $G_{d,k}$. Since for every fixed k, $N_{d,k}$ grows with d, and g is nondecreasing, d_k is well-defined and depends only on k. Let $\mathcal{C}_g = \{G_{d,k} \mid d, k \in \mathbb{N}_+, d > d_k\}$. The rest of the proof is to verify that both $\mathbf{Def}_{\mathcal{L}[\sigma_{gr}]}[\mathcal{P}_g, \mathcal{C}_g]$ and $\mathbf{Sep}_{\mathcal{L}[\sigma_{gr}]}[tc, \mathcal{C}_g]$ hold. To complete the proof for deterministic transitive closure, we just reverse all the edges of $G_{d,k}$, to make all the paths not involving leaves deterministic. □

5 Limitations of the technique

To summarize what has been achieved so far, we know that $FO(C) + \mathcal{O}_1 = TC^0$, and the above results show that for any $k > 1$, DLOG $\not\subseteq FO(C) + \mathcal{O}_k$. Furthermore, DLOG $\not\subseteq FO(C) + \mathcal{P}_g$ for any nondecreasing function g that is not bounded by a constant. Thus, one may ask if the techniques can be pushed further to prove expressivity bounds for $FO(C) + <$.

The lemma below shows that removing the assumption that g is not bounded by a constant is essentially equivalent to having a linear order:

Lemma 5. *Let $g(n)$ be bounded by some constant M for all $n \in \mathbb{N}$. Assume that (deterministic) transitive closure is not in $FO(C) + \mathcal{P}_g$. Then it is not in $FO(C) + <$ either.* □

Thus, a possible avenue for attacking the problem of expressivity with linear order seems to be the following: try to find a class of structures C so that both $\mathbf{Def}_{\mathcal{L}}[\mathcal{P}_M, C]$ and $\mathbf{Sep}_{\mathcal{L}}[q, C]$ would hold, where q is tc, or dtc, or any other query we want to show to be outside of $FO(C) + <$. Here we use $\mathcal{P}_M$ to denote $\mathcal{P}_g$ where $g(n) < M$ for all n.

If we were able to find such a class C, it would show that $q \notin FO(C) + <$. Unfortunately, as the following theorem shows, no such class exists!

Theorem 6. *Let M be a constant, and let q be a query invariant under isomorphism. Let $\mathcal{L}$ be $FO(C)$ or $FO(\mathbf{Q}_u)$. Then there does not exist a class of structures C such that both $\mathbf{Def}_{\mathcal{L}}[\mathcal{P}_M, C]$ and $\mathbf{Sep}_{\mathcal{L}}[q, C]$ hold.*

Proof sketch: Let $M = 1$; then $\mathcal{P}_M$ is a linear order. Assume $\mathbf{Def}_{\mathcal{L}}[\mathcal{P}_M, C]$ and $\mathbf{Sep}_{\mathcal{L}}[q, C]$ hold. Then there is a formula $\varphi(x, y)$ that defines a linear order $<_{\mathcal{A}}$ on each $\mathcal{A}$ with $|A| > n$. Let $r = \mathsf{lr}(\varphi)$ and $d = 3r + 1$. Using $\mathbf{Sep}_{\mathcal{L}}[q, C]$, we can find big enough $\mathcal{A}$ such that $<_{\mathcal{A}}$ is an order and there exist $a, b \in A$ with $a \neq b$ and $a \approx_d^{\mathcal{A}} b$. By [7, 16], there exists a permutation π on A such that $(a, x) \approx_r (b, \pi(x))$ for all $x \in A$. Thus, $a <_{\mathcal{A}} x$ iff $b <_{\mathcal{A}} \pi(x)$ for all x, which is impossible for $a \neq b$, since A is finite. For $M > 1$, $\mathbf{Sep}_{\mathcal{L}}[q, C]$ is used to show the existence of a structure with more than M pairs $a_i \approx_d b_i$, which implies the existence of $a \approx_d b$ with $a <_{\mathcal{A}} b$ and $b \not<_{\mathcal{A}} a$. Then the above proof applies. □

6 On the relative expressive power of auxiliary relations

We give here a few comments about the murky area of expressivity with (pre)orders vs. expressivity without (pre)orders. As was mentioned before, it is known that $FO \subsetneq FO + <$. Note that by $FO + <$ we mean the class of *order-independent* queries in $FO + <$, so this is not a trivial observation. A similar result for $FO(C)$ is implicit in [5]; it also follows from an example due to M. Otto [18].

Proposition 7 (Benedikt-Keisler). $FO(C) \subsetneq FO(C) + <$.

Proof sketch: Consider structures $\mathcal{A} = \langle A, R, U \rangle$ where R is binary and U is unary. The separating query q is the following: If R is an equivalence relation, does the number of distinct sizes of equivalence classes in R equal to the cardinality of U? It was shown to be inexperessible in FO(C) in [5], and we can show that it is definable in FO(C)+ <. □

However, this does not shed any light on what kind of examples might exist (if there are any) in (FO(C)+ <) − (FO(C) + $\mathcal{O}_k$), $k > 1$, as the separating example of Proposition 7 is definable in FO(C) + $\mathcal{O}_k$. Thus, we have

Proposition 8. FO(C) $\subsetneq$ FO(C) + $\mathcal{O}_k$. □

To find a separation from FO(C)+ <, we can only use a small class of preorders which, in a sense, do not have equivalence classes comparable to the size of the universe; this result will be stated in the full version. While this gives us some partial results about expressivity of FO(C) with (pre)orders, it is still not clear how to prove bounds for FO(C)+ < and ultimately for TC^0. We conclude by presenting a query whose inexpressibility in FO(C) (note the absence of order!) would imply bounds on TC^0.

Proposition 9. *a) If there is no* FO(C) *query that defines transitive closure on bushy trees, then* $TC^0 \subsetneq$ NLOG.
b) If there is no FO(C) *query that defines deterministic transitive closure on inverses of bushy trees, then* $TC^0 \subsetneq$ DLOG. □

While we do not know whether queries of Proposition 9 are definable in FO(C), we can give two partial results for *canonical* bushy trees.

Proposition 10. *Transitive closure of canonical bushy trees is definable in* FO(Q_u), *but not in* FO. □

7   Open problems

It still remains an open problem to prove expressivity bounds in the presence of an order relation. We believe that a descriptive complexity approach holds a promise, partly because it does not appear to fit the general scheme of natural proofs of Razborov and Rudich [20]. It is partly the case because we do not know how to translate expressivity bounds with various kinds of auxiliary relations into lower bounds for circuits (if indeed such a translation is possible). Another attempt to interpret such expressivity bounds in terms of circuit complexity is to find different notions of uniformity that will perhaps correspond to different auxiliary relations. We have not explored this yet.

An approach to proving lower bounds for TC^0 circuits, based on arithmetic circuits, was recently proposed in [1]. It may avoid the problems presented in [20]. In [1], the strongest results are obtained in the P-uniform and nonuniform setting, and the weakest for DLOGTIME-uniformity. Our results only apply to a fragment of DLOGTIME-uniform TC^0, so they appear to be of different nature than those in [1].

In conclusion, we believe that recent advances in the study of expressive power of logics with counting and unary quantifiers make it promising to use the tools of finite-model theory and descriptive complexity to attack some of the hard separation problems.

Acknowledgements We thank anonymous reviewers for their comments.

References

1. M. Agrawal, E. Allender and S. Datta. On TC^0, AC^0, and arithmetic circuits. In *Proc. 12th IEEE Conf. on Computational Complexity*, 1997.
2. E. Allender. Circuit complexity before the dawn of the new millennium. In *FST&TCS'96*, Springer LNCS vol. 1180, 1996, 1-18.
3. S. Abiteboul, R. Hull, V. Vianu, *Foundations of Databases*, Addison Wesley, 1995.
4. D.A. Barrington, N. Immerman, H. Straubing. On uniformity within NC^1. *JCSS*, 41:274-306,1990.
5. M. Benedikt, H.J. Keisler. On expressive power of unary counters. *ICDT'97*, Springer LNCS 1186, 1997, pages 291-305.
6. J. Cai, M. Fürer and N. Immerman. An optimal lower bound on the number of variables for graph identification. *Combinatorica* 12 (1992), 389-410.
7. G. Dong, L. Libkin, L. Wong. Local properties of query languages. *Proc. Int. Conf. on Database Theory*, Springer LNCS 1186, 1997, pages 140-154.
8. H.-D. Ebbinghaus. Extended logics: the general framework. In J. Barwise and S. Feferman, editors, *Model-Theoretic Logics*, Springer-Verlag, 1985, pages 25-76.
9. H.-D. Ebbinghaus and J. Flum. *Finite Model Theory*. Springer Verlag, 1995.
10. K. Etessami. Counting quantifiers, successor relations, and logarithmic space, *JCSS*, 54 (1997), 400-411.
11. R. Fagin, L. Stockmeyer, M. Vardi, On monadic NP vs monadic co-NP, *Information and Computation*, 120 (1994), 78-92.
12. H. Gaifman, On local and non-local properties, *in* "Proceedings of the Herbrand Symposium, Logic Colloquium '81," North Holland, 1982.
13. L. Hella. Logical hierarchies in PTIME. *Inform.& Comput.*, 129 (1996), 1-19.
14. N. Immerman. Languages that capture complexity classes. *SIAM Journal of Computing* 16 (1987), 760-778.
15. N. Immerman and E. Lander. Describing graphs: A first order approach to graph canonization. In *"Complexity Theory Retrospective"*, Springer Verlag, Berlin, 1990.
16. L. Libkin. On the forms of locality over finite models. In *LICS'97*, pages 204-215. Full paper "Notions of locality and their logical characterization over finite models" by L. Hella, L. Libkin and J. Nurmonen is available as Bell Labs Technical Memo.
17. J. Nurmonen. On winning strategies with unary quantifiers. *J. Logic and Computation*, 6 (1996), 779-798.
18. M. Otto. Private communication. DIMACS, July 1997.
19. I. Parberry and G. Schnitger. Parallel computation and threshold functions. *JCSS* 36 (1988), 278-302.
20. A. Razborov and S. Rudich. Natural proofs. *JCSS* 55 (1997), 24-35.
21. T. Schwentick. Graph connectivity and monadic NP. *FOCS'94*, pages 614-622.
22. T. Schwentick. Graph connectivity, monadic NP and built-in relations of moderate degree. In *ICALP'95*, Springer LNCS 944, 1995, pages 405-416.

On the Structure of Valiant's Complexity Classes

Peter Bürgisser

Institut für Mathematik, Universität Zürich, Winterthurerstr. 190,
CH-8057 Zürich, Switzerland. E-mail: buerg@amath.unizh.ch

Abstract. In [25, 27] Valiant developed an algebraic analogue of the theory of NP-completeness for computations with polynomials over a field. We further develop this theory in the spirit of structural complexity and obtain analogues of well-known results by Baker, Gill, and Solovay [1], Ladner [18], and Schöning [23, 24].
We show that if Valiant's hypothesis is true, then there is a p-definable family, which is neither p-computable nor VNP-complete. More generally, we define the posets of p-degrees and c-degrees of p-definable families and prove that any countable poset can embedded in either of them, provided Valiant's hypothesis is true. Moreover, we establish the existence of minimal pairs for VP in VNP.
Over finite fields, we give a *specific* example of a family of polynomials which is neither VNP-complete nor p-computable, provided the polynomial hierarchy does not collapse.
We define relativized complexity classes VP^h and VNP^h and construct complete families in these classes. Moreover, we prove that there is a p-family h satisfying $\mathrm{VP}^h = \mathrm{VNP}^h$.

1 Introduction

One of the most important developments in theoretical computer science is the concept of NP-completeness. Recently, initiated by a paper by Blum, Shub, and Smale [5] (BSS-model), there has been a growing interest in investigating such concepts over general algebraic structures, with the purpose of classifying the complexity of continous problems. But already ten years earlier, Valiant [25, 27] had developed a convincing analogue of the theory of NP-completeness in an entirely algebraic framework, in connection with his famous hardness result for the permanent [26]. In fact, the polynomial enumerators of many NP-complete graph problems turn out to be complete in Valiant's sense (cf. [7]). The major differences between the BSS-model and Valiant's model are the absence of uniformity conditions in the latter, and the fact that only straight-line computations are considered (no branching). Both structured models are adapted to the framework of polynomial computations, and we believe that they will be useful for classifing the intrinsic complexity of problems in computer algebra (compare Heintz and Morgenstern [15]).

Our goal is to further develop Valiant's approach along the lines of discrete structural complexity theory.

We show that if Valiant's hypothesis is true, then, over any field, there is a p-definable family which is neither p-computable nor VNP-complete. A similar result due to Ladner [18] in the classical P-NP-setting is well-known. Ladner's proof is a diagonalization argument based on an effective enumeration of all polynomial time Turing machines. However, over uncountable structures, this approach causes problems. Malajovich and Meer [20] were able to carry over Ladner's theorem to the setting of the BSS-model over the complex numbers by employing a transfer principle due to Blum et al. [4], which allows a reduction to the countable field of algebraic numbers. The corresponding question over the reals is still open, but it is known to be true in the nonuniform BSS-setting, cf. Ben-David et al. [2]. (For a detailed treatement of these questions in a general model-theoretic context see Chapuis and Koiran [10].)

In our proof of the analogue of Ladner's theorem, the nonuniformity of Valiant's model is essential, since that allows an enumeration of all polynomial straight-line programs over a possibly uncountable field "in blocks". In Sect. 3 we formalize this idea in a general abstract setting by studying certain compatible quasi-orders on the set $\Omega^{\mathbb{N}}$ of families in a quasi-ordered set $(\Omega, \leq)$, and by proving an analogue of Schöning's uniform diagonalization theorem [23]. Hereby, our notion of a *nice subset* of $\Omega^{\mathbb{N}}$ serves as a substitute for the notion of a recursively presentable set. Based on this theorem, we proceed in Sect. 4 by providing an elegant proof that any countable poset can be embedded in the poset of degrees corresponding to a compatible quasi-order. This is applied in Sect. 5 in Valiant's setting to an analogue of the polynomial Turing reduction (c-reduction), as well as to the p-projection. A similar result in the classical P-NP-setting for polynomial Turing or polynomial many-one degrees was stated by Ladner [18]; however, he presented a proof only in a special case. We further remark that the existence of minimal pairs for VP in VNP can be easily guaranteed by our approach. (See Landweber et al. [19] and Schöning [24] for corresponding results in the classical P-NP setting.)

A striking discovery is that we can describe *specific* families of polynomials which are neither VNP-complete nor p-computable. In fact, the family of cut enumerators over a finite field of characteristic p has this property, provided Mod_pNP is not contained in P/poly. (The latter condition is satisfied if the polynomial hierarchy does not collapse at the second level.) In the classical, as well as in the BSS-setting, only artificial problems are known to have such properties. This is discussed in Section 6.

Finally, in Sect. 7, we define relative versions VP^h and VNP^h of Valiant's complexity classes with respect to a p-family h. For these, we have obtained some results in the spirit of Baker et al. [1]. Over infinite fields, we can construct (artificial) VP^h-complete and VNP^h-complete families with respect to p-projection. Moreover, we can construct a p-family h satisfying $\text{VP}^h = \text{VNP}^h$. We do not know whether there exists a p-family h such that $\text{VP}^h \neq \text{VNP}^h$. Let us remark that Emerson [12] has transfered the results of Baker et al. [1] to the BSS-model.

Acknowledgement. Thanks go to Michael Clausen for encouraging me to investigate these questions.

2 Valiant's Model

We briefly recall the main features of Valiant's algebraic model. For detailed expositions see von zur Gathen [14] and [8, Chap. 21].

In this section $\Omega := k[X_1, X_2, \ldots]$ denotes the polynomial ring over a fixed field k in countably many variables X_i. A *p-family* over k is a sequence $f = (f_n) \in \Omega^{\mathbb{N}}$ of multivariate polynomials such that the number of variables as well as the degree of f_n are polynomially bounded (*p-bounded*) functions of n. An example of a *p*-family is the permanent family PER $= (\text{PER}_n)$, where PER_n is the permanent of an n by n matrix with distinct indeterminate entries.

Let $L(f_n)$ denote the total complexity of f_n, that is, the minimum number of arithmetic operations $+, -, *$ sufficient to compute f_n from the variables X_i and constants in k by a straight-line program. We call a *p*-family f *p-computable* iff $n \mapsto L(f_n)$ is *p*-bounded. The *p*-computable families constitute the complexity class VP. We remark that the restriction to *p*-bounded degrees is a severe one: although X^{2^n} can be computed with only n multiplications, the corresponding sequence is not considered to be *p*-computable, as the degrees grow exponentially.

A *p*-family $f = (f_n)$ is called *p-definable* iff there exists a *p*-computable family $g = (g_n)$, $g_n \in k[X_1, \ldots, X_{u(n)}]$, and a *p*-bounded function $t: \mathbb{N} \to \mathbb{N}$ such that for all n

$$f_n(X_1, \ldots, X_{v(n)}) = \sum_{e \in \{0,1\}^{u(n)-v(n)}} g_n(X_1, \ldots, X_{v(n)}, e_{v(n)+1}, \ldots, e_{u(n)}) \ . \quad (1)$$

The set of *p*-definable families form the complexity class VNP. The class VP is obviously contained in VNP, and *Valiant's hypothesis* claims that this inclusion is strict. We can consider this as an algebraic counterpart of the well-known hypothesis P $\neq$ NP due to Cook [11]. Let us mention the following recent result due to the author, which reveals a close connection between these two hypotheses. The crucial step in its proof is the elimination of constants in the field, which relies on a recent method developed by Koiran [17].

Theorem 2.1 ([6]). *If Valiant's hypothesis were false over the field k, then the nonuniform versions of the complexity classes NC, P, NP, and PH would be equal. In particular, the polynomial hierarchy would collapse to the second level. Hereby, we assume that k is finite or of characteristic zero; in the second case we assume a generalized Riemann hypothesis.*

We define now a quasi-order $\leq_p$ called *p-projection* on the set $\Omega^{\mathbb{N}}$ of families in Ω. Let us call a function $t: \mathbb{N} \to \mathbb{N}$ *p-bounded from above and below* iff there exists some $c > 0$ such that $n^{1/c} - c \leq t(n) \leq n^c + c$ for all n. A polynomial f_n is said to be a *projection* of a polynomial $g_m \in k[X_1, \ldots, X_u]$, for short $f_n \leq g_m$, iff

$$f_n(X_1, \ldots, X_{v(n)}) = g_m(a_1, \ldots, a_u)$$

for some $a_i \in k \cup \{X_1, \ldots, X_{v(n)}\}$. That is, f_n can be derived from g_m through substitution by indeterminates and constants. We call a *p*-family $f = (f_n)$ a *p-projection* of $g = (g_m)$, in symbols $f \leq_p g$, iff there exists a function $t: \mathbb{N} \to \mathbb{N}$ which is *p*-bounded from above and below such that

$$\exists n_0 \ \forall n \geq n_0 : f_n \leq g_{t(n)} . \tag{2}$$

(Our definition of $\leq_p$ differs slightly from the one given in [25] as we also require a lower bound on the growth of the functions t.) Finally, a *p*-family $g \in$ VNP is called *VNP-complete* (with regard to *p*-projection) iff any $f \in$ VNP is a *p*-projection of g.

In [25] Valiant obtained the remarkable result that the permanent family (if char$k \neq 2$) and the family of Hamilton cycle polynomials are VNP-complete. It turns out that the "polynomial enumerators" of several NP-complete graph problems like Clique, H-factors, Hamilton cycles in planar graphs etc. are VNP-complete as well (cf. [7]).

3 An Abstract Diagonalization Theorem

Let a quasi-ordered set $(\Omega, \leq)$ be fixed. Elements of the set $\Omega^{\mathbb{N}}$ of sequences in Ω will be called *families* in the sequel. We may formally define a quasi-order $\leq_p$ (the abstract *p*-projection) on the set $\Omega^{\mathbb{N}}$ of families as in (2). Two families f and g are said to be *p-equivalent* iff $f \leq_p g$ and $g \leq_p f$. We call the equivalence classes *p-degrees* and denote by $\mathcal{D}_p$ the poset of all *p*-degrees with the partial order induced by $\leq_p$. $f <_p g$ shall mean that $f \leq_p g$ but not $g \leq_p f$. The *join* $f \cup g$ of two families $f, g \in \Omega^{\mathbb{N}}$ is defined as

$$f \cup g := (f_0, g_0, f_1, g_1, f_2, g_2, \ldots) .$$

It is easy to see that the join of two *p*-degrees is well-defined and that it is the smallest upper bound of these *p*-degrees in $\mathcal{D}_p$. The poset $\mathcal{D}_p$ of *p*-degrees is thus a join-semilattice.

Definition 3.1. We call a subset of $\Omega^{\mathbb{N}}$ *nice* iff it can be written as a countable union of cartesian products $\prod_{n \in \mathbb{N}} F_n$, where $F_n \subseteq \Omega$.

It is clear that countable unions and finite intersections of nice subsets of $\Omega^{\mathbb{N}}$ are again nice. However, one can show that the nice subsets are not closed under the formation of complements and countable intersections. Every nice set is measurable with respect to the product of σ-algebras $\otimes_n 2^{\Omega}$, but one can show that the converse is not true.

We call two families (f_n) and (g_n) *equal almost everywhere* iff $f_n = g_n$ for all but finitely many n. This is clearly an equivalence relation which is well defined on *p*-degrees. A subset $\mathcal{C} \subseteq \Omega^{\mathbb{N}}$ is called *closed under finite variation* iff $f \in \mathcal{C}$ implies $g \in \mathcal{C}$, provided f and g are equal almost everywhere.

The following theorem is inspired by Schöning's "uniform diagonalization theorem" [23] and can be proved similarly. We note that the nice sets serve as a substitute for the recursively presentable sets appearing there.

Theorem 3.2. *Let $\mathcal{F}, \mathcal{G}$ be nice subsets of $\Omega^{\mathbb{N}}$ which are closed under finite variation. Moreover, let $f, g \in \Omega^{\mathbb{N}}$ such that $f \notin \mathcal{F}$ and $g \notin \mathcal{G}$. Then there exists $h \in \Omega^{\mathbb{N}}$ satisfying $h \leq_p f \cup g$ and $h \notin \mathcal{F} \cup \mathcal{G}$.*

4 An Abstract Embedding Theorem

Again let a quasi-ordered set $(\Omega, \leq)$ be fixed and denote by $\leq_p$ the corresponding abstract p-projection. We extend our discussion to any quasi-order on $\Omega^{\mathbb{N}}$ which satisfies certain some compatibility conditions.

Definition 4.1. A quasi-order $\leq_c$ of $\Omega^{\mathbb{N}}$ is called *compatible*, iff the following conditions are satisfied:

(a) $\forall f, g : f \leq_p g \Rightarrow f \leq_c g$.
(b) $\forall f, g, h : f \leq_c h, g \leq_c h \Rightarrow f \cup g \leq_c h$.
(c) The sets $\{h \mid h \leq_c g\}$ and $\{h \mid f \leq_c h\}$ are nice for all $f, g \in \Omega^{\mathbb{N}}$.

It turns out that the quasi-order $\leq_p$ is compatible.

In the sequel, let a compatible quasi-order $\leq_c$ on $\Omega^{\mathbb{N}}$ be fixed. We call two families f and g *c-equivalent* iff $f \leq_c g$ and $g \leq_c f$. The corresponding equivalence classes are a union of certain p-degrees and called *c-degrees*. We denote by $\mathcal{D}_c$ the poset of all c-degrees with the partial order induced by $\leq_c$. $f <_c g$ means that $f \leq_c g$, but f and g are not c-equivalent. We say that $f <_p g$ *strongly* iff $f \leq_p g$ and $f <_c g$.

Let $(X, \subseteq)$ be a poset. A map $\varphi \colon X \to \Omega^{\mathbb{N}}$ is called an *embedding* of X in $\Omega^{\mathbb{N}}$ (with respect to $\leq_c$) if $x \subseteq y$ implies $\varphi(x) \leq_c \varphi(y)$ and vice versa. We call φ a *strong embedding* iff we have for all $x, y \in X$

$$x \subseteq y \Rightarrow \varphi(x) \leq_p \varphi(y) \text{ and } \varphi(x) \leq_c \varphi(y) \Rightarrow x \subseteq y \ .$$

The following theorem implies immediately that any countable poset can be embedded in both of the posets $\mathcal{D}_p$ and $\mathcal{D}_c$, provided $\mathcal{D}_c$ does not consist of a single point only.

Theorem 4.2. *For any countable poset $(X, \subseteq)$ and elements $f, g \in \Omega^{\mathbb{N}}$ with $f <_c g$ there is an embedding $X \to \{h \mid f <_c h <_c g\}$. If additionally $f \leq_p g$, then there is a strong embedding $X \to \{h \mid f <_p h <_p g\}$.*

It easy to see that any countable poset $(X, \subseteq)$ can be embedded in a countable lattice. For the proof of the above theorem, we may therefore assume that $(X, \subseteq)$ is a lattice.

Lemma 4.3. *Let $(X, \subseteq)$ be a countable lattice. Then there exists an enumeration $x_0, x_1, x_2, \ldots$ of X such each $X_n := \{x_0, \ldots, x_n\}$ is closed under taking meets: that is, $x \cap y \in X_n$ for all $x, y \in X_n$.*

For proving Thm. 4.2 we assume the situation of this lemma and show the following claim by induction on n. There is a map $\varphi_n \colon X_n \to \Omega^{\mathbb{N}}$ satisfying

$$f <_p \cap_{x \in X_n} \varphi_n(x) \text{ strongly }, \quad \cup_{x \in X_n} \varphi_n(x) <_p g \text{ strongly },$$

and for all $x, y, y_1, \ldots, y_s \in X_n$ we have

$$x \subseteq y \Rightarrow \varphi_n(x) \leq_p \varphi_n(y), \quad \varphi_n(x) \leq_c \varphi_n(y_1) \cup \ldots \cup \varphi_n(y_s) \Rightarrow x \subseteq y_1 \cup \ldots \cup y_s .$$

This is done by invoking Thm. 3.2 several times. Due to lack of space, we can not provide more details. To give the reader an idea of how Thm. 3.2 can be applied, we just note that the induction start $n = 0$ is obtained by applying this theorem to the nice sets $\mathcal{F} := \{h \mid g \leq_c h \cup f\}$ and $\mathcal{G} := \{h \mid h \leq_c f\}$. Note that $f \notin \mathcal{F}$ and $g \notin \mathcal{G}$ by our assumption $f <_c g$.

5 Structure of Valiant's Complexity Classes

In this section, we apply our previous results to the setting of Valiant. Let $\Omega := k[X_1, X_2, \ldots]$ denote the polynomial ring over a fixed field k in countably many variables X_i and consider the projection $\leq$, which is a quasi-order on Ω. (Recall that $f \leq g$ iff f can be obtained from g by a substitution of its variables by variables or constants in k.) The corresponding quasi-order $\leq_p$ on $\Omega^{\mathbb{N}}$ is the usual p-projection.

To avoid confusions, we remark that in the future symbols like $f, g, h, \ldots$ will be used to denote either polynomials or sequences of polynomials; it will always be clear from the context what is meant.

We are going to introduce the concept of oracle computations. Let a polynomial $g \in k[X_1, \ldots, X_a]$ be given. We will consider straight-line programs which, beside the usual arithmetic operations, have the ability to evaluate the "oracle polynomial" g at previously computed values at unit cost. This can easily be formalized by considering straight-line programs Γ of type $\{+, -, *, o\}$, where the symbol o stands for the oracle operation of arity a.

Definition 5.1. The *oracle complexity* $L^g(f_1, \ldots, f_s)$ of a set of polynomials $f_1, \ldots, f_s \in \Omega$ with respect to the oracle polynomial g is the minimum number of arithmetic operations $+, -, *$ and evaluations of g (at previously computed values) that are sufficient to compute f from the indeterminates X_i and constants in k.

We introduce next the notion of c-reduction, which can be seen as an analogue of the polynomial Turing reduction for Valiant's setting. (c is an acronym for computation.) One might also interpret the p-projection as an analogue of the polynomial many-one reduction, however, the p-projection is much finer.

Definition 5.2. Let $f = (f_n)$, $g = (g_n) \in \Omega^{\mathbb{N}}$. We call f a *c-reduction* (or polynomial oracle reduction) of g, shortly $f \leq_c g$, iff there is a p-bounded function $t \colon \mathbb{N} \to \mathbb{N}$ such that $n \mapsto L^{g_{t(n)}}(f_n)$ is p-bounded.

It easy to check that $\leq_c$ is a quasi-order of $\Omega^{\mathbb{N}}$. Note that for a p-family f we have $f \leq_c 0$ iff $f \in \text{VP}$. The nonuniformity in the definition of the c-reduction allows to conclude that $\leq_c$ is compatible (compare Def. 4.1(c)).

Lemma 5.3. *The c-reduction $\leq_c$ is a compatible quasi-order on $\Omega^{\mathbb{N}}$.*

This implies for instance that VP and VNP are nice subsets of $\Omega^{\mathbb{N}}$. Let us call a p-degree or a c-degree p-*definable* iff it contains a p-definable family. Note that a p-definable p-degree consists of p-definable families only, whereas a p-definable c-degree might also contain families which are not in VNP. This is because $f \leq_c g$ and $g \in \text{VNP}$ might not imply that $f \in \text{VNP}$. We denote by $\mathcal{PD}_p$ the set of p-degrees of p-definable families and by $\mathcal{PD}_c$ the set of c-degrees of p-definable families.

The main result of this section is analogous to that of Ladner's work [18]. It follows easily from Thm. 4.2.

Theorem 5.4. *Any countable poset can be embedded in either of the posets $\mathcal{PD}_p$ or $\mathcal{PD}_c$, provided Valiant's hypothesis is true.*

Corollary 5.5. *If Valiant's hypothesis is true, then there is a p-definable family which is neither p-computable nor VNP-complete with respect to c-reduction.*

6 A Specific Family neither Complete nor p-Computable

We begin by recalling some facts from discrete complexity theory. For a prime number p the class Mod_pNP is defined as the set of languages $\{x \in \{0,1\}^* \mid \varphi(x) \equiv 1 \bmod p\}$, where $\varphi: \{0,1\}^* \to \mathbb{N}$ is a function in #P (cf. Cai and Hemachandra [9]). This generalizes the class parity polynomial time $\oplus P$, which was introduced by Papadimitriou and Zachos [22]. We remark that if $\varphi: \{0,1\}^* \to \mathbb{N}$ is #P-complete with respect to parsimonious reductions, then the corresponding language $\{x \mid \varphi(x) \equiv 1 \bmod p\}$ is Mod_pNP-complete (with respect to polynomial many-one reductions).

From a well known randomized reduction due to Valiant and Vazirani [28] one can deduce the following inclusion of nonuniform complexity classes

$$\text{NP/poly} \subseteq \text{Mod}_p\text{NP/poly} . \tag{3}$$

(For details see [6]; the notation $\mathcal{C}/\text{poly}$ stands for the nonuniform version of the complexity class $\mathcal{C}$, cf. Karp and Lipton [16].)

Let $K_n = (\underline{n}, E_n)$ denote the complete graph on the set of nodes $\underline{n} := \{1, 2, \ldots, n\}$ and let $w: E_n \to \mathbb{N}$ be a weight function. A *cut* of K_n is a partition $S := \{A, B\}$ of $\underline{n}$ into two nonempty subsets. An edge is said to be separated by S if it connects a node of A with a node of B. The weight $w(S)$ of S is defined as the sum of the weights of all edges separated by S.

The counting problem #CUT is the following: given K_n, a weight function $w: E_n \to \{0, 1, \ldots, n^3\}$, and a natural number $s < D_n := n^3\binom{n}{2}$, what is the number of cuts of K_n of weight s? (The required upper bound n^3 on the weights is just a useful technical assumption.) The related decision problem $\mathrm{Mod}_p\,\mathrm{CUT}$ just asks for the residue class modulo a prime p of the number of cuts of weight s. This problem is clearly in $\mathrm{Mod}_p\mathrm{NP}$.

It is well known that the computation of a cut of maximal weight of a given graph is NP-hard. By a straightforward modification of the proof of this fact given in Papadimitriou [21, p. 191], one can strengthen this as follows.

Proposition 6.1. #CUT *is #P-complete with respect to parsimonious reductions. Thus* $\mathrm{Mod}_p\,\mathrm{CUT}$ *is* $\mathrm{Mod}_p\mathrm{NP}$-*complete.*

The following p-family is related with the #CUT-problem. For $1 \le i < j \le n$ let X_{ij} be distinct indeterminates and set $X_{ji} := X_{ij}$. Let $q \in \mathbb{N}$, $q \ge 2$. The *cut enumerator* Cut_n^q is defined as

$$\mathrm{Cut}_n^q := \sum_S \prod_{i \in A, j \in B} X_{ij}^{q-1} \;,$$

where the sum is over all cuts $S = \{A, B\}$ of K_n. It is easy to see that $\mathrm{Cut}^q := (\mathrm{Cut}_n^q)$ is a p-definable family (over any field).

The connection to the counting problem #CUT is as follows. We may describe a weight function $w\colon E_n \to \{0, 1, \ldots, n^3\}$ of the complete graph K_n by the symmetric matrix $x \in \mathbb{N}^{n \times n}$ defined by $x_{ij} := 2^{nw(\{i,j\})}$. If $c(s)$ denotes the number of cuts in K_n of weight s, then we have

$$\mathrm{Cut}_n^q(x) = \sum_{\mathrm{cut}\, S} 2^{(q-1)nw(S)} = \sum_{s < D_n} c(s) 2^{(q-1)ns} \;.$$

As always $0 \le c(s) < 2^n \le 2^{(q-1)n}$, we can read off all the numbers $c(s)$ from the $2^{(q-1)n}$-ary expansion of $\mathrm{Cut}_n^q(x)$. Moreover, x can be computed from w in polynomial time by a Turing machine. This reasoning, together with Prop. 6.1, shows that the computation of $\mathrm{Cut}_n^q(x)$ for symmetric matrices $x \in \mathbb{N}^{n \times n}$ is #P-hard.

Over finite fields $\mathbb{F}_q$ the situation is different. Let $p := \mathrm{char}\,\mathbb{F}_q$.

Lemma 6.2. *To a symmetric matrix* $x \in \mathbb{F}_q^{n \times n}$ *we assign the graph* $G(x)$ *on the set of nodes* $\underline{n}$ *by requiring that* $\{i, j\}$ *is an edge iff* $x_{ij} = 0$. *Then we have*

$$\mathrm{Cut}_n^q(x) = 2^{N(x)-1} - 1 \bmod p \;,$$

where $N(x)$ *is the number of connected components of* $G(x)$. *In particular, the value* $\mathrm{Cut}_n^q(x)$ *can be computed from a symmetric* $x \in \mathbb{F}_q^{n \times n}$ *in polynomial time by a Turing machine.*

Proof. For any nonzero $\lambda \in \mathbb{F}_q$ we have $\lambda^{q-1} = 1$ by Fermat's theorem. Let $x \in \mathbb{F}_q^{n \times n}$ be symmetric. A partition $\{A, B\}$ of $\underline{n}$ contributes to $\mathrm{Cut}_n^q(x)$ either zero or one. The contribution is one iff $X_{ij} \neq 0$ for all $i \in A$, $j \in B$, that is, none of the nodes of A is connected with any node in B in the graph $G(x)$. This in turn means that A and B are both a union of certain connected components of the graph $G(x)$. The number of such partitions clearly equals $2^{N(x)} - 1$, where $N(x)$ is the number of connected components of $G(x)$. This proves the lemma. $\square$

The main result of this section states that Cut^q is an explicit example of a p-family over the finite field $\mathbb{F}_q$, which is neither p-computable nor complete in VNP.

Theorem 6.3. *The family of cut enumerators Cut^q over a finite field $\mathbb{F}_q$ of characteristic p is neither p-computable nor VNP-complete with respect to c-reduction, provided $\text{Mod}_p\text{NP} \not\subseteq P/\text{poly}$.*

We remark that the inclusion $\text{Mod}_p\text{NP} \subseteq P/\text{poly}$ implied $\text{NP}/\text{poly} \subseteq P/\text{poly}$ by (3) and therefore, by a well-known result by Karp and Lipton [16], the collapse of the polynomial hierarchy at the second level.

Proof. (of Thm. 6.3) Let L be a language in Mod_pNP, say $L = \{x \in \{0,1\}^* \mid \varphi(x) \equiv 1 \bmod p\}$, where $\varphi: \{0,1\}^* \to \mathbb{N}$ is in the class #P. In [6] it is shown that there exists a p-definable family (f_n) over $\mathbb{F}_p$ such that $f_n \in \mathbb{F}_p[X_1, \ldots, X_n]$ and

$$\forall n \; \forall x \in \{0,1\}^n : f_n(x) = \varphi(x) \bmod p \; .$$

Assume now that Cut^q is VNP-complete with respect to c-reduction. Then we have $(f_n) \leq_c \text{Cut}^q$, hence there is a p-bounded function $t: \mathbb{N} \to \mathbb{N}$ such that $L^{\text{Cut}^q_{t(n)}}(f_n)$ is p-bounded. Lemma 6.2 tells us that $\text{Cut}^q_{t(n)}$ can be evaluated in polynomial time on an input over $\mathbb{F}_q$. Hence we may design for each n a boolean circuit C_n of p-bounded size which computes $f_n(x)$ from $x \in \mathbb{F}_q^n$. This implies that the language L is contained in P/poly. We therefore arrive at the conclusion $\text{Mod}_p\text{NP} \subseteq P/\text{poly}$.

Let $K_n = \mathbb{F}_q(\xi_n)$ be a field extension of $\mathbb{F}_q$ of degree $(q-1)D_n$. To an instance $w: E_n \to \{0, 1, \ldots, n^3\}$ of #CUT we assign the symmetric matrix $x \in K_n^{n \times n}$ defined by $x_{ij} := \xi_n^{w(\{i,j\})}$. Then we have

$$\text{Cut}^q_n(x) = \sum_{\text{cut } S} \xi_n^{(q-1)w(S)} = \sum_{s < D_n} (c(s) \bmod p) \, \xi_n^{(q-1)s} \; ,$$

where $c(s)$ is the number of cuts in K_n of weight s. Note that the coefficients $c(s) \bmod p$ are uniquely determined by $\text{Cut}^q_n(x)$.

Assume now that Cut^q is in $\text{VP}_{\mathbb{F}_q}$. Hence for each n there is a straight-line program Γ_n of p-bounded size in n, which computes $\text{Cut}^q_n(X)$ from constants in $\mathbb{F}_q$ and the indeterminates X_{ij} in the polynomial ring $\mathbb{F}_q[X_{ij} \mid 1 \leq i, j \leq n]$. By the universal property of the polynomial ring, Γ_n will compute $\text{Cut}^q_n(x)$ in K_n from the same constants and $x \in K_n^{n \times n}$. We may simulate this computation by a boolean circuit of p-bounded size, since the arithmetic operations in K_n can be simulated by p-bounded circuits (note that D_n is p-bounded). In this way we could solve the Mod_pCUT problem in nonuniform polynomial time. As Mod_pCUT is Mod_pNP-complete by Prop. 6.1, this would imply that $\text{Mod}_p\text{NP} \subseteq P/\text{poly}$. $\square$

Remark 6.4. It would be interesting to find out whether Cut^2 is VNP-complete with respect to c-reduction (or even p-projection) over fields k of characteristic zero.

7 Relativized Complexity Classes

Our investigations here are inspired by the well-known results of Baker, Gill, and Solovay [1] on relativations of the classical P-NP question. Due to lack of space our exposition is very brief. Relative versions of the complexity classes VP and VNP can be defined as follows.

Definition 7.1. Let h be a p-family. VP^h consists of all p-families f such that $f \leq_c h$. VNP^h is the set of all p-families $f = (f_n)$ which can be obtained from some $g = (g_n) \in VP^h$ in the sense of (1).

Note that VP^h and VNP^h specialize to VP and VNP if h is p-computable.

We are able to construct complete families for the complexity classes VP^h and VNP^h. The idea is to use a generalization of the concept of generic computations (cf. [8, Chap. 9]). In order to avoid an exponential growth of degrees, we combine this with an auxiliary result on the computation of homogeneous components, which works by evaluation and interpolation, and requires that k contains sufficiently many points. This way, we can prove the following.

Theorem 7.2. *For any p-family h over an infinite field k there exist VP^h-complete and VNP^h-complete families with respect to p-projection.*

In particular, this gives a new proof for the existence of VNP-complete families, which does not depend on Valiant's intricate reduction for the permanent.

By combining this theorem with some diagonalization argument, we are able to show the following.

Theorem 7.3. *There exists a p-family h such that $VP^h = VNP^h$ over infinite fields.*

Up to now we have not succeeded in establishing a p-family h such that $VP^h \neq VNP^h$. A promising approach for this is as follows (compare Bennett and Gill [3]). For each n choose independently $h_n \in k[X_1, \ldots, X_n]$ of degree most n *at random* according to some probability distribution. Since the classes VP^h and VNP^h are invariant under finite variation of h, the event $\mathcal{E} = \{h \mid VP^h \neq VNP^h\}$ is a so-called tail event. Kolmogorov's zero-one law (cf. Feller [13, Chap. 4]) implies therefore that $\text{Prob}(\mathcal{E}) \in \{0, 1\}$. We conjecture that this probability is one if the h_n are chosen with independent $0, 1$-coefficients.

References

1. T. Baker, J. Gill, and R. Solovay. Relativizations of the $P = ?NP$ question. *SIAM J. Comp.*, 4:431–442, 1975.
2. S. Ben-David, K. Meer, and C. Michaux. A note on non-complete problems in $NP_\mathbb{R}$. Preprint, 1996.
3. C.H. Bennett and J. Gill. Relative to a random oracle A, $P^A \neq NP^A \neq co - NP^A$ with probability 1. *SIAM J. Comp.*, 10:96–113, 1981.

4. L. Blum, F. Cucker, M. Shub, and S. Smale. Algebraic Settings for the Problem "$P \neq NP$?". In *The mathematics of numerical analysis*, number 32 in Lectures in Applied Mathematics, pages 125–144. Amer. Math. Soc., 1996.

5. L. Blum, M. Shub, and S. Smale. On a theory of computation and complexity over the real numbers. *Bull. Amer. Math. Soc.*, 21:1–46, 1989.

6. P. Bürgisser. Cook's versus Valiant's hypothesis. Preprint, University of Zurich, 1997.

7. P. Bürgisser. Some complete families of polynomials. Manuscript, 1997.

8. P. Bürgisser, M. Clausen, and M.A. Shokrollahi. *Algebraic Complexity Theory*. Number 315 in Grundlehren der mathematischen Wissenschaften. Springer Verlag, 1996.

9. J. Cai and L.A. Hemachandra. On the power of parity polynomial time. In *Proc. STACS'89*, number 349 in LNCS, pages 229–239. Springer Verlag, 1989.

10. O. Chapuis and P. Koiran. Saturation and Stability in the Theory of Computation over the Reals. Preprint, 1997.

11. S.A. Cook. The complexity of theorem proving procedures. In *Proc. 3rd ACM STOC*, pages 151–158, 1971.

12. T. Emerson. Relativizations of the P=?NP question over the reals (and other ordered rings). *Theoret. Comp. Sci.*, 133:15–22, 1994.

13. W. Feller. *An introduction to probability theory and its applications*, volume 2. John Wiley & Sons, 1971.

14. J. von zur Gathen. Feasible arithmetic computations: Valiant's hypothesis. *J. Symb. Comp.*, 4:137–172, 1987.

15. J. Heintz and J. Morgenstern. On the intrinsic complexity of elimination theory. *Journal of Complexity*, 9:471–498, 1993.

16. R.M. Karp and R.J. Lipton. Turing machines that take advice. In *Logic and Algorithmic: An international Symposium held in honor of Ernst Specker*, pages 255–273. Monogr. No. 30 de l'Enseign. Math., 1982.

17. P. Koiran. Hilbert's Nullstellensatz is in the polynomial hierarchy. *J. Compl.*, 12:273–286, 1996.

18. R.E. Ladner. On the structure of polynomial time reducibility. *J. ACM*, 22:155–171, 1975.

19. Landweber, Lipton, and Robertson. On the structure of sets in NP and other complexity classes. *Theoret. Comp. Sci.*, 15:181–200, 1981.

20. G. Malajovich and K. Meer. On the structure of $NP_{\mathbb{C}}$. *SIAM J. Comp.* to appear.

21. C.H. Papadimitriou. *Computational Complexity*. Addison-Wesley, 1994.

22. C.H. Papadimitriou and S. Zachos. Two remarks on the power of counting. In *Proc. 6th GI conference in Theoretical Computer Science*, number 145 in LNCS, pages 269–276. Springer Verlag, 1983.

23. U. Schöning. A uniform approach to obtain diagonal sets in complexity classes. *Theoret. Comp. Sci.*, 18:95–103, 1982.

24. U. Schöning. Minimal pairs for *P*. *Theoret. Comp. Sci.*, 31:41–48, 1984.

25. L.G. Valiant. Completeness classes in algebra. In *Proc. 11th ACM STOC*, pages 249–261, 1979.

26. L.G. Valiant. The complexity of computing the permanent. *Theoret. Comp. Sci.*, 8:189–201, 1979.

27. L.G. Valiant. Reducibility by algebraic projections. In *Logic and Algorithmic: an International Symposium held in honor of Ernst Specker*, volume 30, pages 365–380. Monographies de l'Enseignement Mathématique, 1982.

28. L.G. Valiant and V.V. Vazirani. NP is as easy as detecting unique solutions. *Theoret. Comp. Sci.*, 47:85–93, 1986.

On the Existence of Polynomial Time Approximation Schemes for OBDD Minimization (Extended Abstract)

Detlef Sieling*

FB Informatik, LS II, Univ. Dortmund,
44221 Dortmund, Fed. Rep. of Germany
sieling@ls2.cs.uni-dortmund.de

Abstract. The size of Ordered Binary Decision Diagrams (OBDDs) is determined by the chosen variable ordering. A poor choice may cause an OBDD to be too large to fit into the available memory. The decision variant of the variable ordering problem is known to be *NP*-complete. We strengthen this result by showing that there is no polynomial time approximation scheme for the variable ordering problem unless $P = NP$. We also prove a small lower bound on the performance ratio of a polynomial time approximation algorithm under the assumption $P \neq NP$.

1 Introduction

Ordered Binary Decision Diagrams (OBDDs) are the state-of-the-art data structure for Boolean functions in programs for problems like logic synthesis, model checking or circuit verification. The reason is that many functions occurring in such applications can be represented by OBDDs of reasonable size and that for operations on Boolean functions like equivalence test or synthesis with binary operators efficient algorithms on OBDDs are known. Already in the seminal paper of Bryant [4] asymptotic optimal algorithms for many operations are presented. The most important exception is the variable ordering problem for OBDDs, i.e., the task to compute for a given function a variable ordering minimizing the size of the OBDD for this function. The lack of an efficient algorithm for the variable ordering problem affects the applicability of OBDDs because there are important functions for which OBDDs are of reasonable size only for few variable orderings.

Since no efficient algorithms for the variable ordering problem are known, many heuristics for this problem were suggested. Some heuristics assume that the function to be represented is given by a circuit and extract information on the connections between the variables from the circuit description (see, e.g., Fujita, Fujisawa and Kawato [6] or Malik, Wang, Brayton and Sangiovanni-Vincentelli [10]). However, it is an *NP*-hard problem to compute the OBDD

---

* Supported in part by DFG grant We 1066/8.

size for an optimal variable ordering for a function given by a circuit, since the satisfiability problem for circuits is *NP*-complete while the satisfiability problem for OBDDs can be solved in linear time. Hence, we focus on the situation where the given function is described by an OBDD. This is also the situation when applying dynamic variable ordering techniques (see Rudell [15]). Heuristics for this situation optimize the variable ordering, e.g. by a local search or simulated annealing approach (see, e.g., Ishiura, Sawada and Yajima [9], Rudell [15] and Bollig, Löbbing and Wegener [1]).

An exact algorithm for computing optimal variable orderings was presented by Friedman and Supowit [5]. Since this algorithm works on truth tables, it has exponential run time. Also improvements of this algorithm (see e.g. [9]) have exponential worst case run time. The first evidence that the variable ordering problem is hard was given by Tani, Hamaguchi and Yajima [19]. They consider Shared Binary Decision Diagrams (SBDDs), which are the generalization of OBDDs for the representation of more than one function. Their result is the *NP*-completeness of the decision variant of the variable ordering problem for SBDDs. Later on Bollig and Wegener [2] showed that also the decision variant of the minimization problem for OBDDs in *NP*-complete. By these *NP*-completeness results it is justified, also from a theoretical point of view, to use algorithms for the variable ordering problem that do not necessarily compute optimal solutions.

Nevertheless, the *NP*-completeness results do not exclude the existence of good approximation algorithms like polynomial time approximation schemes for the variable ordering problem (even if $P \neq NP$). The *NP*-completeness is proved by a reduction from the problem Optimal Linear Arrangement. The reduction seems not to be approximation preserving and we do not know of any nonapproximability result for Optimal Linear Arrangement. The only indication that good approximation algorithms for the variable ordering problem do not exist is the unsuccessful search for such algorithms. The known heuristics do not provide any guarantee for the quality of their results. Usually, the heuristics are only tested on some set of benchmark circuits.

In this paper we characterize the complexity of the variable ordering problem more precisely by proving that the existence of polynomial time approximation schemes for the variable ordering problem for OBDDs (or SBDDs) implies $P = NP$. Hence, the best we can hope for are approximation algorithms for this problem. We also prove a (small) lower bound on the performance ratio of a polynomial time approximation algorithm for the variable ordering problem (under the assumption $P \neq NP$).

2 Preliminaries

In this section we shortly repeat some definitions and facts concerning OBDDs, SBDDs and the considered problems. For a more detailed introduction into OBDDs see, e.g., Bryant [3] or Wegener [20]. For the definition of notions concerning approximation algorithms, e.g. performance ratio or polynomial time approximation scheme, we refer to the textbook of Garey and Johnson [7].

We start with the definition of OBDDs and SBDDs. An OBDD representing some function $f(x_1, \ldots, x_n)$ is a directed acyclic graph, in which we distinguish sinks and non-sink nodes, also called interior nodes. Sinks are labeled by some Boolean constant 0 or 1. Each interior node v is labeled by some variable x_i and has two outgoing edges, one labeled by 0 and the other one labeled by 1. We say that x_i is tested at v. The ordering condition of OBDDs requires the variables to be tested on each path in the OBDD at most once and according to a prescribed ordering.

With each node v of an OBDD we associate a function f_v, which can be computed in the following way. Let $(a_1, \ldots, a_n)$ be some assignment of the variables. We start the computation at v. If v is labeled by x_i, then we follow that edge leaving v that is labeled by a_i. This is iterated until a sink is reached. The value $f_v(a_1, \ldots, a_n)$ is equal to the label of this sink.

Each OBDD has exactly one source node s and the function represented by the OBDD is the function associated with s. SBDDs are the straightforward generalization of OBDDs for the representation of an arbitrary number of functions (Minato, Ishiura and Yajima [12]). An SBDD for the functions $f_1, \ldots, f_l$ has l distinguished nodes $s_1, \ldots, s_l$ and f_i is equal to the function associated with s_i.

The size of an OBDD or SBDD, resp., is the number of its interior nodes. For the computation of the minimal size of an OBDD for some function f and some fixed variable ordering or the minimal size of an SBDD for some functions $f_1, \ldots, f_l$ and some fixed variable ordering we shall apply the following lemma. This lemma was proved by Sieling and Wegener [18] for the case of OBDDs. The generalization to SBDDs is straightforward.

Lemma 1. *A minimum size SBDD for the functions $f_1, \ldots, f_l$ and the variable ordering $x_1, \ldots, x_n$ contains exactly $|S_i|$ interior nodes labeled by x_i, where*

$$S_i = \{f_{j|x_1=c_1, \ldots, x_{i-1}=c_{i-1}} \mid j \in \{1, \ldots, l\}, c_1, \ldots, c_{i-1} \in \{0, 1\},$$
$$f_{j|x_1=c_1, \ldots, x_{i-1}=c_{i-1}} \text{ essentially depends on } x_i\}.$$

Furthermore, exactly the functions in S_i are the functions associated with the nodes labeled by x_i.

It is well-known that the minimum size SBDD for functions $f_1, \ldots, f_l$ and some fixed variable ordering can be obtained from each SBDD for these functions and this variable ordering by applying two reduction rules bottom-up. By the deletion rule each node whose successors are equal can be eliminated. By the merging rule nodes v and w with the same label, the same 0-successor and the same 1-successor can be merged. Altogether, the effect of the reduction rules is that nodes associated with the same function are replaced by a single node. Hence, nodes cannot be merged if they are associated with different functions.

A function f over some set X of variables is called partially symmetric with respect to the partition $X_1, \ldots, X_l$ of X if the variables in each set X_i can be permuted arbitrarily without changing the function. Then also the OBDD size does not change when permuting the variables in each set X_i. (Totally) symmetric functions are the special case where the partition only consists of one set. Sieling [17] describes a method for the exact computation of the OBDD size for partially symmetric functions. We apply this method in our proof in order to compute the minimum OBDD size for some functions exactly.

Finally, we define the problems considered in this paper.

MinOBDD

Instance: An OBDD H for some function f.

Problem: Compute a variable ordering π minimizing the OBDD size for f and π.

The problem MinSBDD is defined similarly. We remark that from H and the optimal variable ordering π the OBDD for f and π can be computed in polynomial time (Savický and Wegener [16], Meinel and Slobodová [11]). Hence, it suffices to compute a variable ordering instead of the OBDD.

We prove the nonapproximability result by an approximation preserving reduction from MaxCut.

MaxCut

Instance: An undirected graph $G = (V, E)$.

Problem: Compute a partition (V_1, V_2) of V maximizing the number of edges in E with one endpoint in V_1 and the other one in V_2. (The number of such edges is called the size of the cut.)

The following nonapproximability result for MaxCut is due to Håstad [8].

Theorem 2. *For each $\gamma > 0$ the existence of a polynomial time approximation algorithm for MaxCut with a performance ratio of at most $1 + 1/16 - \gamma$ implies $P = NP$.*

3 The Result and a Sketch of its Proof

Our main result is given by the following theorem.

Theorem 3. *For each $\delta > 0$ the existence of a polynomial time approximation algorithm for MinOBDD with a performance ratio of $1 + 1/14943 - \delta$ implies $P = NP$. In particular, the existence of a polynomial time approximation scheme for MinOBDD implies $P = NP$.*

For MinSBDD the slightly larger lower bound $1 + 1/2431 - \delta$ (for each $\delta > 0$) on the performance ratio can be proved.

The proofs of both lower bound are based on approximation preserving reductions from MaxCut. In the following subsections we describe how we obtain an instance for MinOBDD from an instance of MaxCut and how we obtain a solution for the instance of MaxCut from the solution of the constructed instance of MinOBDD. In Section 3.7 we prove that we obtain an approximation preserving reduction in this way. We remark that our reduction is an L-reduction (Papadimitriou and Yannakakis [14]).

Let $G = (V, E)$ be an instance for MaxCut. W.l.o.g. we assume that all nodes in G have a degree of at least two and that $|E| \geq 200$. Let $n = |V|$, and let $m = |E|$. Sometimes, we identify V with the set $\{1, \ldots, n\}$ and E with the set $\{1, \ldots, m\}$. Let $E(v)$ be the set of edges incident to v.

In the following we describe the functions that are represented in the OBDD H, which we construct for the instance G. In a first step we construct an SBDD representing these functions. The variable ordering of this SBDD encodes a cut of G in such a way that the SBDD size corresponds to the size of the cut. The second step is to combine these functions to a single function so that it can be represented by an OBDD. Again we have a correspondence between the variable ordering of the OBDD and the cut as well as a correspondence between the size of the OBDD and the size of the cut. We shall introduce the following functions.

- Functions $f_1, \ldots, f_m$. These functions connect the given graph G and the OBDD H in the following way: If the i-th edge of G is contained in the cut corresponding to the variable ordering of H, then the function f_i has the OBDD size 6. If the i-th edge is not contained in the cut, then the OBDD size is 7.
- Functions $g, h_1, \ldots, h_5$. The functions $f_1, \ldots, f_m$ have the desired property only under further assumptions on the variable ordering. These assumptions include that the variable ordering encodes a cut of G. The functions $g, h_1, \ldots, h_5$ enforce properties of the variable ordering. This means that the size of an SBDD for $f_1, \ldots, f_m, g, h_1, \ldots, h_5$ for variable orderings not encoding any cut for G is much larger than the optimum size.
- Functions h', h''. These functions enforce certain properties of the variable ordering that are helpful when combining all these functions to a single function that is represented by an OBDD.

3.1 The Functions $f_1, \ldots, f_m$

We introduce the variables x_e and y_e for $e \in \{1, \ldots, 2m\}$ and the variables z_e^v for $v \in V$ and $e \in E(v)$. This means that for each edge $e = \{u, v\} \in E$ there are four variables, namely x_e, y_e, z_e^u and z_e^v. The variables x_e and y_e, where $e \in \{m + 1, \ldots, 2m\}$, are not associated with any edge. They are used later on in order to define enforcing components.

For all $e = \{u, v\} \in E$ we define the function $f_e : \{0, 1\}^4 \to \{0, 1\}$ by

$$f_e(x_e, y_e, z_e^u, z_e^v) = \begin{cases} 0 & \text{if } x_e + z_e^u + z_e^v = 0 \text{ or } x_e + z_e^u + z_e^v = 2, \\ 1 & \text{if } x_e + z_e^u + z_e^v = 1, \\ y_e & \text{if } x_e + z_e^u + z_e^v = 3. \end{cases}$$

Note that the functions f_e and $f_{e'}$ for $e \neq e'$ are defined on disjoint sets of variables. Hence, there are no mergings between OBDDs for f_e and $f_{e'}$. The function f_e is partially symmetric with respect to the sets $\{x_e, z_e^u, z_e^v\}$ and $\{y_e\}$. Hence, the OBDD size for f_e is determined only by the position of y_e among $\{x_e, z_e^u, z_e^v, y_e\}$, but not by the relative ordering of x_e, z_e^u and z_e^v. It is easy to check that the OBDD size is 8, if y_e is the first variable among $\{x_e, z_e^u, z_e^v, y_e\}$, that the OBDD size is 7, if y_e is the second or fourth variable, and that the OBDD size is 6, if y_e is the third variable.

Now we explain the relationship between a cut (V_1, V_2) of G and the ordering of the variables. If the variable z_e^u is arranged before y_e in the variable ordering, then the node u is contained in V_1. If z_e^u is arranged after y_e, then the node u is contained in V_2. By introducing the functions h_1, h_2 and h_3 we shall make sure that x_e is always arranged before y_e. Then y_e cannot be the first variable among $\{x_e, z_e^u, z_e^v, y_e\}$ and, hence, it does not occur that the OBDD for f_e has size 8. Now consider the case that $e = \{u, v\}$ is contained in the cut. This means that $u \in V_1$ and $v \in V_2$ or vice versa. In both cases y_e is at the third position among $\{x_e, z_e^u, z_e^v, y_e\}$ and, hence, the OBDD size is 6. If $e = \{u, v\}$ is not contained in the cut, then either $u \in V_1$ and $v \in V_1$ or $u \in V_2$ and $v \in V_2$. In the former case y_e is at the fourth position among $\{x_e, z_e^u, z_e^v, y_e\}$, in the latter case y_e is at the second position. Hence, the OBDD size for f_e is 7.

This construction only works if the classification of u belonging to V_1 or V_2 is consistent for all variables z_e^u and $e \in E(u)$. If there are edges $e = \{u, v\}$ and $e' = \{u, w\}$, it must not happen that z_e^u is arranged before y_e in the variable ordering and $z_{e'}^u$ after $y_{e'}$. Altogether, we shall represent the functions $g, h_1, \ldots, h_5$ in the SBDD in order to enforce the following properties of the variable ordering.

(P1) All x-variables are arranged before all y-variables.

(P2) For each node $u \in V$ the following holds. Either for all $e \in E(u)$ the variable z_e^u is arranged before y_e or for all $e \in E(u)$ the variable z_e^u is arranged after y_e.

Finally, we combine all the functions $f_1, \ldots, f_m$ to one function f. For that purpose we introduce $m - 1$ new variables $\alpha_1, \ldots, \alpha_{m-1}$ and define $f = \bar{\alpha}_1 f_1 \vee (\alpha_1 \bar{\alpha}_2) f_2 \vee \ldots \vee (\alpha_1 \ldots \alpha_{m-2} \bar{\alpha}_{m-1}) f_{m-1} \vee (\alpha_1 \ldots \alpha_{m-1}) f_m$. It can be shown that for each variable ordering for an OBDD for f the OBDD size does not increase when moving the variables $\alpha_1, \ldots, \alpha_{m-1}$ in this order to the beginning of the variable ordering. In this proof we exploit that the functions f_e are defined on disjoint sets of variables. After arranging the α-variables before the x-, y- and z-variables it is easy to see that $m - 1$ nodes labeled by α-variables are necessary and sufficient in each OBDD for f. We obtain the following lemma.

Lemma 4. *The graph G has a cut of size at least C_{cut} iff f has an OBDD of size $8m - 1 - C_{cut}$ with a variable ordering fulfilling (P1) and (P2).*

3.2   The Function g

The function g will make sure that (P2) is fulfilled. We introduce new variables $a_1, \ldots, a_{2m}$, $d_1, \ldots, d_{2m}$ and $\gamma_1, \ldots, \gamma_{3m}$. First we define some components from which g is built up. For each $v \in V$ let $p_v = \bigoplus_{e \in E(v)} z_e^v$. Let

$$p^* = \bigwedge_{v \in V} p_v, \qquad \Gamma = \bigoplus_{i=1}^{3m} \gamma_i, \qquad \text{and} \qquad g^* = \bigvee_{i=1}^{2m} a_1 \ldots a_{i-1} \bar{a}_i y_i d_i \ldots d_{2m}.$$

Finally, we define $g = p^* \wedge g^* \wedge \Gamma$. We may obtain an optimal variable ordering for g in the following way. First for some set $V_1 \subseteq V$ the variables z_e^v, where $v \in V_1$ and $e \in E(v)$, are tested. Here it is important to arrange the z_e^v-variables blockwise, i.e., z_e^v-variables for the same v are arranged adjacently. After these variables we arrange $a_1, \ldots, a_{2m}, y_1, \ldots, y_{2m}, d_1, \ldots, d_{2m}$ in this order. Then the variables z_e^v, where $v \in V - V_1$ and $e \in E(v)$, are arranged blockwise, and finally the γ-variables. By the h-functions defined in the next section we shall make sure that the variable ordering has the following properties.

(P3) All a-variables are arranged before all y-variables.
(P4) All y-variables are arranged before all d-variables.

If these properties are fulfilled, then the OBDD for g^* has size $6m$ and width $2m$. If there is some $v \in V$ such that z_e^v is arranged before y_e and $z_{e'}^v$ is arranged after $y_{e'}$ then in this wide OBDD the value of z_e^v has to be stored. It can be shown that in this case the OBDD for g is much larger than its minimum size. In the following we present this and other properties of g without proof. The purpose of the γ-variables is to control the number of mergings between nodes in the OBDD of g and in the SBDD for h' and h''.

Lemma 5. *For the variable ordering described above the OBDD for g has size $16m - n - 1$.*

Lemma 6. *Each SBDD for f and g has at least $21m - n - 4$ nodes.*

Lemma 7. *Let a variable ordering be given that fulfils (P3) and (P4) but does not fulfil (P2). Then the SBDD size for f and g and this variable ordering is at least $25m - n - 8$.*

3.3   The Functions $h_1, \ldots, h_5$

Our aim is to include functions in the OBDD that ensure (P1), (P3) and (P4). Let us consider, e.g., (P3). A function enforcing (P3) has its optimal variable ordering if all a-variables are arranged before all y-variables. The problem is that an OBDD for this function and the variable ordering $a_1, \ldots, a_{2m-1}, y_1, a_{2m}, y_2, \ldots, y_{2m}$ is not substantially larger. This is the reason why we introduce new variables $b_1, \ldots, b_{2m}$ in order to increase the distance between the a-variables and the y-variables. Similarly, we introduce variables $c_1, \ldots, c_{2m}$ in order to increase the distance between the y-variables and the d-variables. The function $h : \{0,1\}^{4m} \to \{0,1\}$ is defined by

$$
h(p_1, \ldots, p_{2m}, q_1, \ldots, q_{2m}) = \begin{cases} 1 & \text{if } (p_1 + \ldots + p_{2m}) \equiv 0 \bmod 5, \\ q_1 \oplus \ldots \oplus q_{2m} & \text{if } (p_1 + \ldots + p_{2m}) \equiv 1 \bmod 5, \\ 0 & \text{if } (p_1 + \ldots + p_{2m}) \equiv 2 \bmod 5, \\ \overline{q_1 \oplus \ldots \oplus q_{2m}} & \text{if } (p_1 + \ldots + p_{2m}) \equiv 3 \bmod 5, \\ 0 & \text{if } (p_1 + \ldots + p_{2m}) \equiv 4 \bmod 5. \end{cases}
$$

We introduce the functions $h_1, \ldots, h_5$ in order to ensure (P1), (P3) and (P4) to be fulfilled, where $h_1 = h(x_1, \ldots, x_{2m}, a_1, \ldots, a_{2m})$, $h_2 = h(a_1, \ldots, a_{2m}, b_1, \ldots, b_{2m})$, $h_3 = h(b_1, \ldots, b_{2m}, y_1, \ldots, y_{2m})$, $h_4 = h(y_1, \ldots, y_{2m}, c_1, \ldots, c_{2m})$, $h_5 = h(c_1, \ldots, c_{2m}, d_1, \ldots, d_{2m})$.

Note that the function h is partially symmetric with respect to $\{p_1, \ldots, p_{2m}\}$ and $\{q_1, \ldots, q_{2m}\}$. This implies that the functions $h_1, \ldots, h_5$ do not affect the relative ordering of the x-variables (or a-, b-, y-, c- and d-variables, resp.). In the following we present some results on the OBDD size of the h-functions for several variable orderings. These results can be obtained by the method of Sieling [17] for the calculation of the OBDD size for partially symmetric functions.

Lemma 8. *The OBDD size for $h(p_1, \ldots, p_{2m}, q_1, \ldots, q_{2m})$ and the variable orderings where all p-variables are arranged before all q-variables is $14m - 10$. Only such variable orderings are optimal.*

Lemma 9. *The OBDD size for $h(p_1, \ldots, p_{2m}, q_1, \ldots, q_{2m})$ and each variable ordering where after the first q-variable at least m of the p-variables are arranged is at least $19m - 10$.*

Lemma 10. *The OBDD size for $h(p_1, \ldots, p_{2m}, q_1, \ldots, q_{2m})$ and each variable ordering where before the last p-variable at least m of the q-variables are arranged is at least $18m - 10$.*

If, e.g., (P3) is not fulfilled, then there is some a-variable that is tested after at least m of the b-variables, or there is some y-variable that is tested before at least m of the b-variables. In the former case the OBDD for h_2 has size at least $18m - 10$ rather than $14m - 10$, in the latter case the OBDD size for h_3 is at least $19m - 10$ rather than $14m - 10$. In this way it is made sure that (P1), (P3) and (P4) are fulfilled. Here we can see, why we introduced $2m$ rather than only m of the x- and y-variables. The aim is to make the difference between the OBDD size for h and the variable orderings described in Lemma 9 and Lemma 10 and the OBDD size for h and an optimal variable ordering larger.

Finally, we introduce new variables α_m, α_{m+1} and α_{m+2}. Similar to the definition of f we combine the functions h_1, h_3 and h_5 to a single functions h^*, and h_2 and h_4 to a single function h^{**}. Again it can be proved that the OBDD size does not increase when moving the α-variables before the other variables.

3.4 The Functions h' and h''

Up to now we have the following $18m + 2$ variables: x_i, y_i, a_i, b_i, c_i and d_i for $i \in \{1, \ldots, 2m\}$, the variables $\alpha_1, \ldots, \alpha_{m+2}, \gamma_1, \ldots, \gamma_{3m}$ and $2m$ variables z_e^v. In order to simplify the following presentation we rename all these variables to $s_1, \ldots, s_{18m+2}$. The exact correspondence between the old names and the new names is not important. Furthermore, we introduce new variables $s_{18m+3}, \ldots, s_{32m}$. Then we define $h' = s_1 \oplus \ldots \oplus s_{32m}$ and $h'' = \overline{h'}$. Since h' and h'' are symmetric functions the following lemma holds.

Lemma 11. *A minimum size SBDD for h' and h'' consists for all variable orderings of exactly $64m$ interior nodes.*

If $\gamma_1, \ldots, \gamma_{3m}$ are the last variables in the variable ordering, then exactly $6m - 1$ nodes of the SBDD for h' and h'' can be merged with nodes of the OBDD for g.

3.5 The Size of an SBDD for f, g, h^*, h^{**}, h' and h''

We consider the size of an SBDD for f, g, h^*, h^{**}, h' and h'' for two cases. The first case is that the variable ordering fulfils (P1)–(P4). In particular, by the description in Section 3.1 there is a cut of G corresponding to this variable ordering. Furthermore, we assume that the γ-variables are the last variables in the variable ordering. The OBDDs for g, h^* and h^{**} and the SBDD for h' and h'' take their minimum size for some variable ordering fulfilling (P1)–(P4). If we combine an SBDD for f and g, an SBDD for h' and h'' and OBDDs for h^* and h^{**} to an SBDD for the functions f, g, h^*, h^{**}, h' and h'', there are exactly $6m-1$ mergings of pairs of γ-nodes and 6 mergings of pairs of other nodes between the previously separated OBDDs/SBDDs. Together with the upper bounds of the Lemmas 4, 5, 8 and 11, we obtain an upper bound on the size of an SBDD for f, g, h^*, h^{**}, h' and h''. It can be shown that this upper bound is optimal.

Lemma 12. *The graph G has a cut of size at least C_{cut} iff f, g, h^*, h^{**}, h' and h'' have an SBDD of size $152m - n - 54 - C_{cut}$ with a variable ordering fulfilling (P1)–(P4).*

The second case that we consider is the case that at least one of the properties (P1)–(P4) is not fulfilled by the variable ordering. Then the following result can be shown using Lemma 6, Lemma 7, Lemma 9 and Lemma 10.

Lemma 13. *If the variable ordering does not fulfill at least one of the properties (P1)–(P4), then the SBDD size for f, g, h^*, h^{**}, h' and h'' is at least $153m - n - 96$.*

3.6 Construction of an OBDD

The function $\mathcal{F}$ that is represented in the OBDD is defined on the variables $s_1, \ldots, s_{32m}$, which we already introduced, and $32m + 1$ new variables $t, r_1, \ldots, r_{32m}$. The function is defined by

$$\mathcal{F} = \begin{cases} f & \text{if } t = 0 \text{ and } (r_1 + \ldots + r_{32m}) \equiv 0 \bmod 4, \\ g & \text{if } t = 0 \text{ and } (r_1 + \ldots + r_{32m}) \equiv 1 \bmod 4, \\ h^* & \text{if } t = 0 \text{ and } (r_1 + \ldots + r_{32m}) \equiv 2 \bmod 4, \\ h^{**} & \text{if } t = 0 \text{ and } (r_1 + \ldots + r_{32m}) \equiv 3 \bmod 4, \\ 1 & \text{if } t = 1 \text{ and } (r_1 + \ldots + r_{32m}) \equiv 0 \bmod 5, \\ h' & \text{if } t = 1 \text{ and } (r_1 + \ldots + r_{32m}) \equiv 1 \bmod 5, \\ 0 & \text{if } t = 1 \text{ and } (r_1 + \ldots + r_{32m}) \equiv 2 \bmod 5, \\ h'' & \text{if } t = 1 \text{ and } (r_1 + \ldots + r_{32m}) \equiv 3 \bmod 5, \\ 0 & \text{if } t = 1 \text{ and } (r_1 + \ldots + r_{32m}) \equiv 4 \bmod 5. \end{cases}$$

Consider an OBDD for $\mathcal{F}$ and the (nonoptimal) variable ordering t, r-variables, s-variables. For $t = 1$ the function $\mathcal{F}_{|t=1} = h(r_1, \ldots, r_{32m}, s_1, \ldots, s_{32m})$ is computed. As explained above an OBDD for this function has its minimum size if the r-variables are arranged before the s-variables. The r-variables and t determine which of the functions f, g, h^*, h^{**}, h' and h'' is computed. Thus the OBDD contains an SBDD for f, g, h^*, h^{**}, h' and h'' and we have a connection between the OBDD size for $\mathcal{F}$ and the SBDD size for f, g, h^*, h^{**}, h' and h'' given in the following lemma. For the proof of this lemma we use the method of [17].

Lemma 14. *Let τ be some ordering of the s-variables and let S_τ be the minimum size of an SBDD for f, g, h^*, h^{**}, h' and h'' and the variable ordering τ. Let π be the variable ordering seven r-variables, t, the remaining r-variables, the s-variables according to τ. If $S_\tau < 153m$, then among all variable orderings in which the relative ordering of the s-variables is the same as in τ the ordering π leads to minimum OBDD size for $\mathcal{F}$, which is $288m - 27 + S_\tau$. If $S_\tau \geq 153m$, then for each variable ordering in which the relative ordering of the s-variables is the same as in τ the minimum OBDD size for $\mathcal{F}$ is at least $441m - 27$.*

3.7 Proof of Theorem 3

Finally, we sketch how all results of the previous sections imply Theorem 3. The better result for SBDDs can be obtained in a similar way, if we omit all components that are not necessary for the construction of an SBDD.

Let us assume that some polynomial time approximation algorithm A for MinOBDD with a performance ratio of $1 + \varepsilon < 1 + 1/879$ is given. We construct a polynomial time approximation algorithm for MaxCut. Let an instance G of MaxCut be given. We construct an OBDD for $\mathcal{F}$ and a variable ordering leading to a polynomial size OBDD and apply A on this OBDD. Afterwards we reorder the variables of the output to the ordering seven r-variables, t, the remaining r-variables, s-variables without changing the relative ordering of the s-variables. By Lemma 14 the OBDD size does not increase. By Lemma 12 and Lemma 14 the optimum size of the OBDD for $\mathcal{F}$ is $S_{opt} = 440m - n - 81 - C_{opt}$, where C_{opt} is the size of a maximum cut of G. Now we assume that the variable ordering of the computed OBDD does not fulfil at least one of the properties (P1)–(P4). Then by Lemma 13 and Lemma 14 the OBDD size is at least $441m - n - 123$ in contradiction to the upper bound $1 + 1/879$ on the performance ratio of A. Hence, all properties are fulfilled and we can compute a cut of G from the variable ordering of the computed OBDD. Let S_{appr} be the size of the OBDD and C_{appr} be the size of the computed cut. Then by Lemma 12 and Lemma 14 we have $C_{appr} \geq 440m - n - 81 - S_{appr}$. Furthermore, by the definition of the performance ratio $S_{appr} \leq (1+\varepsilon)S_{opt}$. Combining all these formulas for S_{appr} and S_{opt} we can show $C_{appr} \geq C_{opt} - 440\varepsilon m + \varepsilon C_{opt}$. Now we apply the well-known fact that the size of an optimum cut is at least half the number of edges, i.e. $m \leq 2C_{opt}$ (see e.g. Motwani and Raghavan [13]). Then the performance ratio of the constructed algorithm for MaxCut is bounded by $C_{opt}/C_{appr} \leq 1 + \frac{879\varepsilon}{1-879\varepsilon}$. By Theorem 2 the performance ratio is at least $1 + 1/16 - \gamma$ for all $\gamma > 0$, if $P \neq NP$. By solving $1 + \frac{879\varepsilon}{1-879\varepsilon} \geq 1 + 1/16 - \gamma$ for ε and choosing an appropriate γ we obtain Theorem 3.

Acknowledgment This work was inspired by presentations and discussions at the GI Research Seminar on proof verification and approximation algorithms in Dagstuhl in April 1997.

References

1. Bollig, B., Löbbing, M. and Wegener, I.: Simulated annealing to improve variable orderings for OBDDs. In *Proc. of International Workshop on Logic Synthesis IWLS*, 5.1–5.10, 1995.
2. Bollig, B. and Wegener, I.: Improving the variable ordering of OBDDs is *NP*-complete. *IEEE Transactions on Computers* 45, 993–1002, 1996.
3. Bryant, R.E.: Symbolic Boolean manipulation with ordered binary-decision diagrams. *ACM Computing Surveys* 24, 293–318, 1992.
4. Bryant, R.E.: Graph-based algorithms for Boolean function manipulation. *IEEE Transactions on Computers* 35, 677–691, 1986.
5. Friedman, S.J. and Supowit, K.J.: Finding the optimal variable ordering for binary decision diagrams. *IEEE Transactions on Computers* 39, 710–713, 1990.
6. Fujita, M., Fujisawa, H. and Kawato, N.: Evaluation and improvements of Boolean comparison method based on binary decision diagrams. In *Proc. of International Conference on Computer-Aided Design ICCAD*, 2–5, 1988.
7. Garey, M.R. and Johnson, D.S.: Computers and Intractability: A Guide to the Theory of *NP*-Completeness. W.H. Freeman, 1979.
8. Håstad, J.: Some optimal inapproximability results. In *Proc. of 29th Symposium on Theory of Computing STOC*, 1–10, 1997.
9. Ishiura, N., Sawada, H. and Yajima, S.: Minimization of binary decision diagrams based on exchanges of variables. In *Proc. of International Conference on Computer-Aided Design ICCAD*, 472–475, 1991.
10. Malik, S., Wang, A.R., Brayton, R.K. and Sangiovanni-Vincentelli, A.: Logic verification using binary decision diagrams in a logic synthesis environment. In *Proc. of International Conference on Computer-Aided Design ICCAD*, 6–9, 1988.
11. Meinel, C. and Slobodová, A.: On the complexity of constructing optimal ordered binary decision diagrams. In *Proc. of International Symposium on Mathematical Foundations of Computer Science MFCS*, 515–524, 1994.
12. Minato, S., Ishiura, N. and Yajima, S.: Shared binary decision diagram with attributed edges for efficient Boolean function manipulation. In *Proc. of 27th Design Automation Conference DAC*, 52–57, 1990.
13. Motwani, R. and Raghavan, P.: Randomized Algorithms. Cambridge University Press, 1995.
14. Papadimitriou, C.H. and Yannakakis, M.: Optimization, approximation, and complexity classes. *Journal of Computer and System Sciences* 43, 425–440, 1991.
15. Rudell, R.: Dynamic variable ordering for ordered binary decision diagrams. In *Proc. of International Conference on Computer-Aided Design ICCAD*, 42–47, 1993.
16. Savický, P. and Wegener, I.: Efficient algorithms for the transformation between different types of binary decision diagrams. *Acta Informatica* 34, 245–256, 1997.
17. Sieling, D.: Variable orderings and the size of OBDDs for partially symmetric Boolean functions. In *Proc. of the Synthesis and System Integration of Mixed Technologies SASIMI*, 189–196, 1996. Submitted to *Random Structures & Algorithms*.
18. Sieling, D. and Wegener, I.: *NC*-algorithms for operations on binary decision diagrams. *Parallel Processing Letters* 3, 3–12, 1993.
19. Tani, S., Hamaguchi, K. and Yajima, S.: The complexity of the optimal variable ordering problems of shared binary decision diagrams. In *Proc. of 4th International Symposium on Algorithms and Computation ISAAC*, 389–398, 1993.
20. Wegener, I.: Efficient data structures for Boolean functions. *Discrete Mathematics* 136, 347–372, 1994.

Complexity of Problems on Graphs Represented as OBDDs
(Extended Abstract)*

J. Feigenbaum,[1] S. Kannan,[2] ** M. Y. Vardi,[3] *** M. Viswanathan[2] †

[1] AT&T Labs – Research
Room C203, 180 Park Avenue
Florham Park, NJ 07932 USA
jf@research.att.com
[2] Computer and Information Sciences
University of Pennsylvania
Philadelphia, PA 19104 USA
kannan@central.cis.upenn.edu
maheshv@gradient.cis.upenn.edu
[3] Computer Science
Rice University
Houston, TX 77251 USA
vardi@cs.rice.edu

Abstract. To analyze the complexity of decision problems on graphs, one normally assumes that the input size is polynomial in the number of vertices. Galperin and Wigderson [13] and, later, Papadimitriou and Yannakakis [18] investigated the complexity of these problems when the input graph is represented by a polylogarithmically succinct circuit. They showed that, under such a representation, certain trivial problems become intractable and that, in general, there is an exponential blow up in problem complexity.

In this paper, we show that, when the input graph is represented by a small ordered binary decision diagram (OBDD), there is an exponential blow up in the complexity of most graph problems. In particular, we show that the GAP and AGAP problems become complete for PSPACE and EXP, respectively, when the graphs are succinctly represented by OBDDs.

* A full version of this paper is available as AT&T Technical Report 97.1.2. http://www.research.att.com/library/trs/TRs/97/97.1/97.1.2.body.ps

** Work done in part as a consultant to AT&T and supported in part by NSF grant CCR96-19910 and ONR Grant N00014-97-1-0505.

*** Work done as a visitor to DIMACS and Bell Laboratories as part of the DIMACS Special Year on Logic and Algorithms and supported in part by NSF grants CCR-9628400 and CCR-9700061 and by a grant from the Intel Corporation.

† Supported by grants NSF CCR-9415346, NSF CCR-9619910, AFOSR F49620-95-1-0508, ARO DAAH04-95-1-0092, and ONR Grant N00014-97-1-0505.

1 Introduction

The efficiency of algorithms is generally measured as a function of input size [11]. In analyses of graph-theoretic algorithms, graphs are usually assumed to be represented either by adjacency matrices or by adjacency lists. However, many problem domains, most notably computer-aided verification [7,9,16], involve extremely large graphs that have regular, repetitive structure. This regularity can yield very succinct encodings of the input graphs, and hence one expects a change in the time- or space-complexity of the graph problems.

The effect of succinct input representations on the complexity of graph problems was first formalized and studied by Galperin and Wigderson [13]. They discovered that, when adjacency matrices are represented by polylogarithmically-sized circuits, many computationally tractable problems become intractable. Papadimitriou and Yannakakis [18] later showed that such representations generally have the effect of exponentiating the complexity (time or space) of graph problems. Following this line of research, Balcázar, Lozano, and Torán [1–3,22] extended these results to problems whose inputs were structures other than graphs and provided a general technique to compute the complexity of problems with inputs represented by succinct circuits. They characterized the class of problems that become intractable when inputs are represented in this way. Veith [23,24] showed that, even when inputs are represented using Boolean formulae (instead of circuits), a problem's computational complexity can experience an exponential blow-up.

The possibility of representing extremely large graphs succinctly has attracted a lot of attention in the area of computer-aided verification [7,9,16]. In this domain, graphs are represented by *ordered binary decision diagrams* (OBDDs). OBDDs are special kinds of rooted, directed acyclic graphs that are used to represent Boolean circuits. Because of their favorable algorithmic properties, they are widely used in the areas of digital design, verification, and testing [8,9,17]. Experience has shown that OBDD-based algorithmic techniques scale up to industrial-sized designs [10], and tools based on such techniques are gaining acceptance in industry [4]. Although OBDDs provide canonical succinct representations in many practical situations, they are exponentially less powerful than Boolean circuits in the formal sense that there are Boolean functions that have polynomial-sized circuit representations but do not have subexponential-sized OBDD representations [19,20]. (On the other hand, the translation from OBDDs to Boolean circuits is linear [7].) Thus, the results of [2,3,13,18,22–24] do not apply to OBDD-represented graphs. Furthermore, even though Boolean formulae are, in terms of representation size, less powerful than circuits, they are still more succinct than OBDDs. Translation from OBDDs to formulae leads to at most a quasi-polynomial ($n^{\log n}$) blow-up, whereas there are functions (*e.g.*, multiplication of binary integers) that have polynomial-sized formulae but require exponential-sized OBDDs. Indeed, while the satisfiability problem is NP-complete for Boolean formulae, it is in nondeterministic logspace for OBDDs [7]. Therefore, the results in [23,24] do not apply to our case.

In this paper, we show that, despite these theoretical limitations on the power of OBDDs to encode inputs succinctly, using them to represent graphs nonetheless causes an exponential blow-up in problem complexity. That is, the well-studied phenomenon of exponential increase in computational complexity for graph problems with inputs represented by Boolean circuits or formulae [2,3,13,18,22–24] also occurs when the graphs are represented by OBDDs. Graph properties that are ordinarily NP-complete become NEXP-complete. The Graph Accessibility Problem (GAP) and the Alternating Graph Accessibility Problem (AGAP) for OBDD-encoded graphs are PSPACE-complete and EXP-complete, respectively. Both GAP and AGAP are important problems in *model checking*, a domain in which OBDDs are widely used [9,12,14,15].

In section 2, we formally define OBDDs and present some known results about them. In section 3, we discuss the problem in greater detail and compare Papadimitriou and Yannakakis's result to ours. Finally, in sections 4-6, we give our technical results.

Because of space limitations, we have omitted several proofs from this Extended Abstract. They can all be found in our Technical Report:
http://www.research.att.com/library/trs/TRs/97/97.1/97.1.2.body.ps
We thank Ed Clarke for his helpful comments on an earlier version of this paper.

2 Preliminaries

Definition 1. A *Binary Decision Diagram* (BDD) is a single-rooted, directed acyclic graph in which

- Both are external nodes labeled 0, or
- Both are external nodes labeled 1, or
- Each internal node (*i.e.*, a node with nonzero outdegree) is labeled by a Boolean variable.
- Each internal node has outdegree 2. One of the outgoing edges is labeled 1 (the "then-edge") and the other is labeled 0 (the "else-edge").
- Each external node (*i.e.*, a node with zero outdegree) is labeled 0 or 1.

Let $X = \{x_1, x_2, \ldots, x_n\}$ be the set of Boolean variables that occur as labels of nodes in a given BDD B. Each assignment $\alpha = (\alpha_1, \alpha_2, \ldots, \alpha_n)$ of Boolean values to these variables naturally defines a computational path – the one that leads from the root to an external node and has the property that, when it reaches a node labeled x_i, it follows the edge labeled α_i, for any i.

Definition 2. Two nodes u and v of a BDD are *equivalent* if

- u and v are the same node, or
- The label of u is the same as the label of v. Furthermore, u_1 is equivalent to v_1, where $< u, u_1 >$ and $< v, v_1 >$ are the then-edges of u and v, respectively, and u_0 is equivalent to v_0, where $< u, u_0 >$ and $< v, v_0 >$ are the else-edges of u and v, respectively.

A BDD in which no two nodes are equivalent is called *reduced*.

Definition 3. Let $<$ be a total ordering on a set X. An *Ordered Binary Decision Diagram* (OBDD) over $(X, <)$ is a reduced BDD with node-label set X such that, along any path from the root to an external node, there is at most one occurrence of each variable, and the order in which the variables occur along the path is consistent with the order $(X, <)$. The *size* of an OBDD is the number of internal nodes in it.

Definition 4. An OBDD O *represents the Boolean function* $f(x_1, x_2, \ldots, x_n)$ if, for each assignment $\alpha = (\alpha_1, \alpha_2, \ldots, \alpha_n)$ to the variables of f, the computation path defined by α terminates in an external node that is labeled by the value $f(\alpha_1, \alpha_2, \ldots, \alpha_n)$.

An OBDD O *represents the graph* $G = (V, E)$ if O represents the Boolean function adj, where

$$adj(v_1, v_2) = \begin{cases} 1 \text{ if and only if} < v_1, v_2 > \in E \\ 0 \text{ otherwise} \end{cases}$$

Theorem 5 (Bryant [7]). *For each Boolean function f and ordering $(X, <)$ of the set of variables X, there is a unique (up to isomorphism) OBDD over $(X, <)$ that represents f.*

Theorem 6 (Bryant [7]). *Let F and G be OBDDs over $(X, <)$ representing functions f and g, respectively. Let the size of F be m, the size of G be n, and $< op >$ be any Boolean operation. Then there is an OBDD over $(X, <)$ of size at most mn and constructable in time polynomial in m and n that represents $f < op > g$.*

Definition 7. Let $L = (G, <)$ be a linear order on the gates of a circuit, where the inputs and outputs are classified as special instances of gates. We say that the *forward cross section of the circuit at gate g* is the set of wires connected to the output of some gate g_1 and an input of some gate g_2 such that $g_1 \leq g$ and $g < g_2$. The *reverse cross section of the circuit at gate g* is the set of wires connected to an output of some gate g_1 and an input of some gate g_2 such that $g_2 \leq g$ and $g < g_1$.

Definition 8. The *forward width of a circuit under order L*, denoted w_f, is the maximum, over all gates g, of the forward cross section at g. Similarly, *the reverse width of the circuit under order L*, denoted by w_r, is the maximum, over all gates g, of the reverse cross section at g.

Theorem 9 (Berman [5]). *For a circuit and gate-ordering with $w_r = 0$, there exists a variable ordering such that the OBDD size is bounded by $n2^{w_f}$, where n is the number of inputs to the circuit.*

Notation: We will be interested in complexity classes **C** that have universal Turing machines and complete problems. Let $U_\mathbf{C}$ denote the Universal Turing machine for the complexity class **C**. Let $\mathcal{L}(U_\mathbf{C})$ be the language accepted by the machine $U_\mathbf{C}$ *i.e.*, $\mathcal{L}(U_\mathbf{C}) = \{x : x$ encodes a C-bounded Turing machine M and an input y such that M accepts $y\}$.

For an n-bit number x, we will refer to the ith bit by $x^{(i)}$, where $x^{(n)}$ is the most significant bit.

3 Problem Statement

Papadimitriou and Yannakakis [18] show that any NP-complete graph property π to which satisfiability is reducible by a *projection*, in the sense of Skyum and Valiant [21], becomes NEXP-complete when problem instances are encoded as circuits. They do this by first constructing a circuit that computes the clause-literal incidence matrix of a formula $F(x)$ (*i.e.*, given a clause and a literal, the circuit decides whether the literal occurs in the clause in $F(x)$) such that $F(x)$ is satifiable if and only if $x \in \mathcal{L}(U_{\mathbf{NEXP}})$. Then, using the properties of projection, they construct a circuit representing a graph $G(x)$ such that $G(x)$ has a property π if and only if $x \in \mathcal{L}(U_{\mathbf{NEXP}})$.

When graphs are represented by OBDDs, such reductions are not immediately obtainable, for two basic reasons. First, OBDDs are strictly less powerful than Boolean circuits, in the sense that there are Boolean functions that have small circuit representations but no small OBDD representations. In particular, the function that computes the ith bit of the product (or quotient) of two binary numbers cannot be represented by a small OBDD. Second, the size of OBDDs is sensitive to the ordering of the variables of the function, and the OBDD representing the Boolean combination of two OBDDs can be constructed quickly only when the ordering of the variables is consistent in the two OBDDs. Hence, for a result equivalent to Papadimitriou and Yannakakis's to hold for graphs represented by OBDDs, we must construct reductions f such that the jth bit of $f(x)$ can be found by a small OBDD given j as the input, assuming that the ith bit of x can be found by a small OBDD given i as the input. Furthermore, all OBDDs involved must read the bits in consistent order.

4 A Small OBDD for a NEXP Tableau

Applying Cook's theorem to the tableau of a nondeterministic, exponential-time Turing machine produces a Boolean formula with exponentially many clauses and literals such that the formula is satisfiable if and only if the tableau represents a valid, accepting computation. In this section, we show that this formula can be represented succinctly by an OBDD. This means that we can fix an enumeration of the clauses and literals such that, given the indices of a clause and of a literal, we can determine whether the literal occurs in the clause. Our proof exploits the great regularity of the formula in question. Roughly, we use the fact that, for a given input x, there is a small, fixed constant c such that a literal with index l occurs only in clauses with indices between $(l-1)c+1+K$ and $lc+K$, where K is some (not necessarily small) number.

We begin by proving that this range check can be computed by a small OBDD using a fixed ordering consistent with the one in which the bits of the literal index and the bits of the clause index are interleaved starting from the most significant bit of each.

Lemma 10. *There is a circuit with $w_f = log\ Y + 2$ and $w_r = 0$ that, given an ordering $x^{(n)} < x^{(n-1)} < \cdots < x^{(1)}$ on input bits, computes the ith bit of x/Y, for a fixed Y and any i, where "/" denotes integer division. Similarly, there is a circuit with $w_f = log\ Y + 2$ and $w_r = 0$ that, given an ordering $x^{(n)} < x^{(n-1)} < \cdots < x^{(1)}$ on input bits, computes the ith bit of x mod Y, for a fixed Y and any i.*

Lemma 11. *Let f be a function such that $f(x)^{(i)}$ depends only on $x^{(n)}, x^{(n-1)}$, ..., $x^{(i)}$. Given an ordering $x^{(n)} < y^{(n)} < x^{(n-1)} < \cdots < y^{(1)}$ on input bits, the circuit that checks whether $f(x + K) = y$, for a fixed K, has $w_r = 0$ and $w_f = 2W + 4$, where W is the forward width of the circuit that computes $f(x)$ given the same ordering on input bits.*

Notation: Consider the language $\mathcal{L}(U_{\textbf{NEXP}})$. Let $F(x)$ be the CNF Boolean formula obtained by the exponential version of Cook's construction, in which $F(x)$ is satisfiable if and only if $x \in \mathcal{L}(U_{\textbf{NEXP}})$.

Theorem 12. *Let g_x be the Boolean function that decides whether a given literal occurs in a given clause in $F(x)$, i.e., $g_x(Cl, Lt) = 1$ if and only if the literal whose index is Lt occurs in the clause of $F(x)$ whose index is Cl. There is an OBDD of size polynomial in the length of x that represents the function g_x.*

5 NP-Complete Graph Problems

Theorem 12 can be used to prove that most classical NP-complete graph problems are NEXP-complete when graphs are represented by OBDDs. We give one example of such a proof; others are quite similar.

Theorem 13. *The INDEPENDENT SET problem for graphs represented by OBDDs is NEXP-complete.*

Papadimitriou and Yannakakis [18] prove the general theorem that, if the reduction from SAT to an NP-complete problem π is a projection, then π becomes NEXP-complete when the input is represented by a circuit. The fact that the reduction is a projection is a sufficient condition for their result, but it appears far from necessary.

Here we state an analogous result. Let f be a reduction from SAT to an NP-complete problem π. Suppose there is a constant k such that, for all j, $f(x)^{(j)}$ is a function only of the bits $x^{(j_1)}, \ldots, x^{(j_k)}$. Moreover, suppose that there is a finite automaton similar to a Mealy machine that takes the bits of j in some canonical order as input and produces the bits of $j_1, \ldots, j_k$ in most-significant to least-significant order. We will refer to the above class of reductions as NC^0-padding reductions. Note that the class of NC^0-padding reductions is incomparable with the class of projections.

Theorem 14. *Let f be an NC^0-padding reduction from SAT to a problem π in NP. Then π is NEXP-complete when its instances are presented as OBDDs.*

NC^0-padding reductions can be found for a number of graph problems such as CLIQUE, VERTEX COVER, *etc.* Such reductions are obtained by taking the standard reductions and padding the target instances so that each of its indices has sufficient information to allow the reconstruction of the indices on which it depends.

6 The GAP and AGAP Problems

In this section, we examine two graph problems that are crucially important in computer-aided verification. We show that both experience exponential blow-up in worst-case complexity when instances are represented as OBDDs.

Problem 15. (GAP) The Graph Accessibility Problem is:
  Input A directed graph G and vertices s and t in the graph.
  Output Is there a directed path from s to t ?

Theorem 16. *GAP is PSPACE-complete when the graph G is represented by an OBDD.*

Proof. Let p be a polynomial and x be a string that encodes a $p_M(|y|)$-space-bounded Turing machine M and an input y. Without loss of generality, we may assume that the machine M has a single accepting configuration C_f.
  Consider the configuration graph $G_{M,y} = (V_{M,y}, E_{M,y})$ corresponding to the machine M and input y, where

  $V_{M,y} = \{v_C \mid C$ is a possible configuration of machine *i.e.*, C is a string of symbols, of which one is a composite symbol encoding a state of machine M and a tape symbol, and all the rest are tape symbols $\}$, and
  $E_{M,y} = \{< v_{C_1}, v_{C_2} > \mid$ the machine M can go from configuration C_1 to a configuration C_2 in one step $\}$.

  If C_i is the initial configuration of machine M on input y, then M accepts y if and only if the node labeled by C_f is reachable from the node labeled by C_i in $G_{M,y}$. In other words, $x \in \mathcal{L}(U_{\mathbf{PSPACE}})$ if and only if the GAP problem on $G_{M,y}$ has the answer "yes."

Claim: The edge relation $E_{M,y}$ in graph $G_{M,y}$ has a small OBDD representation.

Proof. We need to show that the function e, where,

$$e(C_1, C_2) = \begin{cases} 1 \text{ if and only if } < v_{C_1}, v_{C_2} > \in E_{M,y} \\ 0 \text{ otherwise} \end{cases}$$

can be represented by a small OBDD. Computing the function e entails:

(a) checking that C_1 and C_2 are possible configurations, *i.e.*, there is exactly one composite symbol in each of C_1 and C_2, and
(b) checking whether the configuration C_2 can be reached from C_1 in one step by the machine M.

Checking to see whether a symbol is composite amounts to checking whether the index of the symbol is greater than some constant, because we list all the composite symbols in the end. Hence, check (a) only involves examining the symbols of the configuration in the order in which they occur and thus has a small OBDD representation.

Let $Quad = \{(W, X, Y, Z) |$ if $W, X,$ and Y are the symbols in the $(j-1)^{st}$, j^{th} and $(j+1)^{st}$ cells, respectively, at some time instant, then Z is the symbol in the jth cell at the next time instant $\}$. Checking whether configuration C_2 can be reached from configuration C_1 in one step involves checking whether all the symbols in C_2 arise from the corresponding symbols in C_1. That is, we need to check that $\forall j, (C_1^{(j-1)}, C_1^{(j)}, C_1^{(j+1)}, C_2^{(j)}) \in Quad$. As we saw in the proof of theorem 12, the function that checks whether a given quadruple is in $Quad$ can be represented by a small OBDD. We just read the symbols of C_1 and C_2 alternately and keep checking whether they "conform." Note that we need to "remember" only two symbols of C_1 as we go along. Hence, at any level in the OBDD, there are at most a constant number of nodes, and checking whether one configuration can follow from another is representable by a small OBDD.

□

Because $G_{M,y}$ can be represented by a small OBDD that can be constructed in polynomial time, the GAP problem for graphs represented by OBDDs is PSPACE-complete.

□

Definition 17. An AND-OR graph is a directed graph G with vertices labeled "AND" or "OR." Reachability in such graphs is recursively defined as follows :

(a) Every vertex is reachable from itself.
(b) If u is an AND node, then v is reachable from u if and only if v is reachable from **all** u_i, such that $< u, u_i >$ is an edge in the graph.
(c) If u is an OR node, then v is reachable from u if and only if v is reachable from **any** u_i, such that $< u, u_i >$ is an edge in the graph.

Problem 18. (AGAP)
The Alternating Graph Accessibility Problem is:
Input An AND-OR graph G and vertices s and t in G.
Output Is t reachable from s ?

Theorem 19. *The AGAP problem for graphs represented by OBDDs is EXP-complete.*

Proof. Because the AGAP problem is in P for graphs represented by adjacency matrices, it is in EXP for graphs represented by OBDDs.

Let x be a string that encodes a $2^{p_M(n)}$-time bounded Turing machine M and an input y. We will construct an AND-OR graph with two special vertices s and t, such that t will be reachable from s if and only if $x \in \mathcal{L}(U_{\mathbf{EXP}})$. The construction of the graph is very similar to the construction of the circuit in the proof that CIRCUIT VALUE is P-complete.

Once again let $Quad = \{(W, X, Y, Z) \mid$ if W, X, and Y are the symbols in the $(j-1)^{st}$, j^{th}, and $(j+1)^{st}$ cells, respectively, at some time instant, then Z is the symbol in the jth cell at the next time instant $\}$. Let $<$ be some ordering on the quadruples in $Quad$.

We will construct the graph $G_{M,y}$ in stages, starting with the empty graph.

Stage 0: Add two AND nodes, one labeled 0 and the other 1. These nodes represent "false" and "true," respectively.

Stage 1: For each $j, 0 \le j \le 2^{p_M(n)}$, and each X, where X is either a tape symbol or a composite symbol encoding a state of machine M and a tape symbol, add an OR node labeled $V_{0,j,X}$. Add the edge $< V_{0,j,X}, 1 >$ if the jth symbol in the initial configuration of M on input y is X. Otherwise, add the edge $< V_{0,j,X}, 0 >$.

Stage 2i: For each j and k, add an AND node labeled $N_{i,j,k}$. Add edges $< N_{i,j,k}, V_{i-1,j-1,W} >, < N_{i,j,k}, V_{i-1,j,X} >$, and $< N_{i,j,k}, V_{i-1,j-1,Y} >$, where the kth quadruple in $Quad$ is (W, X, Y, Z), for some Z.

Stage 2i+1: For each j and symbol Z, add an OR node labeled $V_{i,j,Z}$. For each k, if (W, X, Y, Z) is the kth quadruple, for some W, X, and Y, then add the edge $< V_{i,j,Z}, N_{i,j,k} >$.

Stage $2^{p_M(n)} + 2$: Add an OR node s. For all j, add edges $< s, V_{2^{p_M(n)},j,X} >$, where X is a composite symbol encoding a final state and some tape symbol.

The basic idea of the construction is as follows. The node-label $V_{i,j,X}$ means that, during the computation, at time i, the jth tape cell contains the symbol X. From the definition of $Quad$, it can be seen that

$$V_{i,j,Z} = \bigvee_{(W,X,Y,Z) \in Quad} (V_{i-1,j-1,W} \wedge V_{i-1,j,X} \wedge V_{i-1,j+1,Y}).$$

The string y is accepted if, at time $2^{p_M(n)}$, the machine reaches a final state, *i.e.*,

$$\bigvee_j (\bigvee_{X \in F} V_{2^{p_M(n)},j,X}),$$

where F is the set of composite symbols that encode a final state and symbol pair. Hence, the graph $G_{M,y}$ is such that node 1 is reachable from node s if and only if M accepts input y.

Claim: The graph $G_{M,y}$ can be represented by a small OBDD.

Proof. The only interesting edges are those of the form $< N_{i,j,k}, V_{i-1,j',X} >$, where $j' = j$ or $j - 1$ or $j + 1$, and $< V_{i,j,X}, N_{i,j,k} >$. In each case, determining whether nodes $V_{i_1,j_1,X}$ and $N_{I_2,j_2,k}$ are adjacent involves checking whether i_1, i_2 and j_1, j_2 differ by a constant and whether the symbol X occurs in the kth quadruple. That both these checks can be done in by a small OBDD was seen in the proof of theorem 12.

□

Thus the graph $G_{M,y}$ can be represented by a small OBDD that can be constructed in polynomial time. Furthermore, node 1 is reachable from s in $G_{M,y}$ if and only if $x \in \mathcal{L}(U_{\mathbf{EXP}})$. Hence, the AGAP problem is EXP-complete for graphs represented by OBDDs.

□

7 Open Questions

The results that we prove in this paper are all negative; they show that, in the worst case, succinct encoding of instances using OBDDs results in problems that are hard for PSPACE, EXP, or NEXP. However, one of our main motivations for this investigation is the observed good performance of computer-aided verification tools on OBDD-encoded instances. Thus worst-case hardness results do not adequately capture the complexity of the problems on "real-world instances," as is often the case. It would be desirable to have precise characterizations of the special cases that occur in practice and of the special cases that can be solved efficiently.

It would also be desirable to have a general hardness result for OBDDs that is analogous to Papadmitriou and Yannakakis's result for circuits. In other words, is there a class of reductions such that any problem that is complete for NP via a reduction in the class is complete for NEXP when instances are encoded as OBDDs?

References

1. J. L. Balcázar, *The Complexity of Searching Implicit Graphs*, Aritificial Intelligence, 86 (1996), pp. 171-188.
2. J. L. Balcázar and A. Lozano, *The complexity of graph problems for succinctly represented graphs*, in Proc. Graph-Theoretic Concepts in Computer Science, Springer-Verlag, Lecture Notes in Computer Science 411, 1989, pp. 277-285.
3. J. L. Balcázar, A. Lozano, and J. Torán, *The complexity of Algorithmic Problems on Succinct Instances*, Computer Science (R. Baeza-Yates and U. Manber, Eds.), Plenum Press, New York, 1992, pp. 351-377.
4. I. Beer, S. Ben-David, D. Geist, R. Gewirtzman, and M. Yoeli, *Methodology and system for practical formal verification of reactive hardware*, in Proc. 6th Int'l Conf. on Computer-Aided Verification, Springer-Verlag, Lecture Notes in Computer Science 818, 1994, pp. 182–193.

5. C. L. Berman, *Ordered Binary Decision Diagrams and Circuit Structure*, in Proc. IEEE International Conf. on Computer Design, 1989, pp. 392–395.,

6. R. Brayton, A. Emerson, and J. Feigenbaum, *Workshop Summary: Computational and Complexity Issues in Automated Verification*, DIMACS Technical Report 96-15, Rutgers University, Piscataway NJ, June 1996.

7. R. Bryant, *Graph-Based Algorithms for Boolean Function Manipulation*, IEEE Transactions on Computers, C-35 (1986), pp. 677–691.

8. R. Bryant, *Symbolic Manipulation with Ordered Binary Decision Diagrams*, ACM Computing Surveys, 24 (1992), pp. 293–318.

9. J. R. Burch, E. M. Clarke, K. L. McMillan, D. L. Dill, and L. J. Hwang, *Symbolic model checking: 10^{20} states and beyond*, Information and Computation, 98 (1992), pp. 142–170.

10. E. M. Clarke, O. Grumberg, H. Hiraishi, S. Jha, D. E. Long, K. L. McMillan, and L. A. Ness, *Verification of the Futurebus+ Cache Coherence Protocol*, Formal Methods in System Design, 6 (1995), pp. 217–232.

11. T. Cormen, C. Leiserson, and R. Rivest, *Introduction to Algorithms*, The M.I.T. Press, Cambridge, MA (1989).

12. E. A. Emerson and C. L. Lei, *Efficient Model Checking in Fragments of the Propositional mu-Calculus*, Proc. 1st IEEE Symp. on Logic in Computer Science, 1986, pp. 267–278.

13. H. Galperin and A. Wigderson, *Succinct Representations of Graphs*, Information and Control, 56 (1983), pp. 183–198.

14. O. Kupferman and M. Y. Vardi, *Module checking*, Proc. 8th Int'l. Conf. on Computer-Aided Verification, Springer-Verlag, Lecture Notes in Computer Science 1102, 1996, pp. 75–86.

15. R. Kurshan, *Computer-Aided Verification of Coordinating Processes: The Automata-Theoretic Approach*, Princeton University Press, Princeton NJ, 1994.

16. R. Kurshan, *The Complexity of Verification*, Proc. 26th ACM Symp. on Theory of Computing, Montreal, 1994, pp. 365–371.

17. K. McMillan, *Symbolic Model Checking*, Kluwer Academic Publishers, Amsterdam, 1993.

18. C. Papadimitriou and M. Yannakakis, *A Note on Succinct Representations of Graphs*, Information and Control, 71 (1986), pp. 181–185.

19. S. Ponzio, *A lower bound for integer multiplication with read-once branching programs*, Proc. 27th ACM Symp. on Theory of Computation, Las Vegas, 1995, pp. 130–139.

20. S. Ponzio, *Restricted Branching Programs and Hardware Verification*, PhD Thesis, MIT, 1995.

21. S. Skyum and L. Valiant, *A Complexity Theory Based on Boolean Algebra*, in Proc. 23rd IEEE Symp. on Foundations of Computer Science, 1982, pp. 244-253.

22. J. Torán, *Succinct representations of counting problems*, Proc. 6th International Conf. on Applied Algebra, Algebraic Algorithms, and Error-Correcting Codes, Springer-Verlag, Lecture Notes in Computer Science 357, 1988, pp. 415-426.

23. H. Veith, *Languages Represented by Boolean Formulas*, TU Vienna, TR CD 85/95, 1995.

24. H. Veith, *Succinct Representation, Leaf Languages, and Projection Reductions*, in Proc. 11th IEEE Conf. on Computational Complexity, 1996, pp. 118-126.

Equivalence Test and Ordering Transformation for Parity–OBDDs of Different Variable Ordering

Jan Behrens and Stephan Waack

Institut für Numerische
und Angewandte Mathematik
Georg–August–Universität Göttingen
Lotzestr. 16–18, 37083 Göttingen
Germany

Abstract. Ordered binary decision diagrams (OBDDs) have already proved useful for example in the verification of combinational and sequential circuits. But the applications are limited, since the decriptive power of OBDDs is limited. Therefore several more general BDD models are studied. In this paper the so–called Parity–OBDDs are considered. The two polynomial time algorithms given are motivated by the fact that Parity–OBDD representation size essentially depends on the variable ordering. The one decides the equivalence of two Parity–OBDDs of possibly distinct variable ordering. The other one transforms a Parity–OBDD with respect to one variable ordering into an equivalent Parity–OBDD with respect to another preassigned variable ordering.

1 Introduction

In [2] Bryant introduced ordered binary decision diagrams (OBDDs) as a data structure for Boolean functions and circuit verification. See [3] for a survey article on OBDDs, which may be regarded as the state–of–the–art data structure. For convenience, let us briefly recall the basic definition. Let π be a permutation of the set $\{1, 2, \ldots, n\}$. A πOBDD over $\{x_1, x_2, \ldots, x_n\}$ is an acyclic directed graph, where the non-sink nodes are labeled with the Boolean variables from $\{x_1, x_2, \ldots, x_n\}$. The sinks are labeled with Boolean constants. Moreover, on every path from the source to a sink, the variables appear in the ordering given by π. The function represented is canonically defined.

The OBDD representation size decisively depends in many cases on the variable ordering chosen. Unfortunately, it is NP–complete to decide, whether, given a pair $(\mathcal{B}, s)$, where $\mathcal{B}$ is an OBDD and s is a natural number, there is an OBDD $\mathcal{B}'$ of size less than or equal to s, possibly respecting another ordering, which represents the same function as $\mathcal{B}$ (see [4]). Thus the following two problems are motivated.

1. *General equivalence test problem.* Check whether a πOBDD $\mathcal{B}_1$ and a σOBDD $\mathcal{B}_2$ represent the same function.

A polynomial time algorithm which solves this problem has been presented in [6].

2. *Variable ordering transformation problem.* Compute starting from a πOBDD $\mathcal{B}_1$ and a permutation σ a σOBDD $\mathcal{B}_2$ representing the same function as $\mathcal{B}_1$. An efficient solution of this problem is important for the improvement of given variable orderings by simulated annealing or genetic algorithms, and in situations where not compatible representations of functions are made compatible. (See [1] and [9] for more motivation.)

   Savický and Wegener showed in [9] that this problem can be deterministically solved in worst case time $\mathcal{O}\left(\text{SIZE}\left(\mathcal{B}_1\right) \cdot \text{SIZE}\left(\mathcal{B}_2\right) \cdot \log\left(\text{SIZE}\left(\mathcal{B}_2\right)\right)\right)$.

The main drawback of OBDDs is, that their descriptive power is low. Therefore more general BDD models are studied. See [10] for a survey. One possibility is to introduce nondeterminism (inputs can activate many paths) together with parity acception mode. This leads to the so-called Parity-OBDDs ($\oplus$OBDDs). They have been intensively studied in [7] and [11].

We continue this investigations by solving the general equivalence test problem (see Theorem 12) and the variable ordering transformation problem (see Theorem 13) for Parity-OBDDs in time $\mathcal{O}\left(n^2 \cdot \text{SIZE}\left(\mathcal{B}_1\right)^\omega \cdot \text{SIZE}\left(\mathcal{B}_2\right)^\omega\right)$ and in space $\mathcal{O}\left(\text{SIZE}\left(\mathcal{B}_1\right)^2 \cdot \text{SIZE}\left(\mathcal{B}_2\right)^2\right)$.

2 Parity-OBDDs

Let us recall the basic definitions concerning $\oplus$OBDDs as well as the basic results. A permutation σ of the set $\{1, \ldots, n\}$ induces an ordering $\left(x_{\sigma(1)}, \ldots, x_{\sigma(n)}\right)$ of the input variables. A $\oplus$OBDD $\mathcal{B}$ over Boolean variables $\{x_1, \ldots, x_n\}$ with respect to the variable ordering σ, a σ-$\oplus$OBDD for short, is a directed acyclic graph with properties as described in the following. The set of nodes $\mathcal{N}^{\mathcal{B}}$ of $\mathcal{B}$ is partioned into $n+2$ levels denoted by $\mathcal{N}_k^{\mathcal{B}}$, for $k = 0, \ldots, n+1$. Level $\mathcal{N}_{n+1}^{\mathcal{B}}$, which is always not empty, consists of a sequence of nodes, *the sources*. The sources may be connected to any other node by at most one unlabeled arc. The level $\mathcal{N}_0^{\mathcal{B}}$ is always a subset of the singleton set $\{s_1\}$, where s_1 is the *1-sink*. The outdegree of the node s_1 is zero. The level $\mathcal{N}_{n-k+1}^{\mathcal{B}}$, for $k = 1, \ldots, n$, consists of so-called *branching nodes* labeled with the Boolean variable $x_{\sigma(k)}$, for $k = 1, \ldots, n$. A node $v \in \mathcal{N}_{n-k+1}^{\mathcal{B}}$ may be connected to any node on any level $\mathcal{N}_{n-l+1}^{\mathcal{B}}$, for $l > k$, by an arc labeled with 0 or 1. (Parallel arcs must have distinct labels.) Let us denote by level (v) the index of v's level, by $\mathcal{A}^{\mathcal{B}}$ the set of arcs of the diagram $\mathcal{B}$. For any $\oplus$OBDD $\mathcal{B}$, the number of nodes is regarded as *size* of the diagram $\mathcal{B}$, and denoted by SIZE $(\mathcal{B})$. Note, that the storage size of the $\oplus$OBDD $\mathcal{B}$ may be something between SIZE $(\mathcal{B})$ and $\mathcal{O}\left(\text{SIZE}\left(\mathcal{B}\right)^2\right)$.

It is convenient to regard the space $\mathbb{B}_n$ of Boolean functions of n variables as an $\mathbb{F}_2$-algebra, where $\mathbb{F}_2$ is the prime field of characteristic 2, i.e. as an 2^n-dimensional vector space with an additional multiplication operation. The product of two functions $f, g \in \mathbb{B}_n$, which is defined to be the componentwise conjunction, is denoted by $f \wedge g$, or fg. Their sum, which corresponds to the componentwise exclusive-or, by $f \oplus g$. In this context, the variable x_i is taken to represent the projection from $\{0,1\}^n$ to the i-th coordinate, $\bar{x}_i$ as the complement of this function.

Definition 1. Let $\mathcal{B}$ be a $\oplus$OBDD with respect to a permutation σ of the set $\{1, 2, \ldots, n\}$.

1. For each node v of $\mathcal{B}$, we define the *resulting function* $\mathrm{Res}_v^{\mathcal{B}} = \mathrm{Res}_v \in \mathbb{B}_n$ by induction on the index $k = 0, \ldots, n+1$ of the level to which the node v belongs.
   (a) Basis $k = 0$. For v the 1–sink, $\mathrm{Res}_v = 1$.
   (b) If $\mathrm{level}(v) = k$, for $k = 1, \ldots, n$, let $\mathrm{Succ}_\epsilon(v)$ be the set of ϵ–successors: $\mathrm{Succ}_\epsilon(v) := \left\{ u \mid \exists a \in A^{\mathcal{B}} : a = v \stackrel{\epsilon}{\to} u \right\}$. Then $\mathrm{Res}_v := \bar{x}_{\sigma(n-k+1)} \bigoplus_{u \in \mathrm{Succ}_0(v)} \mathrm{Res}_u \oplus x_{\sigma(n-k+1)} \bigoplus_{u \in \mathrm{Succ}_1(v)} \mathrm{Res}_u$. (Since $\mathrm{Res}_v|_{x_{\sigma(n-k+1)}=e} = \bigoplus_{u \in \mathrm{Succ}_e(v)} \mathrm{Res}_u$, for $e = 0, 1$, this is in fact an extension of the well–known Shannon decomposition.)
   (c) If v is a source, and if $\mathrm{Succ}(v)$ is the set of its successors, then $\mathrm{Res}_v := \bigoplus_{u \in \mathrm{Succ}(v)} \mathrm{Res}_u$.
2. The function represented by the whole diagram $\mathcal{B}$, $\mathrm{Res}(\mathcal{B})$, is defined to be $(\mathrm{Res}_{v_1}, \ldots, \mathrm{Res}_{v_m})$, where $(v_1, \ldots, v_m)$ is the sequence of sources of $\mathcal{B}$.

Let $C^\sigma_{\oplus \mathrm{OBDD}}(f)$ be the minimum size of any $\oplus$OBDD that represents the function f using the variable ordering σ.

For describing the main result of [11] in terms of $\oplus$OBDDs representing sequences of functions rather than a single one we need a little more notation. A *subfunction* of a Boolean function $f \in \mathbb{B}_n$ is a function obtained by setting some of the variables of f to constant. A subfunction is formally defined on the whole domain, although it does not depend on the variables set to constant. We say that *it does not essentially depend on these variables*.

Definition 2. 1. We define $\mathbb{B}_k^\sigma$ to be all $g \in \mathbb{B}_n$ not essentially depending on the variables x_i, for $i \in \{\sigma(1), \ldots, \sigma(n-k)\}$. (Clearly, $\mathbb{B}_k^\sigma$ is canonically isomorphic to $\mathbb{B}_k$ via the so–called "diagonal isomorphism" of $\mathbb{F}_2$–algebras.)
   2. For $e \in \{0, 1\}$, and $k = 1, \ldots, n$, let $\pi_k^{\sigma e} : \mathbb{B}_k^\sigma \to \mathbb{B}_{k-1}^\sigma$ be the $\mathbb{F}_2$–linear maps gained by setting $x_{\sigma(n-k+1)}$ to e: $\pi_k^{\sigma e}(\phi) := \phi|_{x_{\sigma(n-k+1)}=e}$.
   3. For $e \in \{0, 1\}$, and $k = 1, \ldots, n$, let $\iota_k^{\sigma e} : \mathbb{B}_{k-1}^\sigma \to \mathbb{B}_k^\sigma$ be the $\mathbb{F}_2$–linear maps defined by
$$\iota_k^{\sigma e}(\phi) := \begin{cases} \phi & \text{if } x_{\sigma(n-k+1)} = e; \\ 0 & \text{otherwise.} \end{cases}$$

Clearly, $\mathbb{B}_k^\sigma$, regarded as a vector space, is the direct sum of two copies of $\mathbb{B}_{k-1}^\sigma$. The maps $\pi_k^{\sigma e}$, $\iota_k^{\sigma e}$, for $e = 0, 1$, are the canonical projections and injections.

Now we can fully describe a $\oplus$OBDD in terms of linear equations in a very easy way.

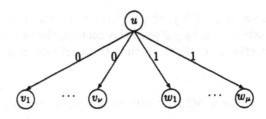

Fig. 1. The local neighbourhood relation in a $\oplus$OBDD

Lemma 3. *Let $\mathcal{B}$ be a $\oplus$OBDD with respect to the permutation σ, let u be a branching node of the diagram $\mathcal{B}$ belonging to the level $\mathcal{N}_l^{\mathcal{B}}$. Then the neighbourhood relation in $\mathcal{B}$ shown in Figure 1 is equivalent to the equation $\text{Res}_u^{\mathcal{B}} = \iota_l^{\sigma 0} \left(\sum_{i=1}^{\nu} \text{Res}_{v_i}^{\mathcal{B}} \right) + \iota_l^{\sigma 1} \left(\sum_{j=1}^{\mu} \text{Res}_{w_j}^{\mathcal{B}} \right)$, or, what is the same, to $\pi_l^{\sigma 1} \left(\text{Res}_u^{\mathcal{B}} \right) = \sum_{j=1}^{\mu} \text{Res}_{w_j}^{\mathcal{B}}$ and $\pi_l^{\sigma 0} \left(\text{Res}_u^{\mathcal{B}} \right) = \sum_{i=1}^{\nu} \text{Res}_{v_i}^{\mathcal{B}}$. (Observe that the sets $\{v_i \mid i = 1, \ldots, \nu\}$ and $\{w_j \mid j = 1, \ldots, \mu\}$ are not necessarily disjoint.)*

Definition 4. 1. Let $f \in \mathbb{B}_n$ be a Boolean function over the variables $\{x_1, \ldots, x_n\}$.
   For $I \subseteq \{1, 2, \ldots, n\}$, let $\mathbb{S}_I(f)$ be the set of all subfunctions obtained from f by setting the variables x_i for all $i \in I$ to all possible constants.
   For σ be a permutation of the set $\{1, 2, \ldots, n\}$, and $k = 0, \ldots, n$, let $\mathbb{S}_k^{\sigma}(f) := \mathbb{S}_I(f)$, where $I := \{\sigma(1), \ldots, \sigma(n - k)\}$, and let $\mathbb{B}_k^{\sigma}(f)$ be the vector space spanned by $\mathbb{S}_k^{\sigma}(f) \cup \ldots \cup \mathbb{S}_0^{\sigma}(f)$. (In particular, $\mathbb{S}_n^{\sigma}(f)$ is the singleton set $\{f\}$, and $\mathbb{S}_0^{\sigma}(f)$ is the set of constant functions defined on $\{0, 1\}^n$ with values in the range of f.)
   2. Let $f = (f_1, \ldots, f_m) \in (\mathbb{B}_n)^m =: \mathbb{B}_n^m$ be a sequence of Boolean functions, and let σ be a permutation of the set $\{1, \ldots, n\}$. If a $\oplus$OBDD $\mathcal{B}$ represents f, then we define $\text{Res}(\mathcal{B})$ to be f, and $(\text{Res}(\mathcal{B}))_i$ to be f_i, for $i = 1, \ldots, m$. Moreover, let us define $\mathbb{S}_I(f) := \bigcup_{i=1}^n \mathbb{S}_I(f_i)$, $\mathbb{S}_k^{\sigma}(f) := \bigcup_{i=1}^n \mathbb{S}_k^{\sigma}(f_i)$, $\mathbb{B}_k^{\sigma}(f) := \text{span}_{\mathbf{F}_2} \mathbb{S}_k^{\sigma}(f) \cup \ldots \cup \mathbb{S}_0^{\sigma}(f)$.

According to [11] we can describe the size–minimal $\oplus$OBDD with respect to a given permutation σ as follows.

Theorem 5. *Let $f \in \mathbb{B}_n^m$ be a sequence of Boolean functions, σ be a permutation. Then $C_{\oplus\text{OBDD}}^{\sigma}(f) = \dim_{\mathbf{F}_2} \left(\mathbb{B}_n^{\sigma}(f) \right) + m$.*
   Moreover, if $\mathcal{B}$ is any $\oplus$OBDD with respect to σ representing f such that $C_{\oplus\text{OBDD}}^{\sigma}(f) = \text{SIZE}(\mathcal{B})$, then $\#\mathcal{N}_k^{\mathcal{B}} = \dim_{\mathbf{F}_2} \left(\mathbb{B}_k^{\sigma}(f) / \mathbb{B}_{k-1}^{\sigma}(f) \right)$, for $k = 1, \ldots, n$, and $\#\mathcal{N}_0^{\mathcal{B}} = \dim_{\mathbf{F}_2} \left(\mathbb{B}_0^{\sigma}(f) \right)$.

The following corollary is easy to prove.

Corollary 6. *Let the σ-$\oplus$OBDD $\mathcal{B}$ have exactly one source. If the σ-$\oplus$OBDD $\mathcal{B}$ is size–minimal, and $\mathrm{Res}\,(\mathcal{B}) \neq 0$, then the source of $\mathcal{B}$ has exactly one successor v, and we have that $\mathrm{Res}_v^{\mathcal{B}} = \mathrm{Res}\,(\mathcal{B})$.*

All that matters now is that there is an algorithm which computes a size–minimal σ-$\oplus$OBDD starting from a arbitrary one.

Let us define *an exponent ω of matrix multiplication* over a field k to be a real number such that multiplication of two square matrices of order h may be algorithmically achieved with $O\,(h^\omega)$ arithmetical operations. (Notice, that this is a nonstandard definition. Usually, *the* exponent of matrix multiplication τ is defined to be the infimum of all ω's.) It is well–known, that matrix multiplication plays a key role in numerical linear algebra. Thus the following problems all have "exponent" ω: matrix inversion, L-R-decomposition, evaluation of the determinant. (For an overview see [8], Chapter 11.) Up to now, the best known ω is 2.376 (see [5]).

Theorem 7 (Minimization of the number of nodes). *There is an algorithm which computes from a σ-$\oplus$OBDD $\mathcal{B}$ representing a Boolean function $f \in \mathbb{B}_n^m$ a size–minimal σ-$\oplus$OBDD representing f in time $\mathcal{O}\,(n \cdot (\mathrm{SIZE}\,(\mathcal{B}))^\omega)$ and space $\mathcal{O}\left(\mathrm{SIZE}\,(\mathcal{B})^2\right)$, for any exponent of matrix multiplication ω.*

Let $\mathcal{B}'$ be a σ'-$\oplus$OBDD and $\mathcal{B}''$ be a σ''-$\oplus$OBDD, both having exactly one source, then we say that $\mathcal{B}'$ and $\mathcal{B}''$ are *equivalent*, abbreviated by $\mathcal{B}' \equiv \mathcal{B}''$, if they represent one and the same function, i. e. $\mathrm{Res}\,(\mathcal{B}') = \mathrm{Res}\,(\mathcal{B}'')$. The equivalence test problem for $\oplus$OBDDs of identical variable ordering is deterministically solved in [11].

Theorem 8. *Let $\mathcal{B}'$ and $\mathcal{B}''$ be two $\oplus$OBDDs with respect to σ. Then we can decide whether $\mathcal{B}' \equiv \mathcal{B}''$ in time $O\,(n \cdot (\mathrm{SIZE}\,(\mathcal{B}') + \mathrm{SIZE}\,(\mathcal{B}''))^\omega)$ and in space $O\left((\mathrm{SIZE}\,(\mathcal{B}') + \mathrm{SIZE}\,(\mathcal{B}''))^2\right)$, where ω is an exponent of matrix multiplication.*

The problem is to extend Theorem 8 to different variable orderings. This is done in Theorem 12.

Definition 9. A $\oplus$OBDD $\mathcal{B}$ over the set of Boolean variables $\{x_1, \ldots, x_n\}$ is defined to be *algebraically reduced* if and only if the vector space $\mathrm{span}_{\mathbb{F}_2}\left\{\mathrm{Res}_v^{\mathcal{B}} \mid v \in \mathcal{N}^{\mathcal{B}}, 0 \leq \mathrm{level}\,(v) \leq n\right\}$ is freely generated by $\{\mathrm{Res}_v^{\mathcal{B}} \mid v \in \mathcal{N}^{\mathcal{B}}, 0 \leq \mathrm{level}\,(v) \leq n\}$.

Since we consider $\oplus$OBDDs representing a sequence $(f_1, \ldots, f_m)$ of Boolean functions, there is an additional problem. The question is to select from $\{f_1, \ldots, f_m\}$ a basis of the space $\mathrm{span}_{\mathbb{F}_2}\{f_1, \ldots, f_m\}$ and to represent the functions not selected in terms of those selected.

Proposition 10. *For each exponent of matrix multiplication ω, there is an algorithm which works as follows. If $\mathcal{B}$ is a algebraically reduced $\oplus$OBDD over $\{x_1, \ldots, x_n\}$ with respect to a permutation σ representing $(f_1, \ldots, f_m) \in \mathbb{B}_n^m$, then*

1. *we can select from the left to the right a basis of the space* $\mathrm{span}_{\mathbb{F}_2}\{f_1,\ldots,f_m\}$ *from the sequence* $(f_1,\ldots,f_m)$ *in time* $\mathcal{O}(\mathrm{SIZE}(B)^\omega)$ *and space* $\mathcal{O}\left(\mathrm{SIZE}(B)^2\right)$,

2. *if* $\{f_1,\ldots,f_\kappa\}$ *is w. l. o. g. the basis selected in Claim 1, then we can compute in additional time* $\mathcal{O}(\mathrm{SIZE}(B)^\omega)$ *the* $\kappa \times (k-\kappa)$*-matrix* M *such that*

$$(f_0,\ldots,f_\kappa)\cdot M = (f_{\kappa+1},f_{\kappa+2},\ldots,f_k).$$

Proof. We know that the functions Res_v^B, for $v \in \mathcal{N}^B$, level $(v) \leq n$, form a basis of the space $\mathbb{B}_n^\sigma(f_1,\ldots,f_m)$. We now represent each f_i in terms of that basis by the help of the $\oplus$OBDD representation on the basis of Lemma 3. Then we apply the well–known methods of numerical linear algebra.

Let B be a $\oplus$OBDD over the set of variables $\{x_1,\ldots,x_n\}$. If $x \in \{x_1,\ldots,x_n\}$, and $e \in \{0,1\}$, then we define the $\oplus$OBDD $B/\{x=e\}$, which is formally regarded as a $\oplus$OBDD over the set of variables $\{x_1,\ldots,x_n\}$, too, as follows. For each node v labeled with x we proceed as follows. First, we lengthen all ingoing arcs of v to all e-successors of v. Second, we delete all outgoing arcs of v. Last, we delete the node itself. Clearly, $B/\{x=e\}$ can be constructed in linear time from B and represents the function $f\mid_{x=e}$. We observe, that $(B/\{x=e\})/\{x'=e'\} = (B/\{x'=e'\})/\{x=e\}$ which we denote by $B/\{x=e,x'=e'\}$. If B is a $\oplus$OBDD with respect to π, then $B/\{x=e\}$ can be regarded as a $\oplus$OBDD with respect to π, too.

Lemma 11. *Let* B_1, B_2 *be two* $\oplus$*OBDDs over* $\{x_1,\ldots,x_n\}$.

1. *If* $B_1 \equiv B_2$, *then* $\forall x \in \{x_1,\ldots,x_n\}\,\forall e \in \{0,1\}: B_1/\{x=e\} \equiv B_2/\{x=e\}$.
2. *If* $\exists x \in \{x_1,\ldots,x_n\}\,\forall e \in \{0,1\}: B_1/\{x=e\} \equiv B_2/\{x=e\}$, *then* $B_1 \equiv B_2$.

Proof. Both assertions follow from Shannon's decomposition and the fact that $B/\{x=e\}$ represents the function $f\mid_{x=e}$.

3 The equivalence test of a π-$\oplus$OBDD and a σ-$\oplus$OBDD

In order to carry out the proof of Theorem 12 we need the following two easy constructions.

First, let B be a $\oplus$OBDD with respect to a permutation σ over the set of variables $\{x_1,\ldots,x_n\}$, and let $V_1,\ldots,V_m$ be subsets of the set of nodes of B. Then the σ-$\oplus$OBDD $B[V_1,\ldots,V_m]$ is constructed from B and $V_1,\ldots,V_m$ as follows. The old sources are removed. Then a sequence of new sources $q_1,\ldots,q_m$ is created. They are connected to the lower level nodes in such a way that $\mathrm{Res}_{q_i}^{B[V_1,\ldots,V_m]} = \sum_{v \in V_i}\mathrm{Res}_v^B$, for $i=1,\ldots,m$. Finally, nodes which cannot be reached from at least one source are superfluous. That's why we delete all nodes from the diagram constructed so far which are not accessible from one of the new sources via a path. The foregoing construction can be done in a straightforward way in time $\mathcal{O}(m\cdot\mathrm{SIZE}(B))$. Moreover, if B is algebraically reduced, then, of course, $\mathcal{B}[V_1, \ldots, V_m]$ is algebraically reduced, too. Finally, $(\text{Res}\,(\mathcal{B}[V_1, \ldots, V_m]))_j = \text{Res}\,(\mathcal{B}[V_j])$, for $j = 1, \ldots, m$.

Second, let $\mathcal{B}_1, \mathcal{B}_2, \ldots, \mathcal{B}_k$ be a sequence of σ-$\oplus$OBDDs for the set of variables $\{x_1, \ldots, x_n\}$, and let $q_1, \ldots, q_k$ be their sequences of sources. We define $\mathcal{B}_1 \curlyvee \mathcal{B}_2 \curlyvee \ldots \curlyvee \mathcal{B}_k$, the *(ordered) bouquet* of the $\oplus$OBDDs $\mathcal{B}_1, \ldots, \mathcal{B}_k$, to be the σ-$\oplus$OBDD gained as follows. We take disjoint copies of $\mathcal{B}_1, \mathcal{B}_2, \ldots, \mathcal{B}_k$, identify their sinks, and commit $(q_1, \ldots, q_k)$ to being the sequence of sources of the new $\oplus$OBDD. Clearly, $\text{Res}\,(\mathcal{B}_1 \curlyvee \mathcal{B}_2 \curlyvee \ldots \curlyvee \mathcal{B}_k) = (\text{Res}\,(\mathcal{B}_1), \ldots, \text{Res}\,(\mathcal{B}_k))$, and $\mathcal{B}_1 \curlyvee \mathcal{B}_2 \curlyvee \ldots \curlyvee \mathcal{B}_k$ can be computed from $\mathcal{B}_1, \mathcal{B}_2, \ldots, \mathcal{B}_k$ in linear time.

Theorem 12 (General equivalence test for $\oplus$OBDDs). *Let $\mathcal{B}_1$ and $\mathcal{B}_2$ be two size–minimal $\oplus$OBDDs over $\{x_1, \ldots, x_n\}$ with respect to π and σ. For each exponent of matrix multiplication ω, there is an algorithm which checks in time $\mathcal{O}\left(n^2 \cdot \text{SIZE}\,(\mathcal{B}_1)^\omega \cdot \text{SIZE}\,(\mathcal{B}_2)^\omega\right)$ and space $\mathcal{O}\left(\text{SIZE}\,(\mathcal{B}_1)^2 \cdot \text{SIZE}\,(\mathcal{B}_2)^2\right)$ whether or not $\mathcal{B}_1 \equiv \mathcal{B}_2$.*

Proof. In order to prove the theorem, we consider the following algorithm. We assume w.l.o.g. that $1 < \text{SIZE}\,(\mathcal{B}_1) \leq \text{SIZE}\,(\mathcal{B}_2)$, and that π is the identity permutation. (The case $1 = \text{SIZE}\,(\mathcal{B}_1) \leq \text{SIZE}\,(\mathcal{B}_2)$ can be trivially decided, since $\text{SIZE}\,(\mathcal{B}_2) = 1$ if and only if $f_2 \equiv 0$.) Let L be the reference to a set of pairs $\rho = \left(\rho^{(left)}, \rho^{(right)}\right)$, where $\rho^{(left)}$ is a set of nodes of $\mathcal{B}_1$, and $\rho^{(right)}$ is a set of equations $\{x_{i_1} = e_{i_1}, \ldots x_{i_\lambda} = e_{i_\lambda}\}$, for $i_1, \ldots i_\lambda$ pairwise distinct indices. (The elements of the set L, which can be canonically regarded as pairs of $\oplus$OBDDs, play an analogous role as in the OBDD-case (see [6]). They must be equivalent in order that $\mathcal{B}_1$ and $\mathcal{B}_2$ be equivalent. But here, moreover, it is necessary to remove "linear dependent" pairs after having checked whether the left and the right components fulfill the same linear equations.)

Algorithm
Inititialization:
Let **var** be an integer variable, let **equiv** be a Boolean variable, let **auxL** be of the same type as L, and let q be the unique successor of the source of $\mathcal{B}_1$ (see Corollary 6).

We assign L $\leftarrow \{\{q\}, \emptyset\}$, **var** $\leftarrow$ level (q) **equiv** $\leftarrow$ **true**, **auxL** $\leftarrow \emptyset$.

Main:
while (var ≥ 1) $\wedge$ (equiv = true) **do**

1. For all $\rho \in$ L having a node $v \in \rho^{(left)}$ such that level $(v) = $ **var do**.
   (a) If $v_1, \ldots, v_{\bar{l}}$ are all nodes $v \in \rho^{(left)}$ such that level $(v) = $ **var**, then we create new objects ρ_e, for $e = 0, 1$, by $\rho_e^{(left)} := \left(\rho^{(left)} \setminus \{v_1, \ldots, v_{\bar{l}}\}\right) \oplus$
   $\bigoplus_{i=1}^{\bar{l}} \text{Succ}_e\,(v_i)$, $\rho_e^{(right)} := \rho^{(right)} \cup \{x_{n-var+1} = e\}$, where "$\oplus$" denotes the symmetric difference of sets in this context. (Clearly, then $\text{Res}\,(\mathcal{B}_1\,[\rho^{(left)}])\,|_{x_{n-var+1}=e} = \text{Res}\,\left(\mathcal{B}_1\,\left[\rho_e^{(left)}\right]\right)$, for $e = 0, 1$, and $\bar{l} = \#\mathcal{N}_{var}^{\mathcal{B}_1}$.)
   (b) We assign L $\leftarrow$ L $\setminus \{\rho\}$, **auxL** $\leftarrow$ **auxL** $\cup \{\rho_0, \rho_1\}$.
od.

2. If auxL $= \emptyset$, then we go to Global Step 7 of this loop.

3. Assume that $L = \{\rho_1, \ldots, \rho_k\}$, and auxL $= \{\rho_{k+1}, \ldots, \rho_{k+l}\}$. (We notice that k as well as l depend on var.) We construct the $\oplus$OBDD $\mathcal{B}_1 \left[\rho_1^{(left)}, \ldots, \rho_{k+l}^{(left)} \right]$, and compute by the help of Proposition 10 an index set I_{var}, $\emptyset \subseteq I_{var} \subseteq \{1, \ldots, l\}$, such that the functions $\mathrm{Res}\left(\mathcal{B}_1 \left[\rho_i^{(left)} \right] \right)$, $\mathrm{Res}\left(\mathcal{B}_1 \left[\rho_{k+j}^{(left)} \right] \right)$, for $i = 1, \ldots, k$, and $j \in I_{var}$, form a basis of the space $\mathrm{span}_{\mathbb{F}_2} \left\{ \mathrm{Res}\left(\mathcal{B}_1 \left[\rho_j^{(left)} \right] \right) \mid j = 1, \ldots, k + l \right\}$.

4. We assume from now on w.l.o.g. that the index set I_{var} computed in Item 3 of this loop to be $\{1, \ldots l'\}$, for $l' \leq l$. Let us abbreviate $\mathrm{Res}\left(\mathcal{B}_1 \left[\rho_j^{(left)} \right] \right)$ for a moment by ϕ_j, for $i = 1, \ldots, k + l$. Compute by Proposition 10 the matrix M_{var} such that $(\phi_1, \phi_2, \ldots, \phi_{k+l'}) \cdot M_{var} = (\phi_{k+l'+1}, \phi_{k+l'+2}, \ldots, \phi_{k+l})$.

5. We construct the bouquet $\mathcal{B}_2/\rho_1^{(right)} \curlyvee \mathcal{B}_2/\rho_2^{(right)} \curlyvee \ldots \curlyvee \mathcal{B}_2/\rho_{k+l'}^{(right)}$, which is a σ-$\oplus$OBDD and reduce it according to Theorem 7. Then we check by the help of Proposition 10 whether the functions $\mathrm{Res}\left(\mathcal{B}_2/\rho_i^{(right)} \right)$, for $i = 1, \ldots, k + l'$, form a basis of the space $\mathrm{span}_{\mathbb{F}_2} \left\{ \mathrm{Res}\left(\mathcal{B}_2/\rho^{(right)} \right) \mid \rho \in L \cup \text{auxL} \right\}$, and, if this question was answered in the affirmative, whether $(\psi_1, \psi_2, \ldots, \psi_{k+l'}) \cdot M_{var} = (\psi_{k+l'+1}, \psi_{k+l'+2}, \ldots, \psi_{k+l})$, where $\psi_i := \mathrm{Res}\left(\mathcal{B}_2/\rho_i^{(right)} \right)$, for $i = 1, \ldots, k + l$. We assign equiv $\leftarrow$ **false**, if one of the foregoing tests has failed.

6. If equiv $=$ **true**, then we assign $L \leftarrow L \cup \{\rho_{k+j} \mid j \in I_{var}\}$, auxL $\leftarrow \emptyset$. (We observe, that if var $= 1$, then $\#L = 1$.)

7. If var $= 1$ and equiv $=$ **true**, then we assign

$$
\text{equiv} \leftarrow \begin{cases} \textbf{true} & \text{if } \mathrm{Res}\left(\mathcal{B}_1 \left[\rho_{fin}^{(left)} \right] \right) \equiv \mathrm{Res}\left(\mathcal{B}_2/\rho_{fin}^{(right)} \right); \\ \textbf{false} & \text{otherwise,} \end{cases}
$$

where ρ_{fin} is the unique element of L.

8. var $\leftarrow$ var $- 1$

od.

Output of the algorithm: equiv.

As for correctness, we consider the following condition Γ_{var}, where var is the variant of the loop, before or after performing the body of the loop.

Γ_{var} is true if and only if the following subconditions are fulfilled.

1. The sequence of functions $\mathrm{Res}\left(\mathcal{B}_1 \left[\rho^{(left)} \right] \right)$, for $\rho \in L$, as well as the sequence of functions $\mathrm{Res}\left(\mathcal{B}_2/\rho^{(right)} \right)$, for $\rho \in L$, are linearly independent.

2. For all $\rho \in L$, and all $v \in \rho^{(left)}$, level$(v) \leq$ var. (In particular, $\mathrm{Res}\left(\mathcal{B}_1 \left[\rho^{(left)} \right] \right) \in \mathbb{B}_{var}^{\pi}(f)$.)

3. $\forall \rho \in L: \mathrm{Res}\left(\mathcal{B}_1 \left[\rho^{(left)} \right] \right) = \mathrm{Res}\left(\mathcal{B}_1/\rho^{(right)} \right)$.

4. $\mathcal{B}_1 \equiv \mathcal{B}_2 \iff \forall \rho \in L : \mathcal{B}_1/\rho^{(right)} \equiv \mathcal{B}_2/\rho^{(right)}$.

Using Lemma 11 it is easy to see that Γ_{var} is a loop invariant. The correctness follows now.

The space bound can be easily verified, since Global Step 5 dominates the space demand of the other global steps. Let us turn to the analysis of the running time.

First, we remark that because of Theorem 5 and the loop invariant $\#L, \#auxL = \mathcal{O}\left(SIZE\left(\mathcal{B}_1\right)\right)$.

Second, we observe, that the running time of Global Step 1 and Global Step 5 of the while–loop dominate those of the other global steps. Clearly, Global Step 5 always takes $\mathcal{O}\left(n \cdot SIZE\left(\mathcal{B}_1\right)^\omega \cdot SIZE\left(\mathcal{B}_2\right)^\omega\right)$ time. It follows that the overall time for Global Step 5 is $\mathcal{O}\left(n^2 \cdot SIZE\left(\mathcal{B}_1\right)^\omega \cdot SIZE\left(\mathcal{B}_2\right)^\omega\right)$.

Third, let us consider Global Step 1. As usual, we assume, that the $\oplus$OBDDs are given in terms of adjacency lists. Moreover, we have precomputed the adjacency matrices of the 1–successor as well as of the 0–successor subgraph of $\mathcal{B}_1$. We implement the sets $\rho^{(left)}$ as arrays of length $SIZE\left(\mathcal{B}_1\right)$. Then Global Step 1 can be done in time $\mathcal{O}\left(\left(\#\mathcal{N}_{var}^{\mathcal{B}_1} + 1\right) \cdot SIZE\left(\mathcal{B}_1\right)^2\right)$, for a fixed var. Thus the overall time for Global Step 1 is $\mathcal{O}\left(SIZE\left(\mathcal{B}_1\right)^3 + n \cdot SIZE\left(\mathcal{B}_1\right)^2\right)$. Since $\omega \geq 2$, and $SIZE\left(\mathcal{B}_1\right) \leq SIZE\left(\mathcal{B}_2\right)$, we have that $SIZE\left(\mathcal{B}_1\right)^3 + n \cdot SIZE\left(\mathcal{B}_1\right)^2 = \mathcal{O}\left(n \cdot SIZE\left(\mathcal{B}_1\right)^\omega \cdot SIZE\left(\mathcal{B}_2\right)^\omega\right)$. The claim on the running time follows.

4 How to transform a π-$\oplus$OBDD into a σ-$\oplus$OBDD

The following theorem introduces an algorithm which transforms an π-$\oplus$OBDD into a σ-$\oplus$OBDD.

Theorem 13 (Variable ordering transformation for $\oplus$OBDDs). *Let P be a size-minimal π-$\oplus$OBDD representation of a Boolean Function f on $X = \{x_1, \ldots, x_n\}$. Then there is for each exponent of matrix multiplication ω an algorithm which computes from P in time $\mathcal{O}\left(n^2 \cdot SIZE\left(Q\right)^\omega \cdot SIZE\left(P\right)^\omega\right)$ and space $\mathcal{O}\left(SIZE\left(Q\right)^2 \cdot SIZE\left(P\right)^2\right)$ a reduced σ-$\oplus$OBDD Q representing f.*

Proof. By renumbering the variables we can assume that σ is the identity permutation. The boolean function f depends w. l. o. g. on all variables $x_1, \ldots, x_n$. (By using the algebraic characterization of f it can be decided whether f depends on x_i or not: The function f depends on x_i if and only if $\mathcal{N}_{n-i+1}^P = \dim_{\mathbb{F}_2}\left(\mathbb{B}_{n-i+1}^\sigma(f)/\mathbb{B}_{n-i}^\sigma(f)\right) \neq 0$.)

The following algorithm constructs the desired $\sigma - \oplus$OBDD top–down in a similar way as in [1] and [9]. But here, moreover, it is necessary to select a "basis" of the nodes constructed so far and replace the other ones according to their linear dependence relations computed before. For each node of Q we store a set of equations $eq := \{x_{i_1} = e_{i_1}, \ldots, x_{i_\lambda} = e_{i_\lambda}\}$, for $i_1, \ldots i_\lambda$ pairwise distinct indices.

Algorithm

Initialization:

Let var be an integer variable and let **Leaves** be a set of nodes, and let v is the only successor of the source of Q. We assign $\mathcal{N}_0^Q \leftarrow \{s_1\}$, $\mathcal{N}_{n+1}^Q \leftarrow \{source\}$, $\mathcal{N}_n^Q \leftarrow \{v\}$, $v.eq := \emptyset$, **Leaves** $\leftarrow \{v\}$, var $\leftarrow n$.

Main:

while (var ≥ 1) **do**

1. For all $v \in$ **Leaves** such that level $(v) = $ var and each $e \in \{0,1\}$ we do the following. We construct and then reduce $P_{v,e} := P/(v.eq \cup \{x_{n-var+1} = e\})$. As to $P_{v,e}$ there are three cases to distinguish now. If $\operatorname{Res}(P_{v,e}) \equiv 0$, then nothing has to be done. If $\operatorname{Res}(P_{v,e}) \equiv 1$, then the e-successor of v is the 1-sink. Otherwise let t be the smallest number, so that $P_{v,e}$ contains an x_t-node. (Remember that we assumed σ as the identity.) Create an x_t-node u, set and $u.eq := v.eq \cup \{x_{n-var+1} = e\}$. Finally, add u to **Leaves** and let u be the e-successor of v.

2. Remove all $v \in$ **Leaves** such that level $(v) = $ var from **Leaves**.

3. Let $v_1, \ldots, v_r$ be all nodes in **Leaves** and let $v_{r+1}, \ldots, v_{r+s} \in \bigcup_{i=var}^n \mathcal{N}_i^Q$ be all nodes complete constructed so far, with the property, that for all $i, j \in \{1, \ldots, r+s\} : i \leq j \Rightarrow$ level $(v_i) \leq$ level (v_j). We establish the bouquet $P_\curlyvee := P/v_1.eq \curlyvee \ldots \curlyvee P/v_r.eq \curlyvee P/v_{r+1}.eq \curlyvee \ldots \curlyvee P/v_{r+s}.eq$. Proposition 10 allows us to compute an index set $I_{var} \subseteq \{1, \ldots, r+s\}$ with the property that the functions $\operatorname{Res}(P/v_i.eq)$, $i \in I_{var}$, form a basis of the space $\operatorname{span}_{\mathbb{F}_2} \{\operatorname{Res}(P/v_i.eq) \mid i = 1, \ldots, r+s\}$.

4. Let the index set I_{var} calculated in Global Step 3 of this loop be $I_{var} = \{\mu_1, \ldots, \mu_t\}$. and $R_{var} = \{\nu_1, \ldots, \nu_{r+s-t}\} = \{1, \ldots, r+s\} - I_{var}$ be the set of the remaining indices. We use Proposition 10 to get the matrix M_{var}, with $(\operatorname{Res}(P/v_{\mu_1}), \operatorname{Res}(P/v_{\mu_2}), \ldots, \operatorname{Res}(P/v_{\mu_t})) \cdot M_{var} = (\operatorname{Res}(P/v_{\nu_1}), \operatorname{Res}(P/v_{\nu_2}), \ldots, \operatorname{Res}(P/v_{\nu_{r+s-t}}))$.

5. We replace the ingoing arcs of $v_{\nu_1}, \ldots, v_{\nu_{r+s-t}}$ according to the representation we have got in Global Step 4. Afterwards the nodes $v_{\nu_1}, \ldots, v_{\nu_{r+s-t}}$ are obsolete and can be removed.

6. var $\leftarrow$ var $- 1$

od.

Output of the algorithm: Q.

To see the correctness of the algorithm we consider the condition Γ_{var}, where var is the variant of the loop, before and after performing the body of the loop. The condition Γ_{var} is true if and only if the following subconditions are fulfilled:

1. All levels $\mathcal{N}_{n+1}^Q, \ldots, \mathcal{N}_{var+1}^Q$ and their outgoing edges are already constructed.
2. For all nodes v so far constructed is valid $\operatorname{Res}(v) = \operatorname{Res}(P/v.eq) \in \mathbb{B}_{level(v)}^\sigma (f)$.
3. $\forall v \in$ **Leaves** : level $(v) \leq$ var.
4. All created nodes represent linear independent subfunctions of f, therefore, $\dim_{\mathbb{F}_2} \left(\operatorname{span}_{\mathbb{F}_2} \left\{ \operatorname{Res}(v) \mid v \in \text{**Leaves**} \cup \bigcup_{i=var}^n \mathcal{N}_i^Q \right\} \right) = \#(\text{**Leaves**} \cup \bigcup_{i=var}^n \mathcal{N}_i^Q) \leq C_{\text{OBDD}}^\sigma (f)$.

It is easy to see that Γ_{var} is a loop invariant and that after performing the algorithm the set Leaves is empty. Thus, Q is size minimal.

In order to show that the algorithm works in time $\mathcal{O}\left(n^2 \cdot \mathrm{SIZE}\,(Q)^\omega \cdot \mathrm{SIZE}\,(P)^\omega\right)$ it is sufficient to prove that Global Step 3 can be performed in time $\mathcal{O}\left(n \cdot \mathrm{SIZE}\,(Q)^\omega \cdot \mathrm{SIZE}\,(P)^\omega\right)$, since Global Step 3 dominates the other ones. The running time for Global Step 3 follows directly from the Subcondition 4 of condition Γ_{var} and the Proposition 10.

Since Global Step 3 dominates the space demand of the other global steps, the space bound can be easily verified.

5 Concluding Remark

The equivalence-test-problem for different orderings described in Section 3 can also be solved by the transformation algorithm presented in Section 4. However, the probability is very high that the equivalence-test-algorithm stops before all levels have been inspected, if the investigated Parity-OBDDs are not equivalent. Therefore, both algorithm are useful.

References

1. J. Bern, C. Meinel, and A. Slobodová. *Global rebuilding of obdds avoiding memory requirement maxima*, IEEE Trans. on Computers 1996, **15**, 131–134.
2. R. E. Bryant, *Graph-based algorithms for Boolean function manipulation*, IEEE Trans. on Computers 1986, **35**, pp. 677–691.
3. R. E. Bryant, *Symbolic Boolean manipulation with ordered binary decision diagrams*, ACM Comp. on Surveys 1992, **24**, pp. 293–318.
4. B. Bollig, I. Wegener, *Improving the variable ordering of OBDDs is NP-complete.*, Forschungsbericht Nr. 542 (1994) des Fachbereichs Informatik der Universität Dortmund.
5. D. Coppersmith, S. Winograd, *Matrix multiplication via arithmetic progressions*, J. Symbolic Computation 1990, **9**, pp. 251–280.
6. St. Fortune, J. Hopcroft, E. M. Schmidt, *The complexity of equivalence and containment for free single variable program schemes*, in: Proc. ICALP 1978, Lecture Notes in Computer Sci. **62**, Springer–Verlag 1978, pp. 227–240.
7. J. Gergov, Ch. Meinel, *Mod-2-OBDDs — a data structure that generalizes EXOR-Sum-of-Products and Ordered Binary Decision Diagrams*, Formal Methods in System Design 1996, **8**, pp. 273–282.
8. J. van Leeuwen (edit.), *Handbook of theoretical computer science, volume A*, Elsevier Science Publishers B.V. 1992.
9. P. Savický and I. Wegener. *Efficient algorithms for the transformation between different types of binary decision diagrams*, Acta Informatica 1997 **34**. pp. 245–256.
10. I. Wegener, *Efficient data structures for Boolean functions*, Discrete Mathematics 1994, **136**, pp. 347–372.
11. St. Waack, *On the decriptive and algorithmic power of parity ordered binary decision diagrams*, in: Proc. 14th STACS 1997, Lecture Notes in Computer Sci. **1200**, Springer–Verlag 1997, pp. 201–212.

Size and Structure of Random Ordered Binary Decision Diagrams

(Extended Abstract)

Clemens Gröpl*, Hans Jürgen Prömel**, and Anand Srivastav**

Humboldt-Universität zu Berlin
Institut für Informatik
Lehrstuhl Algorithmen und Komplexität

Abstract. We investigate the size and structure of ordered binary decision diagrams (OBDDs) for random Boolean functions. Wegener [Weg94] proved that for "most" values of n, the expected OBDD-size of a random Boolean function with n variables equals the worst-case size up to terms of lower order. Our main result is that this phenomenon, also known as strong Shannon effect, shows a threshold behaviour: The strong Shannon effect *does not* hold within intervals of constant width around the values $n = 2^h + h$, but it *does* hold outside these intervals. Also, the oscillation of the expected and the worst-case size is described. Methodical innovations of our approach are a functional equation to locate "critical levels" in OBDDs and the use of Azuma's martingale inequality and Chvátal's large deviation inequality for the hypergeometric distribution. This leads to significant improvements over Wegener's probability bounds.

1 Introduction

A Boolean function is a mapping $f : \{0, 1\}^n \to \{0, 1\}$ depending on Boolean variables. Efficient representation and manipulation of Boolean functions is an important issue in many applications, e. g. formal verification. The state-of-the-art data structure for Boolean functions are ordered binary decision diagrams, abbreviated OBDD (see Bryant's articles [Bry86,Bry92]). For a given variable ordering the OBDD is uniquely determined, and its size usually depends heavily on the chosen variable ordering. While there exist Boolean functions that have exponential OBDD-size for all variable orderings, many functions encountered in practice have polynomially sized OBDDs for *some* variable orderings. Of course, we would like this to be the typical, i. e. average case. Therefore, a thorough investigation of the relation between average-case and worst-case is recommended by theoretical and practical reasons. To do so, we have to make assumptions on the underlying probability distribution. From a structural point of view, the uniform distribution is most natural. In the following, we will briefly call a Boolean function chosen from the uniform distribution a *random* Boolean function.

* Graduate school "Algorithmische Diskrete Mathematik", supported by Deutsche Forschungsgemeinschaft, grant GRK 219/2-97.
** supported by Deutsche Forschungsgemeinschaft, grant Pr 296/3-2.

Previous Work. Recall that the ordered binary decision diagram (OBDD) of a Boolean function can be defined as the result of the application of the merging *and* the deletion rule to its binary decision tree. In the same way, the quasi-reduced ordered binary decision diagram (qOBDD) is the result of applying only the merging rule to its binary decision tree. (See Fig. 1 and 2.) The variables are tested according to some ordering $(x_1, \ldots, x_n)$ along each computation path. A *level* is the set of all nodes testing a particular variable.

The phenomenon that almost all functions have the same OBBD size as the hardest functions (up to a factor of $1 + o(1)$) is called *strong Shannon effect* for random Boolean functions and OBBDs [Weg94]. If almost all functions have the same OBBD size (up to a factor of $1 + o(1)$), but possibly smaller than the worst-case size, we say that the *weak Shannon effect* holds. Wegener proved that the weak Shannon effect holds for random Boolean functions and OBDDs: For arbitrary constants $\epsilon > 0$ the probability that the optimal OBDD size of a random Boolean function differs more than $O(n2^{2n/3})$ from the expected size is at most $O(2^{-n/3+\epsilon n})$. He also observed that the strong Shannon effect holds with probability tending to 1 as $h \nearrow +\infty$ for qOBDDs with a uniformly distributed random number $n \in [2^h .. 2^{h+1} - 1]$ of levels. Since for almost all functions the size of qOBDDs and OBDDs is the same (up to a factor of $1 + o(1)$), this result extends also to OBBDs. Interestingly, these results carry over to read-once branching programs, also called FBDDs.

The proof of Wegener is carried out in two steps. First, the results are proved for an arbitrary, but fixed variable ordering, and in a second step the generalization to arbitrary variable orderings is done using the second moment method. An important methodological innovation in the work of Wegener is the use of *urn experiments* (see [KSC78]) for the estimation of the expectation and the variance of the number of nodes on each OBDD-level.

Results. In Section 3 we show a threshold behaviour for the strong Shannon effect for random Boolean functions and OBDDs with respect to an arbitrary, but fixed variable ordering: The strong Shannon effect *does not* hold within intervals of constant width around the values $n = 2^h + h$, $h \in \mathbb{N}$, but it *does* hold outside these intervals.

In Section 4 we generalize our result to arbitrary variable orderings. This is done by computing large deviations. Unfortunately, the results of [KSC78] on the limit distributions of certain urn experiments cannot be applied in this context. Instead, we derive a special purpose inequality, using Azuma's martingale inequality and invoke Chvatal's bound on the hypergeometric distribution. We show that the probability of "large" de-

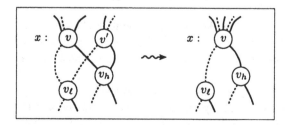 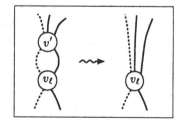

Fig. 1. Merging v and v' **Fig. 2.** Deleting v

viations is doubly exponentially small, which is a substantial improvement over Wegener's probability bound.

As a by-product of our proof we show an oscillation of the worst-case and the expected OBDD size. This generalizes and improves result of Heap and Mercer [HM94] and Liaw and Lin [LL92].

In Section 5, we identify those n for which the gap between the expected and the worst-case qOBDD-size is minimal.

Several important proofs are too complicated, technical and lengthy for an extended abstract. Due to space limitations we had to omit them. A full version of this paper is available electronically [GPS97].

2 Preliminaries

It is straightforward to show that the qOBDD of a Boolean function is uniquely determined for each variable ordering, as is the OBDD. Let f be a Boolean function and let qOBDD(f) be its qOBDD with respect to the variable ordering $(x_1, \ldots, x_n)$. The nodes on level i of qOBDD(f) represent the different subfunctions of f that can be obtained by substituting the first $i - 1$ variables $x_i, \ldots, x_{i-1}$ by constants $c_1, \ldots, c_{i-1}$. Let Y_i resp. Z_i denote the number of nodes on level i of the qOBDD resp. OBDD of f. Define $k_i := 2^{i-1}$, $m_i := 2^{2^{n-i+1}}$, $m_i' := m_i - m_{i+1}$ $w_i := \min\{k_i, m_i\}$ and $w_i' := \min\{k_i, m_i'\}$. Then the inequalities $Y_i \leqslant w_i$ and $Z_i \leqslant w_i'$ are valid (see [Weg94]).

Let us denote the worst-case size of the whole qOBDD by $W(n) := \sum_{i=1}^{n} w_i$. In Section 3.1 we will derive the precise asymptotic values of $W(n)$ for suitable parametrizations of n. For the moment, we only mention that $W(n) = \Theta(2^n/n)$.

We will need the following inequalities, valid for $x \leqslant 0$.

$$\left.\begin{array}{r} 1 + x \\ 1 + x + \dfrac{x^2}{2} + \dfrac{x^3}{6} \end{array}\right\} \leqslant e^x \leqslant 1 + x + \dfrac{x^2}{2}. \tag{2.1}$$

Finally, let us fix some notation. Intervals of integers will be denoted as $[a .. b] := [a, b] \cap \mathbb{Z}$ and $[a] := [1 .. a]$. We write ab/cd for $\frac{ab}{cd}$. The notation $f \sim g$ is equivalent to $f = (1 + o(1))g$.

3 Strong Shannon Effect – Fixed Variable Orderings

In this section f is a random Boolean function. We consider an arbitrary, but *fixed* variable ordering $(x_1, \ldots, x_n)$ and state all results for this case. The extension to arbitrary variable orderings and the proof of the full statement will be carried out in Section 4 via large deviation inequalities.

Define $X_i := w_i - Y_i$. Y_i is the number of nodes at the i-th level of qOBDD(f) while X_i is the number of nodes that are "missing" at the i-th level compared with the worst-case size w_i. Put $X := \sum_i X_i$, $Y := \sum_i Y_i$ and let $E(X)$ resp. $E(Y)$ denote the expectations.

By Wegener's work we already know that the weak Shannon effect holds for random Boolean functions and OBDDs. More precisely, the OBDD-size of almost all functions *is* the expected qOBDD-size $E(Y)$, up to a factor of $1 + o(1)$. (See also Section 4 of this paper.) Thus for the study of the strong Shannon effect it suffices to compute the ratio $E(Y)/W(n)$ or $E(X)/W(n)$.

Theorem 3.1 (Main Theorem). *Let* $B := \bigcup_{h \in \mathbb{N}} [2^h + h - d(h) .. 2^h + h + d(h)]$ *and* $A := \mathbb{N} \setminus B$.

(i) *If* $d(h) \nearrow +\infty$ *and* $n \in A$, *then* $E(X)/W(n) = o(1)$, *i. e. the strong Shannon effect holds for random Boolean functions and OBDDs with respect to the variable ordering* $(x_1, \ldots, x_n)$.

(ii) *If* $d(h) = O(1)$ *and* $n \in B$, *then* $E(X)/W(n) = \Omega(1)$, *i. e. the strong Shannon effect does not hold for random Boolean functions and OBDDs with respect to the variable ordering* $(x_1, \ldots, x_n)$.

In order to prove Theorem 3.1 we compute the worst-case size $W(n)$ and the expectation $E(X)$. The computation of $W(n)$ is done in the next subsection. After this, we analyse $E(X)$, which is much more difficult, and in fact requires new methods, especially the notion of critical levels.

The point i where the two upper bounds k_i and m_i meet turns out to be crucial for all results in this paper. Therefore, let us introduce a concise notation for it. We define the function L by the equation

$$L(n) + \log L(n) = n, \qquad (3.1)$$

and set

$$i_\delta := L(n) + \delta + 1. \qquad (3.2)$$

Then

$$k_{i_\delta} = 2^{\delta + L} \quad \text{and} \quad m_{i_\delta} = 2^{2^{-\delta}L}. \qquad (3.3)$$

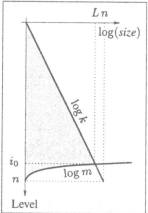

Fig. 3. The worst-case shape of an OBDD

It follows that the critical point is at $i = i_0$, where $k_{i_0} = m_{i_0}$. However, in general i_0 is not integral. It only marks the borderline where the worst-case level width turns from growing exponentially to shrinking doubly exponentially (see Fig. 3). A critical level is an $i_\delta \in \mathbb{N}$, where δ is sufficiently small. We remark that by Definition (3.1), the inverse function of L is $L^{-1}(i) = i + \log i$, and that $L(n) \sim n$.

3.1 The Worst-Case Size of qOBDDs

The behaviour of the worst-case bound $W(n)$ for various parametrizations of n is best expressed in terms of the ratio $W/2^L$. (Previous investigations [LL92,HM94] focused on the ratio $W/(\frac{2^n}{n})$, which does have the same asymptotic behaviour, but leads to more cumbersome computations.) The proofs of the results in this section can be carried out without much difficulties using the functional equation for L, but due to space limitations we have to omit them.

The first theorem gives the asymptotic value of $W/2^L$ for parametrizations of n "close" to $2^h + h$. (See Fig. 4.)

Theorem 3.2. *Assume that $a = o(2^h)$. Then*

$$\left(\frac{W}{2^L}\right)(2^h + h + a) \sim \begin{cases} 2, & a \leqslant 0; \\ 1 + 2^{-a}, & a \geqslant 0. \end{cases}$$

Theorem 3.2 is complemented by Theorem 3.3, which describes how $W/2^L$ develops between $2^h + h$ and $2^{h+1} + h + 1$. (See Fig. 5.)

Theorem 3.3. *Assume that $n \nearrow +\infty$ and let h and $b = b(h)$ be chosen such that $n = b2^h + h$ and $b \in [1, 2]$. Then*

$$\left(\frac{W}{2^L}\right)(b2^h + h) = b\left(1 + 2^{(1-b)2^h}(1 + o(1))\right),$$

which is $\sim b$, if b converges to real number in $]1, 2]$. The convergence of $W/2^L$ is uniform on each interval $[1 + \varepsilon, 2] \ni b$, where $\varepsilon > 0$ is a constant.

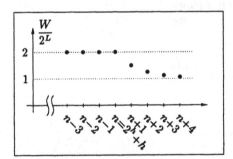

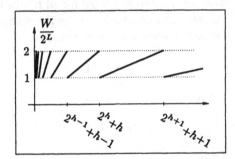

Fig. 4. The worst-case size of OBDDs near $n = 2^h + h$

Fig. 5. The oscillation of the worst-case size of OBDDs

Our refined analysis of the oscillation of the worst-case size also leads to improvements over the upper bound of Liaw and Lin [LL92] and the lower bound of Heap and Mercer [HM94], which are valid for all n.[1]

Corollary 3.4. *Let $\varepsilon > 0$ be an arbitrarily small constant. Then*

$$1 \leqslant \left(\frac{W}{2^L}\right)(n) \leqslant 2 + O\left(2^{-\frac{1-\varepsilon}{2}n}\right).$$

The lower bound is asymptotically tight for any sequence of the form $n = 2^h + h + a$, where $a \nearrow +\infty$, and $a = o(2^h)$. The upper bound is attained e. g. for $n = 2^h + h$.

---

[1] Essentially, the lower bound from [HM94] is $1/2$, and the upper bound from [LL92] is $2 + o(1)$.

3.2 Critical Levels and $E(X)$

Typically, not all nodes that are (numerically) possible at some level will exist. Recall that $X_i = w_i - Y_i$ is the number of nodes that are "missing" at the i-th level of a qOBDD. For each level, $E(X_i)$ can be computed by the following urn experiment (see [KSC78,Weg94]). We think of the subfunctions that result from constant assignments to the first $i - 1$ variables as *balls* and of the possible subfunctions at level i as *urns*. At level i, we are throwing k_i balls into m_i urns, so the expected number of non-empty urns is

$$E(Y_i) = \sum_{j \in [m_i]} \Pr(j\text{-th urn non-empty }) = m_i (1 - q_i),$$

where $q_i := \Pr(\text{ first urn empty }) = \left(1 - \frac{1}{m_i}\right)^{k_i}$. From this we see that

$$E(X_i) = \begin{cases} k_i - m_i (1 - q_i), & i \leqslant i_0; \\ m_i q_i, & i \geqslant i_0. \end{cases} \tag{3.4}$$

The following lemma contains asymptotics for $E(X_i)$. It is a key for the proof of the main theorem.

Lemma 3.5. *(i) If $\delta \leqslant 0$, then $E(X_{i_\delta}) \leqslant 2^{(2-2^{-\delta})L}$.*
(ii) Denote the "middle" part of an qOBDD with n levels by

$$M_\epsilon(n) := \left[L(n) - \varepsilon \mathinner{.\,.} L(n) + 1 + \left(\log L(n) - \log \log e + \varepsilon \right) \log e / L(n) \right],$$

where $\varepsilon > 0$ is an arbitrarily small constant. Then $\sum_{i \in [n] \setminus M(n)} E(X_i) = o(1)$.

So only two levels will be of particular interest for the analysis of the strong Shannon effect.

3.3 Proof of the Main Theorem

There seems to be no way to compute $E(X)/W$ directly, but in view of the results of Section 3.1, we can look at the ratio $E(X)/2^L$ instead. We will show that there is at most *one* "critical" level whose expected width differs significantly from its worst-case width, and that this level must be an $i_\delta \in \mathbb{N}$ for some sufficiently small δ (depending on n). Its existence decides upon whether or not the strong Shannon effect holds. This analysis is summarized in Theorems 3.8 and 3.9, which then lead to the proof of the main theorem.

But before we go into the technicalities, let us demonstrate the idea of the proof in a special case. The next proposition says that the strong Shannon effect does not hold for n of the form $n = 2^h + h$. The ratio of expected and worst-case size is a factor $\leqslant 1 - \frac{1}{2e}$. The tightness of the bounds will be shown in Theorem 3.8.

Proposition 3.6. *It holds*

$$\liminf_{\substack{n=2^h+h \\ h \nearrow +\infty}} \frac{E(X_{i_0})}{2^L} \geqslant \frac{1}{e}, \quad and \quad \liminf_{\substack{n=2^h+h \\ h \nearrow +\infty}} \frac{E(X)}{W} \geqslant \frac{1}{2e}.$$

Proof. Recall that $L(2^h + h) = 2^h$, which implies that $i_0 = 2^h + 1 \in \mathbb{N}$ is a level. Observe that $w_{i_0} = k_{i_0} = m_{i_0} = 2^L = 2^{2^h}$. By (3.4), we have

$$E(X_{i_0}) = m_{i_0} \left(1 - \frac{1}{m_{i_0}}\right)^{k_{i_0}} = 2^{2^h} \left(1 - \frac{1}{2^{2^h}}\right)^{2^{2^h}},$$

and by Theorem 3.2, $W\left(2^h + h\right) = 2 \cdot 2^{2^h}$. $\quad\square$

The proof of Proposition 3.6 was based on the fact that $L\left(2^h + h\right) = 2^h$ is an integer. We will see that taking more than a constant number of steps away from the "bad" values $n = 2^h + h$ is enough to guarantee that the strong Shannon effect holds.

Definition 3.7. Let $\delta'(n)$ denote the gap between $L(n)$ and the next natural number, i.e.

$$\delta'(n) := \ell - L(n), \qquad \text{where } \ell \in \mathbb{N} \text{ is such that } \left[\ell - \tfrac{1}{2}, \ell + \tfrac{1}{2}\right[\ni L(n).$$

So $i_{\delta'(n)} = i_0 + \delta'(n) = \ell + 1$ is the integer nearest to i_0. In case of a tie, we round up. Observe that $\delta'(2^h + h) = 0$. As $|\delta'|$ gets larger, $E(X_{i_{\delta'}})$ becomes more and more negligible compared to 2^L.

Theorem 3.8. *Let $n \in \mathbb{N}$ and choose $h(n) \in \mathbb{N}$ and $a = a(n) \in \mathbb{Z}$ such that $n = 2^h + h + a$ and $|a|$ is minimal.*

(i) *For sequences of n such that $|a(n)| \nearrow +\infty$, $\lim_n \dfrac{E(X_{i_{\delta'}})}{2^L} = 0$.*

(ii) *For sequences of n such that $a(n) = a \in \mathbb{Z}$ is a constant,*

$$\lim_n \frac{E(X_{i_{\delta'}})}{2^L} = \begin{cases} 2^{-a}\left(e^{-2^a} - 1\right) + 1, & a \leqslant 0; \\ 2^{-a}\, e^{-2^a}, & a \geqslant 0. \end{cases}$$

Note that the numbers $h(n)$ and $a(n)$ are well-defined, because $2^h + h + a = 2^{h+1} + h + 1 - a$ would imply that $(2^h + 1)/2 = a \in \mathbb{N}$, a contradiction.

Theorem 3.9. *For all sequences of n,*

$$\lim_n \frac{E(X) - E(X_{i_{\delta'}})}{2^L} = 0.$$

Proof. Take $\varepsilon = \tfrac{1}{5}$ and observe that for large enough n the middle part $M_\varepsilon(n)$ can only contain the levels $i_{\delta'}$ and $i_{\delta'-1}$. By Lemma 3.5 (ii) we have

$$\lim_n \frac{E(X) - E(X_{i_{\delta'}}) - E(X_{i_{\delta'-1}})}{2^L} = 0.$$

Since $\delta' - 1 \leqslant -\tfrac{1}{2}$, Lemma 3.5 (i) gives

$$0 \leqslant \frac{E(X_{i_{\delta'-1}})}{2^L} \leqslant \frac{2^{(2-\sqrt{2})L}}{2^L} = o(1),$$

and the theorem is proved. $\quad\square$

Proof of the main theorem. We write $E(X)/W = \frac{E(X)/2^L}{W/2^L}$. By Corollary 3.4, $1 \leqslant W/2^L \leqslant 3$, so $E(X)/2^L$ gives the asymptotic of $E(X)/W$. By Theorem 3.9, all levels except $i_{\delta'(n)}$ are negligible. We will apply the definition of $a(n)$ from Theorem 3.8.

Assertion (i): Since $d(h) \nearrow +\infty$ and $n \in A$, we have $|a(n)| \nearrow +\infty$. Therefore, $\lim_n E(X_{i_{\delta'}})/2^L = 0$, and we are done.

Assertion (ii): Since $d(h) = O(1)$ and $n \in B$, we have $a(n) = O(1)$. By partitioning the sequence of n into subsequences, we may assume that $a(n) = a$ is a constant. The subsequences may have finite or infinite length, but only a finite number of subsequences can be infinitely long, since $a(n) = O(1)$. For each infinite subsequence, we have proved in Theorem 3.8, Assertion (ii) that $\lim_n E(X_{i_{\delta'}})/2^L > 0$. So the original sequence satisfies $\liminf_n E(X_{i_{\delta'}})/2^L > 0$, i.e., $E(X_{i_{\delta'}})/2^L = \Omega(1)$. □

The main theorem is illustrated by Fig. 6.

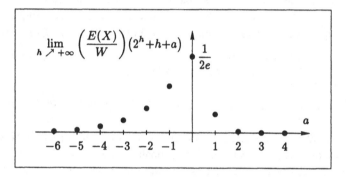

Fig. 6. When the strong Shannon effect does not hold.

4 Strong Shannon Effect – Arbitrary Variable Orderings

For the generalization of Theorem 3.1 to arbitrary variable orderings, we must show that for *almost all* Boolean functions *all* variable orderings lead to an OBDD size which is equal to the expected qOBDD size $E(Y)$ up to a factor of $1+o(1)$. This will be proved in two steps. In Section 4.1 we show that with (very) high probability the qOBDD size for all variable orderings is $E(Y)(1 + o(1))$. In Section 4.2 the effect of the deletion rule is taken into account. We show here that with (very) high probability, the size of qOBDDs and OBDDs for all variable orderings is the same (up to a factor of $1 + o(1)$). Altogether we obtain the desired generalization of Theorem 3.1 to arbitrary variable orderings.

4.1 The Reduction Effect of the Merging Rule

In this subsection, we study large deviations from the expected qOBDD-size which was computed in Section 3. Define $Y_\pi(f)$ to be the qOBDD size with respect to the variable ordering π. Note that the expected size $E(Y_\pi)$ does not depend on π, as we consider the uniform distribution. From Theorems 3.8 and 3.9, a similar argument as in the proof of the main theorem shows that $E(Y) \geqslant \left(1 - \frac{1}{2e} + o(1)\right)2^L = \Omega(2^n/n)$. Wegener [Weg94] proved that

$$\Pr\left(\exists \pi \; |Y_\pi - E(Y)| \geqslant O(2^{2n/3}/n)\right) = O(2^{-n/3+\varepsilon n}),$$

where $\varepsilon > 0$ is an arbitrary constant. Using Azuma's martingale inequality, we will derive the following result.

Theorem 4.1. *For every constant $c > 0$ and large n,*

$$\Pr\left(\exists \pi \; |Y_\pi - E(Y)| \geqslant n2^{\frac{1+c}{2}n}\right) \leqslant e^{-2^{cn}}.$$

Proof. First let us consider the variable ordering $\pi = (x_1, \ldots x_n)$. At each level i, we have an urn experiment where k_i balls are thrown randomly into m_i urns. Y_i is the number of nodes on the x_i-level as well as the number of non-empty urns. Using Azuma's inequality (see [AS91, Chapter 7]), we find that $\Pr\left(|Y_i - E(Y_i)| > \lambda\sqrt{k_i}\right) < 2e^{-\lambda^2/2}$. We apply this bound to all levels $i \leqslant n - 1$ (level n is negligible). At level i, there are $k_i \leqslant 2^{n-2}$ balls. If we set $\lambda := 2^{\frac{c}{2}n+1}$, we get

$$\Pr\left(|Y_i - E(Y_i)| \geqslant 2^{\frac{1+c}{2}n}\right) \leqslant \Pr\left(|Y_i - E(Y_i)| \geqslant 2^{\frac{c}{2}n+1}\sqrt{k_i}\right) \leqslant 2e^{-2^{cn+2}}.$$

Summing over all levels gives the result for any fixed variable ordering. Now we allow arbitrary variable orderings. Summing the probability bound over all $n!$ variable orderings, we get the claimed result, because the additional factor $n!$ is amply absorbed by the factor $e^{-2^{cn+2}}$. $\qquad\square$

In other words, we have shown that the weak Shannon effect holds for arbitrary variable orders and qOBDDs, too.

4.2 The Reduction Effect of the Deletion Rule

Wegener already proved that the deletion rule does not yield much reduction when applied to a random qOBDD. In this section we give a more refined analysis and better probability bounds. Let us denote the difference between $|\text{qOBDD}(f)|$ and $|\text{OBDD}(f)|$ using the canonical variable ordering by $X'(f) := |\text{qOBDD}(f)| - |\text{OBDD}(f)|$. Of course, $X' = \sum_{i=1}^n X'_i$, where $X'_i := Y_i - Z_i$ denotes the number of nodes deleted on level i. With the techniques used in Section 3, the expected reduction by the deletion rule can be estimated as follows.

Theorem 4.2. *We have*

$$E\left(X'_{i_\delta}\right) \leqslant \begin{cases} 2^{\delta+(1-2^{-\delta-1})L}, & \delta \leqslant 0; \\ 2^{2^{-\delta-1}L}, & \delta \geqslant 0. \end{cases}$$

Let $\delta_a := \frac{a\log e}{L(n)}$ and $a^2 = o(L)$. Then $E\left(X'_{i_{\delta_a}}\right) \leqslant \sqrt{2^{L-|a|+o(1)}}$.

As an example for the above theorem, note that $E(X'_{i_{-1}}) \sim 1/2$ and $E(X'_{i_1}) \sim 2^{L/4}$. (The tightness of the above bounds requires an extra argument.) The amount of reduction achieved by the deletion rule changes in a periodic way, too. However, the oscillations are not as distinct as for the merging rule.

Define $X'_*(f) := \max_\pi X'_\pi(f)$. (Again, the additional index π indicates the variable order.) Wegener [Weg94] used the second moment method to prove

$$\Pr\left(X'_* \geqslant O(2^{2n/3}n)\right) = O\left(2^{-n/3+\varepsilon n}\right)$$

for every constant $\varepsilon > 0$. Using Chvátal's inequality [Chv79], we can improve Wegener's result in the following way.

Theorem 4.3. *For every constant $c > 0$ and large n,*

$$\Pr\left(X'_* \geqslant n2^{\frac{1\pm c}{2}n}\right) \leqslant e^{-2^{cn}}.$$

5 Minimizing the Gap

Using the tools developed so far, we can also fix those n for which the gap between expected and worst-case size is *minimal* and show how it develops between both extremes. Somewhat surprisingly, a "typical" random qOBDD with $n = 2^h + 2h$ variables has only $E(X) = O(n^2)$ less than in the worst-case (which is $\Omega(2^n/n)$). But for qOBDD with only *one* level less, the following theorem implies that $E(X) = \Omega\left(2^{0.2786\,n}\right)$, asymptotically. The reason for this jump is that level $i_{\delta'(2^h+2h-1)}$ belongs to the middle part $M(2^h + 2h - 1)$, but $i_{\delta'(2^h+2h)}$ is *not* contained in $M(2^h + 2h)$.

Theorem 5.1.

(i) *Consider OBDDs of random Boolean functions with $n = 2^h + ch$ levels, where $h \nearrow +\infty$ and $c = c(h)$ converges to a real number in $[2, +\infty[$. Then $E(X) = n^{\tilde{c}}$, where $\tilde{c} \sim 2(c-1)$.*

(ii) *Consider OBDDs of random Boolean functions having $n = 2^h + 2h - c$ levels, where Boolean functions $h \nearrow +\infty$ and $c \in \mathbb{N}$ is a positive constant. Then $E(X) = 2^{c'n}$, where $c' = 1 - 2^{-c}\log e + o(1)$.*

6 Discussion

It has been observed by Löbbing, Schröer, and Wegener [SW94,LSW95] that zero-suppressed BDDs, also known as ZBDDs behave quite similar to OBDDs, if random Boolean functions are considered. The analyses of Liaw and Lin [LL92] and Wegener [Weg94] can be carried over without major changes. This is true for our results, too. Recall that ZBDDs can be defined as the result of the application of the merging rule and a modified deletion rule to the decision tree for a given Boolean function. In OBDDs, a node is deleted if both successors represent the same function. In ZBDDs, a node is deleted if its 1-sucessor is the 0-sink. It has been observed [Min93] that ZBDDs are particularly efficient for functions with an on-set consisting mostly of inputs for which many variables are assigned the value 0. However, if random Boolean functions are to be considered, the modified elimination rule for ZBDDs leads to the same probability distribution of X_i^l. So the results of Section 4.2 also hold for ZBDDs.

The situation is more complicated with FBDDs. As yet, we know that for $n = 2^h + h$ the minimal FBDD-size is less than $(1 - 1/2e + o(1))\,W(n)$ with high probability. It is still unknown whether this is true for an even greater gap than $1/2e$. We conjecture that an analogue to Theorem 3.1 holds for FBDDs, too.

References

[AS91]　Noga Alon, Joel H. Spencer: The Probabilistic Method; John Wiley & Sons, New York, 1991.

[Bry86]　Randal E. Bryant: Graph-Based Algorithms for Boolean Function Manipulation; IEEE Transactions on Computers, vol. C-35, no. 8, 1986, 677 – 691.

[Bry92]　Randal E. Bryant: Symbolic Boolean Manipulation with Ordered Binary-Decision Diagrams; ACM Computing Surveys, vol. 24, no. 3, 1992, 293 – 318.

[Chv79]　Vašek Chvátal: The Tail of the Hypergeometric Distribution; Discrete Mathematics, vol. 25, 1979, 285 – 287.

[GPS97]　Clemens Gröpl, Anand Srivastav, Hans Jürgen Prömel: Full version of this extended abstract, available via http://www.informatik.hu-berlin.de/Institut/struktur/algorithmen/forschung/veri/shannon.ps.gz .

[HM94]　Mark A. Heap, Melvin Ray Mercer: Least Upper Bounds on OBDD Sizes; IEEE Transactions on Computers, vol. 43, no. 6, 1994, 764 – 767.

[KSC78]　Valentin F. Kolchin, Boris A. Sevast'îanov, Vladimir P. Chistîakov: Random Allocations; John Wiley & Sons, 1978, Chapter 1.

[LL90]　Heh-Tyan Liaw, Chen-Shang Lin: On the OBDD-Representation of General Boolean Functions; NSC Rep., NSC79-0404-E002-35, 1990.

[LL92]　Heh-Tyan Liaw, Chen-Shang Lin: On the OBDD-Representation of General Boolean Functions; IEEE Transactions on Computers, vol. 41, no. 6, 1992, 661 – 664.

[LSW95]　Martin Löbbing, Olaf Schröer, Ingo Wegener: The Theory of Zero-Suppressed BDDs and the Number of Knights Tours; Proceedings of the IFIP WG 10.5 Workshop on Applications of the Reed-Muller Expansion in Circuit Design (Reed-Muller '95), August 27. – 29. 1995, Makuhari, Chiba, Japan, 1995, 38 – 45.

[Min93]　Shin-Ichi Minato: Zero-Suppressed BDDs for Set Manipulation in Combinatorial Problems; Proceedings of the 30th ACM/IEEE Design Automation Conference, 1993, 272 – 277.

[SW94]　Olaf Schröer, Ingo Wegener: The Theory of Zero-Suppressed BDDs and the Number of Knights Tours; Forschungsbericht Nr. 552/1994, Universität Dortmund, Fachbereich Informatik, 1994, 25 pages.

[Weg94]　Ingo Wegener: The Size of Reduced OBDDs and Optimal Read-Once Branching Programs for Almost All Boolean Functions; IEEE Transactions on Computers, vol. 43, no. 11, 1994, 1262 – 1269.

Provable Security for Block Ciphers by Decorrelation

Serge Vaudenay

Ecole Normale Supérieure — CNRS
Serge.Vaudenay@ens.fr

Abstract

In this presentation we investigate a new way of protecting block ciphers against classes of attacks (including differential and linear cryptanalysis) which is based on the notion of decorrelation which is fairly connected to Carter-Wegman's notion of universal functions. This defines a simple and friendly combinatorial measurement which enables to quantify the security. We show that we can mix provable protections and heuristic protections. We finally propose two new block ciphers family we call COCONUT and PEANUT, which implement these ideas and achieve quite reasonable performances for real-life applications.

Public research on cryptography has been boosted twenty years ago by the discovery of public-key cryptography. In their seminal papers, Diffie and Hellman [8], Merkle and Hellman [23], Rivest, Shamir and Adleman [30] and others public-key pioneers offered some new directions to researchers. While giving a fairly good promotion to the area of computational number theory and complexity theory (it was not a coincidence that it has been invented shortly after the foundations of NP-completeness theory has been set), it concentrated most of the research energy.

Surprisingly, the area of conventional cryptography was quite older but did not received the same boost. Although the U.S. Government adopted the Data Encryption Standard (DES) in 1977 (the development of which was pushed by banking security needs), the real advances on the research of block ciphers used to be quite rare. Security was almost always based on empiristic approaches: block ciphers were developed in a sufficiently enough complicated way to make their cryptanalysis impossible. Implementation cost however needs the complexity to be not so high, but no one has a sharp idea where to make a trade-off in between. Now, this area of research is moving towards a mature state where we can give results on provable security.

Before the second world war, security of encryption used to be based on the secrecy of the algorithm. Mass telecommunication and computer science networking however pushed the development of public algorithms with secret keys. The most important research result on encryption was found for the application to the telegraph by Shannon in the Bell Laboratories in 1949 [31]. It proved the unconditional security of the Vernam's Cipher which had been published in 1926 [37]. Although quite expensive to implement for networking (because the sender and the receiver need to be synchronized, and it needs quite cumbersome huge keys), this cipher was used in the Red Telephone between Moscow and Washington D.C. during the cold war. Shannon's result also proves that unconditional security cannot be achieved in a better (*i.e.* cheaper) way. For this reason, empiristic security seemed to be the only efficient possibility, and all secret key block ciphers which have been publicly developed were considered to be secure until some researcher published an attack on it. Therefore research mostly grew like a ball game between the designers team and the analysts team and treatment on the general security of block ciphers has hardly been done.

Real advances on the security on block ciphers have been made in the early 90's.

One of the most important result has been obtained by Biham and Shamir in performing a *differential cryptanalysis* on DES [2, 3, 4, 5]. The best version of this attack can recover a secret key with a simple 2^{47}-chosen plaintext attack[1]. Although this attack is heuristic, experiments confirmed the results.

Biham and Shamir's attack was based on statistical cryptanalysis idea which have also been used by Gilbert and Chassé against another cipher [11, 10]. Those ideas inspired Matsui who developed a *linear cryptanalysis* on DES [20, 21]. This heuristic attack, which has been implemented, can recover the key with a 2^{43}-known plaintext attack. Since then, many researchers tried to generalize and improve these attacks (see for instance [18, 17, 15, 34, 16, 24, 35]), but the general ideas was quite the same.

The basic idea of differential cryptanalysis is to use properties like "if x and x' are two plaintext blocks such that $x' = x \oplus a$, then it is likely that $C(x') = C(x) \oplus b$". Then the attack is an iterated two-chosen plaintexts attack which consists in getting the encrypted values of two random plaintexts which verify $x' = x \oplus a$ until a special event like $C(x') = C(x) \oplus b$ occurs. Similarly, the linear cryptanalysis consists in using the probability that $C(x)$ is in a given hyperplane when we know that x is in another given hyperplane, when this probability is far from $1/2$. More precisely, linear cryptanalysis is

---

[1]So far, the best known attack was an improvement of exhaustive search which requires on average 2^{54} DES computations.

an incremental one-known plaintext attack where we count how many times this event occurs.

Instead of keeping breaking and proposing new encryption functions, some researchers tried to sit down and understand how to protect ciphers against some classes of attacks. Nyberg first formalized the notion of strength against differential cryptanalysis [25], and similarly, Chabaud and Vaudenay formalized the notion of strength against linear cryptanalysis [6]. With this approach we can study how to make internal computation boxes resistant against both attacks. This can be used in a heuristic way by usual active s-boxes counting tricks (see Heys and Tavares [13] for instance). This has also been used to provide provable security against both attacks by Nyberg and Knudsen [26], but in an unsatisfactory way which introduce some algebraic properties which lead to other attacks as shown by Jakobsen and Knudsen [14].

In this presentation, we introduce a new way to protect block ciphers against various kind of attacks. This approach is based on the notion of universal functions introduced by Carter and Wegman [7, 38] for the purpose of authentication. Protecting block ciphers is so cheap that we call NUT (as for "n-Universal Transformation") the set of ciphers which are protected this way. We finally describe two cipher families we call COCONUT (as for "Cipher Organized with Cute Operations and NUT") and PEANUT (as for "Pretty Encryption Algorithm with NUT") and offer two definite examples as a cryptanalysis challenge.

The paper is organized as follows. First we give some definitions an basic constructions for decorrelation. Then we state Shannon's perfect secrecy notion in term of decorrelation. We show how to express security results in the Luby-Rackoff's security model. Then we compute how Feistel Ciphers are decorrelated. We prove how pairwise decorrelation can protect a cipher against Biham-Shamir's differential cryptanalysis and Matsui's cryptanalysis. Finally, we define the COCONUT and PEANUT construction and show how to generate keys for them.

1   Decorrelation

We first give formal definitions of the notion of decorrelation which plays a crucial role in our treatment.

Definition 1. Given a random function F from a given set $\mathcal{A}$ to a given set $\mathcal{B}$ and an integer d, we define the d-wise distribution matrix $[F]^d$ of F as a $\mathcal{A}^d \times \mathcal{B}^d$-matrix where the (x, y)-entry of $[F]^d$ corresponding to the multipoints

$x = (x_1, \ldots, x_d) \in \mathcal{A}^d$ and $y = (y_1, \ldots, y_d) \in \mathcal{B}^d$ is defined as the probability that we have $F(x_i) = y_i$ for $i = 1, \ldots, d$.

Basically, each row of the d-wise distribution matrix corresponds to the distribution of the d-tuple $(F(x_1), \ldots, F(x_d))$ where $(x_1, \ldots, x_d)$ corresponds to the index of the row.

Definition 2. Given two random functions F and G from a given set $\mathcal{A}$ to a given set $\mathcal{B}$, an integer d and a distance D over the vector space $\mathbf{R}^{\mathcal{A}^d \times \mathcal{B}^d}$, we call $D([F]^d, [G]^d)$ the d-wise D-decorrelation between F and G.

A decorrelation of zero means that for any multipoint $x = (x_1, \ldots, x_d)$ the multipoints $(F(x_1), \ldots, F(x_d))$ and $(G(x_1), \ldots, G(x_d))$ have the same distribution, so that F and G are *correlated* in some sense *to the order d*.

It is also important to study the decorrelation of a given random function F to a reference random function. For instance, let say that R is a random function from $\mathcal{A}$ to $\mathcal{B}$ with a uniform distribution. Saying that F and R have a 1-wise decorrelated of zero (or equivalently that F and R are *correlated*) means that for any x_1 the distribution of $F(x_1)$ is uniform. Saying that F is 2-wise correlated to R means that for any $x_1 \neq x_2$ the points $F(x_1)$ and $F(x_2)$ are uniform and independent.

We note that this notion is fairly similar to the notion of universal functions which was introduced by Carter and Wegman [7, 38]. More precisely, we recall that a random function F from $\mathcal{A}$ to $\mathcal{B}$ is ϵ-almost strongly universal$_d$ if for any pairwise different $(x_1, \ldots, x_d)$ and any $(y_1, \ldots, y_d)$ we have

$$\Pr[F(x_i) = y_i; i = 1, \ldots, d] \leq \frac{1}{\#\mathcal{B}^d} + \epsilon.$$

If we define $\|A\|_\infty = \max_{x,y} |A_{x,y}|$, if F has a d-wise $\|.\|_\infty$-decorrelated of ϵ to a truly random function, then it is ϵ-almost strongly universal$_d$. The converse is true when $\epsilon \geq \frac{1}{\#\mathcal{B}^d}$. Although the notion is fairly similar, we will use our formalism which is adapted to our context.

For a treatment on block cipher, we consider random permutations C on the block-message space $\mathcal{M}$. (Here the randomness comes from the secret key.) In most of cases, we have $\mathcal{M} = \{0, 1\}^m$. If we define the Perfect Cipher C^* as being a random permutation on $\mathcal{M}$ with a uniform distribution, we are interested in the decorrelation between C and C^*.

For the purpose of our treatment, we define the L_2 norm, the infinity weighted norm N_∞ and the L_∞-associated matrix norm $\|\|.\|\|_\infty$ on $\mathbf{R}^{\mathcal{M}^d \times \mathcal{M}^d}$

by:

$$||A||_2 = \sqrt{\sum_{x,y} (A_{x,y})^2} \tag{1}$$

$$N_\infty(A) = \max_{x,y} \frac{|A_{x,y}|}{\Pr[x \overset{C^*}{\mapsto} y]} \tag{2}$$

$$|||A|||_\infty = \max_x \sum_y |A_{x,y}| \tag{3}$$

where C^* is the Perfect Cipher. For the definition of n_∞ we outline that for any d-wise distribution matrix of a permutation, if $\Pr[x \overset{C^*}{\mapsto} y] = 0$, then $A_{x,y} = 0$ and we can assume $0/0 = 0$.

We recall that the $||.||_2$ and $|||.|||_\infty$ norms are matrix norms, which means that $||A \times B|| \leq ||A||.||B||$. Moreover, N_∞ has a similar property. Namely, for a multipoint $x = (x_1, \ldots, x_d)$, let $\equiv_x$ denotes the equivalence relation among $\{1, \ldots, d\}$ defined by

$$i \equiv_x j \iff x_i = x_j.$$

We say that a matrix A has a *support of a permutation* distribution matrix if

$$\forall x, y \quad \equiv_x \neq \equiv_y \Longrightarrow A_{x,y} = 0.$$

For the distribution matrices of any cipher, this property holds. Then, for any such matrices A and B, we have $N_\infty(A \times B) \leq N_\infty(A).N_\infty(B)$. Namely, N_∞ is a norm on the sub-algebra of all matrices which have a support of a permutation distribution matrix.

Multiplicativity of the decorrelation is very useful when we consider product ciphers. Concretely, if C_1 and C_2 are two independent ciphers and C^* is the Perfect Cipher, then we have

$$|||[C_1 \circ C_2]^d - [C^*]^d||| \leq |||[C_1]^d - [C^*]^d|||.|||[C_2]^d - [C^*]^d|||.$$

(This comes from $[C_i]^d \times [C^*]^d = [C^*]^d$.) This property makes the decorrelation a friendly combinatorial measurement.

2 Basic constructions

Perfect 1-wise decorrelation is easy to achieve when the message-block space $\mathcal{M}$ is given a group structure. For instance we can use $C(x) = x + K$ where K has a uniform distribution on $\mathcal{M}$.

We can construct perfect pairwise decorrelated ciphers on a field structure $\mathcal{M}$ by $C(x) = a.x + b$ where $K = (a, b)$ is uniform in $\mathcal{M}^* \times \mathcal{M}$. This requires to consider the special case $a = 0$ when generating K. On the standard space $\mathcal{M} = \{0, 1\}^m$, it also requires to implement arithmetic on the finite field GF(2^m), which may lead to poor encryption rate on software. As an example we can mention the COCONUT Ciphers (see Section 8).

A similar way to construct perfect 3-wise decorrelated ciphers on a field structure $\mathcal{M}$ is by $C(x) = a + b/(x + c)$ where $K = (a, b, c)$ with $b \neq 0$. (By convention we set $1/0 = 0$.)

Perfect decorrelated ciphers with higher orders require dedicated structure. We can for instance use Dickson's Polynomials.

An alternative way consists of using Feistel Ciphers with decorrelated functions [9]. Given a set $\mathcal{M} = \mathcal{M}_0^2$ where $\mathcal{M}_0$ has a group structure and given r random functions $F_1, \ldots, F_r$ on $\mathcal{M}_0$ we denote $C = \Psi(F_1, \ldots, F_r)$ the cipher defined by $C(x^l, x^r) = (y^l, y^r)$ where we iteratively compute a sequence (x_i^l, x_i^r) such that

$$x_0^l = x^l \quad \text{and} \quad x_0^r = x^r$$
$$x_i^l = x_{i-1}^r \quad \text{and} \quad x_i^r = x_{i-1}^l + F_i(x_{i-1}^r)$$
$$y^l = x_r^r \quad \text{and} \quad y^r = x_r^l.$$

(Note that the final exchange between the two halves is cancelled here.) In all the constructions of this paper, $\mathcal{M}_0$ is the group $\mathbf{Z}_2^{\frac{m}{2}}$, so the addition is the bitwise exclusive or which will be denoted $\oplus$.

If $\mathcal{M}_0$ has a field structure, we can use perfect d-wise decorrelated F_i functions by $F_i(x) = a_d.x^{d-1} + \ldots + a_2.x + a_1$ where $(a_1, \ldots, a_d)$ is uniformly distributed on $\mathcal{M}_0^d$.

Decorrelation of Feistel Ciphers depends on the decorrelation of all F_i functions. It will be studied in Section 5.

3 Shannon's unconditional security

In this Section, we consider decorrelation of zero. We note that this decorrelation does not depend on the choice of the distance.

Intuitively, if C is 1-wise correlated to C^*, the encryption $C(x_1)$ contains no information on the plaintext-block x_1, so the cipher C is secure if we use it only once (as one-time pad [37]). This corresponds to Shannon's perfect secrecy theory [31]. Similarly, if C is d-wise correlated to C^*, it is unconditionally secure if we use it only d times (on different plaintexts) as the following Theorem shows.

Theorem 3. *Let C be a cipher d-wise correlated to the Perfect Cipher. If an adversary knows $d - 1$ pairs $(x_i, C(x_i))$ $(i = 1, \ldots, d - 1)$, for any y_d which is different from all $C(x_i)$'s, his knowledge of $C^{-1}(y_d)$ is exactly that it is different from all x_i's. More precisely, for any $x_1, \ldots, x_d$*

$$H(X/C(x_1), \ldots, C(x_d), C(X)) = H(X)$$

where X is a random variable with uniform distribution among all the messages in $\mathcal{M}$ which are different from the x_i's. (Here H denotes Shannon's entropy of random variables.)

We recall that by definition we have $H(X/Y) = H(X, Y) - H(Y)$ and

$$H(X) = -\sum_x \Pr[X = x] \log \Pr[X = x]$$

with the convention that $0 \log 0 = 0$.

Proof. For any x_d which is different from all x_i's, the probability that $C(x_d) = y_d$ knowing all x_i's and $C(x_i)$'s is equal to $\frac{1}{\#\mathcal{M}-r}$ where r is the number of pairwise different x_i's. This is equal to the probability of a uniform X is equal to x_d knowing that it is different from all x_i's. □

4 Security in the Luby-Rackoff model

To illustrate the power of the notion of decorrelation, let us first express the perfect security notion. In the Luby-Rackoff model, an attacker is an infinitely powerful device whose aim is to distinguish a cipher C from the Perfect Cipher C^* by querying an oracle with a limited number n of inputs (see [19]). We assume that we are given an oracle $\mathcal{O}$ which either implements C or C^*, and that the attacker must finally answer 0 ("reject") or 1 ("accept"). We measure the ability to distinguish C from C^* by the advantage $|p - p^*|$ where p (resp. p^*) is the probability of answering 1 if $\mathcal{O}$ implements C (resp. C^*). We have the following Theorem.

Theorem 4. *If C is a cipher, let d be an integer and ϵ be the d-wise N_∞-decorrelation between C and the Perfect Cipher. For any distinguishing attacker which is limited to d queries, the advantage $|p - p^*|$ is at most ϵ.*

In particular, we have unconditional security for $\epsilon = 0$ and we still have a proven quantified security when ϵ is small.

Proof. Each execution of the attack with an oracle which implements C is characterized by a random tape ω and the successive answers $y_1, \ldots, y_d$ of the queries which we denote $x_1, \ldots, x_d$ respectively. More precisely, x_1 depends on ω, x_2 depends on ω and y_1 and so on. The answer thus depends on $(\omega, y_1, \ldots, y_d)$. Let $\mathcal{E}$ be the set of all $(\omega, y_1, \ldots, y_d)$ such that the output of $\mathcal{A}$ is 1. We have

$$
\begin{aligned}
p &= \sum_{(\omega, y_1, \ldots, y_d) \in \mathcal{E}} \Pr[\omega] \Pr[C(x_i(\omega, y_1, \ldots, y_{i-1})) = y_i; i = 1, \ldots, d] \\
&\leq (1 + \epsilon) \sum_{(\omega, y_1, \ldots, y_d) \in \mathcal{E}} \Pr[\omega] \Pr[C^*(x_i(\omega, y_1, \ldots, y_{i-1})) = y_i; i = 1, \ldots, d] \\
&\leq (1 + \epsilon) p^*
\end{aligned}
$$

so we have $p - p^* \leq \epsilon$ for any attacker. We can apply this result to the attacker which acts as the attack does but produces the opposite output to complete the proof. $\qquad\square$

Here is a more precise Theorem in the non adaptive case. (We call an attack "non adaptive" if no x_i queried to the oracle depends on some previous answers y_j.)

Theorem 5. *If C is a cipher, let d be an integer and ϵ be the d-wise $\|||.|||_\infty$-decorrelation between C and the Perfect Cipher. The advantage $|p - p^*|$ of the best non-adaptive distinguishing attacker which is limited to d queries is equal to $\epsilon/2$.*

Proof. For those attacks, with the notations of Theorem 4, we have

$$
p = \sum_x \Pr[x] \sum_y 1_{(x,y) \in \mathcal{E}} \Pr\left[x \overset{C}{\mapsto} y\right]
$$

(where 1_P is defined to be 1 if Predicate P is true and 0 otherwise) thus, for the best attack, we have

$$
|p - p^*| = \max_{\substack{x \mapsto \Pr[x] \\ \mathcal{E}}} \left| \sum_x \Pr[x] \sum_y 1_{(x,y) \in \mathcal{E}} \left(\Pr\left[x \overset{C}{\mapsto} y\right] - \Pr\left[x \overset{C^*}{\mapsto} y\right] \right) \right|.
$$

We can easily see that this maximum is obtained when $x \mapsto \Pr[x]$ is a Dirac distribution on a multipoint $x = x^0$ and $\mathcal{E}$ includes all y's such that $\Pr\left[x_0 \overset{C}{\mapsto} y\right] - \Pr\left[x_0 \overset{C^*}{\mapsto} y\right]$ has the same sign, which gives the result. $\qquad\square$

5 Decorrelation of Feistel Ciphers

In this Section, we assume that $\mathcal{M} = \mathcal{M}_0{}^2$ where $\mathcal{M}_0$ is a group. Thus we can consider Feistel Ciphers on $\mathcal{M}$.

Theorem 5 can be used in a non-natural way. For instance, let us recall the following Theorem.

Theorem 6 (Luby-Rackoff [19]). *Let F_1, F_2, F_3 be three independent uniform random functions on $\mathcal{M}_0$ and d be an integer. For any distinguishing attacker against $\Psi(F_1, F_2, F_3)$ on $\mathcal{M} = \mathcal{M}_0{}^2$ which is limited to d queries, we have*

$$|p - p^*| \le \frac{d^2}{\sqrt{\#\mathcal{M}}}.$$

Thus we can say that

$$|||[\Psi(F_1, F_2, F_3)]^d - [C^*]^d|||_\infty \le 2\frac{d^2}{\sqrt{\#\mathcal{M}}}$$

where C^* is the Perfect Cipher.

For completeness, we also mention some improvements to the previous Theorem due to Patarin [27, 28, 29].

Theorem 7 (Patarin [29]). *Let $F_1, \ldots, F_6$ be six independent uniform random functions on $\mathcal{M}_0$ and d be an integer. For any distinguishing attacker against $\Psi(F_1, \ldots, F_6)$ on $\mathcal{M} = \mathcal{M}_0{}^2$ which is limited to d queries, we have*

$$|p - p^*| \le \frac{37d^4}{(\#\mathcal{M})^{\frac{3}{2}}} + \frac{6d^2}{\#\mathcal{M}}.$$

So, as Theorem 6 guaranties the security of a three-round Feistel Cipher until $d \sim (\#\mathcal{M})^{\frac{1}{4}}$, this one guaranties the security until $d \sim (\#\mathcal{M})^{\frac{3}{8}}$.

The $|||.|||_\infty$-decorrelation of Feistel Ciphers can be estimated with the following Lemma.

Lemma 8. *Let $F_1, \ldots, F_r, R_1, \ldots, R_r$ be $2r$ independent random functions on $\mathcal{M}_0$ such that $|||[F_i]^d - [R_i]^d|||_\infty \le \epsilon_i$ for $i = 1, \ldots, r$. We have*

$$|||[\Psi(F_1, \ldots, F_r)]^d - [\Psi(R_1, \ldots, R_r)]^d|||_\infty \le (1 + \epsilon_1) \ldots (1 + \epsilon_r) - 1.$$

Proof. Let u^i denotes the input of F_i or R_i in $\Psi(F_1, \ldots, F_r)$ or $\Psi(R_1, \ldots, R_r)$. We thus let (u^0, u^1) denote the input of the ciphers, and (u^{r+1}, u^r) denote the output. Here, all u^i's are multipoints, $i.e.$ $u^i = (u_1^i, \ldots, u_d^i)$. We have

$$\Pr_{F_1,\ldots,F_r}[u^0 u^1 \mapsto u^{r+1} u^r] - \Pr_{R_1,\ldots,R_r}[u^0 u^1 \mapsto u^{r+1} u^r]$$

$$= \sum_{u^2,\ldots,u^{r-1}} \left(\prod_{i=1}^r \Pr_{F_i}[u^i \mapsto u^{i-1} \oplus u^{i+1}] - \prod_{i=1}^r \Pr_{R_i}[u^i \mapsto u^{i-1} \oplus u^{i+1}] \right)$$

$$= \sum_{u^2,\ldots,u^{r-1}} \sum_{\substack{I \subseteq \{1,\ldots,r\} \\ I \neq \emptyset}} \prod_{i \in I} (\Pr_{F_i} - \Pr_{R_i})[u^i \mapsto u^{i-1} \oplus u^{i+1}] \prod_{i \notin I} \Pr_{R_i}[u^i \mapsto u^{i-1} \oplus u^{i+1}]$$

hence

$$\sum_{u^{r+1},u^r} \left| \Pr_{F_1,\ldots,F_r}[u^0 u^1 \mapsto u^{r+1} u^r] - \Pr_{R_1,\ldots,R_r}[u^0 u^1 \mapsto u^{r+1} u^r] \right|$$

$$\leq \sum_{u^2,\ldots,u^{r+1}} \sum_{\substack{I \subseteq \{1,\ldots,r\} \\ I \neq \emptyset}} \prod_{i \in I} |\Pr_{F_i} - \Pr_{R_i}|[u^i \mapsto u^{i-1} \oplus u^{i+1}] \prod_{i \notin I} \Pr_{R_i}[u^i \mapsto u^{i-1} \oplus u^{i+1}]$$

$$= \sum_{\substack{I \subseteq \{1,\ldots,r\} \\ I \neq \emptyset}} \prod_{i \in I} \epsilon_i$$

$$= (1 + \epsilon_1) \ldots (1 + \epsilon_r) - 1.$$

$\square$

From this Lemma and the previous observation we obtain the following Theorem.

Theorem 9. *Let $F_1, \ldots, F_r, R$ be r independent random functions on $\mathcal{M}_0$ where R has a uniform distribution and such that $\||[F_i]^d - [R]^d|\|_\infty \leq \epsilon$ for $i = 1, \ldots, r$. For any $k \geq 3$ we have*

$$\||[\Psi(F_1, \ldots, F_r)]^d - [C^*]^d|\|_\infty \leq \left((1+\epsilon)^k - 1 + \frac{2d^2}{\sqrt{\#\mathcal{M}}} \right)^{\lfloor \frac{r}{k} \rfloor}.$$

We can remark that the Lemma remains valid if we replace the group operation used in the Feistel construction by any other (pseudo)group law.

This makes the $\||.|\|_\infty$-decorrelation a friendly tool for constructing Feistel Ciphers.

6 Differential cryptanalysis

In this Section we study the security of pairwise decorrelated ciphers against basic differential cryptanalysis. We study criteria which prove that the attack cannot be better than exhaustive attack. In our model, the exhaustive attack for decrypting a given ciphertext y is an attack which exhaustively request for many random $C(x)$'s until we have $C(x) = y$. The complexity of this attack is obviously within the range of the number of possible text blocks.

We assume that $\mathcal{M} = \{0,1\}^m$ is the set of all bitstrings with length m. Let C be a cipher on $\mathcal{M}$ and let C^* be the Perfect Cipher.

Although differential cryptanalysis has been invented in order to recover a whole key by Biham and Shamir (see [4, 5]), we study here the basic underlying notion which makes it work. We call basic differential cryptanalysis the following distinguisher which is characterized by a pair $(a, b) \in \mathcal{M}^2$ with $a \neq 0$.

1. pick a random x with a uniform distribution and query for $C(x)$ and $C(x \oplus a)$

2. if $C(x \oplus a) = C(x) \oplus b$, stop and output 1, otherwise, start again until n trials has been performed

3. stop and output 0

It is well known that differential cryptanalysis depends on the following $\mathrm{DP}^C(a, b)$ (see for instance [25]). We define

$$\mathrm{DP}^C(a, b) = \Pr[C(X \oplus a) = C(X) \oplus b]$$

where X has a uniform distribution. This quantity thus depends on the key. Here we focus on the average value $E\left(\mathrm{DP}^C(a, b)\right)$ over the distribution of the key. We first mention that it has an interesting linear expression with respect to the pairwise distribution matrix of C. Namely, straightforward computation shows that

$$E(\mathrm{DP}^C(a, b)) = 2^{-m} \sum_{\substack{x_1 \neq x_2 \\ y_1 \neq y_2}} 1_{\substack{x_1 \oplus x_2 = a \\ y_1 \oplus y_2 = b}} \Pr\left[\begin{matrix} x_1 & \overset{C}{\mapsto} & y_1 \\ x_2 & \mapsto & y_2 \end{matrix}\right]. \tag{4}$$

Lemma 10. For the attack above, we have

$$|p - p^*| \leq n \cdot \max\left(\frac{1}{2^m - 1}, E\left(\mathrm{DP}^C(a, b)\right)\right).$$

So, if $E\left(\mathrm{DP}^C(a, b)\right)$ is within the order of 2^{-m}, then the attack above cannot be better than exhaustive attack.

Proof. It is straightforward to see that the probability, for some fixed key, that the attack accepts C is

$$1 - \left(1 - \mathrm{DP}^C(a,b)\right)^n$$

which is less than $n.\mathrm{DP}^C(a,b)$. Hence we have $p \leq n.E\left(\mathrm{DP}^C(a,b)\right)$. Since from Equation (4) we have $E\left(\mathrm{DP}^{C^*}(a,b)\right) = \frac{1}{2^m-1}$, we obtain the result. □

Theorem 11. *Let* $\epsilon = \left\| |[C]^2 - [C^*]^2| \right\|_\infty$ *where* C^* *is the Perfect Cipher. For any basic differential distinguisher between* C *and* C^*, *we have* $|p - p^*| \leq \frac{n}{2^m-1} + n\epsilon$.

Proof. Actually we notice that $E\left(\mathrm{DP}^{C^*}(a,b)\right) = \frac{1}{2^m-1}$ and that

$$\left| E\left(\mathrm{DP}^C(a,b)\right) - \frac{1}{2^m - 1} \right| \leq \epsilon$$

from Equation (4). □

So, if ϵ has the order of 2^{-m}, basic differential cryptanalysis does not work against C, but with a complexity in the scale of 2^m.

7 Linear cryptanalysis

Linear cryptanalysis has been invented by Matsui [20, 21] based on the notion of statistical attacks which are due to Gilbert *et al.* [11, 33, 10]. We study here the simpler version of the original attack. With the notations of the previous Section, we similarly call basic linear cryptanalysis the following distinguisher which is characterized by a pair $(a,b) \in \mathcal{M}^2$ with $b \neq 0$.

1. initialize the counter value c to zero

2. pick a random x with a uniform distribution and query for $C(x)$

3. if $x \cdot a = C(x) \cdot b$, increment the counter

4. go to step 2 until n iterations has been performed

5. stop and give an output which only depends on the counter value c

We notice here that the attack depends on the way it accepts or rejects depending on the final counter c value.

Linear cryptanalysis is based on the following quantity as pointed out by Chabaud and Vaudenay [6].

$$LP^C(a, b) = (2 \Pr[X \cdot a = f(X) \cdot b] - 1)^2$$

where $\cdot$ denotes the inner dot product. (We use Matsui's notations taken from [22].) As for differential cryptanalysis, we focus on $E\left(LP^C(a, b)\right)$, and there is a linear expression of this mean value and the terms of the pairwise distribution matrix $[C]^2$ which comes from straightforward computations :

$$E(LP^C(a, b)) = 1 - 2^{2-2m} \sum_{\substack{x_1 \neq x_2 \\ y_1 \neq y_2}} 1_{\substack{x_1 \cdot a = y_1 \cdot b \\ x_2 \cdot a \neq y_2 \cdot b}} \Pr \begin{bmatrix} x_1 \overset{C}{\mapsto} y_1 \\ x_2 \mapsto y_2 \end{bmatrix}. \tag{5}$$

Lemma 12. For any distinguisher in the above model, we have

$$|p - p^*| \leq 9.3 \left(n \cdot \max \left(\frac{1}{2^m - 1}, E\left(LP^C(a, b)\right) \right) \right)^{\frac{1}{3}}.$$

So, if $E\left(LP^C(a, b)\right)$ is within the order of 2^{-m}, then the attack above cannot be essentially better than exhaustive attack.

Proof. Let N_i be the random variable defined as being 1 or 0 depending on whether or not we have $x \cdot a = C(x) \cdot b$ in the ith iteration. All N_i's are independent and with the same 0-or-1 distribution. Let μ be the probability that $N_i = 1$, for a fixed key K. From the Central Limit Theorem, we can approximate the quantity c/n where c is the counter value to a normal distribution law with mean μ and standard deviation $\sigma = \sqrt{\frac{\mu(1-\mu)}{n}}$. Let A be the set of all accepted c/n quantities. For a fixed key K, the probability that the attack accepts is

$$p^K \approx \int_{t \in A} \frac{e^{-\frac{(t-\mu)^2}{2\sigma^2}}}{\sigma\sqrt{2\pi}} dt.$$

We can compare it to the theoretical expected value p_0 defined by $\mu = \frac{1}{2}$ and $\sigma = \frac{1}{2\sqrt{n}}$. The difference $p^K - p_0$ is maximal when $A = [\tau_1, \tau_2]$ for some values τ_1 and τ_2 which are roots of the Equation

$$\frac{(t - \mu)^2}{\sigma^2} + \log \sigma^2 = 4n \left(t - \frac{1}{2} \right)^2 - \log 4n.$$

Hence, the maximum of the difference $p^K - p_0$ is at most the maximum when $A = [\tau_1, \tau_2]$ over the choice of τ_1 and τ_2, which is the maximum minus the minimum of $p^K - p_0$ when $A =]-\infty, \tau]$. Now we have

$$\int_{-\infty}^{\tau} \frac{e^{-\frac{(t-\mu)^2}{2\sigma^2}}}{\sigma\sqrt{2\pi}} dt = \int_{-\infty}^{\frac{\tau-\mu}{\sigma}} \frac{e^{-\frac{t^2}{2}}}{\sqrt{2\pi}} dt$$

so we have

$$p^K - p_0 \le \left(\max_{\tau} - \min_{\tau}\right) \int_{2\sqrt{n}(\tau-\frac{1}{2})}^{\sqrt{n}\frac{\tau-\mu}{\sqrt{\mu(1-\mu)}}} \frac{e^{-\frac{t^2}{2}}}{\sqrt{2\pi}} dt.$$

We consider the sum as a function $f(\mu)$ on μ. Since we have $f\left(\frac{1}{2}\right) = 0$, we have $|f(\mu)| \le B. \left|\mu - \frac{1}{2}\right|$ where B is the maximum of $|f'(x)|$ when x varies from μ to $\frac{1}{2}$. We have

$$f'(x)\sqrt{\frac{2\pi}{n}} = \left(-\frac{1}{\sqrt{x(1-x)}} - \frac{\frac{1}{2}-x}{x(1-x)}\frac{\tau-x}{\sqrt{x(1-x)}}\right) e^{-\frac{n(\tau-x)^2}{2x(1-x)}}$$

so

$$|f'(x)| \le \sqrt{\frac{n}{2\pi\mu(1-\mu)}} + \frac{\left|\mu-\frac{1}{2}\right|}{\mu(1-\mu)}\frac{e^{-\frac{1}{2}}}{\sqrt{2\pi}} \le \sqrt{\frac{n}{2\pi\mu(1-\mu)}} + \frac{\left|\mu-\frac{1}{2}\right|}{\mu(1-\mu)}.$$

Therefore we have

$$\left|p^K - p_0\right| \le 2\sqrt{\frac{n}{2\pi}}\frac{\left|\mu-\frac{1}{2}\right|}{\sqrt{\mu(1-\mu)}} + 2\frac{\left(\mu-\frac{1}{2}\right)^2}{\mu(1-\mu)}. \tag{6}$$

Let $d = E\left((2\mu-1)^2\right)$ over the distribution of the key. (We recall that μ depends on the key used in the cipher.) Let $\alpha = \frac{1}{8}\left(d\sqrt{\frac{2\pi}{n}}\right)^{\frac{1}{3}}$. Since $d \le 1$ and $n \ge 1$, we have $\alpha \le .17$ so if $\left|\mu - \frac{1}{2}\right| \le \alpha$ we have

$$\left|p^K - p_0\right| \le .55(dn)^{\frac{1}{3}}.$$

Now we have $\left|\mu - \frac{1}{2}\right| \ge \alpha$ with a probability less than $\frac{d}{4\alpha^2}$, which is less than $8.68(dn)^{\frac{1}{3}}$, and in this case we have $\left|p^K - p_0\right| \le 1$. Hence, we have

$$E\left(\left|p^K - p_0\right|\right) \le 9.3(dn)^{\frac{1}{3}}.$$

We note that $d = E\left(LP^C(a,b)\right)$. We also have $E\left(LP^{C^*}(a,b)\right) = \frac{1}{2^m-1}$ from Equation (5), therefore $|p - p^*|$ is too small if $E\left(LP^C(a,b)\right)$ has the order of 2^{-m} and $n \ll 2^m$. $\qquad\square$

Theorem 13. *Let $\epsilon = ||\,|[C]^2 - [C^*]^2|\,||_\infty$ where C^* is the Perfect Cipher. For any basic linear distinguisher between C and C^*, we have $|p - p^*| \leq 9.3 \left(\frac{n}{2^m-1} + 4n\epsilon\right)^{\frac{1}{3}}$.*

Proof. Actually we notice that $E\left(\mathrm{LP}^{C^*}(a, b)\right) = \frac{1}{2^m-1}$ and that

$$\left| E\left(\mathrm{LP}^C(a, b)\right) - \frac{1}{2^m - 1} \right| \leq 4\epsilon$$

from Equation (5). □

So, if ϵ has the order of 2^{-m}, basic linear cryptanalysis does not work against C, but with a complexity in the scale of 2^m.

8 COCONUT: a perfect decorrelation design

In this Section we define the COCONUT Ciphers family which are perfectly decorrelated ciphers to the order two.

The COCONUT Ciphers are characterized by some parameters (m, p). m is the block length, and p is a irreducible polynomial with degree m in GF(2) (which defines a representation of the GF(2^m) Galois Field). A COCONUT Cipher with block length m is simply a product cipher $C_1 \circ C_2 \circ C_3$ where C_1 and C_3 are any (possibly weak) ciphers which can depend from each other, and C_2 is an independent cipher based on a $2m$-bit key which consists of two polynomials A and B with degree at most $m - 1$ over GF(2) such that $A \neq 0$. For a given representation of polynomials into m-bit strings, we simply define

$$C(x) = A.x + B \bmod p.$$

Since C_2 performs perfect decorrelation to the order two and since it is independent from C_1 and C_3, any COCONUT Cipher is obviously perfectly decorrelated to the order two. Therefore Theorems 11 and 13 shows that COCONUT resists to the basic differential and linear cryptanalysis.

One can wonder why C_1 and C_3 are for. Actually, C_2 makes some precise attacks provably impractical, but in a way which makes the cipher obviously weak against other attacks. We believe that all real attacks on any real cipher have an intrinsic *order* d, that is they use the d-wise correlation in the encryption of d messages. Attacks with a large d on real ciphers are impractical, because the d-wise decorrelation can hardly be analyzed since it depends on too many factors. Therefore, the COCONUT approach consists in making the cipher provably resistant against attacks with order at most

2 such as differential or linear cryptanalysis, and heuristically secure against attacks with higher order by real life ciphers as C_1 and C_3. The cipher C_2 alone would actually have been very unsecure since two known plaintexts can recover the key which can be used to decrypt any (third) ciphertext.

In the Appendix we propose a concrete example: the COCONUT98 Cipher with parameters $m = 64$ and $p = x^{64} + x^{11} + x^2 + x + 1$.

9 PEANUT: a partial decorrelation design

In this Section we define the PEANUT Ciphers family, which achieves an example of partial decorrelation. This family is based on a combinatorial tool which has been previously used by Halevi and Krawczyk for authentication in [12].

The PEANUT Ciphers are characterized by some parameters (m, r, d, p). They are Feistel Ciphers with block length of m bits (m even), r rounds. The parameter d is the order of partial decorrelation that the cipher performs, and p must be a prime number greater than $2^{\frac{m}{2}}$.

The cipher is defined by a key of $\frac{mrd}{2}$ bits which consists of a sequence of r lists of d $\frac{m}{2}$-bit numbers, one for each round. In each round, the F function has the form

$$F(x) = g(k_1.x^{d-1} + k_2.x^{d-2} + \ldots + k_{d-1}.x + k_d \bmod p \bmod 2^{\frac{m}{2}})$$

where g is any permutation on the set of all $\frac{m}{2}$-bit numbers.

Let us now estimate the $|||.|||_\infty$-decorrelation of the PEANUT ciphers.

Lemma 14. Let $\mathbf{K} = \mathrm{GF}(q)$ be a finite field, let $r : \{0,1\}^{\frac{m}{2}} \to \mathbf{K}$ be an injective mapping, and let $p : \mathbf{K} \to \{0,1\}^{\frac{m}{2}}$ be a surjective mapping. Let F be a random function defined by

$$F(x) = p(r(A_{d-1}).r(x)^{d-1} + \ldots + r(A_0))$$

where the A_i's are independent and uniformly distributed in $\{0,1\}^{\frac{md}{2}}$. If R is an independent uniformly distributed random function, we have

$$|||[F]^d - [R]^d|||_\infty \leq 2 \left(\left(\frac{q}{2^{\frac{m}{2}}} \right)^d - 1 \right).$$

Proof. Let $x = (x_1, \ldots, x_d)$ be a multipoint in $\{0,1\}^{\frac{m}{2}}$. We want to prove that

$$S = \sum_{y=(y_1,\ldots,y_d)} |[F]^d_{x,y} - [R]^d_{x,y}| \leq 2 \left(\left(\frac{q}{2^{\frac{m}{2}}} \right)^d - 1 \right).$$

Let c be the number of pairwise different x_i's. For any y such that there exists (i,j) such that $y_i \neq y_j$ and $x_i = x_j$, the contribution to the sum is zero. So we can assume that y is defined over the $2^{\frac{cm}{2}}$ choices of y_i's on positions corresponding to pairwise different x_i's. If we let $x_{d+1}, \ldots, x_{2d-c}$ be new fixed points such that we have exactly d pairwise different x_i's, since the probability that F (resp. R) maps x onto y is equal to the sum over all choices of $y_{d+1}, \ldots, y_{d+c}$ that it maps the extended x onto the extended y, we can assume w.l.o.g. that $c = d$.

For any multipoint y we thus have that $\Pr[x \mapsto y] = j.2^{-\frac{md}{2}}$ where j is an integer. Let N_j be the number of multipoints y which verify this property. We have $\sum_j N_j = 2^{\frac{md}{2}}$ and $\sum_j j N_j = 2^{\frac{md}{2}}$. We have

$$S \leq \sum_j N_j \left| \frac{j-1}{2^{\frac{md}{2}}} \right| = 2N_0 . 2^{-\frac{md}{2}}.$$

N_0 is the number of unreachable y's, i.e. the y's which correspond to a polynomial whose coefficients are not all r-images. This number is thus less than the number of missing polynomials which is $q^d - 2^{\frac{md}{2}}$. □

From Theorem 9 with $k = 3$ we thus obtain the following Theorem.

Theorem 15. *Let C be a cipher in the PEANUT family with parameters (m, r, d, p) and let C^* be the Perfect Cipher. We have*

$$||[C]^d - [C^*]^d||_\infty \leq \left(\left(1 + 2 \left(p^d 2^{-\frac{md}{2}} - 1 \right) \right)^3 - 1 + \frac{2d^2}{2^{\frac{m}{2}}} \right)^{\lfloor \frac{r}{3} \rfloor}.$$

We can thus approximate

$$||[C]^d - [C^*]^d||_\infty \approx \left(\frac{6d \left(p - 2^{\frac{m}{2}} \right) + 2d^2}{2^{\frac{m}{2}}} \right)^{\frac{r}{3}}.$$

Example. We can use the parameters $m = 64$, $r = 9$, $d = 2$ and $p = 2^{32} + 15$. We obtain that $||[C]^2 - [C^*]^2||_\infty \leq 2^{-76}$. Therefore from Theorems 11 and 13 no differential or linear cryptanalysis can efficiently distinguish the cipher from the Perfect Cipher. In the Appendix we propose PEANUT98 which is based on those parameters.

In an earlier version of this work [36], we proposed a similar construction (say PEANUT97) which uses prime numbers smaller than $2^{\frac{m}{2}}$. However the result above does not hold with the $|||.|||_\infty$ norm, but rather with the $||.||_2$ one. The drawback is that this norm has less friendly theorems for constructing Feistel ciphers, and in particular we need more rounds to make the cipher provably secure.

10 Note on the key length

One problem with the COCONUT or PEANUT construction is that it requires a long key (in order to make the internal random functions independent). In real-life examples, we can generate this long key by using a pseudorandom generator fed with a short key, but the results on the security based on decorrelation are no longuer valid. However, provided that the pseudorandom generator produces outputs which are undistinguishable from truly random sequences, we can still prove the security.

Actually, let C be the cipher fed with a key spanned with a short key and let C^* be the cipher fed with a truly random key. We assume that we have a result on the security of C^* based on its decorrelation. If there exists an attack against C which would have contradicted the security if it could be applied against C^*, we can use this attack to distinguish the key spanned by the generator from a random key: the distinguisher just give the output of the attack. Hence if the pseudorandom generator is secure, the security results hold on C as they hold on C^*.

11 Conclusion

Decorrelation modules are cheap and friendly tools which can strengthen the security of block ciphers. Actually, we can quantify their security against a class of cryptanalysis which includes differential and linear cryptanalysis. To illustrate this paradigm, we propose two definite prototype ciphers. (see Appendix). Research on other general cryptanalysis is still an open problem, so we strongly encourage research on analyzing the security of those prototype ciphers.

Acknowledgements

I wish to thank Thomas Pornin and Jacques Stern for valuable help. I also thank the CNRS for having strongly motivated this work.

References

[1] E. Biham. A fast new DES implementation in software. In *Fast Software Encryption*, Haifa, Israel, Lectures Notes in Computer Science 1267, pp. 260–272, Springer-Verlag, 1997.

[2] E. Biham, A. Shamir. Differential cryptanalysis of DES-like cryptosystems. In *Advances in Cryptology CRYPTO'90*, Santa Barbara, California, U.S.A., Lectures Notes in Computer Science 537, pp. 2–21, Springer-Verlag, 1991.

[3] E. Biham, A. Shamir. Differential cryptanalysis of DES-like cryptosystems. *Journal of Cryptology*, vol. 4, pp. 3–72, 1991.

[4] E. Biham, A. Shamir. Differential cryptanalysis of the full 16-round DES. In *Advances in Cryptology CRYPTO'92*, Santa Barbara, California, U.S.A., Lectures Notes in Computer Science 740, pp. 487–496, Springer-Verlag, 1993.

[5] E. Biham, A. Shamir. *Differential Cryptanalysis of the Data Encryption Standard*, Springer-Verlag, 1993.

[6] F. Chabaud, S. Vaudenay. Links between differential and linear cryptanalysis. In *Advances in Cryptology EUROCRYPT'94*, Perugia, Italy, Lectures Notes in Computer Science 950, pp. 356–365, Springer-Verlag, 1995.

[7] L. Carter, M. Wegman. Universal clases of hash functions. *Journal of Computer and System Sciences*, vol. 18, pp. 143–154, 1979.

[8] New directions in cryptography. *IEEE Transactions on Information Theory*, vol. IT-22, pp. 644–654, 1976.

[9] H. Feistel. Cryptography and computer privacy. *Scientific american*, vol. 228, pp. 15–23, 1973.

[10] H. Gilbert. *Cryptanalyse Statistique des Algorithmes de Chiffrement et Sécurité des Schémas d'Authentification*, Thèse de Doctorat de l'Université de Paris 11, 1997.

[11] H. Gilbert, G. Chassé. A statistical attack of the FEAL-8 cryptosystem. In *Advances in Cryptology CRYPTO'90*, Santa Barbara, California, U.S.A., Lectures Notes in Computer Science 537, pp. 22–33, Springer-Verlag, 1991.

[12] S. Halevi, H. Krawczyk. MMH: software message authentication in the Gbit/second rates. In *Fast Software Encryption*, Haifa, Israel, Lectures Notes in Computer Science 1267, pp. 172–189, Springer-Verlag, 1997.

[13] H. M. Heys, S. E. Tavares. Substitution-Permutation Networks resistant to differential and linear cryptanalysis. *Journal of Cryptology*, vol. 9, pp. 1–19, 1996.

[14] T. Jakobsen, L. R. Knudsen. The interpolation attack on block ciphers. In *Fast Software Encryption*, Haifa, Israel, Lectures Notes in Computer Science 1267, pp. 28–40, Springer-Verlag, 1997.

[15] L. R. Knudsen. *Block Ciphers — Analysis, Design and Applications*, Aarhus University, 1994.

[16] B. R. Kaliski Jr., M. J. B. Robshaw. Linear cryptanalysis using multiple approximations. In *Advances in Cryptology CRYPTO'94*, Santa Barbara, California, U.S.A., Lectures Notes in Computer Science 839, pp. 26–39, Springer-Verlag, 1994.

[17] X. Lai. *On the Design and Security of Block Ciphers*, ETH Series in Information Processing, vol. 1, Hartung-Gorre Verlag Konstanz, 1992.

[18] X. Lai, J. L. Massey, S. Murphy. Markov ciphers and differential cryptanalysis. In *Advances in Cryptology EUROCRYPT'91*, Brighton, United Kingdom, Lectures Notes in Computer Science 547, pp. 17–38, Springer-Verlag, 1991.

[19] M. Luby, C. Rackoff. How to construct pseudorandom permutations from pseudorandom functions. *SIAM Journal on Computing*, vol. 17, pp. 373–386, 1988.

[20] M. Matsui. Linear cryptanalysis methods for DES cipher. In *Advances in Cryptology EUROCRYPT'93*, Lofthus, Norway, Lectures Notes in Computer Science 765, pp. 386–397, Springer-Verlag, 1994.

[21] M. Matsui. The first experimental cryptanalysis of the Data Encryption Standard. In *Advances in Cryptology CRYPTO'94*, Santa Barbara, California, U.S.A., Lectures Notes in Computer Science 839, pp. 1–11, Springer-Verlag, 1994.

[22] M. Matsui. New structure of block ciphers with provable security against differential and linear crypt-analysis. In *Fast Software Encryption*, Cambridge, United Kingdom, Lectures Notes in Computer Science 1039, pp. 205–218, Springer-Verlag, 1996.

[23] R. Merkle, M. Hellman. Hiding information and signatures in trapdoor knapsacks. *IEEE Transactions on Information Theory*, vol. IT-24, pp. 525–530, 1978.

[24] S. Murphy, F. Piper, M. Walker, P. Wild. Likehood estimation for block cipher keys. Unpublished.

[25] K. Nyberg. Perfect nonlinear S-boxes. In *Advances in Cryptology EUROCRYPT'91*, Brighton, United Kingdom, Lectures Notes in Computer Science 547, pp. 378–385, Springer-Verlag, 1991.

[26] K. Nyberg, L. R. Knudsen. Provable security against a differential cryptanalysis. *Journal of Cryptology*, vol. 8, pp. 27–37, 1995.

[27] J. Patarin. *Etude des Générateurs de Permutations Basés sur le Schéma du D.E.S.*, Thèse de Doctorat de l'Université de Paris 6, 1991.

[28] J. Patarin. In *Advances in Cryptology EUROCRYPT'92*, Balatonfüred, Hungary, Lectures Notes in Computer Science 658, pp. 256–266, Springer-Verlag, 1993.

[29] J. Patarin. About Feistel schemes with six (or more) rounds. To appear in *Fast Software Encryption*, 1998.

[30] R. L. Rivest, A. Shamir, L. M. Adleman. A method for obtaining digital signatures and public-key cryptosystems. *Communications of the ACM*, vol. 21, pp. 120–126, 1978.

[31] C. E. Shannon. Communication theory of secrecy systems. *Bell system technical journal*, vol. 28, pp. 656–715, 1949.

[32] A. Shamir. How to photofinish a cryptosystem? Presented at the Rump Session of Crypto'97.

[33] A. Tardy-Corfdir, H. Gilbert. A known plaintext attack of FEAL-4 and FEAL-6. In *Advances in Cryptology CRYPTO'91*, Santa Barbara, California, U.S.A., Lectures Notes in Computer Science 576, pp. 172–181, Springer-Verlag, 1992.

[34] S. Vaudenay. *La Sécurité des Primitives Cryptographiques*, Thèse de Doctorat de l'Université de Paris 7, Technical Report LIENS-95-10 of the Laboratoire d'Informatique de l'Ecole Normale Supérieure, 1995.

[35] S. Vaudenay. An experiment on DES — Statistical cryptanalysis. In *3rd ACM Conference on Computer and Communications Security*, New Delhi, India, pp. 139–147, ACM Press, 1996.

[36] S. Vaudenay. A cheap paradigm for block cipher security strengthening. Technical Report LIENS-97-3. Unpublished.

[37] G. S. Vernam. Cipher printing telegraph systems for secret wire and radio telegraphic communications. *Journal of the American Institute of Electrical Engineers*, vol. 45, pp. 109–115, 1926.

[38] M. N. Wegman, J. L. Carter. New hash functions and their use in authentication and set equality. *Journal of Computer and System Sciences*, vol. 22, pp. 265–279, 1981.

Appendix: COCONUT98 and PEANUT98

We define here two real-life block ciphers in the COCONUT and PEANUT families: COCONUT98 and PEANUT98. Both are on blocks with length 64 bits. They both use a secret key which is defined as a sequence which can be generated by a secure pseudorandom generator.

Without any mention, we identify the bitstrings in $\{0,1\}^\ell$ and the integers in $\{0,\ldots,2^\ell-1\}$ by the natural binary expansion mapping

$$b_{\ell-1}\ldots b_1 b_0 \longleftrightarrow 2^{\ell-1}b_{\ell-1} + \ldots + 2b_1 + b_0.$$

Let first let φ and g be two functions from $\{0,1\}^{32}$ onto itself. We define

$$\varphi(x) = x + 256.S(x \bmod 256) \bmod 2^{32}$$

and

$$g(x) = R_L^{11}(\varphi(x)) + c \bmod 2^{32}$$

where R_L^{11} is a circular rotation by 11 bits to the left (i.e. multiplication by 2^{11} modulo $2^{32}+1$), c is a constant and S is a lookup table of 256 24-bit integers. We use the hexadecimal expansion of the mathematical constant e to define c and S: given that

$$e = \sum_{i=0}^{\infty} \frac{1}{i!} = 2.\text{b7e15162 8aed2a 6abf71 58809c f4f3c7 62e716}\ldots$$

we define $c = \text{b7e15162}$ and S by Tables 1 and 2.

We observe that $x \bmod 256 = \varphi(x) \bmod 256$. Thus from $\varphi(x)$ we can recover x by $x = \varphi(x) - 256.S(\varphi(x) \bmod 256)$. Since the sum to a constant modulo 2^{32}, R_L and φ are permutations, g is a permutation over the set of 32-bit strings.

The COCONUT98 Cipher

For COCONUT98, we know define

$$f_i(x) = \varphi(g(x \oplus k_i))$$

where $\oplus$ denotes the exclusive or and k_i is a constant defined by the secret key. From the f_i's we can define a Feistel permutation onto the set of 64-bit strings. The COCONUT Cipher consists of two four-round Feistel ciphers $\Psi(f_1, f_2, f_3, f_4)$ and $\Psi(f_5, f_6, f_7, f_8)$ and a decorrelation module in between.

The key space of COCONUT98 is the set of all 256-bit strings K such that if we write the concatenation $K = (K_1, K_2, \ldots, K_8)$ where all K_i's are

	.0	.1	.2	.3	.4	.5	.6	.7
0.	8aed2a	6abf71	58809c	f4f3c7	62e716	0f38b4	da56a7	84d904
1.	bb1185	eb4f7c	7b5757	f59584	90cfd4	7d7c19	bb4215	8d9554
2.	cfbfa1	c877c5	6284da	b79cd4	c2b329	3d20e9	e5eaf0	2ac60a
3.	78e537	d2b95b	b79d8d	caec64	2c1e9f	23b829	b5c278	0bf387
4.	bbca06	0f0ff8	ec6d31	beb5cc	eed7f2	f0bb08	801716	3bc60d
5.	94640d	6ef0d3	d37be6	7008e1	86d1bf	275b9b	241deb	64749a
6.	f10de5	13d3f5	114b8b	5d374d	93cb88	79c7d5	2ffd72	ba0aae
7.	571121	382af3	41afe9	4f77bc	f06c83	b8ff56	75f097	9074ad
8.	5a7db4	61dd8f	3c7554	0d0012	1fd56e	95f8c7	31e9c4	d7221b
9.	c6b400	e024a6	668ccf	2e2de8	6876e4	f5c500	00f0a9	3b3aa7
a.	d1060b	871a28	01f978	376408	2ff592	d9140d	b1e939	9df4b0
b.	c703f5	32ce3a	30cd31	c070eb	36b419	5ff33f	b1c66c	7d70f9
c.	6d8d03	62803b	c248d4	14478c	2afb07	ffe78e	89b9fe	ca7e30
d.	df2be6	4bbaab	008ca8	a06fda	ce9ce7	048984	5a082b	a36d61
e.	558aa1	194177	20b6e1	50ce2b	927d48	d7256e	445e33	3cb757
f.	6b6c79	a58a9a	549b50	c58706	90755c	35e4e3	6b5290	38ca73

Table 1: $S(xy)$ for $y < 8$

	.8	.9	.a	.b	.c	.d	.e	.f
0.	5190cf	ef324e	773892	6cfbe5	f4bf8d	8d8c31	d763da	06c80a
1.	f7b46b	ced55c	4d79fd	5f24d6	613c31	c3839a	2ddf8a	9a276b
2.	cc93ed	874422	a52ecb	238fee	e5ab6a	dd835f	d1a075	3d0a8f
3.	37df8b	b300d0	1334a0	d0bd86	45cbfa	73a616	0ffe39	3c48cb
4.	f45a0e	cb1bcd	289b06	cbbfea	21ad08	e1847f	3f7378	d56ced
5.	47dfdf	b96632	c3eb06	1b6472	bbf84c	26144e	49c2d0	4c324e
6.	7277da	7ba1b4	af1488	d8e836	af1486	5e6c37	ab6876	fe690b
7.	9a787b	c5b9bd	4b0c59	37d3ed	e4c3a7	939621	5edab1	f57d0b
8.	bed0c6	2bb5a8	7804b6	79a0ca	a41d80	2a4604	c311b7	1de3e5
9.	e6342b	302a0a	47373b	25f73e	3b26d5	69fe22	91ad36	d6a147
a.	e14ca8	e88ee9	110b2b	d4fa98	eed150	ca6dd8	932245	ef7592
b.	391810	7ce205	1fed33	f6d1de	9491c7	dea6a5	a442e1	54c8bb
c.	60c08f	0d61f8	e36801	df66d1	d8f939	2e52ca	ef0653	199479
d.	1e99f2	fbe724	246d18	b54e33	5cac0d	d1ab9d	fd7988	a4b0c4
e.	2b3bd0	0fb274	604318	9cac11	6cedc7	e771ae	0358ff	752a3a
f.	3fd1aa	a8dab4	0133d8	0320e0	790968	c76546	b993f6	c8ff3b

Table 2: $S(xy)$ for $y \geq 8$

32-bit strings, not all the 64 bits of (K_7, K_8) are set to zero. We define the k_i's by

i	1	2	3	4
k_i	K_1	$K_1 \oplus K_3$	$K_1 \oplus K_3 \oplus K_4$	$K_1 \oplus K_4$

i	5	6	7	8
k_i	K_2	$K_2 \oplus K_3$	$K_2 \oplus K_3 \oplus K_4$	$K_2 \oplus K_4$

which can easily be performed in hardware: in the first (resp. second) Feistel cipher, we first load K_1 (resp. K_2) in a round-key register and at each round xor it with K_3 or K_4 alternately. The decorrelation module is a function M defined by $(K_5, \ldots, K_8)$ by

$$M(xy) = (xy \oplus K_5 K_6) \times K_7 K_8$$

where the product is performed in $GF(2^{64})$. The Galois Field representation is chosen so that a 64-bit string $b_{63} \ldots b_1 b_0$ represents the polynomial

$$b_{63}.x^{63} + \ldots + b_1.x + b_0$$

modulo $x^{64} + x^{11} + x^2 + x + 1$ modulo 2.

Finally, the COCONUT98 Cipher is defined by

$$C_K(xy) = \Psi(f_5, f_6, f_7, f_8)\left(M\left(\Psi(f_1, f_2, f_3, f_4)(xy)\right)\right)$$

which is illustrated on Figure 1. We assume that the secret key is uniformly distributed among the bitstrings such that $K_7 K_8 \neq 0$.

We note that the decryption can be performed by using a key K^{-1} defined by

$$K^{-1} = (K_2 \oplus K_4, K_1 \oplus K_4, K_3, K_4, K_5', K_6', K_7', K_8')$$

with

$$K_5' K_6' = K_5 K_6 \times K_7 K_8$$
$$K_7' K_8' = 1/K_7 K_8$$

in $GF(2^{64})$.

The PEANUT98 Cipher

PEANUT98 is in the PEANUT family with parameters $(64, 9, 2, 2^{32} + 15)$. It is thus a 9-rounds Feistel Cipher. The secret key is a uniformly distributed 576-bit string which is split into 32-bit strings $K = (K_1, \ldots, K_{18})$. We define the round function in the i-th round as

$$f_i'(x) = g(x.K_{2i-1} + K_{2i} \bmod 2^{32} + 15 \bmod 2^{32}).$$

Decryption consists of inverting the order of the 64-bit blocks in the key string, i.e. flipping K_1 and K_{17}, K_2 and K_{18}, K_3 and K_{15}, etc.

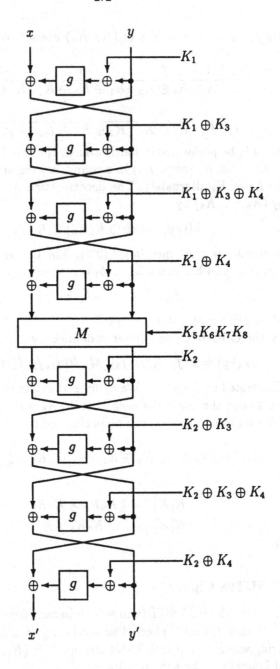

Figure 1: The COCONUT Cipher

Note. We remark that although the PEANUT construction leads to provably secure ciphers against previously mentioned attacks in average among all the keys, it may be possible that some small set of keys are weak. It may be unsecure that there exists some set of keys with density $1/p$ such that there are some simple attacks on when p is not too large. For instance, the PEANUT98 Cipher may accept all the keys such that for some i we have $K_{2i-1} = 0$ as weak keys. Although we did not try to describe an attack in full details, we suggest that those keys shall not be used.

Test Values

As an example, we use the secret key

$$K = \begin{array}{l} \texttt{7c44a4ad56f6bb77220e96e8401694e1} \\ \texttt{6c469dbc516decc517929e9b226ddd64} \end{array}$$

to encrypt the plaintext 6d8779e078ac5f02 with COCONUT98. Here are the intermediate values of the left and right halves in the encryption:

round#1 :	78ac5f02	62b29ee4
round#2 :	62b29ee4	f257ec80
round#3 :	f257ec80	2c554933
round#4 :	6d40ec0e	2c554933
decorrelation :	9c7e2827	751e12b5
round#5 :	751e12b5	ed8378b4
round#6 :	ed8378b4	4f8ba7ff
round#7 :	4f8ba7ff	037910f8
round#8 :	3b2ae895	037910f8

so the ciphertext is 3b2ae895037910f8. For instance, in the first round we have to compute

$$
\begin{aligned}
g(\texttt{78ac5f02} \oplus \texttt{7c44a4ad}) &= g(\texttt{04e8fbaf}) \\
&= \varphi(R_L^{11}(\varphi(\texttt{04e8fbaf})) + c) \\
&= \varphi(R_L^{11}(\texttt{f45e8daf}) + c) \\
&= \varphi(\texttt{f46d7fa2} + c) \\
&= \varphi(\texttt{f46d7fa2} + \texttt{b7e15162}) \\
&= \varphi(\texttt{ac4ed104}) \\
&= \texttt{0f35e704}
\end{aligned}
$$

which is xored onto 6d8779e0 to produce 62b29ee4.

As an example for PEANUT98, we use the secret key

$$k_1 = \texttt{2115e265}$$
$$k_2 = \texttt{9225cb79}$$
$$k_3 = \texttt{cfa1c6fc}$$
$$k_4 = \texttt{bd67eef1}$$
$$k_5 = \texttt{58cb0b8f}$$
$$k_6 = \texttt{fbf151b1}$$
$$k_7 = \texttt{423c41e6}$$
$$k_8 = \texttt{ec11b5d9}$$
$$k_9 = \texttt{b9002c83}$$
$$k_{10} = \texttt{406c0b46}$$
$$k_{11} = \texttt{ba977fbd}$$
$$k_{12} = \texttt{91c0adf4}$$
$$k_{13} = \texttt{5b716ec6}$$
$$k_{14} = \texttt{1533a950}$$
$$k_{15} = \texttt{080b807e}$$
$$k_{16} = \texttt{a1a305e3}$$
$$k_{17} = \texttt{2a0f096e}$$
$$k_{18} = \texttt{4b027140}$$

The encryption of $\texttt{0123456789abcdef}$ is $\texttt{07f141edac6485df}$. In the first round, we have to compute

$$
\begin{aligned}
f_1'(\texttt{89abcdef}) &= g(\texttt{89abcdef.2115e265} + \texttt{9225cb79} \bmod 2^{32} + 15) \\
&= g(\texttt{037724cf7}) \\
&= R_L^{11}(\texttt{703cbff7}) + c \bmod 2^{32} \\
&= \texttt{e5ffbb81} + c \bmod 2^{32} \\
&= \texttt{9de10ce3}
\end{aligned}
$$

which is xored onto $\texttt{01234567}$ to produce $\texttt{9cc24984}$.

Implementation

Because of the finite field multiplication, the COCONUT98 Cipher is adapted to hardware implementations. Software implementations may require dedicated tricks such as partial table look-up for the multiplications. Biham's bit-slice parallel implementation technics may be a good way to solve the problem too (see [1]).

The design of PEANUT98 is well adapted to software implementation on microprocessors enable to perform 32-bit integer multiplications. A software implementation on a Pentium gave a short (non-optimized) code with only 396 clock cycles per block encryption. This leads to a 20.5Mbps encryption rate at 133MHz.

The integer multiplication makes Biham's bit-slice parallel programming method impractical (see [1]). This has however a good consequence on its security because it makes Shamir's photo-finishing attack impossible too [32].

On the Approximation of Finding A(nother) Hamiltonian Cycle in Cubic Hamiltonian Graphs

(Extended abstract) *

Cristina Bazgan[1] Miklos Santha[2] Zsolt Tuza[3]

[1] Université Paris-Sud, LRI, bât.490, F–91405 Orsay, France, bazgan@lri.fr
[2] CNRS, URA 410, Université Paris-Sud, LRI, F–91405 Orsay, France, santha@lri.fr
[3] Computer and Automation Institute, Hungarian Academy of Sciences, H–1111 Budapest, Kende u.13–17, Hungary, tuza@sztaki.hu

Abstract. It is a simple fact that cubic Hamiltonian graphs have at least two Hamiltonian cycles. Finding such a cycle is NP-hard in general, and no polynomial time algorithm is known for the problem of finding a second Hamiltonian cycle when one such cycle is given as part of the input. We investigate the complexity of approximating this problem where by a feasible solution we mean a(nother) cycle in the graph. First we prove a negative result showing that the LONGEST PATH problem is not constant approximable in cubic Hamiltonian graphs unless $P = NP$. No such negative result was previously known for this problem in Hamiltonian graphs. In strong opposition with this result we show that there is a polynomial time approximation scheme for finding another cycle in cubic Hamiltonian graphs if a Hamiltonian cycle is given in the input.

1 Introduction

LONGEST PATH and LONGEST CYCLE are well-known problems in graph theory which were shown to be NP-complete in 1972 by Karp [7]. The approximability of the associated optimization problems is very much open despite considerable efforts in recent years.

Monien [10] gave an algorithm to find a path of length k in time $O(k! \cdot n \cdot m)$ where n and m are respectively the number of vertices and the number of edges of the graph. Karger, Motwani and Ramkumar [8] gave a polynomial time algorithm which finds a path of length $\Omega(\log n)$ in any 1-tough graph. A similar result was obtained also by Fürer and Raghavachari [4]. Since 1-tough graphs include Hamiltonian graphs, these algorithms can be used in particular to find such paths in graphs which contain a Hamiltonian cycle. Alon, Yuster and Zwick [2] generalized this result by giving a polynomial time algorithm which for any

* This research was supported by the ESPRIT Working Group RAND2 n° 21726 and by the bilateral project Balaton, grant numbers 97140 (APAPE, France) and F-36/96 (TéT Alapítvány, Hungary)

$c > 0$, finds a path of length $c \log n$, in a graph which contains such a path. Finding paths of length $\omega(\log n)$ in polynomial time is an open problem even for Hamiltonian graphs.

On the negative side, Karger, Motwani and Ramkumar [8] have proved that unless $P=NP$, LONGEST PATH is not constant approximable in polynomial time. Their proof consists of two parts. First, they have shown that LONGEST PATH doesn't have a polynomial time approximation scheme, unless $P=NP$. They were able to show this even when the input instances are restricted to Hamiltonian graphs. Then they gave a self-improving scheme for the problem, showing that a polynomial time approximating algorithm for *some* constant can be transformed into a polynomial time approximating algorithm for *any* constant. These results remain valid also when the maximum degree of the input graph is bounded by a constant at least four. But their self-improving scheme didn't conserve Hamiltonicity, and they asked if it can be proven also for Hamiltonian graphs that they are not constant approximable in polynomial time, unless $P=NP$.

In this paper we will prove an even stronger negative result. It turns out that we will be able to give a self-improving scheme for LONGEST PATH which preserves Hamiltonicity when the input graphs are further restricted to be also cubic. That LONGEST PATH remains NP-complete even for cubic graphs was shown by Garey, Johnson and Tarjan [6]. In addition we also prove that this problem doesn't have a polynomial time approximation scheme in cubic Hamiltonian graphs, unless $P=NP$. These two results imply that LONGEST PATH is not constant approximable for any constant in cubic Hamiltonian graphs, unless $P=NP$. A similar result follows immediately for LONGEST CYCLE.

The LONGEST CYCLE problem has an interesting variant in cubic Hamiltonian graphs. It is not hard to show [11] that any such graph has at least two Hamiltonian cycles. Therefore if some Hamiltonian cycle is given as part of the input, one can ask to find another Hamiltonian cycle in the graph. We will call this problem SECOND HAMILTONIAN CYCLE. It is a well known instance of what Meggido and Papadimitriou [9] call the class $TFNP$ of total functions. This class contains function problems associated with languages in NP where for every instance of the problem a solution is guaranteed to exist. Other examples in the class are FACTORING and the HAPPYNET problem.

Many functions in $TFNP$ (like the examples quoted above) have a challenging intermediate status between FP and FNP, the function classes associated with P and NP. Although these problems are not NP-hard unless $NP=co\text{-}NP$, no polynomial time algorithm is known for them. We consider here (for the first time up to our knowledge) approximating a problem in $TFNP$. In particular, we show that in striking opposition with the above negative result, SECOND HAMILTONIAN CYCLE admits a polynomial time approximating scheme, where a feasible solution for this problem is a cycle different from the one given in the input.

The paper is organized as follows: In section 2 we give the necessary definitions and reduce LONGEST PATH to approximating the longest path between two fixed vertices in cubic Hamiltonian graphs. In section 3 we prove that this latter problem has no polynomial time approximation scheme, and in section 4 we prove that it is not constant approximable either. In section 5 we describe a ptas for SECOND HAMILTONIAN CYCLE.

2 Preliminaries

In this paper by optimization problem we always mean an NP-optimization problem. Let us recall a few notions about their approximability. Given an instance x of an optimization problem A and a feasible solution y of x, we denote by $m(x, y)$ the value of the solution y, and with $opt_A(x)$ the value of an optimum solution of x. The *performance ratio* of y is

$$R(x, y) = \max\left\{ \frac{m(x, y)}{opt_A(x)}, \frac{opt_A(x)}{m(x, y)} \right\}.$$

For a constant $c > 1$, an algorithm is a *c-approximation* if for any instance x of the problem it returns a solution y such that $R(x, y) \leq c$. We say that an optimization problem is *constant approximable* if for some $c > 1$, there exists a polynomial time c-approximation for it. The set of problems which are constant approximable is denoted by APX. An optimization problem has a *polynomial time approximation scheme* (in short a *ptas*) if for every constant $\varepsilon > 0$, there exists a polynomial time $(1 + \varepsilon)$-approximation for it.

The notion of L-reduction was introduced by Papadimitriou and Yannakakis in [13]. Let A and B be two optimization problems. A is *L-reducible* to B if there are two constants $\alpha, \beta > 0$ such that

1. there exists a polynomial time computable function which transforms an instance x of A into an instance x' of B such that $opt_B(x') \leq \alpha \cdot opt_A(x)$,
2. there exists a polynomial time computable function which transforms any solution y' of x' into a solution y of x such that $|m(x, y) - opt_A(x)| \leq \beta \cdot |m(x', y') - opt_B(x')|$.

For us the important property of this reduction is that it preserves ptas, that is if A is L-reducible to B and B has a ptas then A has also a ptas.

Let $G = (V, E)$ an undirected graph. A *path* of length k in G is a sequence of distinct vertices $v_0, v_1, \ldots, v_k$ such that for $0 \leq i \leq k - 1$, there is an edge between v_i and v_{i+1}. For two vertices s and t, an *s-t path* is a path whose first vertex is s and last vertex is t. A path of length at least three whose first and last vertices coincide is called a *cycle*. A path *covers* a subgraph H if it contains all the vertices of H. A path or a cycle is *Hamiltonian* if it covers G. The graph is called *Hamiltonian* if it has a Hamiltonian cycle, and it is called *cubic* if the degree of all its vertices is three. Finally it is called *cubic with distinguished vertices* s and t if all its vertices have degree three except s and t which have degree two, and there is an edge between s and t.

Our negative result is that there is no constant approximation for the longest path (cycle) problem in cubic Hamiltonian graphs, problems we now define formally.

CH LONGEST PATH (CYCLE)
Input: A cubic Hamiltonian graph G.
Solution: A path (cycle).
Value: The length of the path (cycle).

Since CH LONGEST PATH is trivially L-reducible to CH LONGEST CYCLE, we will prove our non-approximability result for CH LONGEST PATH. For technical reasons it is easier to show it for the following variant of the problem.

CH LONGEST s-t PATH
Input: A cubic Hamiltonian graph G with distinguished vertices s and t .
Solution: An s-t path.
Value: The length of the path.

It is probably standard knowledge (and it was pointed out to us by M. Yannakakis [16]) that these two problems have the same difficulty of approximation. We state here the exact reduction we need.

Lemma 1. *If* CH LONGEST PATH *is constant approximable then* CH LONGEST s-t PATH *is also constant approximable.*

What is particular in these instances of the longest path problem is that the value of the optimum solution is known in advance. Although they remain hard to approximate, this property makes it very unlikely that MAX 3SAT could be L-reduced to them, as we will show it in the next section. Therefore to prove that they still don't have a ptas, we will reduce to them the special case of MAX 3SAT where the value of an optimum solution is also known. Let us define it formally.

SATISFIABLE MAX 3SAT
Input: A formula F with variables $x_1, \ldots, x_n$ and with clauses $C_1, \ldots, C_m$, where F is satisfiable.
Solution: A truth assignment for the variables.
Value: The number of clauses satisfied.

SATISFIABLE MAX 3SAT$(4, \bar{4})$ is the restriction of SATISFIABLE MAX 3SAT in which each variable and its negation appear at most four times in F.

Let us finally state the variant of LONGEST CYCLE for which we will be able to give a ptas.

SECOND HAMILTONIAN CYCLE
Input: A cubic Hamiltonian graph G and a Hamiltonian cycle C.
Solution: A cycle different from C.
Value: The length of the cycle.

3 CH LONGEST s-t PATH has no ptas

The basis of our non-approximability result is the following refinement by Arora et al [1] of Cook's theorem on the NP-hardness of 3SAT.

Theorem 2. *Let L be a language in NP. There exists a polynomial time algorithm and a constant $0 < \varepsilon < 1$ such that, given an input x, the algorithm constructs an instance F_x of 3SAT which satisfies the following properties:*
1. If $x \in L$ then F_x is satisfiable.
2. If $x \notin L$ then no assignment satisfies more than fraction $(1-\varepsilon)$ of the clauses.

The standard way for showing that an optimization problem has no ptas is to show the stronger result that it is hard for APX under L-reduction. But we can not proceed here this way since if $NP{\neq}co\text{-}NP$ then this stronger result doesn't hold for problems where the value of an optimum solution is known. This is somewhat analogous to the result of Megiddo and Papadimitriou [9] showing that an FNP-complete function can not be total unless $NP{=}co\text{-}NP$.

Theorem 3. *If $NP{\neq}co\text{-}NP$ then an optimization problem where the value of an optimum solution is known can not be APX-hard under L-reduction.*

Using Theorem 2 we can prove that SATISFIABLE MAX 3SAT has no ptas.

Lemma 4. SATISFIABLE MAX 3SAT *has no ptas, unless $P{=}NP$.*

Using now the L-reduction of [13] from MAX 3SAT to MAX 3SAT$(4,\bar{4})$, and observing that satisfiable instances are mapped into satisfiable instances, we get the following corollary.

Corollary 5. SATISFIABLE MAX 3SAT$(4,\bar{4})$ *has no ptas, unless $P{=}NP$.*

We now prove the main result of this section.

Theorem 6. CH LONGEST s-t PATH *has no ptas, unless $P{=}NP$.*

Proof. We construct an L-reduction from SATISFIABLE MAX 3SAT$(4,\bar{4})$ to CH LONGEST s-t PATH. The outline of our construction follows the polynomial time reduction given by Papadimitriou and Steiglitz [12] from 3SAT to the Hamiltonian cycle problem. In [14] Papadimitriou and Yannakakis gave an L-reduction from MAX 3SAT$(4,\bar{4})$ to the traveling salesman problem with edges of weight one and two by exploiting the strong connection between this later problem and the Hamiltonian cycle problem. Although we will give an L-reduction which is more constraining than a polynomial time reduction, we basically can avoid the complications in the construction of Papadimitriou and Yannakakis. The reason for that is that (here) we are concerned only with satisfiable instances of MAX 3SAT$(4,\bar{4})$. On the other hand, we have additional difficulties since the graph we construct must be cubic and Hamiltonian. In particular, similarly to both [12] and [14] we will use in our construction so-called variable and clause devices. The variable device will be taken from [12] (which is simpler than the one used in [14]), but for the clause device we will use additional features.

A basic ingredient for both is the modification of the ex-or device from [12] which is shown in Fig.1, where only the edges e_1, e_2, e_3, e_4 are joined with the rest of the graph. The only difference with respect to the original ex-or device is

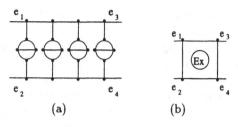

Fig. 1. The Ex-or device and its shorthand representation

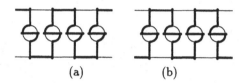

Fig. 2.

that here all vertices have degree three. The ex-or device has the property that any covering path for the device which starts and ends outside it uses either the edge set $\{e_1, e_3\}$, or the edge set $\{e_2, e_4\}$ as connection with the rest of the graph like in Fig. 2(a) and 2(b). Also, it is impossible to have two disjoint paths starting and ending outside the device such that they both contain some vertices of the device and together they cover it. Ex-or devices can be connected in series like in Fig.3(a).

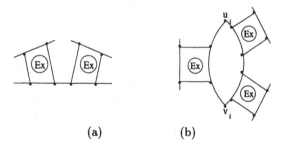

Fig. 3.

Let F be an instance of SATISFIABLE MAX 3SAT$(4, \bar{4})$ with n variables and m clauses. For each variable we will construct a variable device and for each clause a clause device. For $1 \leq i \leq n$, let p_i be the number of positives occurrences of x_i in F and let r_i be the number of its negatives occurrences. For every i, the ith variable device is the following: for two specific vertices u_i and v_i, there are two paths between u_i and v_i. To one of these paths are attached p_i ex-or devices connected like in Fig.3(a), and we say that they are *standing for* x_i. To the other path are attached r_i ex-or devices in series which are standing for $\bar{x}_i$. If $p_i = 0$ or

$r_i = 0$ then the corresponding path consists of just an edge. Figure 3(b) shows the variable device corresponding to a variable with $p_i = 1$ and $r_i = 2$.

The jth clause device corresponding to the clause C_j is shown in Fig.4 where the three ex-or devices stand for the three literals appearing in that clause. If C_j contains the literal x_i then the jth clause device and the ith variable device will share an ex-or device which will stand in the latter for x_i. If C_j contains $\bar{x}_i$ then the same devices share again an ex-or device now standing for $\bar{x}_i$ in the variable device. The specific property satisfied by the clause devices is stated in the next lemma.

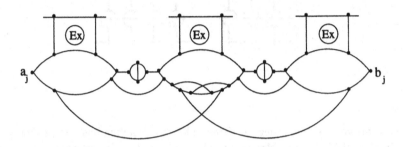

Fig. 4. The clause device C_j

Lemma 7. *For any subset $S \neq \emptyset$ of the three ex-or devices in the jth clause device, there is a path from a_j to b_j which contains exactly those vertices of the clause device which are not in S. On the other hand, there is no path from a_j to b_j which contains all the vertices of the clause device.*

The graph G contains all the variable and clause devices, and two additional vertices s and t. Beside the edges of the devices, there is an edge between s and u_1, between v_i and u_{i+1} for $1 \leq i \leq n-1$, between v_n and a_1, between b_j and a_{j+1} for $1 \leq j \leq m-1$, between b_m and t, and finally between s and t. If there is a satisfying assignment A for F then the path which picks up in each variable device the ex-or devices standing for the literal satisfied by A, and which crosses the clause devices according to Lemma 7 is Hamiltonian. G is also cubic except for vertices s and t which have degree two. We show now that the reduction is indeed an L-reduction.

Let N be the number of vertices in G, then the size of the longest s-t path is $N-1$. The number of clauses m in F is also the value of an optimum assignment for an instance of SATISFIABLE MAX 3SAT$(4,\bar{4})$. Clearly $m = \Theta(n)$ since every literal appears only a constant number of times in the formula. Since the variable and the clause devices have a constant number of vertices, $N = \Theta(m+n) = \Theta(n)$, which shows that the first condition of the L-reduction is satisfied.

For the second condition let us consider an arbitrary s-t path P in G. We will call all the vertices not in this path *missing*.

We construct now from P a partial assignment A_P for the formula F which will give a value to all variables whose corresponding variable device is correctly traversed by P for x_i or $\bar{x}_i$. We say that P *correctly traverses* the ith variable device *for x_i* if it covers all the ex-or devices standing for x_i, these ex-or devices are entered from the variable device, and none of the ex-or devices standing for $\bar{x}_i$ is entered from the variable device. In that case A_P assigns the value **true** for x_i. The definition for correctly traversing the ith variable device for $\bar{x}_i$ is analogous, in which case A_P assigns the value **false** for x_i.

Lemma 8. *If the path P has k missing vertices then the partial assignment A_P satisfies at least $m - 8k$ clauses.*

Proof. Let us suppose that a clause C_j is unsatisfied by A_P. Then either its three literals are made false by A_P or at least one of its literals didn't receive a truth value. In the former case, by the definition of A_P, the variable device of each literal was correctly traversed for the negation of that literal. Therefore the only vertices where P can enter and leave the jth clause device are a_j and b_j, and there must be a missing vertex in that device by Lemma 7. In the latter case there must be a missing vertex in the variable device corresponding to the variable without truth value. Since every variable and its negation appear together at most 8 times in F, the statement follows. □

To finish the proof of Theorem 6 we now show that the second condition in the definition of an L-reduction is also satisfied. Since F is satisfiable, its optimum is m, and since G has a Hamiltonian cycle, its optimum is $N - 1$. Let us given an s-t path P of length $N - 1 - \ell$. Then there are ℓ missing vertices in the graph. Let A be an assignment which extends A_P. By Lemma 8 A satisfies at least $m - 8\ell$ clauses of F. Therefore the second condition is satisfied with $\beta = 8$. □

4 CH s-t Longest Path is not in *APX*

Given an instance $G = (V, E)$ of CH Longest s-t Path with distinguished vertices s and t, we now define the *vertex square* graph G^2 of G which will be an instance of the same problem. The basic idea is to replace in G every vertex v by a copy G_v of G and by a *connector* device C_v. The copy of the connector device for v is shown in Fig.5. This device will connect G_v with the rest of G^2 through the vertices a_v, b_v, c_v which we call *exterior* vertices. The important property of the connector device is stated in the following lemma.

Lemma 9. *For every set $\{x, y\} \subseteq \{a_v, b_v, c_v\}$ there exist two paths P_x starting from x and P_y starting from y such that they are disjoint, together they contain all the vertices of the device, and the other two endpoints of the paths are s_v and t_v in some order.*

G^2 will contain a copy G_v of G and a copy C_v of the connector device for every vertex v except s and t. It will also have two distinguished vertices S and T.

For every v, we identify the distinguished vertices of G_v with the vertices s_v and t_v of C_v, and we delete the edge $\{s_v, t_v\}$. We denote the resulting graph by H_v, and call it the *component* corresponding to v. The components are connected by the following so called *exterior* edges. For every edge $\{v, w\} \in E$, we put an edge between an exterior vertex of C_v and an exterior vertex of C_w. Let s' (respectively t') be the neighbor of s (t) in G different from t (s). We add an edge between S and an exterior vertex of $C_{s'}$ and an edge between T and an exterior vertex of $C_{t'}$. Finally we add an edge between S and T.

Since there is a Hamiltonian s-t path in G, Lemma 9 implies that there is a Hamiltonian S-T path in G^2.

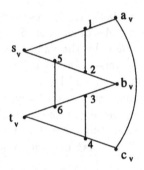

Fig. 5. The connector device C_v

Lemma 10. *Any S-T path of length L in G^2 can be transformed in polynomial time into an s-t path in G of length $\sqrt{L} - 10$.*

This self-improving scheme with Theorem 6 gives using standard arguments

Theorem 11. CH LONGEST s-t PATH *is not constant approximable, unless* $P = NP$.

Our main negative results follow immediately from Lemma 1 and Theorem 11.

Theorem 12. CH LONGEST PATH *and* CH LONGEST CYCLE *are not in APX, unless* $P = NP$.

We can show a stronger non-approximability result under a stronger hypothesis.

Theorem 13. *For any $\varepsilon > 0$, CH LONGEST PATH and CH LONGEST CYCLE are not $2^{O(\log^{1-\varepsilon} n)}$-approximable, unless $NP \subseteq DTIME(2^{O(\log^{1/\varepsilon} n)})$.*

5   SECOND HAMILTONIAN CYCLE has a ptas

In this section we prove that SECOND HAMILTONIAN CYCLE has a ptas in cubic Hamiltonian graphs, which is to our best knowledge the first-ever approximation scheme for a problem in the complexity class TFNP. Actually we are going to prove this result in a much stronger form.

Theorem 14. *Let $G = (V, E)$ be a cubic graph of order n with Hamiltonian cycle $C = v_1 v_2 \cdots v_n$. There is an algorithm that finds a cycle $C' \neq C$ of length at least $n - 4\sqrt{n}$ in $O(n^{3/2} \log n)$ steps.*

We will need the following terminology and notation.
Definitions. We assume throughout that the vertices $v_1, v_2, \ldots, v_n$ follow each other in this order along the given Hamiltonian cycle C of G. The *length* of a chord $e = v_i v_j \in E(G) \setminus E(C)$ $(i < j)$ is defined as $||e|| := \min\{j - i, n + i - j\}$. We denote by P_e the shorter subpath of C with endpoints v_i and v_j if $||e|| < n/2$, and set $P_e := v_i v_{i+1} \cdots v_j$ if $||e|| = n/2$. Two chords e, e' are said to be
–*crossing* if $P_e \cap P_{e'} \neq \emptyset$, $P_e \not\subseteq P_{e'}$, and $P_{e'} \not\subseteq P_e$;
–*incomparable* if $P_e \cap P_{e'} = \emptyset$;
–*parallel* if they do not cross, i.e., either they are incomparable, or $P_e \subset P_{e'}$, or $P_{e'} \subset P_e$.

If $P_e \subset P_{e'}$, we also say that e is *smaller* than e'. The chord e is *minimal* if there is no chord smaller than e.

Proof. Let $k := \lfloor \sqrt{n} \rfloor + 1$. First, we check in $n/2$ steps whether C has a chord of length at most k. If such a chord e exists, then $(E(C) \cup \{e\}) \setminus E(P_e)$ is a cycle of required length. Suppose that all chords are longer than k. We now consider k consecutive chords, say the ones starting from $v_1, \ldots, v_k$. Denoting by z_i the other endpoint of the chord e_i incident to v_i, we can find two subscripts i_1, i_2 such that z_{i_1} and z_{i_2} are at distance less than $(n - k)/(k - 1) < k$ apart on the path $P' := v_{k+1} v_{k+2} \ldots v_n$. Note that the order of the k vertices z_i on P' can be determined in at most $O(k \log k) = O(n^{1/2} \log n)$ steps by any standard sorting algorithm, and then the closest pair can be selected in k steps. If e_{i_1} and e_{i_2} are crossing chords, and say $i_1 < i_2$, then $v_{i_2} v_{i_2+1} \cdots z_{i_1-1} z_{i_1} v_{i_1} v_{i_1-1} \cdots z_{i_2+1} z_{i_2}$ is a cycle of length at least $n - 2k + 2 > n - 2\sqrt{n}$.

Otherwise, if e_{i_1} and e_{i_2} are parallel, we keep them as a starting configuration.

To simplify notation, denote $e_0 := e_{i_1}$, $e'_0 := e_{i_2}$, and assume that $e_0 = v_a v_b$, $e'_0 = v_{a'} v_{b'}$. It may be the case that e_0 and e'_0 are incomparable (i.e., neither of them is smaller than the other), but we may assume without loss of generality (by renumbering the vertices if necessary) that $P_{e_0} = v_a v_{a+1} \cdots v_{b-1} v_b$ and that $P_{e'_0} \not\subseteq P_{e_0}$. We then consider the next k chords $e'_1, \ldots, e'_k$, starting from the vertices $v_{a+1}, \ldots, v_{a+k}$, and select from them two chords f_0 and f'_0 the other endpoints of which are at distance less than k apart. If f_0 and f'_0 are crossing, then a cycle of length at least $n - 2\sqrt{n}$ is easily found as above, therefore we may assume that f_0 and f'_0 do not cross.

If both f_0 and f'_0 are smaller than e_0, and say f_0 is smaller than f'_0, then we rename $e_0 := f_0$, $e'_0 := f'_0$, and do the previous step again. Note that this situation cannot occur more than $O(n)$ times.

Suppose next that f_0 or f'_0 crosses e_0 but it does not cross e'_0. In this situation again, e_0 and the crossing chord create a cycle of length at least $n - 2\sqrt{n}$.

Similarly, if f_0 is smaller than e_0 but f_0' crosses both e_0 and e_0', then f_0' with any one of e_0, e_0' is a suitable choice to construct a cycle of required length.

Finally, suppose that f_0 and f_0' are parallel and they cross both e_0 and e_0'. Remove the two pairs of short arcs (of lengths $< k$) joining the parallel chords (i.e., remove the subpaths of C that join P_{e_0} with $P_{e_0'}$ and also those between P_{f_0} and $P_{f_0'}$) to create four paths of total length at least $n - 4k$. We then obtain a cycle longer than $n - 4\sqrt{n}$ by adjoining the four edges e_0, e_0', f_0, f_0'. □

Remark. By very similar techniques, we can show that if $P \neq NP$ then the traveling salesman problem with weights one and two, restricted to instances where the graph formed by the edges of weight one is cubic and Hamiltonian, has no ptas. On the other hand, when a Hamiltonian cycle is given in the input, the problem has a ptas.

References

1. S. Arora, C. Lund, R. Motwani, M. Sudan, M. Szegedy, *Proof verification and hardness of approximation problems*, Proc. of 33rd FOCS, pages 14-23, 1992.
2. N. Alon, R. Yuster, U. Zwick, *Color-coding: a new method for finding simple paths, cycles and other small subgraphs within large graphs*, Proc. of 26th STOC, pages 326-335, 1994.
3. R. P. Dilworth, *A decomposition theorem for partially ordered sets*, Ann. Math. (2) 51 (1950), 161–166.
4. M. Fürer, B. Raghavachari, *Approximating the Minimum-Degree Steiner Tree to within One of Optimal*, Journal of Algorithms, 17, pages 409-423, 1994.
5. G. Galbiati, A. Morzenti, F. Maffioli, *On the Approximability of some Maximum Spanning Tree Problems*, to appear in Theoretical Computer Science.
6. M. R. Garey, D. S. Johnson, R. E. Tarjan, *The planar Hamiltonian circuit problem is NP-complete*, SIAM J. Comput. 5:4, pages 704-714, 1976.
7. R. M. Karp, *Reducibility among combinatorial problems*. In R. E. Miller and J. W. Thatcher editors, Complexity of Computer Computations, pages 85-103, 1972.
8. D. Karger, R. Motwani, G. Ramkumar, *On Approximating the Longest Path in a Graph*, Proc. of 3rd Workshop on Algorithms and Data Structures, LNCS 709, pages 421-432, 1993.
9. N. Megiddo, C. Papadimitriou, *On total functions, existence theorems and computational complexity*, Theoretical Computer Science, 81, pages 317-324, 1991.
10. B. Monien, *How to find long paths efficiently*, Annals of Discrete Mathematics, 25, pages 239-254, 1984.
11. C. Papadimitriou, *On the Complexity of the Parity Argument and Other Inefficient Proofs of Existence*, Journal of Computer and System Science 48, pages 498-532, 1994.
12. C. Papadimitriou, K. Steiglitz, *Combinatorial Optimization: Algorithms and Complexity*, Prentice-Hall, Englewood Cliffs, NJ, 1982.
13. C. Papadimitriou, M. Yannakakis, *Optimization, Approximation and Complexity Classes*, Journal of Computer and System Science 43, pages 425-440, 1991.
14. C. Papadimitriou, M. Yannakakis, *The traveling salesman problem with distances one and two*, Mathematics of Operations Research, vol. 18, No. 1, pages 1-11, 1993.
15. A. Thomason, *Hamilton cycles and uniquely edge colourable graphs*, Ann. Discrete Math. 3, pages 259-268, 1978.
16. M. Yannakakis, personal communication, 1996.

The Mutual Exclusion Scheduling Problem for Permutation and Comparability Graphs*

Klaus Jansen[1]

Max-Planck Institute for Computer Science, Im Stadtwald, 66 123 Saarbrücken,
Germany
jansen@mpi-sb.mpg.de

Abstract. In this paper, we consider the mutual exclusion scheduling problem for comparability graphs. Given an undirected graph G and a fixed constant m, the problem is to find a minimum coloring of G such that each color is used at most m times. The complexity of this problem for comparability graphs was mentioned as an open problem by Möhring (1985) and for permutation graphs (a subclass of comparability graphs) as an open problem by Lonc (1991). We prove that this problem is already NP-complete for permutation graphs and for each fixed constant $m \geq 6$.

1 Introduction

The following problem arises in scheduling theory: there are n jobs that must be completed on m processors in minimum time t. A processor can execute only one job at a time, and each job requires one time unit for completion. The scheduling is complicated by additional resources (e.g. I-O devices, communication links). A job can only be scheduled onto a processor in a given time unit after it has an exclusive lock on all required resources. Problems of this form have been studied in the Operations Research literature [6, 17]. Another problem arises in load balancing the parallel solution of partial differential equations (pde's) by domain decomposition [5, 2]. The domain for the pde's is decomposed into regions where each region corresponds to a subcomputation. The subcomputations are scheduled on m processors so that subcomputations corresponding to regions that touch at even one point are not performed simultaneously. Other applications are in constructing school course time tables [16] and scheduling in communication systems [14].

1.1 Mutual Exclusion Scheduling

These scheduling problems can be solved by creating an undirected graph $G = (V, E)$ with a vertex for each of the n jobs, and an edge between every pair of conflicting jobs. In each time step, we can execute any subset $U \subset V$ of jobs for

---

* This work was done while the author was associated with the University Trier and supported in part by DIMACS and by EU ESPRIT LTR Project No. 20244 (ALCOM-IT).

which $|U| \leq m$ and U is an independent set in G. A minimum length schedule corresponds to a partition of V into a minimum number t of such independent sets. Baker and Coffman called this graph theoretical problem *Mutual Exclusion Scheduling* (short: MES). Bodlaender and Jansen [7] studied the decision problem of a complementary scheduling problem. Their initial interest was in *compatibility scheduling* (short: CS) which has the same instance and makespan objective function as MES but has a different meaning on the adjacency in G: if two tasks are adjacent in G then they can not be executed on the same processor. Therefore, in MES an independent set is processed in a time unit, whereas in CS an independent set is executed on one processor.

Lonc [18] showed that MES for split graphs can be solved in polynomial time. He proved that MES is NP-complete for complements of comparability graphs and fixed $m \geq 3$ and that MES is polynomial solvable for complements of interval graphs and every m and for cographs and fixed m. However, Bodlaender and Jansen [7] showed that MES is NP-complete when G is restricted to cographs, bipartite or interval graphs. They also proved the following result: if either t or m is a fixed constant, then MES is in P for cographs; if t is a fixed constant, then the problem is in P for interval graphs, and if m is a constant, then it is in P for bipartite graphs. MES remains NP-complete for bipartite graphs and any fixed $t \geq 3$, and for interval graphs and any fixed $m \geq 4$. Independently, Hansen et al. [13] proved that MES restricted to biparite graphs and fixed m is solvable in polynomial time.

Corneil [10] reports that Kirkpatrick has shown the NP-completeness of MES restricted to chordal graphs and fixed $m \geq 3$. For $m = 2$, MES is equivalent to the maximum matching problem in the complement graph and, therefore, in P. On the other hand, for $m = 3$ the complexity of MES is the same as that of the NP-complete problem partition into triangles in the complement graph. Moreover, Baker and Coffman [2] proved that MES is in P for forests and general $t, m \in \mathbb{N}$. A linear time algorithm was proposed in [15] for MES restricted to graphs with constant treewidth and fixed m. Furthermore, MES is NP-complete for complements of line graphs and fixed $m \geq 3$ [9] and polynomial for line graphs and every fixed m [1].

In this paper, we prove the following main result.

Theorem 1. *For each fixed constant $m \geq 6$, the problem MUTUAL EXCLUSION SCHEDULING is NP-complete for permutation graphs (and also comparability graphs).*

1.2 *M*-Machine Scheduling

A *partial order* will be denoted by $P = (V, <_P)$ where V is the set of vertices and $<_P$ is the order relation, i.e. an irreflexive and transitive relation whose pairs $(a, b) \in <_P$ are written as $a <_P b$ (for $a, b \in V$). If the relation is clear, we write $<$ instead of $<_P$. Two elements $a, b \in V$ are *comparable* in P if $a < b$ or $b < a$. A set of pairwise comparable elements is called a *chain*, and a set of pairwise incomparable elements is called an *antichain*. An element a is *minimal* in P, if it has no predecessor. With each partial order $P = (V, <)$, we may associate an undirected graph $G(P)$ as follows. The vertices of $G(P)$ are the elements in V, and two vertices are connected by an edge in $G(P)$ if they are comparable in P. $G(P)$ is called the comparability graph of P. In general, an undirected graph G is called a *comparability graph*, if $G = G(P)$ for some partial order on its vertex set. Algorithmic aspects of comparability graphs are given e.g. in [19].

Let us consider another famous scheduling problem. The *m-machine scheduling* problem with unit times can be modeled by a partition of a partial order $P = (V, <)$ into antichains $A_1, A_2, \ldots, A_t$ such that $|A_i| \leq m$ and

(*) each set A_i consists of some minimal elements of $V \setminus (A_1 \cup \ldots A_{i-1})$ for $i = 1, \ldots, t$.

Inspite of intensive research by several scientists, the complexity status of the problem of finding such a partition with minimum t for each fixed constant $m \geq 3$ still remains unsolved [4]. In fact, this problem is mentioned already in the original list of ten basic open problems in complexity theory [11]. For an overview about different classes of partial orders and complexity results, we refer to [4, 21]. Möhring [20] proposed 1985 a related problem. He asked for the complexity of the problem with condition (*) dropped. This amounts to the mutual exclusion problem restricted to comparability graphs and fixed constant $m \geq 3$. Furthermore, Lonc [18] asked 1991 for the complexity of MES restricted to permutation graphs (a subclass of all comparability graphs) and fixed $m \geq 3$.

In the classical m-machine scheduling problem, the jobs in antichains $A_1, \ldots, A_t$ with $|A_i| \leq m$ corresponding to a partition that satisfies property (*) can be executed on m processors one after another in t time steps. This follows from the fact that each vertex v in an antichain A_i has no predecessor in $V \setminus (A_1 \cup \ldots \cup A_{i-1})$. The makespan of the corresponding schedule is equal to the number of antichains. If we drop condition (*), it is possible that the antichains can not be ordered to form a feasible schedule. Furthermore for each $m \geq 3$, there exists a comparability graph G and a number t such that the answer to the problem MES is *yes* for G and t, but the answer to the m machine scheduling problem is *no* for t and any partial order P (such that G is the comparability graph of P). However, it could be possible to modify the graph in the construction or to extend the ideas in the proof to get a NP-completeness result for the m - machine scheduling problem.

A *linear order* is a partial order without incomparable elements. A linear extension of a partial order $P = (V, <_P)$ is a linear order $L = (V, <_L)$ on the same ground set V that extends P, i.e. $a <_P b$ implies $a <_L b$ for all $a, b \in V$. Linear orders can be written as sequences $L = x_1 \ldots x_n$ defining the order relation $x_1 <_L x_2 <_L \ldots <_L x_n$. The *dimension $dim(P)$* of a partial order P is the smallest number of linear extensions $L_1, \ldots, L_k$ of P, $L_i = (V, <_i)$ whose intersection is P, i.e. $a <_P b$ if and only if $a <_i b$ for $i = 1, \ldots, k$. A partial order is called k - *dimensional* for $k \in \mathbb{N}$, if $dim(P) \leq k$.

Baker et al. [3] have given the following graph theoretic characterization: A partial order P is 2 - dimensional, if and only if the complement graph $G(P)^c$

is also a comparability graph. This implies that a graph G is a comparability graph of a 2 - dimensional partial order if and only if G and its complement graph G^c are comparability graphs. These graphs have been studied also under the name *permutation graphs* [22, 12].

Let $\Pi = (i_1, \ldots, i_n)$ be a permutation of $\{1, \ldots, n\}$. We denote with $\Pi^{-1}(i)$ the position of i in Π. A graph $G = (V, E)$ with $V = \{1, \ldots, n\}$ is a *permutation graph*, if there is a permutation Π such that $\{i, j\} \in E$, if and only if $(i - j) \cdot (\Pi^{-1}(i) - \Pi^{-1}(j)) < 0$. Different techniques for solving algorithmic problems on a permutation graph are given in [8].

1.3 Main Ideas

The main ideas of the NP-completeness proof are the following. First, we choose a restricted SAT problem where each variable occurs three times (once negated and twice unnegated or once unnegated and twice negated) and where each clause contains two or three literals. Let n be the number of variables and r be the number of clauses. We prove that the restricted SAT problem remains NP-complete under a separation condition: for each variable x_i there exists an index $l_i \in \{1 \ldots, r-1\}$ such that x_i appears only negated in the first l_i clauses and unnegated in the other clauses, or vice versa.

Next, we construct a pointset in a rectangle that represents the permutation graph and that simulates the SAT formula. The rectangle can be divided into n horizontal and r vertical stripes such that each horizontal stripe corresponds to a variable and that each vertical stripe corresponds to a clause. If x_i or $\bar{x}_i$ occurs in a clause c_j then we place a point in the square corresponding to the i.th horizontal and j.th vertical stripe. To simulate the variable setting, we differ between two parallel lines in each horizontal stripe and place the points on the upper or lower line in dependence whether the variable is negated or unnegated.

Finally, we place some points (in two chains) around the rectangle such that each independent set in an optimum solution of the restricted coloring problem for the permutation graph covers only points in a horizontal or vertical stripe. Our proof was inspired by the NP-completeness proof of K. Wagner [23].

2 A Restricted Satisfiability Problem

To prove the main theorem, we consider the following restricted satisfiability problem:

Restricted SAT

Given: A set of unnegated variables $X = \{x_1, \ldots, x_n\}$ and negated variables $\bar{X} = \{\bar{x}_1 \ldots, \bar{x}_n\}$, a collection of clauses $c_1, \ldots, c_r$ over $X \cup \bar{X}$ (subsets of $X \cup \bar{X}$) such that

   (i) each clause c_i contains two or three literals $y \in X \cup \bar{X}$,
   (ii) each variable x_j appears either twice unnegated and once negated or twice negated and once unnegated in the clauses,

(iii) no clause c_i contains a pair x_j, $\bar{x}_j$.

Question: Does there exist a truth mapping for the variables such that each clause is satisfied?

The *literals* of a clause c_i are denoted by $y_{i,1}$ and $y_{i,2}$ (and $y_{i,3}$ if we have three literals).

Lemma 2. *The restricted SAT problem is NP-complete.*

Proof. By a reduction from the NP-complete SAT problem where each clause contains exactly three literals [11]. We may assume that each variable appears not only negated or unnegated; otherwise we replace x_j (or $\bar{x}_j$) by the truth values *TRUE* and get a smaller instance. If a clause contains only one literal $y \in \{x_j, \bar{x}_j\}$ (after this reduction), we replace the corresponding variable x_j in all clauses by *TRUE* if $y = x_j$ or by *FALSE* if $y = \bar{x}_j$. If a clause contains a pair x_j and $\bar{x}_j$, we can remove the clause. Then, we have to consider the following remaining cases:

Case 1: A variable x_j appears once unnegated and once negated. In this case, we choose a new variable a_1, insert the clauses $(x_j \vee \bar{a}_1), (\bar{x}_j \vee a_1)$ and replace the second occurrence of x_j in the old clauses by a_1.

Case 2: A variable x_j appears $k \geq 4$ times in the clauses. In this case, we choose new variables $a_1, a_2, \ldots, a_{k-1}$, insert the clauses $(x_j \vee \bar{a}_1), (a_1 \vee \bar{a}_2) \ldots,$ $(a_{k-1} \vee \bar{x}_j)$ and replace the h.th occurrence of x_j in the old clauses by a_{h-1} for $1 < h \leq k$.

In both cases, we obtain that the truth value x_j is equal to $a_1 = \ldots = a_{k-1}$ and that each variable appears now in three different clauses. □

For our reduction, we need a further property for the restricted SAT problem. Let us assume that the clauses are numbered by $1, \ldots, r$ and that the number of the j.th occurrence of variable x_i in the clauses is denoted by $i[j]$. Since each clause contains a variable x_i at most once, we have $i[1] < i[2] < i[3]$. We say that an instance I has the **separation** property if for each variable x_i one of the following cases is true (observe that each variable appears either twice negated and once unnegated or twice unnegated and once negated):

(1) x_i appears negated in clauses $i[1]$ and $i[2]$ and unnegated in $i[3]$,
(2) x_i appears unnegated in clauses $i[1]$ and $i[2]$ and negated in $i[3]$,
(3) x_i appears negated in clauses $i[2]$ and $i[3]$ and unnegated in $i[1]$,
(4) x_i appears unnegated in clauses $i[2]$ and $i[3]$ and negated in $i[1]$.

Lemma 3. *The restricted SAT problem with separation property is also NP-complete (even if there are only case (3) or (4) variables).*

Proof. By a reduction from the restricted SAT problem. For each variable x_i consider the following two possibilities:

Case 1: Variable x_i appears twice unnegated and once negated in the clauses. In this case, we choose two new variables a, b and insert the new clauses $(x_i \vee \bar{a})$,

$(a \vee \bar{b})$, $(b \vee \bar{x}_i)$ at the beginning of the sequence of clauses. Furthermore, we replace the first unnegated occurrence of x_i by a and the second unnegated occurrence by b in the old clauses. Then, we observe that the variables x_i, a and b now have the desired property.

Case 2: Variable x_i appears twice negated and once unnegated in the clauses. We insert the same clauses as in case 1, but we change the order as follows: $(\bar{x}_i \vee b)$, $(\bar{b} \vee a)$, $\bar{a} \vee x_i)$. Next, we replace the first negated occurrence of x_i by $\bar{a}$ and the second negated occurrence by $\bar{b}$. Then, we get the desired property for the variables x_i, a and b. $\qquad\square$

3 The Proof of the Main Theorem

The permutation graph in our proof is represented by a *set of points* $\{p_j = (j, \Pi^{-1}(j)) | 1 \leq j \leq n\}$ in the plane $\mathbb{N}^2$. We notice that an independent set U in G corresponds to a point set in increasing order and that a clique C corresponds to a point set in decreasing order. In the following, we give a proof of the main theorem with fixed constant $m = 6$. The proof can be simply modified such that the result is true for each fixed constant $m \geq 6$.

Proof of Main Theorem: The theorem is proved by a reduction from the restricted SAT problem with separation property (and with only case (3) and (4) variables). Let I_1 be an instance of such problem containing a collection of r clauses $c_1, \ldots, c_r$ over a set of unnegated and negated variables $\{x_1, \bar{x}_1, \ldots, x_n, \bar{x}_n\}$. Let r_2 (and r_3) be the number of 2-clauses (3-clauses) in I_1. The *literals* of clause c_i are denoted by $y_{i,1}$ and $y_{i,2}$ (and $y_{i,3}$ if we have three literals). We denote with $\ell(i)$ the number of literals in clause c_i.

An instance I_2 of the mutual exclusion scheduling problem with bound 6 for each independent set is constructed as follows. The constructed point set for the restricted SAT instance $(x_1 \vee \bar{x}_2 \vee x_i) \wedge (\bar{x}_1 \vee x_2 \vee x_n) \wedge \ldots \wedge (\bar{x}_1 \vee x_2 \vee \bar{x}_n)$ is given in Figure 1. In this example, x_1 and x_n are case (3) variables and x_2 is a case (4) variable. The permutation graph G is given by the corresponding point set

$$P = \bigcup_{i=1}^{n} [A_i \cup B_i] \cup \bigcup_{i=1}^{r} [E_i \cup D_i \cup \{a_{i,1}, a_{i,2}\}] \cup$$
$$\bigcup_{1 \leq i \leq r, \ell(i)=3} [E_i' \cup D_i' \cup \{a_{i,3}\}]$$

where

- A_i consists of one point and B_i consists of three points on the increasing line as marked in Figure 1.
- E_i (and E_i' if clause c_i contains three literals) consists of four points on the increasing line and D_i (and D_i' if c_i contains three literals) consists of one point as marked in Figure 1.
- a_{ik} is a point situated in the crossing of the decreasing line h_i and the increasing line marked by f_ℓ or g_ℓ for $y_{i,k} \in \{x_\ell, \bar{x}_\ell\}$.

We choose the line f_ℓ or g_ℓ in dependence to the cases (3) and (4) for the variable x_ℓ (as constructed in the proof of Lemma 3). We place the points as

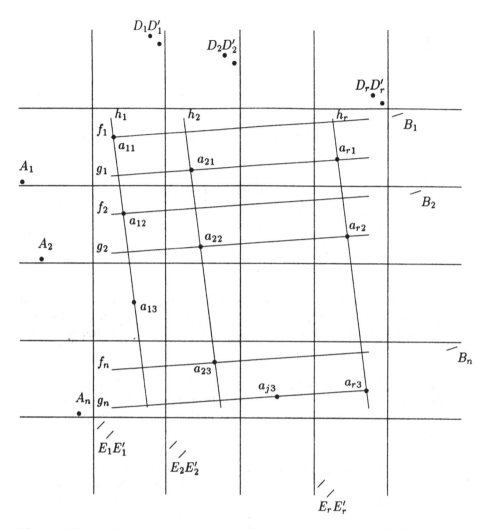

Fig. 1. The point set P constructed for a restricted SAT instance $(x_1 \vee \bar{x}_2 \vee x_i) \wedge (\bar{x}_1 \vee x_2 \vee x_n) \wedge \ldots \wedge (\bar{x}_1 \vee x_2 \vee \bar{x}_n)$.

given in Figure 2. If the first occurrence of variable x_ℓ is unnegated, we have case (3) (otherwise we have case (4)). In case (3), the unnegated literal is placed on line f_ℓ and the negated literals on line g_ℓ. This means for $y_{i,1} = x_\ell$ that $a_{i,1}$ lies in the crossing of the lines h_i and f_ℓ. In case (4), we place the negated literal on line f_ℓ and the unnegated literals on line g_ℓ.

The maximum clique size $\omega(G)$ of the constructed permutation graph is equal to

$$t = n + r_2 + 2 \cdot r_3.$$

One maximum clique is given e.g. by

$$K = A_1 \cup \ldots \cup A_n \cup \{z_i | 1 \leq i \leq r\} \cup \{z_i' | E_i' \neq \emptyset\}$$

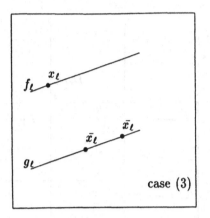 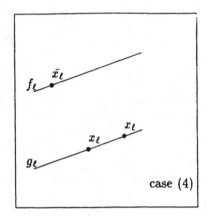

Fig. 2. The way to place the points on the lines f_ℓ or g_ℓ

with arbitrarily choosen $z_i \in E_i$ and $z'_i \in E'_i$ (if c_i is a 3-clause). As illustrated in Figure 1, the point set $\{a_{i,k} | 1 \leq i \leq r, 1 \leq k \leq \ell(i)\}$ can be divided into n horizontal (one stripe for each variable) and r vertical stripes (one stripe for each clause).

We prove the following **claim:** I_1 is a yes - instance of the restricted SAT problem with separation property if and only if there is a coloring of the permutation graph in instance I_2 with at most t colors where each color is used at most 6 times.

Suppose that α is a truth assignment for the variables which makes the clauses $c_1, \ldots, c_r$ true. The first n colors cover the points of $A_j \cup B_j$ and all points (at most two) on f_j if [$\alpha(x_j) = FALSE$ and x_j is a case (4) variable] or [$\alpha(x_j) = TRUE$ and x_j is a case (3) variable] or all points on g_j if [$\alpha(x_j) = FALSE$ and x_j is a case (3) variable] or [$\alpha(x_j) = TRUE$ and x_j is a case (4) variable]. Then, at least one of the points $a_{i,1}, \ldots, a_{i,\ell(i)}$ is colored with one of these n colors for each clause c_i. The remaining $t - n$ colors are used to color the points of D_i, D'_i, E_i, E'_i and the remaining uncolored points in the vertical stripes. We use two colors for a 3-clause and one color for a 2-clause. This gives us a t-coloring of G such that each color is used at most 6-times.

Conversely, suppose that we have a t-**coloring** f of the permutation graph G where each color is used at most 6 times. First, we observe that

$$K = A_1 \cup \ldots \cup A_n \cup \{z_i | 1 \leq i \leq r\} \cup \{z'_i | E'_i \neq \emptyset\}$$

with arbitrarily choosen $z_i \in E_i$ and $z'_i \in E'_i$ and that

$$\bar{K} = D_1 \cup \ldots \cup D_r \cup \{D'_i | 1 \leq i \leq r, \ell(i) = 3\} \cup \{b_1, \ldots b_n\}$$

with arbitrarily choosen $b_i \in B_i$ are cliques in G of size t. This implies that the vertices in the cliques K (and $\bar{K}$) must be colored differently and that points in every chain E_i (and E'_i, B_j) must be colored with the same color. This means e.g. that the number of colors $|\{f(z)|z \in E_i\}|$ is equal 1. Since the cliques K and $\bar{K}$ are disjoint, there must be an $1-1$ assignment between the vertices in K and $\bar{K}$ to obtain a t- coloring.

Feasible Assignments: Since each color can be used at most 6 times, it is not possible to take the same color for a E_i chain (or E'_i chain) and a B_j chain; otherwise we have 7 vertices colored with the same color. This implies that only the following assignments (or colorings) are possible:

(1) the color set $\{f(A_j)|1 \leq j \leq n\}$ is equal to $\{f(b_j)|1 \leq j \leq n, b_j \in B_j\}$,
(2) the color set $\{f(z_i), f(z'_i)|z_i \in E_i, z'_i \in E'_i, 1 \leq i \leq r\}$ is equal to the color set $\{f(D_i), f(D'_i)|1 \leq i \leq r\}$.

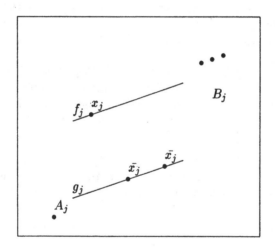

Fig. 3. A horizontal stripe with points A_j and B_j

W.l.o.g. we may assume that $f(A_j) = j$ for $1 \leq j \leq n$. In the next step, we can prove that

(3) $f(A_j) = f(b_j) = j$ with $b_j \in B_j$, $1 \leq j \leq n$,
(4) $f(z_i) = f(D_i)$ with $z_i \in E_i$ and $f(z'_i) = f(D'_i)$ with $z'_i \in E'_i$ for a 3-clause c_i.

For example, consider the points in the sets A_j, B_j. We know already that the color sets

$$\{f(A_j)|1 \leq j \leq n\} = \{f(b_j)|1 \leq j \leq n, b_j \in B_j\}$$
$$= \{1, \ldots, n\}.$$

Since A_1 is in conflict to all points in the chain $B_2, \ldots, B_n$, we get $f(A_1) = f(b_1) = 1$ with $b_1 \in B_1$. For the induction step, observe that A_i is in conflict to the points in $B_{i+1}, \ldots, B_n$ and to the points $A_1, \ldots, A_{i-1}$. Therefore, $f(A_i) \notin \{f(b_j) | 1 \leq j \leq n, j \neq i, b_j \in B_j\}$. This implies that $f(A_i) = f(b_i)$ with $b_i \in B_i$ and proves property (3) above. A similar idea can be used to prove property (4). We notice that it is possible that $f(z_i) = f(D_i')$ and $f(z_i') = f(D_i)$. But we can exchange these colors without problems to obtain a coloring such that property (4) is satisfied.

Horizontal and Vertical Stripes: Finally, we look at a horizontal stripe with vertices A_j and $b_j \in B_j$ colored with the same color j.

Using the separation property for the variable x_j we get a stripe as in Figure 3 (here given for a case (3) variable x_j). Since the point A_j and the points in B_j are colored with j, only the point on f_j or the two points on g_j can be colored with color j. Notice that an independent set must form an increasing chain of points in the point model. Let U_i be the set of points colored with color i, for $1 \leq i \leq t$. We may assume that either the point on f_j or both points on g_j are colored with color j.

We define a truth assignment α for the variables $x_1, \ldots, x_n$ as follows:

$$\alpha(x_j) = \begin{cases} true & \text{if } U_j \text{ contains both points on line } g_j \\ & \text{and if } x_j \text{ is a case (4) variable} \\ true & \text{if } U_j \text{ contains a point on line } f_j \\ & \text{and if } x_j \text{ is a case (3) variable} \\ false & \text{otherwise.} \end{cases}$$

Since $f(z_i) = f(D_i)$ with $z_i \in E_i$ and $f(z_i') = f(D_i')$ with $z_i' \in E_i'$ and since the points $a_{i,k}$ lie on a decreasing line h_i, each independent set U_j with $n + 1 \leq j \leq t$ contains at most one point in a vertical stripe. Therefore, at least one point in each vertical stripe must be covered by the sets $U_1, \ldots, U_n$. This implies that the mapping α is a truth assignment that satisfies all clauses in I_1. □

4 Conclusion

In this paper, we have proved that mutual exclusion scheduling (MES) restricted to permutation (and comparability graphs) is NP-complete for each fixed constant $m \geq 6$. Finding the complexity of MES for smaller constants $m = 3, 4, 5$ could be a step to the solution of the famous m-machine scheduling problem with unit-times.

References

1. N. Alon: A note on the decomposition of graphs into isomorphic matchings, *Acta Mathematica Hungarica* 42 (1983), 221-223.

2. B.S. Baker and E.G. Coffman: Mutual exclusion scheduling, *Theoretical Computer Science* 162 (1996), 225-243.

3. K.A. Baker, P.C. Fishburn and F.S. Roberts: Partial orders of dimension 2, *Networks* 2 (1971), 11-28.

4. M. Bartusch, R.H. Möhring and F.J. Radermacher: M-machine unit time scheduling: A report on ongoing research, in: Optimization, Parallel Processing and Applications (ed. A. Kurzhanski), LNEMS 304, 1988, 165-213.

5. P. Bjorstad, W.M. Coughran and E. Grosse: Parallel domain decomposition applied to coupled transport equations, in: Domain Decomposition Methods in Scientific and Engineering Computing (eds. D.E. Keys, J. Xu), AMS, Providence, 1995, 369-380.

6. J. Blazewicz, K.H. Ecker, E. Pesch, G. Schmidt and J. Weglarz: Scheduling computer and manufacturing processes, Springer, Heidelberg, 1996.

7. H.L. Bodlaender and K. Jansen: On the complexity of scheduling incompatible jobs with unit-times, *Mathematical Foundations of Computer Science*, MFCS 93, LNCS 711, 291-300.

8. A. Brandstädt: On improved time bounds for permutation graph problems, *Graph Theoretic Concepts in Computer Science*, WG 92, LNCS 657, 1-10.

9. E. Cohen and M. Tarsi: NP-completeness of graph decomposition problem, *Journal of Complexity* 7 (1991), 200-212.

10. D.G. Corneil: The complexity of generalized clique packing, *Discrete Applied Mathematics* 12 (1985), 233-239.

11. M.R. Garey and D.S. Johnson: Computers and Intractability: A Guide to the Theorey of NP-Completeness, Freeman, New York, 1979.

12. P.C. Gilmore and A.J. Hoffman: A characterization of comparability graphs and of interval graphs, *Canadian Journal of Mathematics* 16 (1964), 539-548.

13. P. Hansen, A. Hertz and J. Kuplinsky: Bounded vertex colorings of graphs, *Discrete Mathematics* 111 (1993), 305-312.

14. S. Irani and V. Leung: Scheduling with conflicts, and applications to traffic signal control, *SIAM Symposium on Discrete Algorithms*, SODA 96, 85-94.

15. D. Kaller, A. Gupta and T. Shermer: The χ_t - coloring problem, *Symposium on Theoretical Aspects of Computer Science*, STACS 95, LNCS 900, 409-420.

16. F. Kitagawa and H. Ikeda: An existential problem of a weight - controlled subset and its application to school time table construction, *Discrete Mathematics* 72 (1988), 195-211.

17. V. Lofti and S. Sarin: A graph coloring algorithm for large scale scheduling problems, *Computers and Operations Research* 13 (1986), 27-32.

18. Z. Lonc: On complexity of some chain and antichain partition problem, *Graph Theoretical Concepts in Computer Science*, WG 91, LNCS 570, 97-104.

19. R.H. Möhring: Algorithmic aspects of comparability graphs and interval graphs, in: Graphs and Orders (ed. I. Rival), Reidel Publishing, Dordrecht 1985, 41-101.

20. R.H. Möhring: Problem 9.10, in: Graphs and Orders (ed. I. Rival), Reidel Publishing, Dordrecht 1985, 583.

21. R.H. Möhring: Computationally tractable classes of ordered sets, in: Algorithms and Orders (ed. I. Rival), Kluwer Acad. Publishing, Dordrecht 1989, 105-193.

22. A. Pnueli, A. Lempel and W. Even: Transitive orientation of graphs and identification of permutation graphs, *Canadian Journal of Mathematics* 23 (1971), 160-175.

23. K. Wagner: Monotonic coverings of finite sets, *Journal of Information Processing and Cybernetics - EIK* 20 (1984), 633 – 639.

Massaging a Linear Programming Solution to Give a 2-Approximation for a Generalization of the Vertex Cover Problem

Nader H. Bshouty*   Lynn Burroughs**

Department of Computer Science, The University of Calgary, Calgary, Alberta, Canada T2N 1N4, e-mail: {bshouty,lynnb}@cpsc.ucalgary.ca

Abstract. Linear programming relaxations have been used extensively in designing approximation algorithms for optimization problems. For VERTEX COVER, linear programming and a thresholding technique gives a 2-approximate solution, rivaling the best known performance ratio. For a generalization of VERTEX COVER we call vc^t, in which we seek to cover t edges, this technique may not yield a feasible solution at all. We introduce a new method for massaging a linear programming solution to get a good, feasible solution for vc^t. Our technique manipulates the values of the linear programming solution to prepare them for thresholding. We prove that this method achieves a performance ratio of 2 for vc^t with unit weights. A second algorithm extends this result, giving a 2-approximation for vc^t with arbitrary weights. We show that this is tight in the sense that any α-approximation algorithm for vc^t with $\alpha < 2$ implies a breakthrough α-approximation algorithm for VERTEX COVER.

1 Introduction

Linear programming relaxations of integer programs have been used extensively in designing approximation algorithms for optimization problems [5] [11] [3]. A similar idea, relaxing integer programs to semidefinite programs, has recently brought exciting advancements in the approximability of several problems including MAXCUT [2] and k-colourable GRAPH COLOURING [6], although attempts to use semidefinite programming for VERTEX COVER [9] have not provided improvements over existing algorithms.

Given an integer programming description of a problem, we relax some of the constraints to get a linear program which can be solved in polynomial time using the Ellipsoid method [8] or one of many interior point methods, the first of which was given in [7]. The rational solution to the linear program is then interpreted to get an integer approximate solution to the original problem.

VERTEX COVER is one natural application for this tool. The VERTEX COVER problem is, given a graph $G = (V, E)$ and a weight function $\omega : V \to (\mathbb{Z}^+ \cup \infty)$,

---

* This research was supported in part by the NSERC of Canada.

** This research was supported in part by the NSERC of Canada. Part of this work was done while both authors were visiting the Technion, Haifa, Israel.

find a set $C \subseteq V$ with minimum total weight such that all edges in E are 'covered' by having at least one endpoint in C. Numerous approximation algorithms for VERTEX COVER in general graphs exist, including [1]. The best of these achieve a performance ratio of 2. Johan Håstad has proved in [4] that a performance ratio less than $\frac{7}{6}$ is not possible unless P = NP.

Hochbaum [5] gives an approximation algorithm for VERTEX COVER that uses linear programming and a thresholding technique. The linear program is solved to get a value $x_i \in [0, 1]$ for each vertex v_i, and the cover is given by $C = \{v_i \mid x_i \geq \frac{1}{2}\}$. C covers all of the edges and has size at most twice the optimal size. Thus the algorithm has a performance ratio of 2.

For some applications of VERTEX COVER, it may be sufficient to find a set C that covers t edges in E where $t : N \to N$ is a fixed function of $|V|$ or $|E|$. We call C a t-cover of G. This generalization of VERTEX COVER will be called VC^t. A similar variation of VERTEX COVER was considered by E. Petrank [10]. We show that for many functions t, an α-approximation algorithm for VC^t implies an α-approximation algorithm for VERTEX COVER. Therefore it would be difficult to find better than a 2-approximation and highly unlikely that we would find better than a $\frac{7}{6}$-approximation for VC^t in general graphs. One might think that an optimal t-cover is simply a subset of an optimal vertex cover. But this is not true in general. In fact, a feasible t-cover formed from an optimal vertex cover may be arbitrarily larger in size than the optimal t-cover.

Unfortunately Hochbaum's algorithm for VERTEX COVER does not work for VC^t. In fact, the set $\{v_i \mid x_i \geq \frac{1}{2}\}$ may not be feasible for VC^t. We need a different approach. We present a new technique for massaging the LP solution to prepare it for thresholding. This technique guarantees a 2-approximate solution for VC^t for any function t. We present our technique in two algorithms.

Our first algorithm begins by solving the linear program and taking the vertices that exceed a threshold of $\frac{1}{2}$ in the LP solution. We then work with the subgraph induced on the remaining vertices and interpret the remainder of the LP solution as values on the vertices and edges. We massage these values, performing a series of adjustments wherein a fraction of the value of one vertex is shifted to another vertex. While performing these adjustments, we keep the value of the objective function the same, keep the total edge value from decreasing and maintain a certain relationship between the edge and vertex values. Once a vertex assumes a value of $\frac{1}{2}$, we put that vertex in the cover and remove it from the subgraph. When no more adjustments can be made, one final vertex may be placed in the cover. We show that this algorithm runs in polynomial time, covers at least t edges and uses at most $2 + \frac{B-1}{OPT}$ times the optimal number of vertices where B is the largest finite weight in the graph and OPT is the optimal value for the instance. Thus it is a 2-approximation algorithm for VC^t with unit weights. The second algorithm uses the first as a subroutine and improves the bound for VC^t with arbitrary weights to 2 by searching for the largest vertex weight in an optimal cover.

While the use of linear programming relaxations is not new, previous uses of this tool involve direct thresholding [5] or a randomized rounding technique [11]

[3]. We use thresholding, but do not restrict ourselves to using the raw linear program solution. Instead we massage the LP solution to prepare it for the thresholding technique. This allows for a higher threshold and thus a better approximation than is possible from using the raw values.

2 Preliminaries

An instance of vc^t is a graph $G = (V, E)$ and a weight function $\omega : V \to \{1, 2, \ldots, B, \infty\}$ with $|V| = n$. The value of the optimal t-cover for G (i.e., the total weight of the vertices it contains) will be denoted $OPT(G, \omega, t)$ or may be abbreviated to OPT if the instance is clear from the context. For an arbitrary set of vertices C we will use the notation $\omega(C)$ to mean $\sum_{v \in C} \omega(v)$.

The integer program for vc^t is given below. We assign a variable x_i to each vertex v_i and a variable y_{ij} to each edge (v_i, v_j).

$$OPT(G, \omega, t) = \text{Min} \sum_{i=1}^{n} \omega(v_i) x_i$$

$$\text{Subject to} \sum_{(v_i, v_j) \in E} y_{ij} \geq t$$

$$x_i + x_j \geq y_{ij} \quad \forall (v_i, v_j) \in E$$

$$x_i \in \{0, 1\} \quad i = 1, \ldots, n$$

$$x_i = 0 \quad \forall v_i \text{ with } \omega(v_i) = \infty$$

$$y_{ij} \in \{0, 1\} \quad \forall (v_i, v_j) \in E$$

We define the product of zero and ∞ in the objective function to be zero. If the program is infeasible we will define $OPT(G, \omega, t) = \infty$. For feasible programs, an optimal t-cover is given by $C = \{v_i \mid x_i = 1\}$. The edges covered are $\{(v_i, v_j) \in E \mid y_{ij} = 1\}$. Unfortunately there is no known polynomial-time algorithm to solve the integer program for arbitrary t. Such an algorithm would imply that VERTEX COVER can be solved in polynomial time by setting $t = |E|$. We relax the constraints of the integer program to get the linear program below.

$$(P1) \quad LB(G, \omega, t) = \text{Min} \sum_{i=1}^{n} \omega(v_i) x_i$$

$$\text{Subject to} \sum_{(v_i, v_j) \in E} y_{ij} \geq t$$

$$x_i + x_j \geq y_{ij} \quad \forall (v_i, v_j) \in E$$

$$x_i \geq 0 \quad i = 1, \ldots, n$$

$$x_i = 0 \quad \forall v_i \text{ with } \omega(v_i) = \infty$$

$$0 \leq y_{ij} \leq 1 \quad \forall (v_i, v_j) \in E$$

The relaxation can be solved in polynomial time using the Ellipsoid method or an interior point method. Also, since the constraints of the integer program are a subset of the constraints of the linear program, $LB(G, \omega, t) \leq OPT(G, \omega, t)$.

We will assume without loss of generality that for each edge (v_i, v_j), $y_{ij} = \min(x_i + x_j, 1)$. If $y_{ij} < \min(x_i + x_j, 1)$ then we may increase y_{ij} until equality holds. This increase neither violates the constraints of the program, nor does it affect the value of the objective function.

To see why the LP algorithm for VERTEX COVER fails for vc^t, let $t = \frac{2}{3}|E|$ and consider a complete graph on four vertices with unit weights. The optimal solution requires 2 vertices to cover 4 edges. One optimal LP solution is given by $x_1 = x_2 = x_3 = x_4 = \frac{1}{2}$, and $y_{ij} = x_i + x_j = \frac{2}{3}$. Using a threshold of $\frac{1}{2}$ which was sufficient for VERTEX COVER, we form $C = \{v_i \mid x_i \geq \frac{1}{2}\}$. But for this LP solution C is empty and is not a feasible solution to the vc^t instance. With a threshold of $\frac{1}{3}$, we obtain a feasible solution from this LP solution, but we fail with the equally optimal solution given by $x_1 = x_2 = x_3 = \frac{1}{6}$, $x_4 = \frac{5}{6}$ and $y_{ij} = x_i + x_j$. It is not clear that a threshold exists that will always produce a feasible solution. This is the motivation for a multi-phase approach.

3 The First Algorithm

We now present our first algorithm. A discussion follows.

Algorithm ALG1
Input: $G = (V, E)$, $\omega : V \to \{1, 2, \ldots, B, \infty\}$, t
Phase 0:
If $|\{(u, v) \in E \mid \omega(u) \neq \infty \text{ or } \omega(v) \neq \infty\}| < t$ then return V; terminate
Solve the linear program $(P1)$ to get $x_1, \ldots, x_n$ and $\{y_{ij} \mid (v_i, v_j) \in E\}$.
Copy these values to $x_1', \ldots, x_n'$ and $\{y_{ij}' \mid (v_i, v_j) \in E\}$.
Phase 1:
$C_1 \leftarrow \{v_i \mid x_i' \geq \frac{1}{2}\}$
Let (V', E') be the induced subgraph on $V' = V \setminus C_1$
Let $d'(v_i)$ be the degree of vertex v_i in (V', E')
Phase 2:
$C_2 \leftarrow \emptyset$
While $\exists$ distinct v_r and v_g in V' with $x_r' > 0$, $x_g' > 0$ and $\frac{d'(v_r)}{\omega(v_r)} \geq \frac{d'(v_g)}{\omega(v_g)}$ do
$\quad \delta \leftarrow \min\{\frac{1}{2} - x_r', \frac{\omega(v_g)}{\omega(v_r)} x_g'\}$
$\quad x_r' \leftarrow x_r' + \delta$
$\quad x_g' \leftarrow x_g' - \frac{\omega(v_r)}{\omega(v_g)} \delta$
$\quad$ For each edge $(v_\ell, v_k) \in E'$ do $y_{\ell k}' \leftarrow x_\ell' + x_k'$
$\quad$ If $x_r' \geq \frac{1}{2}$ then
$\quad\quad C_2 \leftarrow C_2 \cup \{v_r\}$
$\quad\quad$ Remove v_r and adjacent edges from (V', E')
$\quad\quad$ Update $d'(v_k)$ for all v_k in the new subgraph (V', E')
Phase 3:
If $\exists v_i$ with $x_i' > 0$ and $C_1 \cup C_2$ covers $< t$ edges then $C_3 \leftarrow \{v_i\}$
$\quad$ Else $C_3 \leftarrow \emptyset$
Output: $C_1 \cup C_2 \cup C_3$

ALG1 begins by solving the linear program $(P1)$ to obtain a value x'_i for each vertex v_i and a value y'_{ij} for each edge (v_i, v_j). The next three phases form the sets C_1, C_2, C_3 and together these sets are output as the cover. As we place vertices in these sets, we work with the subgraph induced on the remaining vertices. We use $d'(v_i)$ to denote the degree of v_i in the current subgraph. In phase 1 we form $C_1 = \{v_i \mid x'_i \geq \frac{1}{2}\}$. In phase 2 we work with the subgraph induced on $V \setminus C_1$. At this phase, each vertex has $x'_i < \frac{1}{2}$ and each edge observes $y'_{ij} = x'_i + x'_j$. We perform a series of adjustments to the values x'_ℓ and y'_{ij} as follows. Two distinct vertices v_r and v_g are chosen with $x'_r > 0$ and $x'_g > 0$. We donate a portion of the value of x'_g to x'_r. The values $y'_{\ell k}$ are adjusted to maintain the equality $y'_{\ell k} = x'_\ell + x'_k$ for all edges in the current subgraph. The choice of v_r and v_g is made such that the sum $\sum y'_{\ell k}$ does not decrease as a result of the adjustment. After an adjustment, the value x'_r may be increased to $\frac{1}{2}$. If so, v_r is placed in C_2 and is removed from the subgraph. The reader will note that the adjusted values may no longer form a valid solution to the linear program. It is only necessary that we keep the sums $\sum \omega(v_i)x'_i$ and $\sum y'_{\ell k}$ from decreasing. Phase 2 continues until there are no longer v_r, v_g in the current subgraph with $x'_r > 0$ and $x'_g > 0$. If $C_1 \cup C_2$ covers fewer than t edges, then one vertex is placed in C_3.

3.1 Analysis

We examine each phase separately and compare the size of each C_i with the total vertex value considered in phase i. We will employ the following notation.

x'_i and y'_{ij} will refer to the *final* values in the algorithm.

$$V_1 = C_1 \qquad V_2 = C_2 \qquad V_3 = V - (C_1 \cup C_2)$$

$$W_k = \sum_{v_i \in V_k} w(v_i)x'_i \text{ for } k \in \{1,2,3\}$$

$$E_1 = \{(v_i, v_j) \in E \mid v_i \in V_1 \text{ or } v_j \in V_1\}$$

$$E_2 = \{(v_i, v_j) \in E \mid v_i \in V_2 \text{ or } v_j \in V_2\} \setminus E_1$$

$$E_3 = \{(v_i, v_j) \in E \mid v_i \in V_3 \text{ and } v_j \in V_3\}$$

$$Y_k = \sum_{(v_i, v_j) \in E_k} y'_{ij} \text{ for } k \in \{1,2,3\}.$$

Recall also that

$$LB(G, \omega, t) \leq OPT(G, \omega, t).$$

It should be clear that V_1, V_2 and V_3 are disjoint sets and $V_1 \cup V_2 \cup V_3 = V$. Also, E_1, E_2 and E_3 are disjoint sets with $E_1 \cup E_2 \cup E_3 = E$. We begin with an observation and some lemmas that will be used to establish our first result.

Observation *If no solution covers t edges without using at least one vertex with infinite weight, then $OPT = \infty$ and any infinite solution (such as V) will do. If the finite-weight vertices cover at least t edges then OPT is finite. In this case, for all infinite-weight vertices v_q, the linear program will return $x_q = 0$ and* *nowhere in the algorithm will v_q be included in $C_1 \cup C_2 \cup C_3$. Thus no vertex in $C_1 \cup C_2 \cup C_3$ will have weight larger than B.*

For the remainder of the paper we will assume we are working with instances that have finite optimal solutions.

Lemma 1. *The running time of ALG1 is polynomial in the input size.*

Proof of lemma 1: For each iteration in phase 2, either x'_r becomes $\frac{1}{2}$ and v_r is removed from the graph or x'_g becomes 0 and is not considered again. Thus at most n iterations are performed before phase 2 terminates. The work done in phases 1 through 3 is polynomial and is dominated by the time required to solve the linear program in phase 0. This can be done using an interior point method in time $\mathcal{O}(|V|^{10} + |V|^7 \log B)$. This is the running time of our algorithm.　□

Lemma 2. *With each iteration in phase 2, the total value of the objective function does not change, i.e., $\sum\limits_{i=1}^{n} \omega(v_i)x_i = \sum\limits_{i=1}^{n} \omega(v_i)x'_i = W_1 + W_2 + W_3$.*

Proof of lemma 2: With each iteration, δ is added to x'_r and $\frac{\omega(v_r)}{\omega(v_g)}\delta$ is subtracted from x'_g. The net change to the objective function is

$$\omega(v_r)\delta - \omega(v_g)\left(\frac{\omega(v_r)}{\omega(v_g)}\delta\right) = 0.　□$$

Lemma 3. *For all $(v_i, v_j) \in E$, $y'_{ij} \leq 1$.*

Proof of lemma 3: For edges $(v_i, v_j) \in E_1$, $y'_{ij} = y_{ij}$ and thus $y'_{ij} \leq 1$ by a constraint of the linear program. For edges $(v_\ell, v_k) \in E_2 \cup E_3$, note that at the beginning of phase 2, $y'_{\ell,k} = x'_\ell + x'_k < 1$. This relationship is maintained while (v_ℓ, v_k) remains in the current subgraph. This only changes when and if one endpoint attains a value equal to $\frac{1}{2}$. Then that vertex and edge (v_ℓ, v_k) are removed from the graph and $y'_{\ell,k}$ will never be changed again. Thus the largest value $y'_{\ell,k}$ will attain is strictly less than 1.　□

Lemma 4. $\sum\limits_{(v_i,v_j)\in E} y_{ij} \leq Y_1 + Y_2 + Y_3$.

Proof of lemma 4: We begin with $y'_{ij} = y_{ij}$. There are $d'(v_r)$ edges adjacent to v_r that are increased by δ and $d'(v_g)$ edges adjacent to v_g that are decreased by $\frac{\omega(v_r)}{\omega(v_g)}\delta$. Since $\frac{d'(v_r)}{\omega(v_r)} \geq \frac{d'(v_g)}{\omega(v_g)}$, the net effect to $\sum y'_{ij}$ is

$$d'(v_r)\delta + d'(v_g)\left(-\frac{\omega(v_r)}{\omega(v_g)}\delta\right) \geq d'(v_r)\delta - \frac{d'(v_r)}{\omega(v_r)}\omega(v_r)\delta = 0$$

So the net effect is never a decrease. Thus we always have

$$\sum\limits_{(v_i,v_j)\in E} y_{ij} \leq \sum\limits_{(v_i,v_j)\in E} y'_{ij} = Y_1 + Y_2 + Y_3　□$$

Lemma 5. *If $|C_3| > 0$, then at least Y_3 edges in E_3 are covered by C_3.*

Proof of lemma 5: If $|C_3| > 0$ then $C_3 = \{v_i\}$ for some i and we had $0 < x'_i < \frac{1}{2}$. For each $v_j \in V_3$ with $i \neq j$, we have $x'_j = 0$ as a condition of the termination of phase 2. For edges $(v_k, v_\ell) \in E_3$ not adjacent to v_i, we have $y'_{k\ell} = x'_k + x'_\ell = 0$. For edges $(v_i, v_j) \in E_3$ (there are $d'(v_i)$ of them) we have $y'_{ij} = x'_i < \frac{1}{2}$. Thus we have $Y_3 \leq d'(v_i)/2$. Since C_3 covers $d'(v_i)$ new edges, we have the relationship

$$\text{Edges in } E_3 \text{ covered by } C_3 \geq 2Y_3 \geq Y_3, \qquad \square$$

Lemma 6 establishes the feasibility of solutions produced by the algorithm.

Lemma 6. *Solutions produced by ALG1 cover at least t edges.*

Proof of lemma 6: There are two non-trivial cases. Either $C_3 = \emptyset$ because all remaining v_i have $x'_i = 0$, or $C_3 = \{v_i\}$. For the first case note that all uncovered edges have $y'_{ij} = x'_i + x'_j = 0$. Hence $Y_3 = 0$. Since C_3 is empty, the number of edges in E_3 covered by C_3 is Y_3. For the second case, lemma 5 tells us that the number of edges in E_3 covered by C_3 is at least Y_3. Then for both cases we have

$$\text{Number of edges covered} \geq |E_1| + |E_2| + Y_3$$

Since $y'_{ij} \leq 1$ by lemma 3,

$$\geq \left(\sum_{(v_i, v_j) \in E_1} y'_{ij} \right) + \left(\sum_{(v_i, v_j) \in E_2} y'_{ij} \right) + Y_3$$

$$= Y_1 + Y_2 + Y_3$$

By lemma 4,

$$\geq \sum_{(v_i, v_j) \in E} y_{ij}$$

From a constraint of the linear program,

$$\geq t \qquad \square$$

Lemma 7. *Suppose that $C_3 = \{v_i\}$. Then x'_i has a lower bound of $1/d'(v_i) > 0$.*

Proof of lemma 7: Suppose that $C_3 = \{v_i\}$. This means that $|E_1| + |E_2| < t$. Since t is an integer, we have $|E_1| + |E_2| \leq t - 1$. Then by lemma 3,

$$Y_1 + Y_2 \leq \left(\sum_{(v_i, v_j) \in E_1} 1 \right) + \left(\sum_{(v_i, v_j) \in E_2} 1 \right) = |E_1| + |E_2| \leq t - 1.$$

By lemma 4 and a constraint of the linear program we have $Y_1 + Y_2 + Y_3 \geq t$. This implies that $Y_3 \geq 1$. Now, recall that in V_3, only v_i has $x'_i > 0$. All other vertices $v_j \in V_3$ have $x'_j = 0$. Since $y'_{\ell k} = x'_\ell + x'_k$, we have $y'_{ik} = x'_i$ for all edges adjacent to v_i and $y'_{\ell k} = 0$ for non-adjacent edges. Then $Y_3 = d'(v_i)x'_i \geq 1$. This gives us a lower bound of $x'_i \geq 1/d'(v_i) > 0$. $\qquad \square$

Lemma 8. *The three sets that make up the cover observe the following:*

$$\omega(C_1) \le 2W_1, \qquad \omega(C_2) \le 2W_2, \qquad \omega(C_3) < 2W_3 + B.$$

Proof of lemma 8: The first two inequalities follow from the fact that a vertex v_i is only placed in C_1 or C_2 if $x_i' \ge \frac{1}{2}$. For the last inequality, there are two cases. Either $C_3 = \emptyset$ or $C_3 = \{v_i\}$. In the first case $\omega(C_3) = 0$ and clearly the inequality holds. In the second case we have

$$\omega(C_3) = \omega(v_i) = 2W_3 + \omega(v_i)(1 - 2x_i')$$

By our first observation, $\omega(v_i) \le B$,

$$\le 2W_3 + B(1 - 2x_i')$$

By lemma 7,

$$< 2W_3 + B \qquad \square$$

Now we can determine the performance ratio for ALG1.

Theorem 9. *ALG1 is a $\left(2 + \frac{B-1}{OPT}\right)$-approximation algorithm for vc^t with arbitrary weights and a 2-approximation algorithm for vc^t with unit weights.*

Proof: By lemma 1, ALG1 is polynomial. By lemma 6, it covers t edges. We will now show that $\omega(C_1 \cup C_2 \cup C_3) \le 2OPT(G, \omega, t) + B - 1$.

Since C_1, C_2, C_3 are disjoint,

$$\omega(C_1 \cup C_2 \cup C_3) = \omega(C_1) + \omega(C_2) + \omega(C_3)$$

From lemma 8,

$$< 2(W_1 + W_2 + W_3) + B$$

From lemma 2,

$$= 2\sum_{i=1}^{n} \omega(v_i)x_i + B$$

From the linear program,

$$= 2LB(G, \omega, t) + B$$
$$\le 2OPT(G, \omega, t) + B$$

Now $\omega(C_1 \cup C_2 \cup C_3)$ must be an integer, and thus it must not exceed the greatest integer k with $k < 2OPT(G, \omega, t) + B$. Since $OPT(G, \omega, t)$ and B are both integers, this implies $\omega(C_1 \cup C_2 \cup C_3) \le 2OPT(G, \omega, t) + B - 1$. Therefore ALG1 is a $\left(2 + \frac{B-1}{OPT}\right)$-approximation algorithm for vc^t. When the instance is restricted to unit weights, $B = 1$ and ALG1 is a 2-approximation algorithm. $\square$

4 The Second Algorithm

While ALG1 provides a good approximation for vc^t with unit weights, it may give a poor approximation when B is arbitrarily larger than OPT. In our second algorithm we show how to tighten this bound, giving a 2-approximation algorithm for VC^t with arbitrary weights. To do this, we search for a vertex v_p such that $v_p \in C_{opt}$ for some optimal cover C_{opt} and $\omega(v_p)$ is the largest vertex weight in C_{opt}. It is not necessary to identify v_p, we just create a cover for each potential v_p and then return the cover with the smallest weight.

Algorithm ALG2
Input: $G = (V, E), \omega : V \to \{1, 2, \ldots, B, \infty\}, t$
Order V by decreasing weight: $\omega(v_1) \geq \omega(v_2) \geq \cdots \geq \omega(v_n)$
For $i \leftarrow 1, \ldots, |V|$ do
    $G_i \leftarrow G \setminus \{v_i\}$
    For $j \leftarrow 1, \ldots, i-1$ do $\omega_i(v_j) \leftarrow \infty$
    For $j \leftarrow i, \ldots, |V|$ do $\omega_i(v_j) \leftarrow \omega(v_j)$
    $t_i \leftarrow t - d(v_i)$
    $C_i \leftarrow \text{ALG1}(G_i, \omega_i, t_i) \cup \{v_i\}$
Find k such that $\omega(C_k) = \min\{\omega(C_1), \ldots, \omega(C_n)\}$
Return C_k

4.1 Analysis of ALG2

First we show that ALG2 produces a feasible solution.

Lemma 10. *For all $i \in \{1, \ldots, n\}$, C_i covers at least t edges.*

Proof of lemma 10: ALG1 covers at least t_i edges in G_i and in G. Since $v_i \notin G_i$, adding v_i to the cover increases the number of covered edges in G by $d(v_i)$. Thus there are at least $t_i + d(v_i) = t$ edges in G covered by C_i. $\square$

The next lemma will be used to prove the approximation ratio of ALG2.

Lemma 11. *Let C_{opt} be an optimal t-cover for instance $G; \omega; t$ and let v_p be the vertex of largest weight in C_{opt}. Then*

1. $OPT(G, \omega, t) = OPT(G_p, \omega_p, t_p) + \omega(v_p)$, and
2. $\omega(C_p) \leq 2OPT - 1$.

Proof of lemma 11: Since we are given that C_{opt} contains v_p and covers at least t edges then $C_{opt} \setminus \{v_p\}$ covers at least $t - d(v_p) = t_p$ edges. So $C_{opt} \setminus \{v_p\}$ is feasible for $G_p; \omega_p; t_p$ and has weight $\omega_p(C_{opt}) - \omega_p(v_p)$. Then

$$OPT(G, \omega, t) = \omega(C_{opt}) - \omega(v_p) + \omega(v_p)$$

Since $\omega(v_i) = \omega_p(v_i)$ for all $v_i \in C_{opt}$,

$$= [\omega_p(C_{opt}) - \omega_p(v_p)] + \omega(v_p)$$
$$\geq OPT(G_p, \omega_p, t_p) + \omega(v_p)$$

Now, let $\hat{C}_{opt}$ be optimal for $G_p; \omega_p; t_p$. Since v_p and adjacent edges are not in G_p, $\hat{C}_{opt} \cup \{v_p\}$ covers at least $t_p + d(v_p) = t$ edges in G. So $\hat{C}_{opt} \cup \{v_p\}$ is feasible for $G; \omega; t$. Then

$$OPT(G_p, \omega_p, t_p) + \omega(v_p) = \omega_p(\hat{C}_{opt}) + \omega(v_p)$$

Since $\omega(v) \leq \omega_p(v)$ for all $v \in V$,

$$\geq \omega(\hat{C}_{opt}) + \omega(v_p)$$
$$\geq OPT(G,\omega,t)$$

Thus $OPT(G,\omega,t) = OPT(G_p,\omega_p,t_p) + \omega(v_p)$. For part 2 we have

$$\omega(C_p) = \omega(ALG1(G_p,\omega_p,t_p)) + \omega(v_p)$$

But ALG1 guarantees

$$\leq [2OPT(G_p,\omega_p,t_p) + B - 1] + \omega(v_p)$$

Since $B = \omega(v_p)$,

$$= 2OPT(G_p,\omega_p,t_p) + \omega(v_p) - 1 + \omega(v_p)$$

From the first part of this lemma,

$$= 2(OPT(G,\omega,t) - \omega(v_p)) + \omega(v_p) - 1 + \omega(v_p)$$
$$= 2OPT(G,\omega,t) - 1 \qquad \square$$

We now reach our main result.

Theorem 12. *ALG2 is a 2-approximation algorithm for* vc^t.

Proof: It is clear that ALG2 runs in polynomial time and by lemma 10 covers t edges. For any optimal t-cover C_{opt}, some vertex v_p has maximum weight in C_{opt} and v_p is identified in iteration p of ALG2. For that iteration, lemma 11 states that we produce a cover C_p with $\omega(C_p) \leq 2OPT(G,\omega,t) - 1$. But since $\omega(C_k)$ is minimal we have $\omega(C_k) \leq \omega(C_p)$ and hence $\omega(C_k) \leq 2OPT(G,\omega,t) - 1$. Therefore ALG2 is a 2-approximation algorithm for vc^t. $\qquad \square$

To evaluate our result, we compare vc^t to VERTEX COVER. As theorem 13 shows, their approximability is directly related.

Theorem 13. *For functions t such that $t^{-1}(|V|) = |V|^\beta$ (or $t^{-1}(|E|) = |V|^\beta$) for some constant β, there is an approximation-preserving reduction from* VERTEX COVER *to* vc^t *such that an α-approximation algorithm for* vc^t *implies an α-approximation algorithm for* VERTEX COVER.

Proof: We will prove the case when t is a function of $|E|$. The proof for t a function of $|V|$ follows easily. Let $G = (V,E); \omega$ be any instance of VERTEX COVER and let $OPT(G,\omega)$ be the weight of an optimal solution. We create a vc^t instance G' from G by adding a matching M with s edges on $2s$ new vertices. Each new vertex will have weight equal to the maximum weight of a vertex in G. We choose s such that $t(|E| + s) = |E|$, i.e., $s = t^{-1}(|E|) - |E|$. Then G' has polynomial size if $t^{-1}(|E|) = |V|^\beta$ for some constant β. A vertex cover of G is a t-cover for G' and has the same total weight. Thus $OPT(G',\omega,t) \leq OPT(G,\omega)$. Now, let C be a t-cover of G' that covers $t(|E|+s) = |E|$ edges. Let $m = |C \cap M|$. The m vertices from $C \cap M$ can cover at most m edges in M, leaving at most m edges in the subgraph G uncovered (since the total edges covered is $|E|$). Let $\hat{C} = C \setminus M$ and then add to $\hat{C}$ one endpoint for each edge left uncovered in subgraph G. Since each vertex in G has weight no greater than a vertex in M, we have $\omega(\hat{C}) \leq \omega(C)$. Now, given an α-approximation algorithm for VC^t, we have

$$\omega(\hat{C}) \leq \omega(C) \leq \alpha OPT(G', \omega, t) \leq \alpha OPT(G, \omega).$$

Therefore we have an α-approximation for VERTEX COVER. $\qquad\qquad\Box$

It is a direct consequence of the proof of theorem 13 that any lower bound for VERTEX COVER is also a lower bound for VC^t for these values of t. Thus, following the result of Håstad [4], we have the following.

Corollary *Unless* P $=$ NP, *there can exist no α-approximation algorithm for* VC^t *with* $\alpha < \frac{7}{6}$.

5 Conclusions

We have demonstrated a new technique of massaging a linear programming solution to give good approximate solutions for an interesting generalization of the VERTEX COVER problem. This method may prove useful to other graph problems for which the linear programming solutions may not appear to have a direct interpretation.

References

1. R. Bar-Yehuda and S. Even. A local-ratio theorem for approximating the weighted vertex cover problem. Annals of Discrete Mathematics vol 25, (1985), 27-45.
2. M. Goemans and D. Williamson. .878-Approximation Algorithms for MAX CUT and MAX 2SAT. Proceedings of the Twenty-Sixth Annual ACM Symposium on the Theory of Computing (1994), 422-431.
3. M. Goemans and D. Williamson. New 3/4-Approximation Algorithms for MAX SAT. SIAM Journal of Discrete Mathematics, 7 (1994), 656-666.
4. J. Håstad. Some Optimal Inapproximability Results. To appear in the Proceedings of the Twenty-Ninth Annual ACM Symposium on the Theory of Computing (1997).
5. D. Hochbaum. Approximation Algorithms for Set Covering and Vertex Cover Problems. SIAM Journal on Computing 11 (1982), 555-556.
6. D. Karger, R. Motwani and M. Sudan. Approximate Graph Coloring by Semidefinite Programming. 35th Annual Symposium on Foundations of Computer Science (1994) 2-13.
7. N. Karmarkar. A New Polynomial-Time Algorithm for Linear Programming. Combinatorica vol 4, (1984), 373-395.
8. L. Khachian. A polynomial algorithm for linear programming. Doklady Akad Nauk USSR, vol 244(5) (1979), 1093-1096.
9. J. Klienberg and M. Goemans. The Lovász Theta Function and a Semidefinite Programming Relaxation of Vertex Cover. To appear in SIAM Journal on Discrete Mathematics.
10. E. Petrank. The hardness of approximation: Gap location. Computational Complexity, 4 (1994) 133-157.
11. P. Raghavan and C. Thompson. Randomized Rounding: A Technique for Provably Good Algorithms and Algorithmic Proofs. Combinatorica, 7 (1987) 365-374.

Partially Persistent Search Trees
with Transcript Operations

Kim S. Larsen*

Department of Mathematics and Computer Science, Odense University,
Campusvej 55, DK-5230 Odense M, Denmark. E-mail: kslarsen@imada.ou.dk.

Abstract. When dictionaries are persistent, it is natural to introduce a
transcript operation which reports the status changes for a given key over
time. We discuss when and how a time and space efficient implementation
of this operation can be provided.

1 Introduction

When balanced binary search trees are made partially persistent using the node-
copying method [5], the possibility of searching efficiently for information in the
past is added to the system. The operations of updating the present version and
searching in the present as well as in the past are asymptotically as efficient as
in the corresponding normal (non-persistent) binary search tree.

In database applications, it is sometimes desirable to produce transcripts of
information change over time. If we wish to obtain a transcript of information
related to some key k from version number v_1 to v_2, this can be obtained by
independent search operations in all versions in that interval in time $O(ph)$,
where h is the maximum height of the search tree in that interval, and $p =
v_2 - v_1 + 1$ is the number of versions between v_1 and v_2. We discuss when and
how this can be reduced to $O(h + p)$ by maintaining one extra pointer with a
version number in each node, without changing the asymptotic complexity of
any of the existing operations.

In database applications, search trees are usually leaf-oriented, which means
that all keys reside in the leaves, and internal nodes contain routers guiding
the search to the correct leaf. Leaves are often of another type than the internal nodes, and contain extra pointers or space consuming values associated with
their keys. We remain faithful to this model, and the extra pointer which is introduced only appears in internal nodes. Leaves will contain no extra information
compared to the single version scenario.

* This work was carried out while the author was visiting the Department of Computer
Sciences, University of Wisconsin at Madison. Supported in part by SNF (Denmark),
in part by NSF (U.S.) grant CCR-9510244, and in part by the ESPRIT Long Term
Research Programme of the EU under project number 20244 (ALCOM-IT).

2 Transcript Trees

When search trees are used in database applications, the trees are usually *leaf-oriented*. This means that only the leaves contain keys. The internal nodes contain routers, which are of the same type as the keys and which direct the searches to the correct location as usual in a search tree. However, routers need not be present as keys in the tree. This means that we do not have to update routers whenever a deletion takes place. For an internal node, the keys in its left subtree are smaller than or equal to its router, and the keys in its right subtree are larger. In a leaf-oriented tree, every internal node has two children.

Searching and Updating

To *insert* a key k in the tree, search for k as usual. An unsuccessful search ends up in a leaf, say l. A new internal node u is created in place of l, and l and a new leaf l' containing the key k are made the children nodes of u. The one containing the smaller key will be the left child. The router of u is a copy of the key contained in its left child. Thus, only one pointer in the existing structure is changed. To *delete* a key k from the tree, search for k as usual. If the key is found in the leaf l, its parent is replaced by the sibling node of l. A node and its pointers are deleted, but only one pointer is changed. The insertion and deletion operations are called *update* operations. It is easy to prove that while a router r is present in the tree, no new node with the same router can be created. This is important since otherwise rotations could violate the search tree invariant.

We assume that some scheme for maintaining balanced search trees will be used. Thus, the nodes contain additional fields for registering heights, colors, or other balance information. The update operations may manipulate these fields, and there will be a collection of rebalancing operations, which after each update, by manipulating the fields and applying rotations, will make sure that the balance constraints, if violated by the update, are again fulfilled.

However, we require that any rebalancing operation can be expressed as a constant number of single rotations carried out as described now (see Figure 1).

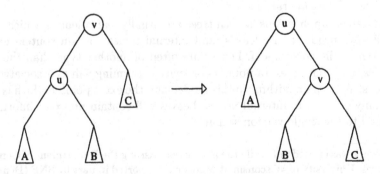

Fig. 1. Single rotation (to the right).

The single rotation can be either *right* (in the direction of the arrow) or *left* (in the opposite direction of the arrow). We give the requirements for a right rotation; the other is similar. Assume that the parent of v is a node w, and assume without loss of generality that it points to v via its left pointer. Then carry out $v.left:=u.right$; $u.right:=v$; $w.left:=u$.

The point is that nodes must keep their identity since we introduce new pointers which should aim at the correct nodes, even when these are moved around by rebalancing operations. We do not know of any scheme which does not conform to this requirement, except the ones which apply some form of global rebuilding as in [2, 6].

Partial Persistence

We now make the search tree partially persistent using the *node-copying method* [5]. A *partially persistent* structure is a structure which supports multiple versions, such that all versions can be accessed, but only the newest version can be modified. We assume that versions are numbered using consecutive integers. All nodes have a field stating under which version they were created. The structure has a number of *entry* pointers (pointers to roots of different versions) which also have version numbers.

For balanced search trees, the node-copying method can be used in a simplified form [5]. We use an extended version of that simplified form. We retain what is referred to as the *copy pointer*, equip it with a version number, and update it as in the original method. We give a very brief description here.

There are three types of information in a balanced search tree: keys, pointers, and balance information. Since we only update the newest version, balance information is not needed for older versions. Thus, it is safe to overwrite old balance information. When we insert a new key, we create a new node for it. So, no key information is being changed. Instead, it is one pointer in one node that is being changed to include the new node. Similarly, when we delete a key, this is done by changing one pointer to cut out the node, in which the key resides. So, we only need to discuss pointer updates.

To avoid copying too many nodes every time a pointer must be changed, nodes have one extra field for a pointer update. So, nodes in the tree have fields: *key, left, right, vn, extra,* and *copy,* where "vn" is short for "version number". The field *extra* is composite, and has the following fields: *ptr, dir,* and *vn* for recording the new pointer, which pointer it replaces (*dir* is *Left* or *Right*), and in which version it was done ("ptr" is short for "pointer" and "dir" is short for "direction"). The field *copy* is also composite with fields *ptr* and *vn*.

An update in the newest version i is handled as described now. We explain the action which must be taken for *one* pointer change. If an update (or a rebalancing operation) involves several pointer changes, the procedure is repeated.

If the node u in which the update is made has version number i, the update overwrites the existing pointer. Otherwise, if the extra field has version number i *and* its direction field indicates the pointer to be updated, the pointer in the extra field is overwritten.

If neither case applies, there are two possibilities depending on whether or not the extra field has already been used. If it has not, the update is made there, also setting the direction field and setting the version number to i.

Otherwise, a new node v must be made. This is the only case, where the procedure does not terminate immediately. The key and the *newest* left and right pointers from u are copied into v. Then v's version number is set to i and its extra and copy fields to nil. Finally, the copy field in u is set to point to v, and its version number is set to i. Thus, copy pointers link together sequences of nodes which are basically the "same" node at different times. Now, the parent of u must be updated to point to v instead. This is done *recursively* using the method just outlined, i.e., the effects of an update continue up in the tree along the update path for as long as there are nodes where the extra field has already been used by an earlier version, or has been used in the opposite direction of the update path by the current version. If the root is copied, then a new entry pointer to the new root with version number i is created. For an example, see Figure 2. Nodes, extra pointers (in the middle), and copy pointers (dashed) have version numbers. Thick lines indicate parts of the tree which was present before the operation. The current version number is 3.

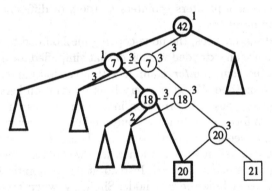

Fig. 2. Inserting 21.

Searching starts by following an entry pointer. When accessing version j, the entry pointer with the largest version number i, $i \leq j$, is used.

At each node, a decision to continue the search to the left or the right must be made. As usual, this decision is made by comparing the key to be found with the router (key) in the node. However, when the decision has been made to proceed to the left, for instance, then the *newest* left pointer no newer than j must be followed. The following result, or more general ones, are proven in [5].

Theorem 1. *When the node-copying method is applied to a pointer structure for which there is an upper bound p on the number of pointers that can point to any one node, then if nodes in the persistent version are equipped with at least p extra pointers, the following holds.*

– *The asymptotic complexity of searching in any version in the persistent struc-*

ture is equal to the asymptotic complexity of searching in the standard structure corresponding to that version.

- *The persistency actions carried out after one pointer change are performed in amortized constant time.*
- *When searching in the newest version, only the last node in a copy sequence can be accessed.*

Note that in connection with the trees we use here, only the parent of a node can point to that node, so one extra field suffices. Also, since we use only a constant amount of new space in each step, space usage per pointer update is also amortized constant.

Transcript Facilitating Additions

Now we extend the updating and rebalancing procedures even further. We first describe our goal informally. If we want to keep track of some key k, then we position ourselves at the internal node under which k is found (or would be inserted). When changing version, from v_1 to v_2 say, the internal node with that property may be deleted, or an insertion or rotation may have the effect that it is now another node which has the given property. Therefore, whenever we make an update or a rotation, we also build a path from the node with the given property in version v_1 to the one with that property in version v_2, such that later, during a transcript operation, it will be easy to get to that new node. We make sure the path is protected in the sense that no later pointer updates can alter it. The result is a tree with protected paths which run through the tree over time (versions, that is) at the levels just above the leaves.

Note first that when an internal node is deleted, no pointer updates can ever be applied to it again. This means that its copy pointer will never be used, so we are free to use it for other purposes (this is safe since the copy pointer is never used in searching). Additionally, when an insertion or a rotation is carried out, we may *trigger* a copying of a node which would not have been made at that time in the original method. This means that in contrast to [5], we can have several copies of a node with the same version number. These are the only changes we are making. Whenever the copy pointer is set, its version number is also set, and it is set to the newest (that is, the current) version.

For deletion, there are two cases. First assume that both of the children of the internal node u to be deleted are leaves. The deletion is performed, and the copy pointer of u is set to point to the newest copy of the parent of u. Now assume that the internal node to be deleted has only one leaf among its two children. After the deletion has been performed, the copy pointer of u is set as follows. If the leaf is its left child, then its copy pointer is set to point to its *in-order internal successor* (the left-most internal node in its right subtree). This case is illustrated in Figure 3. Note that persistency actions will continue above the node with key 7 (this is not shown).

If the leaf is its right child, then its copy pointer is set to point to its *in-order internal predecessor*. The successor (or predecessor) is found by a search in the newest version of the structure starting at the internal node to be deleted.

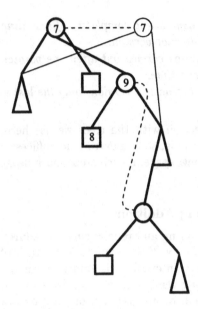

Fig. 3. Deleting the key 8 from a transcript tree.

For insertion, we do the following. As described earlier, we first search for the correct leaf, and the only pointer in the existing structure which is changed is the one in the internal node v which points to the leaf in question. We make the insertion as usual for leaf-oriented trees, applying the necessary persistency actions, i.e., v may be copied. When this is completed, we trigger a new copy of v, or of the copy of v, if v has already been copied during the insertion.

The actions taken for a rotation are similar to those for insertions. First, the rotation is carried out, and all the necessary persistency actions are taken. Then if the middle subtree is a leaf, we trigger a new copy of the newest version of the node which is root of the operation *after* the rotation is carried out.

The Transcript Operation

The transcript operation can now be defined. It takes a key k and two version numbers v_1 and v_2 (with $v_1 \leq v_2$) as parameters and prints the history of k between these two versions. More precisely, a line is printed for every version, stating whether or not k is in the tree. In a practical application, there would most likely be values associated with the keys, and these could also be printed.

The implementation is given in Figure 4. To avoid too many details, we assume that the tree is never empty, and that we have a function Leaf which decides if a node is a leaf, and a function Entry which given a version number, returns the entry pointer to be used.

The function *Find* finds the parent of a given key in a given version. This is simply a search, except that we return the parent of the key. The leaf in the

```
func Go(u: Node, dir: Dir, i: Version): Node
  if u.extra ≠ nil and u.extra.dir = dir and u.extra.vn ≤ i then
    return u.extra.ptr
  else
    return (dir = Left ? u.left : u.right)

func Find(u: Node, k: Key, i: Version): Node
  v := (k ≤ u.key ? Go(u, Left, i) : Go(u, Right, i))
  return (Leaf(v) ? u : Find(v, k, i))

func Status(u: Node, k: Key, i: Version): Bool
  dir := (k ≤ u.key ? Left : Right)
  if u.extra ≠ nil and u.extra.dir = dir and u.extra.vn ≤ i then
    return u.extra.ptr.key = k
  else
    return (dir = Left ? u.left.key : u.right.key) = k

func Advance(k: Key, u: Node, i: Version): ⟨Bool, Node⟩
  u := Find(u, k, i)
  while u.copy ≠ nil and u.copy.vn ≤ i do
    u := Find(u.copy.ptr, k, i)
  return ⟨Status(u, k, i), u⟩

proc Transcript(k: Node, v1, v2: Version)
  u := Entry(v1)
  for i := v1 to v2 do
    ⟨s, u⟩ := Advance(k, u, i)
    print i, k, s
```

Fig. 4. The transcript operation.

direction we would search to find k is referred to as the *leaf possibly containing*
k. *Find* calls *Go* which takes one step down in the tree by going left or right
as appropriate. Assuming that we are located at the parent of the leaf possibly
containing k, *Status* returns the information as to whether or not k is present.

Transcript works by repeatedly calling the *Advance*. The function *Advance* is
called with a key k, a node u, and a version number i. Except for the first call,
the assumption is that the information returned by the previous call to *Advance*
was from u in version $i - 1$, i.e., u was the internal node which could have a leaf
child containing k. *Advance* finds the next similar node in version i, and returns
information on the status of k at that time, as well as the node.

3 Correctness

When searching through version i by starting at the entry pointer, one will see
version i as it appeared when we switched to the next version. However, the

current version can keep changing right until we switch. This complicates the search by *Advance* even after we have switched to a new version because we do not enter via the entry pointer. Thus, we may see parts of version i which were temporary, and which would never be found when entering "correctly". This is the reason for the **while** loop in function *Advance*. Whenever we use *Find*, we are searching in version i as it appeared at some point; not necessarily in its final appearance. For instance, several deletions could be made in the same version so it may be necessary to follow the copy pointer several times.

The reason for triggering a node copying in connection with insertions and rotations is that the parent of the leaf possibly containing k changes. By triggering a copying of the node which used to be the parent, we ensure that no later operations can prevent us from accessing the new parent. Thus, we build a protected path from the parent of the leaf possibly containing k in one version to the (possibly new) parent in the next.

Theorem 2. *If we are at the node u equal to $Find(Entry(i-1), k, i-1)$, $i > 1$, then $Advance(k, u, i)$ will bring us to v equal to $Find(Entry(i), k, i)$.*

Proof. Assuming that *Advance* brings us from u to v, it follows a path in the tree; this is not necessarily a direct path from one node to a descendant since copy pointers can also be followed. We call this the *advance path*. The proof is by induction. However, we strengthen the induction hypothesis by adding that no pointer changes can be made to nodes on the advance path except the last.

We prove by induction in the number of modifications in the tree by version i or greater that *Advance* will find the node v. Actually, since nodes are never physically deleted, and pointers are only overwritten if created by the version performing the update, updates in versions greater than i cannot affect the path, so it is sufficient to consider updates by version i.

The base case is when no modifications have taken place. In that case, the start node u and end node v are identical, so *Advance* will certainly find v (immediately after switching to version i, searching in version i is identical to searching in version $i-1$). Additionally, the advance path consists of a single node (which is then last), so the second part of the hypothesis follows trivially.

Assume that some operation changes the advance path. By induction, the change is made by altering or adding a pointer in the last node on the path. We consider the operations in turn.

Assume that an insertion changed the advance path, and assume without loss of generality that the leaf possibly containing k (before the insertion) is to the left. If the insertion is also made to the left, the advance path will continue first by either following the left pointer (if the node in question is from version i), the extra pointer, or, if that was already used, by following the copy pointer and then the left pointer. This brings us to the new internal node which, by the insertion, has become the new parent of the leaf possibly containing k.

Since a new copy is triggered as the last action taken during an insertion, and since updates are always performed in the newest nodes, none of the pointers described in the above can ever be altered. Thus, it is still the case that no pointer changes can be made to nodes on the advance path except the last.

If the insertion is made to the right, then only the copy pointer will be followed (possibly twice) and we are again at the correct location.

Now, we assume that a deletion makes a change to the last node u on the advance path. If u is the internal node which is deleted, the copy pointer is set to point to the correct next location (since the location is actually found by a search in the newest version), and since a deleted node can never be accessed by an updating operation again, no changes can be made to nodes on the advance path except the last. If u is the parent of the internal node to be deleted, the advance path is either not changed (if the extra pointer was used for the update), or the copy pointer from u to the newest version of u becomes part of this path.

Finally, we must consider rebalancing operations. As required, they consist of a number of single rotation. Thus, we only need to consider one single rotation. Assume that a single rotation (Figure 1) makes a change to the last node on the advance path. If v in Figure 1 is the last node, since it keeps it leaf (C), the path is either not changed or it is extended with one copy pointer. For the node u, the situation is similar if A is the leaf possibly containing k. So, assume that B is the leaf in question. This is exactly the case where a copy of u is triggered after the rotation. Thus, the new pointer in u leading to the new parent of the leaf possibly containing k cannot be altered again. Thus, except for the new last node (v), nodes on the advance path cannot be changed. □

4 Complexity

An interesting case to consider is the one where there are many versions, since then searching from entry pointers would be very time consuming. We assume that we change version no later than after a fixed constant number of updates. We consider the behavior of *Advance* under that restriction.

Complexity of Transcript

Since a copy pointer which is set up during a deletion points to a node which was in the tree at the time (though it may be from an older version), any additional copy pointers from there must be at least as new. This is also the case for copy pointers created by insertions. Since there are only a constant number of updates, the function *Advance* can follow such pointers at most a constant number of times to get from one version to the next.

Rebalancing does not necessarily behave that well. We consider restrictions on the behavior of rebalancing which lead to the best possible complexities. The rotations described earlier which trigger a copying of a node, because the middle subtree is a leaf, are referred to as *expensive* rotations.

Definition 3. A balanced binary search tree scheme is called *neighborhood balanced* if there exists a constant c such that the number of expensive rotations carried out in response to any update is bounded by c.

If a balanced search tree scheme is neighborhood balanced, *Advance* will find the correct node in constant time, since, referring to the proof of Theorem 2, the advance path is extended with only a constant number of edges for each insertion, deletion, and expensive rotation, and we change version after a bounded number of updates. Thus, the complexity of the transcript operation is $O(\log n + p)$, where n is the number of elements in the start version for the reporting.

Red-black trees [7] are neighborhood balanced, since, as it is pointed out in [5], rebalancing after an update consists of at most three rotations (single or double). The remaining rebalancing operations are recolorings. AVL-trees [1] are also neighborhood balanced, since, as it is pointed out in [11], only one rotation (single or double) is necessary after each update. The remaining rebalancing operations consist of height adjustments.

Also treaps [3] are neighborhood balanced. Only single rotations are used in treaps, and in a sequence of rotations following an insertion, it is always the same node which is the bottom-most node of the two internal nodes in the rotation. After it has lost at most two leaf children, its children are going to be internal nodes from that point on. Deletions can be viewed as reverse insertions.

BB[α]-trees [9, 4, 8] are not neighborhood balanced. It is possible to construct a sequence of double rotations where one leaf gets a different parent a logarithmic number of times. However, the number of structural changes after an update is amortized constant. This means that printing the whole history of a key would indeed be done in the same complexity, but printing only a part may not.

Splay trees [10] are not neighborhood balanced either. A sequence similar to the one from BB[α]-trees can be constructed. Note also that, as opposed to the usual case, a leaf-oriented splay tree does not have the property that the most recently accessed keys are close to the root, since it is the last *internal* node on a search path which is rotated up to the root.

Other Complexity Considerations

Searching in a transcript tree is of the same asymptotic complexity as in [5].

For updates, the additions only increase the complexity by a constant factor, since the work carried out in connection with pointer updating and triggering of copies, by Theorem 1, is amortized constant per pointer change. Though setting the copy pointer in connection with a deletion requires a search, this action is taken in connection with an update, so the total complexity of updates remains bounded by the height of the tree.

If the transcript operation is implemented using binary search trees without balance constraints, the complexity of the transcript operation becomes $O(h + p)$, where h is the height of the first version from which we report, and updates become $O(h)$, where h is the height of the tree.

From the discussion above, it also follows that search trees where the number of pointer changes in response to an update is amortized constant will use space only linear in the number of updates. This applies to red-black trees and AVL trees. The expected number of rotations carried out in response to an update in a treap is two, so space consumption for treaps is expected linear.

5 Concluding Remarks

The status of a key is not necessarily changed in every version. Thus, when following a copy pointer, we may skip over a large number of versions. This can be exploited to give a running time which can sometimes be significantly better.

A chronological transcript involving more than one key could be produced by merging separate transcripts. However, *Transcript* has been expressed via *Advance* which reports one change at a time. Thus, the total chronological transcript can be produced directly by maintaining a priority queue. This is more "on-line", i.e., it would start printing out the transcript earlier.

If the number of pointers is not a concern, then there is an easier solution: decide that the standard search tree should have all the parents of leaves connected in a doubly linked list. By [5], we obtain the optimal asymptotic amortized complexities. However, more pointers are needed. The doubly linked list uses additional pointers, but since nodes can now have more than one predecessor (there are now 3), nodes must be equipped with predecessor pointers (see [5]) such that all nodes referring to an updated node can themselves be updated. Additionally, there must be more extra fields corresponding to the number of predecessors (the amortized constant results from [5] depend on this). Though the amortized time complexity would be the same, the worst case complexity of an update would be $\Omega(n)$ instead of $O(\log n)$ obtained by a balanced transcript tree.

References

1. G. M. Adel'son-Vel'skiĭ and E. M. Landis. An Algorithm for the Organisation of Information. *Soviet Math. Doklady*, 3:1259–1263, 1962.
2. A. Andersson. Improving Partial Rebuilding by Using Simple Balance Criteria. In *Lecture Notes in Computer Science, Vol. 382: 1st Wokshop on Algorithms and Data Structures*, pages 393–402. Springer-Verlag, 1989.
3. C. R. Aragon and R. G. Seidel. Randomized Search Trees. In *Proc. 30th Ann. IEEE Symp. on the Foundations of Comp. Sci.*, pages 540–545, 1989.
4. N. Blum and K. Mehlhorn. On the Average Number of Rebalancing Operations in Weight-Balanced Trees. *Theoretical Computer Science*, 11:303–320, 1980.
5. J. R. Driscoll, N. Sarnak, D. D. Sleator, and R. E. Tarjan. Making Data Structures Persistent. *Journal of Computer and System Sciences*, 38:86–124, 1989.
6. I. Galperin and R. L. Rivest. Scapegoat Trees. In *4th ACM-SIAM Symp. on Discrete Algorithms*, pages 165–174, 1993.
7. L. J. Guibas and R. Sedgewick. A Dichromatic Framework for Balanced Trees. In *Proc. 19th Ann. IEEE Symp. on the Foundations of Comp. Sci.*, pages 8–21, 1978.
8. K. Mehlhorn. *Sorting and Searching*. Springer-Verlag, 1986.
9. J. Nievergelt and M. Reingold. Binary Search Trees of Bounded Balance. *SIAM Journal on Computing*, 2(1):33–43, 1973.
10. D. D. Sleator and R. E. Tarjan. Self-Adjusting Binary Trees. In *Proc. 15th Ann. ACM Symp. on the Theory of Computing*, pages 235–245, 1983.
11. M. A. Weiss. *Data Structures and Algorithm Analysis*. The Benjamin/Cummings Publishing Company, Inc., 2nd edition, 1995.

Relating Hierarchies of Word and Tree Automata

Damian Niwiński* and Igor Walukiewicz*

Institute of Informatics, Warsaw University
Banacha 2, 02-097 Warsaw, POLAND
{niwinski,igw}mimuw.edu.pl

Abstract. For an ω-word language L, the derived tree language Path(L) is the language of trees having all their paths in L. We consider the hierarchies of deterministic automata on words and nondeterministic automata on trees with Rabin conditions in chain form. We show that L is on some level of the hierarchy of deterministic word automata iff Path(L) is on the same level of the hierarchy of nondeterministic tree automata.

1 Introduction

For an ω-regular set of infinite words $L \subseteq \Sigma^\omega$, we consider the *derived language* of infinite binary trees, Path(L), consisting of the trees all of whose paths are in L. In this paper, we address the following question: what is the connection between the complexity of L and that of Path(L)? Here, by the complexity of a language of ω-words or trees we understand the minimal complexity (in appropriate sense) of an automaton recognizing this language.

One motivation for considering this question is the following. In application of automata to program verification, a language L usually represents some property of computations of a program, e.g. expressed in a temporal logic of linear time. Now, we are often interested in a more general issue of whether all computations of a program satisfy the given property. Representing all possible computations of a program by a tree we are led to consider the derived tree language Path(L).

In most of the cases our language L will be an ω-regular language. It is easy to transform a deterministic automaton recognising L into a *deterministic* automaton recognising Path(L). However, we may a priori hope that using nondeterminism in some clever way we can simplify the automaton for Path(L).

In [6], Kupferman, Safra and Vardi investigated this question and showed a result that refutes such a hope in a special case of Büchi automata. More specifically, these authors show that a derived tree language is recognizable by a *nondeterministic* Büchi tree automaton if and only if the original language of ω-words is recognizable by a *deterministic* Büchi automaton. Read under the negation, the implication "only if" says that nondeterministic Büchi tree automata do not recognize more languages of the form Path(L) than they must. In yet other words, if an ω-regular language L cannot be recognized by a deterministic Büchi automaton, its inherent difficulty persists in Path(L).

---

* Supported by Polish KBN grant No. 8 T11C 002 11

In the present paper we show that a similar phenomenon occurs for *all* ω-regular languages. To make the statement precise, we need somehow to measure the complexity of languages L and Path(L). This can be done using hierarchies classifying ω-regular languages of trees or words. Our choice here (advocated below) is to consider the hierarchies induced by an index of a Mostowski acceptance condition. This condition, in different phrasing, was introduced by A.W. Mostowski [7]. It was independently introduced by Emerson and Jutla [3]. We use their formulation of the condition.

The Mostowski condition is given by a function $\Omega : Q \to \mathbb{N}$ assigning to each state of the automaton a natural number called its *priority*. Then a computation path is accepting if $\limsup_{n \to \infty}$ of all the priorities occurring along the path is even, in other words, the highest priority persisting infinitely often is even. The index of a condition $\Omega : Q \to \mathbb{N}$ is the pair (ι, n) where ι and n are, respectively, the minimal and maximal values of the function Ω. By the nature of Mostowski condition, one can always scale Ω down so that ι is either 0 or 1. The concept of an index naturally gives rise to the hierarchies of ω-regular languages of words and trees recognized by automata equipped with Mostowski condition. Here we shall focus on the hierarchies induced by *deterministic* automata on ω-words and *nondeterministic* automata on trees. Both hierarchies are known to be infinite and to exhaust the classes of ω-regular languages and regular sets of infinite trees, respectively. Moreover, these hierarchies refine the analogous hierarchies induced by the Rabin acceptance criterion. (Note that a hierarchy of *nondeterministic* automata on ω-words is uninteresting as it collapses at the level of Büchi automata, i.e., automata of the Mostowski index (1,2).)

Our main result is that Path(L) can be recognized by a nondeterministic tree automaton of Mostowski index (ι, n) iff L can be recognized by a deterministic automaton of the same index. That is, L and Path(L) are on the same level in the corresponding hierarchies. In [6] the result was shown for index $(1, 2)$.

Let us comment on the advantages of the Mostowski criterion compared to other acceptance criteria considered in the literature, and on the relevance of the aforementioned hierarchies.

The Mostowski condition is as powerful as Muller condition for nondeterministic automata on trees [7] and deterministic automata on ω-words [16]. Yet, Mostowski condition is much easier to work with. This is mainly because this is the unique condition that admits memoryless winning strategies for both of the players in infinite games with perfect information. This fact allows for a relatively easy proof of Rabin's Complementation Lemma ([8], cf. [14]); it is also used in many other applications of automata. The other reason for considering Mostowski condition is that when translating the μ-calculus into automata one naturally ends up with automata with Mostowski condition [3,4]. Finally, as we have mentioned above, the hierarchies induced by the Mostowski condition are more subtle than those induced by the Rabin condition (cf. [11]); they also enjoy nice symmetry.

The aforementioned hierarchies are interesting for at least two reasons. First, they allow to classify the expressive power of formalisms. For example, one can better understand the expressive power of some logic of programs by translating it into automata of some level of the hierarchy. The second reason comes from considering complexity problems. For example, the general framework for solving satisfiability problems for logics of programs is to translate them into automata and then to do the emptiness check. Up to current knowledge, the deterministic complexity of the emptiness check depends exponentially on the size of the index (in general it is NP-complete for Rabin automata [2] and in NP∩co-NP for Mostowski automata [1]).

Our work was directly inspired by the work of Kupferman, Safra and Vardi [6], but our proof is different from theirs. For our proof we need a characterization of the index of an ω-word language in terms of properties of the graph of a deterministic automaton recognizing the language. This is related with the work of Wagner [16]. One may see our characterization as a simplification of his characterization for the restricted case of Mostowski conditions. Actually our characterization of Mostowski index can be transformed into a polynomial time algorithm computing the index of an ω–regular language (i.e. a minimal index of an automaton recognizing L). This is related with the work of Krishnan, Puri and Brayton [5] who show that computing the Rabin index of a language presented by a deterministic Rabin automaton is NP-complete. On the other hand, solving the same question is polynomial if the winning condition of the automaton is given in Muller form (c.f. Wilke and Yoo [17]).

The plan of the paper is as follows. We start with the preliminary section introducing automata and the hierarchies. In Section 3, we show a simple characterization of the index of a ω-language in terms of properties of the graph of a deterministic Mostowski automaton recognizing the language. In the final section we use this characterization to show the main theorem.

2 Preliminaries

The set of natural numbers is denoted by ω or by N. For a set X, X^* is the set of finite words over X, including the empty word ε, and X^ω is the set of mappings from ω to X usually referred to as *infinite words*. The *length* of a finite word w is denoted by $|w|$; note that $|\varepsilon| = 0$. We write $v \leq w$ to mean that v is an *initial segment* of w. For $u \in X^\omega$, we let $\mathrm{Inf}(u) = \{x \in X : (\forall m \exists n > m)\, u(n) = x\}$ be the set of elements appearing infinitely often in u.

A nonempty subset T of X^* closed under initial segments is called a *tree*. The elements of T are called *nodes*, the $\leq$-maximal nodes are *leaves* and ε is the *root* of T. If $u \in T$, $x \in X$, and $ux \in T$ then ux is an *immediate successor* of u in T. An infinite sequence $P = (w_0, w_1, \ldots)$ such that $w_0 = \varepsilon$ and, for each m, w_{m+1} is an immediate successor of w_m is called a *path* in T.

If Σ is an arbitrary set and T is a tree then a mapping $t : T \to \Sigma$ is called an Σ-*valued* tree or shortly a Σ-*tree*; in this context T is the *domain* of t denoted by $T = \mathrm{dom}(t)$. We say "root of t", "path in t" etc., referring to the corresponding objects in $\mathrm{dom}(t)$. If $P = (w_0, w_1, \ldots)$ is a path in t, we denote by $t(P)$ the *labeling of P*, i.e., the ω–word $t(w_0)t(w_1)\ldots$ in Σ^ω. Note that $\mathrm{Inf}(t(P))$ is the set of values occurring infinitely often on the path P.

We now fix a finite alphabet Σ. For notational convenience, we shall focus in this paper on full binary trees over Σ, i.e., the Σ-trees with $\mathrm{dom}(t) = \{1,2\}^*$. Thus, every node $w \in \mathrm{dom}(t)$ has exactly two successors $w1$ and $w2$. Let T_Σ be the collection of all such trees.

Lemma 1. *The following is the key concept of our paper.*

Definition 2. Let $L \subseteq \Sigma^\omega$. The *path tree language derived from* L is defined by

$$\mathrm{Path}(L) = \{t \in T_\Sigma : \text{for each path } P \text{ in } t, t(P) \in L\}$$

We call a set $M \subseteq T_\Sigma$ a *path tree language* whenever $M = \mathrm{Path}(L)$, for some $L \subseteq \Sigma^\omega$.

Examples. Let $\Sigma = \{a,b\}$, and let $L_1 = \{a,b\}^*a^\omega$, $L_2 = (\{a,b\}^*b)^\omega$ (for simplicity we abbreviate $\{s\}$ by s). Then $\mathrm{Path}(L_1)$ is the set of trees such that on each path, b occurs only finitely many times, while $\mathrm{Path}(L_2)$ is the set of trees such that on each path b occurs infinitely often.

Not every set of trees is a path language. For example, the set of trees t, such that $t(w) = a$ for at least one w, is not.

Automata on words A finite automaton on infinite words can be presented as a tuple $\mathcal{A} = \langle \Sigma, Q, q_0, Tr, Acc \rangle$, where Σ is a finite alphabet, Q is a finite set of *states*, q_0 is an *initial state*, $Tr \subseteq Q \times \Sigma \times Q$ is a set of *transitions*, and $Acc \subseteq Q^\omega$ is an acceptance criterion, usually given in some special finitary form.

A *run* of the automaton $\mathcal{A}$ on an infinite word $u \in \Sigma^\omega$ can be presented as an infinite word $r \in Q^\omega$ such that $r(0) = q_0$, and $\langle r(m), u(m), r(m+1)\rangle \in Tr$ for every $m < \omega$. A run is *accepting* if it belongs to Acc. A word $u \in \Sigma^\omega$ is *accepted* by $\mathcal{A}$ if there exists an accepting run of $\mathcal{A}$ on u. The set of all accepted words is denoted $L(\mathcal{A})$.

An automaton $\mathcal{A}$ as above is *deterministic* whenever Tr is a partial function from $Q \times \Sigma$ to Q, i.e., for every $q \in Q$, $\sigma \in \Sigma$, there is at most one q' such that $\langle q, \sigma, q'\rangle \in Tr$. Note that a deterministic automaton can have at most one run on a given word $u \in \Sigma^\omega$.

Several kinds of automata have been considered in the literature according to the actual form of the acceptance criterion. A *Büchi criterion* is given by a set $F \subseteq Q$, and the corresponding set Acc consists of those r for which $\mathrm{Inf}(r) \cap F \neq \emptyset$. A *Muller criterion* is given by a family $\mathcal{F} \subseteq \wp(Q)$, and $Acc = \{r : \mathrm{Inf}(r) \in \mathcal{F}\}$. A *Rabin criterion* is given by family $\{(L_1, U_1), \ldots, (L_n, U_n)\}$, $L_i, U_i \subseteq Q$, and $Acc = \{r : (\exists i)\ \mathrm{Inf}(r) \cap L_i = \emptyset \text{ and } \mathrm{Inf}(r) \cap U_i \neq \emptyset\}$. Note that Büchi criterion can be presented as a special case of Rabin criterion (namely, $\{\emptyset, F\}$), and the last as a special case of Muller criterion.

In this paper we focus on yet another criterion that we call the *Mostowski acceptance criterion*. (This criterion has been often referred to as Rabin criterion in chain form.) Mostowski criterion can be presented by a *priority function* $\Omega : Q \to \mathbb{N}$, and Acc consists of those r for which $\limsup_{n \to \infty} \Omega(r(n))$ is *even*; in other words, the highest priority repeating infinitely often is even. It is not difficult to see that this can be represented by a Rabin criterion given by a family $(\{q : \Omega(q) \text{ is odd and } \geq i+1\}, \{q : \Omega(q) \text{ is even and } \geq i\})$, where i ranges over even numbers less than or equal to $\max(\Omega(Q))$.

According to the actual form of the acceptance criterion we shall refer to Büchi automata, Mostowski automata, etc.

Two automata $\mathcal{A}_1, \mathcal{A}_2$ are *equivalent* whenever $L(\mathcal{A}_1) = L(\mathcal{A}_2)$. By the remarks above, for every Büchi, Rabin or Mostowski automaton, there is an equivalent Muller automaton.

Moreover, the following facts are well–known (see e.g. [13]).

Theorem 3. *For every Muller automaton there is an equivalent nondeterministic Büchi automaton, and an equivalent deterministic Mostowski automaton, but in general there may be no equivalent deterministic Büchi automaton.*

Automata on trees An automaton over binary trees can be presented by $\mathcal{A} = \langle \Sigma, Q, q_0, Tr, Acc \rangle$, where all the items are as for automata on words except that $Tr \subseteq Q \times \Sigma \times Q \times Q$. A *run* of $\mathcal{A}$ on a tree $t \in T_\Sigma$ is a Q–valued tree $r : \{1, 2\}^* \to Q$ such that $r(\varepsilon) = q_0$, and, for each $w \in \text{dom}(r)$, the tuple $\langle r(w), t(w), r(w1), r(w2) \rangle$ is in Tr.

A path P in a run r is *accepting* if $r(P) \in Acc$; a run is accepting if so are all of its paths. The set of trees *accepted* by $\mathcal{A}$, denoted $L(\mathcal{A})$, consists of those $t \in T_\Sigma$ for which there exists an accepting run of $\mathcal{A}$ on t.

Two automata $\mathcal{A}_1, \mathcal{A}_2$ are *semantically equivalent* whenever $L(\mathcal{A}_1) = L(\mathcal{A}_2)$.

Similarly as for automata on infinite words, we can consider different kinds of tree automata according to the form of the acceptance criterion. Since the concept of acceptance criterion is the same as for automata on words, we have again some trivial inclusions. Moreover, Mostowski [7] established the following.

Theorem 4. *For every Muller tree automaton there is an equivalent Mostowski tree automaton.*

This is the furthest that the analogy with Theorem 3 can go. In contrast to automata on words, tree automata cannot in general be determinized and the expressive power of nondeterministic Büchi tree automata is weaker than that of Muller automata, and incomparable with that of deterministic Muller tree automata [12].

2.1 Index hierarchies

It is an acceptance criterion that gives rise to the expressive power of automata; the criteria also induce several semantical hierarchies. Here we will restrict our considerations to the hierarchy of Mostowski conditions because there is close correlation between this hierarchy and both: the hierarchy of Rabin conditions and the hierarchy of fixpoint alternations (cf. [11]).

Definition 5. An *index* of a Mostowski condition given by the function Ω is a pair (ι, n), where ι and n are the minimal and the maximal values of Ω, respectively.

We say that a language $L \subseteq \Sigma^\omega$ is ι-*n*-*feasible* if L can be recognized by a *deterministic* Mostowski automaton with the condition of index (ι, n). Otherwise, L is ι-*n*-*unfeasible*.

We say that a set of trees $M \subseteq T_\Sigma$ is ι-*n*-*feasible* if it can be recognized by a nondeterministic Mostowski automaton on trees with the condition of index (ι, n), and is ι-*n*-*unfeasible* otherwise.

Note that we express feasibility of sets of words in terms of deterministic automata, and feasibility of sets of trees in terms of general, i.e., nondeterministic automata.

It turns out that the hierarchy of indices induces a strict hierarchy of languages. Let, for $n < \omega$, $\Sigma_n = \{0, 1, \ldots, n\}$. Consider the following families of languages, $M_n, N_n \subseteq \Sigma_n^\omega$:

$$M_n = \{u \in \Sigma_n^\omega : \limsup_{i \to \infty} u(i) \text{ is } even \}$$
$$N_n = \{u \in \Sigma_n^\omega : \limsup_{i \to \infty} u(i) \text{ is } odd \}$$

The following results have been originally stated in different terms but the actual phrasing follows directly from the correspondence between Rabin and Mostowski indices (cf. [11]).

Theorem 6 (Wagner [15]). *For every $n < \omega$: (i) M_n is 0-n-feasible but 1-$(n+1)$-unfeasible; (ii) N_n is 1-$(n+1)$-feasible but 0-n-unfeasible.*

Theorem 7 (Niwiński [9, 11]). *For every $n < \omega$: (i) $Path(M_n)$ is 0-n-feasible but 1-$(n+1)$-unfeasible; (ii) $Path(N_n)$ is 1-$(n+1)$-feasible but 0-n-unfeasible.*

We end this section by an observation which is an easy part of our main result.

Proposition 8. *For a deterministic Mostowski automaton $\mathcal{A}$, $Path(L(\mathcal{A}))$ is recognized by a (deterministic) Mostowski tree automaton of the same index.*

Proof. If $\mathcal{A} = \langle \Sigma, Q, q_0, Tr, Acc \rangle$ is a deterministic automaton recognizing L, then the automaton for $Path(L)$ is $\mathcal{A}' = \langle \Sigma, Q, q_0, Tr', Acc \rangle$, with all the same components as $\mathcal{A}$ but $Tr' = \{\langle q, a, q', q' \rangle : \langle q, a, q' \rangle \in Tr\}$. $\square$

3 Flower Lemma

In this section we will show a connection between the Mostowski index of an ω-word language and the shape of a deterministic Mostowski automaton recognizing the language. Roughly speaking, we will show that in the graph of an automaton recognizing a "hard" language there must be a subgraph, called a flower, "responsible" for this hardness.

Definition 9. Let $\mathcal{A} = \langle \Sigma, Q, q_0, Tr, \Omega \rangle$ be a Mostowski automaton on words. The *graph of* $\mathcal{A}$ is the graph obtained by taking Q as the set of vertices and adding an edge from q to q' whenever $\langle q, a, q' \rangle \in Tr$, for some letter a.

A *path* in a graph is a sequence of vertices $v_1, \ldots, v_j$, such that, for every $i = 1, \ldots, j-1$ there is an edge from v_i to v_{i+1} in the graph. A *maximal strongly connected component* of a graph is a maximal subset of vertices of the graph, such that, for every two vertices v_1, v_2 in the subset, there is a path from v_1 to v_2 and from v_2 to v_1.

For an integer k, a *k-loop* in $\mathcal{A}$ is a path $v_1, \ldots, v_j$ in the graph of $\mathcal{A}$ such that, $v_1 = v_j$, $j > 1$ and $k = \max\{\Omega(v_i) : i = 1, \ldots, j\}$. Please observe that a k-loop must necessarily go through at least one edge.

Given integers m and n, a state $q \in Q$ is an *m-n-flower* in $\mathcal{A}$ if for every $k \in \{m, \ldots, n\}$ there is, in the graph of $\mathcal{A}$, a k-loop containing q.

First, we need an operation allowing us to scale unnecessary indices up.

Definition 10. For a Mostowski automaton $\mathcal{A}$ as above and an integer i we define the automaton $\mathcal{A}{\uparrow}^i$ obtained from $\mathcal{A}$ by *lifting the index i*. The automaton $\mathcal{A}{\uparrow}^i$ has all the same components as $\mathcal{A}$ except for the priority function $\Omega {\uparrow}^i$ defined in the following steps:

1. Take the set $Q_{\leq i} = \{q \in Q : \Omega(q) \leq i\}$. Let $G_{\leq i}$ denote the graph of $\mathcal{A}$ restricted to the states from $Q_{\leq i}$.
2. If S is a trivial maximal strongly connected component of the graph $G_{\leq i}$, i.e., a component consisting of one state without a self loop, then let $\Omega {\uparrow}^i(q) = i + 1$ for q being the unique state in S.
3. If S is a nontrivial maximal strongly connected component of $G_{\leq i}$ and S contains only states of priorities strictly smaller than i then let $\Omega {\uparrow}^i(q) = \Omega(q) + 2$ for every $q \in S$.
4. If a state q is in none of the above components then let $\Omega {\uparrow}^i(q) = \Omega(q)$.

Lemma 11. *For every deterministic Mostowski automaton $\mathcal{A}$ on words and every integer i: $L(\mathcal{A}) = L(\mathcal{A}{\uparrow}^i)$.*

Proof. Let $\mathcal{A} = \langle \Sigma, Q, q_0, Tr, \Omega : Q \to \{0, \ldots, n\} \rangle$ be a deterministic Mostowski automaton on words. Consider an infinite word $w \in \Sigma^\omega$. There is at most one run of $\mathcal{A}$ on w and if it exists it is also a run of $\mathcal{A}{\uparrow}^i$. This run determines the set of states Q_w that are met infinitely often on the run. Let $k = \max\{\Omega(q) : q \in Q_w\}$.

Suppose $k > i$. Every state of priority k in Q_w has still the priority k in $\mathcal{A}{\uparrow}^i$. Observe that if $\Omega {\uparrow}^i(q) \neq \Omega(q)$ then $\Omega {\uparrow}^i(q) \leq i + 1$. Hence $k = \max\{\Omega {\uparrow}^i(q) : q \in Q_w\}$ as well. So, in this case, $\mathcal{A}$ accepts w iff $\mathcal{A}{\uparrow}^i$ does.

If $k \leq i$ then all the states in Q_w belong to the same strongly connected component of the graph $G_{\leq i}$. According to the definition of $\mathcal{A}{\uparrow}^i$, all of the priorities of the states in Q_w are either increased by 2 or left unchanged. In both cases, $\mathcal{A}$ accepts w iff $\mathcal{A}{\uparrow}^i$ does. $\square$

Lemma 12. *Let $\mathcal{A}$ be a deterministic Mostowski automaton and let $i \in \mathbb{N}$. Let $\mathcal{B} = \mathcal{A}{\uparrow}^0{\uparrow}^1 \ldots {\uparrow}^i$. If a state q has priority m in $\mathcal{B}$ then q is a m-i-flower in $\mathcal{B}$.*

Proof. We prove the claim by induction on i. Let us denote by $\Omega_{\leq i}$ the priority function of the automaton $A\uparrow^0\uparrow^1\ldots\uparrow^i$.

First, consider the case $i = 0$. If $\Omega_{\leq 0}(q) > 0$ then the claim is trivial, so suppose $\Omega_{\leq 0}(q) = 0$.

Let, as in Definition 10, $G_{\leq 0}$ be the graph of A restricted to the states that have priority 0. There must be a 0-loop in $G_{\leq 0}$ containing q, otherwise the priority of q would be increased. Call this loop P. Clearly, the operation $\uparrow^0$ increases the priority of no state on this loop, and thus P remains a 0-loop in the automaton $A\uparrow^0$. Therefore, q is a 0-0-flower.

Now, let $i > 0$ and suppose the result holds for $i - 1$. Let $\Omega_{\leq i}(q) = m$. Again, if $m > i$, the claim is trivial, so we can assume $m \leq i$.

Let $G'_{\leq i}$ be the graph of A restricted to the states where priority is not greater than i in the automaton $A\uparrow^0\uparrow^1\ldots\uparrow^{i-1}$. Let $m' = \Omega_{\leq i-1}(q)$. By the induction hypothesis, q is an m'-$(i-1)$-flower in $A\uparrow^0\uparrow^1\ldots\uparrow^{i-1}$.

We consider two cases (clearly $m' \leq m$):

1. Suppose $m' = m$, i.e., the operation $\uparrow^i$ has not changed the priority of q. This is possible only if there existed in $G'_{\leq i}$ an i-loop containing q. Call this loop P. Clearly, the operation $\uparrow^i$ has not shifted the priority of any other node in the maximal strongly connected component of $G'_{\leq i}$ containing q. In particular, P continues to be an i-loop in $A\uparrow^0\uparrow^1\ldots\uparrow^i$. Furthermore every j-loop in $A\uparrow^0\uparrow^1\ldots\uparrow^{i-1}$ containing q, for $j < i$, remains a j-loop in $A\uparrow^0\uparrow^1\ldots\uparrow^i$. Since q is an m-$(i-1)$-flower in $A\uparrow^0\uparrow^1\ldots\uparrow^{i-1}$, it is an m-i-flower in $A\uparrow^0\uparrow^1\ldots\uparrow^i$.

2. $m' < m$. The only possibility is $m = m' + 2$. By the definition of lifting (Definition 10), this means that the operation $\uparrow^i$ has shifted by 2 the priorities of all the states in the maximal strongly connected component of $G'_{\leq i}$ containing q. Thus, every loop containing q that was a j-loop in $A\uparrow^0\uparrow^1\ldots\uparrow^{i-1}$, with $j < i$, has become a $(j+2)$-loop in $A\uparrow^0\uparrow^1\ldots\uparrow^i$. Since q is an m'-$(i-1)$-flower in $A\uparrow^0\uparrow^1\ldots\uparrow^{i-1}$, it is an m-$(i+1)$-flower in $A\uparrow^0\uparrow^1\ldots\uparrow^i$. Hence *a fortiori* an m-i-flower.

□

Definition 13. We say that a language $L \subseteq \Sigma^\omega$ *admits* an m-n-flower if there exists a deterministic Mostowski automaton A, such that $L = L(A)$ and A has an m-n-flower q for some q not useless state in A (i.e., q occurring in some accepting run of A).

Lemma 14 (Flower Lemma). *For every $n \in \mathbb{N}$ and $L \subseteq \Sigma^\omega$: (1) if L is 1-$(n + 1)$-unfeasible then L admits a $2i$-$(2i + n)$-flower, for some i; (2) if L is 0-n-unfeasible then L admits a $(2i + 1)$-$(2i + 1 + n)$-flower, for some i.*

Proof. We prove (1) and leave (2) to the reader.

Let $L = L(A)$ for some deterministic Mostowski automaton A. We can assume without a loss of generality that A has no useless states. Clearly, for arbitrary k, the operation $\uparrow^k$ preserves this property.

Let M be the maximal value of the priority function of $\mathcal{A}$. Choose i such that $2i + n > M$. Consider the automaton $\mathcal{B} = \mathcal{A}\!\uparrow^0\!\uparrow^1 \ldots \uparrow^{2i+n}$. By Lemma 11, $L(\mathcal{B}) = L$. If we can find a state q whose priority in $\mathcal{B}$ is less or equal to $2i$ then we are done since, by Lemma 12, q is a $2i$-$(2i+n)$-flower in $\mathcal{B}$, and by the remark above q is not useless.

Suppose on the contrary that the minimal value of the priorities of all states in $\mathcal{B}$ is $j > 2i$. It should be clear that the maximal priority of a state in $\mathcal{B}$ cannot exceed $2i + n + 1$. (The operation $\uparrow^k$ can change the priority of a state to at most $k + 1$.) By adding, if necessary, some dummy states we obtain that L is $(2i + 1)$-$(2i + 1 + n)$-feasible. By scaling the priorities down we conclude that L is 1-$(n + 1)$-feasible, contradicting the assumption. □

For a given automaton $\mathcal{A}$ and a given integer i one can calculate $\mathcal{A}\!\uparrow^i$ in the time proportional to the number of transitions in $\mathcal{A}$. Hence one can compute $\mathcal{A}\!\uparrow^1 \ldots \uparrow^n$ in time $\mathcal{O}(n|\mathcal{A}|)$; where $|\mathcal{A}|$ denotes the number of transitions in $\mathcal{A}$. As we can easily limit the range of the Ω function to be at most twice as big as the number of states in $\mathcal{A}$ we get:

Corollary 15. *The problem of establishing the index of the language accepted by an automaton with a Mostowski condition can be solved in time $\mathcal{O}(|\mathcal{A}|^2)$.*

4 The main result

Proposition 16. *1. If a language $L \subseteq \Sigma^\omega$ admits a $2m$-$(2m + n)$-flower for some m, then $Path(L)$ is 1-$(n + 1)$-unfeasible.*
2. If a language $L \subseteq \Sigma^\omega$ admits a $(2m + 1)$-$(2m + 1 + n)$-flower for some m, then $Path(L)$ is 0-n-unfeasible.

Proof. We shall prove 1, the case of 2 is analogous.

Let $\mathcal{A}$ be a deterministic Mostowski automaton recognizing L and let q be a $2m$-$(2m + n)$-flower in $\mathcal{A}$, where q is not useless.

Suppose on the contrary that $Path(L)$ is 1-$(n + 1)$-feasible, and let:

$$\mathcal{C} = \langle \Sigma, Q^c, q_0^c, Tr^c, \Omega^c : Q^c \to \{1, \ldots, n + 1\}\rangle$$

be a nondeterministic Mostowski automaton of index $(1, n + 1)$ recognizing $Path(L)$. Without a loss of generality we may assume that there are no useless states in $\mathcal{C}$.

From $\mathcal{C}$ we shall construct a tree automaton $\mathcal{D}$ of the same index, but over the alphabet $\{0, \ldots, n\}$, that will recognize $Path(M_n)$. This, however, will contradict Theorem 7 and will prove that an automaton $\mathcal{C}$ cannot exist.

To this end, we shall use our flower q in $\mathcal{A}$. For every $i = 0, \ldots, n$ fix a word w_i, such that, there is a $(2m + i)$-loop from q to q in $\mathcal{A}$ labelled by w_i. Clearly, this is possible by the definition of a flower.

Apart from $w_0, \ldots, w_n$, fix also a finite word v that is a labeling of a path in $\mathcal{A}$ from the initial state to q. For notational convenience we assume that the length of v as well as all w_i ($i = 1, \ldots, n$) is k. The general case, when the lengths of v and w_i are different, requires more complex indexing.

Before defining formally an automaton for $\text{Path}(M_n)$, we explain its idea. It is based on a transformation of trees over the alphabet $\{0, \ldots, n\}$ into trees over the alphabet Σ. This transformation essentially replaces a node labelled by $i \in \{0, \ldots, n\}$ by a finite path labelled by w_i. Then, when examining a tree t (over the alphabet $\{0, \ldots, n\}$), our automaton $\mathcal{D}$ will mimic the work of the automaton $\mathcal{C}$ on the transformed tree. This will be done in such a way that a computation of $\mathcal{D}$ along a path labelled by $i_0 i_1 i_2 \ldots$ ($i_\ell \in \{0, \ldots, n\}$) will essentially simulate a computation of $\mathcal{C}$ along a path labelled $v w_{i_0} w_{i_1} w_{i_2} \ldots$

Let a function $h : \{1, 2\}^* \to \{1, 2\}^*$ be defined by:

$$h(d_1 \ldots d_i) = 1^{k-1} d_1 1^{k-1} \ldots d_i 1^{k-1}$$

Let us denote by R the smallest tree containing the range of h.

We define an operation transforming a labelled tree $\tau : \{1, 2\}^* \to \{0, \ldots, n\}$ into a labelled tree $\tau^\sharp : R \to \Sigma$. For arbitrary $u \in \{1, 2\}^*$, $d \in \{1, 2\}$ and $i \in \{0, \ldots, k-1\}$ we let:

$$\tau^\sharp(1^i) = \text{the } (i+1)\text{th letter of } v$$
$$\tau^\sharp(h(u)d1^i) = \text{the } (i+1)\text{th letter of } w_{\tau(u)}$$

The key property of this transformation is the following:

for every path $P = d_1 d_2 \ldots$ in $\text{dom}(\tau)$ ($d_\ell \in \{1, 2\}$), there is a path $P^\sharp = 1^{k-1} d_1 1^{k-1} d_2 1^{k-1} \ldots$ in $\text{dom}(\tau^\sharp)$ and this path is labelled by $\tau^\sharp(P^\sharp) = v w_{\tau(\varepsilon)} w_{\tau(d_1)} w_{\tau(d_1 d_2)} \cdots$

Now, it follows form the choice of the words $v, w_0, \ldots, w_n$, that an infinite word of the form $v w_{i_1} w_{i_2} \ldots$ is in $L = L(\mathcal{A})$ iff $\limsup_{\ell \to \infty} i_\ell$ is even. In particular, we note the following.

Observation 17. For every infinite path P in the tree τ we have: $\tau^\sharp(P^\sharp) \in L$ iff $\tau(P) \in M_n$.

Now, we define the automaton:

$$\mathcal{D} = \langle \Sigma^d, Q^d = Q^c \times \{1, \ldots, n+1\} \cup \{q_s\}, q_s, Tr^d, \Omega^d : Q^d \to \{1, \ldots, n+1\}\rangle$$

where $\Sigma^d = \{0, \ldots, n\}$ and the meanings of the other components are defined below.

The state q_s is the initial state and it will be used only in the starting move of $\mathcal{D}$. For the function Ω^d, we let $\Omega^d(q, i) = i$. We let $\Omega^d(q_s) = 1$ but this value does not really matter.

Before defining the transition function let us introduce an auxiliary notion. We say that a pair of states $q_l, q_r \in Q^c$ is i-*reachable* from a state $q_1 \in Q^c$ on a word $w = a_1 \cdots a_k \in \Sigma^*$ if there is a sequence of transitions from Tr^c:

$$\langle q_1, a_1, q_2, q_2'\rangle, \langle q_2, a_2, q_3, q_3'\rangle, \ldots, \langle q_{k-1}, a_{k-1}, q_k, q_k'\rangle, \langle q_k, a_k, q_l, q_r\rangle$$

and $i = \max(\Omega(q_1), \ldots, \Omega(q_k))$.

Now, for every $a \in \{1, \ldots, n\}$, we let $\langle q_s, a, (q_l, i), (q_r, i) \rangle \in Tr^d$ if (q_l, q_r) is i-reachable from the initial state q_0^C of C on the word v. Also, we put a tuple $\langle (q, i), m, (q_l, j), (q_r, j) \rangle \in Tr^d$ if (q_l, q_r) is j-reachable from q on the word w_m.

To prove the claim it is enough to show that $L(\mathcal{D}) = \mathrm{Path}(M_n)$. With the help of the operation $\sharp$ introduced above, we can state the correspondence between C and $\mathcal{D}$.

Observation 18. For every tree $\tau : \{1, 2\}^* \to \{0, \ldots, n\}$ we have: $\tau \in L(\mathcal{D})$ iff $\tau^\sharp$ can be extended to a full binary tree that is accepted by C.

Recall that C accepts a tree iff the labeling of every path of this tree is accepted by $\mathcal{A}$. So, if C accepts an extension of $\tau^\sharp$ then in particular $\tau^\sharp(P^\sharp) \in L(\mathcal{A}) = L$, for each path P in $\mathrm{dom}(\tau)$. By Observation 17, we conclude that $\tau \in \mathrm{Path}(M_n)$. We have proved $L(\mathcal{D}) \subseteq \mathrm{Path}(M_n)$.

For the converse inclusion, let $\tau \in \mathrm{Path}(M_n)$. Then clearly $\tau^\sharp$ can be extended to a tree in $\mathrm{Path}(L)$. The conclusion now follows from the hypothesis that $L(C) = \mathrm{Path}(L)$. $\square$

Theorem 19. *For every $L \subseteq \Sigma^\omega$ and $m, n \in \mathbb{N}$: L is m-n-feasible if and only if $\mathrm{Path}(L)$ is m-n-feasible.*

Proof. The implication "only if" follows from Proposition 8.

Suppose L is m-n-unfeasible. The case of $m > n$ is trivial, so we can assume $m \leq n$.

Suppose m is even. Then clearly L is also 0-$(n-m)$-unfeasible since otherwise we could scale up the priorities in the hypothetical automaton by m. By the Flower Lemma, L admits a $(2i+1)$-$(2i+1+n-m)$-flower, for some i, and then by Proposition 16, $\mathrm{Path}(L)$ is 0-$(n-m)$-unfeasible. Hence clearly it is also m-n-unfeasible, since otherwise we could scale down the hypothetical automaton by m.

If m is odd, we present it as $m = m' + 1$, and the rest of the argument is similar. $\square$

From the connections between indices of Mostowski and Rabin automata (cf. [11]) we obtain:

Corollary 20. *For every $L \subseteq \Sigma^\omega$ and $n \in \mathbb{N}$, L can be recognized by a deterministic Rabin automaton of index n if and only if $\mathrm{Path}(L)$ can be recognized by a nondeterministic Rabin tree automaton of index n.*

Finally let us remark on the problem of constructing a deterministic automaton recognising L from a nondeterministic tree automaton recognizing $\mathrm{Path}(L)$. This can be done by constructing a nondeterministic automaton recognizing L, determinizing it and then applying our index lifting procedure from the flower lemma. The so–obtained deterministic automaton would have the least possible index.

References

1. E. A. Emerson, C. Jutla, and A. Sistla. On model-checking for fragments of μ-calculus. In *CAV'93*, volume 697 of *LNCS*, pages 385–396, 1993.
2. E. A. Emerson and C. S. Jutla. The complexity of tree automata and logics of programs. In *29th FOCS*, 1988.
3. E. A. Emerson and C. S. Jutla. Tree automata, mu-calculus and determinacy. In *Proc. FOCS 91*, 1991.
4. D. Janin and I. Walukiewicz. Automata for the μ-calculus and related results. In *MFCS '95*, volume 969 of *LNCS*, pages 552–562, 1995.
5. S. Krishnan, A. Puri, and R. Brayton. Srtuctural complexity of ω-automata. In *STACS'95*, volume 900 of *LNCS*, 1995.
6. O. Kupferman, S. Safra, and M. Vardi. Relating word and tree automata. In *11th IEEE Symp. on Logic in Comput. Sci.*, pages 322–332, 1996.
7. A. W. Mostowski. Regular expressions for infinite trees and a standard form of automata. In A. Skowron, editor, *Fifth Symposium on Computation Theory*, volume 208 of *LNCS*, pages 157–168, 1984.
8. A. W. Mostowski. Hierarchies of weak automata and week monadic formulas. *Theoretical Computer Science*, 83:323–335, 1991.
9. D. Niwiński. On fixed-point clones. In *Proc. 13th ICALP*, volume 226 of *LNCS*, pages 464–473, 1986.
10. D. Niwiński. Fixed points vs. infinite generation. In *LICS '88*, pages 402–409, 1988.
11. D. Niwiński. Fixed point characterization of infinite behaviour of finite state systems. *Theoretical Computer Science*, 1998. to appear.
12. M. Rabin. Weakly definable relations and special automata. In Y.Bar-Hillel, editor, *Mathematical Logic in Foundations of Set Theory*, pages 1–23. 1970.
13. W. Thomas. Automata on infinite objects. In J. van Leeuven, editor, *Handbook of Theoretical Computer Science Vol.B*, pages 133–192. Elsevier, 1990.
14. W. Thomas. Languages, automata, and logic. In G. Rozenberg and A. Salomaa, editors, *Handbook of Formal Languages*, volume 3. Springer-Verlag, 1997.
15. K. Wagner. Eine topologische Charakterisierung einiger Klassen regulärer Folgenmengen. *J. Inf. Process. Cybern. EIK*, 13:473–487, 1977.
16. K. Wagner. On ω–regular sets. *Information and Control*, 43:123–177, 1979.
17. T. Wilke and H. Yoo. Computing the Rabin index of a regular language of infinite words. *Information and Computation*, 130(1):61–70, 1996.

Languages Defined With Modular Counting Quantifiers
(extended abstract)

Howard Straubing

Computer Science Department
Boston College
Chestnut Hill, MA 02167
Tel.: (617)-552-3977
e-mail:straubin@cs.bc.edu

Topics: Automata and formal languages, logic in computer science (finite model theory), computational complexity (circuit complexity)

1 Introduction

1.1 Statement of the principal result

We prove that a regular language defined by a boolean combination of generalized Σ_1-sentences built from modular counting quantifiers is defined by a boolean combination of generalized Σ_1-sentences in which only regular numerical predicates appear. We will give the precise definitions of all of these terms shortly. This is a special case of an outstanding conjecture in circuit complexity (that the class ACC is properly contained in NC^1) about the power of constant-depth circuits built with modular gates. The techniques used in the present paper may shed some light on the general case.

We use formulas of generalized first-order logic to describe properties of strings over a finite alphabet A. The variables of the formula range over the positions $1, \ldots, |w|$ in the string $w \in A^*$. There are two kinds of atomic formulas: If $a \in A$, then the atomic formula $Q_a x$ is interpreted to mean 'the letter in position x of w is a'. The other kind of atomic formula we call a *numerical predicate*. This is a relation on the positions in the string, such as $x < y$, or 'x is prime' that depends only on $|w|$ (the length of w) and the numerical values of the positions, and not on the letters that appear in those positions. A *regular numerical predicate* is a numerical predicate defined by a first-order formula in which the atomic formulas are all of the form $x < y$ and $x \equiv 0 \pmod{q}$ for some $q > 1$. This terminology is used to suggest an analogy with regular languages, since, in a sense that can be made precise, these are the numerical predicates that can be defined by finite automata.

We use a special kind of quantifier in our sentences. Let $s \geq 0, p > 0$. We say that two integers $m, n \geq 0$ are congruent $\pmod{(s,p)}$ if either

$$m = n < s,$$

or

$$m, n \geq s, \text{and } m \equiv n \pmod{p}.$$

We call s the *stem*, and p the *period*, of the congruence. Let $i \in \mathbf{N}$. We interpret the formula

$$\exists^{i,s,p} x \phi(x)$$

to mean 'the number of x such that $\phi(x)$ is true is congruent to i modulo (s,p)'. We can also apply these generalized quantifiers to k-tuples of individuals. Thus

$$\exists^{i,s,p} (x_1, \ldots, x_k) \phi(x_1, \ldots, x_k)$$

means, 'the number of k-tuples of individuals $(x_1, \ldots, x_k)$ such that $\phi(x_1, \ldots, x_k)$ is true is congruent to i modulo (s,p)'. Sometimes we will abbreviate a k-tuple $(x_1, \ldots, x_k)$ by a single bold-faced letter x. If ϕ is quantifier-free then we call such a formula a *generalized Σ_1-formula*. Observe that if $s = p = 1$ then this is an ordinary existential quantifier and we get the usual Σ_1-formulas. If $s = 0$ and $p > 1$ then these are the modular quantifiers studied in Straubing, Thérien and Thomas [17].

A sentence in this logic thus defines a language in A^*, consisting of all the strings that satisfy the sentence. Our principal result says, for generalized Σ_1-sentences, if the language defined is a regular language, then only regular numerical predicates are needed:

Theorem 1. *Let $L \subseteq A^*$ be a regular language defined by a boolean combination of generalized Σ_1-sentences. Then L is defined by a boolean combination of generalized Σ_1-sentences in which all the numerical predicates are regular numerical predicates. Further, all the quantifiers in the new sentence have stem s and period p, where s is the maximum of the stems, and p the least common multiple of the periods, in the original sentence.*

As consequences, we can prove that a boolean combination of generalized Σ_1-sentences with stem 0 cannot define the language $1^* \subseteq \{0,1\}^*$, and that a boolean combination of generalized Σ_1-sentences with period q cannot define the language consisting of all strings in which the number of 1's is divisible by p, if p is a prime that does not divide q.

We conjecture that this theorem remains true for sentences of any quantifier complexity. A proof of this conjecture would resolve an outstanding question in circuit complexity: whether the class ACC is properly contained in NC^1. We will discuss these connections to circuit complexity in the next subsection.

The techniques used to prove Theorem 1, a combination of semigroup theory and Ramsey theory, draw to some extent on ideas in Barrington, Straubing and Thérien [3] and Barrington and Straubing [5]. We believe these techniques may be useful in settling the general problem as well.

1.2 Background and significance of the principal result

Büchi [6] initiated the study of the description of finite automata in formal logic, by characterizing the regular languages as those languages definable by monadic second-order sentences with successor. McNaughton and Papert [11] showed that the languages defined by first-order sentences with numerical predicates of the form $x < y$ are precisely those whose syntactic monoids contain no nontrivial groups. Straubing, Thérien and Thomas [17] introduced modular quantifiers, a version of the generalized quantifiers considered in the present paper, and characterized the languages (all of them regular) definable by sentences using these quantifiers and numerical predicates of the form $x < y$. Straubing [16] surveys these and many related results.

As long as we restrict the numerical predicates in these sentences to be regular numerical predicates, the languages the sentences define are regular. If we place no restriction on the numerical predicates that can occur, then the classes of languages defined by various kinds of sentences are precisely those belonging to various nonuniform circuit complexity classes. Thus Immerman [10] and Gurevich and Lewis [9] show that a language belongs the the nonuniform circuit complexity class AC^0 (consisting of languages recognized by polynomial-size constant-depth circuit families with unbounded fan-in AND and OR gates) if and only if it is definable by a first-order sentence.

One of the outstanding acheivements of circuit complexity theory is the proof that AC^0 does not contain the language consisting of all bit strings with an even number of 1's (Furst, Saxe and Sipser [8] and Ajtai [1]). Barrington, *et. al.* [2] show that this theorem has a surprisingly neat model-theoretic formulation: A regular language is first-order definable if and only if it is definable by a first-order sentence in which all the numerical predicates are regular numerical predicates. Further, one of the outstanding unsolved problems in this domain concerns the power of constant-depth circuits with modular gates. Razborov[14] and Smolensky[15] introduced the study of families of circuits containing modular gates of a fixed modulus as well as AND and OR gates. The class of languages recognized by polynomial-size constant-depth circuit families of this kind is called ACC. It is conjectured that an ACC circuit family built with MOD_q gates cannot recognize the language consisting of all bit strings in which the number of $1's$ is divisible by p, where p is a prime that does not divide q, and this conjecture is known to hold when q is prime (This was shown in [14] for $q = 2$ and in [15] for arbitrary prime q.) Once again, the conjecture has a model-theoretic interpretation. Barrington, *et. al.*, in [2] show that the conjecture is equivalent to: A regular language is definable by a sentence with generalized quantifiers if and only if it is definable by such a sentence in which only regular numerical predicates are used. These automaton- and model-theoretic formulations of facts and conjectures from circuit complexity are discussed at length in Straubing[16] Our theorem establishes this conjecture for the case of boolean combinations of generalized Σ_1-sentences. There is a simple model-theoretic proof for ordinary Σ_1-sentences (Straubing [16]). Maciel, Péladeau and Thérien [12] extend this to boolean combinations of Σ_1-sentences, using highly nontrivial techniques.

The result in the present paper generalizes theirs, and employs a quite different method.

1.3 Plan of the paper

We will begin by formulating Theorem 1 in a very different form, as a statement concerning congruences on A^* and the behavior of certain kinds of 'programs' over finite commutative monoids. We will then prove this new theorem by a Ramsey-theoretic argument and show how Theorem 1 follows. We discuss the relevance of our methods to the general problem of proving $ACC \neq NC^1$ in a concluding section.

2 Congruences and Programs

2.1 A congruence on A^+.

A congruence on an algebraic structure is an equivalence relation that is compatible with the operations on the structure. The set N of nonnegative integers with the operations of addition and multiplication forms a commutative semiring, and the relation of congruence modulo (s, p) defined above is a congruence on this semiring. In fact, all congruences of finite index on the semiring N have this form. We write $N_{s,p}$ to denote the quotient of N by this congruence. $N_{s,p}$ is itself a commutative semiring, and any commutative monoid M (a monoid is an algebraic structure with an associative operation and an identity element for the operation), with its operation written additively, that satisfies the identity

$$(s + p) \cdot x = s \cdot x$$

for all $x \in M$ is a semimodule over this semiring.

As usual, if A is a finite alphabet, then we denote by A^* the set of all strings over A, and by A^+ the set of all nonempty strings over A. A^* is a monoid, and A^+ a semigroup, with concatenation of strings as the operation.

Let $w \in A^*$, and $v = a_1 \cdots a_k$, with each $a_i \in A$. We say v is a *subword* of w if

$$w \in A^* a_1 A^* \cdots a_k A^*.$$

(This somewhat unusual terminology is from Eilenberg [7]. When $w \in A^* v A^*$, we say v is a *factor* of w.) An occurrence of v as a subword of w is a factorization

$$w = w_0 a_1 w_1 \cdots a_k w_k,$$

with each $w_i \in A^*$. The *signature* of this occurrence is the bit string

$$\sigma = b_0 1 b_1 \cdots 1 b_k,$$

where b_i is the empty string whenever w_i is the empty string, and $b_i = 0$ otherwise.

Example. Consider the boldfaced occurrence of the subword *aba* in the word *aaabbba*. The signature is 01101. In *aaabbaaaaa*, the signature is 0101010.

Let $w \in A^+$, $v \in A^+$, $|v| \leq k$, and let σ be a bit string with $|v|$ 1's and without two consecutive occurrences of 0. We denote by $c(w, v, \sigma)$ the number of occurrences of v as a subword of w with signature σ.

We define an equivalence relation $\theta^k_{s,p}$ on A^+ as follows: $w_1 \theta^k_{s,p} w_2$ if for all v with $|v| \leq k$, and all signatures σ,

$$c(w_1, v, \sigma) \equiv c(w_2, v, \sigma) \quad (\text{mod } (s, p)).$$

This is obviously an equivalence relation on A^+ of finite index. It is easy to see that this is a congruence as well: Let $a \in A$, $w \in A^*$. If v occurs as a subword of aw with signature σ, and $\sigma = 1\tau$, then $v = av'$, and $c(aw, v, \sigma) = c(w, v', \tau)$. If, on the other hand, $\sigma = 01\tau$, then $c(aw, v, \sigma) = c(w, v, 1\tau) + c(w, v, \sigma)$. In either case, we have $w_1 \theta^k_{s,p} w_2$ implies $aw_1 \theta^k_{s,p} aw_2$. Similarly, we conclude $w_1 a \theta^k_{s,p} w_2 a$. Thus $w_1 \theta^k_{s,p} w_2$ and $v_1 \theta^k_{s,p} v_2$ implies $w_1 v_1 \theta^k_{s,p} w_2 v_2$, so $\theta^k_{s,p}$ is a congruence.

We will use the symbol $\theta^k_{s,p}$ to denote not only the congruence, but the homomorphism from A^+ onto the quotient semigroup by this congruence.

2.2 *k*-programs

Let $w \in A^n$, $w = a_1 \cdots a_n$. Let $I = (i_1, \ldots, i_k) \in \{1, \ldots, n\}^k$. We denote by $w(I)$ the *k*-tuple $(a_{i_1}, \ldots, a_{i_k}) \in A^k$. A *k*-program over a finite monoid M associates to each $I \in \{1, \ldots, n\}^k$ a map $f_I : A^k \to M$. The *value* of the program on $w \in A^n$ is

$$\prod f_I(w(I)) \in M,$$

where the *k*-tuples are taken in some fixed but arbitrary order. These are closely related to the programs over finite monoids studied by Barrington and Thérien [5]. In our application, M is commutative, so the ordering doesn't matter. The program thus associates to each word of length n an element of M. Ordinarily we consider a *family* of programs, one for each length $n \geq 1$.

We can also view families of programs as language acceptors. Let $X \subseteq M$. A family of *k*-programs, with X as the accepting set, accepts a string $w \in A^+$ if and only if the value of the $|w|^{th}$ program of the family on w is in X. The language recognized by the program family is the set of all accepted strings.

2.3 A reformulation of the principal theorem

Roughly speaking, this is what our algebraic formulation of the principal theorem says: Suppose we have a family of *k*-programs over a finite monoid M with which we can perform multiplication in a finite semigroup S. If M is commutative, then S is *almost* a commutative semigroup.

Now for the detailed statement: Let $t \geq 0, q > 0$. $\lambda_{t,q}$ maps each $w \in A^*$ to $|w| \bmod (t, q)$. That is, $\lambda_{t,q}$ is a homomorphism onto the additive monoid of $N_{t,q}$. Let M be a finite commutative monoid. We will write the operation in M

additively. By the finiteness of M there exist $s \geq 0, p > 0$, so that $(s+p) \cdot x = s \cdot x$ for all $x \in M$. We will always suppose that M is nontrivial, so in our applications either $s > 0$ or $p > 1$. As noted above, this makes M into a semimodule over $N_{t,q}$.

Let $\phi : A^+ \to S$ be a homomorphism onto a finite semigroup. We will make a technical assumption about the structure of S for now: that $S = S^2 = \{st : s, t \in S\}$, and also that $\phi(A) = S$.

Suppose we have a family of k-programs over M that *simulates* multiplication in S in the following sense: Whenever $|w_1| = |w_2|$, and the program for words of length $|w_1|$ has the same value on w_1 and w_2, then $\phi(w_1) = \phi(w_2)$. That is, $\phi(w)$ is determined by the value of the program on w.

Let $T_r = (\phi \times \theta_{s,p}^k)(A^r)$. (This is a subset of $S \times A^+/\theta_{s,p}^k$.) There exist $t \geq 0, q > 0$ such that $T_t = T_{t+q}$, because of finiteness of $S \times A^+/\theta_{s,p}^k$. It follows that whenever $r_1 \equiv r_2 \pmod{(t,q)}$, then $T_{r_1} = T_{r_2}$.

Theorem 2. ϕ *is refined by* $\theta_{s,p}^k \times \lambda_{t,q}$.

2.4 Step 1 of the proof: Setting the stage.

To prove the theorem, we need to show that if

$$(\theta_{s,p}^k \times \lambda_{t,q})(v) = (\theta_{s,p}^k \times \lambda_{t,q})(w),$$

then

$$\phi(v) = \phi(w).$$

Since $|w| \equiv |v| \pmod{(t,q)}$, we have $T_{|v|} = T_{|w|}$, and thus there exists v' such that $|v| = |v'|$, and

$$(\phi \times \theta_{s,p}^k)(v') = (\phi \times \theta_{s,p}^k)(w).$$

In particular, $v\theta_{s,p}^k v'$. We need to show $\phi(v) = \phi(v')$, from which it will follow that $\phi(v) = \phi(w)$.

Let us write

$$v = a_1 \cdots a_m,$$

and

$$v' = a_1' \cdots a_m'.$$

Our hypothesis about S implies that for each a_i, a_i' and each $k > 0$, there exist strings $\alpha_{i,k}, \alpha_{i,k}'$ of length k such that $\phi(\alpha_{i,k}) = \phi(a_i)$, and $\phi(\alpha_{i,k}') = \phi(a_i')$. We will show how to construct strings

$$V = \alpha_{1,k_1} \cdots \alpha_{m,k_m},$$

and

$$V' = \alpha_{1,k_1}' \cdots \alpha_{m,k_m}',$$

such that the program for words of length $|V| = |V'|$ has the same value on V and V'. This implies $\phi(V) = \phi(V')$, by hypothesis, and thus $\phi(v) = \phi(v')$.

2.5   Step 2: Families of intervals, signatures and colorings.

Let $m = |v| = |v'|$ as in the preceding section. We will choose n much larger than m (exactly how large will be specified later). Subsets of $\{1,\ldots,n\}$ consisting of consecutive integers will be called *intervals*. We will use the usual notation of open and half-open intervals to denote these; for example $[a, b) = \{x : a \le x < b\}$, but we mean by this the set of *integers* x such that $a \le x < b$.

A subset $\{p_1 < \cdots < p_r\}$ of $\{2,\ldots,n\}$ partitions $\{1,\ldots,n\}$ into a set of $r+1$ disjoint nonempty intervals

$$[p_0 = 1, p_1), [p_1, p_2), \ldots, [p_r, p_{r+1} = n+1).$$

We denote a subset of this set of intervals by its *signature;* this is a string of 0's and 1's of length $r+1$ that has 1 in the ith position if and only if $[p_{i-1}, p_i)$ is in the set. We are only interested in subsets in which every one of the elements $p_1,\ldots,p_r$ is included as an endpoint of one of the intervals of the subset. This is the same as requiring that the signature not contain two consecutive 0's, exactly like the signatures of occurrences of subwords introduced in 2.1. Thus each set $U \subseteq \{2,\ldots,n\}$ of cardinality r gives rise to a map $\mathcal{I}_U$ that sends bit strings of length $r+1$ without two consecutive 0's to families of disjoint intervals.

Example. Let $U = \{4, 7\}, n = 8$. The complete set of intervals is $\{[1,4), [4,7), [7,9)\}$. $\mathcal{I}_U$ maps 010 to $\{[4,7)\}$, and 110 to $\{[1,4),[4,7)\}$.

$\mathcal{I}_U$ is a bijection from the set of strings of length $|U|+1$ without consecutive 0's, into the set of families of intervals whose set of endpoints (excluding 1 and $n+1$) is U. In particular, given a family $\mathcal{F}$ of intervals and the value of n, we can deduce the signature $\sigma(\mathcal{F})$ directly.

A k-tuple $I \in \{1,\ldots,n\}^k$ is *compatible* with $\mathcal{F}$ if each component of I is in $\cup\mathcal{F}$, and each interval in $\mathcal{F}$ contains a component of I. Let z be the number of intervals of $\mathcal{F}$. We define a map $\chi_{\mathcal{F}}$ from A^z into M, or from A^{z-1} into M (which of the two possible domains we choose depends on $\mathcal{F}$) as follows.

Case 1. Let t be the smallest endpoint in $\mathcal{F}$, and suppose $\mathcal{F}$ does not contain the interval $[1, t)$. (That is, $\sigma(\mathcal{F})$ begins with 0). Let $(b_1,\ldots,b_z) \in A^z$. For each $i = 1,\ldots,z, k > 0$, there exists a string $\beta_{i,k}$ of length k such that $\phi(b_i) = \phi(\beta_{i,k})$. We build a word w of length n by filling the positions of the the the ith interval in $\mathcal{F}$ by $\beta_{i,k}$, where k is the size of the interval. of the ith interval. The other positions of w can be filled in arbitrarily. We define

$$\chi_{\mathcal{F}}(b_1,\ldots,b_z) = \sum f_I(w(I)),$$

where the f_I are the k-program maps for inputs of length n, and where the sum ranges over all I compatible with $\mathcal{F}$.

Case 2. Again, let t be the smallest endpoint of $\mathcal{F}$. Suppose $\mathcal{F}$ contains $[1, t)$ but not $[t, p_2)$, where p_2 is the second-smallest endpoint. That is, $\sigma(\mathcal{F})$ begins with 10. Let $(b_1,\ldots,b_z) \in A^z$. We define w as in Case 1, except we place $\beta_{1,t-1}$ in the first $t-1$ positions of w, instead of filling the positions in $[t, p_2)$. We then define $\chi_{\mathcal{F}}(b_1,\ldots,b_k)$ by the same equation as above, again taking the sum over all k-tuples compatible with $\mathcal{F}$.

Case 3. The remaining case is when $\mathcal{F}$ contains both $[1,t)$ and $[t,p_2)$, so that $\sigma(\mathcal{F})$ begins with 11. In this case the domain is A^{z-1}. Given $(b_1, \cdots, b_{z-1})$ we build w by placing β_{1,p_2} in the first p_2 positions of w, and fill the remaining positions as in Case 1. We then define $\chi_{\mathcal{F}}(b_1, \ldots, b_{z-1})$ by the same equation as in the preceding two cases.

Thus each set $U \subseteq \{2, \ldots, n\}$ gives rise to a map Δ_U, which maps each binary string σ of length $|U| + 1$ without consecutive 0's to the map $\chi_{\mathcal{F}}$, where $\mathcal{F} = \mathcal{I}_U(\sigma)$. Δ_U is the *color* of U.

2.6 Step 3: Application of Ramsey's Theorem

By Ramsey's theorem we can choose n so large that there is an m-element subset J of $\{2, \ldots, n\}$ such that for each $j \leq k$, all the j-element subsets of J have the same color Δ_j.

$J = \{i_1 < \cdots < i_m\}$ partitions $\{1, \ldots, n\}$ into the sequence of intervals

$$[1, i_1), \ldots, [i_m, n+1).$$

We now construct the words V and V' described in Subsection 2.4. V is formed by placing α_{1,i_2-1} in the first $i_2 - 1$ positions, and, for each $p > 1$, $\alpha_{p,q}$ in the positions of $[i_p, i_{p+1})$, where $q = i_{p+1} = i_p$. We build V' from v' in the analogous fashion. Every k-tuple in $\{1, \ldots, n\}^k$ is compatible with exactly one nonempty subfamily of the above family of intervals. Given a family $\mathcal{F}$ of intervals, let $\{i_{j_1} < \cdots < i_{j_r}\}$ be its set of left endpoints; in the case where $\mathcal{F}$ contains the interval $[1, i_1)$ we take $i_{j_1} = i_1$. Let $I_{\mathcal{F}} = (j_1, \ldots, j_r)$. It follows that the value of the program on V is

$$\sum_{\mathcal{F}} \chi_{\mathcal{F}}(v(I_{\mathcal{F}})),$$

the sum ranging over all subfamilies $\mathcal{F}$ of the $m + 1$ intervals above, such that the number of intervals in $\mathcal{F}$ is no greater than k. We get the same expression, with v' replacing v, for the value of the program on V'. To complete the proof of Theorem 2, we use the monchromaticity together with $v\theta_{s,p}^k v'$ to show that these two sums are equal. The details are given in the full paper.

3 Proof of Theorem 1

To prove Theorem 1, we will first show that if a language is defined by a boolean combination of generalized Σ_1-sentences then it is recognized by a family of k-programs over a finite commutative monoid. We then show that if a regular language L is recognized by a family of k-programs over a monoid M, then we can simulate multiplication in the syntactic monoid of L by a family of programs over a direct product of copies of M. We then apply Theorem 2 to a subsemigroup of the syntactic monoid of M, giving a congruence that refines this semigroup. This will enable us to express the classes of the syntactic congruence by boolean combinations of generalized Σ_1-sentences with regular numerical predicates.

We need the following two lemmas, whose proofs are given in the full paper.

Lemma 3. *If L is defined by a boolean combination of generalized Σ_1-sentences of modulus (s, p), then L is recognized by a family of k-programs over a finite commutative monoid satisfying the identity $(s + p) \cdot x = s \cdot x$.*

The converse of Lemma 3 is also true, but we do not need it here.

Lemma 4. *Let $L \subseteq A^*$, $v \in A^*$. If L is recognized by a family of k-programs over a finite commutative monoid M, then so are the languages $v^{-1}L = \{w : vw \in L\}$, and $Lv^{-1} = \{w : wv \in L\}$.*

We now proceed to the proof of Theorem 1. Let $L \subseteq A^*$ be a regular language defined by a boolean combination of generalized Σ_1-sentences of modulus (s, p). By Lemma 3, L is recognized by a family of k-programs over a finite commutative monoid M that satisfies an identity of the form $(s + p) \cdot x = s \cdot x$. By Lemma 4, each of the languages $u^{-1}Lv^{-1}$ is recognized by a family of k-programs over M. Let $\mu : A^* \to M(L)$ be the syntactic morphism of L. (This is the projection onto the *syntactic monoid* of L, the quotient of A^* by the coarsest congruence for which L is a union of classes. See Eilenberg [7] or Pin [13] for an exposition of the theory.) Each of the sets

$$\{w \in A^* : \mu(w) = m\},$$

where $m \in M(L)$, is a finite boolean combination of languages of the form $u^{-1}Lv^{-1}$, and so is recognized by a family of k-programs over a direct product of copies of M.

Consider the sets $P_t = \mu(A^t)$, for $t > 0$. These form a finite semigroup of subsets of $\mu(A^t)$, and thus there is an idempotent element $P_r = P_{2r} = P_r^2$. That is, $S = P_r$ is a finite semigroup that satisfies $S^2 = S$. Let $B = A^r$, considered as a finite alphabet. We obtain a homomorphism $\nu : B^+ \to S$ simply by restricting μ to the strings of length divisible by r. Each of the sets

$$\{w \in B^+ : \nu(w) = s\},$$

where $s \in S$, is recognized by a family of k-programs over a direct product of copies of M, with B as the input alphabet. This is because each k-tuple of positions in a word over B gives rise to r^k different k-tuples of positions in the corresponding word over A, and we can take the sum of the original program values over these k-tuples of positions. If we now form a direct product of $|S|$ copies of these programs, we obtain a family of programs that simulates ν. We are thus in a position to apply Theorem 2: ν is refined by $\theta_{s,p}^k \times \lambda_{t,q}$. (Keep in mind that the underlying alphabet for this congruence is B, not A.)

We now indicate how the proof is completed; the details are carried out in the full paper. We first show that each class of the congruence $\theta_{s,p}^k \times \lambda_{t,q}$ is defined by a sentence of the appropriate kind, using only regular numerical predicates. Thus the same is true for each of the sets $\nu^{-1}(s)$. We then use this fact to show that the corresponding language over A, namely

$$L_s = \{w \in (A^r)^* : \mu(w) = s\},$$

is a union of intersections of languages defined by generalized Σ_1-sentences with only regular numerical predicates. The language L is a finite union of sets of the form $L_s w$, where $|w| < r$. Since $(L_1 \cap L_2)w = L_1 w \cap L_2 w$ and $(L_1 \cup L_2)w = L_1 w \cup L_2 w$, we need only show that if K is a language defined by a generalized Σ_1-sentence with modulus (s, p) and regular numerical predicates, then so is Kw.

Corollary 5. *If a regular language L is defined by a boolean combination of ordinary Σ_1-sentences, then it is defined by a boolean combination of ordinary Σ_1-sentences with regular numerical predicates.*

Proof. This follows directly from Theorem 1 for quantifiers $\exists^{(1,1,1)}$.

Corollary 6. *The language $1^* \subseteq \{0, 1\}^*$ cannot be defined by a boolean combination of generalized Σ_1-sentences of the form $\exists^{(i,0,p)} x \phi(x)$.*

Proof. Suppose otherwise. This language is regular, so by Theorem 1, it is defined by a boolean combination of sentences with regular numerical predicates and quantifiers of the form $\exists^{(j,0,q)}$, where q is the least common multiple of the moduli in the original sentences. Let U_1 denote the monoid $\{0, 1\}$ with the usual multiplication. By Theorem VII.4.2 of [16], the image of $\{0, 1\}^+$ under the syntactic morphism of the language does not contain a copy of U_1. But the syntactic morphism of 1^* maps $\{0, 1\}$ onto U_1, a contradiction.

The foregoing corollary is equivalent to a theorem in Barrington, Straubing and Thérien [3] (inexpressibility of AND by programs over nilpotent groups).

Corollary 7. *Let q be prime, and let L_q consist of all bit strings in which the number of 1's is divisible by q. Then L_q is not definable by a boolean combination of generalized Σ_1-sentences in which none of the moduli is divisible by q.*

Proof. Suppose otherwise. Since L_q is regular, Theorem 1 implies that it is defined by a boolean combination of sentences with regular numerical predicates and moduli not divisible by q. The syntactic morphism μ of L_q maps 0 to the identity and 1 to the generator of the cyclic group of order q, and thus maps the set of strings of length q onto the group of order q. But by Theorem VII.4.1 of [16], $\mu(\{0, 1\}^q)$ has cardinality dividing a product of the moduli occurring in a defining sentence, a contradiction.

4 Conclusion

The theorem that the circuit complexity class AC^0 does not contain parity, and the conjecture that ACC is strictly contained in NC^1, as well as related conjectures concerning circuits built from modular gates alone, can all be formulated as statements about the expressibility of regular languages in first-order logic and generalized first-order logic. The reformulations all say the same thing: If a regular language L can be defined by a sentence ϕ, then L can be defined by a sentence ϕ', where ϕ' contains the same kinds of quantifiers as ϕ, and has only regular numerical predicates. In the present paper, we have proved this in the case where ϕ is restricted to be a generalized Σ_1-sentence.

Of course, the principal problem left open by this work is that of extending our principal theorem to sentences of higher quantifier complexity. If we set our sights high, we can imagine what a proof of the general result, following along the lines of the present paper, might look like. A language is in the circuit complexity class ACC if and only if it is recognized by a family of k-programs over a finite solvable monoid, that is, a finite monoid that contains only solvable groups. (This follows from a result in Barrington and Thérien [5].) Finite solvable monoids, in turn, are formed, in effect, by iterating commutative monoids. More precisely, congruences on A^* whose quotients are finite solvable monoids are formed by iterating the congruences whose quotients are commutative. The iteration scheme, devised by Thérien [18], works as follows: Let $s \geq 0, p > 0$. We denote by $\alpha_{s,p}^{(0)}$ the congruence on A^* that identifies all strings. Now suppose $r \geq 0$, and $\alpha_{s,p}^{(r)}$ has been defined. Let $w_1, w_2 \in A^*$. We define

$$w_1 \alpha_{s,p}^{(r+1)} w_2$$

if, for all $\alpha_{s,p}^{(r)}$-classes X, Y and all $a \in A$, the number of factorizations

$$w_1 = uau',$$

with $u \in X$ and $u' \in Y$, is congruent, modulo (s,p), to the number of factorizations

$$w_2 = vav',$$

with $v \in X$ and $v' \in Y$. It is not difficult to show that $\alpha_{s,p}^{(r+1)}$ is a congruence of finite index on A^*, and Thérien's results imply that every finite solvable monoid is a quotient of $A^*/\alpha_{s,p}^{(r)}$ for some finite alphabet A and integers $s \geq 0, p > 0$ and $r > 0$. In the case $r = 1$, we get the finite commutative monoids.

We would like to be able to define a sequence of congruences $\theta_{s,p}^{k,(r)}$, $r > 0$, on A^+, with $\theta_{s,p}^{k,(1)} = \theta_{s,p}^k$, such that the analogous result to Theorem 2 holds: If a surjective homomorphism $\phi : A^+ \to S$ is simulated by a family of k-programs over $B^*/\alpha_{s,p}^{(r)}$, for some s, p, r, then ϕ is refined by $\theta_{s,p}^{k,(r)} \times \lambda_{t,q}$ for some t, q. If we can do this in such a manner that the quotient $A^*/\theta_{s,p}^{k,(r)}$ is itself a solvable semigroup, it would follow that ACC is strictly contained in NC^1, for we can simulate any homomorphism onto a finite semigroup in NC^1, so we could pick S to be any nonsolvable finite semigroup and derive a contradiction.

It is not hard to devise plausible candidates for the congruences $\theta_{s,p}^{k,(r)}$, but generalizing the combinatorial arguments in the proof of Theorem 2 presents formidable obstacles. We suspect that the pure group case is somewhat easier to work with, so k-programs over $A^*/\alpha_{0,p}^{(2)}$ might provide a good place to begin.

References

1. M. Ajtai, "Σ_1^1 formulae on finite structures", *Annals of Pure and Applied Logic* **24** (1983) 1–48.
2. D. Mix Barrington, K. Compton, H. Straubing, and D. Thérien, "Regular Languages in NC^1", *J. Comp. Syst. Sci.* **44** (1992) 478–499.
3. D. Mix Barrington, H. Straubing, and D. Thérien, "Nonuniform Automata over Groups", *Information and Computation* **89** (1990) 109-132.
4. D. Mix Barrington and H. Straubing, "Superlinear Lower Bounds for Bounded-width Branching Programs", *J. Comp. Syst. Sci.* **50** (1995) 374-381.
5. D. Mix Barrington and D. Thérien, "Finite Monoids and the Fine Structure of NC^1", *JACM* **35** (1988) 941–952.
6. J. R. Büchi, "Weak Second-order Arithmetic and Finite Automata", *Zeit. Math. Logik. Grund. Math.* **6** (1960) 66-92.
7. S. Eilenberg, *Automata, Languages and Machines*, vol. B, Academic Press, New York, 1976.
8. M. Furst, J. Saxe, and M. Sipser, "Parity, Circuits, and the Polynomial Time Hierarchy", *J. Math Systems Theory* **17** (1984) 13-27.
9. Y. Gurevich and H. Lewis, "A Logic for Constant-Depth Circuits", *Information and Control*, **61** (1984) 65-74.
10. N. Immerman, "Languages That Capture Complexity Classes", *SIAM J. Computing* **16** (1987) 760-778.
11. R. McNaughton and S. Papert, *Counter-Free Automata*, MIT Press, Cambridge, Massachusetts, 1971.
12. A. Maciel, P. Péladeau and D. Thérien, "Programs over Semigroups of Dot-depth One", preprint (1996).
13. J. E. Pin, *Varieties of Formal Languages*, Plenum, London, 1986.
14. A. Razborov, "Lower Bounds for the Size of Circuits of Bounded Depth with Basis $\{\wedge, \oplus\}$" *Math. Notes of the Soviet Academy of Sciences* **41** (1987) 333-338.
15. R. Smolensky, "Algebraic Methods in the Theory of Lower Bounds for Boolean Circuit Complexity", *Proc. 19th ACM STOC* (1987) 77-82.
16. H. Straubing, *Finite Automata, Formal Languages, and Circuit Complexity*, Birkhaüser, Boston, 1994.
17. H. Straubing, D. Thérien, and W. Thomas, "Regular Languages Defined with Generalized Quantifiers", *Information and Computation* **118** (1995) 289-301.
18. D. Thérien, "Classification of finite monoids: the language approach", *Theoret. Comput. Sci.* **14** (1981) 195-208.

Hierarchies of Principal Twist-Closed Trios

Matthias Jantzen

Universität Hamburg FB Informatik, Universität Hamburg
Vogt-Kölln-Straße 30, 22527 Hamburg
jantzen@informatik.uni-hamburg.de

Abstract. The language theoretic operation *twist* from [JaPe 94] is studied in connection with the semiAFLs of languages accepted by reversal bounded multipushdown and multicounter acceptors. It is proved that the least *twist*-closed trio generated by $MIR := \{ww^{rev} \mid w \in \{a,b\}^*\}$ is equal to the family of languages accepted in quasi-realtime by nondeterministic one-way multipushdown acceptors which operate in such a way that in every computation each pushdown makes at most one reversal. Thus, $\mathcal{M}_\cap(MIR) = \mathcal{M}_{twist}(MIR)$ and this family is a principal *twist*-closed semiAFL with a linear context-free generator. This is in contrast to the semiAFL of languages accepted by reversal-bounded multicounter machines in quasi-realtime. This family is a well known semiAFL which is principal as an *intersection*-closed semiAFL with generator $B_1 := \{a_1^n \bar{a}_1^n \mid n \in I\!N\}$, see [Grei 78], but is not principal as a semiAFL. It is here shown that it forms a hierarchy of *twist*-closed semiAFLs and therefore is not principal as *twist*-closed semiAFL.

1   Introduction

In connection with a representation of Petri net languages by Dyck-reductions of (linear) context-free sets the operation *twist* was defined and used for the first time, see [JaPe 91,JaPe 94]. The definition of this new language theoretic operation is based upon a mapping from strings to strings which rearranges letters: for a string $w := x_1 x_2 \cdots x_{n-1} x_n$ the unique new string is $twist(w) := x_1 x_n x_2 x_{n-1} \cdots x_{\lfloor \frac{n}{2} \rfloor + 1}$. Observe, that $twist : \Sigma^* \longrightarrow \Sigma^*$ is a bijection and it's inverse mapping $twist^{-1}(w)$ yields a unique string v with $twist(v) = w$. It follows that for each string w there exists a non-negative integer $k \in I\!N$ such that $twist^k(w) = w$. The mapping *twist* can also be regarded as a permutation of the n distinct positions for the symbols of a string of length n. As language theoretic operation *twist* is generalized to languages and families of languages in the obvious way, see Def.4.

It was shown in [JaPe 94], Th.2.10, that the family $\mathcal{R}eg$ of regular sets is closed with respect to *twist*. The inclusion $twist(\mathcal{R}eg) \subsetneq \mathcal{R}eg$ must be proper since $twist(MIR) = \{a^2, b^2\}^*$, where $MIR := \{ww^{rev} \mid w \in \{a,b\}^*\}$ is the non-regular context-free set of palindromes of even length. This observation means, that the regular set $\{a^2, b^2\}^*$ will never appear as $twist(R)$ for any regular set $R \in \mathcal{R}eg$. Notice, $twist^{-1}(MIR) = COPY := \{ww \mid w \in \{a,b\}^*\}$. In [JaPe 94],

Th.2.11, it was also proved that the family $\mathcal{L}_0 := \mathcal{M}_\cap(D_1)$is closed with respect to the operation *twist*. Again, $twist(\mathcal{L}_0) \subsetneq \mathcal{L}_0$, since $twist(MIR) \in \mathcal{L}_0$ but $MIR \notin \mathcal{L}_0$ follows from [Grei 78,Jant 79a] using [Mayr 84,Kosa 82]. In [Jant 97] a new morphic characterization of the recursively enumerable sets was given by $\mathcal{R}e = \hat{\mathcal{H}}(\mathcal{H}^{-1}(twist(lin\,Cf))) = \hat{\mathcal{M}}_{twist}(dMIR)$. Similar results are known for principal *intersection*-closed full trios (see [BaBo 74]) and for full principal trios, the generator of which is as rich in structure as the twinshuffle language L_{TS} (see [Salo 81], Chapt. 6, for a condensed presentation. L_{TS} was there abbreviated by L_Σ). A slight improvement of Theorem 4.5 of [Bran 87] has been shown by Engelfriet in [Enge 96] for the reverse twin shuffle language L_{RTS}. However, neither L_{TS} nor L_{RTS} is context-free.

It was proved in[BoNP 74] that the *intersection*-closed trio $\mathcal{M}_\cap(MIR)$ equals the class of languages accepted by one-way reversal-bounded multipushdown machines in quasi-realtime and that three pushdown stores suffice. Hence, this family is a principal semi-AFL. We will show here that $twist(lin\,Cf) \subseteq \mathcal{M}_\cap(MIR)$ and $\mathcal{M}_\cap(MIR) = \mathcal{M}_{twist}(MIR)$.

The situation becomes different for the semiAFL of languages accepted by one-way reversal-bounded multicounter machines in quasi-realtime. This family is a well known semiAFL which is principal as an *intersection*-closed semiAFL with generator $B_1 := \{a_1^n \bar{a}_1^n \mid n \in \mathbb{N}\}$, which is not a principal semiAFL, see [FiMR 68,Grei 78]. (For the definition of the languages B_i, C_1, and C_i see Definition 3 below). $\bigcup_{i \geq 0} \mathcal{M}(C_i) = \mathcal{M}_\cap(C_1) = BLIND = \bigcup_{i \geq 0} \mathcal{M}(B_i) = \mathcal{M}_\cap(B_1) = RBC$ is known from [Grei 78]. The known situation for these hierarchies is as follows: For all $i \geq 1$ we have $\mathcal{M}(B_i) \subsetneq \mathcal{M}(B_{i+1})$, see [Gins 75], $\mathcal{M}(C_i) \subsetneq \mathcal{M}(C_{i+1})$, and $\mathcal{M}(B_i) \subseteq \mathcal{M}(C_i)$, shown in [Grei 76,Grei 78], [Latt 77], and [Latt 78,Latt 79]. We will in addition show here that $\bigcup_{i \geq 0} \mathcal{M}(C_i) = \mathcal{M}_\cap(C_1)$ forms a hierarchy of *twist*-closed semiAFLs and therefore is not principal as *twist*-closed semiAFL. Each such semiAFL $\mathcal{M}(C_k)$ can be characterized as family of languages acceptable by blind k-counter machines in quasi-realtime, [Grei 78]. Slightly improving a construction by Greibach we will prove $\mathcal{M}(B_i) \subseteq \mathcal{M}_{twist}(B_i) = \mathcal{M}(C_i) = \mathcal{M}_{twist}(C_i) \subseteq \mathcal{M}(B_{i+1})$ for each $i \geq 1$.

2 Basic Definitions

Definition 1. Let $\mathcal{R}eg$ (resp. $lin\,Cf$, Cf, Cs, $\mathcal{R}ec$, $\mathcal{R}e$) denote the families of regular sets (linear context-free, context-free, context sensitive, recursive, and recursively enumerable languages, respectively).

Definition 2. Let $w_1, w_2 \in \Sigma^*$, $w_1 := x_1 x_2 \cdots x_m$, and $w_2 := y_1 y_2 \cdots y_n$ where $x_i, y_j \in \Sigma$ for $1 \leq i \leq m$ and $1 \leq j \leq n$.

Then the *shuffle* $\sqcup\!\sqcup$ and the *literal shuffle* $\sqcup\!\sqcup_{lit}$ are defined as follows:

$$w_1 \sqcup\!\sqcup w_2 := \left\{ u_1 v_1 u_2 v_2 \cdots u_n v_n \;\middle|\; \begin{array}{l} n \in \mathbb{N}, u_i, v_i \in \Sigma^*, w_1 = u_1 u_2 \cdots u_n, \\ w_2 = v_1 v_2 \cdots v_n \end{array} \right\},$$

$$w_1 \sqcup\!\sqcup_{lit} w_2 := \begin{cases} x_1 y_1 x_2 y_2 \cdots x_m y_m y_{m+1} \cdots y_n \,, & \text{if } m \leq n \\ x_1 y_1 x_2 y_2 \cdots x_n y_n x_{n+1} \cdots x_m \,, & \text{if } n < m \end{cases}$$

Definition 3. Specific languages we consider are constructed using the alphabets $\Gamma := \{a, b\}$, $\overline{\Gamma} := \{\overline{a}, \overline{b}\}$ and $\Gamma_n, \overline{\Gamma}_n$ specified for each $n \in I\!\!N, n \geq 1$ by: $\Gamma_n := \{a_i, b_i \mid 1 \leq i \leq n\}$, $\overline{\Gamma}_n := \{\overline{a}_i, \overline{b}_i \mid 1 \leq i \leq n\}$, and the homomorphisms $^-, h, \overline{h}$, and h_i defined by:

$$\overline{x} := \begin{cases} \overline{x}, & \text{if } x \in \Gamma \\ x, & \text{if } x \in \overline{\Gamma} \end{cases}, \quad h(x) := \begin{cases} x, & \text{if } x \in \Gamma \\ \lambda, & \text{if } x \in \overline{\Gamma} \end{cases}, \quad \overline{h}(x) := \begin{cases} \lambda, & \text{if } x \in \Gamma \\ x, & \text{if } x \in \overline{\Gamma} \end{cases},$$

and $h_i(a_1) := a_i, h_i(b_1) := b_i$ for $i \in I\!\!N \setminus \{0\}$,

By $|w|_x$ we denote the number of occurences of the symbol $x \in \Sigma$ within the string $w \in \Sigma^*$ and $|w| := \Sigma_{x \in \Sigma} |w|_x$ is the length of w.

$$B_i := \begin{cases} B_{i-1} \sqcup \{a_i^n b_i^n \mid n \in I\!\!N\}, & \text{if } i \geq 2 \\ \{a_1^n b_1^n \mid n \in I\!\!N\}, & \text{if } i = 1 \end{cases}$$

$$C_i := \begin{cases} C_{i-1} \sqcup h_i(C_1), & \text{if } i \geq 2 \\ \{w \in \{a_1, b_1\}^* \mid |w|_{a_1} = |w|_{b_1}\}, & \text{if } i = 1 \end{cases}$$

$$D_i := \begin{cases} D_{i-1} \sqcup h_i(D_1), & \text{if } i \geq 2 \\ \{w \in \{a_1, b_1\}^* \mid |w|_{a_1} = |w|_{b_1}, \\ \text{and } \forall w = uv : |u|_{a_1} \geq |u|_{b_1}\}, & \text{if } i = 1 \end{cases}$$

$$dMIR := \{wcw^{rev} \mid w \in \Gamma^*\}.$$

$$MIR := \{ww^{rev} \mid w \in \Gamma^*\}.$$

$$PAL := \{w \mid w = w^{rev}, w \in \Gamma^*\}.$$

$$dCOPY := \{wcw \mid w \in \{a, b\}^*\}$$

$$COPY := \{ww \mid w \in \Gamma^*\}$$

$$L_{TS} := \{w \in (\Gamma \cup \overline{\Gamma})^* \mid \overline{h(w)} = \overline{h}(w)\}$$

$$L_{RTS} := \{w \in (\Gamma \cup \overline{\Gamma})^* \mid \overline{h(w)} = \overline{h}(w^{rev})\}$$

$$twinPAL := \{w \in (\Gamma \cup \overline{\Gamma})^* \mid h(w) \in MIR \text{ and } \overline{h}(w) \in \overline{MIR}\}$$

$\square$

Let us repeat the basic notions and results from AFL-theory details of which are to be found in the textbooks of Ginsburg, [Gins 75], and Berstel, [Bers 80].

A family of languages $\mathcal{L}$ is called trio if it is closed under inverse homomorphisms, intersection with regular sets, and nonerasing homomorphisms. The least trio containing the family $\mathcal{L}$ is written $\mathcal{M}(\mathcal{L})$. If $\mathcal{L} := \{L\}$, then L is a generator of the trio $\mathcal{M}(\mathcal{L})$, shortly written as $\mathcal{M}(L)$ and then called principal. A union-closed trio is called semiAFL. Any principal trio is closed with respect to union and thus forms a semiAFL. If a trio is closed under arbitray homomorphisms, then it is called a full trio, written $\hat{\mathcal{M}}(\mathcal{L})$.

A family of languages $\mathcal{L}$ is called an AFL (or full AFL) if it is a trio (full trio, resp.) which is closed under the operations union, product and Kleene plus. The smallest AFL (or full AFL) containing the family $\mathcal{L}$ is written $\mathcal{F}(\mathcal{L})$ ($\hat{\mathcal{F}}(\mathcal{L})$, resp.). Each full AFL is closed with respect to Kleene star.

If a trio $\mathcal{M}(\mathcal{L})$ (or an AFL $\mathcal{F}(\mathcal{L})$) is in addition closed with respect to one further operation $\circledast$ then this family will be called $\circledast$-closed and we use $\mathcal{M}_\circledast(\mathcal{L})$ (resp. $\mathcal{F}_\circledast(\mathcal{L})$) to denote the smallest trio (AFL, resp.) containing $\mathcal{L}$ and closed with respect to $\circledast$.

dMIR, *MIR* and *PAL* are well-known context-free generators of the family *lin Cf* of linear context-free languages: $lin\,Cf = \mathcal{M}(dMIR) = \mathcal{M}(MIR) = \mathcal{M}(PAL)$. These languages are precisely the languages accepted by nondeterministic on-line single pushdown acceptors which operate in such a way that in every accepting computation the pusdown store makes at most one reversal. And this family is not closed with respect to product or Kleene plus.

Similarly, the *intersection*-closed semiAFL $\mathcal{M}_\cap(MIR)$ can be identified with the family of languages accepted by nondeterministic on-line multipushdown acceptors which operate in such a way that in every computation each pushdown makes at most one reversal and that work in quasi-realtime, see [BoGr 70]. This family, however, becomes the set of recursively enumerable languages if erasing is allowed and was characterized in [BaBo 74] by $\mathcal{R}e = \hat{\mathcal{M}}_\cap(MIR) = \hat{\mathcal{M}}(twinPAL)$.

The language D_1 defined above is the so-called semi-Dyck language on one pair of brackets which is often abbreviated by $D_1'^*$, see e.g. [Latt 77,Latt 79] or [Bers 80]. D_n here denotes the n-fold shuffle of disjoint copies of the semi-Dyck language D_1 and it is known [Grei 78] that $\bigcup_{i\geq 0}\mathcal{M}(D_i) = \mathcal{M}_\cap(D_1) = PBLIND(n)$. The latter family is the family of languages accepted in quasi-realtime by nondeterministic one-way multicounter acceptors which operate in such a way that in every computation no counter can store a negative value, and whether or not the value stored in a counter is *zero* is not used for deciding the next move.

The languages C_n are the Dyck languages on n pairs of brackets $a_i, \bar{a}_i$, often abbreviated by D_n^*, see again [Latt 77,Latt 79] or [Bers 80]. Greibach, [Grei 78], has shown that $\bigcup_{i\geq 0}\mathcal{M}(C_i) = \mathcal{M}_\cap(C_1) = BLIND = BLIND(lin) = BLIND(n) = \bigcup_{i\geq 0}\mathcal{M}(B_i) = \mathcal{M}_\cap(B_1) = RBC(n) = RBC \subsetneqq PBLIND$.

Here *BLIND* (resp. *BLIND(n)*) denotes the family of languages accepted (in quasi-real time, resp.) by nondeterministic one-way multicounter acceptors which operate in such a way that in every computation all counters may store arbitrary integers, and the information on the contents of the counters is not used for deciding the next move. The family $RBC(n)$ is the family of languages accepted by nondeterministic one-way multicounter acceptors performing at most one reversal in each accepting computation.

The least *intersection*-closed full semiAFL $\hat{\mathcal{M}}_\cap(B_1)$ has been characterized in [BaBo 74] as the family of languages accepted by nondeterministic on-line multicounter acceptors which operate in such a way that in every computation each counter makes at most one reversal. It was there shown that this class contains only recursive sets, i.e. $\hat{\mathcal{M}}_\cap(B_1) \subseteq \mathcal{R}ec$.

Definition 4. Let Σ be an alphabet, then *twist* : $\Sigma^* \longrightarrow \Sigma^*$ is recursively defined for any $w \in \Sigma^*$ and $a \in \Sigma$ by: $twist(aw) := a \cdot twist(w^{rev})$, and $twist(\lambda) := \lambda$.

For sets of strings L and families of languages $\mathcal{L}$ the operation *twist* is generalized as usual: $twist(L):= \{twist(w) \mid w \in L\}$ and $twist(\mathcal{L}):= \{twist(L) \mid L \in \mathcal{L}\}$.

$\square$

3 Results

We see that $twist(w) = x_1 x_n x_2 x_{n-1} \cdots x_{\lfloor \frac{n}{2} \rfloor + 1}$ for any string $w \in \Sigma^*, w :=$ $x_1 x_2 \cdots x_{n-1} x_n$ where $x_i \in \Sigma$ for all $i \in \{1, \cdots, n\}$.

Viewed as the permutation π_{twist} of the n subscripts $1, 2, \ldots, n$, i.e. the positions of the symbols that form the string $w := x_1 x_2 \cdots x_{n-1} x_n$ this yields

$$\pi_{twist}(i) := \begin{cases} 2 \cdot i - 1 \,, & \text{if } 0 \leq i \leq \lceil \frac{n}{2} \rceil \\ 2(n+1-i) \,, & \text{otherwise} \end{cases}$$

Twisting a context-free language obviously yields a context-sensitive language. We have $twist(\mathcal{C}f) \subsetneq \mathcal{C}s$ and the inclusion must be proper since $twist(L)$ has a semilinear Parikh image whenever L has this property. Note that $twist(L)$ may not be context-free even for a linear context-free language $L := L_{lin}$ or a one-counter language $L := L_{count}$.

In order to use the operation $\frac{1}{2}$ in connection with $twist$ we shall define a slightly generalized version of this operation, compare [HoUl 79]:

Definition 5. For any string $w := x_1 x_2 \cdots x_n, x_i \in \Sigma$, let $\frac{1}{2}(w) := x_1 x_2 \cdots x_{\lceil \frac{n}{2} \rceil}$. $\square$

Hence, $\frac{1}{2}(abaab) = \frac{1}{2}(abaabb) = aba$.

Lemma 6. *Any trio which is closed with respect to twist is also closed under reversal and $\frac{1}{2}$.*

Proof: Let $L \subseteq \Sigma^*, \$ \notin \Sigma$ be a new symbol and $f : (\Sigma \cup \{\$\})^* \longrightarrow \Sigma^*$ a homomorphism defined by $f(\$) := \lambda$ and $\forall x \in \Sigma : f(x) := x$. Then $L^{rev} = g^{-1}(twist(f^{-1}(L) \cap \{\$\}^* \Sigma^*))$ where $g : \Sigma^* \longrightarrow (\Sigma \cup \{\$\})^*$ is a homomorphism given by $\forall x \in \Sigma : g(x) := \$x$. Thus any $twist$-closed trio $\mathcal{M}(\mathcal{L})$ is closed with respect to reversal.

To express the operation $\frac{1}{2}$ by trio operations and $twist$ that works for strings of both even and odd length we have to insert a dummy symbol for those of odd length and then mark half of the symbols. To do this we use an inverse homomorphism h_1^{-1}. By intersection with a suitable regular set we then can fix the position of the dummy symbol and the marked symbols.

In detail we define: $\overline{\Sigma} := \{\overline{x} \mid x \in \Sigma\}$ as a disjoint copy of Σ and the homomorphism $h_1 : (\Sigma \cup \overline{\Sigma} \cup \{\$\})^* \longrightarrow \Sigma^*$ by: $h_1(x) := h_1(\overline{x}) := x$ for all $x \in \Sigma$ and $h_1(\$) := \lambda$. Now, for any string $w \in \Sigma^*$, $h_1^{-1}(w)$ may contain an arbitrary number of extra $\$$-symbols and likewise barred symbols from $\overline{\Sigma}$ at any position. Then $K_1 := h_1^{-1}(L) \cap \Sigma^* \{\$, \lambda\} \overline{\Sigma}^*$ contains at most one extra symbol $\$$ and all and only the barred symbols at the right hand side. Define new alphabets $\Gamma := \{\langle x, y \rangle \mid x \in \Sigma, y \in \overline{\Sigma}\}$, $\Gamma_\$:= \{\langle x, \$ \rangle \mid x \in \Sigma\}$ and a homomorphism $h_2 : (\Gamma_\$ \cup \Gamma)^* \longrightarrow (\Sigma \cup \overline{\Sigma} \cup \{\$\})^*$ by $h_2(\langle x, y \rangle) := xy$. Now $K_2 := h_2^{-1}(twist(K_1)) \cap (\Gamma^* \cup \Gamma^* \Gamma_\$)$ is a set of strings, each of which describes the $twist$ of a string from K_1 in the projection of both components of the new symbols from $\Gamma \cup \Gamma_\$$. Since the first $\lceil \frac{|w|}{2} \rceil$ symbols of the original string w are put into the first component of the corresponding string from K_2 a simple coding will retrieve the string $\frac{1}{2}(w)$. With $h_3 : (\Gamma \cup \Gamma_\$) \longrightarrow \Sigma$ defined by $h_3(\langle x, y \rangle) := x$ one obtains $\frac{1}{2}(L) := h_3(K_2)$. The only operations we used to define $\frac{1}{2}(L)$ were trio operations and *twist* so that the lemma was proved completely.

□

Lemma 7. *Each twist-closed trio $\mathcal{L}$ that is in addition closed under product is also closed with respect to intersection.*

Proof: Let $L_1, L_2 \subseteq \Sigma^*, L_1, L_2 \in \mathcal{L}$ and Γ a copy of Σ with $h : \Sigma \longrightarrow \Gamma$ being the bijection between the alphabets. By Lemma 6 $L_2^{rev} \in \mathcal{L}$ and then also $L_3 := g^{-1}(twist(L_1 \cdot h(L_2^{rev}))) \in \mathcal{L}$ where $g : \Sigma^* \longrightarrow (\Sigma\Gamma)^*$ is defined by $g(x) = xh(x)$ for all $x \in \Sigma$. Obviously $L_3 = L_1 \cap L_2$, and this proves the lemma.

□

There exist families of languages that are closed with respect to the operations *twist* and *product* but not under intersection! The family $\mathcal{L}_{slip}$ of languages having a semi-linear Parikh image, i.e. being letter equivalent to regular sets, is such a family. This is because this family is not a trio since it is not even closed with respect to intersection by regular sets! To see this, consider the language $L := \{ab^{2^n} \mid n \in \mathbb{N}\} \cup \{b\}^*\{a\}^* \in \mathcal{L}_{slip}$, where one has $L \cap \{a\}\{b\}^* \notin \mathcal{L}_{slip}$.

This observation indicates that it might not be easy to express the operation *twist* by means of known operations in abstract formal language theory.

Using simple and standard techniques we can show that the languages *MIR*, *COPY* and their deterministic variants all are generators of the same *twist*-closed trio $\mathcal{M}_{twist}(MIR)$.

Theorem 8. $\mathcal{M}_{twist}(dCOPY \cup \{\lambda\}) = \mathcal{M}_{twist}(COPY) = \mathcal{M}_{twist}(MIR) = \mathcal{M}_{twist}(PAL) = \mathcal{M}_{twist}(dMIR \cup \{\lambda\}) = \mathcal{M}_{twist}(L_{RTS}) = \mathcal{M}_{twist}(L_{TS})$

Proof:

(a) $COPY \in \mathcal{M}_{twist}(dCOPY \cup \{\lambda\})$ follows since *COPY* is obtained from *dCOPY* by limited erasing of the symbol c and it is well known that every trio is closed with respect to this operation.

(b) $MIR \in \mathcal{M}_{twist}(COPY)$ follows by observing that $MIR = twist(COPY)$. This can be shown by induction on the length and structure of the strings involved.

(c) $dMIR \in \mathcal{M}_{twist}(MIR)$, (d) $PAL \in \mathcal{M}_{twist}(dMIR \cup \{\lambda\})$, and (e) $MIR \in \mathcal{M}_{twist}(PAL)$ follow from the well known: $\mathcal{M}(dMIR \cup \{\lambda\}) = \mathcal{M}(MIR) = \mathcal{M}(PAL)$.

(f) $dCOPY \in \mathcal{M}_{twist}(dMIR)$: $K_2 := \{w\$^i w^{rev} \rlap{/}{\mathfrak{c}}^j \mid w \in \{a, b\}^*, i, j \in \mathbb{N} \setminus \{0\}\} \in \mathcal{M}(dMIR)$ is easily proved. Likewise, $K_3 := twist(K_2) \cap (\{a, b\}^*\{\rlap{/}{\mathfrak{c}}\})^* \cdot \{\$\rlap{/}{\mathfrak{c}}\} \cdot (\{\$\}\{a, b\})^* \in \mathcal{M}(dMIR)$ and then $dCOPY = f(h^{-1}(K_3))$ follows with $h : \{a, b, c, \bar{a}, \bar{b}\}^* \longrightarrow \{a, b, \$, \rlap{/}{\mathfrak{c}}\}^*$ defined by $h(a) := a\rlap{/}{\mathfrak{c}}, h(b) := b\rlap{/}{\mathfrak{c}}, h(c) := \$\rlap{/}{\mathfrak{c}}, h(\bar{a}) := \$a, h(\bar{b}) := \$b$, and $f(a) := f(\bar{a}) := a, f(b) := f(\bar{b}) := b, f(c) := c$. Consequently, $dCOPY \in \mathcal{M}_{twist}(dMIR)$.

(g) Obviously, $L_{RTS} \cap \Gamma^* \overline{\Gamma}^* = \{w\overline{w}^{rev} \mid w \in \Gamma^*\}$, hence $MIR \in \mathcal{M}(L_{RTS})$. Let $\rlap{/}{\mathfrak{c}} \notin \Gamma$ and the homomorphism $h_{\rlap{/}{\mathfrak{c}}} : (\Gamma \cup \overline{\Gamma})^* \longrightarrow (\Gamma \cup \overline{\Gamma} \cup \{\rlap{/}{\mathfrak{c}}\})^*$ be defined by

$h_{\phi}(x) := x$, if $x \in \Gamma \cup \overline{\Gamma}$ and $h_{\phi}(\phi) := \lambda$. Then $L := twist(h_{\phi}^{-1}(\{w\overline{w} \mid w \in \Gamma^*\})) \cap$ $\{x\phi, \phi y \mid x \in \Gamma, y \in \overline{\Gamma}\}^*$ is an element of $\mathcal{M}_{twist}(COPY)$. Now, $L_{RTS} = g^{-1}(L) \in$ $\mathcal{M}_{twist}(COPY)$ for g being defined by $g : (\Gamma \cup \overline{\Gamma})^* \longrightarrow (\Gamma \cup \overline{\Gamma} \cup \{\phi\})^*$ with $g(x) := x\phi$, if $x \in \Gamma$ and $g(y) := \phi y$, if $x \in \overline{\Gamma}$.

(h)Similar to **g** we have $L_{TS} \cap \Gamma^* \overline{\Gamma}^* = \{w\overline{w} \mid w \in \Gamma^*\}$, hence $COPY \in$ $\mathcal{M}(L_{TS})$. As in **(g)** one shows $L_{RTS} = g^{-1}(L) \in \mathcal{M}_{twist}(MIR)$.

$\square$

Since the mapping $twist$ only performs a permutation of the symbols that form a string it is easily seen that $\mathcal{R}e$, $\mathcal{R}ec$, and $\mathcal{C}s$ are $twist$-closed families. The family $\mathcal{M}_{\cap}(MIR)$ of quasi-realtime multipushdown languages [BoNP 74,BoGr 78] will be shown to be another $twist$-closed family.

Lemma 9. *The family $\mathcal{M}_{\cap}(MIR)$ is closed with respect to the operation twist.*

Proof: Let $L \in \mathcal{M}_{\cap}(MIR)$ be accepted by some nondeterministic on-line reversal-bounded multipushdown machine M_L which operates in such a way that in every computation each pushdown makes at most one reversal and runs in linear time, see [BoNP 74]. In order to accept $K := twist(L)$ we use machine M_L and add one further pushdown store to obtain machine M_K that accepts K as follows: M_K reads the symbols at odd positions of an input string $w \in K$, beginning with the first symbol of w and behaves on them exactly as the machine M_L. Beginning with the second symbol of w the symbols at even positions alternatively are pushed onto the new pushdown. After having read the last symbol of the input string the symbols from the pushdown are popped and now treated as input for the machine M_L. M_K accepts if the new pusdown is emptied and M_L accepts its input $twist^{-1}(w)$. Hence, M_K accepts $twist(L)$ and operates on each pushdown with only one reversal. It must be observed that M_K works in linear but not in quasi-realtime. That this is not a loss follows from a result in [BoNP 74] stating that the class $\mathcal{M}_{\cap}(dMIR \cap \{\lambda\})$ is closed with respect to linear erasing homomorphisms.

$\square$

Lemma 9 showed $\mathcal{M}_{twist}(dMIR \cup \{\lambda\}) \subseteq \mathcal{M}_{\cap}(dMIR \cup \{\lambda\})$ and by the following results we will be able to prove equality of these two classes. Since the family $lin\, Cf = \mathcal{M}(dMIR \cup \{\lambda\}) = \mathcal{M}(MIR)$ is not closed with respect to product we cannot apply Lemma 7 for proving that $\mathcal{M}_{twist}(dMIR \cup \{\lambda\})$ is *intersection*-closed.

Lemma 10. $\overline{dMIR} \cdot dMIR \in \mathcal{M}_{twist}(dMIR \cup \{\lambda\})$ for $\overline{dMIR} := \{\overline{w} \mid w \in dMIR\}$.

Proof: Let $L_1 := \{c_1^{k_1} w\, c_2^{k_2}\, v\, c_3\, v^{rev}\, c_4^{k_4}\, w^{rev}\, c_5 \mid w \in \{a,b\}^*, v \in \{\overline{a}, \overline{b}\}^*, k_1,$ $k_2, k_4 \in I\!N \setminus \{0\}\} \in lin\, Cf = \mathcal{M}(dMIR \cup \{\lambda\}) \subset \mathcal{M}_{twist}(dMIR \cup \{\lambda\})$. Then let $L_2 \in \mathcal{M}_{twist}(dMIR \cup \{\lambda\})$ be defined by $L_2 := twist(L_1) \cap R_1$, where $R_1 :=$ $\{c_1c_5\}(\{c_1\}\{a,b\})^*\{c_1c_4\}(\{a,b\}\{c_4\})^*\{c_2c_4\}(\{c_2\}\{\overline{a}, \overline{b}, c_3\})^*\{c_2c_4\} \in \mathcal{R}eg$.

One observes $L_2 := \{c_1 c_5 c_1 w_1 c_1 w_2 \cdots c_1 w_n c_1 c_4 w_1 c_4 w_2 \cdots c_4 w_n c_4 c_2 c_4 c_2 v_1 c_2$
$\cdots v_m c_2 c_3 c_2 v_m c_2 v_{m-1} \cdots c_2 v_1 \mid w_i \in \{a,b\} \text{ and } v_j \in \{\overline{a}, \overline{b}\}\}$ and from this
$L_3 := \{c_1 w c_2 w c_3 v c_4 v^{rev} c_5 \mid w \in \{a,b\}^* \text{ and } v \in \{\overline{a}, \overline{b}\}^*\} \in \mathcal{M}_{twist}(dMIR \cup \{\lambda\})$
follows easily. By a similar technique we will finally get the stated result: Let
$L_4 := \{c_1^{k_1} w c_2^{k_2} w c_3^{k_3} v c_4 v^{rev} c_5 \mid w \in \{a,b\}^* \text{ and } v \in \{\overline{a}, \overline{b}\}^*, k_1, k_2, k_3 \in \mathbb{N} \setminus \{0\}\} \in \mathcal{M}(L_3)$ and $L_5 := twist(L_4) \cap R_2$ where $R_2 \in \mathcal{Reg}$ is given by $R_2 := \{c_1 c_5\}(\{c_1\}\{\overline{a}, \overline{b}, c_4\})^* \{c_1 c_3\}(\{a,b\}\{c_3\})^* \{c_2 c_3\}(\{c_2\}\{a,b\})^* \{c_2 c_3\}$. One gets
$L_6 := \{\$_1 v \$_2 v^{rev} \$_3 w \$_4 w^{rev} \$_5 \mid v \in \{\overline{a}, \overline{b}\}^* \text{ and } w \in \{a,b\}^*\} \in \mathcal{M}(L_5)$ and finally $\overline{dMIR} \cdot dMIR \in \mathcal{M}(L_6) \subseteq \mathcal{M}_{twist}(dMIR \cup \{\lambda\}) = \mathcal{M}_{twist}(MIR)$.

$\square$

Now, given two languages $L_1, L_2 \in \mathcal{M}_{twist}(dMIR \cup \{\lambda\})$ we know that each of them is obtained by finitely many applications of a-*transducer* mappings (each represented by trio operations) and the operation *twist* in any order. Our goal is to show, that each of this sequences op_1 and op_2 that are applied to the generator *dMIR* can be replaced by one sequence of operations which simulates these sequences on each single component of the generator $\overline{dMIR} \cdot dMIR \subset \{\overline{a}, \overline{b}\}^* \cdot \{a,b\}^*$. Since this is easy as long as only a-*transducer* mappings are used we have to show that also *twist* can be applied separately to each single component. This will be proved in Lemma 11 below.

Lemma 11. *Let* $L \in \mathcal{M}_{twist}(\mathcal{L})$ *such that* $L := L_1 \cdot L_2$ *for* $L_1 \subseteq \Sigma_1^*, L_2 \subseteq \Sigma_2^*$ *with* $\Sigma_1 \cap \Sigma_2 = \emptyset$. *Then* $K_1 \cdot K_2 \in \mathcal{M}_{twist}(\mathcal{L})$ *for each choice of* $K_i \in \{L_i, L_i^{rev}, twist(L_i)\}, i \in \{1, 2\}$.

Proof: With $L_1 \cdot L_2 \in \mathcal{M}_{twist}(\mathcal{L})$ also $L_3 := \{c_1\}^* L_1 \{c_2\}^* L_2 \in \mathcal{M}_{twist}(\mathcal{L})$.
Then $L_2^{rev} \cdot L_1 = f^{-1}(twist(L_3) \cap (c_1 \Sigma_2)^* (\Sigma_1 c_2)^*)$ for $f(x) := \begin{cases} c_1 x, & \text{if } x \in \Sigma_2 \\ x c_2, & \text{if } x \in \Sigma_1 \end{cases}$.
Now, by Lemma 6 also
(a) $L_1^{rev} \cdot L_2 = (L_2^{rev} \cdot L_1)^{rev} \in \mathcal{M}_{twist}(\mathcal{L})$.
    Starting with $L_4 := L_1 \{c_2\}^* L_2 \{c_3\}^* \in \mathcal{M}_{twist}(\mathcal{L})$ one easily shows
(b) $L_1 \cdot L_2^{rev} \in \mathcal{M}_{twist}(\mathcal{L})$ by the same technique.
    $L_5 := twist(L_1 \cdot L_2 \cdot \{c_3\}^*) \cap (\Sigma_1 c_3)^* \Sigma_2^*$ is the basis for proving
(c) $L_1 \cdot twist(L_2) \in \mathcal{M}_{twist}(\mathcal{L})$.
    With $L_1 \cdot L_2 \in \mathcal{M}_{twist}(\mathcal{L})$ also $L_2^{rev} \cdot L_1^{rev} \in \mathcal{M}_{twist}(\mathcal{L})$ by Lemma 6. Applying (b) and then (c) yields $L_2^{rev} \cdot twist(L_1) \in \mathcal{M}_{twist}(\mathcal{L})$, and a reversal followed by application of (a) finally yields
(d) $twist(L_1) \cdot L_2 \in \mathcal{M}_{twist}(\mathcal{L})$.
    All other combinations stated in the lemma are now obtainable from combinations of cases (a) to (d).

$\square$

As described in the motivation before Lemma 11 we can now apply any finite sequence of applications of trio operations (a-*transducer* mappings) and/or *twist* to the two components specified by the two generators from $\mathcal{L}$ for a language $L = L_1 \cdot L_2 \in \mathcal{M}_{twist}(\mathcal{L})$ in order to verify Theorem 12 with the help of Lemma 11:

Theorem 12. *If* $\mathcal{L}$ *is family of languages that is closed with respect to product then* $\mathcal{M}_{twist}(\mathcal{L})$ *is closed under product, too.*

Since we know $\overline{dMIR} \cdot dMIR \in \mathcal{M}_{twist}(dMIR \cup \{\lambda\}) = \mathcal{M}_{twist}(MIR)$ by Lemma 10 we obtain immediately:

Corollary 13. *The family $\mathcal{M}_{twist}(MIR)$ is closed with respect to product.*

Combining Lemma 7, Corollary 13, and Lemma 9 we get the main result:

Theorem 14. $\mathcal{M}_{twist}(MIR) = \mathcal{M}_\cap(MIR)$

A consequence of this new characterization of the family of languages accepted in quasi-realtime by reversal-bounded on-line multipushdown machines we find a new characterization of the recursively enumerable languages:

Corollary 15. $\mathcal{R}e = \hat{\mathcal{M}}_{twist}(MIR)$

Corollary 15 which is similar to the characterizations presented in [BaBo 74], [Salo 81] and [Enge 96] can now be obtained from Theorem ?? by a more direct construction instead of using the new characterization of the family $\mathcal{M}_\cap(MIR)$.

In [JaPe 94] it was proved that the family $\mathcal{M}_\cap(D_1) = PBLIND(n)$ is closed with respect to *twist*. We will now show that for each $k \geq 1$ the family $\mathcal{M}(C_k)$ of languages accepted by blind k-counter machines in quasi-realtime is *twist*-closed, too. To do this let us recall that each blind k-counter machine M can easily be described by a finite state transition diagram in which a directed arc from state p_1 to p_2 is inscribed by the input symbol x to be processed and a vector $\Delta \in \{+1, 0, -1\}^k$ used for updating the counters by adding Δ to the current contents $C_1 \in \mathbf{Z}^k$ of the counters. This will be written as $p_1 \xrightarrow[\Delta]{x} p_2$. A string $w = x_1 x_2 x_3 \cdots x_{n-1} x_n$, $x_i \in \Sigma$ is accepted by M, iff there exists a path in the transition diagram of the form $q_0 \xrightarrow[\Delta_1]{x_1} q_1 \xrightarrow[\Delta_2]{x_2} q_2 \xrightarrow[\Delta_3]{x_3} q_3 \cdots q_{n-1} \xrightarrow[\Delta_n]{x_n} q_n$, where q_0 (q_n) is an initial (resp. final) state and $\Sigma_{i=1}^n \Delta_i = 0$ in each component. The machine starts with empty counters and accepts only when all counters are zero again. It is easy to construct a blind k-counter machine M_2 that accepts L^{rev} from the machine M_1 that accepts $L \in \mathcal{M}(C_k)$: One just has to revert the arcs in the state transition diagram of M_1 and exchange the sets of final and initial states. Now it is not difficult to show that $\mathcal{M}(C_k)$ is *twist*-closed for each $k \geq 1$.

Lemma 16. $\forall k \in \mathbb{N}, k \geq 1 : \mathcal{M}_{twist}(C_k) = \mathcal{M}(C_k)$.

Proof: Let $L \in \mathcal{M}(C_k), L \subseteq \Sigma^*$, then there exists a blind k-counter machine M which accepts $L = L(M)$ in quasi-realtime. In order to accept the set $twist(L)$ we modify the machine M to a new machine M_{twist} as follows: Let Q (Q_0, and Q_f) be the set of states (initial and final states, resp.) of the machine M, then $Q_{twist} := Q^2 \times \{o, e\}$ is the set of states of M_{twist}. The sets of initial (and final) states $Q_{0,twist}$ ($Q_{f,twist}$, resp.) of M_{twist} are given by $Q_{0,twist} := \{(p_0, p_f, o) \mid p_0 \in Q_0, p_f \in Q_f\}$ and $Q_{f,twist} := \{(p, p, o), (p, p, e) \mid p \in Q\}$. Let $w \in L$, $w = x_1 x_2 x_3 \cdots x_{n-1} x_n$, $x_i \in \Sigma$ be a string of length $|w| = n$ which is accepted by

M in a sequence of transitions $q_0 \xrightarrow[\Delta_1]{x_1} q_1 \xrightarrow[\Delta_2]{x_2} q_2 \xrightarrow[\Delta_3]{x_3} q_3 \cdots q_{n-1} \xrightarrow[\Delta_n]{x_n} q_n$. The new machine M_{twist} now uses the finite control of M in the first components of the elements in Q_{twist} in each odd step beginning with the first move, while it is used every second (even) step in the second components backwards. The sequence of transitions of M_{twist} accepting $twist(w)$ now would be

$$\begin{pmatrix} q_0 \\ q_n \\ o \end{pmatrix} \xrightarrow[\Delta_1]{x_1} \begin{pmatrix} q_1 \\ q_n \\ e \end{pmatrix} \xrightarrow[\Delta_n]{x_n} \begin{pmatrix} q_1 \\ q_{n-1} \\ o \end{pmatrix} \xrightarrow[\Delta_2]{x_2} \begin{pmatrix} q_2 \\ q_{n-1} \\ e \end{pmatrix} \xrightarrow[\Delta_{n-1}]{x_{n-1}} \cdots \xrightarrow[\Delta_{\lfloor \frac{n}{2} \rfloor +1}]{x_{\lfloor \frac{n}{2} \rfloor +1}} \begin{pmatrix} q_{\lceil \frac{n}{2} \rceil} \\ q_{\lceil \frac{n}{2} \rceil} \\ x \end{pmatrix},$$

where $x \in \{o, e\}$ depends on the length of the input string:

The last step in this computation is $\begin{pmatrix} q_{\frac{n}{2}} \\ q_{\frac{n}{2}+1} \\ e \end{pmatrix} \xrightarrow[\Delta_{\frac{n}{2}+1}]{x_{\frac{n}{2}+1}} \begin{pmatrix} q_{\frac{n}{2}} \\ q_{\frac{n}{2}} \\ o \end{pmatrix}$, if n is even, and

is $\begin{pmatrix} q_{\lfloor \frac{n}{2} \rfloor} \\ q_{\lceil \frac{n}{2} \rceil} \\ o \end{pmatrix} \xrightarrow[\Delta_{\lceil \frac{n}{2} \rceil}]{x_{\lceil \frac{n}{2} \rceil}} \begin{pmatrix} q_{\lceil \frac{n}{2} \rceil} \\ q_{\lceil \frac{n}{2} \rceil} \\ e \end{pmatrix}$, otherwise.

Conversely, every accepting computation in M_{twist} can be unfolded to yield a valid computation in M showing that only strings of the form $twist(w), w \in L(M)$ are accepted by M_{twist}.

□

Theorem 17. $\forall k \in \mathbb{N}, k \geq 1: M_{twist}(B_k) = \mathcal{M}(C_k)$.

Proof: From Lemma 16 and $B_k \in \mathcal{M}(C_k)$ we see $M_{twist}(B_k) \subseteq \mathcal{M}(C_k)$. To show the converse we have to verify $C_k \in M_{twist}(B_k)$. Let $\not\in \Gamma_k$ and the homorphism $h_{\not\in} : (\Gamma_k \cup \{\not\in\})^* \longrightarrow \Gamma_k^*$ be defined by $h_{\not\in}(x) := x$, if $x \in \Gamma_k$ and $h_{\not\in}(\not\in) := \lambda$. Then $L_k := twist(h_{\not\in}^{-1}(B_k)) \cap \{\not\in b_i, a_j \not\in \mid b_i, a_j \in \Gamma_k\}^*$ is an element of $M_{twist}(B_k)$ and the pairs $\not\in b_i, a_j \not\in$ may appear in any order for all $1 \leq i, j \leq k$. Hence, applying the inverse homomorphism $g : \Gamma_k^* \longrightarrow (\Gamma_k \cup \{\not\in\})^*$ with $g(a_i) := a_i \not\in$ and $g(b_i) := \not\in b_i$ we see $C_k = g^{-1}(L_k) \in M_{twist}(B_k)$.

□

Greibach showed $C_1 \in \mathcal{M}(B_3)$, Lemma 1 in [Grei 78]. We want to show that it is sufficient to use only one more counter to accept C_k using only k+1 partially blind reversal-bounded counters.

Theorem 18. $\forall k \in \mathbb{N}, k \geq 1: \mathcal{M}(C_k) \subsetneq \mathcal{M}(B_{k+1})$.

Proof: "$\subseteq$". Let C_k be accepted by some blind k-counter machine working in quasi-realtime. The new reversal-bounded (k+1)-counter machine M_{k+1} is given as follows: One counter, call it z_0, is used to non-deterministically find the middle of each string $w \in C_k$ by adding 1 in each step when reading a prefix u of $w = uv$. We call this the first phase of the work of M_{k+1}. Then, non-deterministically this phase is stopped and in the following, second, phase the counter z_0 is decreased by 1 in each and every step. One has $|u| = |v| = \frac{|w|}{2}$ if and only if this counter reached zero after reading the last symbol of w. All other counters z_1 to z_k are treated differently in the first and the second phase of M_{k+1}'s computation: If $\Delta = (\delta_1, \ldots, \delta_k)$ is a counter-update used in the first phase of M_k, then $\Delta' := (+1, \delta_1 + 1, \ldots, \delta_k + 1) \in \{2, 1, 0\}^{k+1}$ is the non-decreasing counter-update in M_{k+1} to be used instead. Likewise, if $\Delta = (\delta_1, \ldots, \delta_k)$ is a counter-update used in the second phase of M_k, then $\Delta' := (-1, \delta_1 - 1, \ldots, \delta_k - 1) \in \{-2, -1, 0\}^{k+1}$ is the non-increasing counter-update to be used in M_{k+1} instead. Since the first and the second phase consist of equally many steps, the overall change of the counters is zero again, and exactly the strings form C_k are accepted using k+1 reversal-bounded partially blind counters. Since the new counter-updates now are elements of $\{-2, -1, 0, 1, 2\}^{k+1}$ and not of $\{-1, 0, 1\}^{k+1}$ we have to modify the machine M_{k+1} by splitting moves that increase (or decrease) a counter by 2 into two moves that increase (or decrease) this counter by 1 and all other counters are treated as before in the first step and stay stationary in the second step. This modification gives a partially blind (k+1)-counter machine that accepts C_k in quasi-realtime and with one reversal on each of its counters, hence $\mathcal{M}(C_k) \subseteq \mathcal{M}(B_{k+1})$.

In order to see the strictness of this inclusion we use known results from Ginsburg, [Gins 75], Greibach, [Grei 76], or Latteux, [Latt 78,Latt 79], where it was shown that $B_{k+1} \cap \{a_1\}^* \cdots \{a_{k+1}\}^* \{b_{k+1}\}^* \cdots \{b_1\}^*$ is not an element of $\mathcal{M}(C_k)$, hence $\mathcal{M}(C_k) \subsetneq \mathcal{M}(B_{k+1})$ for each $k \geq 1$.

$\square$

Consequently, $\bigcup_{i \geq 0} \mathcal{M}(C_i) = \mathcal{M}_\cap(C_1)$ forms a strict hierarchy of *twist*-closed semiAFLs and therefore is not principal as *twist*-closed semiAFL.

Ginsburg, [Gins 75] Example 4.5.2, has shown $\mathcal{M}(B_k) \subsetneq \mathcal{M}(B_{k+1})$ and in [Grei 76,Jant 79a] it was shown that $B_{k+1} \notin \mathcal{M}(D_k)$.

We conjecture the following sharpening of the above results: $\forall k \in \mathbb{N}, k \geq 1$: $\mathcal{M}(B_k) \subsetneq \mathcal{M}(C_k)$.

The strictness of the inclusion $\mathcal{M}(B_1) \subsetneq \mathcal{M}(C_1)$ can be shown easily: $\mathcal{M}(B_1) \subseteq \mathcal{M}(lin\,Cf)$ but $\mathcal{M}(C_1) \not\subseteq \mathcal{M}(lin\,Cf)$, and only $\mathcal{M}(B_k) \subseteq \mathcal{M}(C_k)$ is obvious.

Acknowledgement: I thank Julia Maas for discussing various aspects of the twist operation, Olaf Kummer for Theorems 16 and 18and especially one referee for her or his very competent and detailed comments.

References

[BaBo 74]   B.S. Baker and R.V. Book. Reversal-bounded multipushdown machines, J. Comput. Syst. Sci., **8** (1974) 315–332.

[Bers 80]   J. Berstel, Transductions and Context-free Languages, Teubner Stuttgart (1980).

[BoGr 70]   R.V. Book and S. Greibach. Quasi-realtime languages, Math. Syst. Theory **19** (1970) 97–111.

[BoGr 78]   R.V. Book and S. Greibach. The independence of certain operations on semiAFLs, RAIRO Informatique Théorique, **19** (1978) 369–385.

[BoNP 74]   R.V. Book, M. Nivat, and M. Paterson. Reversal-bounded acceptors and intersections of linear languages, Siam J. on Computing, **3** (1974) 283–295.

[Bran 87]   F.J. Brandenburg. Representations of language families by homomorphic equality operations and generalized equality sets, Theoretical Computer Science, **55** (1987) 183–263.

[Chan 81]   T.-H. Chan. Reversal complexity of counter machines, in: Proc. 13th annual ACM Sympos. on Theory of Computing, Milwaukee, Wisconsin, (1981) 146–157.

[EnRo 79]   J. Engelfriet and G. Rozenberg. Equality languages and fixed point languages, Information and Control, **43** (1979) 20–49.

[Enge 96]   J. Engelfriet. Reverse twin shuffles, Bulletin of the EATCS, vol. **60** (1996) 144.

[FiMR 68]   P.C. Fischer, A.R. Meyer, and A.L. Rosenberg. Counter machines and counter languages, Math. Syst. Theory, **2** 1968 265–283.

[Gins 75]   S. Ginsburg, Algebraic and Automata Theoretic Properties of Formal Languages, North Holland Publ. Comp. Amsterdam (1975).

[Grei 76]   S. Greibach. Remarks on the complexity of nondeterministic counter languages, Theoretical Computer Science, **1** (1976) 269–288.

[Grei 78]   S. Greibach. Remarks on blind and partially blind one-way multicounter machines, Theoretical Computer Science, **7** (1978) 311–236.

[HoUl 79]   J.E. Hopcroft and J.D. Ullman, Introduction to Automata Theory, Languages, and Computation, Addison-Wesley Publ. Comp. (1997).

[Jant 79a]   M. Jantzen. On the hierarchy of Petri net languages, R.A.I.R.O., Informatique Théorique, **13** (1979) 19–30.

[Jant 79b]   M. Jantzen. On zerotesting-bounded multicounter machines. In: Proc. 4th GI-Conf. Lecture Notes in Compuer Science, vol. **67**, Springer, Berlin, Heidelberg, New York (1979) 158–169.

[Jant 97]   M. Jantzen. On twist-closed trios: A new morphic characterization of the r.e. sets. In: Lecture Notes of Computer Science vol ?, Springer, Berlin, Heidelberg, New York (1997), to appear.

[JaPe 91]   M. Jantzen and H. Petersen. Twisting Petri net languages and how to obtain them by reducing linear context-free sets, in: Proc. 12th Internat. Conf. on Petri Nets, Gjern (1991) 228–236.

[JaPe 94]   M. Jantzen and H. Petersen. Cancellation in context-free languages: enrichment by reduction. Theoretical Computer Science, **127** (1994) 149–170.

[Kosa 82]   S.R. Kosaraju. Decidability of reachability of vector addition systems, 14th Annual ACM Symp. on Theory of Computing, San Francisco, (1982) 267–281.

[Latt 77]   M. Latteux. Cônes rationnels commutativement clos. R.A.I.R.O., Informatique Théorique, **11** (1977) 29–51.

[Latt 78]   M. Latteux.Langages commutatifs, Thesè Sciences Mathematiques, Univ. Lille (1978).

[Latt 79]   M. Latteux. Cônes rationnels commutatifs. J. Comput. Syst. Sci., **18** (3) (1979) 307–333.

[Mayr 84]   E.W. Mayr. An algorithm for the general Petri net reachability problem, SIAM J. of Computing, **13** (1984) 441–459.

[Salo 78]   A. Salomaa. Equality sets for homomorphisms and free monoids, Acta Cybernetica **4** (1978) 127–139.

[Salo 81]   A. Salomaa. Jewels of formal Language Theory, Computer Science Press, Rockville (1981).

Radix Representations of Algebraic Number Fields and Finite Automata

Taoufik SAFER

LIAFA, Université Paris 7, 2 place Jussieu,
75252 Paris Cedex 05, France
safer@liafa.jussieu.fr

Abstract. Let L be the set of algebraic integers of a number field $\mathbf{Q}[\gamma]$. Let $\beta \in L$ and let A and D be two finite subsets of L with $0 \in D$. Assume that β and all its conjugates have moduli greater than one. Denote by $\nu : A^* \longrightarrow D^*$ the normalization relation which maps any representation of an algebraic integer in base β with digits in A onto the ones of the same number with digits in D. In this case, the relation ν is shown to be computable by a right finite state automaton. If (β, D) is a valid number system, then the normalization ν is a right sub-sequential function. We also prove that the question whether (β, D) does or does not give a valid number system for L can be decided by executing a finite number of arithmetical operations.

1 Introduction

In this work we are interested in the representation of numbers as strings of digits taken from a finite alphabet. This type of symbolic representation allows using discrete techniques to perform computations.

The usual way of representing positive real numbers by their decimal or binary expansions can be generalized to represent complex numbers. Using a fixed *base* $\beta \in \mathbf{C}$ and a fixed *set of digits* $D \subset \mathbf{C}$, every complex number may have a *positional radix expansion* in the *number system* (β, D).

For instance, using the complex base $-1 + i$ with digits from the set 0 and 1, every complex number can be written in the form

$$z = \sum_{k \leq l} d_k (-1 + i)^k,$$

where $l \in \mathbf{Z}$ and $d_k \in D = \{0, 1\}$, for $k \leq l$ (see Knuth [14]). This representation is denoted by

$$z = (d_l d_{l-1} \cdots d_0 . d_{-1} \cdots)_{-1+i},$$

which will be simply noted by the *infinite word* $d_l d_{l-1} \cdots d_0 . d_{-1} \cdots$. The digits to the left of the radix point, $(d_l d_{l-1} \cdots d_0)_{-1+i}$, constitute the *integer part* of the representation.

So, the base $-1 + i$ provides a binary representation of complex numbers using 0 and 1 as digits and integer parts of these representations correspond to *Gaussian integers*, that are elements of $\mathbf{Z}[i] = \{x + iy \mid x, y \in \mathbf{Z}\}$. For example

$$2 + 3i = (-1 + i)^3 + (-1 + i) + 1 = 1011,$$
$$\tfrac{-1+2i}{5} = \sum_{k \leq 1} (-1 + i)^{2k} = 0.01010 \cdots.$$

I. Kátai and J. Szabó [13] proved that this system gives a valid number system for the Gaussian integers. A number system (β, D) is said to be *valid* if the digit set is a complete residue system modulo β for the set L of algebraic integers of $\mathbf{Q}[\beta]$ (*i.e.* $d - d'$ is not multiple of β in L, for all $d, d' \in D$), and if integer parts of representations correspond exactly to elements of L. This implies that a representation of an algebraic integer in a valid number system is unique.

To do arithmetic (addition, subtraction and multiplication) in a number system, two methods can be checked out. The first one is algorithmic; it was used for the decimal system by Al Kashi (1427, see [11] page 311) and for the binary one by G.W. Leibniz [16]. It states for example, that when adding two integers in the binary system the carry is 0 or 1 as $2 = (10)_2$ and have to be carried to the left column. In [7, 8], the same method is used by W. Gilbert to study arithmetic in base $-1 + i$. Addition and multiplication of two numbers written in this base can be performed in the same way as real arithmetic in base 2, except for a change in the carry digits as in this case $2 = (1100)_{-1+i}$. However, division is more complicated and is not to be discussed here (see [9] for division in base $-1 + i$).

We focus our interest in the second method which uses automata. It is a well known example that addition in the usual binary system can be computed by a right sub-sequential finite state automaton (see Eilenberg [5], Chap. V and XI). Its states correspond exactly to the two only possible carries 0 and 1.

The normalization is the function which maps any representation of a positive integer onto the normal one, obtained by the greedy algorithm. Addition is a particular case of normalization. It is known (see [3, 6, 17]) that normalization in base β, where $\beta \in \mathbf{N} \setminus \{0, 1\}$, is a right sub-sequential function.

In the case where the base β is a real number > 1, it is known that the normalization function, which maps any representation of a positive real number onto its greedy representation, is computable by a finite automaton if and only if β is a Pisot number, *i.e.* an algebraic integer such that all its conjugates are smaller than 1 in modulus [1].

A less standard example of normalization function is given by addition in base $-1 + i$ with digits 0 and 1 (see [19]). As in the binary system, there is only a finite number of possible carries while processing addition. This fact implies that only a finite memory is required to store them to do this computation. The automaton given by Figure 1 simulates this process.

This result is generalized to algebraic integer base and digits of a number field $\mathbf{Q}[\gamma]$, seen as a subset of the vector space $\mathbf{Q}[\gamma] \otimes \mathbf{R}$. An appropriate norm is used, instead of the modulus in $\mathbf{Q}[i] \otimes \mathbf{R} \cong \mathbf{C}$, to keep the desirable topological property of the set L of algebraic integers of $\mathbf{Q}[\gamma]$.

More precisely, let us assume the base β to be an algebraic integer such that all its conjugates have moduli greater than one, and let A and D be two finite subsets of L. We show (Theorem 1) that in this case, the normalization relation $\nu : A^* \longrightarrow D^*$ can be computed by a right finite state automaton.

If additionally (β, D) is a valid number system, the normalization ν is shown (Corollary 1) to be a right sub-sequential function.

The question whether a number system is a valid one is slightly different. W. Gilbert [10] shows that if β is an algebraic integer and D is a complete residue system for $\mathbf{Z}[\beta]$ modulo β, that contains 0, then (β, D) is a valid number system if and only if β and all its conjugates have moduli greater than one and there is no positive integer t for which

$$d_{t-1}\beta^{t-1} + \cdots + d_1\beta + d_0 \equiv 0 \pmod{\beta^t - 1} \text{ with } d_{t-1}, \cdots, d_1, d_0 \in D.$$

The same arguments used for normalization functions allow us to prove a slightly different result from Gilbert one (Theorem 2): Assume that the base β and all its conjugates have moduli greater than one and that $D \subset L$ is a complete residue system modulo β. The question whether (β, D) does or does not give a valid number system for L can be decided by executing a finite number of arithmetical operations.

Our result generalize the one given by S. Körmendi [12] in the cubic number field $\mathbf{Q}[\sqrt[3]{2}]$.

2 Preliminaries

2.1 Automata

Let us recall some definitions on automata. More details can be found in [2, 5]. Let A and D be two finite sets. A *finite state automaton* with inputs in A and outputs in D, $\mathcal{A} = (Q, A \times D, E, q_0, \varphi)$, is a directed labeled graph with a finite set of edges, labeled by elements of $A \times D$, and where $\varphi : Q \longrightarrow D^*$ is the *terminal function*. Here, D^* denotes the set of finite sequences of elements of D which are called *words*. The empty word is denoted ε.

We will be interested only in *right finite state automaton*. That is one which processes words from right to left. Such an automaton is said to be *sub-sequential* if it is deterministic on inputs, that is to say, if $p \xrightarrow{a/d} q$ and $p \xrightarrow{a/d'} q'$ are two edges of E, then necessarily $q = q'$ and $d = d'$. In such a case, the automaton outputs serially one digit for each input digit. The image by the terminal function φ of the last state read comes as a prefix of the output.

Sub-sequentiality of the automaton is very important for algorithmic efficiency reasons. In this paper, the automaton may be not sequential and φ may be just a relation and not necessarily a function. So, the input word may have several output words corresponding to different paths in the automaton and different images by φ of the last read states.

The language of the right finite state automaton $\mathcal{A}$ is the set

$$\Big\{(x,y) \in A^* \times D^* \mid y = y''y', q_0 \xrightarrow{x/y'} q \text{ is a path in } \mathcal{A} \text{ and } y'' \in \varphi(q)\Big\}.$$

A function $f : A^* \longrightarrow D^*$ is *computable by a right finite state automaton* (respectively *by a sub-sequential one*), if there exists such an automaton $\mathcal{A}$ the language of which is equal to the graph of f.

2.2   Number system

Let us be given an algebraic number field $\mathbf{Q}[\gamma]$ of degree n and denote by L the ring of its algebraic integers. The field $\mathbf{Q}[\gamma]$ can be seen as a subset of the $\mathbf{R}$-vector space $\mathbf{Q}[\gamma] \otimes \mathbf{R}$.

Let β be an algebraic integer of $\mathbf{Q}[\gamma]$. The multiplication by β in $\mathbf{Q}[\gamma]$ extends to $\mathbf{Q}[\gamma] \otimes \mathbf{R}$ as a linear application denoted by ϕ. The norm of β is defined as the *determinant of ϕ*: $N(\beta) = det(\phi)$.

In the vector space $\mathbf{Q}[\gamma] \otimes \mathbf{R}$, the canonical basis $\mathcal{B} = \{1 \otimes 1, \gamma \otimes 1, \cdots, \gamma^{n-1} \otimes 1\}$ is fixed. A vector $\alpha \in \mathbf{Q}[\gamma] \otimes \mathbf{R}$ will be indifferently denoted by its coordinate colon vector in $\mathcal{B}$:

$$\alpha = \sum_{j=0}^{n-1} \alpha_j(\gamma^j \otimes 1) \cong \begin{pmatrix} \alpha_0 \\ \alpha_1 \\ \vdots \\ \alpha_{n-1} \end{pmatrix}$$

The Euclidean vector norm of α is $\|\alpha\| = \sqrt{\sum_{j=0}^{n-1} \alpha_j^2}$. The corresponding matrix norm is defined, for a matrix U with real coefficients, by

$$\|U\| = \max_{\alpha \neq 0} \frac{\|U\alpha\|}{\|\alpha\|}.$$

This norm is called the *spectral norm*, since $\|U\|$ is equal to the spectral radius of U.

Denote by P_β the minimum polynomial of β and by M the matrix of ϕ in $\mathcal{B}$. The eigenvalues of M are exactly the Galois conjugates of β, *i.e.* the roots $\lambda_1, \cdots, \lambda_n$ of its minimum polynomial. The *spectral radius* of M is defined by

$$\rho(M) = \max_{1 \leq j \leq n} |\lambda_j|,$$

where $|.|$ is the usual modulus on $\mathbf{C}$.

The value homomorphism v is defined on $\mathbf{Q}[\gamma] \otimes \mathbf{R}$ by

$$v\Big(\sum_{i=0}^{n-1} y_i \gamma^i \otimes 1\Big) = \sum_{i=0}^{n-1} y_i \gamma^i \in \mathbf{C}.$$

Denote by L the ring of algebraic integers of $\mathbf{Q}[\gamma] \cong \mathbf{Q}[\gamma] \otimes 1$. Let D be a finite subset of L, with $0 \in D$. The set D is a *complete residue system modulo β* if for any d and $d' \in D$, there is no algebraic integer $m \in L$ such that $d - d' = m\beta$. The number system (β, D) is said to be a *valid number system* for L if the set D is a complete residue system modulo β and if every $\alpha \in L$ can be written in the form

$$\alpha = \sum_{k=0}^{m} d_k \beta^k,$$

where $m \in \mathbf{N}$ and where the digits d_k are in D. When the digits chosen to represent numbers are $D = \{0, 1, \cdots, |N(\beta)| - 1\}$, then (β, D) is said to be a *canonical number system*.

2.3 Normalization Relations and Functions

Let π be the numerical value function of finite words on the (infinite) alphabet $\mathbf{C}$,

$$\pi : \quad \mathbf{C}^* \quad \longrightarrow \mathbf{C}$$
$$u_l u_{l-1} \cdots u_0 \longmapsto \sum_{j=0}^{l} u_j \beta^j .$$

Let A be a finite set of L. The normalization relation

$$\nu : A^* \longrightarrow D^*$$
$$f \longmapsto \nu(f),$$

is defined by $\nu(f) = \{g \in D^* \mid \pi(f) = \pi(g)\}$. It associates to a representation $f = u_l u_{l-1} \cdots u_0$ of a complex number $z = \pi(u) = \sum_{j=0}^{l} u_j \beta^j$ on the alphabet A, all its representations on the alphabet D.

Remark *From now on we will only deal with $\mathbf{Q}[\gamma]$, not with the whole of $\mathbf{Q}[\gamma] \otimes \mathbf{R}$. We will keep denoting by ϕ and v the restriction to $\mathbf{Q}[\gamma]$ of the corresponding objects in $\mathbf{Q}[\gamma] \otimes \mathbf{R}$. The base β and the two finite sets A and D are assumed to be in L, with $0 \in D$. In particular the elements of A and D are vectors, not numbers: the corresponding numbers are obtained by applying the value homomorphism v.*

3 Main results

The main results of this paper are in two parts. In the first one is given a characterization of algebraic integers for which the normalization relation is computed by a right finite state automaton (Theorem 1), or a sub-sequential one (Corollary 1). In the second is given a characterization of those which may give a valid number system (Theorem 2).

Theorem 1 *If the algebraic integer β and all its conjugates have moduli greater than one and if the two alphabets A and D are finite subsets of L, then the normalization relation $\nu : A^* \longrightarrow D^*$ can be computed by a right finite state automaton.*

Proof. Let M be the matrix associated with β and let τ be the spectral radius of the matrix M^{-1}. Its spectral norm is $\|M^{-1}\| = \tau$. We set

$$m = \max_{a \in A, d \in D} \|a - d\| \text{ and } c = \frac{m\tau}{1 - \tau}.$$

As the algebraic integer β and all its conjugates have moduli greater than one, we have $0 < \tau < 1$, and $0 < c$.

To normalize a word of A^*, the latter is read digit by digit from right to left. The first carry is $s = 0$. The algorithm consists in adding the carry to the input digit $a \in A$ and searching (if there exist some)

$$d \in D \text{ and } s' \in L, \text{ such that } s + a = Ms' + d.$$

The new carry is s'. We set $s := s'$.

It is easily seen that all the carries s have an Euclidean norm $\|s\| < c$, since the first one is zero ($c > 0$) and recursively: if $\|s\| < c$, the next carry $s' = M^{-1}(s + a - d)$ has a norm

$$\|s'\| \leq \|M^{-1}\|(\|s\| + \|a - d\|) < \tau(c + m) = c.$$

The finite state automaton $\mathcal{T} = (Q, A \times D, E, 0, \varphi)$ is defined as follows: the states are the possible carries. So, any $s \in Q$ is an algebraic integer of modulus smaller than c. The initial state is zero; for every $s, s' \in Q$ an edge is directed from s to s' and labeled by $a/d \in A \times D$ (*i.e.* $s \xrightarrow{a/d} s'$ is an edge of $\mathcal{T}$), if and only if

$$s + a = Ms' + d.$$

The terminal function φ associates to a state $s \in Q$ all its representations in base β with digits from D.

As the state norm is bounded by the constant c, $\mathcal{T}$ has only a finite number of states from the next well-known fact (see [15]):

Lemma 1 *The set L of algebraic integers in $\mathbf{Q}[\gamma]$ forms a lattice.*

To end the proof, it is easily shown that the language of $\mathcal{M}$ is exactly the graph of ν. $\square$

Corollary 1 *If the algebraic integer β and all its conjugates have moduli greater than one and if (β, D) is a number system, then $\nu : A^* \longrightarrow D^*$ can be computed by a right sub-sequential finite state automaton.*

Proof. If two edges of the form

$$s \xrightarrow{a/d_1} s_1 \text{ and } s \xrightarrow{a/d_2} s_2$$

are in $\mathcal{T}$, then we have $s + a = Ms_1 + d_1 = Ms_2 + d_2$. So, d_1 and d_2 are congruent modulo β. We deduce that $d_1 = d_2$ et $s_1 = s_2$. $\square$

In [7, 9], W. Gilbert shows that: *if β or any of its conjugates have modulus smaller than one, it cannot give a number system using digits from $\{0, 1, \cdots, |N(\beta)| - 1\}$ (i.e. canonical number system).* This fact can easily be generalized to digit sets $D \subset L$.

Thus, in Corollary 1 the hypothesis (β, D) *"is a number system"* implies that β and all its conjugates have moduli greater than or equal to one (*i.e.* $\tau \leq 1$). The reciprocal assertion is given separately for the cases: $\tau < 1$ and $\tau = 1$.

For the first case, we have the following

Theorem 2 *Assume that β and all its conjugates have moduli greater than one* (i.e. $\tau < 1$) *and let $D \subset L$ be a complete residue system modulo β. The question whether (β, D) does or does not give a number system for L can be decided by executing a finite number of arithmetical operations.*

Proof. Let M, τ and c be defined as before with now $m = \max_{d \in D} \|d\|$. To represent an algebraic integer $s \in L$ in base β with digits in D, we proceed as follows: set $s_0 := s$ and for all $j \in \mathbf{N}$

$$s_j = M s_{j+1} + d_j, \text{ with } d_j \in D.$$

So, (β, D) is a number system if and only if the sequence $(s_i)_{i \geq 0}$ converges to zero for all integral initial values.

It is easily seen that for all initial values, the sequence $(s_i)_{i \geq 0}$ is strictly increasing until having norm smaller than or equal to c: if $\|s_j\| > c$ for some j, then

$$\|s_{j+1}\| = \|M^{-1}(s_j - d_j)\| \leq \tau(\|s_j\| + m) < \|s_j\|,$$

so

$$\limsup_j \|s_j\| \leq c.$$

From Lemma 1, the bounded domain $F = \{s \in L \mid \|s\| \leq c\}$ contains only a finite number of vectors. Let t be the cardinal of F.

Thus, (β, D) is a number system if and only if the sequence $(s_i)_{i \geq 0}$ converges to zero in less than t steps for all initial values $s_0 \in F$. $\qquad\square$

Example 1 In the space $\mathbf{Q}[i] \otimes \mathbf{R}$, the canonical basis $\{1 \otimes 1, i \otimes 1\}$ is fixed. The matrix associated with the multiplication by $-1 + i$ in this space is

$$M_{-1+i} = \begin{pmatrix} -1 & -1 \\ 1 & -1 \end{pmatrix}$$

It is well known that this base gives a canonical number system using the digit set $D = \{0, 1\}$ (see [14, 13, 7]). This fact is proved here again using Theorem 2. The set D is a complete residue system modulo $-1 + i$ of $L = \mathbf{Z}[i]$. The spectral radius of the matrix M_{-1+i}^{-1} is $\tau = \max\{|(-1 + i)^{-1}|, |(-1 - i)^{-1}|\} = \frac{\sqrt{2}}{2}$, and $m = \max\{\|d\| \mid d \in D\} = 1$. So, we have $c = \frac{m\tau}{1-\tau} = 1 + \sqrt{2}$, thus

$$F = \{s \in \mathbf{Z}[i] \otimes 1 \mid \|s\| \leq 1 + \sqrt{2}\}$$
$$\cong \{0, 1, -1, i, -i, 1 - i, -1 + i, -1 - i, 1 + i, 2, -2, 2i, -2i, 1 + 2i, 1 - 2i, 2 + i,$$
$$2 - i, -2 + i, -2 - i, -1 + 2i, -1 - 2i\}.$$

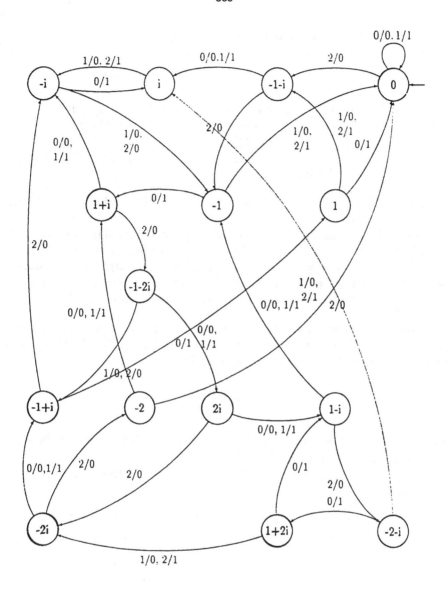

Figure 1: Addition in base -1 + _i._

Each element of F can be represented in base $-1 + i$ with digits 0 and 1 as follows:

$$-1 = 11101, \qquad -1 + i = 10, \qquad -2 = 11100,$$
$$1 - 2i = 101, \qquad -2 - i = 11101011, \quad -1 - i = 110,$$
$$2i = 1110100, \qquad 2 + i = 1111, \qquad 1 - i = 111010,$$
$$1 + i = 1110, \qquad -2i = 100, \qquad 2 - i = 111011,$$
$$-i = 111, \qquad 2 = 1100, \qquad 1 + 2i = 1110101,$$
$$-2 + i = 11111, \qquad i = 11, \qquad -1 + 2i = 11001,$$
$$-1 - 2i = 11101001.$$

From Theorem 2, we deduce that $(-1 + i, D)$ is a valid number system for the Gaussian integers.

By Corollary 1, addition and multiplication by a fixed Gaussian integer in $(-1 + i, D)$ can be computed by a right sub-sequential finite state automaton. In Figure 1 is given the automaton computing addition. Its state set Q is formed by Gaussian integers which are the possible carries. Its terminal function $\varphi :$ $Q \longrightarrow \{0, 1\}^*$ associates to a carry its representation in base $-1 + i$.

Example 2 Let us now consider the space $\mathbf{Q}[i\sqrt{3}] \otimes \mathbf{R}$ with the canonical basis $\{1 \otimes 1, i\sqrt{3} \otimes 1\}$. Let us consider the number system $(i\sqrt{3}, D)$, where $D = \{0, 1, \frac{1 + i\sqrt{3}}{2}\}$. The matrix associated with $i\sqrt{3}$ is

$$M_{i\sqrt{3}} = \begin{pmatrix} 0 & -3 \\ 1 & 0 \end{pmatrix}.$$

As in Example 1, we show that $(i\sqrt{3}, D)$ is a valid number system for the set of Eisenstein integers $L = \mathbf{Z}[\frac{1 + i\sqrt{3}}{2}]$ (see [18]).

By Corollary 1, addition and multiplication by a fixed Eisenstein integer are sub-sequential.

We end this paper by conjecturing that *if the base β or any of its conjugates have modulus one (i.e. $\tau = 1$), it cannot give a number system for any digit set $D \subset L$.*

Acknowledgment

Dominique Bernardi greatly contributed to the improvement of a preliminary version of this work. I would like also to thank Jean-Paul Allouche, Gilles Christol and Christiane Frougny for interesting discussions. By their careful reading, the referees helped to the achievement of this work.

References

1. D. Berend and Ch. Frougny, Computability by finite automata and Pisot bases. *Math. Systems Theory* **27** (1994), 274–282.

2. J. Berstel, *Transductions and context-free languages*, Teubner, 1979.

3. K. Čulik II et A. Salomaa, Ambiguity and decision problems concerning number systems. *Inform. and Control* **56** (1983), 139–153.

4. J. Duprat, Y. Herreros and S. Kla, New redundant representations of complex numbers and vectors, *IEEE Transactions on Computers*, vol. **42**, no. **7**, Jul. 1993, 817–824.

5. S. Eilenberg, *Automata, Languages and Machines*, vol. A, Academic Press, 1974.

6. Ch. Frougny, Representation of numbers and finite automata. *Math. Systems Theory* **25** (1992), 37–60.

7. W.J. Gilbert, Radix representations of quadratic fields. *J. Math. Anal. Appl.* **83** (1981), 264–274.

8. W.J. Gilbert, Arithmetic in complex bases. *Math. Mag* **57** (1984), 77–81.

9. W.J. Gilbert, The Division algorithm in complex bases. *Can. Math. Bul.* **39** (1996), 47–54.

10. W.J. Gilbert, Gaussian Integers as Bases for Exotic Number Systems. Unpublished manuscript (1994).

11. G. Ifrah, *Histoire Universelle des Chiffres, tome 2*, Robert Laffont, 1994.

12. S. Körmendi, Canonical number systems in $\mathbf{Q}[\sqrt[3]{2}]$. *Acta Sci. Math.* **50** (1986), 351–357.

13. I. Kátai and J. Szabó, Canonical number systems for complex integers. *Acta Sci. Math. Hung.* **37** (1975), 255–260.

14. D.E. Knuth, *The Art of computer programming, vol. 2: Seminumerical algorithms*, 2nd ed., Addison-Wesley, 1988.

15. S. Lang, *Algebra*, Addison-Wesley, 1965.

16. G.W. Leibniz, *Mémoires de l'Académie Royale des Sciences*, Paris (1703), 110–116.

17. G. Rauzy, Systèmes de numération. *Journées S.M.F.* Théorie élémentaire et analytique des nombres, 1982, 137–145.

18. A. Robert, A good basis for the computing with complex numbers. *Elemente der Mathematik* **49** (1994), 111–117.

19. T. Safer, *Représentation des nombres complexes et automates finis*, Ph.D dissertation, Université Paris 6, 1997.

Sorting and Searching on the Word RAM*

Torben Hagerup

Max-Planck-Institut für Informatik, D–66123 Saarbrücken, Germany
torben@mpi-sb.mpg.de

Abstract. A word RAM is a unit-cost random-access machine with a word length of w bits, for some w, and with an instruction repertoire similar to that found in present-day computers. The simple lower bounds for the problems of sorting and searching valid in the comparison-based model do not hold for the word RAM, so that the well-known algorithms for these tasks are not optimal for the word RAM. This paper gives an overview of faster algorithms known for sorting and searching on the word RAM, many of which were developed within the last few years.

Key words. Sorting, searching, dictionaries, word RAM, word-level parallelism, conservative algorithms, exponential range reduction, tries, fusion trees, exponential search trees, multiplication, AC^0, network flow.

1 Introduction

The traditional model of computation used in the study of such problems as sorting is the comparison-based model, which counts only comparisons, but on the other hand insists that the output of an algorithm must be deducible from the outcomes of the comparisons between input keys carried out by the algorithm, other steps in the algorithm serving only to figure out which comparisons to make. It is well-known that the complexity of sorting n keys in the comparison-based model is $\Theta(n \log n)$. The comparison-based model is simple, fundamental, and intellectually appealing, but it fails to account for very practical and successful algorithms such as bucket sorting, which sorts small integers in linear time by using them as addresses, i.e., by interpreting each input key directly as a pointer to the "bucket" to which it belongs. Bucket sorting does not carry out any comparisons between input keys and cannot be formulated in the comparison-based model, but is certainly a viable sorting method for RAMs and real, commercial computers. This seems to point to a serious gap between "theory" (the comparison-based model) and "practice" (the RAM).

For a long time, bucket sorting was treated as an odd bird that "cheats" in a seemingly well-understood and therefore harmless manner, but the question of the complexity of (integer) sorting on the RAM was occasionally addressed. Paul and Simon [33] showed that a unit-cost RAM with addition, subtraction,

---

* This work was carried out while the author held a visiting position at the Department of Computer Science, University of Copenhagen, Denmark.

and multiplication as its only arithmetic instructions cannot sort faster than in $\Theta(n \log n)$ time. On the other hand, they and Kirkpatrick and Reisch [30] demonstrated that with a richer instruction repertoire, the unit-cost RAM becomes capable of linear-time (integer) sorting. The latter results did not bring about a revolution in the theory and practice of sorting because the corresponding algorithms were felt to cheat in a very real manner. In order to sort n integers of w bits each, they use intermediate results of nw or more bits, which makes them practically irrelevant and theoretically less interesting—the unit-cost assumption had never been intended to be stretched that far. Kirkpatrick and Reisch were aware of this issue and suggested as a more reasonable model that of *conservative* algorithms, algorithms that work on w-bit input integers using only operations on integers of $O(w)$ bits each. This suggestion was embraced by later work, beginning with that of Fredman and Willard [24, 25], who blasted through the $\Theta(n \log n)$ barrier of comparison-based sorting by showing that n integers can be sorted in $O(n\sqrt{\log n})$ time by a conservative RAM algorithm. Although the work of Fredman and Willard represented an exciting breakthrough, it is probably fair to say that a subsequent flurry of research activity took its inspiration mainly from the demonstration by Andersson et al. [5] of a very simple conservative sorting algorithm with a running time of $O(n \log \log n)$, the arguments of Fredman and Willard having been too complicated to invite much follow-up work. The aim of this paper is to give simple descriptions of some of the algorithms and data structures that have been developed and to draw connections between them and discuss their importance. We will focus on sorting and the problem sometimes known as *searching*, namely maintaining a set of objects, each identified by an integer key, under insertion, deletion, and queries for the objects with keys closest on either side to a given key value. The development begins slowly, but picks up speed to cover also some new results of interest to the specialist. A certain familiarity with standard algorithms and data structures will be assumed. The aim is not a meticulous assignment of credit; it will be stated who did what, but the algorithms will not be described in the chronological order of their discoveries, for which reason incremental work may appear more significant than it was. We will not deal in any detail with lower bounds or with randomized or parallel algorithms, but provide pointers to related work on these subjects.

Rather than speaking about conservative algorithms, we will use the possibly more suggestive term *word-based algorithms*, and the corresponding model will be called the *word model*. An illuminating view is that a word-based algorithm potentially may exploit *word-level parallelism*: We can think of each bit position as having a tiny processor that, during the execution of each instruction, operates in parallel with the processors of other bit positions and to some extent independently of them. The independence among the bit processors is perfect in the case of bitwise Boolean operations, where each bit processor simply functions as a Boolean gate. In the case of addition and subtraction, carry bits cause "communication" between the bit processors, but of a very restricted kind, whereas the notion of bit processors essentially breaks down for multiplication. At any rate, word-level parallelism is the only source from which a word-based algorithm may derive increased efficiency relative to more traditional algorithms. On a 64-bit machine, e.g., we should therefore not expect a word-based algorithm to be more than 64 times faster than its traditional counterparts (except, perhaps, if it makes heavy use of multiplication). However, this is good news: Although 64 is a "constant factor", it is a pretty large one, and in practical terms a speedup of 64 would be fabulous. Unfortunately, known word-based algorithms do not offer that kind of speedup, even theoretically, because they view fields of several bit positions as the smallest unit, not exploiting parallelism within fields (except as embodied in the usual arithmetic operations). In the author's opinion, it is not clear at the time of writing whether word-based algorithms will exert a direct practical influence; for positive evidence, see [8]. As concerns the theory of algorithms, on the other hand, there can be little doubt that the word-based algorithms are with us to stay. The word model is natural and could even be claimed to be "*the* right model", and some of the word-based algorithms and data structures are so simple and appealing that they will gain entry into textbooks.

What should be considered a word-based algorithm or data structure is largely a matter of taste. Algorithmic techniques such as the "four Russians' trick" in its many variations and algorithms and data structures for "bounded universes" such as the van Emde Boas tree [43] or plain bucket sorting have been around for a long time. Nonetheless, the recent results do seem to have a distinctive flavor, one characteristic feature being the emphasis on bounds that depend on n, the problem size, but not on w, the word length (compare this with the van Emde Boas tree, e.g., which offers execution times of $O(\log w)$ per operation).

As suggested by the word-parallelism view, the existing body of knowledge concerning parallel algorithms is a rich source of ideas for the design of word-based algorithms. However, word-based algorithms are not just parallel algorithms in a new guise. For one thing, the relevant parallel model is not precisely like any of the traditional parallel models, the closest relative, somewhat unexpectedly, appearing to be the concurrent-read concurrent-write parallel RAM (CRCW PRAM). Moreover, the design of word-based algorithms calls for a unique and interesting blend of sequential and parallel algorithmic techniques, as the reader is invited to discover.

After a definition and discussion of the word model of computation in Sect. 2, we investigate the problem of searching, formalized in the definition of a *neighbor dictionary*, under the assumption that the operation of multiplication is not available. Section 3 considers a simple special case of the problem, where only a small number of objects with small keys are to be maintained, while Sect. 4 deals with the general case. Section 5 describes a very simple algorithm for sorting in $o(n \log n)$ time. In Sect. 6 we return to the searching problem, now allowing multiplication, but reducing the amount of space needed.

2 The Word Model of Computation

2.1 Definition

The word model of computation is a variation of the classic RAM model of Cook and Reckhow [17], the main differences being that the contents of all memory cells are assumed to be integers in the range $\{0, \ldots, 2^w - 1\}$, that some additional instructions are available, and that some of the new instructions are natural only if integers are assumed to be represented as strings of w bits, an assumption that we will therefore make.

Just as a usual RAM, a *word RAM* is assumed to have an infinite memory consisting of cells with addresses $0, 1, 2, \ldots$, not assumed to be initialized in any particular way. We adopt usual conventions for presenting input to the machine and obtaining output from it and assume that the machine operates on its memory according to a program that is a finite list of possibly labeled instructions with operands. The instruction set is assumed to contain the usual range of load and store instructions and conditional and unconditional jumps, and the possible forms of operands are assumed to allow for immediate operands (i.e., literal constants) as well as direct and indirect addressing. We shall not elaborate on any of this, but concentrate on the differences to the usual RAM.

As mentioned above, a parameter of the word RAM is its word length w, a positive integer. The content of every memory cell is an integer x in the range $\{0, \ldots, 2^w - 1\}$, often called a *word* and identified with the bit string $x_{w-1} \cdots x_0$, where $x = \sum_{i=0}^{w-1} x_i \cdot 2^i$. The arithmetic instructions take two operand words x and y and compute a result word z equal to $(x \oplus y) \bmod 2^w$, where $\oplus \in \{+, -, \cdot\}$ in the case of addition, subtraction, and multiplication, $x \oplus y = x \cdot 2^y$ in the case of left shift, and $x \oplus y = \lfloor x \cdot 2^{-y} \rfloor$ in the case of right shift; the latter operations can also (and more naturally) be viewed as (noncyclic) shifts of x by y bit positions with zero filling. The binary Boolean operations AND and OR take two operand words $x_{w-1} \cdots x_0$ and $y_{w-1} \cdots y_0$ and compute $z_{w-1} \cdots z_0$, where $z_i = x_i \bowtie y_i$, for $i = 0, \ldots, w-1$, and $\bowtie \in \{\wedge, \vee\}$. The operation NOT, finally, takes a single word operand $x_{w-1} \cdots x_0$ and produces the word $z_{w-1} \cdots z_0$ with $z_i = 1 - x_i$, for $i = 0, \ldots, w-1$. The execution of every instruction is assumed to take constant time.

Rather unfortunately, perhaps, the word RAM actually is not a single model, but a family of related models differing in the exact arithmetic instruction set assumed to be available. We have to deal with this issue explicitly, since much of the effort in word-based algorithms in fact has been directed towards translating algorithms between instruction sets. At the very least, we will assume the availability of a *restricted instruction set* consisting of addition, subtraction, left and right shifts, and the bitwise Boolean operations AND, OR, and NOT, and speak of the *restricted model*. It is frequently necessary to augment the restricted instruction set one way or another. Augmenting it with just the multiplication instruction, we arrive at the *multiplication instruction set* and the *multiplication model*. Another possibility is to allow precisely all AC^0 *instructions*, those instructions that implement functions computable by unbounded-fanin circuits of constant depth and polynomial size (here "polynomial" means polynomial in the word length w). This convention is known to exclude the multiplication instruction [26, 28], but to allow all of the other instructions discussed above as well as a host of other instructions. We will call the corresponding model the AC^0 model.

It turns out to be necessary to supply most of the interesting word-based algorithms with a small number of integer constants that depend (only) on the word length w—if nothing else, the algorithms need to know the number w itself. All of the relevant constants can be computed in $O(\log w)$ time; but since we do not assume w to be bounded in terms of the problem size n, $\Theta(\log w)$ could far exceed the running time of the remaining parts of an algorithm. As a concession to reality, where the computation of such universal constants is hardly the biggest challenge, we therefore allow a fixed number of constants to be available free of charge; one can imagine, e.g., that each is loaded into a register by its own new instruction. We will call such constants *native constants*, since one can think of them as having been hard-wired into a machine by its manufacturer.

Issues less of the model of computation than of the kinds of results desired are space requirements and randomization. We may insist that the space used to solve problems of size n be $O(n)$—measured, of course, in w-bit words—or we may not restrict the space used, in which case the space used by many of the algorithms described here is around 2^w ("around" in a very vague sense). Likewise, we may insist that algorithms be deterministic, or we may allow randomization and be content with expected rather than worst-case bounds. In the latter case, the instruction repertoire is assumed to include one more instruction, *Random*, that takes a word operand x and produces an integer drawn from the uniform distribution over the set $\{0, \ldots, x\}$ and independent of all other such integers.

2.2 Discussion

How realistic and practically relevant is the word RAM? First of all, the basic assumption of a binary, word-oriented computer certainly fits present-day computers very well. The restriction to integer data is a restriction, but one that still leaves many important application areas. Moreover, as has been pointed out repeatedly, the IEEE 754 floating-point standard [29, p. 228] contains provisions that may allow algorithms developed for integer data to run without change also on floating-point data. Thus algorithms working on integers are well-motivated. Our model does not have any provisions for representing negative numbers. This design decision was made in the interest of simplicity and carries no real significance. It is easy to translate all algorithms discussed here to a machine that represents negative numbers using, e.g., the two's-complement system.

The unit-cost assumption is a good approximation in many situations, and is a bad approximation in others. Since any shortcomings of the model in this regard appear to be as serious for the usual RAM as for the word RAM, we will ignore them here.

Practically all modern computers have instruction sets that include addition, subtraction, and bitwise Boolean and shift operations, and multiplication is also widely available. Thus our restricted and multiplication instruction sets appear to be quite realistic; if anything, they underestimate the capabilities of modern computers (which may, e.g., have hardware instructions for integer division), but not as badly as the comparison-based model. The urge, nonetheless, to avoid the use of multiplication stems partly from the fact that alone among the instructions considered, it is not an AC^0 instruction, i.e., it cannot be realized by unbounded-fanin circuits of constant depth and size $w^{O(1)}$, which may make the unit-cost assumption appear untenable for large word lengths. The validity of this argument in practical terms should not be overvalued. Even AC^0 instructions such as addition are not realized in practice through bounded-depth circuits, except in so far as everything on a chip is "bounded-depth", and hardware designers are seldom particularly interested in the complexity class AC^0, one reason being that gates with a large fanin, essential in all nontrivial constant-depth circuits, are considered expensive (in terms of area, power consumption, and operational speed) or unfeasible. Still, the fact remains that multiplication on real computers is typically somewhat slower than addition.

In practical terms, shift operations with a variable shift distance are also somewhat more problematic than addition, and one might be tempted to exclude them from the model. However, it seems intuitively plausible, and was shown by Ben-Amram and Galil [12] in the case of sorting, that an efficient algorithm for many problems must employ at least one instruction that allows the higher-order bits of words used in a computation to influence the lower-order bits. Among the instructions in the multiplication instruction set, only right shift has this property, all other operations moving information only towards the more significant bits. Thus we cannot hope to make do even without shifts (unless we introduce other "right-moving" operations such as double-precision multiplication, where lower-order bits of the more significant result word are influenced by higher-order bits of the operand words).

Instruction sets of real computers, of course, do not contain arbitrary AC^0 instructions. In future they might, however, provided that the general opinion converges on a very small number of nonstandard AC^0 instructions considered to be truly useful. This is one possible motivation for studying the AC^0 model.

The space issue is very important in practical terms. The algorithms discussed here are of interest only if w is considerably larger than $\log n$, where n is the problem size, and then an algorithm with space requirements around 2^w is not practical. The issue of randomization is also important, although probably less so. Randomized algorithms usually perform very well in practice. On the other hand, randomized algorithms are invariably implemented with pseudo-randomness as opposed to true randomness, which is unsatisfactory in theory and occasionally leads to problems in practice. One should note that the space and randomization issues are related. Given a (deterministic or randomized) algorithm for the multiplication model that needs a lot of space, we can always use universal hashing [15] to derive a randomized algorithm with the same expected running time, up to a constant factor, that uses space at most proportional to the running time. The method is simply to replace accesses to a large (but sparsely used) memory by accesses via a hash function chosen at random from a universal class to a table of size proportional to the maximum number of memory cells in use at any one time. The hashing schemes of [15] use instructions not present in the multiplication instruction set, but Dietzfelbinger et al. [20] describe a universal class of hash functions that can be evaluated in constant time in the multiplication model.

We finally discuss two minor technical issues. First, whenever we deal with an input of size n or with a data structure storing n keys, we will assume that $n < 2^w$. This is natural, since otherwise n is not a representable integer. Some of the algorithms described here, however, may need more than 2^w memory cells, or it may take additional tricks to bring them below 2^w. We will not worry about this, but assume that the machine architecture provides a way to address sufficient memory (e.g., double-word addresses, which can be shown to suffice).

Second, although the word length is exactly w bits, standard algorithms for arithmetic on multiple-precision integers (e.g., the school methods for addition, multiplication, etc., generalized to w-bit numbers) allow us to treat a constant number of words (of altogether $O(w)$ bits) as a single word, the only penalty being a constant factor in the time and space requirements. We will therefore assume that intermediate results may have $O(w)$ bits, rather than exactly w bits. In particular, we can simulate a double-precision multiplication in constant time.

3 Ranking Dictionaries

3.1 Static Ranking

In this subsection we consider the problem of computing the *rank* of an integer x in a multiset S of k integers, i.e., the number of elements in S smaller than or equal to x. A parallel algorithm for this problem might comprise the following steps:

1. Broadcast x to each of k processors $P_1, \dots, P_k$.
2. For $i = 1, \dots, k$, let P_i compare x with the ith element y_i of S, setting $b_i = 1$ if $x \geq y_i$ and setting $b_i = 0$ otherwise.
3. Compute and output $\sum_{i=1}^{k} b_i$.

Our next task is to translate this algorithm to the word-based setting. Of course, we will assume that the numbers x and $y_1, \dots, y_k$, which will be called *keys*, are representable, i.e., lie in the range $\{0, \dots, 2^w - 1\}$. In order to make use of the word-level parallelism, we will in fact assume more, namely that the keys $y_1, \dots, y_k$ are so small that we can store all of them together in a single word of w bits. Thus each key resides in its own *field* of f bits, where $f = w/k$ (here and in the following, we will ignore issues of integrality, thus writing $f = w/k$ instead of $f = \lfloor w/k \rfloor$; no essential ideas will be lost). The fields containing $y_1, \dots, y_k$ are placed next to each other in a single word Y, as shown in Fig. 1.

In the context of word-based algorithms, Step 1 of the parallel algorithm above obviously means to create the word X shown in Fig. 2. In the multiplication model, this can be done simply by multiplying x by the integer $\sum_{i=0}^{k-1} 2^{if}$, which we can consider to be a native constant if f and therefore $k = w/f$ are

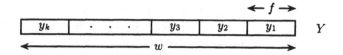

Fig. 1. The entire multiset $S = \{y_1, \ldots, y_k\}$ stored in a single word Y.

fixed. The reason why this works is illustrated in Fig. 3. Although multiplication is a commutative operation, it is often useful to view it as shifting several
copies of one operand, as determined by the bits set (i.e., equal to 1) in the other
operand, and then adding the shifted copies. In Fig. 3, copies of x are shifted
left by $0, f, 2f, \ldots, (k-1)f$ bit positions, i.e., by $0, 1, 2, \ldots, k-1$ field widths,
which precisely puts a copy of x in each field. Frequently, as in this example,
there will be no carries, so that the addition of the copies has the same effect as
ORing them together.

Fig. 2. The integer x replicated k times.

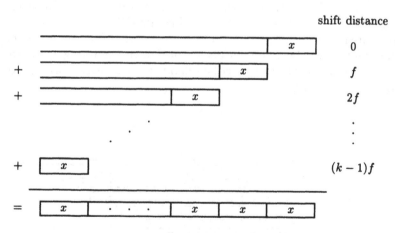

Fig. 3. Replicating x by means of a multiplication.

Consider now Step 2, which requires us to compare x and y_i for all i. A
subtraction seems to be what is called for, since $x - y_i$ is nonnegative if and only
if $x \geq y_i$. We intend subtractions to take place simultaneously in every field,
so that the word-level parallelism can come into play. However, we need a way
to prevent a carry from one field from corrupting the computation in a field to
its left. This can be done by ending each field on the left with a sentinel that
stops any carry. In concrete terms, before Y is subtracted from X, the most

significant bit, called the *test bit*, of each field in X is set to 1, and the test bit of each field in Y is set to 0 (see Fig. 4), which is easy to do by means of bitwise Boolean operations and native constants. This also gives us a nice way to recognize the outcome of each comparison, since the ith test bit b_i in $X - Y$ is 1 if and only if $x \geq y_i$, for $i = 1, \ldots, k$. Because of the test bits, one bit is lost for the representation of keys, but we can compensate for this by storing S in two words, rather than one.

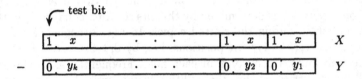

Fig. 4. A single subtraction carries out k comparisons.

In order to carry out the final Step 3, we must sum the test bits of $X - Y$ after clearing all the other bits by means of bitwise Boolean instructions and native constants (i.e., with a "mask"). In the multiplication model, this can be done with a single multiplication, again with the constant $\sum_{i=0}^{k-1} 2^{if}$, which lines up all test bits in a common column and sums them (see Fig. 5). Subsequently, $b = \sum_{i=1}^{k} b_i$, which is the desired rank of x in S, can be shifted to the right word boundary, and irrelevant bits can be eliminated. The conscientious reader will have noted that the algorithm works correctly only if b fits in a single field, a condition that will be automatically satisfied in all interesting applications. Most ideas in the algorithm just described derive from an early sorting algorithm of Paul and Simon [33].

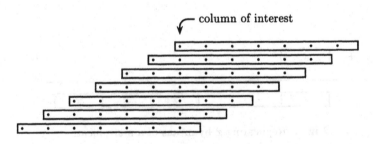

Fig. 5. A multiplication sums the test bits, shown as dots.

If we move from the multiplication model to the AC^0 model, we can no longer carry out Step 1 with a multiplication. However, Step 1 is entirely trivial in the AC^0 model, as it can be realized by a circuit that has no gates at all and consists only of "fanout". Step 2 did not use multiplication, but in order to carry out Step 3 in constant time, we need to change the representation of S

to make $y_1 \leq y_2 \leq \cdots \leq y_k$ (i.e., we choose to store S in sorted order). Then $b_1 \geq b_2 \geq \cdots \geq b_k$, so that the problem essentially becomes one of determining the *most significant bit (MSB)* of the word obtained from $X - Y$ by clearing all bits except the test bits. The MSB of a word Z is the largest bit position in Z holding a 1, i.e., equal to $\lceil \log_2 Z \rceil$ if $Z > 0$, and (say) equal to -1 if $Z = 0$. For $Z > 0$, the MSB of Z can be computed in two steps. The first step clears all bits except for the leftmost bit equal to 1 and can be formulated as an AC^0 operation by noting that an output bit should be set if and only if the corresponding input bit is set and no input bit to its left is, a description that translates directly into a Boolean circuit of depth 2. The second step replaces the single surviving bit by its position (represented in binary) and can be realized by a single level of OR gates; e.g., the OR gate whose output is the rightmost output bit has as its inputs all the odd-numbered input bits. In the present setting, a trivial modification of the second step allows us to compute the field number of the leftmost test bit equal to 1 rather than its bit position, or we can compute its bit position, add 1, and divide by f (which we should then choose as a power of two, so that the division is just a right shift).

If we have only the restricted instruction set at our disposal, we cannot replicate x in constant time, but we can create the word X in $O(\log k)$ time by $\log k$ applications of a shift-and-OR step that doubles the number of copies of x. The summation of the test bits $b_1, \ldots, b_k$ can be done by table lookup, using the word containing (only) the test bits (all other bits having been cleared with a mask) as an index into an (enormous) table. If we keep S sorted, as for the AC^0 model, then $b_1 \geq b_2 \geq \cdots \geq b_k$, as noted above, so that at most $k + 1$ entries of the table will ever be needed. Thus the table can be precomputed in $O(k)$ time.

3.2 Dynamic Ranking

It is not difficult to turn the static algorithm developed above for ranking in a multiset into a data structure that maintains an initially empty multiset S under the powerful operations *insert* (insert (one copy of) a key in S), *delete* (delete (one copy of) a key from S), *rank* (return the rank in S of a key), and *select* (for given r, return the rth smallest element in S, where elements of S are counted with their multiplicities). For lack of a standard name, we will call a data structure with these operations a *ranking dictionary*. The *capacity* of a ranking dictionary is the maximum number of keys that it can hold at any one time.

Lemma 1. *A ranking dictionary with a capacity of w/f f-bit keys can be implemented so that when m keys are stored, each operation takes constant time in the restricted, multiplication, and AC^0 models, except that each operation in the restricted model other than select needs $O(\log m)$ initial processing that depends only on the key of the operation, and that the capacity is also bounded by 2^f in the multiplication model. The space needed is $O(m)$ for the multiplication and AC^0 models, and $O(2^w)$ for the restricted model.*

Proof. We store the multiset S of the dictionary in sorted order. We have already seen how to execute *rank*. The condition $m \leq 2^f$ for the multiplication model ensures that every rank fits in an f-bit field (not quite, but we can enlarge the fields by two bits). Since $S = \emptyset$ initially, we can compute all masks and the "summation table" of the restricted model as S grows, i.e., without having to appeal to a nonconstant number of native constants, and save space by deleting them again when S shrinks. The operation *select* is trivial—we just use a mask to pick out the key in question and shift it right. In order to insert a key, we first use *rank* to determine where it should go, then isolate all larger keys in S by means of a mask and move them one field width to the left, and finally shift the new key appropriately and OR it together with the other keys. In order to delete a key, we again determine its rank, then isolate the two pieces to its left and right and move them together to exclude the key in question. □

The space used by the ranking dictionary, except for a constant number of words, is taken up by tables and masks that can be shared among several instances of the dictionary. Another observation is that we can implement a (simple) set, rather than a multiset, by ignoring all requests to insert a key that is already present, a condition that can be tested by means of *rank* and *select*.

In most applications of dictionary data structures, keys have associated information that, following Cormen et al. [18], we will call *satellite data*. If nothing else is stated, we will always assume that the satellite data of a key fits in a constant number of words. After a brief prelude, satellite data, although essential, is often ignored in the treatment of traditional data structures. The reason is that it usually is obvious that satellite data can follow its key in the sense that it can be stored together with the key and accessed effortlessly once the key has been located. This, however, is no longer obvious and not always the case for word-based data structures, since several keys may be handled together at unit cost. One should therefore pay more than usual attention to whether a particular data structure supports satellite data. Once satellite data is present, there should be an operation *lookup* that returns the satellite data associated with a key or an indication that the key is not present. If multiple copies of a key can be present, the definition of *lookup* and other operations such as *select* may need elaboration; we leave the details to the reader. When nothing else is stated, the intention is that each data structure discussed in the following allows each key to be accompanied by $O(w)$ bits of satellite data.

The ranking dictionary described in Lemma 1 does not immediately support satellite data. However, it is easy to see that we can support satellite data of $O(f)$ bits per key: For a key stored in a particular field, store its satellite data in corresponding fields of a constant number of other words and operate on these words "in the same way" as on the word containing the keys. Provided that the number of keys stored is bounded by 2^f (a condition that was already imposed for the multiplication model), we can use this to support w-bit satellite data in linear space: The satellite data words are stored in a contiguous block of memory. When a key is inserted, the block expands by one word to accommodate its satellite data, and an f-bit pointer to the data is stored directly with the key. When a key x is deleted, the last word of satellite data is copied to the cell previously containing the satellite data of x (this also involves resetting the pointer to that word), and the block of satellite data shrinks by one word. Finally, it is trivial to go from w-bit satellite data to full $O(w)$-bit satellite data.

4 Neighbor Dictionaries without Multiplication

Despite its rather complicated side conditions, the essence of Lemma 1 is simply that few short keys can be maintained efficiently. We would like to extend this result, to the extent possible, to many long keys, where "many" means a number that does not depend on w, and "long" means w-bit. Depending on the order in which we attend to the two aspects "many" and "long", we obtain algorithms with different properties. In this section, we achieve first "many" and then "long", while in Sect. 6 we go the other way. By a lower bound due to Fredman and Saks [23, Theorem 1], a ranking dictionary is too much to hope for, and we will instead implement what we, again for lack of a standard name, will call a *neighbor dictionary*. A neighbor dictionary for a set S of distinct keys with satellite data supports the operations usually associated with search trees, namely *insert*, *delete*, *lookup*, *min* and *max* (return min S or max S), and *succ* and *pred* (return the successor or the predecessor in S of a given value x, i.e., the smallest key in S larger than x or the largest key in S smaller than x). A search tree supports additional operations, some of which can also be implemented efficiently for the data structures described here. If objects with keys that are not necessarily distinct are to be accommodated, one can generally store all objects with a given key in a linked list and store only a pointer to this list in the dictionary proper; for this reason, we will assume in the following that dictionaries need deal only with distinct keys.

4.1 Short Keys: B-Trees

There is a standard way to go from maintaining few keys to maintaining many keys, namely by means of a *B-tree* [10]. Recall that a (leaf-oriented) B-tree with parameter $k \geq 2$ is an ordered search tree with all leaves of the same depth, in which each internal node has between k and $2k - 1$ children, except that the root has between 2 and $2k - 1$ children (unless the tree contains fewer than three nodes). Each leaf stores exactly one key of the set represented by the tree, and the keys appear in increasing order from left to right and are linked together in this order in both directions. An internal node with d children stores $d - 1$ *splitter values* that can guide a (top-down) search for a key value to the correct child.

Lemma 2. *Suppose that there is a ranking dictionary $\mathcal{D}$ with a capacity of $2k$ f-bit keys that supports $(\log N)$-bit satellite data, executes each operation in time at most t and uses space s. Then there is a neighbor dictionary with a capacity of N f-bit keys that supports w-bit satellite data and, when the number of keys stored is n, executes each operation in $O(t \log n / \log k)$ time and uses $O(n(1 + s/k))$ space.*

Proof. The basic idea is to implement each node of a B-tree via an instance of $\mathcal{D}$. Each splitter value of a node is stored as a key in its ranking dictionary with a pointer to the corresponding child as its satellite data; this leaves one child pointer, which is easily handled separately. Suppose first that only insertions and queries (*lookup*, *min*, *max*, *succ*, and *pred*) are to be supported. The implementation of queries is straightforward and left to the reader; the time bound follows from the observation that the height of the tree is $O(\log n/\log k)$. Insertions are handled as follows: As soon as a node v acquires its $(k+1)$st child, it starts separating the leftmost half and the rightmost half of its children and storing each half in a new instance of $\mathcal{D}$. What this means is that each time v acquires a new child, starting with the $(k+1)$st child, it inserts into its lower-half (upper-half, respectively) dictionary the smallest (largest, respectively) splitter value not yet there (together with its satellite pointer); it is easy to keep track of the ranks of these splitter values, so that they can be located. As a result, when v acquires its $(2k)$th child and thus has one child too many, the lower-half and upper-half dictionaries are ready to "take over", and v splits into two nodes in constant time. As usual in a B-tree, the parent of v, which now has one more child, may also need to split, and so on. If the root needs to split, the height of the tree grows by 1, a new root with two children being created.

We handle deletions by means of *lazy deletion*, i.e., a key to be deleted is simply marked as deleted (in its satellite data) and taken out of the doubly-linked list of leaves. The only problem with this is that space is wasted. In order to keep the wasted space $O(n)$, where n is the number of keys currently stored, we resort to the standard technique of global rebuilding, i.e., we continuously migrate from one data structure to another, getting rid of deleted elements in the process. Assume that once we inaugurate a particular instance D of $\mathcal{D}$, the proportion of deleted keys in D is at most $1/4$; moreover, assume that we have an exact copy of D called the *spare copy*. We immediately freeze the spare copy and, while continuing to execute operations using D, step through the elements in the spare copy and insert those that are not marked as deleted into a new data structure D' and its spare copy. This "background" process should work 16 times faster than the "foreground" process, i.e., each time we execute an operation on D, we copy 16 keys from its spare copy to D' and its spare copy (or discard them). Once the copying has been completed, we execute all of the operations that were executed on D since its spare copy was frozen (including lazy deletions) also on D' and its spare copy, still 16 times faster than the foreground process. Subsequently D' is up-to-date, and we inaugurate it and discard D and its spare copy and reclaim their storage. Since $\sum_{i=1}^{\infty}(1/16)^i < 1/8$, the proportion of deleted keys in D never exceeds $1/2$, and the proportion of deleted keys in D' does not exceed $1/4$ at the time of its inauguration, so that the process is back where it started. A key is stored in at most four different dictionaries, and the proportion of deleted keys never exceeds $1/2$ in any dictionary. A space bound of $O(n)$ follows. □

Note the following advantage of lazy deletion: Parameters of a data structure that depend on the current number n of keys can be recomputed each time we migrate to a new instance of the data structure, allowing the parameters to adapt to changes in n. Since all such parameters will be more than robust enough to tolerate changes in n by a constant factor, we can choose the parameters pretending n to be known and fixed.

4.2 Long Keys: Exponential Range Reduction

Combining Lemmas 1 and 2, we see that there is a neighbor dictionary for n f-bit keys in the restricted model with operation times of $O(\log(w/f) + \log n/\log(w/(f + \log n)))$. Here $O(\log n/\log(w/(f + \log n)))$ is the depth of the B-tree (we increase the field width sufficiently to hold $O(\log n)$-bit pointers), and $O(\log(w/f))$ accounts for the duplication of the argument key, which must be carried out only once. E.g., if $w/f = 2^{\sqrt{\log n}}$, then the operations run in $O(\sqrt{\log n})$ time. While this is not terribly interesting in itself, it turns out that there is a way of reducing the (very interesting) case of full w-bit keys to the case of f-bit keys, where f satisfies the relation above, spending only $O(\sqrt{\log n})$ additional time. The way to achieve this is a staple of word-based algorithms, *exponential range reduction*. In general, range reductions transform a problem on long keys into a problem on shorter keys, possibly in a series of stages. Exponential range reduction takes us in one stage from dealing with b-bit keys to dealing with $(b/2)$-bit keys and comes in different flavors for different problems. The exponential range reduction needed here, which we next describe, was invented by van Emde Boas et al. [43].

Suppose that we have a neighbor dictionary $\mathcal{D}$ that can deal with b-bit keys. We want to construct from $\mathcal{D}$ a neighbor dictionary $\mathcal{D}'$ to maintain a set S of $(2b)$-bit keys. As is natural, we view each $(2b)$-bit key z as a pair (x, y), where x consists of the most significant b bits and y consists of the least significant b bits of z. One component of $\mathcal{D}'$ is the *top-level structure*, an instance of $\mathcal{D}$ that maintains the set X of current first components of keys in S (see Fig. 6). For each $x \in X$, let Y_x be the set of those second components y with $(x, y) \in S$. X is also maintained as a bit vector of size 2^b, and in yet another array of size 2^b, for each $x \in X$ we store $\min Y_x$, $\max Y_x$, and a pointer to a *bottom-level structure*, an instance D_x of $\mathcal{D}$ that maintains the set Y_x as well as any satellite data of the corresponding keys.

In order to find the successor of (x, y), we first use the bit vector to test whether $x \in X$. If this is the case and $y \neq \max Y_x$, we query D_x to find the successor y' of y and return (x, y'). In the opposite case ($x \notin X$ or $y = \max Y_x$), we instead query the top-level structure to find the successor x' of x and return $(x', \min Y_{x'})$. In either case, we use constant time plus one query in an instance of $\mathcal{D}$. Similarly, each of the other operations *pred*, *min*, *max*, *lookup*, *insert*, and *delete* can be realized using constant time plus one operation in an instance of $\mathcal{D}$ (plus, possibly, the creation of and insertion of one key into an empty instance of $\mathcal{D}$ or the deletion from and subsequent destruction of an instance of $\mathcal{D}$ containing exactly one key, operations that we will assume to take constant time). A standard trick of storing all initialized positions of the bit vector compactly in a (steadily expanding) block of memory and letting each purportedly initialized position point to the position in the block where it is mentioned lets us

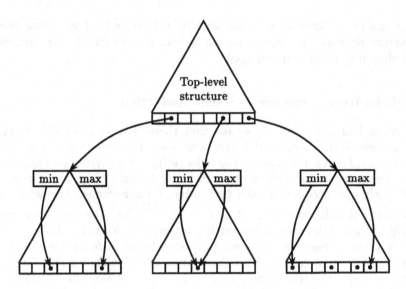

Fig. 6. Exponential range reduction for neighbor dictionaries.

dispense with any need to initialize the bit vector before the first operation. The conclusion is that spending constant extra time per operation, we can handle keys twice as long. Put differently, spending constant extra time per operation reduces the length of the keys that need to be maintained by a factor of two.

In the context of the combination of Lemmas 1 and 2 discussed above, we can use $\sqrt{\log n}$ stages of exponential range reduction, which either take us down to maintaining 1-bit keys, a trivial task, or ensure that $w/f = 2^{\sqrt{\log n}}$. This proves the following result, shown by Fredman and Willard [24] for the multiplication model (in a different way) and by Andersson [3] for the restricted model (essentially as described here). The theorem makes no mention of the space requirements, which are very bad, due to the many large arrays.

Theorem 3. *There is a neighbor dictionary (for w-bit keys) in the restricted model that uses $O(\sqrt{\log n})$ time per operation when n keys are present.*

In the sequel, when the key length of a data structure is not explicitly mentioned, full w-bit keys should be understood to be supported.

5 Sorting

Before continuing the development of neighbor dictionaries, we discuss the problem of sorting. The reasons for this interlude are that the sorting algorithm to be described is simpler and probably of more interest to some readers, and that it gives us an opportunity to develop a few more programming tricks that will come in handy later.

An important feature of a parallel machine, even if it is of the single-instruction multiple-data (SIMD) type, is the ability to let processors selectively execute the next instruction or not depending on some local condition. Of course, the word model does not directly offer anything of the kind, but we can nevertheless simulate it in many cases. Consider a word conceptually divided into f-bit fields including a test bit and assume that a field signals that it should participate in some operation by setting its test bit to 1. We have already seen that a fieldwise $\leq$ comparison can be made to set the test bits in this way, and the same clearly holds also for the other standard arithmetic comparisons $=$, $<$, etc. From the word X consisting only of the test bits (all other bits having been cleared with a mask), we would like to compute a mask A in which every test bit is 0, and every other bit of a field has the same value as the test bit of the corresponding field of X, since such a mask allows us to "extract" the fields of interest and operate on them as we choose, after which we may OR them back together with the remaining fields. A can be obtained simply by multiplying X by $1 - 2^{-(f-1)}$, i.e., by a shift and a subtraction.

5.1 Sorting Short Keys: Independent Slices

As a simple application of conditional operation, we can create a constant-time procedure $MinMax$ that replaces two words divided into the same number of f-bit fields by the word containing the minima and the word containing the maxima of the keys in corresponding fields— in other words, $MinMax$ separates the minima and the maxima. We will use this procedure as the workhorse of a sorting algorithm and again take our inspiration from parallel computing. The famous *bitonic-sorting algorithm* of Batcher [9] for sorting n keys, where n is a power of two, works in $\log n$ stages, in the ith of which sorted sequences of length 2^{i-1} each are merged in pairs to create sorted sequences of length 2^i each, for $i = 1, \ldots, \log n$. In the ith stage, a sequence $(x_1, \ldots, x_{2m})$ of length $2m = 2^i$ is first obtained by concatenating one sequence of length m sorted in the previous stage with the reverse of another, and then the sequence is sorted by a procedure that first executes $(a_i, b_i) := MinMax(x_i, x_{m+i})$ for $i = 1, \ldots, m$, where here $MinMax$ is assumed to operate on single pairs of keys and not on pairs of words, and then calls itself recursively to sort $(a_1, \ldots, a_m)$ and $(b_1, \ldots, b_m)$, the resulting sequences subsequently being concatenated and returned. As can be gleaned from this (rather brief) description, sequential bitonic sorting of n keys works in $O(n(\log n)^2)$ time, the keys are touched only in calls of the form $MinMax(x_i, x_j)$, and in each such call, the indices i and j are entirely independent of the input (except that they depend on n). It is this latter *obliviousness* of the algorithm that makes it suitable both for its original formulation as a sorting network and for its use in a word-based algorithm. The latter possibility was pointed out (for a different oblivious sorting algorithm) by Dessmark and Lingas [19], who take the word-parallelism to an extreme, in a certain sense, by suggesting to solve completely independent subproblems in different word "slices". In the present context of sorting, if we can accommodate k keys per word, we simply split the input sequence into k sequences, each of length n/k, store the sequences in n/k words, with each sequence occupying a separate slice, and then sort all k sequences in parallel using (sequential) bitonic sorting, but modified to use $MinMax$ on words rather than on single keys. In $O(n + n(\log n)^2/k)$ time, this yields k sorted sequences, which are finally merged (without resorting to word-level parallelism) in a binary-tree fashion in $O(n \log k)$ time to obtain the final sorted output sequence. With $k = (\log n)^2$, the total time is $O(n \log \log n)$.

5.2 Sorting Long Keys: Exponential Range Reduction

As in Sect. 4, at this point we have an efficient algorithm for short keys, and we would like to derive from it an efficient algorithm for full w-bit keys. The solution, now as then, is exponential range reduction. Exponential range reduction in the context of sorting was described by Kirkpatrick and Reisch [30]. Referring back to Fig. 6 and its context, suppose that we implement the top-level data structure and each bottom-level data structure simply as a linked list. Stepping once through the input and spending $O(n)$ time, we process each pair (x, y) by inserting x into the top-level list, if it is not already there (as we can tell, using the bit vector), and by inserting y into the appropriate bottom-level list. It suffices to sort the top-level list and each bottom-level list, since subsequently we can step through the bottom-level lists, guided by the top-level list, and output the original keys in sorted order in $O(n)$ time. The total size of the lists exceeds n by the number of distinct first components. However, we can reduce the total size to n by finding the minimum in each nonempty bottom list (again spending $O(n)$ time) and not letting it participate in the sorting. In linear time, hence, we can reduce the problem of sorting n b-bit keys to that of sorting n $b/2$-bit keys within disjoint groups. Note that the sorting subroutine used need not support satellite data and that, indeed, the algorithm of the previous subsection does not have this capability.

After the reduction has been used $2 \log \log n$ times, $(\log n)^2$ keys will fit in a word. We have already seen that such keys can be sorted in $O(\log \log n)$ time per key, which leads to the following theorem.

Theorem 4. *n keys can be sorted in $O(n \log \log n)$ time in the restricted model.*

The theorem was proved by Andersson et al. [5]. The proof given here is probably the simplest one known. Because of the use of exponential range reduction, the space requirements are high.

Thorup has shown that using just $O(n)$ space, n keys can be sorted in the restricted model either in $O(n \log \log n)$ expected time using randomization [38] or in $O(n(\log \log n)^2)$ time deterministically [40]. Andersson et al. [5] describe a randomized sorting algorithm for the multiplication model that can sort n keys in $O(n)$ expected time using $O(n)$ space if w is sufficiently large relative to n, namely if $w \geq (\log n)^{2+\epsilon}$ for some fixed $\epsilon > 0$. Various parallel word-based sorting algorithms can be found in [2, 5, 19].

6 Neighbor Dictionaries Using Linear Space

In this section we carry out the program announced in Sect. 4 of providing first ranking dictionaries for few general keys and then neighbor dictionaries for arbitrarily many general keys. We discuss the AC^0 and multiplication models separately.

6.1 Linear-Space Neighbor Dictionaries in the AC^0 Model

The main goal of this subsection is to prove the following theorem.

Theorem 5. *A linear-space ranking dictionary with a capacity of $w^{1/3}$ (w-bit) keys can be implemented in the AC^0 model so that each operation takes constant time.*

The exponent of $1/3$ appearing in Theorem 5 is immaterial and can be replaced by any other positive constant c. To see this, it essentially suffices to appeal to the construction of Lemma 2 of a B-tree of constant height for w^c keys (not to the lemma itself, which promises only a neighbor dictionary). The only new issue is that for a node in the tree of degree d, we must maintain the number of keys stored in its i leftmost subtrees, for $i = 1, \ldots, d$, which is quite easy. The same remark applies to the exponent $1/9$ of Theorem 7 and to both of the exponents $1/8$ and 5 of Theorem 9.

Theorem 5 is new. Related earlier results will be discussed in the next subsection. Combining Lemma 2 and Theorem 5, we obtain the following result.

Theorem 6. *There is a linear-space neighbor dictionary in the AC^0 model that executes each operation in $O(1 + \log n/\log w)$ time when n keys are present.*

Theorem 6 has as an interesting corollary the fact, not previously known, that if w is sufficiently large relative to n, then deterministic sorting in linear time and space is possible, at least in the AC^0 model. This is because $O(n)$ neighbor-dictionary operations clearly suffice to sort.

We now come to the rather involved proof of Theorem 5, which combines ideas of Ajtai et al. [1], Fredman and Willard [24, 25], Raman [34], and Andersson et al. [7] with some new ideas.

Let $1 \leq m \leq w^{1/3}$ and let $\mathcal{Y}$ be a set of m keys $y_1, \ldots, y_m$ with $y_1 < y_2 < \cdots < y_m$. We view the elements of $\mathcal{Y}$ as strings of w bits each, beginning at the leftmost and most significant bit, and define V as the set of strings that are prefixes of at least one string in $\mathcal{Y}$. The *digital search tree* or *trie* T of $\mathcal{Y}$ is illustrated in Fig. 7. Its node set is V, the root is the empty string, and for each nonempty string $v \in V$, the parent of v in T is the string u obtained from v by deleting its last bit b, and the edge between u and v is labeled b.

Given an edge e in a rooted tree, we define the *parent endpoint* of e as the endpoint of e of smaller depth, its *child endpoint* as its other endpoint, and its *depth* as the depth of its child endpoint. We call two edges *siblings* if they have the same parent endpoint, and an edge in the trie T will be called *distinguishing* if it has a sibling, i.e., if its parent endpoint is of degree 2. We consider each pair of siblings in T to be ordered such that the sibling labeled 0 is to the left of the sibling labeled 1. Then the m leaves of T, in the order from left to right, are precisely $y_1, \ldots, y_m$.

Given two w-bit strings x and y, we define the *similarity* of x and y, $Sim(x, y)$, as one more than the length of the longest common prefix of x and y. For $i = 1, \ldots, w$, call i a *distinguishing position* for $\mathcal{Y}$ if at least one distinguishing edge in T is of depth i. Equivalently, $i = Sim(y_j, y_{j+1})$ for some j with $1 \leq j \leq m-1$.

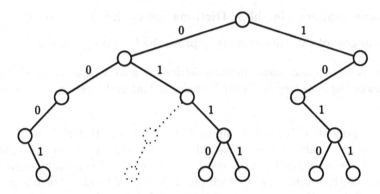

Fig. 7. The digital search tree T of the strings 0001, 0110, 0111, 1010, and 1011. Dotted lines show changes that would be caused by the insertion of $x = 0100$.

The *semi-compressed trie* T' for $\mathcal{Y}$ (this is not a standard concept) is obtained from T by contracting each edge whose depth is not a distinguishing position (see Fig 8); informally, we get rid of all levels in the tree at which "nothing happens". Finally, the (fully) compressed trie T'' for $\mathcal{Y}$ is obtained from either T or T' by contracting each nondistinguishing edge (see Fig. 9); informally, we get rid of all nodes at which "nothing happens".

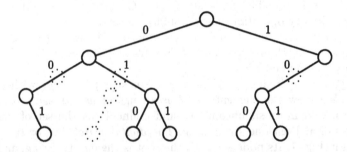

Fig. 8. The semi-compressed trie T' for the strings of Fig. 7. Dotted lines show changes that would be caused by the insertion of $x = 0100$.

We now describe the data structures used to represent the set $\mathcal{Y}$ in the ranking dictionary. First of all, the elements of $\mathcal{Y}$ are stored in a block FullKeys of m memory cells in no particular order. There is also a sorting permutation SortPerm stored in a single word, the jth field of which, for $j = 1, \ldots, m$, points to the position in FullKeys of the element of $\mathcal{Y}$ of rank j (m is sufficiently small for this to be possible). Next, $\{Sim(y_j, y_{j+1}) \mid 1 \le j \le m - 1\}$ is stored as a multiset (with exactly $m - 1$ elements) in a ranking dictionary MultiDistPos and as the corresponding simple set S (with duplicates removed) in a ranking dictionary DistPos. Both of the ranking dictionaries are realized according to Lemma 1.

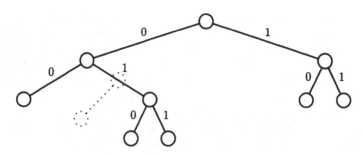

Fig. 9. The compressed trie T'' for the strings of Fig. 7. Dotted lines show changes that would be caused by the insertion of $x = 0100$.

The semicompressed trie T' is represented through two arrays Dist and Label. For $j = 1, \ldots, m$ and $i = 1, \ldots, |S|$, if the ith edge e on the path from the root of T' to the leaf y_j is distinguishing, then $\text{Dist}[i, j] = 1$ and $\text{Label}[i, j]$ is the label of e. If e is not distinguishing, then $\text{Dist}[i, j] = 0$ and $\text{Label}[i, j]$ is arbitrary. Although Dist and Label are thought of as twodimensional arrays, each is stored in column-major order in successive fields of a single word. We store in the same way an array PrefixDist containing the columnwise prefix sums of Dist, i.e., for $j = 1, \ldots, m$ and $i = 1, \ldots, |S|$, $\text{PrefixDist}[i, j] = \sum_{l=1}^{i} \text{Dist}[l, j]$.

T'' is represented using the so-called *Euler-tour technique*, well-known in parallel computation and introduced by Tarjan and Vishkin [36]. An *Euler tour p* of T'' is a path in T'' that moves around the contour of T'' from left to right, using each edge once in each direction and starting and ending at the root. Each leaf occurs exactly once on p, each internal node occurs exactly three times on p, and the total number of nodes on p is $4m - 3$. For $i = 1, \ldots, 4m - 3$, let v_i be the ith node on p. We store T'' in the form of two arrays Depth and LeafRank, indexed from 1 to $4m - 3$. For $i = 1, \ldots, 4m - 3$, $\text{Depth}[i]$ is the depth of v_i in T'', and $\text{LeafRank}[i]$ is the number of leaves encountered on the initial part of p until and including v_i. E.g., for the compressed trie of Fig. 9, the entries in Depth and LeafRank would be $0, 1, 2, 1, 2, 3, 2, 3, 2, 1, 0, 1, 2, 1, 2, 1, 0$ and $0, 0, 1, 1, 1, 2, 2, 3, 3, 3, 3, 3, 4, 4, 5, 5, 5$, respectively. Each of Depth and LeafRank is stored as consecutive fields of a single word. Finally, there is an array LeafPos, also stored in a single word, whose jth element, for $j = 1, \ldots, m$, is a pointer to the location in Depth and LeafRank that corresponds to the (single) occurrence of the leaf y_j, i.e., to the jth local maximum of the sequence stored in Depth. E.g., in the example of Fig. 9, the entries of LeafPos would be $3, 6, 8, 13, 15$.

Let us now see how to determine the rank in $\mathcal{Y}$ of a w-bit string x. The reader may wish to visualize claims and algorithmic steps in what follows as they apply to the example worked in Figs. 7–9. Assume first that we know a string y_l in $\mathcal{Y}$ whose similarity r with x is maximal. If $r = w + 1$, then $x = y_l$, and we simply use LeafPos and LeafRank to look up the rank of y_l in $\mathcal{Y}$. If x is either smaller or larger than all strings in $\mathcal{Y}$, we can easily discover this fact and output the rank of x. Assume that these easy special cases do not occur. Then $r - 1$ is the depth in T of the node at which the path from the root to x would branch off from the rest of the tree if we were to insert x. Determining the rank of $r - 1$ in the set S of distinguishing positions, we find the depth r' of the edge in T' "on" or at a node immediately below which the new path to x would branch off. Inspecting $\texttt{PrefixDist}[r', l]$, we translate r' to the depth r'' of the edge e in T'' "on" which the path to x would branch off. We want to determine the parent endpoint u of e. This can be done by searching left and right in $\texttt{Depth}$, starting from the position of y_l (located with the aid of $\texttt{LeafPos}$), for the first occurrence of the integer $r'' - 1$, an operation that can be realized with the aid of an MSB and an analogous *least significant bit* (*LSB*) computation. Assume that the two nearest occurrences of $r'' - 1$ in $\texttt{Depth}$ are encountered in positions t_1 and t_2, where $t_1 < t_2$. The two positions t_1 and t_2 correspond to the two visits to u that "enclose" y_l and x. All leaves encountered on p before position t_1 are smaller than x, all leaves after position t_2 are larger than x, and the leaves between t_1 and t_2 are larger than x if and only if y_l is. Hence either $\texttt{LeafRank}[t_1]$ or $\texttt{LeafRank}[t_2]$ is the desired rank of x in $\mathcal{Y}$, and the choice between the two can be made on the basis of a comparison between x and y_l.

So far we assumed a string y_l in $\mathcal{Y}$ of maximum similarity with x to be known. In order to understand the algorithm for determining y_l, imagine a procedure for this task that searches for x in T, starting at the root and maintaining the set A of strings that still agree with x. Initially, $A = \mathcal{Y}$, and whenever the search crosses a distinguishing edge, we remove from A those strings that "take the wrong turn". When the path to x branches off from T altogether, we return as y_l any string that remained in A until the end. In light of this procedure, the following can be seen to be true: Whenever all strings in A agree on the next bit to be "tested", we may assume that in fact they also agree with x on that bit. The reason is that if the assumption is wrong, as it may very well be, then every string in the current set A is of maximum similarity with x, and we will eventually return one such string as the answer. This observation justifies testing only bits that correspond to edges in T'', which we implement as follows: First, we create a string x_S of $|S|$ fields, called the *compressed key* of x, whose ith field, for $i = 1, \ldots, |S|$, has the value of the bit of x in the ith distinguishing position, called the ith *significant bit* of x; in other words, we extract from x the significant bits and store them, each in a separate field. We will call each such field a *target field* and the $|S|$ target fields collectively the *target area*. Next we make a copy $\texttt{Label}_x$ of $\texttt{Label}$ and, using $\texttt{Dist}$ as a mask, change it to agree with x in every entry that corresponds to a nondistinguishing edge. More precisely, for $j = 1, \ldots, m$ and $i = 1, \ldots, |S|$, if $\texttt{Dist}[i, j] = 1$, then $\texttt{Label}_x[i, j] = \texttt{Label}[i, j]$, and otherwise $\texttt{Label}_x[i, j]$ equals the value of the ith field of x_S (see Fig. 10). The observation above implies that it suffices to find a string in $\texttt{Label}_x$, with each column considered as a string, of maximum similarity with x_S. And for this it suffices to compute the rank of x_S in $\texttt{Label}_x$, since the string of maximum similarity with x_S in $\texttt{Label}_x$, if different from x_S, is either the predecessor or the successor of x_S in $\texttt{Label}_x$. We compute the rank of x_S in $\texttt{Label}_x$ using the ranking algorithm of the proof of Lemma 1—note that $\texttt{Label}_x$ is sorted.

What remains is to describe how to compute the compressed key x_S of x. What we need is a *multiselect* instruction that picks out those bits of a "data

Dist						Label						x_S		Label_x				
1	1	1	1	1		0	0	0	1	1		0		0	0	0	1	1
1	1	1	0	0		0	1	1				1		0	1	1	1	1
0	1	1	1	1		0	1	0	1			0		0	0	1	0	1

Fig. 10. Data structures used in a search for $x = 0100$ in the example of Fig. 8. Blank positions indicate don't-care values.

word", x, that are specified in an "address word", the word storing the set S. In the AC^0 model, a multiselect circuit can be formed out of $M - 1$ copies, where M is the capacity of the dictionary, of a *select circuit* that picks out a single bit based on a single address. Such a circuit is simple to describe (see Fig. 11): The address is compared in parallel for equality with all possible addresses (which are hardwired into the circuit), the conjunction of the output of each comparison with the data bit in the corresponding position is formed, and the disjunction of all the resulting bits is the output bit. The equality test is not the function of a standard gate, but can clearly be realized by a circuit of linear size and bounded depth.

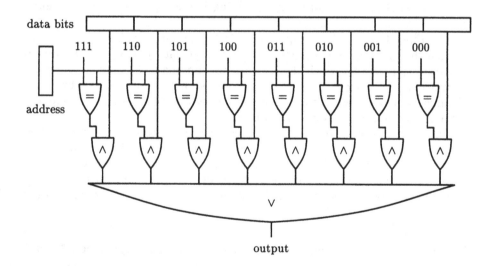

Fig. 11. A select circuit of bounded depth.

Consider now the remaining operations of a ranking dictionary. Due to the availability of SortPerm, *select* is trivial. In order to insert a new key x, we begin by computing its rank in $\mathcal{Y}$ as described above, after which updating the various data structures of the representation of $\mathcal{Y}$ to record the presence of x is largely a routine matter. x is inserted into FullKeys, and a pointer to its position in FullKeys is stored in SortPerm after making room for it by moving the pointers to larger keys one field width to the left. One new similarity is inserted into MultiDistPos; if it is not present in DistPos (i.e., it is the first occurrence of a particular value), it is inserted there as well.

Room for a new column for x is made in Label, Dist, and PrefixDist by moving the columns for larger keys one position to the right. The entries in rows $1, \ldots, r'$ of the new column for x in each of the three arrays are copied from the column of a key y_l in $\mathcal{Y}$ of maximum similarity with x, the remaining entries of the new column of Dist are set to zero, and the remaining entries of the new column of PrefixDist are set to r''. If the insertion of x makes a new position distinguishing (i.e., if a new value is inserted into DistPos), room for a new row below the previous row r' is made in Label, Dist, and PrefixDist by moving rows $r' + 1, r' + 2, \ldots$ down by one position (for this to be easily possible, we should reserve space for $M - 1$ rows from the outset), and the new row is initialized to contain zeros in the case of Dist and to be a copy of the old row r' in the case of PrefixDist. Whether or not a new position becomes distinguishing, the arrays are then changed in accordance with the two new distinguishing edges of depth $r' + 1$. This entails setting $\text{Dist}[r' + 1, j]$ to 1 for $j = \text{LeafRank}[t_1] + 1, \ldots, \text{LeafRank}[t_2] + 1$, where t_1 and t_2 are the positions found during the ranking of x, setting the corresponding entries of Label to the appropriate values (e.g., if x "branches off" to the right, then $\text{Label}[r' + 1, t_2 + 1] = 1$ and $\text{Label}[r' + 1, j] = 0$ for $j = \text{LeafRank}[t_1] + 1, \ldots, \text{LeafRank}[t_2]$), and modifying PrefixDist accordingly by adding 1 to every entry in a rectangular area.

The Euler tour of T'', as represented in Depth and LeafRank, is expanded through the insertion of two easily computed segments of lengths 1 and 3 near the positions t_1 and t_2, which also necessitates an update of LeafPos. Moreover, all values stored in Depth between the two new segments are increased by 1, and all values stored in LeafRank after the larger new segment (which corresponds to the actual new leaf) are increased by 1.

The updates needed in the case of a deletion of a key are similar, and their design will be left to the reader. The maximum number M of keys that can be supported is determined by PrefixDist, which is stored in a single word and should be able to contain $M(M - 1)$ fields of $O(\log M)$ bits each. For $M = w^{1/3}$, the relation $M^2 \log M = O(w)$ is clearly satisfied. This ends the proof of Theorem 5.

6.2 Linear-Space Ranking Dictionaries in the Multiplication Model

A deterministic result corresponding to Theorem 5 is not currently known for the multiplication model, a randomized such result having been provided by Raman [34]. The best that we can do deterministically is represented by the two results proved in this subsection. First (Theorem 7) we will see a *static ranking dictionary*, a dictionary that supports only the query operations *rank* and *select*, but not the update operations *insert* and *delete* (*delete* can be supported, but without *insert* this is of little interest). Such a data structure was invented and described for the multiplication model by Fredman and Willard [24], who used it as the central piece of their *fusion trees*, and ported to the AC^0 model by

Andersson et al. [7] (the latter result is subsumed by Theorem 5). Subsequently (Theorem 9) we develop a (dynamic) ranking dictionary that has essentially the performance bounds of Theorem 5, but relies on a table that needs space and computation time $2^{M^{\Theta(1)}}$, where M is the maximum number of keys to be supported. Because of the excessive precomputation time, this dictionary can normally be used for only very small numbers of keys, even if w is large. The following subsection shows how to obtain a dynamic neighbor dictionary from the static ranking dictionary of the present subsection.

Theorem 7. *A linear-space static ranking dictionary for $m \leq w^{1/9}$ keys can be constructed in $O(m)$ time using $O(m)$ space in the multiplication model so that each query takes constant time.*

Proof. We need to change two parts of the construction described in the proof of Theorem 5, namely the MSB and LSB computations and the multiselect operation. We begin with the former and first consider the following simpler task: Suppose that we are given a word B of k f-bit fields, where $f \geq 2 + \log k$, and that the ith field (counted from the right) contains $b_i \in \{0, 1\}$, for $i = 1, \ldots, k$. We want to determine the largest (in the case of MSB) or smallest (in the case of LSB) index of a nonzero field, if any. In order to solve this task, we first multiply B by the word I with 1 in each field, which gives us a word C in which the ith field contains $c_i = \sum_{j=1}^{i} b_j$, for $i = 1, \ldots, k$ (see Fig. 5). If $B \neq 0$, we can now clear all fields of C except the unique field i with $b_i = 1$ and either $c_i = c_k$ (for MSB) or $c_i = 1$ (for LSB). Converting the resulting word to a mask and using the native constant whose ith field contains i, for $i = 1, \ldots, k$, we obtain a word that contains the desired answer in some field and zero in all other fields. A multiplication by I transmits the answer to field k, from where it can be extracted.

In order to use this procedure to compute the MSB or LSB of a given w-bit word, we divide the word into fields of $f = 2 + \log w$ bits each, carry out a constant-time procedure that replaces each nonzero field value by 1, and use the procedure to locate the leftmost (for MSB) or rightmost (for LSB) nonzero field. Now the answer is narrowed down to a single field, and we create a word consisting of f copies of this field. After using a mask to clear all except the ith bit (counted from the right, starting at 1) in the ith field, for $i = 1, \ldots, f$, we can use the procedure again, now with a field width of $f + 1$ bits, and calculate the final answer.

Consider next the problem of extracting the significant bits of a word x to create the compressed key x_S. Recall that a multiplication can be used to place certain bits of an argument word "on top of" each other (see Fig. 5). The basic idea now is to use a multiplication with a word A instead to place the significant bits in the rightmost positions of their respective fields of the target area, which will be located in the rightmost part of the more significant word of a double-word product. The most straightforward realization of this idea meets with the obstacle that we cannot prevent a bit of A from "acting on" a bit of x for which it was not intended, which may corrupt the bits of interest. For this reason, we buy flexibility by allowing A to place each significant bit of x anywhere in its corresponding target field, relying on a (by now familiar) subsequent computation to move the bit to the rightmost position within its field.

Write the set of distinguishing positions as $S = \{s_1, \ldots, s_r\}$ with $s_1 < s_2 < \cdots < s_r$ and let B be the word whose 1 bits are precisely in the positions $s_1, \ldots, s_r$. The multiplier A will have exactly one 1 bit for each 1 bit in B, say, in the positions $a_1, \ldots, a_r$. We want to choose A in such a way that (a) for $i = 1, \ldots, r$, position $a_i + s_i$ is within the ith target field; (b) for $1 \le i, j, k, l \le r$, if $(i, j) \ne (k, l)$, then $a_i + s_j \ne a_k + s_l$. Given the word x, we first clear all bits with positions outside of S, then multiply the resulting word with a word A satisfying (a) and (b), and finally clear all bits in the product with positions outside of $\{a_i + s_i \mid 1 \le i \le r\}$. This places the ith significant bit of x in the ith target field for $i = 1, \ldots, r$, as desired, and the two masks needed in the process will be seen below to be easy to compute.

We can visualize the selection of the multiplier A as follows: Suppose that we have drawn a copy of B on each of r strips of transparent film, with each bit equal to 1 drawn as a circle and the ith of these circles marked specially on the ith strip, say, by a cross, for $i = 1, \ldots, r$. The task then is to slide the strips, placed over each other, horizontally and independently of each other over a ruler divided into bit positions and with the target fields indicated in order to find a placement for which (a) the cross on the ith strip falls within the ith target field, for $i = 1, \ldots, r$; (b) no two circles, with or without crosses, collide, i.e., are over each other. It turns out that we can place the strips one by one in any order (see Fig. 12). With a field width of f bits, condition (a) allows f positions for the next strip to be placed, and since each of the r circles on this strip can collide in at most one position with each of the at most $r(r-1)$ circles on strips already placed, condition (b) rules out at most $r^2(r-1)$ positions. A multiplier with properties (a) and (b) thus exists if $f \ge r^3$.

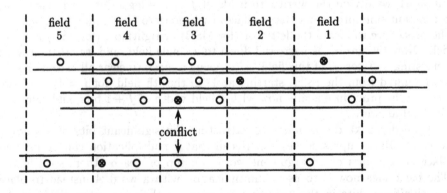

Fig. 12. Finding a legal position for the next strip, shown at the bottom.

We change the construction slightly to allow for an efficient computation of A. Observe first that we can permit arbitrary collisions, provided that no cross collides, and provided that the r^2 positions to the right of each cross are free, i.e., no circles from strips other than the one holding the cross are placed there. The reason is that in a multiplication of two numbers, each with r bits set, no carry can "run" for $r^2 + 1$ positions. If we choose A in this manner, the fact that each cross has a "protected interval" of r^2 positions to its right rules out another at most $r(r-1) \cdot 2r^2 \leq 2r^4$ of the f positions allowed by condition (a), so that we must now require $f \geq 3r^4$. Since $r \leq m-1$, a field width of $3m^4$ suffices. More importantly, in the selection of the next bit of A we now need to keep track of only the target area and r^2 positions to its right. Based on this observation, it is not difficult and left to the reader to devise a way to use word-level parallelism to test all possibilities for the next bit of A simultaneously, provided that the word length is sufficient to allow $3m^4$ copies of the target area (of $3m^5$ bits) to be placed next to each other. By assumption, $m \leq w^{1/9}$, so that this requirement is satisfied, and the construction can be carried out in constant time per element of S, i.e., in $O(m)$ time altogether. □

Corollary 8. *A linear-space static ranking dictionary for n keys given in sorted order can be constructed in $O(n)$ time using $O(n)$ space in the multiplication model so that each query takes $O(1 + \log n / \log w)$ time.*

As anticipated in the beginning of this subsection, we next attempt to dynamize the static ranking dictionary of Theorem 7 as far as possible. Theorem 9 below was proved by Fredman and Willard [25].

Theorem 9. *There is a linear-space ranking dictionary for the multiplication model with a capacity of $M \leq w^{1/8}$ keys that executes each operation in constant time, but uses a table that needs space and computation time $2^{\Theta(M^5)}$.*

Proof. The fact that only a static dictionary was to be supported was used in the preceding proof only in the computation of compressed keys, and there only in the calculation of the multiplier A. Suppose that we were to compute A as in the static case, setting a new bit of A to 1 whenever a new element appears in S and clearing it again when that element of S disappears. The problem is that while we can select each bit of A so as to provide the corresponding cross with an empty protected interval and respecting the protected intervals of crosses already present, we cannot prevent new elements of S from violating the protected interval of a cross that arrived earlier, creating "noise" that may "mask" the bit of x represented by the cross. Figuratively, it is as though new circles suddenly appear in unforeseeable places on the transparent strips already placed, possibly invalidating the chosen placement. The solution is to deduce the significant bits of x in the reverse order of the time of the inclusion of their indices in S (i.e., latest arrivals first), and to carefully subtract off the noise caused by each significant bit of x as it is computed, leaving the next significant bit of x "unmasked". While this procedure would be quite easy to carry out in $O(r)$ time, where $r = |S|$, what we need is a constant-time solution.

It does not seem very likely that the computation in question can be executed in constant time in the multiplication model, and we have to resort to the "table-lookup" method of precomputing and storing the results of all possible computations. We thus store the target area as well as any auxiliary information needed to interpret the target area correctly into a single word z and obtain the compressed key x_S as $F[z]$, where F is a precomputed table. If we can manage to pack the information in z into q bits and each entry in F can be computed in polynomial or even exponential time (as is the case here), F needs 2^q space and can be computed in $2^{O(q)}$ time. Our remaining task is simply to design auxiliary information to be stored in z that does not make q too large. As can be seen from the statement of Theorem 9, the requirement is that the auxiliary information should fit in $O(M^5)$ bits; in particular, information whose size grows unboundedly with w is out of the question.

In order to interpret the target area, we certainly need to know, for $i = 1, \ldots, r$, the location in the target area of the cross that corresponds to the ith last element s'_i of S to arrive. This information fits in $O(M \log M)$ bits, and maintaining it (as we must) is routine. At the price of another $O(M \log M)$ bits, we can also include information about the rank of s'_i in S, which allows us to do away with having a separate target field for each rank. As a consequence, a target area consisting of a single target field of $3M^4$ bits suffices, and we need not worry about changing the multiplier A in response to rank changes within S.

All that is still missing is information that allows us to subtract off the "noise" caused by a particular bit of B (which, recall, is a bit-vector representation of S) in the product $A \cdot B$. The word A would certainly suffice for the purpose, but A is too large to be included in the auxiliary information. Note, however, that only a small piece of A is relevant to the "noise" caused in the target area by a single bit of B, namely a piece whose length h is the length of the target area plus the length of a protected interval. It would thus suffice to include such a piece of A in the auxiliary information for each element of S. It is not clear, however, whether concatenated pieces of this type can be maintained efficiently under changes to A. The problem is essentially one of misalignment: The various pieces are drawn from A at arbitrary positions, and an update of A may need to be reflected by an update in a different position of each piece, something that is difficult to handle in the word model. We solve this last problem by enforcing alignment. A is divided into *blocks* of h bits each, and we store for each element of S a piece consisting of the one or (usually) two relevant blocks of A. An update of A happens in a particular block, and the representation of S stored in the ranking dictionary DistPos makes it easy to determine which elements of S hold copies of that block, after which the update of the copies can be carried out in constant time. The auxiliary information is dominated by the at most M copies of blocks of size $O(M^4)$, so that the space and preprocessing time needed by the ranking dictionary are $2^{O(M^5)}$. □

6.3 From Static to Dynamic Dictionaries: Exponential search trees

We have already seen one example of a static dictionary (Corollary 8), one whose performance improves as w grows relative to n. In this section we develop first a second static dictionary (Theorem 11), one that performs best for small w, and then a general method for converting static dictionaries to dynamic ones.

The static dictionary for small w is essentially the construction of Theorem 3, but with the space requirements reduced by means of perfect hashing (which both prevents the dictionary from being dynamic and needs multiplication) and with the exponential range reduction continued for a full $\log w$ stages, rather than $\sqrt{\log n}$ stages (because this leads to the more interesting result, in view of Corollary 8). What we need to know about perfect hashing is encapsuled in the following lemma.

Lemma 10 ([4]). *Given a set S of n (w-bit) keys, a function that maps S injectively into a range of size $O(n)$ and that can be evaluated in constant time and stored in $O(n)$ space can be computed in $O(n^3 w^2)$ time in the multiplication model.*

Proof. Fredman et al. [22] proved that the lemma holds with a construction time of $O(n^3 w)$ if the available instruction set is augmented to include integer division (with remainder). A division can be simulated in $O(w)$ time, which is sufficiently fast for the construction phase, given that we allow a factor of $\Theta(w)$ more time. The function evaluation also uses division, but only by $O(n)$ numbers fixed during the construction phase, and we can replace division by such a number q by multiplication with $1/q$, suitably scaled and rounded and precomputed in $O(w)$ time to a sufficient precision during the construction phase. $\qquad\square$

The following lemma, due to Willard [44] and Andersson [4], describes the static dictionary for small w.

Theorem 11. *A static neighbor dictionary for m keys that uses $O(m^4)$ space and supports queries in $O(\log w)$ time can be constructed in $O(m^4)$ time in the multiplication model.*

Proof. We can assume that $w \leq m^{1/3}$, since otherwise (sorting plus) Corollary 8 provides what we need. We store the keys using $\log w$ stages of the exponential range reduction illustrated in Fig. 6, but storing each array of pointers to bottom-level structures and to satellite data in a hash table according to Lemma 10, which still allows constant-time access. This is possible because all the elements to be stored in one of these arrays are known before the array needs to be accessed for the first time, a consequence of the dictionary being static. Each key is stored in $O(\log w)$ hash tables, so the total time and space needed are $O(m^3 w^2 \log w) = O(m^4)$. $\qquad\square$

When the cost of adding a new child of a node in a search tree is an increasing function of the node degree, as is the case when the node is implemented via a ranking dictionary according to Corollary 8 or Theorem 11, it is not optimal to let the node degrees be uniform all over the tree, as in a B-tree. Instead degrees should be constant near the leaves, where changes are frequent, and increase towards the root, where changes are rare. Such trees form a standard paradigm in parallel computation. They are exploited in Valiant's doubly-logarithmic merging algorithm [42] and called "balanced doubly logarithmic height trees" by

Berkman et al. [13]. Here we describe a variant of the *exponential search trees* of Andersson [4].

Suppose that we change the ground rules of insertion-only leaf-oriented B-trees as follows: Instead of splitting a node when its degree has doubled, we split a node when its *weight*, the number of its leaf descendants, has doubled. Moreover, each of the two nodes resulting from the split takes over precisely half of the leaf descendants and constructs from these a new search tree, perfectly balanced in the sense that any two siblings in the tree have weights that differ by at most a constant factor; we do not require all leaves to have the same depth. It can be verified that with these rules, no node degree ever increases by more than a constant factor before the node is split. This is no different from B-trees, but we have the added property that the number of operations that must "pass through" a node u from it is created until it is split is at least half its weight at any time. Imagine that each such operation places one "credit" at u, where a credit represents the ability to "pay for" a constant amount of computation. When a child v of u is split, we would like the credits that have accumulated at v to be sufficient to pay for the necessary reconstruction of u. Assume therefore that we choose the nodes degrees such that the cost of reconstructing a node u to accommodate an additional child is within a constant factor of the weight of any child of u. Provided that the degree of each internal node is at least 2 and that the degree of each node is at least twice that of each of its children, the credits are also able to pay for global rebuildings of entire subtrees. To see this, suppose that the maximal subtree T rooted at a node v needs to be rebuilt and recall that the number of credits available is at least half the weight m of v. Consider a random experiment in which each internal node of T selects one of its subtrees uniformly at random and each leaf selects itself. By assumption, the cost of the global rebuilding is within a constant factor of the minimum total weight (over all random choices) of all selected subtrees; however, the expected total weight of the selected subtrees is $\leq 2m$, since each leaf has probability at most 2^{-i} of belonging to a selected subtree of height i, for $i = 0, 1, \ldots$

These considerations imply that the worst-case cost of a query and the amortized cost of an insertion will both be within a constant factor of the worst-case cost of carrying out one query at each node on a root-to-leaf path in the search tree. Deletions can be supported at the same worst-case cost, as described in the proof of Lemma 2. In order to estimate the operation cost, note that for each of our candidate node dictionaries (Corollary 8 and Theorem 11), the cost of reconstructing a node u of degree d is polynomial in d, say, $O(d^k)$. If the weight of u is m, we can then choose $d = \Theta(m^{1/(k+1)})$ in order to make the cost of reconstructing u be $O(m/d)$, as required above. With this choice, the other degree restrictions imposed above are automatically satisfied outside of subtrees of weights bounded by a constant.

The choice $d = \Theta(m^{1/(k+1)})$ means that as we pass from an internal node in the tree to one of its children, the logarithm of the weight drops by a constant factor. In particular, the height of a tree storing n keys is $O(\log \log n)$. If we implement each node in the search tree according to Corollary 8, the search costs along a search path form a geometric series, and the operation count becomes

$O(\log n/\log w + \log \log n)$. Another possibility is, for a constant $c > 0$, to implement the upper $c \log(2 + \log n/(\log w)^2)$ levels (if there are that many) according to Theorem 11, at a search cost of $O(\log w)$ per level, and the remaining levels according to Corollary 8. If c is chosen sufficiently large, the upper levels will reduce the logarithm of the weight from $\log n$ to at most $(\log w)^2$, so that the search cost spent on the bottom levels is just $O(\log w + \log \log n) = O(\log w)$. Thus we have proved the following result of Andersson [4]:

Theorem 12. *There is a linear-space neighbor dictionary in the multiplication model that executes each operation in*

$$
O\left(\min\left\{\begin{array}{l} \log n/\log w + \log \log n \\ \log(2 + \log n/(\log w)^2)\log w \end{array}\right\}\right)
$$

time when n keys are present, with the bound for insertion applying only in the amortized sense. In particular, for every combination of n and w, the operation times are $O(\sqrt{\log n})$.

It is instructive to compare Theorem 12 with Theorems 3 and 6. Disregarding the differences in the models of computation, Theorem 12 subsumes Theorem 3, although it does not subsume the bound of $O(\log w)$ of [43] obtained by using just exponential range reduction. Theorems 6 and 12 are incomparable. The bound of $O(\sqrt{\log n})$ of Theorem 12 is nearly matched by a lower bound of $\Omega(\sqrt{\log n/\log \log n})$ due to Beame and Fich [11].

7 Conclusions

As we have seen, the area of word-based algorithms and data structures already possesses a fairly rich body of knowledge, especially in view of the many additional results that have not been discussed here. The latter include work on static or dynamic plain dictionaries (data structures supporting *lookup* and possibly *insert* and *delete*) [6, 14, 21, 31, 32] and on priority queues (data structures supporting, as a minimum, *insert* and *extractmin*, inspection and removal of an object with minimal key value) and closely related graph problems [25, 34, 35, 37–41]. The most pressing issue, from a practical point of view, is surely whether any of the algorithms and data structures are or can be made competitive with or even superior to traditional methods. One argument sometimes heard against this possibility goes along the following lines: Real word lengths are fairly modest, and once keys become so small that you can put several of them into a single word, it is actually trivial to deal with them by traditional means. However, a proponent of this line of reasoning, so it seems, must also believe radix sorting to be a panacea for all sorting problems, a point of view that is not widespread. On the other hand, it should be recognized that many of the algorithms described here are not practical in their current form. Others appear reasonable, and it would be interesting to carry out practical experiments and comparisons. In particular, are there very efficient hashing schemes that would allow the fastest algorithms described here to be implemented with reasonable space requirements without adding an unacceptable overhead to the running time?

Theoretically, there are several bigger and smaller open problems that one might try to tackle. What is the complexity of sorting in any of the word models with no holds barred, i.e., allowing arbitrary amounts of space and perhaps randomization? Is it $\Theta(n)$ for all combinations of n and w, or does it sometimes go up to $\Theta(n \log \log n)$ (cf. Theorem 4)? Of course, one might also ask for, e.g., the complexity of deterministic sorting in linear space. Is there an analogue of Theorem 5 for the multiplication model? And is $\Theta(\sqrt{\log n})$ the correct answer for the problem of maintaining a neighbor dictionary in linear space in the multiplication model (cf. Theorem 12)?

To date, with few exceptions, only the most straightforward parallel algorithms and concepts have been employed in the design of word-based algorithms. It remains to be seen whether more sophisticated parallel techniques will also find applications. Algorithms and data structures for problems outside of the realm of very basic ordering questions so far have received scant attention. As an example of a simple but still interesting new result for a more complex task, we observe that even in the restricted model, a maximum flow in an n-vertex network with w-bit integer capacities can be computed in $O(n^{17/6})$ time. This follows by combining a recent algorithm of Goldberg and Rao [27], with a running time of $O(n^{8/3}w)$, with an algorithm of Cheriyan et al. [16], with a running time of $O(n^{8/3} \log n \log w + n^3/w)$ (see below). If $w \leq n^{1/6}$, the first algorithm runs in $O(n^{17/6})$ time, and otherwise the second algorithm runs in $O(n^{17/6})$ time (using only $n^{1/6}$ bits of the word length). A running time of $O(n^{8/3} \log n \log w + n^3/w)$ is not mentioned explicitly by Cheriyan et al., but can be obtained by combining their Lemmas 7.5, 8.3, and 8.4(b) (used with $h = n$) with the remark at the end of their Sect. 6, whose quantity x corresponds to what is called w here, and noting that the bound of $O(n^3/x + qx + n^2)$ of the remark is overly conservative and can be replaced by $O(n^3/x + q \log x + n^2)$ for a machine with the restricted instruction set.

References

1. M. Ajtai, M. Fredman, and J. Komlós, Hash functions for priority queues, *Inform. and Control* **63** (1984), pp. 217–225.
2. S. Albers and T. Hagerup, Improved parallel integer sorting without concurrent writing, *Inform. and Comput.* **136** (1997), pp. 25–51.
3. A. Andersson, Sublogarithmic searching without multiplications, in Proc. 36th Annual IEEE Symposium on Foundations of Computer Science (FOCS 1995), pp. 655–663.
4. A. Andersson, Faster deterministic sorting and searching in linear space, in Proc. 37th Annual IEEE Symposium on Foundations of Computer Science (FOCS 1996), pp. 135–141.
5. A. Andersson, T. Hagerup, S. Nilsson, and R. Raman, Sorting in linear time?, *J. Comput. System Sci.*, to appear. A preliminary version appeared in Proc. 27th Annual ACM Symposium on the Theory of Computing (STOC 1995), pp. 427–436.

6. A. Andersson, P. B. Miltersen, S. Riis, and M. Thorup, Static dictionaries on AC^0 RAMs: Query time $\Theta(\sqrt{\log n/\log\log n})$ is necessary and sufficient, in Proc. 37th Annual IEEE Symposium on Foundations of Computer Science (FOCS 1996), pp. 441–450.

7. A. Andersson, P. B. Miltersen, and M. Thorup, Fusion trees can be implemented with AC^0 instructions only, *Theoret. Comput. Sci.*, to appear.

8. A. Andersson and M. Thorup, Implementing monotone priority queues, Proc. 1996 DIMACS Challenge, to appear.

9. K. E. Batcher, Sorting networks and their applications, in Proc. 32nd AFIPS Spring Joint Computer Conference (1968), pp. 307–314.

10. R. Bayer and E. McCreight, Organization and maintenance of large ordered indexes, *Acta Inform.* **1** (1972), pp. 173–189.

11. P. Beame and F. Fich, On searching sorted lists: A near-optimal lower bound, manuscript, 1997.

12. A. M. Ben-Amram and Z. Galil, When can we sort in $o(n \log n)$ time?, in Proc. 34th Annual IEEE Symposium on Foundations of Computer Science (FOCS 1993), pp. 538–546.

13. O. Berkman, B. Schieber, and U. Vishkin, Optimal doubly logarithmic parallel algorithms based on finding all nearest smaller values, *J. Algorithms* **14** (1993), pp. 344–370.

14. A. Brodnik, P. B. Miltersen, and J. I. Munro, Trans-dichotomous algorithms without multiplication — some upper and lower bounds, in Proc. 5th International Workshop on Algorithms and Data Structures (WADS 1997), Lecture Notes in Computer Science, Vol. 1272, Springer, Berlin, pp. 426–439.

15. J. L. Carter and M. N. Wegman, Universal classes of hash functions, *J. Comput. System Sci.* **18** (1979), pp. 143–154.

16. J. Cheriyan, T. Hagerup, and K. Mehlhorn, An $o(n^3)$-time maximum-flow algorithm, *SIAM J. Comput.* **25** (1996), pp. 1144–1170.

17. S. A. Cook and R. A. Reckhow, Time bounded random access machines, *J. Comput. System Sci.* **7** (1973), pp. 354–375.

18. T. H. Cormen, C. E. Leiserson, and R. L. Rivest, *Introduction to Algorithms*, The MIT Press, Cambridge, MA, 1990.

19. A. Dessmark and A. Lingas, On the power of nonconservative PRAM, in Proc. 21st International Symposium on Mathematical Foundations of Computer Science (MFCS 1996), Lecture Notes in Computer Science, Vol. 1113, Springer, Berlin, pp. 303–311.

20. M. Dietzfelbinger, T. Hagerup, J. Katajainen, and M. Penttonen, A reliable randomized algorithm for the closest-pair problem, *J. Algorithms* **25** (1997), pp. 19–51.

21. F. Fich and P. B. Miltersen, Tables should be sorted (on random access machines), in Proc. 4th International Workshop on Algorithms and Data Structures (WADS 1995), Lecture Notes in Computer Science, Vol. 955, Springer, Berlin, pp. 482–493.

22. M. L. Fredman, J. Komlós, and E. Szemerédi, Storing a sparse table with $O(1)$ worst case access time, *J. Assoc. Comput. Mach.* **31** (1984), pp. 538–544.

23. M. L. Fredman and M. E. Saks, The cell probe complexity of dynamic data structures, Proc. 21st Annual ACM Symposium on Theory of Computing (STOC 1989), pp. 345–354.

24. M. L. Fredman and D. E. Willard, Surpassing the information theoretic bound with fusion trees, *J. Comput. System Sci.* **47** (1993), pp. 424–436.

25. M. L. Fredman and D. E. Willard, Trans-dichotomous algorithms for minimum spanning trees and shortest paths, *J. Comput. System Sci.* **48** (1994), pp. 533–551.

26. M. Furst, J. B. Saxe, and M. Sipser, Parity, circuits, and the polynomial-time hierarchy, *Math. Syst. Theory* **17** (1984), pp. 13–27.

27. A. V. Goldberg and S. Rao, Beyond the flow decomposition barrier, in Proc. 38th Annual IEEE Symposium on Foundations of Computer Science (FOCS 1997), pp. 2–11.

28. J. Hastad, Almost optimal lower bounds for small depth circuits, Proc. 18th Annual ACM Symposium on Theory of Computing (STOC 1986), pp. 6–20.

29. J. L. Hennessy and D. A. Patterson, *Computer Organization and Design: The Hardware/Software Interface*, Morgan Kaufmann Publ., San Mateo, CA, 1994.

30. D. Kirkpatrick and S. Reisch, Upper bounds for sorting integers on random access machines, *Theoret. Comput. Sci.* **28** (1984), pp. 263–276.

31. P. B. Miltersen, Lower bounds for static dictionaries on RAMs with bit operations but no multiplication, in Proc. 23rd International Colloquium on Automata, Languages and Programming (ICALP 1996), Lecture Notes in Computer Science, Vol. 1099, Springer, Berlin, pp. 442–453.

32. P. B. Miltersen, Error correcting codes, perfect hashing circuits, and deterministic dynamic dictionaries, in Proc. 9th Annual ACM-SIAM Symposium on Discrete Algorithms (SODA 1998).

33. W. J. Paul and J. Simon, Decision trees and random access machines, in Proc. International Symposium on Logic and Algorithmic, Zürich, 1980, pp. 331–340.

34. R. Raman, Priority queues: Small, monotone and trans-dichotomous, in Proc. 4th Annual European Symposium on Algorithms (ESA 1996), Lecture Notes in Computer Science, Vol. 1136, Springer, Berlin, pp. 121–137.

35. R. Raman, Recent results on the single-source shortest paths problem, *SIGACT News* **28**:2 (1997), pp. 81–87.

36. R. E. Tarjan and U. Vishkin, An efficient parallel biconnectivity algorithm, *SIAM J. Comput.* **14** (1985), pp. 862–874.

37. M. Thorup, On RAM priority queues, in Proc. 7th Annual ACM–SIAM Symposium on Discrete Algorithms (SODA 1996), pp. 59–67.

38. M. Thorup, Randomized sorting in $O(n \log \log n)$ time and linear space using addition, shift, and bit-wise boolean operations, in Proc. 8th Annual ACM–SIAM Symposium on Discrete Algorithms (SODA 1997), pp. 352–359.

39. M. Thorup, Undirected single source shortest paths in linear time, in Proc. 38th Annual IEEE Symposium on Foundations of Computer Science (FOCS 1997), pp. 12–21.

40. M. Thorup, Faster deterministic sorting and priority queues in linear space, Proc. 9th Annual ACM-SIAM Symposium on Discrete Algorithms (SODA 1998).

41. M. Thorup, Floats, integers, and single source shortest paths, Proc. 15th Symposium on Theoretical Aspects of Computer Science (STACS 1998), Lecture Notes in Computer Science, Springer, Berlin.

42. L. G. Valiant, Parallelism in comparison problems, *SIAM J. Comput.* **4** (1975), pp. 348–355.

43. P. van Emde Boas, R. Kaas, and E. Zijlstra, Design and implementation of an efficient priority queue, *Math. Syst. Theory* **10** (1977), pp. 99–127.

44. D. E. Willard, Log-logarithmic worst-case range queries are possible in space $\Theta(n)$, *Inform. Process. Lett.* **17** (1983), pp. 81–84.

Communication-Efficient Deterministic Parallel Algorithms for Planar Point Location and 2d Voronoi Diagram*

Mohamadou Diallo[1], Afonso Ferreira[2] and Andrew Rau-Chaplin[3]

[1] LIMOS, IFMA, Campus des Cézeaux, BP 265, F-63175 Aubière Cedex, France.
E-mail: mdiallo@ifma.fr.
[2] CNRS, INRIA, Projet SLOOP, BP 93, 06902 Sophia Antipolis, France.
E-mail: ferreira@sophia.inria.fr.
[3] Faculty of Computer Science, Dalhousie University, P.O. Box 1000, Halifax, Nova
Scotia, Canada B3J 2X4. E-mail: arc@tuns.ca. Partially supported by NSERC.

Abstract. In this paper we describe deterministic parallel algorithms
for planar point location and for building the Voronoi Diagram of n
co-planar points. These algorithms are designed for BSP-like models
of computation, where p processors, with $O(\frac{n}{p}) \gg O(1)$ local memory each, communicate through some arbitrary interconnection network.
They are communication-efficient since they require, respectively, $O(1)$
and $O(\log p)$ communication steps and $O(\frac{n \log n}{p})$ local computation per
step. Both algorithms require $O(\frac{n}{p}) = \Omega(p)$ local memory.

1 Introduction

High performance computing systems nowadays, be either multicomputers or
networks of workstations, consist of a set of p *state-of-the-art* processors, each
with considerable local memory, connected to some interconnection network.
These systems are usually *coarse grained*, i.e. the size of each local memory is
"considerably larger" than $O(1)$.

Recently, there has been a growing interest in coarse grained computational
models [4, 7, 16] and the design of coarse grained algorithms [10, 5, 8, 6, 9]. The
work on computational models has tended to be motivated by the observation
that "fast algorithms" for fine-grained models rarely translate to fast code running on coarse grained machines. The BSP model ([16]) was proposed in order
to benefit from slackness in the number of processors and memory mapping via
hash functions to hide communication latency and provide for the efficient execution of fine grained PRAM algorithms on coarse grained hardware. Other coarse
grained models focus more on utilizing existing sequential code and minimizing
global communication operations. These include the Coarse Grained Multicomputer (CGM) from [7] used in this paper and to be described below. In this

* Part of this work was done while the second author was with the LIP at the ENS
Lyon and while the authors visited each other in Lyon and in Halifax. Support from
the respective Institutions is acknowledged.

mixed sequential/parallel setting, there are three important measures of any coarse grained algorithm, namely, the amount of local computation, the number and type of global communication phases required and the scalability of the algorithm, that is, the range of values for the ratio $\frac{n}{p}$ for which the algorithm is efficient and applicable.

This paper describes efficient scalable parallel algorithms for the planar point location and the 2d Voronoi diagram problems within the coarse grained multicomputer context. The planar point location algorithm requires local storage $\frac{n}{p} = \Omega(p)$ and is optimal with respect to local computation ($O(\frac{n \log n}{p})$) and communication phases ($O(1)$). This algorithm is then used as a procedure in the Voronoi Diagram algorithm, which also requires local storage $\frac{n}{p} = \Omega(p)$, but uses $\lceil \log p \rceil$ communication phases with $O(\frac{n \log n}{p})$ local computation per phase.

The Model

The *Coarse Grained Multicomputer* model, or $CGM(n, p)$ for short, can be seen as a weak-CREW BSP machine ([7, 10]). On a $CGM(n, p)$ a problem of size n is solved using p processors each with a local memory of size $O(\frac{n}{p})$. The processors communicate through some arbitrary interconnection network or a shared memory. The term "*coarse grained*" refers to the fact that (as in practice) the number of words of each local memory $O(\frac{n}{p})$ is defined to be "considerably larger" than $O(1)$. This is clearly true for all currently available coarse grained parallel machines. In the following, when determining time complexities both local computation time and inter-processor communication time are considered in the standard way. Also note that we assume, for clarity of explanation, that $p = 2^k$ for some fixed integer k.

In this model, all global communications are performed by a small set of standard communications operations - Segmented broadcast, Segmented gather, All-to-All broadcast, Personalized All-to-All broadcast, Partial sum and Sort, which are typically efficiently realized in hardware or system level code. If a parallel machine does not provide these operations they can be, in the worst case, implemented in terms of a constant number of sorting operations [7].

Furthermore, it was shown that, given $n^{1-\frac{1}{c}} > p$ ($c \geq 1$), sorting $O(n)$ elements distributed evenly over p processors in the CGM, BSP or LogP models can be achieved in $O(\log n / \log(h + 1))$ communication rounds and $O(n \log n / p)$ local computation time, for $h = \Theta(\frac{n}{p})$, i.e. with optimal local computation and $O(1)$ h-relations, when $\frac{n}{p} = \Omega(p)$ [10]. Therefore, using this sort, the communication operations of the $CGM(s, p)$ can be realized in the BSP or LogP models in a constant number of h-relations, where $h = \Theta(\frac{s}{p})$.

Hence, finding an optimal algorithm in the CGM model is equivalent to minimizing the number of global communication rounds as well as the local computation time. It has also been shown that minimizing the number of rounds also results in improved portability across different parallel architectures [16].

Previous Work

Many algorithms (sequential or parallel) have been proposed for solving the multi-planar point location problem [1, 13], where $O(n)$ query points are located in a planar convex subdivision with n vertices. The sequential complexity of the problem is $\Theta(n \log n)$ time with $O(n)$ space. In the fine grained parallel setting, algorithms have been described for many architectures including the CREW PRAM [3], the Hypercube [15] and the Mesh [12]. Except for the PRAM, these algorithms are not work-optimal (time in $O(\sqrt{n})$ and $O(\log^2 n)$ for the Mesh and the Hypercube, respectively).

For the Voronoi diagram (see Figure 1), whose sequential complexity is $\Theta(n \log n)$, the only time-optimal parallel algorithm (although not work-optimal since it runs in $O(\sqrt{n})$ time with n processors) was proposed in [12] for the Mesh. The same technique (to be explored further in this text) was used in [15] to design a $O(\log^3 n)$ time algorithm for the Hypercube. Finally, the best existing PRAM algorithm requires $O(\log n \log \log n)$ time. With respect to the CGM, no efficient deterministic algorithm exist. The *randomized* algorithm from [5] builds the Voronoi diagram in time $O(\frac{n \log n}{p})$, *with high probability*, and requires $n/p = \Omega(p^2)$.

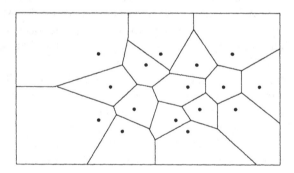

Fig. 1. Voronoi diagram of a set of points.

Our Results

In this paper we first describe a scalable coarse-grained **deterministic** algorithm for the Convex Planar Multi-Point Location problem. Our algorithm requires time $O(\frac{n \log n}{p})$ in the worst case. Furthermore, it requires only a constant number of global communication rounds and local memory space $\frac{n}{p} = \Omega(p)$ to locate $O(n)$ query points in a planar convex subdivision with n vertices.

Using this algorithm as a subprocedure, we propose an algorithm for solving the Voronoi diagram problem for points in $\mathcal{R}^2$ for the same model, with the same space complexity, $\lceil \log p \rceil$ communication rounds, and $O(\frac{n \log n}{p})$ local computation per round. Our algorithm is deterministic and is also more scalable than the algorithm given in [5] in that it is efficient and applicable for a larger range of values for the ratio n/p.

Our approach (which is very different from the one presented in [2, 11]) presents two particular strengths. First, all inter-processor communications are restricted to usages of a small set of simple communication operations. This has the effect of making the algorithms both easy to implement, in that all communications are performed by calls to a standard highly optimized communication library, and very fast in practice. Second, most of the local computation is done through well known algorithms designed for the very same problems. Therefore, costs associated with software development are largely reduced.

2 Planar Point Location

The problem of planar multi-point location on a convex subdivision is stated as follows: Locate $O(n)$ points in a planar convex subdivision defined by $O(n)$ edges. Each edge is labeled with the regions to its left and its right, and regions are defined by coordinates of one interior point (called the center of the region).

To locate a point in the planar subdivision, we design a coarse-grained algorithm based on the chain method originally described in the sequential setting ([13]) and then utilized in the fine-grained parallel setting for MCC ([12]) and hypercubes ([15]). The idea is to perform planar point location via a binary search on a balanced binary tree whose nodes represent a chain of edges of the planar subdivision. The tree is built as follows.

First the regions are sorted by x-coordinate of their centers. There is a chain of edges which share half regions to left and half to right (left and right regions correspond to centers lying to left or right of the chain). The same is applied to left and right half of regions recursively and a monotone complete set of chains is obtained (i.e. the set of chains so that for any two chains c_1 and c_2 the vertices of c_1 that are not on c_2 are on the same side of c_2). These chains are the nodes of the balanced binary tree mentioned above.

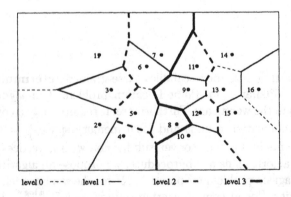

level 0 - - -   level 1 ——   level 2 — —   level 3 ▬▬

Fig. 2. Construction of the chains.

The leaves of this tree correspond to regions of the subdivision (see Figure 2). Chains may share common edges. If an edge e belongs to more than one chain then it belongs to all members of a set of consecutive chains. There is a unique member c of this set which, in the binary search tree, is a common ascendant of all the other members of the set (the highest chain, in the hierarchy, to which e belongs). In order to avoid duplication of edges, we assign e to such a member c. By $O(\log n)$ discriminations (deciding on which side of chain c a query point lies) each query point can be located.

Each chain has a level and an index. The level of a chain is the height were the chain is located in the tree (the root has the highest level). The index is the rank of the chain in the chains of a given level, ranked from left to right. And as described above, each edge is assigned to exactly one chain. The level and the index of an edge are those of the chain to which it belongs to. The levels and indices of the edges can be determined in constant time using the rules described in [12]: for a given edge e, find the "bit exclusive or", say ψ, of the binary indices of centers of e. The level of e, say l_e, is $l_e = \lfloor \log \psi \rfloor$. The index of e is $((2's\ complement(2^{l_c}) - 2^{l_c}) \wedge (index\ of\ center\ of\ e))/2^{l_c+1}$.

2.1 Coarse-Grained Planar Multi-Point Location

We describe in this subsection a planar multi-point location algorithm that requires a constant number of communication rounds. The entire data for a given problem is assumed to be initially distributed across the local memories and remains there until the problem is solved. Given a set Q of n query points, a planar convex subdivision of the plane into n regions (e.g. a Voronoi diagram) and a p processor coarse grained multicomputer we show how to locate the query points into the subdivision. The basic approach is as follows:

1. Divide the plane into the p regions or vertical slabs (Figure. 3) $V_1, V_2, \ldots V_p$ defined by the $p-1$ highest level chains.
2. For each point $q \in Q$ determine $vs(q) \in \{V_1, V_2, \ldots V_p\}$ the vertical slab q is located in. This is done by forming horizontal slabs (Figure. 3) from the chains computed in Step 1 and performing a point location within these horizontal slabs after first having load balanced the points and slabs.
3. Finally, load balance the vertical slabs and the points such that each processor stores $O(1)$ vertical slabs of total size $O(n/p)$ and $O(n/p)$ points that must be located in them. Locally execute planar multi-point location on all processors.

The main challenge lies in computing for each point which vertical slab it is in (Step 2) in a constant number of communication phases and under the constraint given by the memory size. The idea will be to partition the vertical slabs into p horizontal slabs that are bounded by lines rather than polygonal chains. Our Planar Multi-Point Location algorithm is described in detail below.

CGM's Planar Multi-Point Location(Q, $Vor(S)$)

Input: A set Q of $O(n)$ query points and a planar subdivision defined by $O(n)$ edges.
Output: The $O(n)$ query points labeled by the center of the region to which they belong.

1. For each edge, determine to which chain it belongs using the method by Jeong [12]. The method involves sorting the regions' centers by their x-coordinate. Recall that using this method, each edge belongs to only one chain. Note that we are only interested in the $p-1$ higher level chains, these chains partition the plane into p "vertical" slabs $V_1, V_2, \ldots V_p$ (Figure. 3). Let C denotes the set of the edges that define the $p-1$ chains.

2. Sort the edges in C by their largest y-coordinates. Each processor i receives $O(n/p)$ edges denoted H_i and can determine a horizontal line that defines its upper boundary by looking for the largest received y-coordinate (Figure. 3). Perform an all-to-all broadcast of these horizontal lines so that every processor stores a copy of H, the set of these p horizontal lines.

3. Each processor determines for each edge $c \in C$ it stores the elements of H it intersects, denoted $range(c)$. Note that, because the chains are y-monotonic, $range(c)$ is a (contiguous) interval that can be computed by binary search in H, each edge is intersected by at most p horizontal lines and each element of H intersects at most p elements of C. Perform a personalized all-to-all broadcast such that each edge c, for which $range(c) = [i, j]$ is not empty, is broadcast to processors i through j.

4. For each point $q \in Q$ determine $hs(q) \in \{H_1, H_2, \ldots H_p\}$ the horizontal slab q is located in and for each horizontal slab H_i, compute $C(H_i) = \lceil \frac{|\{q \in Q : hs(q) = H_i\}|}{\frac{n}{p}} \rceil$, for $1 \leq i \leq p$. Create $C(H_i)$ copies of H_i and distribute them such that each processor stores at most two horizontal slabs. Redistribute Q such that each point $q \in Q$ is stored on a processor that also stores a copy of $hs(q)$.

5. Each processor locally executes Kirkpatrick's planar multi-point location algorithm ([14]). When a point is located to the right or the left of an edge, the vertical slab to which it belongs, $vs(q)$ is obtained by consulting the rank of the center of the region associated to the edge, in the sorted list.

6. For each vertical slab V_i, compute $C(V_i) = \lceil \frac{|\{q \in Q : vs(q) = V_i\}|}{\frac{n}{p}} \rceil$, for $1 \leq i \leq p$. Create $C(V_i)$ copies of V_i and distribute them such that each processor stores at most two vertical slabs. Redistribute Q such that each point $q \in Q$ is stored on a processor that also stores a copy of $vs(q)$.

7. All processors now locally execute Kirkpatrick's planar multi-point location ([14]). The location is done in the vertical slab into which the points are located and each point is now precisely located.

Theorem 1. *Algorithm* **CGM's Planar Multi-Point Location()** *locates $O(n)$ query points in a planar convex subdivision defined by $O(n)$ edges in $O(\frac{n \log n}{p})$ time. It requires $\frac{n}{p} = \Omega(p)$ local memory space and a constant number of communication rounds.*

Proof. The correctness follows from the correctness of the chain method described in [12], the correctness of Kirkpatrick's sequential planar multi-point

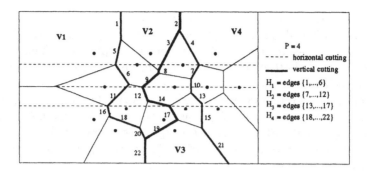

Fig. 3. Horizontal and vertical cuttings.

location method [14, pages 56-58], and the following observations. (1) Both the vertical and horizontal slabs have a size of $O(n/p)$. (2) The total number of slabs created in Steps 4 and 6 is $O(p)$. (3) The total number of queries moved in steps 4 and 6 is $O(n/p)$. The space requirement is thus $O(\frac{n}{p} + p) = O(\frac{n}{p})$ per processor. In each step, the local computation time is at most $O(\frac{n}{p} \log n)$. The global communication in each step reduces to a constant number global sorts and communications operations.

3   Building a 2D-Voronoi Diagram on a CGM

In this section, we first present an algorithm for merging two Voronoi diagrams on a $CGM(n,p)$ which requires only $O(1)$ communication phases and then show how this algorithm can be used to help build the Voronoi diagram of a set of 2d-points through a divide-and-conquer approach. The merge algorithm in turn uses the planar multi-point location algorithm described in the previous section as a basic subprocedure.

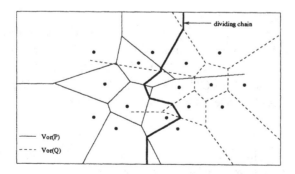

Fig. 4. The dividing chain.

Let a set S of n points (the center of each region) in the plane be given and

P and Q be two disjoint subsets of S, of size $\frac{n}{2}$ each, such that all points of P are located to the left of all points of Q. Suppose that the Voronoi diagrams of P and Q are known and denoted by $\text{Vor}(P)$ and $\text{Vor}(Q)$, respectively. Finally, suppose that $\text{Vor}(P)$ and $\text{Vor}(Q)$ are each represented by a set of edges distributed evenly over $p/2$ processors.

Our merging algorithm follows the scheme described in [12] in which a dividing chain between two Voronoi diagrams is computed (see Figure 4). Since the problem is analogous with respect to P or Q, we will describe the details of the merging from only the point of view of P. The following are the main steps of the algorithm.

Two-Way Merging($\text{Vor}(P)$, $\text{Vor}(Q)$)

Input: A distributed representation of $\text{Vor}(P)$ and $\text{Vor}(Q)$, each over $\frac{p}{2}$ processors.
Output: A distributed representation of $\text{Vor}(P \cup Q)$ over p processors.

1. Partition the edges of $\text{Vor}(P)$ into three sets:
   (a) PP, those that have both their endpoints closer to P than to Q,
   (b) PQ, those that have one of their endpoints closer to P than to Q, and the other one closer to Q than to P.
   (c) QQ those that have both their endpoints closer to Q than to P.
2. For each of the sets found above, decide which edges are intersected by the dividing chain (actually the problem is just for QQ).
3. Compute the new endpoints for the edges that are intersected by the dividing chain (intersection point with the dividing chain) and discard the portion of the edge laying in the wrong side.
4. Globally sort all the newly generated endpoints (of the edges of $\text{Vor}(P)$ and $\text{Vor}(Q)$) in order to obtain the edges of the dividing chain (for the infinite rays, it suffices to look at the two points, one in P and the other one in Q, that are closer to their finite endpoint to find their slope).
5. Perform Steps 1 through 4, analogously, with respect to $\text{Vor}(Q)$.
6. All the current edges form $\text{Vor}(S)$. Distribute them over the p processors.

Theorem 2. *Given two sets P and Q of $\frac{n}{p}$ points in the plane, $P \cup Q = S$, such that all points in P are on the left of all points in Q, and a distributed representation of the two Voronoi diagrams $\text{Vor}(P)$ and $\text{Vor}(Q)$, each distributed over $\frac{p}{2}$ processors, then algorithm* **Two-Way Merging()** *merges $\text{Vor}(P)$ and $\text{Vor}(Q)$ to form $\text{Vor}(S)$ in $O(\frac{n \log n}{p})$ time. It requires $\frac{n}{p} = \Omega(p)$ local memory space and a constant number of communication rounds.*

Proof. We now consider the correctness and complexity of each step of the algorithm **Two-Way Merging($\text{Vor}(P)$, $\text{Vor}(Q)$)**.

- Step 1: Partitioning of the edges into the sets PP, PQ and QQ can be computed for the finite edges by performing a planar point location of the endpoints of the edges. For the semi-infinite edges, it has been established [12] that, if all the semi-infinite edges of $\text{Vor}(Q)$ are sorted by their slope δ, then for the infinite endpoint v_i and the semi-infinite edges e_i of $\text{Vor}(P)$, and two consecutive semi-infinite edges e_j and e_{j+1} of $\text{Vor}(Q)$, v_i is laying in the unbounded region bordered by e_j and e_{j+1} if and only if $\delta_{e_j} \leq \delta_{e_i} \leq \delta_{e_{j+1}}$.

Using this result, we can find the center of the region, in $\text{Vor}(Q)$, containing the endpoint at infinity and thus see to which set it is closer to by just computing the bisector between the closest point in P and the closest one in Q and then see if the semi-infinite edge crosses this bisector.

Hence, the time complexity of this step is also dominated by calls to the planar point location algorithm, that is $O(\frac{n \log n}{p})$.

- Step 2: In [12] it was also shown that the edges in PP do not cross the dividing chain, the edges in PQ cross it once, and for the edges in QQ we have two cases: if they cross the dividing chain they cross it twice, or else they do not cross it at all (see Figure 5). A simple technique to distinguish these two cases involves again a planar point location. The point location concerns, for each edge of QQ, a unique and precise point X on the concerned edge. Each edge which is determined to be intersected twice is split into two edges of type PQ at the point X. For and edge e in QQ, X is the intersection point between e and the horizontal line passing through one of the centers of the two regions associated to e. The chosen center is the one with the greatest x-coordinate. Here again, the time complexity of this step is also dominated by calls to the planar point location algorithm, that is $O(\frac{n \log n}{p})$.

- Step 3: Compute one intersection point per edge since the edges that are intersected twice are now split into two edges of type PQ. The computation of the intersection point can be done in constant time by computing the bisector between the point in P (the one with the greatest x-coordinate) closest to the first endpoint and the point in Q closest to the second endpoint, and then compute the intersection of the edge with this bisector.

- Step 4: A global sort. Once the new endpoints are sorted (using their y-coordinate as principal key), the dividing chain is built. Recall that this chain is y-monotonic that it is crossed at most once by all horizontal lines. The time complexity of this step is thus $T_s(n, p)$.

- Step 5: A communication phase in which the newly built dividing chain is distributed over the appropriate processors.

Note that all of the steps consist of at most $O(\frac{n \log n}{p})$ local computation and a constant number of calls to the planar point location algorithm, therefore the time complexity follows. The correctness follows from [12].

Using **Two-Way Merging()** we can now easily describe a CGM algorithm for building the Voronoi diagram. Recall that $p = 2^k$ for some integer k.

408

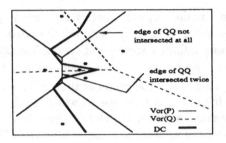

Fig. 5. Example of an edge that is intersected twice.

Voronoi diagram(S)

Input: Each processor stores a set of $\frac{n}{p}$ points drawn arbitrarily from S.
Output: A distributed representation of the Voronoi diagram of S.

1. Globally sort the points in S by x-coordinate. Let S_i denote the set of $\frac{n}{p}$ sorted points now stored on processor i.
2. Independently and in parallel, each processor i computes the Voronoi diagram of the set S_i. Let $\text{Vor}(S_i^1)$ denote the result on processor i.
3. For $j = 1$ to $\log p$ in parallel do
   $\text{Vor}(S_i^{j+1}) \leftarrow$ **Two-Way Merging**($\text{Vor}(S_{2i}^j)$, $\text{Vor}(S_{2i+1}^j)$), (i from 0 to $\frac{p}{2^j} - 1$).

We have therefore proved the second main result of this paper:

Theorem 3. *Algorithm* **Voronoi diagram**() *computes the Voronoi diagram of a set S of n points in the plane, Vor(S) , on a $CGM(n,p)$. It requires $\frac{n}{p} = \Omega(p)$ local memory space, $\lceil \log p \rceil$ communication rounds, and $O(\frac{n \log n}{p})$ local computation time per round.*

References

1. S. Akl and K. Lyons. *Parallel Computational Geometry*. Prentice Hall, 1993.
2. M. Atallah and J. Tsay. On the parallel-decomposability of geometric problems. In *Proceedings of the 5th Annual ACM Symposium on Computational Geometry*, pages 104–113, 1989.
3. R. Cole, M. Goodrich, and C. Dunlaing. Merging free trees in parallel for efficient voronoï diagram construction. In *17th ICALP, England*, July 1990.
4. D. Culler, R. Karp, D. Patterson, A. Sahay, K. Schauser, E. Santos, R. Subrarnonian, and T. von Eicken. LogP: Towards a realistic model of parallel computation. In *Fifth ACM SIGPLAN Symposium on the Principles and Practice of Parallel Programming*, 1993.
5. F. Dehne, X. Deng, P. Dymond, A. Fabri, and A. Khokhar. A randomized parallel 3d convex hull algorithm for coarse grained multicomputers. In *Proc. 7th ACM Symp. on Parallel Algorithms and Architectures*, pages 27–33, 1995.
6. F. Dehne, A. Fabri, and C. Kenyon. Scalable and archtecture independent parallel geometric algorithms with high probability optimal time. In *Proceedings of the 6th IEEE SPDP*, pages 586–593. IEEE Press, 1994.

7. F. Dehne, A. Fabri, and A. Rau-Chaplin. Scalable parallel geometric algorithms for coarse grained multicomputers. In *ACM 9th Symposium on Computational Geometry*, pages 298–307, 1993.

8. X. Deng and N. Gu. Good algorithm design style for multiprocessors. In *Proc. of the 6th IEEE Symposium on Parallel and Distributed Processing, Dallas, USA*, pages 538–543, October 1994.

9. A. Ferreira, A. Rau-Chaplin, and S. Ubéda. Scalable 2d convex hull and triangulation for coarse grained multicomputers. In *Proc. of the 6th IEEE Symposium on Parallel and Distributed Processing, San Antonio, USA*, pages 561–569. IEEE Press, October 1995.

10. M. Goodrich. Communication-efficient parallel sorting. In *Proc. of the 28th annual ACM Symposium on Theory of Computing Philadephia, USA*, May 1996.

11. M. Goodrich, J. Tsay, D. Vengroff, and J. Vitter. External-memory computational geometry. *Proceedings of the Symposium on Foundations of Computer Science*, 1993.

12. C. Jeong. An improved parallel algorithm for constructing voronoï diagram on a mesh-connected computer. *Parallel Computing*, 17:505–514, 1991.

13. D.T. Lee and F. Preparata. Location of a point in a planar subdivision and its applications. *SIAM Journal on Computing*, 6(3):594–606, 1977.

14. F. Preparata and M. Shamos. *Computational Geometry: An Introduction*. Springer Verlag, 1985.

15. I. Stojmenovic. Computational geometry on a hypercube. Technical report, Computer Science Dpt., Washington State University, Pullman, Washington 99164–1210, 1987.

16. L. Valiant. A bridging model for parallel computation. *Communication of ACM*, 38(8):103–111, 1990.

On Batcher's Merge Sorts as Parallel Sorting Algorithms

Christine Rüb

Max-Planck-Institut für Informatik, Im Stadtwald
D-66123 Saarbrücken

Abstract. We examine the average running times of Batcher's bitonic merge and Batcher's odd-even merge when they are used as parallel merging algorithms. It has been shown previously that the running time of odd-even merge can be upper bounded by a function of the maximal rank difference for elements in the two input sequences. Here we give an almost matching lower bound for odd-even merge as well as a similar upper bound for (a special version of) bitonic merge. From this follows that the average running time of odd-even merge (bitonic merge) is $\Theta((n/p)(1+\log(1+p^2/n)))$ $(O((n/p)(1+\log(1+p^2/n)))$, resp.) where n is the size of the input and p is the number of processors. Using these results we then show that the average running times of odd-even merge sort and bitonic merge sort are $O((n/p)(\log n + (\log(1+p^2/n))^2))$, that is, the two algorithms are optimal on the average if $n \geq p^2/2^{\sqrt{\log p}}$.

1 Introduction

Batcher's bitonic merge sort and odd-even merge sort are two well known comparator networks for sorting [1,6]. The two networks can easily be converted into parallel sorting algorithms where each processor holds more than one element by replacing comparisons between input elements by a split–and–merge procedure [6]. The running time of a straightforward implementation will then be $O((n/p)(\log n + \log^2 p))$, where n is the size of the input and p is the number of processors. Although this is not optimal, the constants involved are small and the two algorithms can be the fastest available for small input sizes.

In contrast to odd-even merge sort, bitonic merge sort has often been used in comparative studies of sorting algorithms for parallel computers. Because of the small constant factors, here bitonic merge sort proved to be the fastest sorting algorithm for small input sizes.

In [8] it has been shown that the average running time of the odd-even merge (odd-even merge sort) algorithm can be improved much by keeping the amount of communication to a minimum. The resulting running times are $O((n/p)(1 + \log(1+p^2/n)))$ for merging and $O((n/p)(\log n + \log p \log(1+p^2/n)))$ for sorting. (In the case of merging (sorting) we assume that each outcome of the merging (each permutation of the input elements, resp.) is equally likely.) In the meantime this version of odd-even merge sort has been used in two comparative studies of sorting [3,9]. In both cases, odd-even merge sort was the fastest sorting algorithm among those considered for some input size.

However, the run time predictions in [3] showed that the upper bounds from [8] were too pessimistic. In fact, as we will show in this paper, the average running time of the odd-even merge sort algorithm can be bounded by $O((n/p)(\log n + (\log(1 + p^2/n))^2))$.

In [8] it was pointed out that the average running time of bitonic merge sort can be improved by storing the input elements such that the smaller indexed processor always receives the smaller elements and again keeping the amount of communication at a minimum. This means that an already established ordering among elements will be preserved. In this paper we prove that the average running time of the order-preserving bitonic merge (bitonic merge sort, resp.) algorithm can be upper bounded by $O((n/p)(1 + \log(1 + p^2/n)))$ $(O((n/p)(\log n + (\log(1 + p^2/n))^2)))$, resp.).

We also report shortly on experimental results that we have obtained using a simulation program (more results can be found in [7]). For these experiments we do not use a specific computer model; rather, we assume that the processors form a complete graph and that the number of exchanged elements determines the running time. The actual average running time of an implementation depends strongly on the parallel machine used and such an investigation is beyond the scope of this paper. However, we hope the results obtained here will help to estimate actual average running times in the future.

This paper is organized as follows. Section 2 contains some preliminaries. Section 3 is concerned with odd-even merge, Section 4 with bitonic merge, and Section 5 with the two merge sort algorithms. Section 6 gives some experimental results and Section 7 contains some conclusions.

2   Preliminaries

The algorithms considered here are based on comparator networks. A comparator network for merging (sorting) p elements can be turned into a parallel algorithm for merging (sorting, resp.) $n = pr$ elements using p processors by replacing each comparator by a split–and–merge procedure [6]. To improve the average running time of the algorithm, the exchange procedure can be implemented such that only the minimal number of elements necessary is exchanged.

In the remainder of this paper we mean by odd-even merge (sort) and bitonic merge (sort) parallel procedures that use such an exchange procedure. Additionally we imply that the input elements are stored such that the smaller elements are always sent to the processor with the smaller index. (In the case of bitonic merge the corresponding merging network is known as the balanced merger [4]. In [2] it was shown that the bitonic merge network and the balanced merger are essentially the same networks.) We will also assume that the number p of processors is a power of two.

We will make statements of the following kind. Let *alg* stand for an algorithm, let n be the size of the input and let p be the number of processors. Each processor stores $r = n/p$ elements of the input. The algorithm consists of T steps where in each step pairs of processors communicate and possibly exchange elements. We define the functions t_{alg} and R_{alg} as follows.

Definition 1. $t_{alg}(T+1, j) = 0$, $0 \leq j \leq p - 1$.
$t_{alg}(i, j) = \max\{t_{alg}(i + 1, j), t_{alg}(i + 1, k)\} + x(j, k)$, $1 \leq i \leq T$, where P_k is the processor that communicates with P_j in step i and $x(j, k)$ is the number of elements P_j and P_k exchange. If P_j does not communicate with any processor in step i, $t_{alg}(i, j) = t_{alg}(i + 1, j)$.
Finally, $R_{alg}(r, p) = \max\{t_{alg}(1, j); 0 \leq j \leq p - 1\}$.

A lower bound for R_{alg} will always be a lower bound for the running time of the algorithm. For our upper bounds on R_{alg} we assume that $x(i, j) = r$ whenever the two communicating processors exchange an element. By doing this, we neglect the communication network but we account for the local computation time.

Due to space limitations, we state several lemmas and theorems without proof. The proofs can be found in [7].

3   Odd-even merge

In this section we show that the running time of odd-even merge is closely related to the maximal rank difference of elements in the two input sequences. Namely, we show the following. Let A and B be the two sorted sequences to be merged and let $d_{max} = \max\{| \text{ rank of } x \text{ in } A - \text{ rank of } x \text{ in } B|; x \in A \cup B\}$. Then the running time of odd-even merge is $\Theta((n/p)(1 + \log(1 + d_{max}p/n)))$ where n is the size of the input. From these bounds upper and lower bounds for the average running time of odd-even merge will be derived.

From the odd-even merge network we can derive the following parallel merge procedure. Assume we want to merge two sorted lists A and B of length m such that the even indexed processors hold subsequences of A and the odd indexed processors hold subsequences of B. Each processor holds $2m/p =: r$ elements. The following procedure will merge the two lists.

procedure Odd_Even_Merge(p);
for all i, $0 \leq i < p/2$, **pardo**
    compare–exchange(P_{2i}, P_{2i+1});
for $i = \log p - 1$ **downto** 1 **do**
    for all $j, 1 \leq j \leq (p - 2^i)/2$, **pardo**
        compare–exchange(P_{2j-1}, P_{2j+2^i-2});

Compare-exchange(P_i, P_j) denotes a procedure where P_i gets the smallest r elements stored at P_i and P_j, and P_j gets the largest r elements.
First we will analyze the running time of the for-loop of the odd-even merge algorithm. For ease of notation, we call the step of the for-loop where the index i has a certain value k, step k of the for-loop or step k of the algorithm. Correspondingly, the first step of odd-even merge will also be called step $\log p$.

Let $A^j = A_{jr}, A_{jr+1}, ..., A_{(j+1)r-1}$ and $B^j = B_{jr}, B_{jr+1}, ..., B_{(j+1)r-1}$, $0 \le j \le p-1$. We define i_{max} as the first (or maximal) i where at least two elements are exchanged in the for-loop of the algorithm and e_{max} as the maximal number of elements that any processor exchanges in step i_{max}. The following lemma gives an upper bound on d_{max} that depends on i_{max} and e_{max}.

Lemma 1. $d_{max} \ge (2^{i_{max}-1} - 1)r + 2e_{max}$.

Lemma 1 gives immediately rise to an upper bound on the running time that depends on d_{max}: since i_{max} is the first step of the for-loop that is executed, no processor can exchange more than $(i_{max} - 1)r + e_{max}$ elements during the execution of the for-loop.

With the help of the following four lemmas it is possible to prove a lower bound for the running time of odd–even merge that depends on d_{max}. The four lemmas examine which paths an element takes during the execution of odd-even merge, and which paths are particularly expensive.

Lemma 2. Let x be stored at P_j at the beginning of the for-loop and at $P_{j+\delta}$ at the end of the for-loop where $\delta = \pm(2/3)(2^i - 1 + \alpha)$, $\alpha \in \{0, 1/2\}$ and $i \in N$. Then x will be moved to another processor exactly in the last i steps of the for-loop.

The following two lemmas show that if one element is moved a certain number z of processors forward during the for-loop, we can always find a subsequence Z of the input (i.e., $Z = A^j$ or $Z = B^j$) where all elements of Z are moved y or $1 + y$ processors forward for all $y \le z - 2$. From this we can then derive a lower bound for the running time of odd-even merge.

Lemma 3. Let B_j have rank $j + \delta$ in A and let B_k have rank $k + \gamma$ in A, $j < k$ and $\delta > \gamma$. Let $\delta > \beta > \gamma$. Then there exists an index l, $j < l < k$, such that B_l has rank $l + \beta$ in A, that is, every rank difference between δ and γ occurs between B_j and B_k. A similar claim holds for A and B interchanged.

Lemma 4. Let x be an element that is moved from processor P_{2j+1} to processor $P_{2j+1+\alpha}$ during the execution of the for-loop, $\alpha \ge 2$. Let $0 \le \beta \le \alpha - 2$. Then there exists a processor P_{2k+1}, $k > j$, where all elements stored at processor P_{2k+1} before the execution of the for-loop will be stored at processor $P_{2k+1+\beta}$ or at processor $P_{2k+2+\beta}$ at the end of the for-loop.

By combining Lemma 2 and Lemma 4, we can show a relationship between d_{max} and the maximal distance a subsequence of A or B has to travel.

Lemma 5. If $d_{max} \ge (2/3)(2^i + 7/2)r$, $i \ge 2$, there exists a subsequence A^j of A or a subsequence B^j of B, where A^j (B^j, resp.) is moved to a different processor during steps i through 2 of the for-loop.

Next we put together the results shown above to prove lower and upper bounds on the running time of the for-loop of odd-even merge that depend on d_{max}. The subscript "for" stands for the for–loop of odd–even merge.

Theorem 6. Let $T = (\lfloor \log(\max\{2, d_{max}/r \cdot (3/2) - 7/2\}) \rfloor - 1)r$, and let $S = (\lfloor \log(d_{max}/r + 1) \rfloor + 1)r$.
1. $T \leq R_{\text{for}}(r, p) \leq S$.
2. If $d_{max} \leq r$, then $R_{\text{for}}(r, p) \leq d_{max}/2$.
3. Let $d_{max}/r = 2^x y - 1$, $1 \leq y < 2$ (thus $S = (x+1)r$). Let $d_{max}/r \cdot 3/2 - 7/2 \geq 0$. If $(3/2)y - 5/2^x \geq 2$, then $S - T = r$. If $(3/2)y - 5/2^x < 2$, then $S - T = 2r$.
4. $S - T \in \{r, 2r\}$.

The following theorem also considers the first step of odd–even merge. The subscript "odd" stands for odd-even merge.

Definition 2. Let $P(\Delta)$ be the probability that there exists an element x in $A \cup B$ where the rank of x in A differs from the rank of x in B by Δ. For $\Delta > 0$, let $Q_A(\Delta)$ be the probability that there exists an element x in A where the rank of x in B minus the rank of x in A is Δ.

Theorem 7.

$$r \left(Q_A(2r - 1) + \sum_{i=2}^{\lfloor \log((3/2)p - 7) \rfloor - 1} Q_A \left((2^{i+1} + 7) \, r/3 \right) \right)$$

$$\leq \text{Exp}(R_{\text{odd}}) \leq r \left(1 + \sum_{i=1}^{\log p - 1} P \left((2^{i-1} - 1) \, r \right) \right)$$

Closed formulations for the average running time will be derived in Section 5.

4 Bitonic merge

As odd–even merge, bitonic merge can be defined as a recursive procedure. To obtain a network from this recursion we have to decide how to store the input. Fig. 1 shows the network we use here where A and B are stored alternating. This corresponds to the following merging algorithm.

```
procedure Bitonic_Merge(p);
for i = log p downto 1 do
    { mask = 2^i - 1;
      for all j, j&2^{i-1} = 0, pardo
          compare-exchange(P_j, P_j^mask);
    }
```

Here, & denotes bitwise AND and ^ denotes bitwise XOR. That is, in step i each processor P_j communicates with the processor whose index is obtained by flipping the rightmost i bits of the binary representation of j. (Similar to odd-even merge, we denote the step of bitonic merge where the index i has value k by step k.)

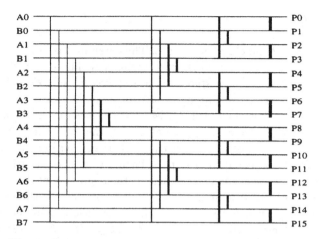

Fig. 1. Order–preserving bitonic merge with 16 processors

Note that the distances between communicating processors in a given step of Bitonic_Merge differ much: the distances lie between 1 and $2^i - 1$ in step i. Since we want to establish a relationship between d_{max} (see Section 3) and the running time of Bitonic_Merge, we divide the communicating pairs of processors into $\log p$ batches. The membership to a batch is determined by the difference between the indexes of the subsequences of A and B the two processors hold at the beginning of the algorithm. Batch i covers all (odd) differences δ where $2^{i-2} \le \delta < 2^{i-1}$ if $i \ge 2$ and where $\delta = 0$ if $i = 1$. Thus, Batch 1 contains all communicating pairs of step 1, and Batch i, $i \ge 2$, contains all communicating pairs (P_j, P_k) of steps $\log p$ through i where $2^{i-1} < k - j < 2^i$ ($0 < k - j < 4$ if $i = 2$). Fig. 1 shows the balanced merger of size 16 where membership to a batch is indicated by the thickness of the corresponding vertical line.

The following two lemmas can be proved by examining which bits are flipped in each step.

Lemma 8. *Each processor P_j can belong to at most one communicating pair of Batch i if $i \ge 3$ and to at most two communicating pairs of Batch 2.*

To derive an upper bound for the running time of bitonic merge Lemma 8 does not suffice: it might happen that a processor has to wait for other processors that belong to batches with smaller numbers. The following lemma shows that no processor has to wait for a long time.

Lemma 9. *Let $P_{j_{\log p+1}}, P_{j_{\log p}}, ..., P_{j_{i+1}}, P_{j_i}$ be a path an element x can take during steps $\log p$ through i of the algorithms, $i \ge 2$.*
Then there exists at most one index t, $\log p \ge t > i$, where P_{j_i} and $P_{j_{i+1}}$ belong to Batch i or smaller in step t.

Definition 3. Let i_{max} be the maximal i where a communicating pair of Batch i exchanges an element and let e_{max} be the maximal number of elements exchanged in Batch i_{max}. Let the subscript "bit" stand for bitonic merge.

Lemma 10. $d_{max} \geq 2^{i_{max}-2}r + 2e_{max}$ if $i_{max} \geq 3$ ($d_{max} \geq 1$ if $i_{max} = 2$).

Proof. For $i_{max} = 2$ the claim is obvious. Thus assume that $i_{max} > 2$.
Let P_g and P_h be two processors that exchange elements in step $v > i_{max}$ and that belong to Batch d_{max} in step v. If processors P_g and P_h hold the same elements they held at the beginning of the for-loop the claim follows through simple calculations. Thus assume otherwise. Then at least one of them exchanged an element in step w where $w > v$. Let P_h be this processor. According to Lemmas 8 and 9, P_h belongs to Batch $i_{max} - 1$ or smaller in step w. Let P_j be the processor P_h is connected to in step w. Then, before the execution of step w, P_h and P_j both hold the elements they held at the beginning of the for-loop (this follows from Lemma 9). Assume that $h \& 2^{v-1} = 0$, that is, h is the smaller of the two communicating processors in step v (the other case is symmetric). Assume further that P_h belongs to Batch l in step w, $l < i_{max}$. Then $h = h_1 2^w + 2^{w-1} + h_2$ or $h = h_1 2^w + 2^{w-1} - 2^{l-1} + h_2$ where $0 \leq h_2 < 2^{l-1}$. Since $h \& 2^{v-1} = 0$ and $l < v < w$, $h = h_1 2^w + 2^{w-1} + h_2$ where $0 \leq h_2 < 2^l$, that is, P_h is the larger of the two processors in step w. Accordingly, P_j is the smaller of the two processors and after the execution of step w P_h can only store elements that have an even larger rank difference. □

Lemma 11. $R_{bit}(r,p) \leq (i_{max} + 1)r$.

Comment. The above bound on R_{bit} cannot be improved easily: if a processor has to wait before it can start with the execution of step i_{max}, this might be because of a rank difference some distance away and we cannot argue for a larger d_{max} than in Lemma 11.

Now we can prove the main results of this section.

Theorem 12.
1. If $d_{max} \leq r$, $R_{bit}(r,p) \leq d_{max} + \min\{r, r/2 + d_{max}\}$.
2. $R_{bit}(r,p) \leq (3 + \max\{0, \lfloor \log((d_{max} - 1)/r) \rfloor\})r$.

Theorem 13.

$$\text{Exp}\left(R_{bit}(r,p)\right) \leq r\left(3 + \sum_{i=1}^{\log p - 3} P\left(2^i r\right)\right).$$

5 Asymptotic Bounds

In this section we derive closed formulations for the bounds on the average running times of the merging and sorting algorithms. We will use the following lemma.

Lemma 14. *Let A and B be two sorted sequences of length m each, and let each outcome of merging A and B be equally likely. Then*

$$\frac{\sqrt{\pi}}{2}\sqrt{\frac{m^2}{(m^2-\Delta^2)}}e^{-\left(\frac{\Delta^2}{m}+(2\ln 2-1)\frac{\Delta^4}{m^3}\right)} \leq P(\Delta) \leq \frac{4}{\sqrt{\pi}}\sqrt{\frac{m^2}{(m^2-\Delta^2)}}e^{-\frac{\Delta^2}{m}}$$

and

$$\frac{\sqrt{\pi}}{2}\sqrt{\frac{m^2}{(m^2-\Delta^2)}}e^{-\left(\frac{\Delta^2}{m}+(2\ln 2-1)\frac{\Delta^4}{m^3}\right)} \leq Q_A(\Delta).$$

This follows from the repeated reflection principle (see [5]) and Stirling's approximation for $n!$.

Theorem 15.

$$\frac{2m}{p}\left(0.84\mathrm{max}\left\{0,\left(\left\lfloor\log\left(\mathrm{max}\left\{1,0.339\sqrt{p^2/m}-7\right\}\right)\right\rfloor-2\right)\right\}+$$

$$0.886e^{-\left(\frac{4m}{p^2}+(2\ln 2-1)\frac{16m}{p^4}\right)}\right) \leq \mathrm{Exp}\left(R_{odd}(r,p)\right) \leq$$

$$\frac{2m}{p}\left(1.17+\left\lceil\log\left(1+0.84\sqrt{p^2/m}\right)\right\rceil\right)$$

Theorem 16.

$$\mathrm{Exp}\left(R_{bit}(r,p)\right) \leq \frac{2m}{p}\left(3.17+\mathrm{max}\left\{0,\left\lceil\log\left(0.42\sqrt{p^2/m}\right)\right\rceil\right\}\right)$$

Proof of Theorem 15 and Theorem 16.

First we proof the upper bounds. In Theorem 7 and Theorem 12 we gave upper bounds for $R_{odd}(r,p)$ and $R_{bit}(r,p)$ that contain summands of the form $P(\Delta_i)$ where $\Delta_i = (2^{i-1}-1)r$, $1 \leq i \leq \log p - 1$, for R_{odd} and $\Delta_i = 2^i r$, $1 \leq i \leq \log p - 3$, for R_{bit}. By Lemma 14 we know that $P(\Delta_i) \leq 8/\sqrt{3\pi}e^{-\Delta_i^2/m}$ (here we made use of the fact that $\Delta_i \leq m/2$). Note that $\Delta_{i+1}/\Delta_i \geq 2$ (for odd-even merge this is true if $i \geq 2$). Let i_α be the smallest i such that $e^{-\Delta_i^2/m} \leq \alpha$, $\alpha < 1$. Then $e^{-\Delta_{i_\alpha+j}^2/m} \leq \alpha^{(2^j)^2}$ where $j \geq 1$. In the case of bitonic merge we choose $\alpha = 0.5$ and bound the first i_α terms of the sum by 1, and in the case of odd-even merge we choose $\alpha = 0.5^4 = 0.0625$ and bound the first $i_\alpha - 1$ terms of the sum by 1.

A similar technique can be used to proof the lower bound. □

Let the subscript "oSort" stand for odd-even merge sort and let the subscript "bSort" stand for bitonic merge sort.

Theorem 17.

$$\mathrm{Exp}\left(R_{oSort}(r,p)\right) \leq 2\log p - 1 + 0.195 \qquad \text{if } n \geq 2p^2\ln 2, \quad \text{and}$$

$$\mathrm{Exp}\left(R_{oSort}(r,p)\right) \leq 2\log p + 0.195 + \left\lceil\frac{1}{2}\log\left(\frac{1.39p^2}{n}\right)\right\rceil\left(1+\left\lceil\frac{1}{2}\log\left(\frac{2.78p^2}{n}\right)\right\rceil\right)$$

if $n < 2p^2 \ln 2$.

$$\text{Exp}\,(R_{bSort}(r,p)) \le 3 \log p - 3 + 0.195 \qquad \text{if } n \ge 0.5p^2 \ln 2, \qquad \text{and}$$

$$\text{Exp}\,(R_{bSort}(r,p)) \le 3 \log p - 1.805 + \left[\frac{1}{2} \log \left(\frac{0.35p^2}{n}\right)\right] \left(1 + \left[\frac{1}{2} \log \left(\frac{0.7p^2}{n}\right)\right]\right)$$

if $n > 0.5p^2 \ln 2$.

Proof. Both algorithms work by dividing the input into two sets, sorting the two sets recursively and then merging the resulting lists. Before the final merging step can start, both recursive calls must be finished.

Let R_{merge} stand for either R_{odd} or R_{bit} and let R_{sort} stand for either R_{oSort} or R_{bSort}. Above we have shown that $\text{Exp}(R_{merge}(r,p)/r)$ can be upper bounded by a function $f(r,p)$ and that the probability that $R_{merge}(r,p)/r$ is at least $f(r,p) + j - 1 + 1/r$ is bounded by $(8/\sqrt{3\pi})2^{-(2^j)^2}$, $j \ge 1$. We treat the contribution of $f(r,p)$ and the running times exceeding $f(r,p)$ separately. The contribution of the second part can be represented as follows.

Given is a complete binary tree with $p/2$ leaves where each node stands for a call of the merging procedure. The nodes are labeled with numbers in $\{0, 1, ..., \log p\}$ where a label of 0 denotes a running time of at most $f(r,p')$ (p' the appropriate number of processors) and where a label of j denotes a running time between $f(r,p') + j - 1 + 1/r$ and $f(r,p') + j$. The running time of the sorting algorithm corresponds to the "heaviest" path from the root to a leaf in this tree, where the weight of a path is the sum of the labels of its nodes.

Let us examine the labels of the nodes and their probabilities more closely. (We will use the notation from the proof of Theorem 15 and Theorem 16.) Firstly, for odd-even merge sort (bitonic merge sort) the labels of nodes corresponding to 2 or 4 (2 to 8, resp.) processors are always 0. Secondly, the probability that a node v has a certain label depends on the relationship between r and the number of processors used in the corresponding merge. Suppose that the number of processors p_α is such that $i_\alpha - 1 \le 1$ for odd-even merge ($i_\alpha \le 1$ for bitonic merge sort), and suppose that p_α is maximal with this property. Let $\beta = 0.5$. Then the probability that v has label j, $j \ge 1$, is bounded by $(8/\sqrt{3\pi})\beta^{(2^j)^2}$. This is also true for all nodes on the same level as or on a higher level than v. On the other hand, consider a node w that corresponds to a merge with $p' = p_\alpha/2^k$ processors. Then the probability that w is labeled with j, $j \ge 1$, is bounded by $(8/\sqrt{3\pi})\beta^{(2^j)^2 2^k}$. Consider the subtree with root v and suppose that v_1 and v_2 are the children of v. With the help of induction one can show that it is possible to replace v_1 (v_2, resp.) by one node with the same probabilities as the nodes on higher levels. This leads to a complete binary tree of depth $\log p - \log p_\alpha + 2$ where for any node the probability that its label is j, $j \ge 1$, is $(8/\sqrt{3\pi})0.5^{(2^j)^2}$.

Let $T(d,i)$ be the probability that the weight of the root of the tree with depth d is at least i. We can bound $\text{Exp}(R_{sort}(r,p)/r)$ by

$$\sum_{i=1}^{\log p} f(r, 2^i) + \sum_{i=1}^{\log p(\log p + 1)/2} T\left(\min\left(\log p - c, \log(p/p_\alpha) + 2, i\right), i\right),$$

where $c = 2$ and $p_\alpha = 2^{\lfloor \log(r/(2\ln 2))\rfloor}$ for odd-even merge sort, and $c = 3$ and $p_\alpha = 2^{\lfloor \log(2r/\ln 2)\rfloor}$ for bitonic merge sort. Thus it suffices to upper bound $T(d, i)$. Instead of doing this directly we will lower bound $S(d, i) = 1 - T(d, i+1)$.

For ease of calculation we assume that labels range from 0 to ∞ and replace y_1 by $q_1 = (8/\sqrt{3\pi})2^{-4}$, y_j by $q_j = q_1/2^{12(j-1)}$ for $j \geq 2$, and y_0 by $1 - \sum_{j=1}^{\infty} q_j$. Let $\delta = 2^{-12}$, $\gamma = q_1/(1-\delta) \approx 0.163$. We will show that $S(d, d-1+i) \geq 1 - \gamma^{i+1}$ for $i \geq 0$. This is obvious for $d = 1$. For larger p's we have

$$S(d, d-1+i) = \sum_{j=0}^{d-1+i} q_j(S(d-1, d-1+i-j))^2 \geq 1 - \gamma^{i+1}$$

since $\gamma \leq 1/2 - q_1 - \delta$. Thus we arrive at

$$\sum_{i=1}^{\log p(\log p+1)/2} T(d, i) \leq d - 2 + \frac{1}{1-\gamma}.$$

This leaves the contribution of $f(r, 2^i)$. Here we only prove the appropriate upper bound for odd-even merge sort. We have

$$\sum_{i=1}^{\log p} f(r, 2^i) = \sum_{i=1}^{\log p} \left(1 + \left\lceil \log\left(1 + \sqrt{\ln 2}\sqrt{2^{i+1}/r}\right)\right\rceil\right).$$

Let $y = \log(r/(2\ln 2)) - 1$. Note that $\log p_\alpha = \lfloor y \rfloor$. As long as $i \leq y$, each term of the sum will be bounded by 2. Assume that $\log p_\alpha < \log p$. Then we get

$$\sum_{i=\log 2p_\alpha}^{\log p} \left\lceil \log\left(1 + \sqrt{\frac{2^{i+1}\ln 2}{r}}\right)\right\rceil \leq 1 + \log\frac{p}{p_\alpha} + \left\lceil\frac{1}{2}\log\left(\frac{1.39p^2}{n}\right)\right\rceil\left\lceil\frac{1}{2}\log\left(\frac{2.78p^2}{n}\right)\right\rceil.$$

By taking into into account that for 2 and 4 processors only 1 and 2, resp., steps are executed, we arrive at the claimed upper bounds. □

Comment. The upper bound for bitonic merge sort is in most cases much larger than the upper bound for odd-even merge sort. However, if we use the number of elements actually exchanged instead of assuming that always r elements are exchanged, the 3 in the above bounds can be replaced by 2 for large n.

6 Experimental results

The upper bounds derived in this paper are asymptotically similar and do not allow an actual comparison of odd-even merge and bitonic merge. In this section we report shortly on some simulation results (obtained by a sequential simulation algorithm) for the algorithms.

When comparing R_{bSort}/r and R_{oSort}/r for 1024 processor with r ranging from 10 to 200, one can see that the average number of exchanges per element drops rapidly towards a value of slightly larger than $10 = \log 1024$. Bitonic merge sort performs always a bit better than odd-even merge sort. If we count the number of rounds where elements are exchanged, odd-even merge sort performs in most cases better than bitonic merge sort. Thus here the relationship between the running times is the same as that of the bounds derived in Section 5.

When comparing $R_{bSort}/100$ and $R_{oSort}/100$ for 100 elements per processor and various number of processor, one can observe that the average number of exchanges per element grows very slowly and is for 16384 processors still smaller than $2\log p$. In most cases, the two values are almost identical. If we consider the average maximum number of rounds, odd-even merge sort performs better in many cases than bitonic merge sort.

7 Conclusions

In this paper we have derived upper bounds for the average running time of special variants of odd-even merge sort and bitonic merge sort. We have shown that for large sizes of input each element will be exchanged at most twice (three times, resp.) with high probability. For several reasons, the derived constant factors are not tight. Firstly, we wanted to derive closed formulations. By evaluating the formulas used in Sections 3 to 5 numerically, better bounds are possible. Secondly, we assumed that each time one element has to be exchanged all elements are exchanged. If only the minimal number of elements is exchanged, for large sizes of input each element will be sent not much more often than $\log p$ times.

References

1. K. E. Batcher. Sorting networks and their applications. *Proceedings of AFIPS Spring Joint Computer Conference*, pages 307–314, 1968.
2. G. Bilardi. Merging and sorting networks with the topology of the omega network. *IEEE trans. on comp.*, C-38, 10:1396–1403, 1989.
3. K. Brockmann and R. Wanka. Efficient oblivious parallel sorting on the MasPar MP-1. In *Proc. 30th Hawaii International Conference on System Sciences (HICSS)*. IEEE, 1997.
4. M. Dowd, Y. Perl, L. Rudolph, and M. Saks. The periodic balanced sorting network. *Journal of the ACM*, 36(4):738–757, 1989.
5. W. Feller. *An Introduction to Probability Theory and Its Applications I*. John Wiley, New York, second edition, 1950.
6. D. E. Knuth. *Sorting and Searching*, volume 3 of *The Art of Computer Programming*. Addison-Wesley, Reading, MA, USA, 1973.
7. C. Rüb. On batcher's merge sorts as parallel sorting algorithms. Technical Report MPI-I-97-1-012, Max-Planck-Institut für Informatik, 1997.
8. C. Rüb. On the average running time of odd–even merge sort. *Journal of Algorithms*, 22(2):329–346, 1997.
9. A. Wachsmann and R. Wanka. Sorting on a massively parallel system using a library of basic primitives: Modeling and experimental results. In *European Conference in Parallel Processing (EURO-PAR)*, 1997.

Minimum Spanning Trees for Minor-Closed Graph Classes in Parallel

Jens Gustedt

TU Berlin, Sekr. MA 6-1, D-10623 Berlin – Germany
gustedt@math.TU-Berlin.DE

Abstract. For each minor-closed graph class we show that a simple variant of Borůvka's algorithm computes a MST for any input graph belonging to that class with linear costs. Among minor-closed graph classes are *e.g* planar graphs, graphs of bounded genus, partial k-trees for fixed k, and linkless or knotless embedable graphs. The algorithm can be implemented on a CRCW PRAM to run in logarithmic time with a work load that is linear in the size of the graph. We develop a new technique to find multiple edges in such a graph that might have applications in other parallel reduction algorithms as well.

Keywords: graph algorithms, parallel algorithms, minimum spanning tree, graph minors

1 Introduction and Overview

Minor-closed classes of graphs, *i.e* classes of graphs that are stable under contraction of edges and deletion of vertices and edges, have found a lot of attention in recent years because they cover the topological intuition behind many definitions of graph classes quite well and because of the stimulating work on structure and algorithms for these classes by Robertson and Seymour, see the book of Diestel (1997) for an introduction. Examples for minor-closed graph classes are:

Planar graphs.

Graphs of bounded genus g, *i.e* graphs that can be embedded without crossing edges into some surface of genus g.

Partial k-trees, *i.e* graphs that are intersection graphs of a collection C of subtrees of a tree T such that each vertex v of T is found in at most $k+1$ subtrees in C.

Linkless embedable graphs, *i.e* graphs that can be embedded into 3-space such that no two cycles of the graph are linked.

Knotless embedable graphs, *i.e* graphs that can be embedded into 3-space such that no cycle of the graph forms a non-trivial knot under the embedding.

A **minimum spanning tree** (forest), *MST* for short, in an edge-weighted graph is a tree (forest) that spans (all components of) the graph and whose sum of edge weights is minimal among all spanning trees (forests). Minimum spanning trees and forests are among the basic tools in graph algorithms: it is a classical problem to compute such a spanning tree in a graph in the most efficient way. The currently best known deterministic sequential algorithm runs in $O(m\alpha \log \alpha)$, Chazelle (1997), the best randomized algorithm has expected linear time, Karger et al. (1995).

At a first look it might be surprising that such an efficient computation can be done in almost linear time, even if only the comparison model on the edge weights is assumed. The book of Tarjan (1983) gives a good overview about the classical approaches to this problem. It classifies the different approaches into how they apply to specific rules to some edge e, which in the terminology used in this article look as follows:

blue rule: If e is of minimum weight in some cut it is in the *MST* and is contracted.

red rule: If e is of maximum weight on some cycle it is not in the *MST* and is deleted.

In general arbitrary cuts and cycles are not so easy to maintain, so to be efficient we restrict ourselves to the following two rules:

ultraviolet rule: If e is of minimum weight for one endpoint it is in the *MST* and is contracted.

infrared rule: If e has a sibling e' with the same endpoints and a higher weight than e' it doesn't belong to the *MST* and is deleted.

Applying the ultraviolet rule simultaneously on as many edges as possible is often called a **Borůvka step**, see Borůvka (1926). It is easy to see that the number of vertices halves in each such step, so after a logarithmic number of steps the graph is contracted to a single vertex and the *MST* is known. This general scheme has been widely used to compute the *MST* efficiently and forming the base to linear and almost linear time and work algorithms as well as in sequential, Yao (1975), Cheriton and Tarjan (1976), Karger et al. (1995), and parallel, Halperin and Zwick (1996), Poon and Ramachandran (1997). Linear time resp. work algorithms when there is no restriction on the input graphs are only known in a *randomized* setting. The main obstacle in the deterministic case being that after applying a certain number of Borůvka steps the graph gets more and more dense. To reduce the number of edges as well as the number of vertices usually an efficient variant of the red rule is applied, see Dixon et al. (1992), King (1997). This leads to linear expected time or work algorithms, but in a deterministic setting some involved data structures are required that introduce some sub-logarithmic factor.

The picture changes a bit when the input graphs are restricted to some graph classes that guarantee sparseness: Cheriton and Tarjan (1976) also give a deterministic linear time algorithm for *planar* graphs that generalizes to –in todays vocabulary– minor-closed graph classes, see Tarjan (1983). The later fact doesn't seem to be too well known in the community, we will discuss an approach that achieves this below.

The key observation here is that simple graphs in such a minor-closed graph class are sparse. So all that an algorithm must do after a Borůvka step is to eliminate duplicate edges, *i.e* to apply the ultraviolet rule to all possible edges. In a sequential setting this is relatively simple, sorting the edges with a two-fold bucketsort easily finds duplicates. But in parallel, no efficient equivalent to bucketsort is known. The aim of this paper is to circumvent this problem and to show how at least a substantial fraction of the duplicate edges can be found. By that we are able to prove the following theorem.

Main Theorem. *On any non-trivial minor closed graph class C the MST problem can be solved with a linear work load and a time of $O(\log n \log^* n)$ on a EREW PRAM or a time of $O(\log n)$ on a CRCW PRAM.*

This paper is organized as follows. The next section provides the necessary notations and facts for graphs in general, minor-closed graph classes and *MST*s. Section 3 then revisits the optimal sequential algorithm and proves the linear time bound for all minor-closed graph classes. Section 4 develops the approach that helps to deal with multiple edges. In particular it shows that if the graph has a certain amount of edges per vertex, for high degree vertices the edge lists always contains duplicate edges that are only of constant distance in the list. Section 5 then plugs all the tools together into the final PRAM algorithm.

2 Graph-Theoretic Background

In general, graphs in this paper are finite undirected multi-graphs with loops, *i.e* may contain several edges between the same one or two-element set of vertices. If e and e' are edges with the same endpoints they are called **siblings** of each other. If a graph has no loops or multiple edges it is called **simple**. If G is a multigraph its **simplification** G^0 is the graph obtained by deleting loops and multiple copies of edges. For a graph G we will denote the number of vertices by n_G and the number of edges by m_G. A graph is **connected** if for any partition V_1, V_2 of the vertex set there is an edge e with one endpoint in V_1 and the other in V_2. A connected graph is a **tree** if the removal of any edge disconnects it. Clearly that a tree T has no loops and multiple edges and that m_T is always equal to $n_T - 1$.

Let G be a graph and $v \in V_G$ and $e \in E_G$ be a vertex resp. edge of the graph. We will use the following standard notations for certain graphs obtained from G:

$G \setminus \{e\}$, a **subgraph**, obtained by deleting e from E_G.

$G \setminus \{v\}$, an **induced subgraph**, obtained by deleting v and its incident edges.

$G/\{e\}$, a **contraction**, obtained by identifying the endpoints of e.

Observe, that contraction of an edge may create multiple edges between vertices or loops at a vertex. Since the result of a sequence consisting of operations of one type only does not depend on the order in which these operations are performed, we extend the above notation to sets of vertices and edges.

A subgraph H of G is a **reduction** of G if $V_H = V_G$. A reduction H is **spanning** if it has the same number of components as G and for all $v \in V_G$ there is an edge $e \in E_H$ such that $v \in e$. It is a **spanning tree** if it is a tree and spanning.

A graph H obtained from another graph G by a sequence of the above operations is called a **minor** of G. We denote this relation among graphs by $H \preceq G$. Observe that $\preceq$ is a transitive relation. A property P of graphs is called **minor-closed** if for any pair of graphs with $H \preceq G$ and such that P holds for G, P also holds for H.

If $H \preceq G$ the vertices $V(H)$ correspond to pairwise disjoint subgraphs of G. If not stated otherwise, we keep loops and multiple copies of edges that may appear after contractions. So if we are explicitly given given a series of contractions and deletion, for any edge $e \in E_H$ we may uniquely identify a pre-image $e' \in E_G$ under these operations. Usually we will not distinguish e and its corresponding pre-image e'.

2.1 Minors and Minimum Spanning Trees

In the rest of the paper all graphs will be edge weighted, *i.e* a graph G is considered as being a triple (V_G, E_G, ω_G) where $\omega_G : E_G \to R$, the **weight function**. The weight $\omega_G(H)$

of a minor $H \preceq G$ is just the sum $\sum_{e \in E_H} \omega_G(e)$ where the edges of H are identified with edges of G as stated above. For convenience we assume that all weights of edges are pairwise distinct. This can be achieved by adding small values on the weights that differ for each edge, $a\varepsilon$ where a is the unique "number" of the edge and $\varepsilon > 0$ is a value that is suitable small. Given a weighted connected graph G, the **minimum spanning tree**, *MST*, problem asks for a spanning tree T of minimum weight over all spanning trees. It is well known that under the uniqueness condition on ω_G the minimum spanning tree is unique. We denote it by T_G.

If G is not connected the *minimum spanning forest* of G is the union of the minimum spanning trees of its components. In general the distinction between looking for a spanning forest versus a spanning tree will not be in the focus of our attention. We will denote such a spanning forest also with the symbol T_G.

The applications of the ultraviolet and infrared rules as presented in the introduction are essential for our discussion. It is already known in the early literature on *MST*, see Borůvka (1926): contracting an edge of the minimum spanning tree and deleting multiple copies of edges doesn't influence the rest of the spanning tree.

2.2 Minor Closed Graph Classes are Sparse

The following theorem is a crucial prerequisite for the discussion in this paper.

Theorem 1 (Mader (1967)). *There is a function h such that for every r > 0 every simple graph G with average degree at least h(r) contains K_r as a (topological) minor.*

A *topological minor* is a certain restriction of the minor relation, which by itself is not important for our discussion here. We use this theorem in the form of following corollary which basically states that minor-closed graph classes are sparse.

Corollary 2. *For any non-trivial minor closed graph class C there is a value μ_C such that all simple graphs in C have average degree at most μ_C.*

Proof. Since C is non-trivial there is some graph $H \notin C$. Choose r in Thm. 1 as being n_H and $\mu_C = h(r)$. Then every simple graph G of average degree more than μ_C has K_r as a minor which in turn has H as a subgraph. So $H \preceq G$ and so it can't belong to C. □

In the following we will assume that for any minor-closed graph class C a threshold μ_C as in Cor. 2 is given and fixed. For convenience we will also assume that $\mu_C \geq 1$.

Corollary 3. *For any non-trivial minor closed graph class C and for any simple graph G in C there are at least $n_G/2$ vertices that have degree at most $2\mu_C$.*

Proof. Suppose the contrary, i.e there are more than $n_G/2$ vertices of degree $> 2\mu_C$. Then the average degree is larger than $2\mu_C n_G/(2n_G) = \mu_C$, a contradiction to Cor. 2. □

In fact it may seem annoying that only existence of the constant μ_C is known for a particular class C. But whenever we know some obstruction, i.e a graph that is not in the class, we may estimate μ_C with the following theorem.

Theorem 4 (Kostochka (1982), Thomason (1984)). *There is a c > 0 such that, for every positive integer r, every graph G of average degree $d(G) \geq cr\sqrt{\log r}$ has a K_r minor. Up to the value of c, this bound is best possible as a function of r.*

3   An Optimal Sequential Algorithm

Now we are ready to formulate our prototype of an algorithms that will solve the *MST* problem for any minor-closed graph class. It is rather a prototype than a complete algorithm, since it depends on a certain parameter u, and it leaves the parts that depend on the machine model (*i.e* sequential or parallel) open.

Theorem 5. *Let C be any fixed minor-closed graph class and $u = \mu_C$ the degree threshold for C as given above. Then, given as input a member G of C with weight function ω_G, Alg. 1 with parameter u, MST^u, can be implemented on a sequential machine such that it outputs the MST of G in linear time.*

Proof. Correctness follows immediately from the validity of the ultraviolet and infrared rules. For the complexity it suffices to show, that each performance of the while-loop needs linear time in the number of edges m of the graph, and that this number is reduced by a factor $0 < \varepsilon_u < 1$ in each round. The total time then is $O(\frac{1}{1-\varepsilon_u}m) = O(m)$.

To see that the time for each round is linear, observe first that line 1 poses no problem at all and that 2 can be easily done if we assume that the edges are sorted. To re-insure this invariant after every round we just have to rename all vertices with consecutive values, propagate these values to the edges, and to apply a two-fold bucketsort to newly order the edge lists. Because the remaining statements basically perform operations on vertices of bounded degree, in total they also do need no more than linear time.

After line 2 the graph has no more than $m \leq \mu_C n/2 = un/2$ edges (Cor. 2). Because of Cor. 3 we know that $|L| \geq n/2$. Every edge might be chosen at most twice, so at least $n/4$ edges are contracted in line 6. So we reduced their number by a factor of $\varepsilon_u \leq (m - n/4)/m \leq (m - m/2u)/m = 1 - 1/2u$. □

Observe that the dependency on $u = \mu_C$ in the running time is $\frac{1}{1-\varepsilon_u} = 2u$. This is so since the running time in the sequential setting of each round does not depend on u. E.g finding the minimum weight edge in line 4 only gives a total work that is linear in $m_G + n_G$.

Algorithm 1: The Prototype of Algorithms *MST* for Parameter u, MST^u.

Input: A graph G with edge weight ω_G.
Output: A minimum spanning tree T.
while *G is not trivial* **do**

1   Delete loops;
2   Replace duplicate edges by the one with the least weight;
3   Find the set L of all vertices of degree less than $2u$;
     foreach $v \in L$ **do**
4   Find the edge e_v of v with least weight;
5   Add e_v to T;
6   Contract e_v;

---

4 High Degree Vertices: A Challenge for Parallelism

When heading to parallelize Alg. 1, most other parts but line 2 can be treated with some standard tricks. Line 2, *i.e* the application of the infrared rule, makes problems, since no way to achieve the linear work load of bucket sort in a parallel setting is known up to now.

So for our *MST* problem finding duplicates of edge is the real problem. Low degree vertices can be detected with a time of $O(2u) = O(1)$ each, and whether their edge lists have duplicates can be checked in $O(u \log u) = O(1)$. So for them we can circumvent the problem by raising the constants for the work to be done.

The problem on the high degree vertices is much harder. Sorting their edges would introduce a log-factor (or similar, depending on the model of parallelism) into the overall work that is performed.

Our main idea here, is not to find all duplicate edge in one round, but to find sufficiently many, such that the size of the graph is reduced substantially. Our method owes a lot to Bodlaender and de Fluiter (1997) and de Fluiter (1997) where a parallel recognition algorithm for partial 2-trees is given. Their idea was to let each edge look ahead a constant number of edges in the edge list, whether it finds a sibling of itself. The argument then uses an estimation of a certain kind of degree in the graphs in question. This estimation in turn is done by arguing with a potential tree-decomposition of the graph.

To be able to estimate the number of matches here, we can not rely on something like a tree-decomposition. Instead, we generalize the idea of Cor. 3, namely we control the number of vertices in a *simple* graph that have low degree.

4.1 Grades in Simple Graphs

For a simple graph G of class C we say that a vertex v has **grade** 0, denoted by $g_G(v) = 0$, if it has degree at most $2\mu_C$ and that it is of **higher grade** otherwise. The subgraph G^+ of G is the subgraph of G induced by the vertices of higher grade. If we iterate this process of deleting the low degree vertices we get a kind of *shelling* of our graph G. A vertex v is of higher grade $g(v)$ is recursively defined as $g(v) = g_{G^+}(v) + 1$. The **grade** of G, denoted by $g(G)$, is the maximum grade over all its vertices. So $g(G)$ reflects the number of levels in the shelling of G.

The following two observations follow easily from Cor. 3.

Remark 6. Let G be a simple graph of class C.

(a) $g(G) \leq \log n_G$.

(b) Let $k \leq g(G)$. Then there are at most $2^{-k} n_G$ vertices in G of grade $\geq k$.

The grade $g_G(e)$ of an edge e is the minimum grade of its endpoints. It has **higher grade** if $g_G(e) > 0$. So the grade of an edge reflects the level in the shelling of G for which one or both of its endpoints are of low degree. As an easy consequence of Cor. 3 we get:

Remark 7. For a simple graph $G \in C$ there are at most $n_G \mu_C / 2$ edges of higher grade.

Proof. Indeed, $n_{G^+} \leq n_G / 2$ and since $G^+ \in C$ we get $m_{G^+} \leq 2\mu_C n_{G^+} \leq \mu_C n_G$. $\qquad \square$

It follows directly from the definition that a vertex v is never incident to an edge e such that $g(e) > g(v)$. We say that an edge e is **accounted** to one of its endpoints v if $g(e) = g(v)$. The **grade-degree** $d_G^+(v)$ of a vertex v is the amount of edges e that are accounted to it. The following easy observation is crucial for our discussion.

Remark 8. Let G be a simple graph in class C. Then $d^+(v) \leq 2\mu_C$ for all $v \in V_G$.

4.2 Bad Edges in a Multigraph

We switch back to *multigraphs*. For such a graph G in class C the values and terms **grade, grade-degree, accounted...** of a vertex (edge, the graph itself...) are the values and terms of the corresponding objects in the simplification G^0.

Let us suppose that a graph G of class C is given in the conventional data structure as cyclic list of edges for each vertex. For what follows we assume that one out of many possible representations of G in such a data structure is given and fixed. We also assume that a "magic" constant $\zeta > 2$ is fixed for the moment. A concrete value for it will be chosen later. Let e be an edge that is accounted to v. We call e **bad** for v if it has some sibling, but it has no sibling in the edge list of v that is closer than distance $\zeta \cdot \mu_C$ in the list. e is bad if it is bad for one of its endpoints.

Lemma 9. *Let C be a non-trivial minor-closed graph class and $G \in C$ be non-trivial given as above. Then G has at most $2m_G/\zeta$ bad edges.*

Proof. For any vertex v there are at most $2\mu_C$ different neighbors w such that v and w can be the endpoints of a bad edge (Rem. 8). Since bad edges that have the same w as other endpoint must have distance at least $\zeta\mu_C$ in the list of v there are at most $d(v)/\zeta\mu_C$ such bad edges incident to v and w. So in total there are at most $2\mu_C d(v)/\zeta\mu_C = 2d(v)/\zeta$ bad edges incident to v. The claim follows by summation over all vertices. $\square$

Call an edge **good for contraction** if it has grade 0 and is the one with lowest weight among all edge incident to the same vertex v of grade 0. Call an edge **good for deletion** if has a sibling and is not bad. Call an edge **good** if it is so for deletion or contraction and call it **indifferent** if it is neither good nor bad. Observe that an edge can be only indifferent if it is has no sibling. The following is an easy consequence of Cor. 7.

Remark 10. There are at most $n_G\mu_C$ indifferent edges in G.

The main result in this section is the following:

Theorem 11. *Let C be a non-trivial minor-closed graph class and $G \in C$ be non-trivial given as above. Then G has at least $m_G/8\mu_C$ good edges.*

Proof. We distinguish two cases:

$m_G \leq 2\mu_C n_G$: In this case we know by a little variation of Cor. 3 that we always find $n/4 \geq m/8u$ edges that are good for contraction.

$m_G > 2\mu_C n_G$: If we plug all our estimations together we see that we have at least the following number of good edges

$$m_G - 2m_G/\zeta - n_G\mu_C > m_G - 2m_G/\zeta - m_G/2 = m_G((\zeta-4)/2\zeta) \qquad (1)$$

So by choosing $\zeta = \left\lceil \frac{16\mu_C}{4\mu_C-1} \right\rceil$, a value between 4 and 6, we easily get the claim. $\square$

5 Designing a PRAM Algorithm

To build an efficient PRAM algorithm we will use the technique of Bodlaender and Hagerup (1995) for *parallel reduction algorithms*. In fact the **reduction rules** that are of interest for us are among the most simple ones – we only have the infrared and the ultraviolet rules, *i.e* deletions and contractions of good edges.

The main idea for parallel reduction algorithms is to choose a non-interfering set of graph reductions that still is at least of size εn. This can be done by observing, that any of the above reduction rules, *i.e* good edges, can only interfere with a constant number of such other rules. If we consider the cyclic order of edges as being fixed, good edges have at most 4 neighbors in the edge lists of their endpoints.

We need some further technical definitions. Two deletable edges e and f are **interfering** if they share a common end point v and they are neighbors in the circular edge list of v. Clearly that every such edge may interfere with at most 4 other edges.

Unfortunately for contractable edges we need a definition that is slightly more complicated. Two contractable edges e and f are **interfering** if they share a common end point v and

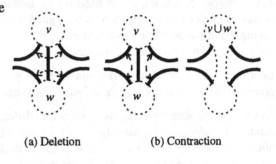

- v is a low degree vertex, or
- v is a high degree vertex and they are neighbors in the circular edge list of v.

    (a) Deletion   (b) Contraction

Fig. 1. Applying Rules Locally

Observe that if we have a set S of good edges that are pairwise non-interfering, we may perform the necessary deletions or contractions of these edges in parallel in constant time. For a deletion a processor may just link the two neighboring edges of each endpoint without running into conflicts with other good edges in S. For a deletion we have to link the four neighboring edges across, see Fig. 1.

Now we have all utilities in hand to build a PRAM algorithm, see Alg. 2. Observe that the only substantial changes when comparing to Alg. 1 are in lines 2, +4 and +6. Lines 2 and +4 take care that we only delete or contract a set of non-interfering edges. Line +6 ensures that after the contractions all edges still know about their endpoints – in the contraction phase this information might get corrupted since edge lists of several vertices might be linked into one large cyclic list.

We will only prove the EREW part of the Main Theorem. The derandomization tricks for symmetry breaking that are needed to avoid the $\log^* n$ factor on a CRCW PRAM are quite the same as given in Bodlaender and Hagerup (1995) and don't give new insights into the *MST* problem itself. So we content ourselves by proving the following proposition.

Proposition 12. *Algorithm 2 is correct and can be implemented in such a way such that it runs with a linear work load and a time of $O(\log n \log^* n)$ on a EREW PRAM.*

Algorithm 2: The Algorithms *MST* for Parameter *u* on a PRAM.

Input: A graph G with edge weight ω_G.

Output: A minimum spanning tree T.

while G *is not trivial* **do**

1	Delete loops;
2	Find a non-interfering set of edges good for deletion and eliminate these;
3	Find the set L of all vertices of degree less than $2u$;
	foreach $v \in L$ **do**
4	$\quad$ Find the edge e_v of v with least weight;
+4	Find a non-interfering set L' of contractable edges;
	foreach $e_v \in L'$ **do**
5	$\quad$ Add e_v to T;
6	$\quad$ Contract e_v;
+6	Update the sets of endpoints for all edges;

Proof. Correctness follows immediately from the correctness of Alg. 1. For the complexity, we will show that

(a) We always find a substantial fraction of the edges to be deleted or to contracted to obtain a logarithmical number of rounds of the algorithm.
(b) There are not more than $O(\log^* n)$ rounds that need $O(\log n)$ time.
(c) For the remaining rounds the time is not more than $O(\log^* n)$ per round.
(d) The overall work is linear.

The first tool to guarantee that we always have sufficiently many good edges is Thm. 11. Not every edge that is good for deletion can in fact be deleted. To be deleted it must have a weight that is higher than the one of a sibling that is close in one of the adjacency lists of the endpoints. But it is easy to see that at least every third edge that is good for deletion can in fact be deleted.

But these remaining good edges may interfere. If we build the conflict graph of these edges, see Bodlaender and Hagerup (1995), we notice that this graph has bounded degree, d say. For every constant d there is a constant $\varepsilon_d > 0$ such that a graph of bounded degree d has an independent set that consists of at least a ε_d-fraction of the vertices. In our conflict graph, an independent set of vertices corresponds to a set of rules that do not interfere, and so if we are also able to find such an independent set we get (a).

A fractional independent set, *i.e* an independent set of size at least a ε_d fraction, in a graph of bounded degree d can be found by two different methods:

Method A In $O(\log n)$ time and linear work (Hagerup (1990)).
Method B In $O(\log^* n)$ time and a work of $O(n \log^* n)$ by symmetry breaking (Goldberg et al. (1987)).

None of these methods alone would allow us to achieve the right complexity. As we will see below they lead to implementations of Alg. 2:

Variant A with $O(\log^2 n)$ time and linear work,

Variant B with $O(\log^* n \log n)$ time and $O(n \log^* n)$ work.

We use an argument of Bodlaender and Hagerup (1995) to combine the two methods, using Variant A for the first $\log^* n$ rounds, **phase A**, and Variant B for the remaining ones, **phase B**. By that combination we easily achieve the desired running time of $O(\log n \log^* n)$.

For the total work observe that after the $\log^* n_G$ rounds of phase A the size of the graph G is already reduced by a factor of $(1 - \varepsilon)^{\log^* n_G}$ and the overall work done in phase B does not exceed

$$O\left(\log^*\left((1-\varepsilon)^{\log^* n_G} n_G\right)(1-\varepsilon)^{\log^* n_G} n_G\right) \leq O\left(\log^*(n_G)(1-\varepsilon)^{\log^* n_G} n_G\right). \quad (2)$$

But since $1 - \varepsilon$ is a constant that is strictly smaller than 1 the factor $\log^*(n_G)(1 - \varepsilon)^{\log^* n_G}$ is itself globaly bounded by a constant and phase B does a work of $O(n_G)$. So if we are able to show that phases A and B can be implemented as claimed above we are done.

Work of Phase X, X equal to A or B: We use Method X for line 2 and +4. For line +6, *i.e* the update of the endpoints for each edge, we will see below how this can in fact be done in constant time and a linear work load.

So in total we achieve a running time and work load as desired *per round*. But then by the same type of argument as in the proof of Thm. 5 the overall work for all rounds is only worsened by a constant factor of $\frac{1}{1-\varepsilon}$.

Updating the endpoints of edges: After having chosen a set of non-interfering edges for contraction consider a set S of vertices that are identified into one. Only two cases remain.

Case 1: If all vertices in S are of low degree then $|S| = 2$. This holds, since at most one edge per low degree vertex is chosen. We arbitrarily choose one vertex $v \in S$ and update all edges to have v as the other endpoint.

Case 2: Otherwise there is a unique vertex v_0 of high degree in S. This is so, since at least one endpoint of a contractable edge is of low degree. We update all edges of the low degree vertices to have v_0 as a new endpoint.

Since these updates only must be done for vertices of low degree we can easily avoid read conflicts that would occur when some edges of the same vertex are asking for the new endpoint at the same time. The updates are simply performed sequentially for each vertex.

So this update procedure concludes the proof that phases A and B have the complexities as claimed. □

Observe that the reduction factor of $1/8\mu_C$ from Thm. 11 for the number of good edges in each round depends on μ_C. In the presentation of the proof above, this factor gets much worsened because the maximum degree of the conflict graph also depends on μ_C. This can be avoided by a more involved argument. In fact we may change the definition of interference of contractable edges to be the same as for deletables. Then, in a more careful analysis of the symmetry breaking step it can be shown that this interference graph allows an independent set of size of about $1/4$ that still allows an $O(1)$ update of the endpoint information of low degree vertices that are contracted.

References

Bodlaender, H. and de Fluiter, B. (1997). Parallel algorithms for treewidth two. In Möhring et al., editors, *Graph-Theoretic Concepts in Computer Science*, Lecture Notes in Computer Science. Springer-Verlag. 23rd International Workshop WG '97, to appear.

Bodlaender, H. L. and Hagerup, T. (1995). Parallel algorithms with optimal speedup for bounded treewidth. In Fülöp, Z. and Gécseg, F., editors, *Automata, Languages and Programming*, volume 944 of *Lecture Notes in Comp. Sci.*, pages 268–279. Springer-Verlag. Proceedings of the 22nd International Colloquium ICALP'95.

Borůvka, O. (1926). O jistém problému minimálním. *Práca Moravské Přírodovědecké Společnosti*, 3:37–58. In Czech, cited after Tarjan (1983).

Chazelle, B. (1997). A faster deterministic algorithm for minimum spanning trees. In *38th Annual Symposion On Foundations of Computer Science*. IEEE, The Institute of Electrical and Electronics Engineers, IEEE Computer Society Press.

Cheriton, D. and Tarjan, R. E. (1976). Finding minimum spanning trees. *SIAM J. Computing*, 5:724–742.

de Fluiter, B. (1997). *Algorithms for Graphs of Small Treewidth* (Algoritmen voor grafen met kleine boombredte). PhD thesis, Universiteit Utrecht.

Diestel, R. (1997). *Graph Theory*, volume 173 of *Graduate Texts in Mathematics*. Springer-Verlag.

Dixon, B., Rauch, M., and Tarjan, R. E. (1992). Verification and sensitivity analysis of minimum spanning trees in linear time. *SIAM Journal on Computing*, 21(6):1184–1192.

Goldberg, A. V., Plotkin, S. A., and Shannon, G. E. (1987). Parallel symmetry-breaking in sparse graphs. In *Proceedings of the Nineteenth Anual ACM Symposion on Theory of Computing*, pages 315–324. ACM, Assoc. for Comp. Machinery.

Hagerup, T. (1990). Optimal parallel algorithms on planar graphs. *Information and Computation*, 84:71–96.

Halperin, S. and Zwick, U. (1996). An optimal randomised logarithmic time connectivity algorithm for the EREW PRAM. *J. Comput. System Sci.*, 53(3):395–416.

Karger, D. R., Klein, P. N., and Tarjan, R. E. (1995). A randomized linear-time algorithm to find minimum spanning trees. *Journal of the ACM*, 42(2):321–328.

King, V. (1997). A simpler minimum spanning tree verification algorithm. *Algorithmica*, 18(2):263–270.

Klein, P. N. and Tarjan, R. E. (1994). A randomized linear-time algorithm for finding minimum spanning trees. In *Proceedings of the Twenty Sixth Anual ACM Symposion on Theory of Computing*, pages 9–15. ACM, Assoc. for Comp. Machinery.

Kostochka, A. V. (1982). On the minimum of the Hadwiger number for graphs with given mean degree of vertices. *Metody Diskretn. Anal.*, 38:37–58.

Mader, W. (1967). Homomorphieeigenschaften und mittlere Kantendichte von Graphen. *Math. Ann.*, 174:265–268.

Poon, C. and Ramachandran, V. (1997). A randomized linear work EREW PRAM algorithm to find a minimum spanning tree. In Jaffar, J. and Leong, H. W., editors, *Algorithms and Computation*. Springer-Verlag. Proceedings of the Eighth Annual International Symposium ISAAC'97, to appear.

Tarjan, R. E. (1983). *Data structures and network algorithms*. Society of Industrial and Applied Mathematics (SIAM), Philadelphia.

Thomason, A. (1984). An extremal function for contractions of graphs. *Math. Proc. Cambridge Philos. Soc.*, 95:261–265.

Yao, A. C.-c. (1975). An $O(|E|\log\log|V|)$ algorithm for finding minimum spanning trees. *Inform. Processing Letters*, 4:21–23.

Optimal Broadcasting in Almost Trees and Partial k-trees

Anders Dessmark [*]   Andrzej Lingas [*]   Hans Olsson [*]   Hiroaki Yamamoto [†]

Abstract

We consider message broadcasting in networks that have almost tree topology. The source node of the input network has a single message which has to be broadcasted to all nodes of the network. In every time unit each node that has already received the message can send it to one of its neighbors. A broadcasting scheme prescribes in which time unit a given node should send a message to which neighbor. It is minimum if it achieves the smallest possible time for broadcasting the message from the source to all nodes. We give the following algorithms to construct a minimum broadcasting scheme for different types of weakly cyclic networks:

A linear-time algorithm for networks whose cycles are node-disjoint and in which any simple path intersects at most $O(1)$ cycles.

An $O(n \log n)$-time algorithm for networks whose cycles are edge-disjoint and in which a node can belong to at most $O(1)$ cycles.

An $O(n^k \log n)$-time algorithm for networks whose each edge-biconnected component is convertible to a tree by removal of at most k edges.

We also present an $O(n^{4k+5})$-time algorithm for constructing a minimum broadcasting scheme for partial k-trees.

1 Introduction

A communication network is usually modeled by a connected graph whose nodes correspond to the processors and whose edges correspond to the communication lines between the processors. The following *broadcasting* problem is an important component of many network protocols: *The source node has a message which must be transmitted to all other nodes.* It often occurs in distributed computing, e.g., in global processor synchronization and updating distributed databases. Also, it is implicit in many parallel computation problems, where data and results are distributed among processors (e.g., in matrix multiplication, parallel solving of linear systems, parallel computing of Discrete Fourier Transform, or parallel sorting, cf. [6], [10] and [17]).

The efficiency of network broadcasting algorithms is usually measured by the number of rounds (*parallel time*) required. Various communication models impose different constraints on actions that can be executed simultaneously, i.e., in the same round. In the *whispering* model pairs of nodes communicating in the same round have to be disjoint, while in the *shouting* model a node can simultaneously broadcast to all its neighbors (cf. surveys [11],

---

[*]Department of Computer Science, Lund University, Box 118, S-22100 Lund, Sweden. E-mail: { Anders.Dessmark, Andrzej.Lingas, Hans.Olsson}@dna.lth.se
Research supported in part by TFR grant 96-278.

[†]Department of Information Engineering, Shinshu University, 500 Wakasato, Nagano-shi, 380 Japan. E-mail: yamamoto@cs.shinshu-u.ac.jp

[13] and [14]). In case of one-message broadcasting, the whispering model is equivalent the so called simultaneous send/receive postal model discussed by Bar-Noy and Kipnis in [3, 4].

Finding a minimum broadcasting scheme for a given source in an arbitrary network in the whispering model is an NP-hard problem [12]. It remains NP-hard even when restricted to planar networks [15]. Hence it is important to investigate natural classes of networks in which a minimum broadcasting scheme can be constructed efficiently. Slater *et al.* [22] gave a linear-time algorithm constructing such a scheme in any tree. From [15] a polynomial time algorithm can be deduced for finding a minimum broadcasting sceme in partial k-trees, only in the case of bounded degree (explicitly or implicitly by requiring their cut parameter to be $O(1)$). An approximation heuristic for partial k-trees of unbounded degree yielding solutions within a logarithmic factor from the minimum is known [19]. Suprisingly no other polynomial-time algorithm for optimal broadcasting in any non-trivial superclass of acyclic networks is known in the litterature.

In this paper we consider message broadcasting in weakly cyclic networks and partial k-trees (see [2]) in the whispering model. We consider four different types of weakly cyclic networks:

Type 1: networks whose cycles are node-disjoint and in which any simple path intersects at most $O(1)$ cycles.

Type 2: networks whose cycles are edge-disjoint and in which a node can belong to at most $O(1)$ cycles.

Type 3: networks whose each edge-biconnected component is convertible to a tree by removal of at most k edges and in which any simple path intersects at most $O(1)$ cycles $(k = O(1))$.

Type 4: networks whose each edge-biconnected component is convertible to a tree by removal of at most k edges $(k = O(1))$.

A *k-tree* is a network which can be reduced to a k-clique (i.e., a complete network on k nodes) by a sequence of eliminations of degree k vertices whose neighbors induce a k-clique. A **partial k-tree** is a sourced subnetwork of a k-tree.

It is natural to ask if the optimal broadcasting algorithm of Slater *et al.* [22] for trees can be modified to meet the needs of weakly cyclic networks or partial k-trees, e.g., networks of type 4. A straightforward approach is to enumerate all removals of minimal number of edges resulting in tree networks and running the algorithm of Slater *et al.* on these tree networks. Note here that the minimum number of edges necessary to remove from a network of type 4 in order to convert it into a tree is generally not bounded by any function in k. Thus, just the enumeration of the possible minimal edge removals could take exponential or super-exponential time. Nevertheless the tree-like structure of networks of type 4 can be easily combined with the idea of the algorithm of Slater *et al.* [22] to yield an $O(n^{k+1})$-time solution.

Our first main results are more sophisticated algorithms for constructing a minimum broadcasting scheme for networks of type 1 and 2 running in time $O(n)$ and $O(n \log n)$, respectively. As a corollary, we show that a minimum broadcasting scheme for networks of type 3 and 4 can be constructed in time $O(n^k)$ and $O(n^k \log n)$, respectively. Our final main result is an $O(n^{4k+5})$-time algorithm for constructing a minimum broadcasting scheme for a partial k-tree, obtained by reduction to network flow.

The rest of the paper is organized as follows. In section 2 we give a precise definition of the communication model and introduce terminology. In section 3 through 5 we outline our algorithms for optimal broadcasting in networks of type 1 through 4 respectively. In section 6, we present our algorithm for partial k-trees. In section 7, we shortly discuss some open problems.

2 Preliminaries and model description

Let $G = (V, E, r)$ be an undirected network with the distinguished source r of broadcasting. If u, v are adjacent nodes in G then $\{u, v\}$ denotes the edge connecting u with v whereas $[u, v]$ denotes the same edge additionally directed from u to v. For a node v of G, $deg(v)$ denotes the degree of v in G. For a subnetwork S of G, $|S|$ denotes the size of S, i.e., the number of nodes and edges in S.

An *edge-biconnected component* of G is a subnetwork of G induced by a maximal subset of its nodes such that any two nodes in the subset are connected by at least two edge-disjoint paths. The tree of edge-biconnected components of G is the tree whose vertices correspond to the edge-biconnected components of G and whose edges correspond to the edges of G connecting the corresponding edge-biconnected components; the tree is rooted at the node corresponding to the component containing r. By running the linear-time algorithm for the standard biconnected components (see [1]) on the graph induced by the edges of G, we obtain the following fact.

Proposition 2.1 *The tree of the edge-biconnected components of a network on n nodes can be computed in time $O(n)$.*

Proposition 2.2 *A network of type 1 through 4 on n nodes has $O(n)$ edges.*

Proof: Each non-singleton biconnected component of such a network can be converted to a tree by removing $O(1)$ edges.   □

The *local source* r_C of broadcasting for a biconnected or edge-biconnected component C of G not containing the source of broadcasting r is the endpoint of the edge of G in C or the articulation point connecting the parent component of C to C. For the component C containing r, $r_C = r$ for convention. G_{r_C} is the minimal subnetwork of G including r_C and all nodes of G reachable from r_C without the use of any edge from the parent component of C (if any). For a component C' and a node v outside $G_{r_{C'}}$, the subnetwork $G_{r_{C'}}$ is said to be *attached* to v if $r_{C'}$ and v are adjacent in G.

A *directed spanning tree* of G is a spanning tree of G directed in such a way that each its node is reachable by a directed path from the source node r.

A *message broadcasting scheme* for the network G is a pair (T, s), where $T = (V, E_T)$ is any directed spanning tree of G and s is any function $s : E_T \to Z^+$ which satisfies the following condition:

1. For no two edges $[v, u]$, $[v, w]$,

$$s([v, u]) = s([v, w]).$$

2. If u is a parent of v and v is a parent of w then

$$s([u, v]) < s([v, w]).$$

The positive integer $s([u, v])$ is the number of the round in which the message is sent along edge $[u, v]$. Condition 1 says that a node cannot send two copies of the message in the same round. (The requirement that a node cannot receive two copies of the message in the same round follows from the fact that T is a tree and messages travel only in the direction of edges of T.) Condition 2 says that each node can send only the message previously obtained from the parent to its children in T. Since the message is sent along all edges of T, all nodes will get it.

Note that in any broadcasting scheme (T, s) of G, for any biconnected or edge-biconnected component C of G, the message has to reach r_C before reaching any other node in G_{r_C}.

The execution *time* of a scheme s is the maximum of integers $s(e)$ over all edges $e \in E$. A message broadcasting scheme for a network G is *optimal* if its time does not exceed the time of any other message broadcasting scheme for G. The time of an optimal scheme is denoted by $B(G, r)$ and is called the *optimal broadcasting time* for the network G with the source r. Often, when the source of broadcasting is obvious from the context, we shall omit the parameter r.

Proposition 2.3 *The optimal broadcasting time for an n-node connected network is at most* $n - 1$.

Note that the upper bound given in Proposition 2.3 is tight for the tree of height 1 whose root has $n - 1$ children or for a path of length $n - 1$.

Consider an acyclic network T on n nodes with the source of broadcasting r. It is rather obvious that a minimum broadcasting scheme for T can be found as follows.

For each non-leaf node v of T with children v_1, v_2, ..., v_l recursively construct a minimum broadcasting scheme for T_v by sorting v_i's by the key $B(T_{v_i})$ in a decreasing order and sending the message from r to v_i in the sorted order.

Note here that in case $deg(v)$ is very small in comparison with $B(T_{v_i})$ one cannot use the linear-time radix sorting here [1]. For this reason, Slater et al. proposed an $n - 1$ phase incremental implementation of the above method [22]. In the j-th phase, each node w for which $B(T_w)$ is already determined and equal to j updates the not yet determined $B(T_v)$ for its parent by $B(T_v) \leftarrow \max\{B(T_v) + 1, B(T_w) + 1\}$ and appends itself to the front of the list of children of v with the already determined optimal broadcasting time. Note that when the list of children of v is finally completed it gives the desired sorted order and $B(T_v)$ is also correct. Hence, we have the following fact due to Slater et al. [22].

Proposition 2.4 *A minimum broadcasting scheme for an acyclic network on n nodes can be constructed in time* $O(n)$.

3 Optimal broadcasting in networks of type 1

Let r_C be the local source of broadcasting for an edge-biconnected component C of a network of type 1. By the node-disjointness of the cycles in G, C is a simple cycle. We want to compute $B(G_{r_C}, r_C)$ on the basis of the minimum broadcasting times for the subnetworks $G_1, ..., G_l$ of G_{r_C} disjoint from C and attached to r_C and the other subnetworks of G_{r_C} attached to the nodes of C different from r_C.

There exist an optimal broadcasting scheme that uses $|C| - 1$ edges in C. The idea is to find an edge $e \in C$ whose removal will not change the optimal broadcasting time $B(G_{r_C}, r_C)$. The removal of such an edge e disconnects the subnetwork G_C of G induced by C and all subnetworks of G attached to the nodes of C different from r_C into two subnetworks $G_C^1(e)$ and $G_C^2(e)$. Let $v_0, v_2, ..., v_m$ be the nodes on the cycle C where $r_C = v_0$. To begin with, we compute $M(i,j) = B(G_C^i(\{v_j, v_{j+1 \bmod m+1}\}), r_C)$ for $i = 1\ 2$, $j = 0, ..., m$ by running the procedure Compute Cuts (see the next page).

In step 1 the procedure Compute cuts preprocess the nodes v_l of the cycle C different from r_C by computing two values $(x_l^1$ and $x_l^2)$ that will help determine the optimal broadcasting time at its later steps. The vector $Jump$ is initialized with the lower numbered node. Step 2 simply states that the time taken to broadcast a message in the ascending direction is 0 when $\{v_0, v_1\}$ is removed. In the third step we compute all the remaining broadcasting times in the ascending direction. The cut is made below the node v_l and x is the optimal broadcasting time for the node v_j. In the steps 3b and 3c (iv), j is updated by the help of the jump vector and the jump vector is in turn updated to, in the future, ignore nodes that simply will add 1 to the broadcasting time. The broadcasting times in the descending direction is similarly computed in steps 5 and 6.

Lemma 3.1 *The procedure Compute cuts computes the matrix $M(i,j)$ on the basis of the optimal broadcasting times for the subnetworks attached to the nodes of the cycle C different from r_C in time $O(|C| + s)$ where s is the total number of the nodes in the aforementioned subnetworks.*

Proof: sketch. The correctness of the procedure follows from the discussion preceding the lemma. By Proposition 2.3, the optimal broadcasting times for each of the subnetworks is bounded by its size. Hence, if we use a bucket sort, step 1 takes $O(s)$ time totally (in fact, we shall use the incremental sorting method in phases of Slater *et al.* outlined at the end of Section 2) in our complete algorithm for networks of type 1, which will make the bucket sort superfluous). Steps 2 and 5 take $O(1)$ time whereas step 4 takes $O(|C|)$ time. For a node v_l of C different from r_C let s_l be the number of nodes in the subnetworks attached to v_l. Note that whenever an edge $\{v_{l+r}, v_{l+r+1 \bmod m}\}$, where $s_l < r \leq m$, is removed then the optimal broadcasting time at v_l is dominated by that at v_{l+1} plus one. It follows that v_l can be considered (i.e., not jumped over) only $O(s_l)$ times in step 3. Hence, step 3 takes $O(|C| + s)$ time. Symmetrically, step 6 takes also $O(|C| + s)$ time. $\square$

Given the matrix $M(i,j)$, the procedure Reduce computes $B(G_{r_C}, r_C)$ and such an edge e whose deletion doesn't affect $B(G_{r_C}, r_C)$ on the basis of optimal broadcasting times for the subnetworks $G_1, G_2...G_q$ attached to r_C.

procedure: *Compute cuts*

Input: A cycle C on nodes $v_0, v_1, ..., v_m$ with $v_1 = r_C$ the optimal broadcasting times for all subnetworks attached to $v_1, v_3, ..., v_m$.

Output: the matrix $M(i, j)$

1. **For** $l \leftarrow 1$ **to** m **do**

   (a) Sort the subnetworks G_1, $G_2, ...$ attached to v_l in descending order according to their broadcasting times $B(G_1) \geq B(G_2) \geq ...$

   (b) $x_l^1 \leftarrow max(B(T_j) + j)$

   (c) $x_l^2 \leftarrow min(B(T_j) | B(T_j) + j = x_l^1)$

   (d) $Jump(l) \leftarrow l - 1$

2. $M(1, 0) \leftarrow -1$

3. **For** $l \leftarrow 1$ **to** m **do**

   (a) $x \leftarrow x_l^1, j \leftarrow l$

   (b) $x \leftarrow x + j - Jump(j) - 1, pre \leftarrow j, j \leftarrow Jump(j)$

   (c) **While** $j > 1$ **do**

      i. **If** $x < x_j^2$ **then** $x \leftarrow x_j^1$

      ii. **Else if** $x < x_j^1$ **then** $x \leftarrow x_j^1 + 1$

      iii. **Else** $x \leftarrow x + 1, Jump(pre) \leftarrow Jump(j)$

      iv. $x \leftarrow x + j - Jump(j) - 1, pre \leftarrow j, j \leftarrow Jump(j)$

   (d) $M(1, l) \leftarrow x$

4. **For** $l \leftarrow m$ **downto** 1 **do** $Jump(l) \leftarrow l + 1$

5. $M(2, m) \leftarrow -1$

6. **For** $i \leftarrow m$ **downto** 2 **do**

   (a) $x \leftarrow x_l^1, j \leftarrow l$

   (b) $x \leftarrow x + Jump(j) - j - 1, pre \leftarrow j, j \leftarrow Jump(j)$

   (c) **While** $j < m + 1$ **do**

      i. **If** $x < x_j^2$ **then** $x \leftarrow x_j^1$

      ii. **Else if** $x < x_j^1$ **then** $x \leftarrow x_j^1 + 1$

      iii. **Else** $x \leftarrow x + 1, Jump(pre) \leftarrow Jump(j)$

      iv. $x \leftarrow x + Jump(j) - j - 1, pre \leftarrow j, j \leftarrow Jump(j)$

   (d) $M(2, l) \leftarrow x$

Lemma 3.2 *The procedure Reduce computes $B(G_{r_C}, r_C)$ and an edge e of C whose removal doesn't affect $B(G_{r_C}, r_C)$ on the basis of the sorted list $G_{i_1}, ..., G_{i_l}$ and the matrix $M(i, j)$ in time $O(\log |C|)$.*

Proof: sketch. The optimal broadcasting time for G_{r_C} is first computed as B_{temp}. The two critical time-thresholds t_1 and t_2 in the procedure are the optimal broadcasting times for the subnetworks that have to be beaten by both or one respectively of the two parts of the cycle separated by the removed edge e to avoid any increase in the total broadcasting time. The procedure in turn tries to find an edge e whose removal will give no, one and two time steps of increase to B_{temp}. If this is not successful the procedure minimizes the time by minimizing the maximum of the two parts and output the corresponding separating edge e.

As for the time complexity, the result follows by the assumption and the fact that only a constant number of binary searches is performed by the procedure Reduce, each in time

procedure: *Reduce*

1. $G_{i_1}, ..., G_{i_l} \leftarrow$ a sorted list of G_i's by the keys $B(G_i)$ in descending order

2. $B_{temp} \leftarrow max_{1 \leq j \leq q}(B(G_{i_j}) + j)$

3. $t_1 \leftarrow max_{1 \leq j \leq q}(B(G_{i_j})|B(G_{i_j}) + j = B_{temp})$

4. $t_2 \leftarrow min((max_{1 \leq j \leq q}(B(G_{i_j})|B(G_{i_j}) + j + 1 \geq B_{temp})), t_1 - 1)$

5. Find (by two binary searches on $M(i, j)$) an edge e of C such that $B(G_C^1(e)) < t_2$ and $B(G_C^2(e))$ is minimal (observing the first condition).

6. If $B(G_C^1(e)) < t_1$ then return B_{temp} and e

7. Find an edge e of C such that $B(G_C^2(e)) < t_1$ and $B(G_C^1(e))$ is minimal (observing the first condition).

8. If $B(G_C^1(e)) \leq B_{temp} + 1$ then return $B_{temp} + 1$ and e

9. Find $e \in C$ (assume $B(G_C^1(e)) \leq B(G_C^2(e))$) such that $max(B(G_C^1(e)), B(G_C^2(e)))(= B(G_C^2(e))$ is minimal and $B(G_C^2(e))$ is minimal (while still observing the first condition).

10. If $B(G_C^1(e)) \leq B_{temp} + 2$ and $B(G_C^2(e)) \leq B_{temp} + 1$ then return $B_{temp} + 2$ and e

11. If $B(G_C^1(e)) = B(G_C^2(e))$ then return $B(G_C^1(e)) + 1$ and e else return $B(G_C^1(e))$ and e

$O(\log |C|)$. □

Now it is sufficient to run the procedures Compute cuts and Reduce for the local sources of broadcasting of the biconnected components of G in a bottom up fashion in the tree of biconnected components rooted at the component containing r to obtain the main result of this section (note that biconnected components coincide with edge-biconnected components in case of networks of type 1).

Theorem 3.1 *A minimum broadcasting scheme and optimal broadcasting time for a network of type 1 on n nodes can be computed in time $O(n)$.*

Proof: Let G be a network of type 1 on n nodes. First, we compute the tree of biconnected components of the input network G rooted at the component containing r and its local sources of broadcasting in time $O(n)$ [1]. For each trivial (i.e., singleton) leaf-component we set the broadcasting time to zero. For the local source of broadcasting r_C of each non-trivial leaf-component, we run the procedures Compute cuts and Reduce to determine $B(G_{r_C}, r_C)$ and the edge e to remove. We proceed analogously with the intermediate components in the tree of components whenever their children components have been already processed. In case of trivial intermediate components we need solely to use the sorting part of Reduce proceeding analogously as in case of trees (see Proposition 2.4). Eventually, we can process the root component in this way and compute $B(G, r)$. We get a minimum broadcasting scheme by backtracking or just by removing the determined edges and running the known algorithm for tree networks [22].

The total taken by running Reduce is $O(n)$ plus the total cost of the sorting by Lemma 3.2. In order to ensure a linear upper time-bound on the cost of sorting, we simply implement our algorithm in phases corresponding to the optimal broadcasting times following the idea of Slator *et al.* given in the proof of Proposition 2.4. When all nodes v of a cycle C contain a complete sorted list of the optimal broadcasting times of the subnetworks attached to v

then we can run the procedures Compute cuts and Reduce for C. By the definition of a network of type 1, each node of G can be a node of at most $O(1)$ subnetworks of the form G_{r_C} where C is a non-trivial biconnected component, i.e., a non-degenerate cycle. Hence, the total time taken by the procedure Compute cuts is $O(n)$ by Lemma 3.1. $\qquad\Box$

4 Optimal broadcasting in networks of type 2

For the sake of presentation, let us consider first the case of a network G of type 2 whose cycles are mutually node-disjoint. Let r_C be the local source of broadcasting for its cycle C.

As in the previous section, we compute $B(G_{r_C}, r_C)$ and an edge e of C whose removal doesn't affect $B(G_{r_C}, r_C)$ on the basis of the minimum broadcasting times for the subnetworks $G_1, ..., G_l$ of G_{r_C} disjoint from C and attached to r_C and the other subnetworks of G_{r_C} attached to nodes of C different from r_C.

Since the tree of biconnected components of G restricted to the non-singleton ones (i.e., cycles) might have a non-constant radius, the recursive application of the procedure Compute cuts could lead to a quadratic complexity. For this reason, we simply run the procedure Reduce for r_C without the preprocessing in the form of the matrix $M(i, j)$. Instead, at each step of the binary searches, we recompute $B(G^1_{r_C}(e), r_C)$ and $B(G^2_{r_C}(e), r_C)$ on the basis of the sorted optimal broadcasting times for the subnetworks attached to the nodes of C different from r_c. We additionally precompute the two critical values for each node on C different from r_C analogously as in the procedure Compute cuts, in total time $O(\sum_{v \in G} deg(v).)$ Having these values, each computation or recomputation of $B(G^1_{r_C}(e), r_C)$ and $B(G^2_{r_C}(e), r_C)$ will take $O(|C|)$ time. We conclude that the so modified procedure Reduce runs in time $O(\sum_{v \in C} deg(v) + |C| \log |C|)$. Hence, arguing similarly as in the proof of Theorem 3.1, we obtain the following theorem by Proposition 2.2.

Theorem 4.1 *A minimum broadcasting scheme and the optimal broadcasting time for a network of type 2 on n nodes whose cycles are mutually node-disjoint can be found in time $O(n \log n)$.*

Due to space considerations, the reader is refered to [8] for an extension of Theorem 4.1 to include general networks of type 2.

Theorem 4.2 *A minimum broadcasting scheme and the optimal broadcasting time for a network of type 2 on n nodes can be found in time $O(n \log n)$.*

5 Optimal broadcasting in networks of type 3 and 4

The general case of networks of type 3 and 4 where the cycles can edge overlap is much more computationally difficult. Here we can offer only moderate improvements over the rather straightforward upper bound of $O(n^{k+1})$ by a reduction to the case network of type 1 or 2, respectively.

Let G be a network of type 4 with a source node r. To begin with, we form a tree of edge-biconnected components of G rooted at the component containing r and find the local source

r_C of broadcasting for each edge-biconnected component C of G. All these operations take linear time by Proposition 2.1. Next, for each edge-connected component C we compute $B(G_{r_C}, r_C)$ on the basis of $B(G_{r_{C'}}, r_{C'})$ where C' ranges over the children edge-biconnected components of C and the optimal broadcasting times for the other subnetworks of G_{r_C} attached to r_C, as follows.

First, we augment C with the edges leaving to the children components of C and with their endpoints to a subgraph C_1. Next, for each subset of at most $k-1$ edges of C, we test whether their removal from C_1 results in a subgraph C_2 containing at most one cycle. If so, we treat the source nodes $r_{C'}$ for the children components C' of C in C_2 as the roots of virtual subnetworks H of type 2 satisfying $B(H, r_{C'}) = B(G_{r_{C'}}, r_{C'})$ and compute the minimum broadcasting time for so virtually extended C_2 using the method of Section 4. By Theorem 4.1, it takes time $O(n_{C_1} \log n_{C_1})$ where n_{C_1} is the number of nodes of C_1. Now, to obtain $B(G_{r_C}, r_C)$ it is sufficient to take the minimum of the aforementioned minimum broadcasting times over all the removals of at most $k-1$ edges of C that leave a single cycle in C_1.

We conclude that we can compute $B(G, r)$ in time $O(n^k \log n)$. Also, a minimum broadcasting scheme can be found within time $O(n^k \log n)$ by backtracking.

Theorem 5.1 *For a network of type 4 on n nodes, the optimal broadcasting time and a minimum broadcasting scheme can be found in time $O(n^k \log n)$.*

In case of a network of type 3, we proceed analogously just replacing the method of Section 4 with that of Section 3. Then the time taken by the computation of $B(G_{r_C}, r_C)$ is $O(|C|^{k-1} n_{G_{r_C}})$, where $n_{G_{r_C}}$ is the number of nodes in G_{r_C} by Theorem 3.1. Since the tree of edge-biconnected components of G restricted to non-singleton ones has a constant radius in G, we obtain the following theorem.

Theorem 5.2 *For a network of type 3 on n nodes, the optimal broadcasting time and a minimum broadcasting scheme can be found in time $O(n^k)$.*

6 Optimal broadcasting in partial k-trees

Following [21], a *tree-decomposition* of a network G is a pair (T, χ) where T is a tree and χ is a mapping from $V(T)$ to the set of subsets of $V(G)$ satisfying:

1. for each $(u, v) \in E(G)$ there is an $x \in V(T)$ s.t. $\{u, v\} \subset \chi(x)$;

2. for any x, y, $z \in V(T)$, if y is on the path from x to z in T then $\chi(x) \cap \chi(z) \subset \chi(y)$.

For $v \in V(T)$, the subset $\chi(v)$ of $V(G)$ is often identified with v and called a *bag*. The maximum size of a bag in (T, χ), minus one is called the *width* of (T, χ). If all bags in (T, χ) have the same size and for each pair of bags, the symmetric difference is exactly two, then (T, χ) is called a *normalized tree-decomposition*. The *treewidth* of a graph is the minimum width of its tree-decomposition. The following facts are well known (see, for instance, [18].)

procedure: *k-tree broadcasting*
Input: a partial k-tree G and a source node $r \in V(G)$
Output: the minimum broadcasting number

1. Compute a width-k normalized tree-decomposition $TD(G) = (T, \chi)$ of G and root it at a vertex x where $r \in \chi(x)$;

2. For each vertex v of T, generate the set $NCS(v)$ of non-contradictory states;

3. For each leaf v of T, set $Reachable(v)$ to $NCS(v)$;

4. For each non-leaf v of T, set $Reachable(v)$ to $\emptyset$;

5. **For** $h \leftarrow 2$ **to** $height(T)$ **do**

    For each vertex v of T of height h **do**

        For each state s in $NCS(v)$ **do**

            If for each child w of v, there is a state in s_w in $Reachable(w)$ such that the members of $\{s\} \cup \bigcup_w \{s_w\}$ are pairwise compatible **then** augment $Reachable(v)$ by s

6. $m \leftarrow \max_{s \in Reachable(x)} s_2(r)$

7. **Return** $n - m - 1$

Lemma 6.1 *Every graph of treewidth k has a normalized k-width tree-decomposition.*

Lemma 6.2 *The class of partial k-trees is exactly the class of graphs of treewidth at most k.*

For a rooted T, the subgraph of G induced by the union of the bags of the maximal subtree of T rooted at v will be denoted by $G_{(v)}$.

By a *(broadcasting) state* of a bag $\chi(v)$, we shall mean a function $s : \chi(v) \rightarrow V(G) \times \{0, ..., n - 1\}$. For a node $u \in \chi(v)$, $s(u) = (s_1(u), s_2(u))$ is interpreted as follows: u receives (for the first time) the message from the node $s_1(u)$ in round $s_2(u)$. The state s is *non-contradictory* iff for all pairs of nodes $u, w \in \chi(v)$ $s(u) \neq s(w)$, and for each node $u \in \chi(v) \setminus \{r\}$, if $s(u) = (s_1(u), s_2(u))$, $s_1(u) \in \chi(v)$ and $s(s_1(u)) = (s_1(s_1(u)), s_2(s_1(u)))$ then $(u, s_1(u))$ is an edge of G and the inequality $s_2(s_1(u)) < s_2(u)$ holds.

Two states a, b of two different bags $\chi(v_1), \chi(v_2)$ are *compatible* if a, b coincide on $\chi(v_1) \cap \chi(v_2)$ and the common extension s of a, b to $\chi(v_1) \cup \chi(v_2)$ is not contradictory, i.e., for all pairs of nodes $u, w \in \chi(v_1) \cap \chi(v_2)$ $s(u) \neq s(w)$, and for each node $u \in \chi(v_1) \cup \chi(v_2) \setminus \{r\}$ if $s(u) = (s_1(u), s_2(u))$ where $s_1(u) \in \chi(v_1) \cup \chi(v_2) \setminus \{r\}$ and $s(s_1(u)) = (s_1(s_1(u)), s_2(s_1(u)))$ then $(u, s_1(u))$ is an edge of G and the inequality $s_2(s_1(u)) < s_2(u)$ holds.

The procedure *k-tree broadcasting* determines the minimum broadcasting number for a partial k-tree G with the source node r.

Theorem 6.1 *The minimum broadcasting number of a partial k-tree can be determined in time $O(n^{4k+5})$.*

Proof: sketch. The proof of correctness of Procedure k-tree broadcasting consists in proving by induction on the height of v, that $Reachable(v)$ is exactly the set of non-contradictory states s such that if the nodes in $\chi(v)$ that should receive the message from a node outside the subgraph $G_{(v)}$ in rounds prespecified by s then all the nodes in $G_{(v)}$ can receive the message within $n-1$ rounds by broadcasting consistent with the state s on $\chi(v)$.

The first step of the algorithm can be easily implemented within the time-bound of the lemma by [7]. Let u be a member of a bag $\chi(v)$. We can generate the superset of the set of non-contradictory states s of v where the first coordinate of $s(u)$ is a neighbor of u in time $O(n^{2k+2})$. Each member of the superset can be checked for whether or not it satisfies the other requirements of non-contradictness in time $O(k^2)$. Since the are $O(nk)$ vertices in T and k is a constant, we conclude that the second step of the algorithm can be implemented in time $O(n^{2k+3})$.

Clearly, any member in $Reachable(v)$ is a non-contradictory state of v. Thus, $|Reachable(v)| = O(n^{2k+2})$ holds. There are $O((n^{2k+2})^{deg_T(v)})$ possible sequences of states $s_w \in Reachable(w)$ where w ranges over children w of v. To avoid the costly enumeration of them in order to find a sequence whose elements are mutually compatible and compatible with s, we observe the following.

If each member of such a sequence is compatible with s then two elements s_{w_1}, s_{w_2} of such a list are incompatible if for the unique nodes $u_1 \in \chi(w_1) \setminus \chi(w)$ and $u_2 \in \chi(w_2) \setminus \chi(w)$ the equality $s_{w_1}(u_1) = s_{w_2}(u_2)$ holds. The equality means that u_1 and u_2 should receive the message from the same node z in the same round which is impossible. It follows from the structure of partial k-trees that the common sender z is in $\chi(v) \cap \chi(w_1) \cap \chi(w_2)$.

These observations lead to the reduction of the problem of determining the membership of s in $Reachable(v)$ to the problem of finding a maximum $0-1$ flow in the following network H.

H has a source node, a sink node and two layers of nodes in between. The nodes of the first layer are in one-to-one correspondence with the children of v. The nodes in the second layer are in one-to-one correspondence with pairs (z, t) where z is a node in $\chi(v)$ and $t \in \{1, ..., n-1\}$. The source node is adjacent to each node of the first layer whereas the sink node is adjacent to each node of the second layer. A node of the first layer corresponding to w is adjacent to a node of the second layer corresponding to (z, t) iff there is a state $s' \in Reachable(w)$ compatible with s such that for the unique node $u \in \chi(w) \setminus \chi(v)$ the equality $s'(u) = (z, t)$ holds. Also, the node of the first layer is directly adjacent to the sink iff there is a state $s' \in Reachable(w)$ compatible with s such that for the unique node $u \in \chi(w) \setminus \chi(v)$, $s'_1(u) \notin \chi(v)$ holds.

It follows from the definition of H that s is in $Reachable(v)$ iff there is a $0-1$ flow from the source to the sink of H of value equal to the number children of v in T. Since H has $O(deg_T(v) + n)$ nodes, a maximum $0-1$ flow in H can be found in time $O((deg_T(v) + n)^3)$ by Dinic's algorithm [9]. The construction of H can be easily performed in time $O(deg_T(v)n^{2k+2})$.

Since there are $O(n^{2k+2})$ states for testing for membership in $Reachability(w)$, we conclude that these tests take $O(deg_T(v)n^{4k+4})$ time for $k \geq 2$. Consequently, the whole algorithm takes $O(n^{4k+5})$ time for $k \geq 2$. $\qquad\Box$

A minimum broadcasting scheme for a sourced partial k-tree can be also produced in time $O(n^{4k+5})$ by backtracking .

References

[1] A.V. Aho, J.E. Hopcroft and J.D. Ullman, The Design and Analysis of Computer Algorithms, Addison-Wesley, Reading, MA, 1974.

[2] S. Arnborg, A. Proskurowski, Linear time algorithms for NP-hard problems on graphs embedded in k-trees. Discrete Applied Mathematics 23 (1989), pp. 11-24.

[3] A. Bar-Noy and S. Kipnis, Designing broadcasting algorithms in the postal model for message passing systems, Proc. 5th Ann. ACM Symp. on Par. Alg. and Arch. (1992), 11-22.

[4] A. Bar-Noy and S. Kipnis, Broadcasting multiple messages in simultaneous send/receive systems, Discr. Appl. Math. 53 (1994), 95-105.

[5] D. Barth and P. Fraigniaud, Approximation algorithms for Structured communication problems, Proc. 9th Ann. ACM Symp. on Par. Alg. and Arch. (1997), 180-188.

[6] D.P. Bertsekas and J.N. Tsitsiklis, Parallel and Distributed Computation: Numerical Methods, Prentice-Hall, Englewood Cliffs, NJ, (1989).

[7] H. Bodlaender. A linear time algorithm for finding tree-decompositions of small treewidth. Proc. 33rd ACM STOC, pp. 226-234.

[8] A. Dessmark, A. Lingas, H. Olsson and H. Yamamoto, Optimal broadcasting in almost trees and partial k-trees, Technical Report, LU-EX-CS:97-193, Dept. Computer Science, Lund University, Sweden, 1997.

[9] S. Even. Graph Algorithms. Computer Science Press, 1979.

[10] G. Fox, M. Johnsson, G. Lyzenga, S. Otto, J. Salmon and D. Walker, Solving Problems on Concurrent Processors, Volume I, Prentice Hall, (1988).

[11] P. Fraigniaud and E. Lazard, Methods and problems of communication in usual networks, Disc. Appl. Math. 53 (1994), 79-133.

[12] M. Garey and D. Johnson, Computers and Intractability: a guide to the theory of NP-completeness, Freeman and Co., San Francisco (1979).

[13] S.M. Hedetniemi, S.T. Hedetniemi and A.L. Liestman, A survey of gossiping and broadcasting in communication networks, Networks 18 (1988), 319-349.

[14] J. Hromkovič, R. Klasing, B. Monien and R. Peine, Dissemination of information in interconnection networks (broadcasting and gossiping), in: F. Hsu and D.-Z. Du (Eds.), Combinatorial Network Theory, Kluwer Academic Publishers, 1995, 125-212.

[15] A. Jacoby, R. Reischuk, Ch. Schindelhaner, The complexity of broadcasting in planar and decomposable graphs, Proc. 20th International Workshop WG'94, June 1994, LNCS 903, 219-231.

[16] K. Jansen, H. Müller, The Minimum Broadcast Time Problem, Proc. 1st Canada-France Conference on Parallel Computing, Montreal, LNCS 805, 1994, 219-234.

[17] S.L. Johnsson and C.T. Ho, Matrix multiplication on Boolean cubes using generic communication primitives, in: Parallel Processing and Medium-Scale Multiprocessors, A. Wouk (Ed.), SIAM, (1989), 108-156.

[18] J. van Leeuwen. Graph Algorithms. Handbook of Theoretical Computer Science A, North Holland, Amsterdam 1990 pp. 527-631.

[19] G. Kortsarz and D. Peleg, Approximation Algorithms for Minimum Time Broadcast, Proc. of the Israel Symposium on Theoretical Computer Science, LNCS 601, 1992.

[20] R. Ravi, Rapid Rumour Ramification: Approximating tha minimum broadcast time, Proc. 35th IEEE Symposium on Foundations of Computer Science (1994), 202-213.

[21] N. Robertson and P. Seymour. Graph Minors II. Algorithmic aspects of tree-width. J. Algorithms No. 7 (1986), pp. 309-322.

[22] P.J. Slater, E. Cockayne and S.T. Hedetniemi, Information dissemination in trees, SIAM J. Comput. 10 (1981), 692-701.

Local Normal Forms for First-Order Logic with Applications to Games and Automata

Thomas Schwentick   Klaus Barthelmann

Johannes Gutenberg-Universität Mainz

Abstract. Building on work of Gaifman [Gai82] it is shown that every first-order formula is logically equivalent to a formula of the form $\exists x_1, \dots, x_l \forall y \varphi$ where φ is r-local around y, i.e. quantification in φ is restricted to elements of the universe of distance at most r from y.
From this and related normal forms, variants of the Ehrenfeucht game for first-order and existential monadic second-order logic are developed that restrict the possible strategies for the spoiler, one of the two players. This makes proofs of the existence of a winning strategy for the duplicator, the other player, easier and can thus simplify inexpressibility proofs.
As another application, automata models are defined that have, on arbitrary classes of relational structures, exactly the expressive power of first-order logic and existential monadic second-order logic, respectively.

1   Introduction

First-order (FO) logic and its extensions play an important role in many branches of (theoretical) computer science. Examples that will be considered in this paper are automata theory and descriptive complexity. Since Büchi's and Elgot's famous characterization of the regular string languages as the sets of models of (existential) monadic second-order (MSO) sentences, (existential) MSO logic has been used as a guideline in the search for reasonable automata models for other kinds of structures like trees or graphs. In descriptive complexity, since Fagin [Fag74] showed that the complexity class **NP** coincides with the sets of models of existential second-order (Σ_1^1) sentences, many complexity classes have been characterized by extensions of FO logic (for references, see [EF95]) and there is still hope that separations of complexity classes might be possible by separating the expressive power of the respective logics. For a recent result in this direction see the paper of Libkin and Wong in this volume.

Despite its importance as an ingredient for more expressive logics, it is well-known that the expressive power of FO logic is rather limited. It can only express properties that depend on the local appearance of a structure. This intuition has been formalized in different ways by Hanf [Han65] and Gaifman [Gai82]. Hanf showed that, for every first-order formula ψ, there is an r such that whether ψ holds in a structure A ("$A \models \psi$") only depends on the multiset of isomorphism types of all r-spheres in A. Here an r-sphere is a substructure of A which is induced by all elements of A that have distance at most r from a fixed element of A. On the other hand, Gaifman showed that, for every first-order formula ψ, there are r and d such that whether $A \models \psi$ holds depends only on how many elements with pairwise disjoint r-neighbourhoods exist that fulfil θ, for every formula θ of quantifier depth at most d.

The starting question for the present investigation was to which extent Hanf's and Gaifman's conditions could be combined. The goal was to replace the isomorphism types in Hanf's condition by something weaker and to get rid of the "disjoint r-neighbourhoods" constraint in Gaifman's condition. (For very interesting recent results concerning Hanf's and Gaifman's theorems from a different point of view see [Lib97,DLW97].) It is easy to see that the straightforward attempt to replace the isomorphism type of a sphere S in Hanf's condition by its Hintikka type for some d (i. e. by the set of formulas of quantifier depth at most d that hold in S) does not work. A counterexample is given by the set of clique graphs. For every d, the spheres of a graph consisting of one $2d$-clique fulfil exactly the same formulas of quantifier depth at most d as those of a graph which consists of two disjoint d-cliques. Nevertheless, it turns out that it is indeed possible to combine the two approaches in the following sense. For every FO formula ψ there are l and r such that $A \models \psi$ if and only if it is possible to put l pebbles onto A such that in the resulting structure all r-spheres fulfil the same FO formula φ. Put in another way, every FO formula is logically equivalent to a formula of the form $\exists x_1, \ldots, x_l \forall y \varphi$ where φ is r-local around y, i. e. quantification in φ is restricted to elements of the universe with distance at most r from y. From this normal form one can easily derive normal forms for other logics like monadic second-order logic. For existential monadic second-order logic we can show even more restricted normal forms.

As one application of the normal forms we get variants of the Ehrenfeucht game [Ehr61] for first-order logic and existential monadic second-order logic in which the spoiler has only restricted global access to the structures that are played. Another application concerns a form of automata on relational structures. It is well-known that regular sets of strings can be obtained as projections of locally testable sets, namely the sets of transition sequences of a (nondeterministic) automaton. Thomas [Tho91] used this idea to extend the notion of recognizability to other objects like grids and graphs of uniformly bounded degree. The normal forms allow to generalize the method of local testing further to sets of arbitrary relational structures. Moreover, they maintain the correspondence to definability in a natural logic.

In Section 2 we give basic definitions and fix some notation. In Section 3 we show the normal form theorems. In Section 4 we define the simplified games for first-order logic and monadic Σ_1^1-logic. The analogous results for automata are presented in Section 5. Section 6 contains a conclusion.

2 Definitions and Notations

A *relational signature* σ is a finite set of relation symbols R, each with a fixed arity $a(R)$, and constant symbols c. We do not use function symbols. A σ-*structure*

A consists of a universe U^A (the *vertices* of A), an $a(R)$-ary relation R^A on U^A, for every relation symbol R of σ and a constant c^A, for every constant symbol c of σ. All theorems in this paper are valid for infinite and finite structures. We assume that all structures contain at least two elements.

Let the *Gaifman graph* of a σ-structure A have universe U^A and edges between vertices a and b whenever a and b occur in a common tuple of a relation of A. The *distance* $\delta(a, b)$ of the vertices a and b of A is given by their (standard graph) distance in the Gaifman graph of A. For a vertex a of A and $r \geq 0$ we define $S^r(a) := \{b \in U^A \mid \delta(a, b) \leq r\}$. For a tuple $\mathbf{a} = a_1, \ldots, a_l$ of vertices of A we define $S^r(\mathbf{a}) := \bigcup_{i=1}^{l} S^r(a_i)$. $N^r(b)$ (resp. $N^r(\mathbf{a})$, for $\mathbf{a} = a_1, \ldots, a_l$) is the substructure of A which is induced by the vertices of $S^r(b)$ (resp. $S^r(\mathbf{a})$) and has b (the a_i) as distinguished elements.

Let $H^d(A)$ denote the *depth-d-Hintikka-type* of A, i.e., the set of all FO formulas of quantifier-depth at most d that hold in A [EF95]. Recall that $H^d(A)$ contains only finitely many different formulas w.r.t. logical equivalence. The *elementary type* of a tuple $a_1, \ldots, a_k$ of vertices of A is the conjunction of all atomic σ-formulas with variables from $x_1, \ldots, x_k$ that hold in $\langle A, a_1, \ldots, a_k \rangle$.

Now we are going to define our notions of locality, r-locality and basic locality. Informally, a formula is r-local around its free variables $\mathbf{x}$, if its truth depends only on $S^r(\mathbf{x})$. More formally, a FO formula $\varphi(\mathbf{x}, \mathbf{u})$ with free variables from $\mathbf{x} = x_1, \ldots, x_l$ and $\mathbf{u} = u_1, \ldots, u_m$ is r-*local around* $\mathbf{x}$ if all variables that are quantified in φ are bounded to $S^r(\mathbf{x})$. I.e., if $\exists y \psi$ (resp. $\forall y \psi$) is a subformula of φ then ψ is of the form $(\delta(y, \mathbf{x}) \leq r) \wedge \chi$ (resp. $(\delta(y, \mathbf{x}) \leq r) \to \chi$), for some χ. A formula is *local around* $\mathbf{x}$ if it is r-local around $\mathbf{x}$ for some r. A formula φ is *basic local* around $\mathbf{x}$ if it is a Boolean combination of formulas each of which is local around some variable x_i. I.e., properties that are expressed by basic formulas depend only on combinations of the properties of spheres. A formula is $\exists^* \forall$-*local* if it is of the form $\exists x_1, \ldots, x_l \forall y \varphi$ where φ is basic local around $x_1, \ldots, x_l, y$. It should be noted that in Gaifman's terminology [Gai82] a formula is local around a variable x only if x is its single free variable.

3 Normal Forms

By definition, every basic local formula around $\mathbf{x}$ is logically equivalent to a local formula around $\mathbf{x}$. The following lemma shows that the converse is also true. This will be an important tool in the proof of the normal form theorem.

Lemma 1. *Every first-order formula φ which is local around variables $\mathbf{x} = x_1, \ldots, x_l$ is logically equivalent to a formula which is basic local around $\mathbf{x}$.*

Proof. Let φ be a first-order formula with free variables $\mathbf{x} = x_1, \ldots, x_l$ and $\mathbf{u} = u_1, \ldots, u_m$ that is local around $\mathbf{x}$. The proof is by induction on the structure of φ. If φ is an atomic formula, it is basic local around $\mathbf{x}$ by definition. If φ is of the form $\neg \psi$ or $\psi_1 \wedge \psi_2$ then the statement holds by induction.

In the only remaining case, φ is of the form $\exists y ((\delta(y, \mathbf{x}) \leq r) \wedge \chi)$ for some r and some formula χ which is local around $\mathbf{x}$. By induction, χ is logically equivalent to a formula which is basic local around $\mathbf{x}$. By writing χ in disjunctive normal form, we get that χ is logically equivalent to a formula of the form $\bigvee_j \bigwedge_i \chi_{ji}$, where, for some r', every χ_{ji} is r'-local around x_i. Hence, φ is logically equivalent to a disjunction of formulas

$$\exists y((\delta(y, \mathbf{x}) \le r) \wedge \bigwedge_i \chi_i) , \qquad (*)$$

where every χ_i is r'-local around x_i. In the following, we assume w. l. o. g. that φ is of the form (*). In a sense, we have to distribute the quantification of y over the χ_i. The problem which arises is that some of the x_i might be close to each other, so that y might play a role for several χ_i simultaneously. To get around this problem we choose, for every "cluster" C of vertices x_i that are close to each other, a representative $v(C)$ such that quantification around any x_i of C can be replaced by quantification around $v(C)$. More formally, we proceed as follows. We associate with every structure A and tuple $\mathbf{a} = a_1, \dots, a_l$ of vertices of A a graph $G(A, \mathbf{a})$ with vertex set $V = \{1, \dots, l\}$ and edge set E, which contains the edge (i, j) whenever $\delta(a_i, a_j) \le R := r + r' + 1$. We are going to construct, for every graph G on $\{1, \dots, l\}$, formulas θ_G and φ_G each of which is basic local around $\mathbf{x}$, such that, for every A and $\mathbf{a}$ it holds that

- $A \models \theta_G$ iff $G(A, \mathbf{a}) = G$, and
- if $G = G(A, \mathbf{a})$ then $[A \models \varphi_G \Longleftrightarrow A \models \varphi]$.

Once we have established the existence of such formulas the statement of the lemma follows immediately because $\varphi \equiv \bigvee_G (\theta_G \wedge \varphi_G)$, where the disjunction is over all graphs G on $\{1, \dots, l\}$.

Let in the following a graph $G = (V, E)$ be fixed. The definition of θ_G is straightforward:

$$\theta_G := \bigwedge_{i \in V} (\bigwedge_{\substack{j \in V \\ (i,j) \in E}} (\delta(x_i, x_j) \le R) \wedge \bigwedge_{\substack{j \in V \\ (i,j) \notin E}} (\delta(x_i, x_j) > R)) .$$

It is easy to see that, for every i, the i-th conjunct can be made R-local around x_i. In order to construct φ_G we choose from each connected component C of G a vertex $v(C)$. Then φ_G can be defined as

$$\bigvee_{C, \alpha} (\exists y[(\delta(y, x_{v(C)}) \le R|C|) \wedge \alpha \wedge \bigvee_{i \in C} (\delta(y, x_i) \le r) \wedge \bigwedge_{i \in C} \chi'_{i,C}] \wedge \bigwedge_{i \notin C} \chi'_{i,C,\alpha}) .$$

Here, the disjunction is over all connected components C of G and all (finitely many) elementary types α of $\mathbf{x}, \mathbf{u}, y$. If $i \in C$ then $\chi'_{i,C}$ is obtained from χ_i by making all quantifications R-local around $x_{v(C)}$. If $i \notin C$ then $\chi'_{i,C,\alpha}$ is obtained from χ_i by

- replacing every atomic subformula that contains the variable y and a variable that is bound in χ_i with <u>false</u>, and

– rewriting all other atomic formulas that refer to y according to α.

These replacements assure that y does not occur in $\chi'_{i,C,\alpha}$ and that every $\chi'_{i,C}$ and every $\chi'_{i,C,\alpha}$ is local around x_i. It should be pointed out that the subformula $[\cdots]$ can only become true if $y \in S^r(x_i)$, for some $i \in C$, hence if $y \notin S^{r'+1}(x_j)$, for $j \notin C$. This justifies the replacement of atomic subformulas that contain the variable y and a variable that is bound in χ_i with false.

Theorem 2. *Every FO formula is logically equivalent to a $\exists^*\forall$–local formula.*

Proof. Let Ψ be a first-order formula with free variables $\mathbf{u} = u_1, \ldots, u_m$. First we are going to show that Ψ is logically equivalent to a positive Boolean combination of $\exists^*\forall$–local formulas. In the following, if φ is r-local around its single variable, we write $\exists_l^r x\varphi(x)$ as an abbreviation for

$$\exists x_1, \ldots, x_l[\bigwedge_{i=1}^{l} \varphi(x_i) \wedge \bigwedge_{i \neq j} (\delta(x_i, x_j) > 2r)] .$$

Gaifman's theorem implies that Ψ is logically equivalent to a positive Boolean combination of formulas of the following three types.

1. $\exists_l^r x\varphi(x)$,
2. $\neg\exists_l^r x\varphi(x)$, and
3. local formulas around $\mathbf{u}$.

We are going to show that formulas of each of these types are logically equivalent to positive Boolean combinations of $\exists^*\forall$–local formulas. For formulas of type 1 this is obvious, for formulas of type 3 it follows with Lemma 1. We still have to consider formulas of type 2. Let $\psi \equiv \neg\exists_l^r x\varphi(x)$. ψ is logically equivalent to

$$\neg\exists x\varphi(x) \vee \bigvee_{i=1}^{l-1} [\exists_i^r x\varphi(x) \wedge \neg\exists_{i+1}^r x\varphi(x)] .$$

Of course, $\neg\exists x\varphi(x)$ is equivalent to $\forall x\neg\varphi(x)$, which is of the required form. For every i, $\exists_i^r x\varphi(x) \wedge \neg\exists_{i+1}^r x\varphi(x)$ is logically equivalent to

$$\exists x_1, \ldots, x_i \forall y[\bigwedge_i \varphi(x_i) \wedge \bigwedge_{i \neq j} (\delta(x_i, x_j) > 2r) \wedge \neg((\delta(y, \mathbf{x}) > 2r) \wedge \varphi(y)) \wedge$$

$$\forall z_1, \ldots, z_{i+1} \neg(\bigwedge_j (\delta(z_j, \mathbf{x}) \leq 2r) \wedge \bigwedge_{i \neq j} (\delta(z_i, z_j) > 2r) \wedge \bigwedge_j \varphi(z_j))] ,$$

expressing that there are $x_1, \ldots, x_i$ that fulfil φ, but

– neither there is a y which fulfils φ and is far from all the x_j,
– nor there exist different $z_1, \ldots, z_{i+1}$ all of which fulfil φ and are close to the x_j.

As the $[\cdots]$ part of this formula can be made local around $x_1, \ldots, x_i, y$, it is logically equivalent to a basic local formula around $x_1, \ldots, x_i, y$ by Lemma 1. We conclude that ψ is logically equivalent to a $\exists^*\forall$–local formula.

It remains to show that every positive Boolean combination of $\exists^*\forall$–local formulas is logically equivalent to a $\exists^*\forall$–local formula. Consider two $\exists^*\forall$–local formulas $\psi_1 \equiv \exists x_1, \ldots, x_k \forall y \varphi_1(\mathbf{x}, y)$ and $\psi_2 \equiv \exists x_1', \ldots, x_m' \forall y' \varphi_2(\mathbf{x}', y')$. $\psi_1 \wedge \psi_2$ is equivalent to $\exists \mathbf{x}, \mathbf{x}' \forall y [\varphi_1(\mathbf{x}, y) \wedge \varphi_2(\mathbf{x}', y)]$ and $\psi_1 \vee \psi_2$ is equivalent to

$$\exists z, z', \mathbf{x}, \mathbf{x}' \forall y (z = z' \wedge \varphi_1(\mathbf{x}, y)) \vee (z \neq z' \wedge \varphi_2(\mathbf{x}', y)) .$$

In fact, by a closer inspection of Theorem 2, we can go one step further and show that local quantification around one single variable is enough.

Theorem 3. *Every first-order formula is logically equivalent to a formula of the form $\exists x_1, \ldots, x_l \forall y \varphi(\mathbf{x}, y)$, where φ is local around y.*

There are straightforward analogues of Theorem 2 for other logics. Of special interest is the case of monadic Σ_1^1-logic, as quantification of unary relations does not change the locality properties of a structure. We get immediately that every monadic Σ_1^1-formula is logically equivalent to a formula of the type $\exists X_1, \ldots, X_l \exists x_1, \ldots, x_m \forall y \varphi$, where φ is local around y. For classes of structures that have a connected Gaifman graph we can show even stronger normal forms. The basic idea is that global information about the structure can be transported along the relations and collected in a designated place. This generalizes a similar procedure in [Tho97], where the transport is much more deterministic.

Theorem 4. *Let C be a class of structures with a connected Gaifman graph. Then the following hold.*

(a) *On C every monadic Σ_1^1 formula is equivalent to a formula of the form $\exists X_1, \ldots, X_l \exists x \forall y \varphi$, where φ is local around y.*

(b) *If there exists a formula ρ that is local around its one free variable such that, for every structure A of C, $A \models \exists! x \rho(x)$ then on C every monadic Σ_1^1 formula is equivalent to a formula $\exists X_1, \ldots, X_l \forall y \varphi$, where φ is local around y.*

Proof. We only give sketches of the proofs.

(a) Let $\psi \equiv \exists X_1, \ldots, X_l \exists x_1, \ldots, x_m \forall y \varphi$, where φ is basic r-local around y. We explain how the x_i can be eliminated in favour of one x. We have to find a way to distinguish m vertices in a structure A. The idea is to guess a vertex x and m minimal paths $p_1, \ldots, p_m$ in the Gaifman graph of A such that, for every i, p_i starts in x and ends in x_i. Note that we view these graph as directed although the Gaifman graph is an undirected graph. Every p_i is encoded by two unary relations Y_i, Z_i. Y_i contains all vertices of p_i. As p_i is a shortest path we can conclude that x and x_i have degree 1 w.r.t. p_i and all other vertices of p_i have degree 2 w.r.t. p_i. Z_i is used to give p_i an orientation. $v \in Z_i$ just in case v is the j-th vertex of p_i, for some j, (where x is the 0-th vertex) and j is congruent to one of 0,1,3 modulo 6.

It is straightforward that there exists a formula $\rho(x,y)$ that is local around y such that $\langle A, \mathbf{Y}, \mathbf{Z}, x \rangle \models \forall y \rho$ if and only if $\mathbf{Y}$ and $\mathbf{Z}$ encode m paths that have their starting point in x. ($\mathbf{Y}$ and $\mathbf{Z}$ might also encode some directed cycles but this does not matter.) It remains to show how in φ all references to variables x_j can be replaced. All atomic formulas θ in which a x_j occurs together with y or with a variable that is bound in φ (around y!) can be easily replaced by a formula $\exists z((\delta(z,y) \le r+1) \wedge \chi_i(z) \wedge \theta')$, where $\chi_i(z)$ tests that z is the sink of p_i (this can be checked 5-locally around z, hence $(r+6)$-locally around y) and in θ' every occurrence of x_j is replaced by z.

To replace atomic formulas that only refer to variables of $\mathbf{x}$ (and $\mathbf{u}$, the free variables) we proceed as follows. For every atomic formula α which only contains variables from $\mathbf{x}$ and $\mathbf{u}$ we introduce a unary relation T_α. The intention is that $T_\alpha = U^A$ in the case that $\langle A, \mathbf{x}, \mathbf{u} \rangle \models \alpha$ and $T_\alpha = \emptyset$ otherwise. Because A is connected, the formula

$$\forall y[y \in T_\alpha \leftrightarrow \forall z((\delta(z,y) \le 1) \rightarrow z \in T_\alpha)]$$

checks that T_α contains either all or no vertices. It is also easy to check by a 6-local formula around y (using some χ_i as needed) that T_α contains all vertices just in case the endpoints of (some of) the m paths behave according to α.

(b) In the construction of part (a), x only occurs in θ. Instead of guessing x and paths that start in x we can guess paths that start in the vertex that is distinguished by ρ. □

Of course, Theorem 2 is also true with $\forall^*\exists$-local formulas in place of $\exists^*\forall$-local formulas. Although it is easy to see that $\exists^*$-local formulas do not capture all first-order properties (e. g. they cannot express the property "every vertex is coloured black"), it follows from results of Compton [Com83] that Boolean combinations (including negations!) of $\exists^*$-formulas do.

From Theorem 2 we can conclude that every first-order (and monadic Σ_1^1) formula on graphs (suitably represented by adjacency lists) can be evaluated by a nondeterministic Random-Access-Machine with unit-cost measure in time $O(nd^{O(1)})$, where n denotes the number of vertices and d the maximal vertex-degree.

4 Games

Ehrenfeucht games – invented in [Ehr61] building on work of Fraïssé [Fra54] – are an important tool for proving inexpressibility results in Mathematical Logic. In fact, in Finite Model Theory, where only finite structures are considered, they are the major tool available (cf. [Fag97]). To show that a given property P of finite structures is not expressible in FO logic it is enough to prove that the duplicator, one of two players, has a winning strategy in the ordinary FO Ehrenfeucht game for P. Variants of Ehrenfeucht games are available for proving inexpressibility results for many other logics, For a definition of Ehrenfeucht games and many of its varaints see [EF95].

Proving the existence of a winning strategy for the duplicator is often very difficult. To simplify such proofs the following approaches have been taken. There have been developed several conditions that assure that the duplicator has a winning strategy on two given structures: the Hanf-condition [FSV95], the Arora-Fagin condition [AF97] and the condition of Schwentick [Sch96] (for a survey see [Fag97]). All of these conditions exploit the fact that FO logic can only express combinations of local properties, i.e. properties of regions of bounded size. On the other hand there have been attempts to make the game easier to play for the duplicator. An important example is the invention of the Ajtai-Fagin game for existential second-order logic which allows the duplicator to choose the second structure *after* the spoiler (the duplicator's opponent) has selected relations for the first structure. The idea behind that game can be used whenever all formulas of a logic have an existential quantifier-prefix.

In this section we introduce simplified Ehrenfeucht games for FO logic and monadic Σ_1^1 logic. Additionally, we characterize the mentioned logics in terms of Hintikka-types (cf. Section 2). First of all, we describe the *simplified first-order Ehrenfeucht game* for a class C of σ-structures. Like the ordinary Ehrenfeucht game it is played by two players, called the spoiler and the duplicator. It has three parameters, l, r and d and consists of three stages.

Stage 1 The duplicator chooses a σ-structure $A \in C$. The spoiler selects vertices $x_1, \ldots, x_l$ from A. Then the duplicator chooses a σ-structure $A' \notin C$ and vertices $x'_1, \ldots, x'_l$ from A'.

Stage 2 The spoiler chooses a vertex y' from A', then the duplicator chooses a vertex y from A.

Stage 3 The spoiler and the duplicator play an ordinary d-round Ehrenfeucht game on the structures $\langle N^r(y), x_1, \ldots, x_l \rangle$ and $\langle N^r(y'), x'_1, \ldots, x'_l \rangle$.

Here $\langle N^r(y), x_1, \ldots, x_l \rangle$ denotes the structure that is induced by $S^r(y)$ and $\mathbf{x}$ and has $x_1, \ldots, x_l$ and y as distinguished elements. The spoiler wins the game if he wins the game of stage 3 in the usual sense. Otherwise, the duplicator wins.

Theorem 5. *Let C be a class of σ-structures. The following are equivalent.*

1. *C is first-order definable.*
2. *For some l, r and d, the spoiler has a winning strategy in the simplified FO Ehrenfeucht game on C with parameters l, r and d.*
3. *There exists a set H of Hintikka-types such that, for some l, r and d, for every σ-structure A it holds that $A \in C$, if and only if there exist $x_1, \ldots, x_l \in A$ with $\{H^d(\langle N^r(y), x_1, \ldots, x_l \rangle) \mid y \in U^A\} \subseteq H$.*

The proof is straightforward. The simplified FO Ehrenfeucht game can be easily adapted for the case of monadic Σ_1^1-logic. The resulting game has one additional parameter m. In stage 1 of the *simplified monadic Σ_1^1 Ehrenfeucht game* the spoiler chooses, before the duplicator chooses A', unary relations $X_1, \ldots, X_m$ and vertices $x_1, \ldots, x_l$ in A and the duplicator has to choose corresponding relations $X'_1, \ldots, X'_m$ and vertices $x'_1, \ldots, x'_l$ in A'.

Theorem 6. *Let C be a class of σ-structures. The following are equivalent.*

1. *C is monadic Σ_1^1 definable.*
2. *For some m, l, r and d, the spoiler has a winning strategy in the simplified monadic Σ_1^1 Ehrenfeucht game on C with parameters m, l, r and d.*
3. *There exists a set H of Hintikka-types such that, for some m, l, r and d, for every σ-structure A it holds that*

$$A \in C \iff \text{there exist unary relations } X_1, \ldots, X_m \text{ and elements } x_1, \ldots, x_l$$
$$\text{with } \{H^d(\langle N^r(y), X_1, \ldots, X_m, x_1, \ldots, x_l \rangle) \mid y \in U^A\} \subseteq H .$$

In a similar manner one can derive respective games from Theorem 4. Although we have not used Theorems 5 and 6 to derive any new inexpressibility results we are optimistic that the simplified Ehrenfeucht games will turn out to be a useful tool to get such results.

5 Automata

The conditions (3) of Theorems 5 and 6 give rise to a generalized form of automata. In this section we are going to introduce automata models for FO logic and monadic Σ_1^1 logic, respectively. Informally the FO automaton works as follows. First it nondeterministically pebbles vertices $b_1, \ldots, b_g$ of its input structure A, for some g. Then, for every vertex a of A, it inspects in a constant number of steps (alternating between nondeterminism and parallelism) the neighbourhood of a. Navigation through the neighbourhood is only along edges.

We now define the model more formally. Let σ be a relational signature. A first-order σ-automaton M consists of a tuple (g, l, I, φ), where $g \geq 0$ is the size of the *global read-only store*, $l \geq 0$ is the size of the *local store*, I is a finite sequence of *instructions* and φ is a *test*. The store will hold a vector of (pointers to) elements of the structure. Instructions are of the form $\langle \text{any } i, j \rangle$ or $\langle \text{all } i, j \rangle$, where $1 \leq i, j \leq l$. The test is a quantifier-free formula with variables from $x_1, \ldots, x_l, y_1, \ldots, y_g$. A *configuration* $(J, \mathbf{b}, \mathbf{a})$ consists of a (possibly empty) sequence J of instructions yet to be executed and the contents $\mathbf{b}$ and $\mathbf{a}$ of the global and local store, respectively. Let $\mathbf{b} \in A^l$ and $v \in A$. A *configuration tree of M for $\mathbf{b}$ and v* is defined as follows. It is a rooted tree, directed from the root to the leaves, and has configurations of M as vertices. It has the *start configuration* $(I, \mathbf{b}, v^l)$ as its root. The leafs are *terminal configurations* $(\varepsilon, \mathbf{b}, \mathbf{a})$. An inner vertex $(\iota J, \mathbf{b}, \mathbf{a})$ has

- one child $(J, \mathbf{b}, \mathbf{a}')$, where a'_j is a_i or a neighbour of a_i, and $a'_k = a_k$, for every other k, if ι is $\langle \text{any } i, j \rangle$, and
- for every vertex a that is a_i or a neighbour of a_i a child $(J, \mathbf{b}, \mathbf{a}')$, with $a'_j = a$ and $a'_k = a_k$ for every other k, if ι is $\langle \text{all } i, j \rangle$.

A terminal configuration $(\varepsilon, \mathbf{b}, \mathbf{a})$ is *accepting*, if $\langle A, \mathbf{a}, \mathbf{b} \rangle \models \varphi(\mathbf{x}, \mathbf{y})$, otherwise *rejecting*. A configuration tree is *accepting*, if all its leaves are accepting. We say that M *accepts* A if there is a $\mathbf{b} \in A^g$ such that for every $v \in A$ there is an accepting configuration tree of M with $\mathbf{b}$ and a.

A monadic Σ_1^1 σ-automaton similarly consists of a tuple (Q, g, l, I, φ), where g, l, I, φ are as before and Q is a finite set of *states*. It starts by nondeterministically selecting a mapping $f : U^A \to Q$ (represented by unary relations F_q, one for each $q \in Q$). Afterwards it continues like the FO automaton (where the input structure is extended by the F_q).

Theorem 7. *A class of σ-structures is first-order definable if and only if it is accepted by a first-order σ-automaton.*

The proof is straightforward.

Corollary 8. *A class of σ-structures is monadic Σ_1^1 definable if and only if it is accepted by a monadic Σ_1^1 automaton.*

For structures of bounded degree, the monadic Σ_1^1 automata generalize those of Thomas [Tho91,Tho97]. In this case a Hintikka-type of a small neighbourhood boils down to an isomorphism type, called *tile*. The automata of Thomas check that each vertex possesses a neighbourhood of one among a finite number of allowed isomorphism types. Moreover, some of them must occur *at least* a certain number of times and others *at most* a certain number of times. (This is the remainder of Hanf's condition, from which it is actually derived.) Inspecting Theorem 3 again with this in mind, one finds that φ describes all admissible tiles, while the center y runs over all vertices. The variables $\mathbf{x}$ are used to distinguish a finite number of vertices. It is therefore clear that it suffices to require that some tiles (namely those with a center in $\mathbf{x}$) occur *exactly* once. Moreover, for structures with a connected Gaifman graph, Theorem 4 implies that *one* occurrence constraint of this form suffices.

6 Conclusion

We introduced a normal form for first-order logic that formalizes the intuition that this logic is only able to express properties of the form "there are some important parts of the structure that fulfil given conditions and everywhere else nothing forbidden happens". Although our normal forms are mainly useful when structures of unbounded diameter are considered (otherwise the whole structure is contained in the neighbourhood of all of its vertices and local formulas are just general first-order formulas) their translation into the language of automata gives uniform means of evaluating first-order properties "along the edges" of structures. On the other hand it is applicable to structures of unbounded degree, even in situations where the Hanf argument does not work.

Our main open question is whether the normal forms have other meaningful applications and to find new inexpressibility results with the help of the simplified games. Another question is whether automata models can be designed that inspect the neighbourhoods of vertices in a more deterministic fashion.

This investigation was inspired by a talk that was given by Wolfgang Thomas in Mainz in December 1996. We would like to thank him, Clemens Lautemann, Juha Nurmonen, Ron Fagin for many fruitful discussions and suggestions. Thanks also to the anonymous referees.

References

[AF97]   Sanjeev Arora and Ronald Fagin. On winning strategies in Ehrenfeucht–Fraïssé games. *Theoretical Computer Science*, 174(1–2):97–121, 1997.

[Com83] K. Compton. Some useful preservation theorems. *Journal of Symbolic Logic*, 48:427–440, 1983.

[DLW97] G. Dong, L. Libkin, and L. Wong. Local properties of query languages. In *Proc. Int. Conf. on Database Theory*, LNCS, pages 140–154. Springer-Verlag, 1997.

[EF95]   H.-D. Ebbinghaus and J. Flum. *Finite Model Theory*. Springer-Verlag, 1995.

[Ehr61] A. Ehrenfeucht. An application of games to the completeness problem for formalized theories. *Fund. Math.*, 49:129–141, 1961.

[Fag74] R. Fagin. Generalized first–order spectra and polynomial–time recognizable sets. In R. M. Karp, editor, *Complexity of Computation, SIAM-AMS Proceedings, Vol. 7*, pages 43–73, 1974.

[Fag97] R. Fagin. Easier ways to win logical games. In *Proceedings of the DIMACS Workshop on Finite Models and Descriptive Complexity*. American Mathematical Society, 1997.

[Fra54] R. Fraïssé. Sur quelques classifications des systèmes de relations. *Publ. Sci. Univ. Alger. Sér. A*, 1:35–182, 1954.

[FSV95] R. Fagin, L. Stockmeyer, and M. Vardi. On monadic NP vs. monadic co-NP. *Information and Computation*, 120:78–92, 1995. Preliminary version appeared in 1993 IEEE Conference on Structure in Complexity Theory, pp. 19-30.

[Gai82] H. Gaifman. On local and nonlocal properties. In J. Stern, editor, *Logic Colloquium '81*, pages 105–135. North Holland, 1982.

[Han65] W. Hanf. Model-theoretic methods in the study of elementary logic. In J. Addison, L. Henkin, and A. Tarski, editors, *The Theory of Models*, pages 132–145. North Holland, 1965.

[Lib97] L. Libkin. On the forms of locality over finite models. In *Proc. 12th IEEE Symp. on Logic in Computer Science*, 1997.

[Sch96] T. Schwentick. On winning Ehrenfeucht games and monadic NP. *Annals of Pure and Applied Logic*, 79:61–92, 1996.

[Tho91] W. Thomas. On logics, tilings and automata. In *Proc. ICALP'91*, number 510 in Lecture Notes in Computer Science, pages 441–453. Springer, 1991.

[Tho97] W. Thomas. Automata theory on trees and partial orders. In *Proc. TAPSOFT'97*, number 1214 in Lecture Notes in Computer Science, pages 20–38. Springer, 1997.

Axiomatizing the Equational Theory of Regular Tree Languages Extended Abstract

Z. Ésik[*]

Department of Computer Science
József Attila University
6720 Szeged
Árpád tér 2
Hungary
esik@inf.u-szeged.hu

Abstract. We show that a finite set of equation schemes together with the least fixed point rule gives a complete axiomatization of the valid identities of regular tree languages. This result is a generalization of Kozen's axiomatization of the equational theory of regular word languages.

1   Introduction

The theory of tree automata is by now a well-studied generalization of the classical framework. Regular languages were extended to tree languages in the 1960's, see [36, 30]. For varieties of regular tree languages we refer to [1, 2, 34, 35, 39, 18]. Trees and tree automata in connection with formal logic have been studied in a number of papers, see [37, 38] for overviews. The texts [21, 22] provide comprehensive treatment of tree automata theory.

In this paper, we address the axiomatization of the equational theory of (regular) tree languages. Suppose Σ is a ranked signature of function symbols and A is a set. Subsets of the absolutely free Σ-algebra $T_\Sigma(A)$ of Σ-trees over A are called *tree languages*. Tree languages form the carrier of an algebra $P_\Sigma(A)$, equipped with the following operations:

1. A *complex operation* associated with each symbol in Σ.
2. A binary *sum operation* + modeled by set union, and the constant 0 denoting the empty set.
3. A *fixed point* or *iteration operation* providing least solutions to polynomial fixed point equations.

---

[*] Partially supported by grant T22423 of the National Foundation of Hungary for Scientific Research, the US-Hungarian Joint Fund under grant no. 351, the Fulbright Commission, and by the French–Hungarian Science and Technology Bilateral Governmental Program financed by the OMFB and its French partner: "Le Ministére des Affaires Entragéres".

By the fixed point operation, each "μ-term" $\mu x.t$ induces a function in the algebra $P_\Sigma(A)$.

More exactly, the structure $P_\Sigma(A)$ is an example of an *iteration Δ-algebra* [15, 6, 9], where the signature $\Delta = \Sigma_{\{+,0\}}$ is obtained by adding to Σ the binary symbol $+$ and the nullary symbol 0. In a similar way, if B is a Σ-algebra, or a non-deterministic Σ-algebra, see below, then the *complex algebra* $P(B)$ of all subsets of B is also an iteration Δ-algebra. In fact, the iteration algebras $P_\Sigma(A)$ and $P(B)$, and the iteration algebras $R_\Sigma(A)$ of *regular tree languages* in $P_\Sigma(A)$ satisfy the same set of identities between μ-terms.

It is not difficult to prove that a finite set of identities and the usual infinitary implication derived from continuity, expressing that $\mu x.t$ is the sup of the corresponding Kleene approximation sequence, is complete for the valid identities of the complex algebras $P(B)$. But we can do a lot better than that. In our main result, we give a simple *finitary* implicational axiomatization. We show that a finite set of equation schemes and the *least fixed point rule* is complete:

$$t[y/x] = y \Rightarrow \mu x.t \leq y,$$

where t is any μ-term and $\mu x.t \leq y$ is an abbreviation for $y + \mu x.t = y$. This result is a generalization of Kozen's axiomatization [28] of the equational theory of regular word languages. Our completeness proof depends on recent advances [16, 17] in the study of iteration theories.

The study of the valid identities of regular word languages has a long history. Redko [32] and Conway [12] proved that the valid identities have no finite complete equational axiomatization. In contrast, Salomaa [33] proved that there is a finite complete first order axiomatization based on the *unique fixed point rule*. Salomaa's result has been further refined recently by Archangelsky and Gorshkov [3], Krob [29], Boffa [11], Kozen [28], and by Bernátsky et al. [5], who gave several finite complete implicational theories. (The book [12] also contains a finite complete implicational theory of this sort, but the first proof of its completeness appears to be that given in [29, 11].) Kozen's system [28] contains both the least fixed point rule and its dual. The fact that the least fixed point rule suffices alone follows from the results of [29, 11]. Infinite complete equational axiomatizations were given in [29, 7]. The system of Krob [29] confirms a classical conjecture of Conway [12].

The axiomatization of the equational theory of regular tree languages has received less attention. The only works addressing this question that the author is aware of are the paper [26] by Ito and Ando and the extended abstract [14]. The system of Ito and Ando is based on the unique fixed point rule and is thus an extension of Salomaa's system. It has the disadvantage that the unique fixed point rule fails in most natural models such as the complex algebras $P(A)$. The result of [14], formulated in the framework of iteration theories, presents without proof an infinite and somewhat complicated complete set of equation schemes.

All proofs are omitted from this extended abstract.

2 Iteration algebras

Iteration algebras were introduced in [15] and subsequently studied in [6]. The concept originates in the study of iteration theories [8, 10]. Our presentation follows [9]. The simple language of μ-terms originates in the μ-calculus [27].

Suppose that $\Sigma = \cup_{n \geq 0} \Sigma_n$ is a signature, where the sets Σ_n, $n \geq 0$ are pairwise disjoint. Let X denote a countably infinite set disjoint from Σ. The set of μ-terms on Σ, denoted μT_Σ, is defined to be the smallest set of expressions satisfying the following:

- $\Sigma_0 \cup X \subseteq \mu T_\Sigma$,
- $\sigma \in \Sigma_n$, $t_1, \ldots, t_n \in \mu T_\Sigma$, $n > 0$ $\Rightarrow$ $\sigma(t_1, \ldots, t_n) \in \mu T_\Sigma$,
- $t \in \mu T_\Sigma$, $x \in X$ $\Rightarrow$ $\mu x.t \in \mu T_\Sigma$.

The variable x is bound in $\mu x.t$. We identify μ-terms which differ only in their bound variables (α-conversion). Hence, when needed, we may tacitly assume that a variable occurring bound in a μ-term is different from any other variable under consideration. Note that the set of μ-terms containing no occurrence of the symbol μ is just the set $T_\Sigma(X)$ of Σ-trees (or terms) over X mentioned in the Introduction. The set $FV(t)$ of free variables in the term t is defined as usual.

Suppose that t is a μ-term, $\mathbf{x} = [x_1, \ldots, x_n]$ is a vector of variables with distinct components, and $\mathbf{s} = [s_1, \ldots, s_n]$ is a vector of μ-terms. We write

$$t[s_1/x_1, \ldots, s_n/x_n] \text{ or } t[\mathbf{s}/\mathbf{x}]$$

to denote the term obtained by substituting s_i for each free occurrence of the variable x_i in t, for each $i \in [n] = \{1, \ldots, n\}$. By our convention about the bound variables, no free variable may become bound as the result of the substitution. Thus, if $t = \sigma(x_1, \ldots, x_n)$, the term $\sigma(s_1, \ldots, s_n)$ is $t[s_1/x_1, \ldots, s_n/x_n]$.

Definition 1. Suppose that A is a nonempty set and for each $t \in \mu T_\Sigma$, t_A is a function $A^X \to A$ depending at most on the arguments that correspond to the variables in $FV(t)$. We call the system consisting of the set A and the functions t_A, where t is a μ-term, a **preiteration Σ-algebra**, if the following hold:

- For each $x \in X$ and $\rho \in A^X$,

$$x_A(\rho) = \rho(x). \tag{1}$$

- For each μ-term $t[t_1/x_1, \ldots, t_n/x_n]$ and for each $\rho \in A^X$,

$$(t[t_1/x_1, \ldots, t_n/x_n])_A(\rho) = t_A(\rho'), \tag{2}$$

where

$$\rho'(x) = \begin{cases} (t_i)_A(\rho), & \text{if } x = x_i, \ i \in [n]; \\ \rho(x), & \text{otherwise.} \end{cases}$$

– For each $t, t' \in \mu T_\Sigma$ and $x \in X$, and for each $\rho \in A^X$, if the maps

$$a \mapsto t_A(\rho_a^x)$$
$$a \mapsto t'_A(\rho_a^x), \quad a \in A$$

are equal, where $\rho_a^x(x) = a$ and $\rho_a^x(y) = \rho(y)$, for all $y \neq x$, then

$$(\mu x.t)_A(\rho) = (\mu x.t')_A(\rho). \tag{3}$$

It follows from condition (2) that

$$t_A = t'_A, \ (t_i)_A = (t'_i)_A, \ i \in [n] \Rightarrow (t[t_1/x_1, \ldots, t_n/x_n])_A = (t'[t'_1/x_1, \ldots, t'_n/x_n])_A.$$

Remark 2. Each preiteration Σ-algebra A determines an ordinary Σ-algebra. When $\sigma \in \Sigma_n$, define the operation $\sigma_A : A^n \to A$ by

$$\sigma_A(a_1, \ldots, a_n) = (\sigma(x_1, \ldots, x_n))_A(\rho),$$

where $a_1, \ldots, a_n$ are in A, $x_1, \ldots, x_n$ are distinct variables, and $\rho : X \to A$ is any function with $x_i \mapsto a_i$, $i \in [n]$. It is usually not possible to recover in a natural way a μ-operation, or a family of μx-operations on the set A.[2] However, it is possible to define these operations on the *polynomials* $A^X \to A$. Suppose that $f : A^X \to A$ is a polynomial, so that $f = t_A$, for some term t. Then we define

$$\mu x.f = (\mu x.t)_A.$$

This definition makes sense by condition (3) above.

Example 1. The set μT_Σ of μ-terms on Σ may be turned into a preiteration Σ-algebra in a natural way. If t is in μT_Σ with free variables in $\{x_1, \ldots, x_n\}$, and when ρ is a function $X \to \mu T_\Sigma$, define

$$t_{\mu T_\Sigma}(\rho) = t',$$

where t' is the term obtained by substituting the term $\rho(x_i)$ for x_i in t, for all $i \in [n]$.

Example 2. Suppose that A is an ω-cpo with a bottom element and suppose that for each $\sigma \in \Sigma_n$, the function $\sigma_A : A^n \to A$ is ω-continuous. The poset A, equipped with the operations σ_A, is a *continuous Σ-algebra*, see [23, 24]. Each continuous Σ-algebra A determines a preiteration algebra. When t is a μ-term and $\rho \in A^X$, define $(\mu x.t)_A(\rho)$ to be the *least (pre-)fixed point* of the map $A \to A$, $a \mapsto t_A(\rho_a^x)$.

Definition 3. Suppose that A and B are preiteration Σ-algebras. A **preiteration Σ-algebra homomorphism** (or homomorphism, for short) from A to B is a function $h : A \to B$ such that for each term t in μT_Σ the diagram

---

[2] These operations can be defined on the set A whenever the algebra is free in some variety of preiteration algebras. See below.

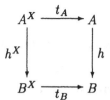

commutes.

Here, h^X denotes the function $\rho \mapsto \rho'$ such that $\rho'(x) = h(\rho(x))$, for all $x \in X$. When A is a subset of B and h is the inclusion, we call A a *sub preiteration algebra* of B.

Example 3. The preiteration Σ-algebra μT_Σ is freely generated by the set X. Any function $\rho : X \to A$, where A is a preiteration Σ-algebra, extends to a unique homomorphism $\rho^\sharp : \mu T_\Sigma \to A$. In fact, for any μ-term t, $t\rho^\sharp = t_A(\rho)$.

An *atomic formula* is a formal equality $t = t'$ between μ-terms t and t'. First-order formulas are constructed from atomic formulas in the usual way. We call a closed formula of the form

$$\forall \mathbf{x}\ t = t'$$

an *identity* or *equation*. Here, we assume that the vector $\mathbf{x}$ contains all of the free variables of t and t'. An *implication* is a closed formula

$$\forall \mathbf{x}\,[f_1 = f_1' \wedge \ldots \wedge f_k = f_k' \Rightarrow g = g'],$$

where f_i, f_i', g, g', $i \in [k]$, are in μT_Σ. We will usually omit the universal quantifiers in an implication or equation.

Suppose that φ is a formula, A is a preiteration Σ-algebra and ρ is a function $X \to A$. Satisfaction of φ by (A, ρ), denoted $(A, \rho) \models \varphi$, is defined as usual. When φ is a closed formula, we define $A \models \varphi$ if $(A, \rho) \models \varphi$ for all (or some) $\rho : X \to A$. In this case we also say φ holds in A, or is satisfied by A. Moreover, we say that φ holds in a class K of preiteration Σ-algebras, denoted $K \models \varphi$, if φ holds in all algebras in K.

Definition 4. A preiteration Σ-algebra is an **iteration Σ-algebra** if it satisfies each identity that holds in all continuous Σ-algebras. A morphism of iteration Σ-algebras is a preiteration Σ-algebra homomorphism.

Iteration Σ-algebras form a variety [6] of preiteration Σ-algebras. For (equational) axiomatizations of iteration algebras we refer the reader to [13, 17], or the survey paper [10]. When $\Sigma \neq \Sigma_0$, any equational axiomatization must contain an infinite number of equation schemes. It is known that when Σ is finite, there is a polynomial time algorithm to decide whether a given equation holds in all iteration Σ-algebras.

3 The main result

For the balance of this section, let Σ denote a signature. The signature Δ was defined in the Introduction.

Definition 5. A **non-deterministic Σ-algebra** is a set A equipped with an operation

$$\sigma_A : A^n \rightarrow P(A),$$

for each $\sigma \in \Sigma_n$.

Any Σ-algebra may be identified with a non-deterministic Σ-algebra.

Definition 6. Suppose that A is a non-deterministic Σ-algebra. Then the **complex Δ-algebra** or **power set algebra** $P(A)$ has as its carrier the power set of A, and operations

$$\sigma_{P(A)} : P(A)^n \rightarrow P(A)$$

$$(A_1, \ldots, A_n) \mapsto \bigcup_{a_i \in A_i} \sigma_A(a_1, \ldots, a_n), \quad \sigma \in \Sigma_n, \ n \geq 0,$$

and

$$A_1 + A_2 = A_1 \cup A_2$$
$$0 = \emptyset,$$

for all $A_1, A_2 \in P(A)$.

Note that if A is a set and $T_\Sigma(A)$ denotes the absolutely free Σ-algebra on A, then $P_\Sigma(A)$ is just the complex algebra $P(T_\Sigma(A))$.

Any complex algebra $P(A)$ is partially ordered by set inclusion. Moreover, $P(A)$ is a complete lattice and the operations are completely additive, i.e., preserve arbitrary sups. Thus, $P(A)$ is a continuous Δ-algebra, hence an iteration Δ-algebra.

For the definition and basic properties of regular (or recognizable) tree languages we refer to [21]. Suppose that A is a set. If t is μ-term in μT_Δ and if $\rho : X \rightarrow P_\Sigma(A)$ such that $\rho(x)$ is regular, for each $x \in X$, then the language $t_{P_\Sigma(A)}(\rho)$ is also regular. It follows that the regular tree languages determine a sub preiteration Δ-algebra of $P_\Sigma(A)$ that we denote by $R_\Sigma(A)$. Clearly, $R_\Sigma(A)$ is also an iteration algebra.

Proposition 7. *The following three conditions are equivalent for an identity $t = t'$, for any $t, t' \in \mu T_\Delta$.*

1. *$t = t'$ holds in all algebras $P_\Sigma(A)$, where A is any set.*
2. *$t = t'$ holds in all algebras $R_\Sigma(A)$, where A is any set.*
3. *$t = t'$ holds in all algebras $P(A)$, where A is any (non-deterministic) Σ-algebra.*

Thus, the following three classes of preiteration Δ-algebras, in fact iteration Δ-algebras, see below, coincide: the variety generated by the algebras $P_\Sigma(A)$, where A is any set; the variety generated by the algebras $R_\Sigma(A)$, where A is any set; the variety generated by the algebras $P(A)$, where A is any (non-deterministic) Σ-algebra. Other models with the same property include complete semilattices equipped with a Σ-algebra structure such that each Σ-operation is completely additive.

Below, when t and t' are μ-terms in μT_Δ, we write $t \leq t'$ as an abbreviation for $t + t' = t'$. The *least fixed point rule* is the implication

$$t[y/x] = y \Rightarrow \mu x.t \leq y, \tag{4}$$

where t is any term in μT_Δ and x, y are variables.

Theorem 8. *An identity holds in all algebras $P_\Sigma(A)$, where A is any set, iff it holds in all preiteration Δ-algebras satisfying the least fixed point rule (4) as well as the identities (5)–(12):*

$$x + (y + z) = (x + y) + z \tag{5}$$

$$x + y = y + x \tag{6}$$

$$x + 0 = x \tag{7}$$

$$x + x = x \tag{8}$$

$$\sigma(x_1 + y_1, \ldots, x_n + y_n) = \sum_{z_i \in \{x_i, y_i\}} \sigma(z_1, \ldots, z_n) \tag{9}$$

$$\sigma(x_1, \ldots, 0, \ldots, x_n) = 0 \tag{10}$$

where $\sigma \in \Sigma_n$, $n > 0$,

$$\mu x.\mu y.t = \mu z.t[z/x, z/y] \tag{11}$$

$$t[\mu x.t/x] = \mu x.t, \tag{12}$$

where t is any μ-term in μT_Δ.[3]

The meaning of the equations (9) and (10) is that the Σ-operations distribute over finite sums. It follows from the axioms that the two constants $\mu x.x$, usually denoted $\perp$, and 0 are the same. Thus, we could spare one constant. Equation (11) is the *double iteration identity* and (12) is the *fixed point identity*. Note that these identities are equation schemes rather than single equations. Also, the least fixed point rule is an implication scheme.

By Theorem 8, the identities (5)–(12) as axioms, the rules of equational logic modified to μ-terms, and the least fixed point rule give a complete inference system for proving the valid identities of power set algebras, or (regular) tree languages.

---

[3] When writing μ-terms, we assume that the scope of a prefix μx extends to the right as far as possible.

Remark 9. The axioms appearing in Theorem 8 are not complete for the implicational theory of the power set algebras.

Remark 10. In the presence of the rest of the axioms, the least fixed point rule is equivalent to the *least pre-fixed point rule*

$$t[y/x] \leq y \Rightarrow \mu x.t \leq y, \tag{13}$$

for all terms t in μT_Δ and variables x, y. This rule is sometimes called the *fixed point induction*, or *Park induction rule*, cf. [31, 16]. It is known that the least pre-fixed point rule and the fixed point identity imply the double iteration identity (11), see, e.g., [16]. Thus, Theorem 8 remains valid if we remove the double iteration identity and replace the least fixed point rule (4) by the least pre-fixed point rule (13). This version of the result is particularly simple, since it says that the identities axiomatizing the equational theory of the finite tree languages together with the fixed point identity (12) and the least pre-fixed point rule (13) are complete.

Notation Let K denote the class of all iteration Δ-algebras satisfying the axioms of Theorem 8, i.e., the least fixed point rule and the identities (5)–(12).

Thus, Theorem 8 can be reformulated as follows: The variety of preiteration Δ-algebras generated by K is the same as that generated by the complex algebras $P(A)$, where A is any non-deterministic Σ-algebra. In the course of proving Theorem 8, we also establish the following result.

Theorem 11. *For each set Z, the Δ-iteration algebra $R_\Sigma(Z)$ is freely generated in K by the set Z. More exactly, let η denote the function $Z \to R_\Sigma(Z)$ defined by $z \mapsto \{z\}$, for all $z \in Z$. Let B be an algebra in K and h a mapping $Z \to B$. Then there is a unique iteration Δ-algebra homomorphism $h^\| : R_\Sigma(Z) \to B$ such that $h^\|(\eta(z)) = h(z)$, for all $z \in Z$.*

Theorem 8 follows from Theorem 11 immediately. Indeed, the equations (5)–(12) and the least fixed point rule (4) hold in any complex algebra $P(A)$. Moreover, if an equation holds in all complex algebras $P(A)$, then this equation holds in all algebras $R_\Sigma(A)$. Hence, by Theorem 11, it holds in all algebras in K.

The proof of Theorem 11 requires more argument. It can be found in the research report [19]. Here we only sketch the main ideas that are based on the formalization of a part of tree automata theory within equational logic. (See also [28] and [7] for related proofs.) The first observation is that $R_\Sigma(Z)$ is in K, for any set Z. Moreover, it is enough to prove the freeness property of $R_\Sigma(Z)$ for the set $Z = X$, the set of variables used to construct μ-terms. Second, by the main result in [16], each algebra in K is an iteration Δ-algebra, so that all the identities of iteration algebras hold in K. Then we assign an essentially unique normal form term in μT_Δ to any non-deterministic tree automaton which denotes the regular set in $R_\Sigma(X)$ accepted by the tree automaton, and show that any μ-term is provably equal to a term in normal form. This part of the argument is rather technical but quite standard, see also [33, 26], or [8] for a very general normal form theorem of this sort. Next, using standard constructions from tree automata theory, we show that two non-deterministic tree automata are equivalent if and only if they can be connected by a finite chain of functional and "dual" functional simulations (morphisms). The concept of simulation originates in [13], see also [8, 7]. Then, we prove that if two non-deterministic tree automata are connected by a simulation or a dual simulation, then the corresponding normal form terms are provably equal. For this, we employ the commutative identity [13, 8] that holds in all iteration algebras, and the "dual" functorial implication that we establish in K. This ends the proof, for it follows now that for any two μ-terms t and t', the equation $t = t'$ holds in K iff t and t' denote equal regular sets in $R_\Sigma(X)$.

4 Remarks

Each class V of continuous Σ-algebras generates both a Birkhoff-variety V_0 of Σ-algebras, and a variety $\widehat{V}$ of iteration Σ-algebras. In general, $\widehat{V}$ has no finite axiomatization over V_0 (in terms of equation schemes between μ-terms, say). This holds for instance for the class V of all continuous Σ-algebras, whenever Σ contains a non-nullary symbol.

But sometimes the equational theory of $\widehat{V}$ does have a simple axiomatization consisting of equational axioms for V_0 and the least pre-fixed point rule. This has been shown recently for the class V of all continuous Σ-algebras [16], and this is shown here for the class V of all complex Σ-algebras. It seems to be very hard to achieve a characterization of all varieties with this property. (See also [25].)

The proof of Theorem 8 gives rise to several additional completeness results and the conjecture that the iteration Δ-algebra identities and the equations (5)–(7), (9), (10), as well as the equation

$$\mu x . x + y = y$$

give a complete equational axiomatization of the valid identities of the algebras $P(A)$, where A is any non-deterministic Σ-algebra. Since the class of iteration algebras can be axiomatized by the Conway identities and the group-identities associated with the finite (simple) groups [17], this conjecture corresponds to Krob's result [29].

The results of the paper may be expressed in several different formalisms. These include the "where expressions" of [25] and the "regular expressions" of [21] that use $\cdot_x$-products and $*_x$-iterations, where x is any variable. In fact, the axiomatization of Ito and Ando [26] is formalized using this latter syntax. In the language of categorical algebra, iteration theories [8] provide yet another framework. The translation between these formalisms is routine.

5 Acknowledgment

The results of this paper were obtained during the author's visit at the Department of Computer Science of Stevens Institute of Technology, during the summer of 1997. The author would like to thank Stephen L. Bloom and the members of the department for their hospitality.

References

1. J. Almeida, On pseudovarieties, varieties of languages, filters of congruences, pseudoidentities and related topics, *Algebra Universalis*, 27(1990), 333–350.
2. J. Almeida, *Finite Semigroups and Universal Algebra*, Word Scientific, 1994.
3. K.V. Archangelsky and P.V. Gorshkov, Implicational axioms for the algebra of regular events, *Dokl. Akad. Nauk USSR Ser A, 10(1987), 67–69* (in Russian).
4. H. Bekič, Definable operations in general algebras, and the theory of automata and flowcharts, *Technical Report*, IBM Laboratory, Vienna, 1969.
5. L. Bernátsky, S.L. Bloom, Z. Ésik, Gh. Stefanescu, Equational theories of relations and regular sets, in: *Words, Languages, and Combinatorics, Kyoto, 1992*, Word Scientific, 1994, 40–48.
6. S.L. Bloom and Z. Ésik, Iteration algebras, *Int. J. Foundations of Computer Science*, 3(1992), 245-302.
7. S.L. Bloom and Z. Ésik, Equational axioms for regular sets, *Mathematical Structures in Computer Science*, 3(1993), 1–24.
8. S.L. Bloom and Z. Ésik, *Iteration Theories: The Equational Logic of Iterative Processes*, EATCS Monographs on Theoretical Computer Science, Springer–Verlag, 1993.
9. S.L. Bloom and Z. Ésik, Solving polynomial fixed point equations, in: *Mathematical Foundations of Computer Science '94*. LNCS 841, Springer–Verlag, 1994, 52-67.
10. S.L. Bloom and Z. Ésik, The logic of fixed points, *Theoretical Computer Science*, 179(1997), 1–60.
11. M. Boffa, Une remarque sur les systemes complets d'identites rationelles, *Theoret. Inform. Appl.*, 24(1990), 419–423.
12. J.C. Conway, *Regular Algebra and Finite Machines*, Chapman and Hall, London, 1971.
13. Z. Ésik, Identities in iterative and rational algebraic theories, *Computational Linguistics and Computer Languages*, 14(1980), 183–207.
14. Z. Ésik, An axiomatization of regular forests in the language of algebraic theories with iteration, in: *Fundamentals of Computation Theory, Szeged, 1981*, LNCS 117, Springer-Verlag, 130-136.
15. Z. Ésik, Algebras of iteration theories, *J. of Computer and System Sciences*, 27(1983), 291–303.
16. Z. Ésik, Completeness of Park induction, *Theoretical Computer Science*, 177(1997), 217–283.
17. Z. Ésik, Group axioms for iteration, to appear.
18. Z. Ésik, A variety theorem for trees and theories, *LITP Report Series*, 97/15, May 1997.
19. Z. Ésik, Axiomatizing the equational theory of regular tree languages, *Stevens Institute of Technology Research Reports*, No. 9703, September 1997.
20. Z. Ésik and A. Labella, Equational properties of iteration in algebraically complete categories, Theoretical Computer Science, to appear. Extended abstract in: *Proc. MFCS '96*, LNCS 1113, Springer–Verlag, 1996, 336–247.
21. F. Gécseg and M. Steinby, *Tree Automata*, Akadémiai Kiadó, Budapest, 1984.

22. F. Gécseg and M. Steinby, Tree languages, in: *Handbook of Formal Languages*, vol. 3, eds.: G. Rozenberg and A. Salomaa, Springer–Verlag, 1997, 1–68.
23. J. Goguen, J. Thatcher, E. Wagner, and J. Wright, Initial algebra semantics and continuous algebras, *J. ACM*, 24(1977), 68–95.
24. I. Guessarian, *Algebraic Semantics*, LNCS 99, Springer–Verlag, 1981.
25. A.J.C. Hurkens, M. McArthur, Y.N. Moschovakis, L.S. Moss and G. Whitney, The logic of recursive equations, *J. of Symbolic Logic*, to appear.
26. T. Ito and S. Ando, A complete axiom system of super-regular expressions, in: *Information Processing* 74, Amagasaki, Japan, North-Holland, 1974, 661–665.
27. D. Kozen, Results on the propositional μ-calculus, *Theoretical Computer Science*, 27(1983), 333–354.
28. D. Kozen, A completeness theorem for Kleene algebras and the algebra of regular events, *Information and Computation*, 110(1994), 366–390.
29. D. Krob, Complete systems of B-rational identities, *Theoretical Computer Science*, 89(1991), 207–343.
30. M. Magidor and G. Moran, Finite automata over infinite trees, *Technical report No. 30, Hebrew University*, 1969.
31. D.M.R. Park, Fixpoint induction and proofs of program properties, in: *Machine Intelligence* 5, D. Michie and B. Meltzer, Eds., Edinburgh Univ. Press, 1970, 59–78.
32. V.N. Redko, On the determining totality of relations of an algebra of regular events, *Ukrain. Mat. Z.*, 16(1964), 120–126 (in Russian).
33. A. Salomaa, Two complete axioms systems for the algebra of regular events, *J. of the ACM*, 13(1966), 158–169.
34. M. Steinby, Syntactic algebras and varieties of recognizable sets, in: *Proc. Coll. Lille*, 1979, 226–240.
35. M. Steinby, A theory of tree language varieties, in: *Tree Automata and Languages*, North-Holland, Amsterdam, 1992, 57–81.
36. J.W. Thatcher and J.B. Wright, Generalized finite automata theory with an application to a decision problem of second order logic, *Mathematical System Theory*, 2(1968), 57–81.
37. W. Thomas, Logical aspects in the study of tree languages, *Ninth Colloq. Trees in Algebra and Programming*, Cambridge Univ. Press, 1984, 31–49.
38. W. Thomas, Automata on infinite objects, in: *Handbook of Theoretical Computer Science*, vol. B, North-Holland, 1992, 133–191.
39. Th. Wilke, An algebraic characterization of frontier testable languages, *Theoretical Computer Science*, 154(1996), 85–106.

A Logical Characterization of Systolic Languages

Angelo Monti[1] and Adriano Peron[2]

[1] Dipartimento Scienze dell'Informazione, Università di Roma (La Sapienza), 00198,
Via Salaria 113, Italy, E-mail: monti@dsi.uniroma1.it

[2] Dipartimento di Matematica e Informatica, Università di Udine, 33100,
Via Delle Scienze 206, Italy, E-mail: peron@dimi.uniud.it

Abstract. In this paper we study, in the framework of mathematical logic, $\mathcal{L}(SBTA)$ i.e. the class of languages accepted by Systolic Binary Tree Automata. We set a correspondence (in the style of Büchi Theorem for regular languages) between $\mathcal{L}(SBTA)$ and $MSO[Sig]$, i.e. a decidable Monadic Second Order logic over a suitable infinite signature Sig. We also introduce a natural subclass of $\mathcal{L}(SBTA)$ which still properly contains the class of regular languages and which is proved to be characterized by Monadic Second Order logic over a finite signature $Sig' \subset Sig$. Finally, in the style of McNaughton Theorem for star free regular languages, we introduce an expression language which precisely denotes the class of languages defined by the first order fragment of $MSO[Sig']$.
(Keywords: Automata and formal languages, Logic in computer science)

1 Introduction

The connection between finite state automata and logic was established in the sixties by Büchi, Elgot and Trakhtenbrot (see [1,3,13]). They showed that finite automata and monadic second-order logic (interpreted over finite words) have the same expressive power, and that the transformations from formulas to automata and vice-versa are effective. Later, such an equivalence was established also between finite state automata and monadic second order logic over infinite words and trees. (For a complete survey on the relation between finite state automata and monadic second order logic we refer to [11,12].) The results have been fundamental for a number of areas of computer science: the decidability of those theories has been the basic engine for proving the decidability of other interesting mathematical theories and logics of programs; the correspondence between finite state automata and monadic second order logic has been refined and extended to allow practical use (e.g. the verification of finite state programs by model checking); the classification theory of formal languages was deepened by including logical notions and techniques.

This paper sets a correspondence between classes of languages accepted by systolic binary tree automata ($SBTA$ for short) and Monadic Second Order logic over suitable signatures (for a signature Sig, we write $MSO[Sig]$). Systolic automata (see [5] for a survey) are synchronous networks of (memoryless) parallel processors working in discrete time. These automata, which extend the expres-

Regular Lang.	$\subset$	$\mathcal{L}(SBTA)$	$= MSO[<, pow, (\approx_{p,q}, \equiv_{p,q})_{p,q \geq 0}]$

Regular Lang. $\subset$ $\mathcal{L}(SBTA)$ $= MSO[<, pow, (\approx_{p,q}, \equiv_{p,q})_{p,q \geq 0}]$

$\cup$ $\cup$

Regular Lang. $\subset$ $\mathcal{L}(SBTA - \#)$ $=$ $MSO[<, pow]$

$\cup$ $\cup$ $\cup$

star free Regular Lang. $\subset \mathcal{L}(star\ free\ SBTA - \#) =$ $FO[<, pow]$

Fig. 1. The hierarchy of systolic languages and their logic characterizations.

sive power of finite state automata though preserving their closure and decidability properties, have been the main theoretical models used to study various basic problems concerning systolic architectures, systems and computations (e.g. the power of various interconnections, the power of different communication modes with the environment, etc.).

Finitary systolic languages have been widely studied from an operational viewpoint (i.e. as languages recognized by automata) but they have not been studied yet from a declarative viewpoint (i.e. as languages defined by logical formulas speaking about properties of words). Only a logical characterization of languages of infinite words accepted by systolic automata (introduced in [8]) has been recently proposed (see [9]), which, however, cannot be trivially rephrased in the finitary case. In this paper we provide the declarative viewpoint of the class of (finitary) languages recognized by $SBTA$ ($\mathcal{L}(SBTA)$, for short). We prove that there is a correspondence - in the style of Büchi Theorem for regular languages- between $\mathcal{L}(SBTA)$ and $MSO[Sig]$, i.e. the Monadic Second Order logic over a suitable infinite signature Sig containing two relational symbols $<$ and pow and a pair of families of unary predicates $(\approx_{p,q}, \equiv_{p,q})_{p,q \geq 0}$. The theory $MSO[<, pow, (\approx_{p,q}, \equiv_{p,q})_{p,q \geq 0}]$ is a proper decidable extension of the theory $MSO[<]$ which characterizes regular languages. We show also that the finite fragment of the signature Sig consisting of the two relational symbols pow and $<$ is sufficient to characterize a wide and relevant subclass of $\mathcal{L}(SBTA)$ (it still properly contains the class of regular languages). From the operational viewpoint, the class of languages corresponding to $MSO[<, pow]$ is obtained by imposing a natural constraint on the definition of the transition function of $SBTA$ (that subclass of $SBTA$ is denoted by $SBTA - \#$). Finally we state, for languages recognized by $SBTA - \#$, a result which is the analogous of McNaughton theorem for regular languages (see [7]). McNaughton theorem establishes a correspondence between star free regular languages and the first order fragment of $MSO[<]$. Similarly, we define an expression language which denotes the first order fragment of $MSO[<, pow]$.

We believe that investigating logical characterizations of systolic languages is meaningful for both theroretical and practical reasons. The first advantage of the proposed logical characterization of systolic languages is that it allows to introduce a new classification of systolic languages. The hierarchy of systolic languages and the corresponding hierarchy of monadic second order characterizations are summarized in Fig.1. From the practical viewpoint, since all of the proposed characterizations allow to effectively construct automata from logical formulas, we expect that our investigation could contribute in an original way to the research oriented to developing systematic and sufficiently automatizable methods to synthesize systolic systems from high level specifications. Due to the restriction on the length of the paper, we are forced to skip theorem proofs.

2 Preliminaries

Throughout this paper Σ denotes an alphabet, Σ^* (resp.: Σ^+) denotes the set of words (resp.: the set of words devoid of the empty word ϵ) on Σ. Words are indicated by $u, v, w, \ldots$, $|u|$ denotes the length of the word u and $u(i)$, with $0 \le i \le |u| - 1$ denotes the i-th element of u. The symbol $\cdot$ denotes concatenation on words of Σ^*. In the subsection 2.1, we recall the standard notions of logical definability of languages and in the subsection 2.2 we recall the notions about operational definability (i.e. automata acceptance) of systolic languages.

2.1 Logical definability

Since logical formulas will be interpreted over words, it is more convenient to identify a word $w \in \Sigma^+$ with the structure $\underline{w} = \langle dom(w), <, (Q_a)_{a \in \Sigma} \rangle$, where

- $dom(w) = \{0, \ldots, |w| - 1\}$;
- $<$ is the restriction to $dom(w)$ of the usual ordering of natural numbers;
- $Q_a \subseteq dom(w) = \{i : w(i) = a\}$, for $a \in \Sigma$.

Properties of words will be described by first order and monadic second order formulas, using individual variables $x, y, z, \ldots$ for elements of $dom(w)$ (i.e., positions of the word w) and set variables $X, Y, Z, \ldots$, for subsets of $dom(w)$.

In the sequel we shall consider a second order language over a signature Sig consisting of a family of unary relation symbols $(U_s)_{s \in S}$ and a family of binary relation symbols $(B_t)_{t \in T}$, for some S and T.

Atomic formulas are of the form $x = y$, $U_s(x)$, xB_ty, $X(x)$ and $Q_a(x)$ interpreted as equality between x and y, $x \in U_s$, $\langle x, y \rangle \in B_t$, $x \in X$ and $x \in Q_a$, respectively.

Formulas are built up from atomic formulas by means of the boolean connectives $\neg$, $\wedge$, $\vee$, $\rightarrow$ and the quantifiers $\forall$ and $\exists$ ranging over both individual and set variables.

A formula without free variables is called a *sentence*. If $\phi(X_1, \ldots, X_n)$ is a formula with at most $X_1, \ldots, X_n$ occurring free in ϕ, w is a word and $P_1, \ldots, P_n$ are subsets of $dom(w)$, we write $\langle \underline{w}, P_1, \ldots, P_n \rangle \models \phi(X_1, \ldots, X_n)$ if w satisfies ϕ under the above mentioned interpretation, where P_i is taken as interpretation of X_i. If ϕ is a sentence we write $\underline{w} \models \phi$. The language $L(\phi)$ defined by a sentence ϕ is the set of all words $w \in \Sigma^+$ such that $\underline{w} \models \phi$.

Given the signature Sig, a language L is *Monadic Second Order definable* w.r. to Sig (written $L \in MSO[Sig]$) if there is a monadic second order sentence ϕ with $L = L(\phi)$; L is *Existential Monadic Second Order definable* w.r. to Sig

(written $L \in EMSO[Sig]$) if there is a sentence $\phi = \exists X_1 \ldots \exists X_n \psi(X_1, \ldots, X_n)$ such that $L = L(\phi)$ with ψ containing only first order quantifiers (i.e. ψ ranges only over individual variables); L is *First Order definable* w.r. to *Sig* (written $L \in FO[Sig]$) if there is a sentence ϕ containing only first order quantifiers such that $L = L(\phi)$.

2.2 Recognizability by Systolic Binary Tree Automata

Systolic languages are sets of words accepted by systolic automata (see [5]). A systolic binary tree automaton (see [2]) consists of an infinite number of nodes which can be interpreted as memoryless processors. Nodes are linked among them and the resulting structure is an (infinite) leafless perfectly balanced binary tree. In order to process a word w, the first level m of the tree is chosen which has at least $|w|$ nodes. Now, the automaton is fed in such a way that adjacent processors at level m are fed with adjacent symbols of w, and that the leftmost processor is fed with the first symbol of w. If the number of processors is greater than the word length, then exceeding processors (i.e. each i-th processor, for $|w| \le i < 2^m$) are fed with a special symbol #. In Fig. 2 an example is given for an input word of length five. Now, all the processors at level m synchronously output, according to the *input relation*, a symbol belonging to the *state alphabet* Q. Each processor at level $m-1$ receives the couple of states output by its pair of sons and it synchronously (with respect to processors at the same level) outputs a symbol belonging to Q according to the *transition relation*. Therefore, information flows bottom-up, in parallel and synchronously, level by level. The word is accepted whenever the root of the tree outputs a symbol belonging to the given set of *final states*.

Definition 1. A *systolic automaton* is a tuple $\mathcal{A} = \langle \Sigma, Q, in, f, F \rangle$, where

- $Q \supseteq \{\#\}$ is the finite set of *states*;
- $in \subseteq (\Sigma \cup \{\#\}) \times Q$ is the *input relation* such that $\langle x, \# \rangle \in in$ iff $x = \#$;
- $f \subseteq Q \times Q \times Q$ is the *transition relation* s.t. $\langle p, q, \# \rangle \in f$ implies $p = q = \#$;
- $F \subseteq Q - \{\#\}$ is the set of *final states*.

The relation $O_{\mathcal{A}} \subseteq (\Sigma \cup \{\#\})^+ \times Q$ is recursively defined as follows:

- if $|w| = 1$, then $\langle w, q \rangle \in O_{\mathcal{A}}$ iff $\langle w, q \rangle \in in$;
- if $2^{m-1} < |w| \le 2^m$, with $m > 0$, then $\langle w, q \rangle \in O_{\mathcal{A}}$ iff $\langle q_1, q_2, q \rangle \in f$ where q_1, q_2 are such that $\langle w_1, q_1 \rangle, \langle w_2, q_2 \rangle \in O_{\mathcal{A}}$ with $|w_1| = |w_2| = 2^{m-1}$ and $w_1 \cdot w_2 = w \cdot \#^{2^m - |w|}$.

The language recognized by $\mathcal{A}$ is the set $\mathcal{L}(\mathcal{A}) = \{w \in \Sigma^+ : \langle w, q \rangle \in O_{\mathcal{A}}, q \in F\}$. The class of languages recognized by *SBTA* is denoted by $\mathcal{L}(SBTA)$.

Example 2. The automaton $\mathcal{A}$ with $\Sigma = \{a\}$, $Q = \{a, \#\}$, *in* the identity function, $f = \{\langle a, a, a \rangle\}$ and $F = \{a\}$, recognizes the (non-regular) language of powers of two, i.e., $\mathcal{L}(\mathcal{A}) = \{a^{2^n} : n \ge 0\}$.

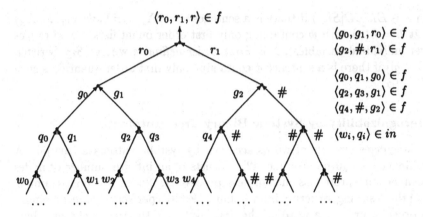

Fig. 2. A computation of a $SBTA$ fed with $w_0 \cdot w_1 \cdot w_2 \cdot w_3 \cdot w_4$

Properties (see [5]): The class $\mathcal{L}(SBTA)$ is effectively closed under union, intersection, complementation and projection. Moreover, this class properly contains the class of regular languages and the emptiness problem for $L \in \mathcal{L}(SBTA)$ is decidable.

3 The logic characterization of $\mathcal{L}(SBTA)$

In this section we define a signature Sig such that the class of $MSO[Sig]$-definable languages is precisely $\mathcal{L}(SBTA)$. The logic characterization of regular languages is given by Büchi Theorem which provides also (in its proof) an effective way for converting finite state automata to MSO-formulas and vice-versa.

Büchi Theorem([1,3]) A language $L \subseteq \Sigma^*$ is $MSO[<]$-definable iff L is regular.

Since the class of regular languages is strictly contained in $\mathcal{L}(SBTA)$, it is natural to consider a proper extension of the signature containing only the relational symbol $<$. The considered extension contains a binary relation symbol pow (actually a function on positive natural numbers) and two (infinite) families of unary predicate symbols $(\approx_{p,q})_{p,q \geq 0}$ and $(\equiv_{p,q})_{p,q \geq 0}$. In the second part of this section we show that the finite signature containing only the two binary relation symbols $<$ and pow is powerful enough to describe a meaningful subclass of $\mathcal{L}(SBTA)$. We start with introducing a function $rm : \mathbb{N}^+ \to \mathbb{N}$, which, applied to a natural positive number, gives the position of the rightmost 1 in its binary representation, i.e.

if $x = 2^{k_n} + 2^{k_{n-1}} + \ldots + 2^{k_0}$ with $k_n > \ldots > k_1 > k_0$, then $rm(x) = k_0$.

The function pow, when applied to a positive number, removes the rightmost 1 in its binary representation.

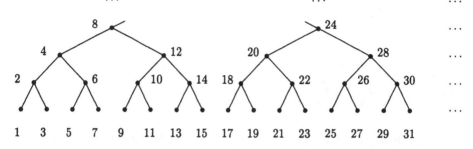

Fig. 3. The upward unbounded binary tree structure.

Definition 3. The *power function pow* : $\mathbb{N}^+ \to \mathbb{N}$ is such that

$$pow(x) = x - 2^{rm(x)}, \text{ for } x > 0.$$

The power function allows to impose on the flat domain of natural numbers a structure which is useful for simulating a systolic binary tree computation, namely the (upward) *unbounded tree structure* depicted in Fig. 3.
For a number z, with $rm(z) > 0$, the left and the right son of z are the numbers

$$z - 2^{rm(z)-1} \text{ and } z + 2^{rm(z)-1}, \text{ respectively.}$$

In particular, it is easy to see that y is the right (resp.: left) son of z iff

$$y = max\{k : pow(k) = z\}(\text{resp.}: y = max\{k : k < z, pow(k) = pow(z)\}). \quad (1)$$

We can define also a notion of *bounded tree structure* over a set $\{0,\ldots,n\}$ by exploiting the same definitions of right and left son of a number z given in Eq.1 (variables are instantiated now over the finite domain $\{0,\ldots,n\}$). As an example, consider the bounded tree structure over $\{0,\ldots,4\}$ depicted in Fig.4.b.
The unbounded structure could be easily exploited to simulate a systolic computation which is, as it can be seen in Fig. 2, a binary tree labelled over the set of states Q. The idea of the simulation for the word $w_0 \cdot w_1 \cdot w_2 \cdot w_3 \cdot w_4$ of Fig. 2 is suggested in Fig. 4.a, where state names have the same meaning as in Fig. 2. The leaves of the structure (i.e. odd numbers) record the first step of the computation on the input word and the other layers of the structure record the state output in the other bottom-up computation steps. A power of two (4 in this case) collects the result of the complete computation.
Unfortunately, the simulation of a systolic computation over a (finite) word w cannot be performed by using the unbounded structure but has to be performed on the bounded structure over the finite domain $dom(w)$. Since, in general, the length of w is not a power of two, the resulting tree structure is not guaranteed to be a perfectly balanced binary tree. For instance, consider the binary tree structure for the domain $\{0,\ldots,4\} = dom(w_0 \cdot w_1 \cdot w_2 \cdot w_3 \cdot w_4)$ depicted in Fig. 4.b. In that structure the nodes 5, 6 and 7 which are used in Fig. 4.a to

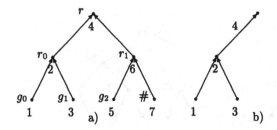

Fig. 4. a) Simulation of a systolic computation. b) The bounded tree structure over $\{0, \ldots, 4\}$.

record the states g_2, r_1 and #, respectively, are missing. In general, for a word w whose length is not a power of two, the bounded tree structure built on $dom(w)$ lacks exactly all of the nodes for recording the output of left-right input pairs of the form $\langle q, \# \rangle$ with $q \in Q$. (As a matter of fact, only the output of left-right input pairs $\langle q, \# \rangle$ with $q \neq \#$ carries a relevant information since the output for $\langle \#, \# \rangle$ is forced to be #.) For solving that problem we need some auxiliary notions. For a number j, $height(j)$ is the number i such that $j = 2^i + m2^{i+1}$ for some $m \geq 0$ (i.e. j belongs to the i-th layer of the unbounded tree structure). Let us consider now the bounded tree structure over a domain $\{0, \ldots, n\}$. For $x, y \in \{0, \ldots, n\}$ with y the right son of x, we say that the *depth* of x is the number $height(x) - height(y)$. Otherwise, if $x \in \{0, \ldots, n\}$ has not a right son (i.e. either x is odd or $x = n$), the *depth* of x is $height(x) + 1$. For instance, the depth of each number in the binary unbounded structure is 1 and the depth of 4 in the structure over the domain $\{0, \ldots, 4\}$ is 3 (see Fig. 4.b).

Let us came back to the simulation of a systolic computation over a bounded tree structure. Assume that the left son and the right son of a node z are labeled by states q_1 and q_2, respectively. If the depth of z is 1, then the structure is in that point complete and z can be labeled by a state q such that $\langle q_1, q_2, q \rangle \in f$. On the other hand, if the depth is $s > 1$, then some of the nodes in the right subtree of z are missing. In that case z can be labelled by a symbol q such that $\langle q_1, \overline{q_2}, q \rangle \in f$ and $\overline{q_2}$ is obtained by combining q_2 with # (according to f) $s - 1$ times, namely $\overline{q_2} \in S_{q_2}(s - 1)$ with $S_p(i) \subset Q$ (with $i \geq 0$) the sequence of sets recursively defined as follows:

$$S_p(0) = \{p\} \text{ and } S_p(i) = \{q : \langle q', \#, q \rangle \in f, q' \in S_p(i - 1)\}.$$

If z has not a right son, z can be labelled by a symbol q computed as in the previous case by taking q_2 in such a way that $\langle w_n, q_2 \rangle \in in$, with w_n the last symbol of the considered input word.

The finiteness of Q implies that the sequence $S_p(0), S_p(1), \ldots$ is ultimately periodic. As a consequence, in order to label correctly a node of the binary structure having right son labelled by state q, it is necessary to know its exact depth only if it is less than or equal to the preperiod of S_q. If the depth is greater than the preperiod, it is enough to know only its relative position within the period. The relevant consequence is that only a finite number of different depth values have to be considered in the simulation of the computation of a systolic automaton. The required numbering of the depth of a node "modulo" a preperiod p and a period q is accomplished by two families of unary predicates $(\approx_{p,q})_{p,q\geq0}$ and $(\equiv_{p,q})_{p,q\geq0}$ defined as follows.

Definition 4. The predicate $\approx_{p,q}$ holds true on $n \in \mathbb{N}$ whenever the rightmost occurrence of a 1 in the binary representation of n precedes a sequence of zeros of length $mq + p$, i.e., if $rm(n) = mq + p$ for some $m \geq 0$.
The predicate $\equiv_{p,q}$ holds true on $n \in \mathbb{N}$ whenever the two rightmost occurrences of a 1 bound a sequence of 0 of length $mq + p$, i.e.,
if $rm(pow(n)) - rm(n) - 1 = mq + p$ for some $m \geq 0$.

For instance, $\approx_{0,0} (n)$ holds true iff n is odd.

We consider now the extension of MSO[<] by the *pow* function and the two families of predicates $(\approx_{p,q})_{p,q\geq0}$ and $(\equiv_{p,q})_{p,q\geq0}$.

Example 5. In this example and in the following we shall use the predicate *Power*(x) (i.e. x is a power of two) as a short-hand for the formula $\forall y(pow(x) \leq y)$ (note that if x is a power of two, then $pow(x) = 0$).
A formula of $MSO[<, pow, (\approx_{p,q}, \equiv_{p,q})_{p,q\geq0}]$ defining the language of words having length equal to a power of two, i.e. the set $\{w : |w| = 2^n, n \geq 0\}$ is

$$\exists X, x(\approx_{0,0} (x) \wedge Last(x) \wedge X(x)\wedge$$
$$\forall y((X(y) \wedge \neg Power(y)) \rightarrow \equiv_{0,0} (y) \wedge \exists z(pow(y) = z \wedge X(z)))).$$

where $Last(x)$ (i.e. x is the greatest number) is a short-hand for $\forall y(y \leq x)$.
A first order formula defining the language $\{w : |w| = 3(2^n) + 1, n \geq 0\}$ is

$$\exists x, y(Last(x) \wedge pow(x) = y \wedge \equiv_{0,0} (x) \wedge Power(y)).$$

Theorem 6. *A language $L \in \mathcal{L}(SBTA)$ iff L is $MSO[<, pow, (\approx_{p,q}, \equiv_{p,q})_{p,q\geq0}]$ definable.*

The theorem above is proved by exploiting the same technique used for Büchi Theorem in [11]. In particular, the computation of a systolic automaton can be described in $MSO[<, pow, (\approx_{p,q}, \equiv_{p,q})_{p,q\geq0}]$ by using the algorithm for labelling a bounded binary tree structure (definable as in Eq.1) sketched in this section. The converse implication is proved by showing that the languages induced by *pow*, $\approx_{p,q}$ and $\equiv_{p,q}$ (for any p,q) belong to $\mathcal{L}(SBTA)$ and proceeding then by induction on the structure.
The characterization of $\mathcal{L}(SBTA)$ given in Th.6 cannot be obtained by rephrasing the characterization of systolic languages on infinite words given in [9]. In fact, that class of languages can be defined by a Monadic Second Order Logic over a finite signature (consisting of the binary relation symbols < and *pow*, only) whereas the finiteness of word models forces in our case the introduction of an infinite signature. As in the case of Büchi Theorem, the translation of a systolic automaton into a $MSO[<, pow, (\approx_{p,q}, \equiv_{p,q})_{p,q\geq0}]$-formula

(and vice-versa) is effective. Since the emptiness problem for $\mathcal{L}(SBTA)$ is decidable, also the theory $MSO[<, pow, (\approx_{p,q}, \equiv_{p,q})_{p,q\geq 0}]$ is decidable. (Note that $MSO[<, pow, (\approx_{p,q}, \equiv_{p,q})_{p,q\geq 0}]$ is a proper extension of the well known decidable theory $MSO[<]$).

Corollary 7. *The theory* $MSO[<, pow, (\approx_{p,q}, \equiv_{p,q})_{p,q\geq 0}]$ *is decidable.*

Moreover, the logical description of acceptance of a systolic binary tree automaton can be given in terms of an $EMSO[<, pow, (\approx_{p,q}, \equiv_{p,q})_{p,q\geq 0}]$-formula. So, in the line of the classical result stating that $MSO[<]$ and $EMSO[<]$ have the same expressive power, we have the following corollary:

Corollary 8. *Any* $MSO[<, pow, (\approx_{p,q}, \equiv_{p,q})_{p,q\geq 0}]$*-definable language is* $EMSO[<, pow, (\approx_{p,q}, \equiv_{p,q})_{p,q\geq 0}]$*-definable.*

The characterization of $SBTA$ languages of Theorem 6 is a monadic second order logic over an infinite signature. We shall show in the remaining part of this section that a finite fragment of that signature is sufficient to characterize a relevant subclass of $\mathcal{L}(SBTA)$. As previously seen, the two families of unary predicates $(\approx_{p,q})_{p,q\geq 0}$ and $(\equiv_{p,q})_{p,q\geq 0}$ have been introduced to deal with the problem of handling, in the simulation of a systolic computation, the output of nodes having input pairs of the form $\langle q, \# \rangle$, for $q \in Q$. These nodes do not carry relevant information in the subclass of $SBTA$ where states are not affected by composition with $\#$, namely in the subclass of automata whose transition function f fulfills the restriction that $\langle q, \#, q' \rangle \in f$ implies $q = q'$.
(This restriction is not new in the literature, see for instance [6].) We denote this subclass of $SBTA$ as $SBTA - \#$ and by $\mathcal{L}(SBTA - \#)$ the corresponding class of recognized languages. The class of languages $\mathcal{L}(SBTA - \#)$ is strictly contained in the class $\mathcal{L}(SBTA)$. For instance, the systolic language of powers of two defined in Ex.2, does not belong to $\mathcal{L}(SBTA - \#)$. However, it is interesting to see that the class $\mathcal{L}(SBTA - \#)$ enjoy the same properties as $\mathcal{L}(SBTA)$.

Proposition 9. *The class* $\mathcal{L}(SBTA - \#)$ *is effectively closed under union, intersection, complementation and projection. This class strictly includes the class of regular languages and the emptiness problem is decidable.*

The finite signature containing $<$ and pow allows to logically characterize the class of systolic languages $\mathcal{L}(SBTA - \#)$.

Theorem 10. *A language* $L \in \mathcal{L}(SBTA - \#)$ *iff* L *is* $MSO[<, pow]$*-definable.*

4 First Order definable $\mathcal{L}(SBTA - \#)$

The classical result of McNaughton (see [7]) states that the class of languages denoted by star free expressions is the subclass of regular languages which is definable in the first order fragment of $MSO[<]$.

McNaughton Theorem ([7]): A language L is denoted by a star free expression iff it is $FO[<]$-definable.

In this section we define an expression language which denotes the subclass of systolic languages definable in the first order fragment of $MSO[<, pow]$. By analogy to the case of regular languages, we call that class of languages *star free* $\mathcal{L}(SBTA - \#)$. (In [4] it can be found a Kleene-like characterization of the entire class $\mathcal{L}(SBTA)$ based on an signature which is infinite and uncomparable with respect to that we are introducing here.) The class of star free $\mathcal{L}(SBTA - \#)$ contains the class of star free regular languages since $FO[<]$ is a fragment of $FO[<, pow]$. Moreover, the containment is strict. For instance, it is well known (see [12]) that the regular language of words having even length is not $FO[<]$-definable (i.e. it is not regular star-free). That language is $FO[<, pow]$-definable. The formula defining it is

$$\exists x \forall z (z \leq x \wedge (z < x \rightarrow z \leq pow(x))).$$

The class of $FO[<, pow]$-definable languages contains also non-regular languages. For instance, the set $\{a^{2^n} \cdot b^+ : n \geq 0\}$ is non regular and it is defined by

$$\exists x (Power(x) \wedge (\forall z(z \geq x \rightarrow Q_b(z)) \wedge z < x \rightarrow Q_a(z))).$$

Systolic star free expressions are built up from two symbols $\diamond$ and $\odot$ denoting two operations of restricted concatenation and from a symbol $\triangleright$ denoting an operation of prefix replacement. The corresponding operations on languages are defined as follows.

Definition 11. Let $U, V \subseteq \Sigma^+$ be languages:
$U \diamond V$ denotes the set $\{u \cdot v : u \in U, v \in V, |v| \leq 2^{rm(|u|)}\}$;
$U \odot V$ denotes $\{u \cdot v : u \in U, v \in V, |v| = 2^m, \text{ with } 0 \leq m < rm(|u| - 1)\}$;
$U \triangleright V$ denotes the set $\{u \cdot v : u \in U, u' \cdot v \in V, |u| = |u'|\}$.

Note that the operation of restricted concatenation $\odot$ acts on the length of its operands as the *pow* function acts on natural numbers. More precisely, if $u \in U$ with $|u| = n$, $v \in V$ and $u \cdot v \in U \odot V$ with $|u \cdot v| = m$, then $pow(m - 1) = n - 1$.

Definition 12. *Systolic star free expressions* over the alphabet Σ are built up from language symbols a (with $a \in \Sigma$) and the language symbol $\Sigma^\wedge$ by means of operations $\cup$, $\cap$, $\sim$, $\diamond$, $\odot$ and $\triangleright$. The expression Σ^* is also admitted as abbreviation of $\Sigma^\wedge \cup \sim \Sigma^\wedge$.
Every systolic star free expression denotes a language over Σ: a denotes the language $\{a\}$, $\Sigma^\wedge$ denotes the language $\{u : u \in \Sigma^+, |u| = 2^n + 1, n \in \mathbb{N}\}$, $\cup$, $\cap$ and $\sim$ denote union, intersection and complement of languages, respectively, and $\diamond$, $\odot$ and $\triangleright$ denote the operations on languages in Def.11. A language L is *systolic star free* if there exists a systolic star free expression denoting it.

Example 13. Given $\Sigma = \{a, b\}$ and $s \in \Sigma$, the systolic expression $E_s = ((\Sigma^* \diamond s) \triangleright \Sigma^*) \cup s$ denotes the language of words in Σ^+ having at least an occurrence of the symbol s.

The language $\{a^{2^n} \cdot b^+ : n \geq 0\}$ is denoted by $(((\sim E_b) \diamond b) \cap \Sigma^\wedge) \triangleright (\sim E_a)$.
The language of even length words is denoted by $(\Sigma^* \odot (a \cup b)) \cup (a \cup b \diamond a \cup b)$.

Theorem 14. *A language L is systolic star free iff it is $FO[<, pow]$-definable.*

It is easy to prove that each star free expression is first order definable. The proof of the converse implication, i.e. the translation of first order formulas into systolic star free expressions, follows the approach (with non-trivial changes) of [10,12] by applying the Ehrenfeucht-Fraïsseé technique.

References

1. J. R. Büchi, Weak Second-Order Arithmetic and Finite Automata, *Z. Math. Logik Grundl. Math.* **6** (1960), 66-92.
2. K. Culik II, A. Salomaa, D. Wood, Systolic Tree Acceptors, R.A.I.R.O Informatique Théorique **18** (1984) 53–69.
3. C. C. Elgot, Decision Problems of Finite Automata Design and Related Arithmetic, *Trans. Amer. Math. Soc.* **98** (1961), 21– 52.
4. E. Fachini and A. Monti, A Kleene-like Characterization of Languages Accepted by Systolic Tree Automata, *Journal of Computer and System Science*, **49** (1994), 133–147.
5. J. Gruska, Synthesis, Structure and Power of Systolic Computations, *Theoretical Computer Science* **71** (1990), 47–78.
6. J. Gruska, M. Napoli, D. Parente, On the Minimization and Succinctness of Systolic Binary Tree Automata. In G. Ciobanu, V. Felea Eds., *Proceedings of the 9th Romanian Symposium on Computer Science*, **ROSYCS '93**, *Iasi, Romania, 12 – 16 Novembre, 1993*, pp. 206–219. "A.I. Cuza" Università di Iasi, Dipartimento di Informatica, 1994.
7. R. McNaughton, S. Papert: Counter-Free Automata, MIT Press, Cambridge, MA, 1971.
8. A. Monti and A. Peron, Systolic tree ω-languages, in: *Proc. STACS'95*, Lecture Notes in Computer Science Vol. 900 (Springer, Berlin,1995), 131–142.
9. A. Monti and A. Peron, Systolic Tree ω-Languages: The Operational and the Logical View, *Theoretical Computer Science*, to appear.
10. D. Perrin and J.-E. Pin, First-Order Logic and Star-Free sets, *Journal of Computation System Science* **32** (1986), 393–406.
11. W. Thomas, Automata on Infinite Objects, in: J. van Leeuwen, ed., *Handbook of Theoretical Computer Science, Vol. B*, (Elsevier, Amsterdam, 1990) 133–191.
12. W. Thomas, Languages, Automata and Logic, in: G. Rozenberg and A. Salomaa, eds., *Handbook of formal languages, Vol. 3*, Springer, 1997, 389–455.
13. B.A. Trakhtenbrot, Finite Automata and Monadic Second Order Logic, *Siberian Math. J* **3** (1962), 103–131. (Russian; English translation in: AMS Transl. 59 (1966), 23-55).

Optimal Proof Systems for Propositional Logic and Complete Sets

Jochen Messner and Jacobo Torán

Universität Ulm
Theoretische Informatik
D-89069 Ulm, Germany
{messner,toran}@informatik.uni-ulm.de

Abstract. A polynomial time computable function $h : \Sigma^* \to \Sigma^*$ whose range is the set of tautologies in Propositional Logic (TAUT), is called a proof system. Cook and Reckhow defined this concept in [5] and in order to compare the relative strength of different proof systems, they considered the notion of p-simulation. Intuitively a proof system h p-simulates a second one h' if there is a polynomial time computable function γ translating proofs in h' into proofs in h. A proof system is called optimal if it p-simulates every other proof system. The question of whether p-optimal proof systems exist is an important one in the field. Krajíček and Pudlák [13, 12] proved a sufficient condition for the existence of such optimal systems, showing that if the deterministic and nondeterministic exponential time classes coincide, then p-optimal proof systems exist. They also gave a condition implying the existence of optimal proof systems (a related concept to the one of p-optimal systems). In this paper we improve this result obtaining a weaker sufficient condition for this fact. We show that if a particular class of sets with low information content in nondeterministic double exponential time is included in the corresponding deterministic class, then p-optimal proof systems exist. We also show some complexity theoretical consequences that follow from the assumption of the existence of p-optimal systems. We prove that if p-optimal systems exist then the class UP (and some other related complexity classes) have many-one complete languages, and that many-one complete sets for NP ∩ SPARSE follow from the existence of optimal proof systems.

1 Introduction

A systematic study of the complexity of proof systems for Propositional Logic, was started some time ago by Cook and Reckhow [5]. They were interested in studying the shortest proofs of propositional tautologies in different proof systems, and defined the abstract notion of proof system in the following way:

Definition 1. Let TAUT be the set of all Boolean tautologies (written in a fixed alphabet Σ). A *propositional proof system* (or just *proof system*) is a polynomial time computable function $h : \Sigma^* \to \Sigma^*$ whose range is TAUT.[1]

For example the function h defined as

$$h(w) = \begin{cases} \varphi & \text{if } w = \langle \varphi, v \rangle \text{ and } v \text{ is a resolution proof of } \varphi, \\ x \vee \overline{x} & \text{otherwise.} \end{cases}$$

is a proof system.

If $h(w) = \varphi$ we say that w is a proof of φ in h. Observe that in the given definition a proof system h is not required to be polynomially honest. For a tautology φ, the shortest proof of φ in h, can be much longer than φ.

A *polynomially bounded proof system* h is a proof system in which every tautology has a short proof. More formally, there is a polynomial q such that for every $\varphi \in$ TAUT, there is a string w of length bounded by $q(|\varphi|)$ with $h(w) = \varphi$.

It is not known whether polynomially bounded proof systems exist, on the other hand, many concrete proof systems have been shown not to be polynomially bounded (see for example [5],[16]). There are different motivations for studying the complexity of proof systmes. On the one hand, there are close relations between proof-complexity and Bounded Arithmetic (see e.g. [11]), also concrete proof systems like for example resolution or Frege systems are interesting in their own right, and recently, important connections between these systems and Boolean circuit complexity have been established. Another main motivation for the study of proof systems comes in fact from the following relation between the NP versus co-NP question and the existence of polynomially bounded systems.

Theorem 2. [5] NP $=$ co-NP *if and only if polynomially bounded proof systems exist.*

This result started the so called *Cook-Reckhow Program*: a way to prove that NP is different from co-NP might be to study more and more powerful concrete proof systems, showing that they are not polynomially bounded, until hopefully we have gained enough knowledge to be able to separate NP from co-NP (see [3]).

In order to compare the relative powers of two different proof systems, the notion of polynomial simulation (or p-simulation) was introduced in [5].

Definition 3. Let h and h' be two propositional proof system. We say that h simulates h' if there is a function γ that is polynomially bounded in length and translates proofs in h' into proofs in h. In other words, there is a polynomial p such that for every x $|\gamma(x)| \leq p(|x|)$, and for every tautology φ and every proof w of φ in h', $\gamma(w)$ is a proof of φ in h. If in addition γ is computable in polynomial time, we say that h p-simulates h'.

---

[1] The original definition allows in fact the use of different alphabets for the domain and different languages for the range of h, but for the purposes of this paper the given definition suffices.

Observe that p-simulation is a stronger notion than simulation. It is easy to see that simulation and p-simulation are reflexive and transitive relations. It is also clear that if a proof system h is not polynomially bounded, and h simulates another system h', then h' cannot be polynomially bounded. Cook and Reckhow used p-simulation in order to classify proof systems in different classes with polynomially related derivation strength.

The notion of simulation between proof systems is closely related to the notion of reducibility between problems. Continuing with this analogy, the notion of a complete problem would correspond to the notion of an optimal proof system.

Definition 4. A proof system is optimal (p-optimal) if it simulates (p-simulates) every other proof system.

An important open problem is whether optimal proof systems exist [3]. Observe that if this were the case, then in order to separate NP from co-NP it would suffice to prove that a concrete proof system is not polynomially bounded.

Krajíček and Pudlák have given sufficient conditions for the existence of p-optimal and optimal proof systems.

Theorem 5. [13, 12]

If NE $= co$-NE *then optimal proof systems exist.*
If E $=$ NE *then p-optimal proof systems exist.*

On the other hand, to our knowledge, only weak complexity-theoretic consequences of the existence of optimal proof systems were known.[2]

In the present paper we improve the mentioned result from [13, 12] by weakening the conditions that are sufficient for the existence of optimal and p-optimal proof systems. We show in Section 3 that if the deterministic and nondeterministic double exponential time complexity classes coincide (EE $=$ NEE) then p-optimal proof systems exist, and that NEE $= co$-NEE is sufficient for the existence of optimal proof systems. In fact we give a probably weaker sufficient condition, using a special kind of sets with small information content. Let us say that a set is *almost tally*, if its words belong to the set 0^*10^*. We show that if the class of almost tally sets in NEE is included in EE, then p-optimal proof systems exist, and that optimal proof systems exist if almost tally sets in NEE belong also to co-NEE.

On the other side, we also show some consequences from the existence of optimal and p-optimal proof systems, proving completeness results for the complexity classes UP and NP $\cap$ SPARSE, that follow from the existence of such proof systems. Since complete problems for these classes have been unsuccessfully searched for in the past, the results give some evidence of the fact that optimal proof systems might not exist. At the same time they strengthen the connection between the notions of optimal proof systems and complete sets.

---

[2] From [14] follows that if optimal proof-systems exist, then the class of disjoint pairs in NP has a complete pair under a weak reduction.

The presentation is organized as follows: In Section 4 we show that if p-optimal proof systems exist, then the class UP (unambiguous NP) of problems in NP that can be accepted by nondeterministic polynomial time machines with at most one accepting path for every input [17], has complete problems under the logarithmic space many-one reductions. The existence of complete problems for UP has been studied in [7], where the authors show the existence of a relativization under which this is not possible. Considering that p-optimal proof systems exist, we also show the existence of complete sets for related *promise* classes like FewP. We also consider the weaker hypothesis of the existence of optimal proof systems, and show that the completeness results for the mentioned classes still hold for nonuniform many-one reductions. Finally in Section 5 we prove that optimal proof systems imply the existence of complete problems for the class NP ∩ SPARSE for many-one logarithmic space reductions. The question of the existence of such sets is subtle and has been intensively investigated. Although many-one complete set in NP ∩ SPARSE are not known, Hartmanis and Yesha prove in [8] that there is a sparse set in NP that is Turing complete for NP ∩ SPARSE, (in fact the given set is tally) and ask whether the result can be improved to the many-one case. Hartmanis has also shown that the set of satisfiable formulas with small Kolmogorov complexity SAT ∩ $K[\log, n^2]$ is Turing complete for NP ∩ SPARSE [6], but the completeness of this set under many-one reductions would imply unexpected consequences in the exponential time hierarchy. More recently Schöning has proven that there are sets that are complete for this class under many-one randomized reductions [15].

2 Basic Notions

We assume some familiarity with the standard results and notions about deterministic and nondeterministic complexity classes. For undefined complexity theory notions, and the definition of standard complexity classes, we refer the reader to the standard books in the area like [2]. We will use a pairing function $\langle \, , \, \rangle$ that is polynomial time computable and invertible. For a set A, $\mathfrak{P}(A) = \{L \mid L \subseteq A\}$ represents the power set of A. Let E and EE, denote the time complexity classes DTIME($2^{O(n)}$) and DTIME($2^{O(2^n)}$), respectively, and NE, NEE their nondeterministic counterparts.

In the introduction we defined the notion of proof system. The next lemma shows that for every set of tautologies that is polynomial time computable, there is a proof system in which the tautologies of the set have short proofs, and moreover, the proof can be found easily.

Lemma 6. *If $T \subseteq$ TAUT and $T \in$ P, then there exists a proof system h and a function $t \in$ FP that produces proofs in h for every tautology in T. That is, for every $\varphi \in T$, $h(t(\varphi)) = \varphi$.*

Proof. Let $h' \in$ FP be a proof system. We can define a new proof system h as follows:

$$h(w) = \begin{cases} h'(v) & \text{if } w = 0v, \\ v & \text{if } w = 1v \text{ and } v \in T, \\ x \vee \overline{x} & \text{otherwise.} \end{cases}$$

Clearly, h is a proof system. The function t producing proofs in h for the elements of T is just $t(v) = 1v$.

In the definitions of proof systems and p-optimality we allowed both functions, the proof system h and the translation function, to be computable in polynomial time. The following lemma shows that the most part of the computational complexity of both functions can be concentrated in one of them, whereas the other function may be computed, for example, in logarithmic space. To formulate the lemma, let us say that a proof system h is *logspace-optimal* if it simulates in logarithmic space every other proof system h', which means that there is a logspace computable function γ such that $h(\gamma(w)) = h'(w)$ for every word w.

Lemma 7. *The following statements are equivalent*

1. *A p-optimal polynomial time computable proof system exists.*
2. *A logspace-optimal polynomial time computable proof system exists.*
3. *A p-optimal logspace computable proof system exists.*

Proof. Clearly, *2* and *3* imply *1*.

To obtain *2* from *1* let h be a p-optimal polynomial time computable proof system, and let g be defined by

$$g(w) = \begin{cases} h(w') & \text{if } w = \langle M, 0^l, v \rangle, \text{ and } M \text{ is a deterministic Turing trans-} \\ & \text{ducer which on input } v \text{ outputs } w' \text{ in at most } l \text{ steps,} \\ x \vee \neg x & \text{else.} \end{cases}$$

Clearly, g is polynomial time computable. We show that g is logspace-optimal. Let h' be a proof system. By assumption, there is a polynomial time computable translation function γ such that $h(\gamma(w)) = h'(w)$ for any w. Let M be a deterministic Turing transducer computing γ with time bounded by a polynomial $p(n)$. It's easy to see that the logspace computable function γ' with $\gamma'(w) = \langle M, 0^{p(|w|)}, w \rangle$ translates proofs in h' into proofs in g.

We now show that the existence of a p-optimal proof system h implies a logspace computable p-optimal proof system f. Let M be a polynomial time machine computing h. Let $f(w) = \varphi$ if w encodes a complete computation of M (given by the sequence of configurations) with output φ, let $f(w) = x \vee \neg x$ otherwise. Choosing a suitable encoding f is logspace computable. Also f p-simulates h. Therefore f p-simulates every proof system.

3 A sufficient condition

We give now a sufficient condition for the existence of optimal proof systems, based on almost tally sets in double exponential time.

Theorem 8.

If $\text{NEE} \cap \mathfrak{P}(0^*10^*) \subseteq \text{EE}$ *then there exists a p-optimal proof system.*
If $\text{NEE} \cap \mathfrak{P}(0^*10^*) \subseteq \text{co-NEE}$ *then there exists an optimal proof system.*

Proof. Let M_1, M_2, M_3, ... be some standard enumeration of deterministic Turing transducers with binary input alphabet such that there is an universal Turing machine which is able to simulate k steps of M_i in $(ik)^2$ time for any $k \geq 0$ (clearly, such enumerations exist). Now define the language

$$T = \{0^j 10^i \mid \text{for any word } w \text{ of length at most } 2^{2^n}, \text{ where } n = i + j + 1: \text{ if } M_i \text{ stops on input } w \text{ in at most } 2^{2^n} \text{ steps, } M_i \text{ outputs a tautology}\}.$$

It is not hard to see that $T \in \text{co-NEE}$. Assuming $\text{NEE} \cap \mathfrak{P}(0^*10^*) \subseteq \text{EE}$ we also have $\text{co-NEE} \cap \mathfrak{P}(0^*10^*) \subseteq \text{EE}$ and therefore $T \in \text{EE}$. Therefore there is a deterministic Turing machine M_T which decides T in time $2^{c \cdot 2^n}$ for some $c > 0$.

We describe a p-optimal proof system h:

On input $\langle 0^j 10^i, 0^s, w \rangle$ examine if $s \geq 2^{2^l}$ and $|w| \leq 2^{2^l}$, where $l = i + j + 1$, and test whether M_T accepts $0^j 10^i$ in at most s steps. If this is the case, output $M_i(w)$ if M_i stops after at most 2^{2^l} steps on input w. (If some other case applies, output some fixed tautology).

Clearly, $0^j 10^i \in T$ implies that the Turing machine M_i on input w outputs a tautology if the computation needs at most 2^{2^l} steps. Therefore $h(\Sigma^*) \subseteq$ TAUT. Also, h is computable in polynomial time. We have to show that h p-simulates every other proof system. Let g be a proof system computed by the deterministic Turing transducer M_i with time bound $n^k + k$. A proof w for g is translated into the proof $w' = \langle 0^j 10^i, 0^s, w \rangle$ where $s = 2^{c \cdot 2^{i+j+1}}$, and $j = \max(0, \lceil \log\log |w|^k + k \rceil - i - 1)$. By the construction of h, we have $h(w') = g(w)$. We just have to show that the translation $w \mapsto w'$ is computable in polynomial time. Clearly, this is the case if the length of 0^s is polynomially bounded by $|w|$. Now observe that $s = 2^{c \cdot 2^{i+j+1}} \leq 2^{c \cdot 2^{i+1} \cdot 2^{\log\log |w|^k + k}} = (|w|^k + k)^{c \cdot 2^{i+1}}$ which is polynomial in $|w|$.

Similar considerations can be used to show the second part of the theorem. $\qquad \square$

If in the previous proof we replace each occurrence of the number 2 by some arbitrary constant d (and log by $\log_d$) we obtain that already $\text{NTIME}(2^{O(d^n)}) \cap \mathfrak{P}(0^*10^*) \subseteq \text{DTIME}(2^{O(d^n)})$ for some $d > 0$ implies the existence of a p-optimal proof system (a similar result follows for optimal proof systems).[3] However, the proof seems not to translate directly to a further exponential level as $2^{2^{2^{c + \log\log\log n}}} \notin n^{O(1)}$ for any real $c > 0$.

---

[3] This improvement emerged from an email discussion with S. Ben-David.

4 Complete problems for UP

In this and the next sections we prove that the existence of optimal proof systems imply the existence of complete sets in certain *promise* complexity classes. The machines computing the sets in these classes can not be guaranteed to keep the condition of the class for all possible inputs, and because of this, complete problems for these classes are not known. We will first prove the existence of complete sets for UP under many-one polynomial time reductions, considering the existence of p-optimal proof systems. Later we will strengthen this result to logarithmic space reductions.

Define the set CAT containing descriptions of machines that are categorical, i.e., have at most one accepting path for all inputs up to a given length.

$$\text{CAT} = \{\langle M, 0^l, 0^n \rangle \mid M \text{ is a nondeterministic Turing machine and for every}$$
$$\text{input } x, |x| \leq n, M \text{ has at most one accepting path of length} \leq l\}.$$

Clearly, CAT $\in$ *co*-NP and since TAUT is *co*-NP complete for polynomial time many-one length-increasing reductions, there is such a function $f \in$ FP reducing CAT to TAUT.

For every fixed nondeterministic Turing machine M and every fixed monotone polynomial q, the set

$$\text{CAT}_{M,q} = \{\langle M, 0^{q(n)}, 0^n \rangle \mid n \geq 1\} \cap \text{CAT},$$

is in P because it is either finite or the set of all such triples. Also the image of $\text{CAT}_{M,q}$ under f, is in P; in order to test whether a given formula ϕ belongs to $f(\text{CAT}_{M,q})$ it suffices to generate the words of $\text{CAT}_{M,q}$ up to a given length, and check whether the image of one of these words after applying f coincides with ϕ.

Let $\varphi_{M,l,n}$ denote the formula $f(\langle M, 0^l, 0^n \rangle)$. Clearly, if the machine M is categorical and its running time is bounded by q, then for every n, $\varphi_{M,q(n),n}$ is a tautology.

The following lemma intuitively says that under the hypothesis of the existence of p-optimal proof systems, for every categorical machine M there is a polynomial time computable function producing proofs of the categoricity of M.

Lemma 9. *Let* $h \in$ FP *be a p-optimal proof system. For every categorical machine* M *with running time bounded by a polynomial* q*, there is a function* $g_{M,q} \in$ FP *such that for every* $l \in \mathbb{N}$*,* $g_{M,q}(0^l)$ *produces an output* w *and* $h(w) = \varphi_{M,q(l),l}$*.*

Proof. Let M be a categorical machine polynomially time bounded by a polynomial q. Every formula in the set $f(\text{CAT}_{M,q})$ is a tautology. As we have seen this set is in P and by Lemma 6 there is a proof system h' and a function $t \in$ FP that produces short proofs in h' for the tautologies in $f(\text{CAT}_{M,q})$. Formally, for every tautology $\varphi_{M,q(l),l} \in f(\text{CAT}_{M,q})$, $h'(t(\varphi_{M,q(l),l})) = \varphi_{M,q(l),l}$.

Since h is p-optimal, it p-simulates h'. This means that there is a function $\gamma \in$ FP, translating proofs in h' into proofs in h, and for every tautology $\varphi_{M,q(l),l} \in f(\mathrm{CAT}_{M,q})$ we have $h(\gamma(t(\varphi_{M,q(l),l}))) = \varphi_{M,q(l),l}$.

The claimed function $g_{M,q}$ on input 0^l computes the formula $\varphi_{M,q(l),l} = f(\langle M, 0^{q(l)}, 0^l \rangle)$, and applies functions t and γ to it. Clearly $g_{M,q} \in$ FP.

We can now prove that p-optimal proof systems imply the existence of complete sets for UP.

Theorem 10. *If p-optimal proof systems exist then there are sets that are complete for* UP *under polynomial time many-one reductions.*

Proof. Let h be a p-optimal proof system and consider the set

$$A = \{\langle M, 0^l, w, x \rangle \mid M \text{ is the description of a NDTM and } h(w) = \varphi_{M,l,|x|}$$
$$\text{and } M \text{ accepts } x \text{ in } l \text{ steps or less}\}.$$

The set A is clearly in UP since $h(w) = \varphi_{M,l,|x|}$ means that this formula is a tautology, and therefore for every input of length smaller or equal than $|x|$ (and in particular for x), M has at most one accepting path of length l.

For the hardness part, let B be a set in UP, accepted by a machine M in time bounded by a polynomial q. W.l.o.g. we can suppose that for every n, $q(n) \le q(n+1)$, and that on any input x, every computation path of M halts after exactly $q(|x|)$ steps. Consider the function $g_{M,q} \in$ FP whose existence was proved in the above lemma. The function $\lambda \in$ FP defined for every $x \in \Sigma^*$ as

$$\lambda(x) = \langle M, 0^{q(|x|)}, g_{M,q}(0^{|x|}), x \rangle$$

many-one reduces B to A since $h(g_{M,q}(0^{|x|})) = \varphi_{M,q(|x|),|x|}$.

To see that p-optimal proof systems also imply the existence of logspace many-one complete problem for UP, we show that the construction in the proof of Lemma 9 can be modified so that $g_{M,q}$ is logspace computable. In fact, by the same observations it can be seen that even weaker reductions are possible. Remember $g_{M,q}(0^l) = \gamma(t(f(\langle M, 0^{q(l)}, 0^l \rangle)))$ where f is a length-increasing reduction from CAT to TAUT, $t(x) = 1x$ is the function from the proof of Lemma 6, and γ is a function translating proofs in a proof system h' into proofs in h, whose existence is guaranteed by the p-optimality of system h. Clearly, $\langle M, 0^{q(l)}, 0^l \rangle$ can be computed in logarithmic space from 0^l, and the function t is logspace computable. Also f can be chosen to be a logspace computable (and length-increasing) many-one reduction from CAT to TAUT. By Lemma 7 we can assume h to be logspace-optimal, and therefore we can also assume γ to be logspace computable. As the composition of logspace computable functions is again logspace computable we obtain:

Theorem 11. *If p-optimal proof systems exist then there are sets that are complete for* UP *under logarithmic space many-one reductions.*

The completeness result for UP can be extended to the related complexity classes FewP and Few as stated in the next theorem. The classes FewP and Few were defined in [1] and [4] as a generalizations of the class UP. For space reasons we omit the definition of these classes and the proof of the next theorem.

Theorem 12. *If p-optimal proof systems exist then there are sets that are complete under logarithmic space many-one reductions for the classes FewP and Few.*

Let us mention at this point that in [7] an oracle is constructed under which the class UP does not have many-one complete sets. In [9] this result is improved from many-one to Turing reducibility, and it is also shown that under certain relativizations the class FewP does not have Turing complete sets. Since the proofs in this paper relativize, we can state the following corollary:

Corollary 13. *There exists a relativization under which p-optimal proof systems do not exist.*

If we only consider the existence of optimal proof systems (instead of p-optimal systems), then we can prove a version of the above result for nonuniform reductions. Intuitively a set is polynomial time nonuniformly many-one reducible to a second one, if the reduction function is not necessarily in FP, as in the usual polynomial time reductions, but it is computed by a family of polynomial size circuits [10].

Due to space reasons we omit the formal definition of non-uniform reductions as well as the proof of the next theorem.

Theorem 14. *If optimal proof systems exist then there are sets that are complete for UP under nonuniform polynomial time many-one reductions.*

As expected, the many-one completeness results for Few and FewP from Theorem 12 become completeness results for nonuniform many-one reductions if only the existence of optimal proof systems is considered.

5 Complete sets for NP ∩ SPARSE

We prove now that there are many-one complete sets for NP ∩ SPARSE under the hypothesis of the existence of optimal proof systems. The proof follows the same lines as the previous one for complete sets in UP, but in this case we do not need p-optimality, and the existence of optimal proof systems suffices.

Let us define the set SP containing descriptions of nondeterministic machines that do not accept too many strings up to a given length:

$$SP = \{\langle M, 0^l, 0^n \rangle \mid M \text{ is a nondeterministic Turing machine and there are}$$
$$\text{at most } l \text{ pairs } (x_i, y_i), |x_i| \leq n, |y_i| \leq l, \text{ such that } x_i \neq x_j \text{ for } i \neq j,$$
$$\text{and } y_i \text{ is an accepting path of } M \text{ on input } x_i\}.$$

It is not hard to see that SP $\in$ *co*-NP, and therefore SP is polynomial-time many-one reducible to TAUT. Let $f \in$ FP be a length increasing function that reduces SP to TAUT.

Let M be a fixed nondeterministic Turing machine with running time bounded by a polynomial q, that for every length l accepts at most $q(l)$ words of length l. The set

$$SP_{M,q} = \{\langle M, 0^{q(n)}, 0^n\rangle \mid n \geq 1\} \cap SP,$$

is in P, and the image of $SP_{M,q}$ under f, is also in P;

Let $\zeta_{M,l,n}$ denote the formula $f(\langle M, 0^l, 0^n\rangle)$. Clearly, if the machine M runs in time bounded by q and accepts a q-sparse set of inputs, then for every n, $\varphi_{M,q(n),n}$ is a tautology.

The following lemma is analogous to Lemma 9, and says that in an optimal proof system, a proof of the fact that a machine accepts a sparse language up to a given length, can be polynomially bounded. The proof of the lemma follows the same lines as the one for Lemma 9 and it is omitted.

Lemma 15. *Let* $h \in$ FP *be a p-optimal proof system. For every nondeterministic Turing machine* M *with running time bounded by a polynomial* q, *and such that for every* $n \in \mathbb{N}$, M *accepts at most* $q(n)$ *words of length* n, *there is a polynomial* r *such that for every* $l \in \mathbb{N}$, *there is a string* $w \in \Sigma^*$, *with* $|w| \leq r(l)$ *and* $h(w) = \zeta_{M,q(l),l}$.

We can now prove that optimal proof systems imply the existence of complete sets for NP $\cap$ SPARSE.

Theorem 16. *If optimal proof systems exist then there are sets that are complete for* NP $\cap$ SPARSE *under logarithmic space many-one reductions.*

Proof. Let h be an optimal proof system, and let S be the set

$$S = \{0^{\langle M,l,j,n\rangle}1x \mid M \text{ is the description of a NDTM which accepts } x \text{ in } l \text{ steps}$$
$$\text{or less, } |x| = n, \text{ and there is a string } w, |w| \leq j, \text{ such that } h(w) = \zeta_{M,l,n}\}.$$

S belongs clearly to NP. Also, the number of string x such that $0^{\langle M,l,j,n\rangle}1x \in S$ is bounded by l, since $0^{\langle M,l,j,n\rangle}1x \in S$ implies that $\zeta_{M,l,|x|}$ is a tautology. Therefore for every length n there are at most n words of this length in S. This proves that S is sparse.

In order to see that S is hard for the class, let S' be a set in NP $\cap$ SPARSE, accepted by a nondeterministic Turing machine M with time bounded by a polynomial q, and with density also bounded by q. By Lemma 15 there is a polynomial r such that for every $l \in \mathbb{N}$, there is a string w with $|w| \leq r(l)$ and $h(w) = \zeta_{M,q(l),l}$. The reduction from S' to S is given by the function

$$\lambda(x) = 0^{\langle M,q(|x|),r(|x|),|x|\rangle}1x.$$

Observe that this function is computable in logarithmic space, one-to-one, length increasing and also invertible in logarithmic space.

Let us mention at this point that contrary to the UP case, there is no known relativization under which the class NP ∩ SPARSE does not have many-one complete sets. For this reason, and considering the existing results on sparse sets mentioned in the introduction, we feel that Theorem 16 only provides a weak consequence of the existence of optimal proof systems.

References

1. E. Allender. Invertible functions. Ph.D. dissertation, Georgia Institute of Technology, 1985.
2. J. L. Balcázar, J. Díaz, and J. Gabarró. *Structural Complexity I*, volume 11 of *EATCS Monographs on Theoretical Computer Science*. Springer-Verlag, 1988.
3. S. Buss. Lectures on Proof Theory. Tech Report No. SOCS-96.1, McGill University, 1996. (http://www.cs.mcgill.ca/~denis/TR.96.1.ps.gz)
4. J. Cai and L. Hemachandra. On the power of parity polynomial time. *Mathematical Systems Theory* **23**, pp. 95–106, 1990.
5. S. Cook and R. Reckhow. The relative efficiency of propositional proof systems. *Journal of Symbolic Logic* **44**, pp. 36–50, 1979.
6. J. Hartmanis. Generalized Kolmogorov complexity and the structure of feasible computations. In *Proceedings of the 24th IEEE Symposium on Foundations of Computer Science* (FOCS'83), pp. 439–445, 1983
7. J. Hartmanis and L. Hemachandra Complexity classes without machines: On complete languages for UP. *Theoretical Computer Science*, **58**, pp. 129–142, 1988.
8. J. Hartmanis and J. Yesha. Computation times of NP sets of different densities. *Theoretical Computer Science*, **34**, pp. 17–32, 1984.
9. L. Hemaspaandra, S. Jain and N. Vereshchagin. Banishing robust Turing completeness. *Int. Journal of Foundations of Computer Science*, 4, pp. 245–265, 1993.
10. R. Karp and R. Lipton. Some connections between nonuniform and uniform complexity classes. In *Proceedings of the 12th ACM Symposium on Theory of Computing*, pp. 302–309, 1980.
11. J. Krajíček. *Bounded Arithmetic, Propositional Logic and Complexity Theory* Cambridge University Press 1995.
12. J. Krajíček and P. Pudlák. Propositional proof systems, the consistency of first order theories and the complexity of computations. *Journal of Symbolic Logic* **54**, pp. 1063–1079, 1989.
13. P. Pudlák. On the length of proofs of finitistic consistency statements in first order theories. Logic Colloquium'84 (J. B. Paris et al., editors), North-Holland, Amsterdam, pp. 165–196, 1986
14. A. A. Razborov. On provably disjoint NP-pairs. Technical Report RS-94-36, Basic Research in Computer Science Center, Aarhus, 1994.
15. U. Schöning. On random reductions from sparse sets to tally sets. *Information Processing Letters*, **46**, pp. 239–241, 1993.
16. A. Urquhart. The complexity of propositional proofs. *Bulletin of Symbolic Logic* **1**, pp. 425–467, 1995.
17. L. Valiant. The relative complexity of checking and evaluating. *Information Processing Letters*, **5**, pp. 20–23, 1976.

The (Parallel) Approximability of Non-Boolean Satisfiability Problems and Restricted Integer Programming

Maria Serna[1]*, Luca Trevisan[2] and Fatos Xhafa[1]*

[1] Departament de Llenguatges i Sistemes Informàtics
Universitat Politècnica de Catalunya
Módul C6 - Campus Nord, Jordi Girona Salgado, 1-3
08034-Barcelona, Spain
E-mail: {mjserna,fatos}@lsi.upc.es

[2] MIT Laboratory for Computer Science
545 Technology Square, Room NE43-371, Cambridge, MA 02139, USA
E-mail: luca@theory.lcs.mit.edu

Abstract. We present parallel approximation algorithms for maximization problems expressible by integer linear programs of a restricted syntactic form introduced by Barland et al. [BKT96]. One of our motivations was to show whether the approximation results in the framework of Barland et al. holds in the parallel setting. Our results are a confirmation of this, and thus we have a new common framework for both computational settings. Also, we prove almost tight non-approximability results, thus solving a main open question of Barland et al.

We obtain the results through the constraint satisfaction problem over multi-valued domains, for which we show non-approximability results and develop parallel approximation algorithms.

Our parallel approximation algorithms are based on linear programming and random rounding; they are better than previously known sequential algorithms. The non-approximability results are based on new recent progress in the fields of Probabilistically Checkable Proofs and Multi-Prover One-Round Proof Systems [Raz95, Hås97, AS97, RS97].

1 Introduction

Expressing combinatorial optimization problems as integer linear programs (ILP) has several applications. In particular, several approximation algorithms start from the linear programming relaxation of the ILP formulation, and then use randomized rounding [RT87, GW94], primal-dual methods [GW96], or more sophisticated methods [LLR95, ENRS95].

An interesting new *structural* use of Integer Linear Programming has been taken in a recent paper of Barland, Kolaitis and Thakur [BKT96], where *syntactic classes* of maximization problems are introduced. A problem belongs to

* Research supported by the ESPRIT Long Term Research Project No. 20244 - ALCOM IT and the CICYT Project TIC97-1475-CE.

one such class if it can be expressed by an ILP with a certain restricted format. The approximability properties of the problem in a class are then implied by the approximability of the respective prototypical ILP. The main goal of [BKT96] was to overcome some limitations of the standard way of defining syntactic classes, namely the approach of logical definability [PY91, PR93, KT94, KT95]. The latter approach, indeed, fails to explain why problems with similar logical definability, such as Max k-dimensional Matching and Max Clique have very different approximability properties. Furthermore, using ILP, classes are defined in terms of a single parameter that determines the hardness of the problems. This parameter is either the maximum number of occurrences of any variable or the maximum size of the domain of the variables. The latter kind of restriction gives rise to a family of classes that Barland et al. call MAX FSBLIP (for Maximum Feasible Subsystem of Bounded Layered Integer Program). Letting the variables to take values in a constant, logarithmic, or polynomial range allowed them to capture syntactic maximization classes that are constant-approximable, polylog-approximable and poly-approximable, respectively. An interesting question is whether these three classes form a proper hierarchy. Barland et al. did not completely resolve this point and left improved non-approximability results as an open question.

In this paper our interest is twofold. In one hand, we use the integer programming as a framework for parallel approximability, aiming to obtain improved parallel approximation results. It is known that all the problems contained in logically defined syntactic classes that are constant-factor approximable, are also constant-factor approximable[3] in NC. This feature of logically defined syntactic classes is desirable for at least two reasons: it reduces the study of sequential and parallel approximability to the same framework, and is in accordance with the fact that almost all the constant factor approximation algorithms that are known also admit a parallel version with a comparable approximation ratio. The issue of parallel approximability is not raised in the paper of Barland et al. Our parallel results state that in the new framework of integer programming the sequential results holds as well as in the parallel setting thus, again, we have a common framework for both computational settings. Having this outcome, the second question that we consider is what are the limits of parallel, as well as sequential, approximability for these problems. We show that our approximation factors are nearly the best possible by providing some new non-approximability results (the non-approximability results will also hold for sequential algorithms.) In both cases, our main results will be expressed in terms of the multi-valued constraint satisfaction problem, and then translated, by means of reductions, in terms of the model of Barland et al.

In the rest of this section, we state our results and we discuss their relation with previously known ones.

In this paper, a crucial role is played by the constraint satisfaction problem over multi-valued domains. In an instance of this problem, we are given a set of

---

[3] An NC algorithm is an algorithm that runs in poly-logarithmic time on a parallel shared-memory machine with a polynomial number of processors. See e.g. [DSST97].

constraints of arity at most k over multi-values variables where a constraint is a boolean valued function over $\{0, 1, \ldots, d-1\}^k$ and is given a positive weight. We can think of a k-ary domain-d constraint as a set of k-tuples values (i.e. a relation over $\{0, 1, \ldots, d-1\}^k$) and say that an assignment satisfies the constraint if the corresponding values to the variables of the constraint form a k-tuple belonging to the relation. The goal is to find an assignment to the variables that maximizes the total weight of satisfied constraints. This problem is a common generalization of several known and well-studied problems. To begin with, it is a natural generalization of the boolean constraint satisfaction problem MAX kCSP, introduced by Khanna et al. [KMSV94] and then studied in [Cre95, Tre96, KSW97] (in the boolean case, the domain is $\{0, 1\}$, that is, $d = 2$.) It also generalizes Multi-Prover One-Round Proof Systems and the MAX CAPACITY REPRESENTATIVES problem (introduced by Bellare et al. [Bel93] and further considered by Barland et al.). The version over multi-valued domain has been studied in the restricted case of binary constraints [LW96] and that of "planar instances" [KM96]. In this paper we address, for the first time, the approximability of the problem in its full generality. We present a parallel approximation, based on linear programming and random rounding, that achieves an approximation factor $1/d^{k-1}$. The algorithm can be efficiently parallelized and de-randomized. Our major contribution here is the definition and the analysis of an appropriate random rounding scheme. The parallelization mimics a similar proof in [Tre96], but is not entirely straightforward. For the special case of binary constraint ($k=2$), our approximation guarantee is twice better than the $1/2d$-approximate algorithm of [LW96].

We also prove several non-approximability results under different complexity assumptions. Such results follow from recent advances in the fields of Probabilistically Checkable Proofs [Hås97] and of Multi-Prover One-Round Proof Systems [Raz95, RS97, AS97] and from the fact that multi-valued constraint satisfaction problems generalize both models. We use reductions from the multi-valued constraint satisfaction problem to derive negative approximation results for the rest of the problems of interest. In terms of the class FSBLIP, our result states that the classes MAX FSBLIP(2), MAX FSBLIP(log) and MAX FSBLIP(poly) form a proper hierarchy (the separation of the two last classes derives from a result of Bellare [Bel93] stating that MAX CAPACITY REPRESENTATIVES which belongs to MAX FSBLIP(poly) is not log-approximable; we separate the first two classes by proving that MAX CAPACITY REPRESENTATIVES(log) is not constant-approximable.)

We also consider the class of integer programs MAX FMIP (for Maximum Feasible Majority Integer Program) for which MAX MAJORITY SAT is a canonical problem. [BKT96] showed that this class contains only constant-approximable problems. For the general MAX FMIP problem, we present a slight improvement and simplification over their approximation result. The latter result does not depend on the constraint satisfaction problem. We also prove an almost tight non-approximability result for the problems of this class by reducing from the boolean constraint satisfaction problem.

2 Preliminaries

For an integer n, we denote by $[n]$ the set $\{0, \ldots, n-1\}$. A combinatorial optimization problem is characterized by the set of *instances*, by the finite set of *feasible solutions* associated to any instance, and by a *measure* function that associates a non-negative *cost* to any feasible solution of a given instance. We refer e.g. to [BC93] for the formal definition of NP Optimization problem.

Definition 1 (MAX CAPACITY REPRESENTATIVES-d). For a function $d : \mathcal{Z}^+ \to \mathcal{Z}^+$, MAX CAPACITY REPRESENTATIVES-$(d(n))$ problem is defined as follows:

Instance: A partition of $\{1, \ldots, n\}$ into sets $S_1, \ldots, S_m$, each of cardinality at most d; and weights $w_{i,j} \geq 0$ for any two elements belonging to different sets of the partition.
Solution: The choice of a representative in any set.
Measure: The sum of the weights $w_{i,j}$ for any i and j that are representatives in different sets of the partition.

Definition 2 (Constraint). A k-ary, domain-d constraint over $x_1, \ldots, x_n$ is a pair $(f, (i_1, \ldots, i_k))$ where $f : [d]^k \to \{0, 1\}$ and $i_j \in \{1, \ldots, n\}$ for $j = 1, \ldots, k$. A constraint $C = (f, (i_1, \ldots, i_k))$ is *satisfied* by an assignment $\mathbf{a} = a_1, \ldots, a_n$ to $x_1, \ldots, x_n$ if $C(\mathbf{a}) \overset{\text{def}}{=} f(a_{i_1}, \ldots, a_{i_k}) = 1$.

We say that a function $f : [d]^k \to \{0, 1\}$ is *conjunctive* if it can be expressed as a conjunction of equations, i.e. there are values $v_1, \ldots, v_k \in [d]$,

$$f(x_1, \ldots, x_k) = 1 \text{ if and only if } [x_1 = v_1] \wedge \ldots \wedge [x_k = v_k] .$$

When this will not cause confusion, we will sometimes blur the important difference between a constraint $(f, (i_1, \ldots, i_k))$ and the function f. For example we say that a constraint $(f, (i_1, \ldots, i_k))$ is conjunctive if function f is, and so on.

Definition 3 (MAX kCSP-d and MAX kCONJ-d). For any integer $k \geq 1$ and function $d = d(n)$, the MAX kCSP-d is defined as follows:

Instance: A set $\{C_1, \ldots, C_m\}$ of domain-d constraints of arity at most k over $x_1, \ldots, x_n$, and associated non-negative weights $w_1, \ldots, w_m$.
Solution: An assignment $\mathbf{a} = (a_1, \ldots, a_n) \in [d]^n$ to the variables $x_1, \ldots, x_n$.
Measure: The total weight of satisfied constraints.

MAX kCONJ-d is the restriction of MAX kCSP-d to instances where all the constraints are conjunctive.

Definition 4 (Integer Linear Programming (ILP)). The ILP is as follows:

Instance: A matrix $A \in \mathcal{Z}^{m \times n}$ and two vectors $\mathbf{c} \in \mathcal{Z}^n$ and $\mathbf{b} \in \mathcal{Z}^m$.
Solution: A vector $\mathbf{x} \in \mathcal{Z}^n$ satisfying $A\mathbf{x} \leq \mathbf{b}$.
Measure: $\mathbf{c} \cdot \mathbf{x}$.

Note that in this formulation, the goal is to maximize the measure $\mathbf{c} \cdot \mathbf{x}$. The variables appearing (with non-zero coefficients) in the objective function are called *objective variables* and those appearing only in the linear constraints are *program variables*. The *width* of a constraint is equal to the number of its variables.

Definition 5 (Constraint Dominance). Given a linear constraint of the form $g(1-t) + \mathbf{a} \cdot \mathbf{q} \geq b$, where t is $0/1$ variable, it is said that t dominates the constraint if (a) for $t = 0$ the constraint is satisfied whatever is the assignment to the rest of variables; (b) if an assignment satisfies $\mathbf{a} \cdot \mathbf{q} \geq b$, then the constraint is satisfied for any value of t.

Definition 6 (MAX FSBLIP($d(n)$) [BKT96]). For a given function $d(n)$, the class MAX FSBLIP($d(n)$) contains all the optimization problems A for which there are positive integer constants l, m, k (that only depend on A) such that every instance of A can be expressed as an ILP with the following structure:

- The program variables can take values in $\{0, 1, \ldots d(n) - 1\}$.
- Each objective variable t_i occurs only in constraints of the form $(1 - t_i) + q_{i,1} + \cdots + q_{i,z} \geq 1$, where $z \in \mathbf{N}$ can be polynomial in n, and each $q_{i,j}$, $1 \leq j \leq z$ is a $0/1$ program variable associated with the objective variable t_i. These constraints are referred to as objective constraints.
- Each variable $q_{i,j}$ appearing in an objective constraint occurs in at most l other constraints and dominates each of them.
- All constraints that are not objective ones have width m and are dominated by some $q_{i,j}$ associated with some objective variable t_i.
- Each objective variable t_i appears in at most k objective constraints.

Definition 7 (MAX FMIP [BKT96]). An optimization problem Π belongs to the class MAX FEASIBLE MAJORITY IP (in short, MAX FMIP) if there exist positive constants k, σ and a polynomial p such that for any instance I of Π we can find a set of linear inequalities over the integers

$$Ax \geq b$$
$$\mathbf{x} \in \{-k, -k+1, \ldots, k-1, k\}^n$$

where $b_j \leq \sigma$, the entries of A are integers of absolute value at most $p(n)$, and the optimum of I is precisely the maximum number of inequalities that are simultaneously satisfiable.

3 Reductions Among Problems

Theorem 8. *For any constant k and function $d(n)$, MAX kCONJ-$d(n)$ belongs to MAX FSBLIP($d(n)$).*

Proof. Our formulation is similar to that of MAX CAPACITY REPRESENTATIVES given in [BKT96, Section 3]. Let $\{C_1, \ldots, C_m\}$ be a set of k-ary domain-d conjunctive constraints over $x_1, \ldots, x_n$, and $w_1, \ldots, w_m$ be associated non-negative weights. We use two 0/1 variables t_j and f_j for any constraint, and we use a d-valued variable y_i for any variable x_i. The integer linear program is

$$\max \sum_j w_j t_j$$
s.t.
$$(1 - t_j) + f_j \geq 1 \quad \forall j = 1, \ldots, m$$
$$d(1 - f_j) + y_i \geq v \quad \forall j = 1, \ldots, m, \ \forall [x_i = v] \in C_j$$
$$d(1 - f_j) - y_i \geq -v \ \forall j = 1, \ldots, m, \ \forall [x_i = v] \in C_j$$

Notice that each objective variable t_j appears in a unique objective constraint, each variable f_j in an objective constraints occurs in at most $2k$ other constraints dominating each of them, and, finally, any constraints has width 2. □

Theorem 9. *If* Max kConj-d *is r-approximable (in* NC*) and $k^d = \text{poly}(n)$, then* Max kCSP-d *is r-approximable (in* NC*).*

Proof. For any constraint C_j of weight w_j, let s be the number of satisfying assignments to its variables (note that $s \leq k^d$). Then we can express C_j as the disjunction of s conjunctive constraints $K_j^1, \ldots, K_j^s$, each one enforcing one of the satisfying assignments of C_j. Observe that any (global) assignment, satisfies at most one of the K_j^i constraints and satisfies one if and only if satisfies C_j. Let us substitute C_j with the $K_j^1, \ldots, K_j^s$ constraints, and give weight w_j to all of them. We repeat the same substitution for any constraint. The new instance is equivalent to the former, in the sense that they share the same set of feasible solutions, and the cost of each solution is always the same. Observe that the substitution process can be done also in parallel for all the constraints. □

Theorem 10. Max 2Conj-d *is r-approximable (in* NC*) if and only if* Max Capacity Representatives-d *is r-approximable (in* NC*).*

Proof. It is easy to see that the two problems are equal. Without loss of generality we can assume that any set in a Max Capacity Representatives-d instance has exactly d elements (add dummy elements and give weight zero to the pairs corresponding to such elements) and that in a Max 2Conj-d instance with n variables there are all the possible $\binom{n}{2} d^2$ conjunctive constraints (add the missing constraints with weight zero). Now, the equivalence is immediate: every set S_i in Max Capacity Representatives-d corresponds to a d-valued variable $s_i = a$, $a = 0, 1, \ldots, d - 1$, meaning that the representative of set S_i is a; to a pair of representatives in different sets S_i, S_j corresponds a conjunctive constraint $s_i = a \wedge s_j = b$; the weight of a constraint is that of the edge from which it was derived. Clearly, starting from an instance of Max Capacity Representatives-d we construct (in NC) an instance of Max 2Conj-d such that its feasible solutions are also feasible solutions of the same cost for Max Capacity Representatives-d and vice-versa. The theorem thus readily follows. □

Theorem 11. Max kConj-2 *can be expressed as a* Max FMIP *problem with* $p(n) = 1$, $k' = 2$ *and* $\sigma = k$.

Proof. Let φ be an instance of MAX kCONJ-2. We have a variable $y_i \in \{-1, 0, 1\}$ for any variable x_i of φ. For any constraint C_j, let P_j (resp. N_j) be the set of indices of variables that are assigned to 1 (resp. 0) in C_j. Let k_j be the arity of C_j. Then C_j is expressible as

$$\bigwedge_{i \in P_j} [x_i = 1] \wedge \bigwedge_{i \in N_j} [x_i = 0] .$$

We translate C_j into the constraint $\sum_{i \in P_j} y_i + \sum_{i \in N_j} -y_i \geq k_j$. Under the understanding that $\{-1, 1\}$ assignments to y_i should be mapped to $\{0, 1\}$ assignments for x_i (i.e. $x_i = (1 + y_i)/2$), the two constraints are equivalent. We repeat the translation for any constraint, and the theorem thus follows. $\square$

4 Positive Results: Algorithms

We now consider a linear programming relaxation of MAX kCONJ-d. We have a variable z_j for any constraint C_j, with the intended meaning that $z_j = 1$ when C_j is satisfied and $z_j = 0$ otherwise. We also have a variable $t_{i,v}$ for any variable x_i and any value $v \in [d]$, meaning that $t_{i,v} = 1$ if $x_i = v$ and $t_{i,v} = 0$ otherwise.

$$\max \sum_j w_j z_j$$
$$\text{s.t.}$$
$$z_j \leq t_{i,v} \qquad \forall i, v, [x_i = v] \in C_j$$
$$\sum_{v \in [d]} t_{i,v} = 1$$
$$0 \leq t_{i,v} \leq 1 \qquad \forall i \in [n], \forall v \in [d]$$

$$\text{(CONJ)}$$

Lemma 12. *The linear program* (CONJ) *is* $(1 - o(1))$-*approximable in* NC.

Proof. Generalization of a result of [Tre96]. The proof is omitted from this extended abstract. $\square$

Lemma 13 (Random Rounding for MAX kCSP-d). *Let* $(\mathbf{z}, \mathbf{t})$ *be a feasible solution for* (CONJ). *Consider the random assignment obtained by setting, for any* i, v

$$\Pr[x_i = v] = (k - 1)/dk + t_{i,v}/k .$$

Then such an assignment has an average cost at least $\frac{1}{d^{k-1}} \sum_j w_j z_j$. *The analysis only assumes that the distribution is* k-*wise independent.*

Proof. It is sufficient to prove that any constraint C_j is satisfied with probability at least $\frac{1}{d^{k-1}} z_j$; the lemma will then follow by the linearity of expectation. Observe that if the atom $[x_i = v]$ occurs in C_j then $z_j \leq t_{i,v}$. Then

$$\Pr[C_j \text{ is satisfied}] \geq \left(\frac{k-1}{dk} + \frac{1}{k} z_j \right)^k \geq \frac{1}{d^{k-1}} z_j . \qquad (1)$$

For the last inequality, we consider the function

$$f(z) = \frac{\left(\frac{k-1}{dk} + \frac{1}{k}z\right)^k}{z}$$

in the interval $0 \leq z \leq 1$, compute its first derivative, and show that f has a minimum in $z = 1/d$, that is $f(z) \geq f(1/d) = 1/d^{k-1}$, $\forall z, 0 \leq z \leq 1$. In the first inequality of Eq. (1) we have assumed that the random variables induced by the clause C_j are independent. □

Remark. The above analysis is tight and establishes that the integrality gap of (CONJ) is d^{k-1}. The bound is achieved e.g. by the instance consisting of clauses $C_1, C_2, \ldots, C_{d^k}$ that are all possible size k (domain-d) conjunctions of $\{x_1, \ldots, x_k\}$.

Theorem 14. *For any $d = d(n)$ and $k = k(n)$ such that $d^k = n^{O(1)}$, there is an* NC *$(1/d^{k-1} - o(1))$-approximate algorithm for* MAX *kCSP-d. In particular, there is a $(1/d - o(1))$-approximate* NC *algorithm for* MAX CAPACITY REPRESENTATIVES-*d.*

4.1 The MAX FMIP Problems

A prototypical problem in MAX FMIP is MAX MAJORITY SAT, which is the variation of MAX SAT where a clause is satisfied if at least half the literals (rather than at least one) are satisfied. Baralnd et al. [BKT96] showed that this class contains only constant-approximable problems (using, once more, the syntactic structure of integer programs) and gave a structural explanation of this result.

It is easy to find a 2-approximate solution for MAX MAJORITY SAT. Any clause is either satisfied by the assignment $x_i = 0$, $\forall i$, or by the assignment $x_i = 1$, $\forall i$. Thus one of the two assignments satisfies at least half the clauses.[4]

For the general MAX FMIP problem, we present a slight improvement and simplification over the approximation result of Barland et al. [BKT96].

Theorem 15. *Given an instance of a* MAX FMIP *problem, the random assignment where each variable is set to $-k$ or to k with probability $1/2$ independently at random satisfies each constraint with probability at least $1/2^{1+\lceil \sigma/k \rceil}$, provided that the constraint is satisfiable.*

Proof. Consider a constraint $\sum_i a_i x_i \geq b$. If the constraint is satisfiable, then $\sum_i |a_i| k \geq b$. Since the a_i are integers, there must be a set J of at most $\lceil b/k \rceil$ indices such that $\sum_{i \in J} |a_i| k \geq b$. Under the uniform distribution, with probability at least $1/2^{|J|} \geq 1/2^{\lceil b/k \rceil}$ we will have $\sum_{i \in J} a_i x_i \geq b$. It is also easy to see that, by symmetry, with probability at least $1/2$ we have $\sum_{i \notin J} a_i x_i \geq 0$. The theorem thus follows since for the whole set of constraints, $b_j \leq \sigma$, $\forall j$. □

The above theorem can be derandomized in NC through the techniques of Karger and Kholler [KK94].

---

[4] This nice idea is due to Michel Goemans.

5 Negative Results: Hardness of Approximation

We first define Probabilistically Checkable Proof Systems and Multi-Prover One-Round Proof Systems. Our notation merges the notations of [BGLR93] and [BGS96]. For an integer d, we denote by $[d]^*$ the set of all strings over $[d]$.

Definition 16 (Verifier). A verifier V for a language L is a randomized polynomial time oracle Turing machine. V receives in input a string x and has oracle access to a string π that is an alleged proof that $x \in L$.

Definition 17 (PCP and MIP). Let $c, s, r, q, d : \mathcal{Z}^+ \to \mathcal{Z}^+$ such that $0 \leq s(n) < c(n) \leq 1$ for any n; we say that a language L belongs to $\text{PCP}_{c,s}[r, q, d]$ if there exists a verifier V such that

1. For any input string x and oracle proof $\pi \in [d(n)]^*$, V queries at most $q(n)$ entries of π and uses at most $O(r(n))$ random bits;
2. For any $x \in L$, there exists a $\pi \in [d(n)]^*$ such that the probability that V accepts x with oracle π is at least $c(n)$;
3. For any $x \notin L$, for any $\pi \in [d(n)]^*$, the probability that V accepts x with oracle π is at most $s(n)$.

The class $\text{MIP}_{c,s}[r, q, d]$ is similar, with the only difference that π is presented as a sequence of q strings $\pi_1, \ldots, \pi_q$, where $\pi_i \in [d]^*$, and V has the further restriction that it can read at most one entry of any π_i.

From the above definition it follows that $\text{MIP}_{c,s}[r, q, d] \subseteq \text{PCP}_{c,s}[r, q, d]$ for any choice of the parameters. The following result is folklore.

Theorem 18. *If* MAX kCSP-$(d(n))$ *is* $\rho(n)$-*approximable, then, for any* $c(n)$ *and* $s(n)$ *such that* $s(n)/c(n) < \rho\left(n^{O(1)} 2^{O(r(n))}\right)$, *it holds*

$$\text{PCP}_{c(n),s(n)}[r(n), k(n), d(n)] \subseteq \text{DTIME}\left(2^{O(r(n)+k(n)\log d(n))}\right) .$$

Theorem 19. *The following statements hold (n is the size of the input):*

(1) *A constant $c > 0$ exists such that, for any constant $d \geq 2$, it is NP-hard to approximate* MAX 2CSP-d *within $1/d^c$. Furthermore, for any $\varepsilon > 0$, it is infeasible to approximate* MAX 2CSP-$(\log n)$ *within $2^{\log^{1-\varepsilon} n}$ unless* $\text{NP} \subseteq \text{DTIME}\left(n^{\log^{O(1/\varepsilon)} n}\right)$.

(2) *For any constant d, for any $k \geq 3$, for any $\varepsilon > 0$, it is NP-hard to approximate* MAX kCSP-d *within $1/d^{\lfloor k/3 \rfloor} + \varepsilon$.*

(3) *Constants k and c exist such that it is NP-hard to approximate* MAX kCSP-$(\log n)$ *within $1/\log n^c$.*

(4) *For any $k \geq 5$, any $\varepsilon > 0$, it is NP-hard to approximate* MAX kCSP *within $2^{\log^{1/3-\varepsilon} n}$.*

(5) *For any $\varepsilon > 0$, a constant $k = O(1/\varepsilon)$ exists such that it is NP-hard to approximate* MAX kCSP *within $2^{\log^{1-\varepsilon} n}$.*

(6) *For any* $\varepsilon > 0$, MAX FMIP *problems are hard to approximate within* $1/2^{\lfloor \sigma/3 \rfloor} + \varepsilon$.

Proof. (Sketch) For **(1)**, Raz [Raz95] has shown that a constant $c' > 0$ exists such that, for any $k : \mathcal{Z}^+ \to \mathcal{Z}^+$, $\mathsf{NP} \subseteq \mathsf{MIP}_{1,2^{-ck(n)}}[k(n)\log n, 2, 3^{k(n)}]$. The first part of the claim follows by setting $k(n) = \lfloor \log_3 d(n) \rfloor$; the second part by setting $k(n) = \log^{O(1/\varepsilon)}(n)$. Next, for **(2)**, Håstad [Hås97] has shown that for any $\varepsilon > 0$, for any fixed prime p, $\mathsf{NP} = \mathsf{PCP}_{1-\varepsilon,1/p+\varepsilon}[\log, 3, p]$. The claim follows by choosing $p = k/3$. Further, **(3)**, **(4)** and **(5)** are re-statements of the results of Raz and Safra [RS97], and Arora and Sudan [AS97] using Theorem 18. Finally, **(6)** follows from the hardness of MAX kCSP-2 and from Theorem 11. $\square$

Barland et al. asked in [BKT96] whether the problem MAX CAPACITY REPRESENTATIVES($\log n$) is constant-approximable. Part **(1)** of Theorem 19 and Theorem 10 imply a negative answer to such question. Finally, it is worth to mention the almost tight non-approximability result for the problems of class MAX FMIP.

Acknowledgment

We thank an anonymous referee for useful comments.

References

[AS97] Arora, S., and Sudan, M. Improved low degree testing and its applications. In *Proc. of 29th ACM STOC* (1997) 485–495

[BC93] Bovet, D.P. and Crescenzi, P. *Introduction to the Theory of Complexity.* Prentice Hall (1993)

[Bel93] Bellare, M. Interactive proofs and approximation: Reductions from two provers in one round. In *Proc. of 2nd IEEE ISTCS* (1993)

[BGLR93] Bellare, M., Goldwasser, S., Lund, C., and Russell, A. Efficient probabilistically checkable proofs and applications to approximation. In *Proc. of the 25th ACM STOC* (1993) 294–304

[BGS96] Bellare, M., Goldreich, O., and Sudan, M. Free bits, PCP's and non-approximability – towards tight results (4th version). Technical Report TR95-24, ECCC (1996) Preliminary version in *Proc. of FOCS'95.*

[BKT96] Barland, I., Kolaitis, P.G., and Thakur, M.N. Integer programming as a framework for optimization and approximability. In *Proc. of 11th IEEE CCC* (1996) 249–259

[Cre95] Creignou, N. A dichotomy theorem for maximum generalized satisfiability problems. *J. of Comp. and Sys. Sci.*, (1995) 51(3):511–522

[DSST97] Díaz, J., Serna, M., Spirakis, P., and Torán, J. *Paradigms for fast parallel approximability.* Camb. Univ. Press (1997)

[ENRS95] Even, G., Naor, J., Rao, S., and Schieber, B. Divide-and-conquer approximation algorithms via spreading metrics. In *Proc. of 36th IEEE FOCS* (1995) 62–71

[GW94]   Goemans, M.X., and Williamson, D.P. New 3/4-approximation algorithms for the maximum satisfiability problem. *SIAM J. of Disc. Math.* (1994) 7:656–666

[GW96]   Goemans, M.X., and Williamson, D.P. The primal-dual method for approximation algorithms and its application to the network design problems. In *Approximation Algorithms for NP-hard Problems*. PWS Pub. (1996)

[Hås97]   Håstad, J. Some optimal inapproximability results. In *Proc. of 29th ACM STOC* (1997) 1–10

[KK94]   Karger, D.R. and Koller, D. (De)randomized construction of small sample spaces in $\mathcal{NC}$. In *Proc. of 35th IEEE FOCS* (1994) 252–263

[KM96]   Khanna, S., and Motwani, R. Towards a syntactic characterization of PTAS. In *Proc. of 28th ACM STOC* (1996) 329–337

[KMSV94] Khanna, S., Motwani, R., Sudan, M., and Vazirani, U. On syntactic versus computational views of approximability. In *Proc. of 35th IEEE FOCS* (1994) 819–830

[KSW97]   Khanna, S., Sudan, M., and Williamson, D.P. A complete classification of the approximability of maximization problems derived from boolean constraint satisfaction. In *Proc. of 29th ACM STOC* (1997) 11–20

[KT94]   Kolaitis, P.G. and Thakur, M.N. Logical definability of NP optimization problems. *Inf. and Comp.* (1994) 115(2):321–353

[KT95]   Kolaitis, P.G. and Thakur, M.N. Approximation properties of NP minimization classes. *J. of Comp. and Syst. Sci.* (1995) 50:391–411

[LLR95]   Linial, N., London, E., and Rabinovich, Y. The geometry of graphs and some of its algorithmic applications. *Combinatorica* (1995) 15(2):215–245

[LN93]   Luby, M., and Nisan, N. A parallel approximation algorithm for positive linear programming. In *Proc. of 25th ACM STOC* (1993) 448–457

[LW96]   Lau, H.C., and Watanabe, O. Randomized approximation of the constraint satisfaction problem. In *Proc. of 5th SWAT*, LNCS 1097, Springer-Verlag (1996) 76–87

[PR93]   Panconesi, A., and Ranjan, D. Quantifiers and approximations. *Theoret. Comp. Sci.* (1993) 107:145–163

[PY91]   Papadimitriou, C.H., and Yannakakis, M. Optimization, approximation, and complexity classes. *J. of Comp. and Syst. Sci.* (1991) 43:425–440

[Raz95]   R. Raz. A parallel repetition theorem. In *Proc. of 27th ACM STOC* (1995) 447–456

[RS97]   Raz, R., and Safra, S. A sub-constant error-probability low-degree test, and a sub-constant error-probability PCP characterization of NP. In *Proc. of 29th ACM STOC* (1997) 475–484

[RT87]   Raghavan, P., and Thompson, C.D. Randomized rounding: a technique for provably good algorithms and algorithmic proofs. *Combinatorica* (1987) 7:365–374

[Tre96]   Trevisan, L. Positive linear programming, parallel approximation, and PCP's. In *Proc. of 4th ESA*, LNCS 1136, Springer-Verlag (1996) 62–75

Interactive Protocols on the Reals

Sergei Ivanov[1] and Michel de Rougemont[2]

[1] Steklov Institute, St. Petersburg, Russia
[2] Université Paris-II & LRI Bâtiment 490,
F-91405 Orsay Cedex, France

Abstract. We introduce the classes $IP_{\mathcal{R}_+}$ (resp. $IP_{\mathcal{R}_\times}$) as the class of languages that admit an interactive protocol on the reals when the verifier is a BSS-machine with addition (resp. addition and multiplication). Let $BIP_{\mathcal{R}_+}$ (resp. $BIP_{\mathcal{R}_\times}$) its restriction when only boolean messages can be exchanged between the prover and the verifier. We prove that the classes $BIP_{\mathcal{R}_+}$ and $PAR_{\mathcal{R}_+}$, the class of languages accepted in parallel polynomial time coincide. In the case of multiplicative machines, we show that $BIP_{\mathcal{R}_\times} \subseteq PAR_{\mathcal{R}_\times}$.

We also separate $BIP_{\mathcal{R}}$ from $IP_{\mathcal{R}}$ in both models by exhibiting a language L which is not in $PAR_{\mathcal{R}_\times}$ but in $IP_{\mathcal{R}_+}$. As a consequence we show that additive quantifier elimination can't be solved in $PAR_{\mathcal{R}_\times}$ and that all boolean languages are in $IP_{\mathcal{R}_+}$.

1 Introduction

The complexity class IP [GMR89, Bab85] is the class of languages L for which there exists an interactive proof. A *random verifier* in such interactive proofs can be far more efficient than a classical one for discrete combinatorial problems. In this paper, we consider continuous problems, define the notion of an interactive proof in this context, and show similar results.

Let us suppose that computations take place over the set $\mathbf{R}$ of reals, i.e that the verifier is a BSS-machine [BSS89] and that the prover and the verifier exchange real numbers at a unit cost. A BSS-program is a sequence of instructions built from a finite set of parameters with the operations $+, -, <$ in the case of additive machines and $+, -, \times, <$ in the case of multiplicative machines. Natural classes P and NP generalize to $P_{\mathcal{R}}$ and $NP_{\mathcal{R}}$ but the class $PSPACE$ identical to PAR (the class of P-uniform polynomial depths circuits) in the classical theory has only $PAR_{\mathcal{R}}$ as a generalization [Cuc93, CK95]. One of the goals of this paper is to study the generalization of Shamir's classical result $IP = PSPACE$ [Sha90] in the case of real computations. We write $IP_{\mathcal{R}}$ to denote both classes $IP_{\mathcal{R}_+}$ (additive machines) and $IP_{\mathcal{R}_\times}$ (multiplicative machines) in the case of general properties that hold for both classes.

In the class $IP_{\mathcal{R}}$, the verifier is a $BPP_{\mathcal{R}}$ algorithm where random reals can be selected and reals are exchanged between the prover and the verifier. We study the influence of boolean exchanges vs. real exchanges and define the class BIP_R (Boolean IP) as the class of languages that admit IP protocols where exchanges between the prover and the verifier are limited to boolean values and the verifier uses boolean tossings as in [CKK+95]. Our main results are :

Theorem 1 : *On additive machines, $BIP_{R_+} = PAR_{R_+}$ whereas on multiplicative machines $BIP_{R_\times} \subseteq PAR_{R_\times}$.*

We then study a natural integral problem INT : given a vector $(x, 1, 1, ..., 1) \in \mathbf{R}^n$, decide if x in an integer less than 2^{2^n}. We show that INT is not in the class $PAR_{R_\times}$, yet admits an IP_{R_+} protocol. We then obtain:

Theorem 2 : *On both additive and multiplicative machines, the class BIP_R is strictly included in the class IP_R. The class $PAR_{R_\times}$ does not contain IP_{R_+}.*

As a consequence, we obtain that additive quantifier elimination is not in the class $PAR_{R_\times}$. We also show that all boolean languages are in $IP_{R_\times}$. In the second section, we set the notations for IP_R and BIP_R and recall some key properties of real computations that we will use. In section 3, we show that BIP_R is always contained in PAR_R. In section 4, we show that $BIP_{R_+} = PAR_{R_+}$ for additive machines by giving a BIP_{R_+} protocol for a PAR_{R_+}-complete problem. In section 5, we show that an integral problem INT is not in $PAR_{R_\times}$ and give an IP_{R_+} protocol for this problem. We then show how to use this protocol to solve any boolean problem.

2 Interactive proofs and computations on the reals

Classical interactive proofs were designed for boolean inputs, i.e. $S \subset \{0, 1\}^*$. As complexity theory can be generalized to arbitrary structures [Poi95], we study such interactive proofs on the structures $\mathcal{R}_+ = (\mathbf{R}, +, <, 0, 1)$ and $\mathcal{R}_{+,\times} = (\mathbf{R}, +, \times, <, 0, 1)$ where $\mathbf{R}$ is the set of real numbers. In this case computations are made as in the BSS model [BSS89], where the prover P and the verifier V exchange real numbers at a unit cost.

In an interactive proof, a prover P exchanges information with a limited verifier V. On an input x of length n, V accepts after m rounds ($P.V(x) = 1$) or rejects ($P.V(x) = 0$). A *protocol* is a BPP-algorithm that the verifier follows and a set S has an interactive proof [GMR89, Bab85] if there exists a protocol such that if $x \in S$, then there exists a prover P such that $\mathbb{P}rob[P.V(x) = 1] = 1$ and if $x \notin S$, then for all prover P', $\mathbb{P}rob[P'.V(x) = 1] \leq \frac{1}{2}$. Notice that the first condition (the completeness condition) was in [GMR89] : if $x \in S$, then there exists a prover P such that $\mathbb{P}rob[P.V(x) = 1] > 1 - |x|^{-k}$. It is however equivalent to our definition, as in [Bab94].

2.1 Boolean Interactive Proofs : $BIP_{\mathcal{R}}$

Let V be a verifier in an interactive proof system deciding a set $S \subset \bigcup_n \mathbf{R}^n$. Denote $S_n = S \cap \mathbf{R}^n$. For an input $x \in \mathbf{R}^n$, V uses a probabilistic sample $\sigma \in \{0,1\}^k$ (which may be unknown to P) and finishes after $m-1$ rounds of exchanges. Here k and m are polynomial functions of n. In a $BIP_{\mathcal{R}}$ protocol, we require that the prover P and the verifier V only exchange bits.

We denote by $(w_1, p_1, \ldots, w_{m-1}, p_{m-1}, w_m) \in \{0,1\}^{2m-1}$ a sequence of questions and responses in a protocol. In such a protocol, $w_1 \in \{0,1\}$ is the first question of V, $p_1 \in \{0,1\}$ is the first answer of the prover, etc, and $w_m \in \{0,1\}$ is the result of computations (1 if V accepts, 0 otherwise). The questions w_j are obtained as $w_j = V_j(\sigma, x, p_1, \ldots, p_{j-1})$ where V_j are functions computed by V (in polynomial time). The responses p_j are $p_j = P_j(x, w_1, \ldots, w_j)$ where $\{P_j\}$ are arbitrary functions (computed by P). The behavior of P is completely determined by the functions $\{P_j\}$. We say "for any prover P" instead of "for any family of functions P_j". The notation $w_m = 1$ is equivalent to $P.V(x) = 1$.

2.2 Computations on the reals

We will need some basic computations on the structures $\mathcal{R}_+ = (\mathbf{R}, +, -, <, 0, 1)$ and $\mathcal{R}_\times = (\mathbf{R}, +, -, \times, <, 0, 1)$. We can multiply a real by any small (polynomial size) integer and solve a linear system $A \cdot x = y$ where A is a matrix with integer coefficients and y is a real vector, i.e. find a pair (x', m) where x' is a real vector, m is an integer and $\frac{x'}{m}$ is a solution of the system. Let $v_1, \ldots, v_k \in \mathbf{Z}^n \subset \mathbf{R}^n$ be vectors with integer coordinates given in binary notation. Let $x \in \mathbf{R}^n$ be any vector. There is a polynomial algorithm for additive BSS-machine which does the following :

1. It checks if the vectors $\{v_i\}$ are linearly independent, and if so, if x is a linear combination of them.
2. If x is such a linear combination, it finds the (unique) coefficients of the linear combination of the form "real over integer".

The problem reduces to solving a linear system with integer coefficients and real constants in the right-hand part. We now describe more specific properties of the structure of polyhedral cones. A detailed exposition of the subject can be found in [SW70]. We say that a vector $x \in \mathbf{R}^n$ is a positive linear combination of vectors $v_1, \ldots, v_k$ if $x = \sum \lambda_i v_i$ for some positive real numbers $\lambda_1, \ldots, \lambda_k$. We allow the collection $\{v_i\}$ to be empty in which case we assign zero value for the sum. We will use *Caratheodory's theorem* : If a vector x is a positive linear combination of a family $\{v_i\}_{i \in I}$, then $\{v_i\}$ contains a linearly independent sub-family $\{v_j\}_{j \in J \subset I}$ such that x is a positive linear combination of $\{v_j\}$. Note that the sub-family $\{v_j\}$ depends on both $\{v_i\}$ and x.

A set $C \subset \mathbf{R}^n$ is a (convex) *polyhedral cone* if it is the set of solutions of a finite system of homogeneous linear inequalities, $C = \{x \in \mathbf{R}^n : \forall j \in J \quad L_j x \geq 0\}$

where J is a finite set and L_j are linear functions from $\mathbf{R}^n$ to $\mathbf{R}$. Such a cone is called *line-free* if it contains no straight lines. This condition is trivially equivalent to that $\cap_{j \in J} \ker L_j = \{0\}$. For a non-zero vector $v \in \mathbf{R}^n$ we denote by $R(v)$ the ray (half-line) $\{\lambda v \; : \; \lambda \geq 0\} \subset \mathbf{R}^n$. A ray $R(v)$ is an *edge* of C if $v \in C$ and v is a solution of a nondegenerated $(n-1) \times n$ linear system of the form $\{L_{j_1}(v) = 0, \ldots, L_{j_{n-1}}(v) = 0\}$ where $j_r \in J$ for $r = 1, \ldots, n-1$. Clearly there are only finitely many of edges of C.

The *Minkowski sum* of sets X and Y in $\mathbf{R}^n$ is the set $X + Y = \{x + y \; : \; x \in X, y \in Y\}$. Note that the sum of several rays emanating from the origin coincide with their *convex hull*, i.e. the minimal convex set containing them. We will use the following result which is a special case of a classical Minkowski theorem ([SW70], Theorem 2.8.6) :

Proposition 1 *Any line-free polyhedral cone $C \subset \mathbf{R}^n$ is the Minkowski sum of its edges, i.e.,*

$$C = \{0\} + R(v_1) + \ldots + R(v_N) = \left\{ \sum \lambda_i v_i \; : \; \lambda_i \geq 0 \right\}$$

where $\{v_i\}$ are vectors generating the edges of C.

Proof : Every convex polyhedral set is the convex hull of its primitive faces, i.e. those faces which are congruent to linear subspaces or half-subspaces ([SW70], 2.5.2, 2.5.4). Since C contains no lines, all its primitive faces are half-lines and points (vertices). The proposition follows since the only vertex of a cone is the origin $\{0\}$. (See also [Kle57] for generalizations of this statement to arbitrary convex sets.)

3 BIP_R is contained in PAR_R

In this section, all the results concern both additive and multiplicative machines. Denote by $Pass(\sigma, x, w_1, p_1, \ldots, p_{j-1}, w_j)$ the relation

$$V_1(\sigma, x) = w_1 \; \& \; V_2(\sigma, x, p_1) = w_2 \; \& \ldots \& \; V_j(\sigma, x, p_1, \ldots, p_{j-1}) = w_j.$$

This relation means that, for given x and σ, a sequence of questions $w_1, \ldots, w_j$ is generated by V if the answers of P are $p_1, \ldots, p_{j-1}$. For a fixed language S and verifier V, we define the integer-valued functions for $j = 1, \ldots, m$:

$$Q_j(x, w_1, p_1, w_2, p_2 \ldots, p_{j-1}) = \max \#\{\sigma \; : \; Pass(\sigma, x, w_1, \ldots, w_{j-1}) \; \& \; w_m = 1\}$$
$$W_j(x, w_1, p_1, \ldots, p_{j-1}, w_j) = \max \#\{\sigma \; : \; Pass(\sigma, x, w_1, p_1, \ldots, p_{j-1}, w_j) \; \& \; w_m = 1\}$$

where the maxima are taken over all provers that give responses $p_1, \ldots, p_{j-1}$ to the questions $w_1, \ldots, w_{j-1}$ on the input x. In other words, these functions calculate 2^k times the probability that V starts with questions $w_1, \ldots, w_{j-1}$ (resp. $w_1, \ldots, w_j$) and accepts at the end, provided that P answers $p_1, \ldots, p_{j-1}$ in the first $j-1$ rounds and behaves optimally after that. Notice that the function Q_j depends on p_{j-1} because of the maximum and that $w_m = 1$ (V accepts) depends on the prover.

Lemma 1

(1) $x \in S_n \iff Q_1(x) > 2^{k-1}$.

(2) $Q_j(x, w_1, \ldots, p_{j-1}) = \sum_{w_j \in \{0,1\}} W_j(x, w_1, \ldots, p_{j-1}, w_j)$ for $j = 1, \ldots, m$.

(3) $W_j(x, w_1, \ldots, w_j) = \max_{p_j \in \{0,1\}} Q_{j+1}(x, w_1, \ldots, w_j, p_j)$ for $j = 1, \ldots, m-1$.

(4) $W_m(x, w_1, \ldots, w_m) = w_m \cdot \#\{\sigma \; : \; Pass(\sigma, x, w_1, p_1, \ldots, w_m)\}$.

Proof : (1) is an immediate consequence of the definitions.
The right-hand side of (4) is equal to the expression after "max" in the definition of W_m. This expression does not actually depends on prover, so "max" can be omitted.
To prove (2), note that the equality $\#\{\sigma \; : \; Pass(\sigma, x, w_1, \ldots, w_{j-1}) \; \& \; w_m = 1\} = \sum_{w_j \in \{0,1\}} \#\{\sigma \; : \; Pass(\sigma, x, w_1, \ldots, p_{j-1}, w_j) \; \& \; w_m = 1\}$ holds for any fixed prover P.
So it suffices to check that there is a prover which gives the maxima for both summands in the right-hand side. Let P_0 and P_1 be provers realizing the maxima for $w_j = 0$ and $w_j = 1$, respectively. Since the strategy of a prover can be chosen arbitrarily, one can define a new prover P which acts as P_0 if $w_j = 1$ and as P_1 if $w_j = 0$. Clearly P gives the maxima for both cases.
The equality (3) simply expresses the fact that all the provers involved in the definition of $W_j(x, w_1, \ldots, w_j)$ are divided into two disjoint classes : those for which $p_j = P_j(x, w_1, \ldots, w_j) = 0$, and those for which $p_j = 1$. The left-hand side of (3) is the maximum of a certain function over provers of both classes, while in the right-hand side one takes the maxima of the same function for two classes separately and let the larger one be the result. So the right- and left-hand sides of (3) are equal.

Note that the function Q_m can be easily computed in parallel polynomial time by means of the equality (4) from Lemma 1. Other statements of Lemma 1 can be thought of as recursive rules for computing the characteristic function of S_n. Since the depth of recursion is equal to $m = poly(n)$, the computation can be done in parallel polynomial time. It follows that the original problem S belongs to PAR and we obtain:

Lemma 2 *The classes $BIP_{\mathcal{R}_+}$ (resp $BIP_{\mathcal{R}_\times}$) are included in the classes $PAR_{\mathcal{R}_+}$ (resp. $PAR_{\mathcal{R}_\times}$).*

4 $PAR_{\mathcal{R}_+}$ is in $BIP_{\mathcal{R}_+}$ for additive machines

In [CK95] the problem called DTRAO (digital theory of reals with addition and order) is proved to be $PAR_{\mathcal{R}_+}$-complete. It is the problem of evaluating a first-order sentence built from real constants, variables, additions, subtractions, comparisons and quantifiers of the form Qx where $Q \in \{\exists, \forall\}$ and $x \in \{0, 1\}$. More precisely, the set DTRAO can be defined as the set of all true formulae of the form

$$\Phi = Q_1 y_1 \in \{0, 1\} \ldots Q_m y_m \in \{0, 1\} \; F(y_1, \ldots, y_m)$$

where F is a quantifier-free formula. We may assume that the literals of F are of the form $S \geq 0$ where S is constructed by additions and subtractions from the variables x_i and real constants. We will give a $BIP_{\mathcal{R}_+}$ protocol for this problem.

We first transform the above formula Φ into a homogeneous (constant-free) form. Let F contain $n-1$ nonzero constants. Introduce n new variables $x_1, \ldots, x_n$ and let Ψ be the formula

$$\Psi(x_1, \ldots, x_n) = Q_1 y_1 \in \{0, x_1\} \ldots Q_m y_m \in \{0, x_1\} \ G(x_1, \ldots, x_n, y_1, \ldots, y_m)$$

where G is obtained from F by replacing each nonzero constant by one of the variables $x_2, \ldots, x_n$. The problem DTRAO reduces to a new one, HDTRAO (H stands for "homogeneous"), whose set of solutions is the set of pairs (Ψ, x) in which Ψ is a constant-free formula in the above form and $x = (x_1, \ldots, x_n)$ is a vector such that $\Psi(x)$ is true. Let h denote the length of Ψ and n the number of its free variables. The size of input (Ψ, x) is then equal to $h + n$.

4.1 h-decomposition of a real

The idea of our protocol is to replace an input $x \in \mathbf{R}^n$ by a small integer vector v such that $\Psi(v)$ is equivalent to $\Psi(x)$. The formula $\Psi(v)$, to which $\Psi(x)$ is equivalent, reduces to an instance of QBF, so a known IP protocol for QBF [Sha90] can be used to evaluate it.

Lemma 3 Let $x, x' \in \mathbf{R}^n$ be vectors such that $\operatorname{sign}(L(x)) = \operatorname{sign}(L(x'))$ for every linear function $L : \mathbf{R}^n \to \mathbf{R}$ with integral coefficients of size $\leq h$. Then the values $\Psi(x)$ and $\Psi(x')$ are equal.

Proof : By a standard procedure, Ψ can be transformed into a quantifier-free formula (of exponential length) composed of terms of the form $G_{\alpha_1, \ldots, \alpha_m}(x_1, \ldots, x_n) = G(x_1, \ldots, x_n, \alpha_1, \ldots, \alpha_m)$ where $\alpha_i \in \{0, x_1\}$. This formula is also constant-free and all of its terms can be written in the form $L(x_1, \ldots, x_n) \geq 0$ where L is a linear function with small integral coefficients (the absolute values of these coefficients are bounded by h). Therefore the value $\Psi(x)$ is determined by the signs of those linear functions at x, and the lemma follows.

To be able to obtain an integral vector v which is equivalent to a given $x \in \mathbf{R}^n$ in the sense of Lemma 3, as well as to verify this property, we use the decomposition technique described below.

Definition 1 For $h \in \mathbf{N}$ and $x \in \mathbf{R}^n$, we say that a finite (possibly empty) collection of vectors $v_1, \ldots, v_k \in \mathbf{Z}^n$ is an h-decomposition of x, if :

1. $size(v_i) \leq hn^2$ for all i.
2. $v_1, \ldots, v_k$ are linearly independent.
3. x is a positive linear combination of vectors $\{v_i\}$.
4. For all linear functions $L : \mathbf{R}^n \to \mathbf{R}$ with integral coefficients of size $\leq h$, either $L(v_i) \geq 0$ for all i, or $L(v_i) \leq 0$ for all i.

Basic properties of convex polyhedral cones lead to :

Proposition 2 For every $h \in \mathbf{N}$ and $x \in \mathbf{R}^n$ there exists an h-decomposition of x.

Proof : For $x \in \mathbf{R}^n$ consider the polyhedral cone $C_h(x)$ obtained as follows. Take all the homogeneous linear inequalities whose coefficients are integers of size $\leq h$ and for which x is a solution, and consider them as one large system. Let $C_h(x)$ be the set of solutions of that system. Note that the polyhedral cone $C_h(x)$ is line-free. Indeed, for each $i = 1, \ldots, n$ at least one of the inequalities $x_i \geq 0$ and $-x_i \geq 0$ is among those defining $C_h(x)$, and therefore $C_h(x)$ cannot contain a line generated by a vector whose ith coordinate is nonzero.

Let $v_1, \ldots, v_N$ be vectors generating the edges of $C_h(x)$. These vectors are obtained as solutions of non-degenerated homogeneous $(n-1) \times n$ linear systems with integer coefficients of size $\leq h$. So we may choose $\{v_i\}$ to have integer coordinates of size $\leq hn^2$ (e.g., those given by Cramer's rule), and therefore satisfying the requirement (1) of Definition 1. By the construction of $C_h(x)$ and the fact that $v_i \in C_h(x)$, they also satisfy (4) of Definition 1. By Proposition 1,

$$C_h(x) = \left\{ \sum \lambda_i v_i \ : \ \lambda_i \geq 0 \right\}.$$

In particular, $x = \sum_{i=1}^{N} \lambda_i v_i$ for some nonnegative values λ_i. Removing those v_i with $\lambda_i = 0$ from the collection, we obtain a decomposition with positive coefficients. By Caratheodory's theorem, a linearly independent sub-family of vectors can be extracted from $\{v_i\}$ such that x is a positive linear combination of vectors of this sub-family. Then this sub-family is a desired h-decomposition.

Lemma 4 *Let $\{v_1, \ldots, v_k\}$ be an h-decomposition of x, and let $v = \sum v_i$. Then $\Psi(x)$ is equal to $\Psi(v)$.*

Proof : By Lemma 3, it suffices to prove that for any linear function $L : \mathbf{R}^n \to \mathbf{R}$ with integer coefficients of size $\leq h$, one has $sign(L(v)) = sign(L(x))$. By the definition of h-decomposition, all nonzero values $L(v_i)$ have the same sign. We may assume that $L(v_i) \geq 0$ for all i (otherwise take $-L$). If $L(v_i) = 0$ for all i, then $L(x) = L(v) = 0$. Otherwise $L(v_i) > 0$ for at least one i, and therefore $L(v) > 0$ and $L(x) > 0$. In both cases, $sign(L(v)) = sign(L(x))$.

4.2 A $BIP_{\mathcal{R}_+}$ protocol for HDTRAO

Given a formula Ψ of length h and $x \in \mathbf{R}^n$, the protocol is :

HDTRAO-Protocol:

1. The verifier asks the prover for an integer $k \geq 0$ and integer vectors $v_1, \ldots, v_k \in \mathbf{Z}^n$ (in binary representation) that form an h-decomposition of x.

2. The verifier checks that $\{v_i\}$ are linearly independent and solves the system $\sum \lambda_i v_i = x$ in unknowns λ_i. (If the solution exists, it is unique and can be found as fractions of the form "real over integer" : see 2.2.) The verifier then checks that its components λ_i are positive.

3. The verifier uses a standard IP protocol to check that $\{v_i\}$ satisfy the condition (4) from Definition 1.

4. The verifier computes the (binary representation of) vector $v = v_1 + \ldots + v_k \in Z^n$ and constructs a quantified boolean formula which is equivalent to $\Psi(v)$.

5. The verifier uses an IP protocol for QBF [Sha90] to check that the formula $\Psi(v)$ is true. The verifier accepts (Ψ, x) if the QBF protocol accepts.

Lemma 5 *The HDTRAO-Protocol is a $BIP_{\mathcal{R}_+}$ protocol that verifies if $(\Psi, x) \in HDTRAO$.*

Proof : In step 3, there is a standard protocol because the condition is co-NP for the standard Turing machines and the v_i are integer vectors of polynomial size. In step 4, the construction of the QBF formula is as follows: write

$$\Psi(v) = Q_1 y_1 \in \{0, v^1\} \ldots Q_m y_m \in \{0, v^1\} \ G(v^1, \ldots, v^n, y_1, \ldots, y_m)$$

(where v^i are the coordinates of v) in the form

$$Q_1 b_1 \in \{0, 1\} \ldots Q_m b_m \in \{0, 1\} \ G(v^1, \ldots, v^n, b_1 \cdot v^1, \ldots, b_m \cdot v^1)$$

and then express the arithmetics of $G(\ldots)$ in terms of the boolean variables b_i and the binary digits of v^i. The hypotheses of the lemma 4 are checked in the steps 1,2 and 3 and assure that $\Psi(v)$ is true iff $(\Psi, x) \in HDTRAO$. The protocols used in steps 3 and 5 can have an arbitrary small probability of error and the total error can be made less than $\frac{1}{2}$.

With the inclusion proved in the previous section (lemma 2), we obtain :

Theorem 1 *The classes $BIP_{\mathcal{R}_+}$ and $PAR_{\mathcal{R}_+}$ coincide.*

5 The interactive complexity of integral problems

In this section we prove some separation results. Although a (multiplicative) BSS-machine can compute large integers (e.g., the number 2^{2^n} can be obtained by n squarings), it is a hard problem to decide whether a given large real *is* an integer. We consider a problem of this kind (INT, see below) which is not in $PAR_{\mathcal{R}_x}$ and give an $IP_{\mathcal{R}}$ protocol for it. The protocol only uses multiplication to multiply reals by large powers of two (like 2^{2^n}). We show that this kind of multiplication (the problem $PROD_n$, below) can be verified by an IP protocol for an *additive* machine, so the protocol for INT is an $IP_{\mathcal{R}_+}$ one. This separates IP from BIP in both additive and multiplicative cases. As an application of the INT protocol, we show that all subsets of $\{0, 1\}^*$ are in $IP_{\mathcal{R}_+}$.

We denote by $[x]$ the integral part of an $x \in R$. For $0 \leq x \leq 2^n$, $[x]$ can be computed by an additive machine in time $O(n^2)$: just compute the numbers $2^0, 2^1, \ldots, 2^n$ (each is obtained by adding the previous one to itself) and subsequently decrease x by the maximal possible one of them so that x remains nonnegative. The process stops in at most n steps and the sum of the subtracted numbers is the desired integral part. Below we consider numbers that are too large for such algorithms.

Define the following decision problems:

- INT_n, deciding if an input $(x, 1, ...1) \in \mathbf{R}^n$ satisfies $x \in \mathbf{Z} \cap [0, 2^{2^n}]$.
- $PROD_n$, deciding if an input $(x, y, 1, ..., 1) \in \mathbf{R}^n$ satisfies $x = y \cdot 2^{2^n}$.

We will also denote by $INT_n(x)$ and $PROD_n(x, y)$ the respective boolean values for given $x, y \in \mathbf{R}$, and by INT and $PROD$ the unions of the affirmative sets over all n.

Remark : These problems are in the class "quantifier elimination" as they can be defined by the following formulae : $PROD_0(x, y) \Leftarrow x = y + y$,
$INT_0(x) \Leftarrow (x = 0 \vee x = 1 \vee x = 2)$,
$INT_n(x) \Leftarrow \exists y, y', z \, (x = y' + z \wedge PROD_{n-1}(y', y) \wedge INT_{n-1}(y) \wedge INT_{n-1}(z))$ and $PROD_n(x, y) \Leftarrow \exists z \, (PROD_{n-1}(x, z) \wedge PROD_{n-1}(z, y))$. With new quantified variables, a standard procedure expands these definitions to quantified formulae of polynomial length. Notice that the above relations contain additions and equality tests only. Hence the following negative result for the class $PAR_{\mathcal{R}_x}$ also applies to the problem of *quantifier elimination* in the theory of reals with addition.

Proposition 3 *The problem INT does not belong to $PAR_{\mathcal{R}_x}$.*

Proof : It is known ([Cuc93]) that if $S \subset \mathbf{R}^n$ is recognizable in parallel time t, then

$$t = \Omega \left(\sqrt{\frac{\log_2(\text{number of connected components of } S)}{n}} \right).$$

For the set $S = \{(x_1, ..., x_n) : x_1 \in \mathbf{Z} \cap [0, 2^{2^n}]\}$, the number of connected components is $2^{2^n} + 1$, so $t = \Omega(2^{n/2}/\sqrt{n})$.

5.1 Deciding $PROD$ with an $IP_{\mathcal{R}_+}$ protocol on additive machines

The following interactive protocol for additive machines checks $PROD_n(x, y)$ in time $O(n^2)$.

PROD-protocol :
1. Let $m := n$.
2. If $m = 0$, accept if $x = y + y$ and reject otherwise.
3. Get a $z \in \mathbf{R}$ from the prover (a correct prover must give $z = 2^{2^{m-1}}$). Pick a random $k \in \{1, ..., 2n\}$ and compute $x' = x + kz$ and $y' = z + ky$.
4. Go to step 2 with $m := m - 1$, $x := x'$, $y := y'$.

Proposition 4 *The PROD-protocol above is an $IP_{\mathcal{R}_+}$ protocol for the problem PROD.*

Proof : At step 2, $PROD_m(x,y)$ is implied by $PROD_{m-1}(x,z) \wedge PROD_{m-1}(z,y)$, and these expressions are equivalent for a correct prover. Since the relation $PROD_{m-1}$ is linear, it is either true for (x,z), (z,y) and all pairs $(x+kz, z+ky)$, or false for all these pairs except at most one. Thus the probability to get a true $PROD_{m-1}(x',y')$ from a false $PROD_m(x,y)$ is at most $1/2n$, so finally the probability of error is $\leq 1/2$.

Corollary 1 *There is an $IP_{\mathcal{R}_+}$ protocol for the following problem: for given $x,y \in \mathbf{R}$ and a binary representation of an integer $N \geq 0$, decide whether $x = y \cdot 2^N$.*

Proof : Let n be the size of N. The binary digits of N determine its representation in the form $2^{n_1} + \ldots + 2^{n_k}$ where $0 \leq n_1 < \ldots < n_k < n$. Multiplying by N then reduces to several (at most n) subsequent multiplications by numbers of the form 2^{2^m}, $m < n$. For each of these operations, the verifier may get the result from the prover and verify it by the $PROD_m$ protocol (repeated several times so that the error probability becomes $\leq 1/2n$).

5.2 Checking INT with an $IP_{\mathcal{R}_+}$ protocol

We describe an $IP_{\mathcal{R}_+}$ protocol checking $INT_n(x)$ where the prover and verifier exchange nonnegative reals. When we say that the verifier "asks to divide a by b", we mean that it gets two reals x and y from the prover and verifies that $bx + y = a$, $x \geq 0$, and $0 \leq y < b$. (The correct prover must give integer values x and y if a and b are integers.) When numbers of the form 2^{2^m} need to be computed or multiplied, the verifier gets the result from the prover and verifies it by means of the $PROD$-protocol given in the previous section.

INT-protocol :
1. Get the value $N = 2^{2^n}$ from the prover and verify it by the PROD-protocol. If $x = N$, accept immediately, if $x > N$ or $x < 0$, reject.
2. Calculate $q = (4n)!$. The binary size of q is polynomial.
3. Let $m := n$ and iterate the following procedure.
4. If $m = 0$, check that $x < 2q$ and that x is an integer, using a trivial algorithm for the integral part mentioned at the beginning of section 5. Accept or reject according to these tests.
5. Ask the prover to divide x by q. Get x' and z such that $x = q \cdot x' + z$, $x' \geq 0$, $0 \leq z < q$. Check that z is an integer, using the algorithm as in step 4.
7. Ask the prover to divide x' by $2^{2^{m-1}}$. Get nonnegative reals u and v, and verify that $x' = 2^{2^{m-1}} u + v$ by means of the PROD-protocol.
8. Pick a random integer $k \in \{1, \ldots, 4n\}$ and let $x'' = u + kv$.
9. Let $m := m - 1$, $x := x''$, and go to step 4.

Proposition 5 *The INT-protocol is an $IP_{\mathcal{R}_+}$ protocol for the problem INT.*

Proof : The cycle 4–9 is repeated at most n times. The multiplications used in the protocol all have either a polynomial-size integer or a number of the form 2^{2^m} as one of two arguments. The multiplications of the first kind can be reduced to polynomially many additions, and those of the second kind are made by means of the $IP_{\mathcal{R}_+}$ PROD-protocol. Hence the INT-protocol can be performed by an additive machine in polynomial time.

Assume first that $INT_n(x)$ is true and the prover is correct. In this case, at step 4 one always has $x \in \mathbf{Z}$ and $x < q \cdot 2^{2^m}$. This can be proved by induction: if $x < q \cdot 2^{2^m}$, then $x' \leq x/q < 2^{2^m}$, $u \leq x'/2^{2^{m-1}} < 2^{2^{m-1}}$, and $v < 2^{2^{m-1}}$ as a remainder in the division operation, therefore $x'' = u + kv \leq (k+1) \cdot 2^{2^{m-1}} < q \cdot 2^{2^{m-1}}$. Since the prover is correct, all numbers involved are integers. It follows that at the final iteration (when $m = 0$), both tests $x < 2q$ and $x \in \mathbf{Z}$ pass and the verifier accepts.

Now suppose that $x \notin \mathbf{Z}$ at step 4 and consider further computations up to step 9. In the PROD-protocol involved, the probability of error can be made $< 1/4n$ by $O(\log n)$ repetitions. We will then estimate the probability that $x'' \notin \mathbf{Z}$ assuming that the relation $x' = 2^{2^{m-1}} u + v$ is true. At step 5, $x' \notin \mathbf{Z}/q := \{y : qy \in \mathbf{Z}\}$ since $x \notin \mathbf{Z}$, $x = qx' + z$ and $z \in \mathbf{Z}$. Consider all the possible values $x'' = u + kv$ that can be obtained at step 8. Suppose two of them, $u + kv$ and $u + k'v$, $0 < k < k' \leq 4n$, are integers. Then $(k' - k)v \in \mathbf{Z}$, and since $(k' - k)$ is a divisor of q, we obtain $qv \in \mathbf{Z}$. Since $v \in \mathbf{Z}/q$ and $u + kv \in \mathbf{Z}$, we have $u \in \mathbf{Z}/q$ and therefore $x' = 2^{2^{m-1}} u + v \in \mathbf{Z}/q$ with a contradiction. It follows that all the numbers $\{u + kv : 0 < k \leq 4n\}$, except at most one, are not integers, so $\mathbb{P}rob[x'' \in \mathbf{Z}] \leq 1/4n$. Taking into account the probability of error in the PROD-protocol we obtain that $\mathbb{P}rob[x'' \in \mathbf{Z}] \leq 1/2n$ if $x \notin \mathbf{Z}$. So if the original x is not an integer, it remains non integer until the end with probability at least $(1 - 1/2n)^n \geq 1/2$.

From the propositions 5 and 3, we obtain :

Theorem 2 $PAR_{\mathcal{R}_\times}$ *does not contain* $IP_{\mathcal{R}_+}$. *Therefore the classes* $BIP_{\mathcal{R}_+} = PAR_{\mathcal{R}_+}$ *and* $BIP_{\mathcal{R}_\times}$ *are strictly included in the classes* $IP_{\mathcal{R}_+}$ *and* $IP_{\mathcal{R}_\times}$, *respectively.*

5.3 Deciding all boolean problems

Consider a boolean decision problem $S \subset \{0,1\}^*$. Every word $w \in \{0,1\}^n$ can be encoded in a natural number $N = N(w)$, $2^n \leq N(w) < 2^{n+1}$, such that w is obtained from the binary representation of $N(w)$ by stripping the first digit (which is always 1). This is a standard 1-1 correspondence between $\{0,1\}^*$ and the natural numbers. If S contains all but finitely many elements of $\{0,1\}^*$, it can be trivially decided by a Turing machine in time $O(1)$. Otherwise, S is represented by the binary expansion of the real number $\alpha = \sum_{w \in S} 2^{-N(w)} \in [0,1)$: the Nth digit of α is 1 or 0 according to whether the corresponding word $w \in \{0,1\}^*$ (such that $N = N(w)$) belongs to S or not. Consider a BSS machine having α as a built-in constant and let it perform the following computations :

1. *Given a word $w \in \{0,1\}^n \subset \{0,1\}^*$, compute $N = N(w)$.*
2. *Let $\beta = \alpha \cdot 2^{N-1}$.*
3. *Let $\gamma = \beta - [\beta]$.*
4. *If $2\gamma \geq 1$, return "yes", otherwise return "no".*

This algorithm solves the set S. Indeed, $w \in S$ if and only if the Nth digit in the binary expansion of α is 1. Multiplying by 2^{N-1} shifts the binary expansion by $N - 1$ positions, so the digit representing w in β is the first one after the integral part. It is also the first digit of the binary expansion of γ ($0 \leq \gamma < 1$), so $\gamma \geq 1/2$ if and only if $w \in S$.

A verifier in an interactive proof system may get β from the prover and verify that $\beta = \alpha \cdot 2^{N-1}$ by means of Corollary 1. Since $\beta < 2^N < 2^{2^{n+1}}$, it may also get γ from the prover, check that $0 \leq \gamma < 1$ and verify the relation $INT_{n+1}(\beta - \gamma)$, which is equivalent to $\gamma = \beta - [\beta]$, with the INT-protocol. This makes the above algorithm an $IP_{\mathcal{R}_+}$ protocol and we obtain :

Corollary 2 *All boolean problems are in $IP_{\mathcal{R}_+}$.*

References

[Bab85]   L. Babai. Trading group theory for randomness. *Symposium on the Theory of Computing*, pages 421–429, 1985.

[Bab94]   L. Babai. Transparent proofs and limits to approximation. *Proceedings of the first European Congress of Mathematics*, pages 31–91, 1994.

[BSS89]   L. Blum, M. Shub, and S. Smale. On a theory of computation and complexity over the real numbers: NP-completeness, recursive functions and universal machines. *Bulletin of the American Mathematical Society*, (21(1)):1–46, 1989.

[CK95]   F. Cucker and P. Koiran. Computing on the reals with addition and order. *Journal of Complexity*, pages 358–376, 1995.

[CKK+95] F. Cucker, M. Karpinski, P. Koiran, Lickteig T., and K. Werther. On real turing machines that toss coins. *Proceedings of the 37-thACM Symposium on Theory of Computing*, pages 335–342, 1995.

[Cuc93]   F. Cucker. On the complexity of quantifier elimination : the structural approach. *The Computer Journal*, 36:400–408, 1993.

[GMR89]   S. Goldwasser, S. Micali, and C. Rackoff. The knowledge complexity of interactive proof systems. *SIAM Journal of Computing*, pages 186–208, 1989.

[Kle57]   V. Klee. Extremal structure of convex sets. *Arch. Math.*, pages 234–240, 1957.

[Poi95]   B. Poizat. *Les petits cailloux : une approche modèle-théorique de l'algorithmie.* Aléas, 1995.

[Sha90]   A. Shamir. IP = PSPACE. *IEEE Symposium on Foundations of Computer Science*, 1990.

[SW70]   J. Stoer and C. Witzgall. *Complexity and Optimization in Finite Dimensions I.* Springer-Verlag, 1970.

Result-Indistinguishable Zero-Knowledge Proofs: Increased Power and Constant-Round Protocols

Giovanni Di Crescenzo*   Kouichi Sakurai**   Moti Yung***

Abstract. We investigate result-indistinguishable perfect zero-knowledge proof systems [8] for "transferring the decision of whether the membership of an input in a language is true or not". Previously only a single number-theoretic language was known to have such a proof system and possible extensions were left as an open question. We show that all known random self-reducible languages (e.g., graph isomorphism, quadratic residuosity, discrete log) and compositions over them have such systems. We also consider techniques for constant-round protocols for these languages in this model, and obtain a 5 round protocol scheme.

1 Introduction

Zero-Knowledge Proofs. These proofs, introduced by Goldwasser, Micali and Rackoff [12], are a method a prover to convince a polynomial-time verifier that a certain assertion (a membership of a string x in a language L) is true, without giving any additional computational advantage. Applications of this important notion include: identification schemes [5], public-key encryption schemes [14], and multi-party protocols [11].

Result-Indistinguishable Zero-Knowledge Proofs. After zero-knowledge proofs of membership had been introduced, some related and more elaborated notions were investigated, such as result-indistinguishable zero-knowledge transfers of decision, introduced in [8]. Informally, a transfer of decision is a method for a prover to convince a polynomial-time bounded verifier of whether a certain assertion is true or false, enjoying strong security properties. Specifically, at the end of the interaction, any verifier will receive no other advantage but the bit denoting whether the statement is true or not (that is, the protocol is perfect zero-knowledge); and any observer of the communication between prover and verifier will receive no advantage at all (that is, the protocol is perfect result-indistinguishable). Proof systems in this model have been shown to have applications to the construction of cryptosystems (see [8]).

In the case of *computational zero-knowledge*, using results in [10, 13], all languages in PSPACE can be shown to have a transfer of decision, assuming some

---

* Computer Science and Engineering Dep., University of California San Diego, La Jolla, CA, 92093-0114. E-mail: giovanni@cs.ucsd.edu

** Dept. of Computer Science and Comm. Eng., Kyushu Univ., Fukuoka 812-81, Japan. E-mail:sakurai@csce.kyushu-u.ac.jp

*** CertCo, New York NY, USA. E-mail: moti@certco.com, moti@cs.columbia.edu

cryptographic assumption. The resulting proof system, however, is merely an extension of the proof systems of membership for the same languages. However, in the case of *perfect zero-knowledge*, i.e., when the property of the protocol being zero-knowledge is proved without resorting to unproven assumptions, only one language, namely, quadratic residuosity modulo Blum integers, was shown to have such a protocol (also assuming that the integer was previously verified). Indeed, the original paper [8] left open the existence of other languages having a zero-knowledge and result-indistinguishable transfer of decision.

Our results. We solve the open question posed in [8] by significantly enlarging the class of languages that are known to have perfect zero-knowledge and perfectly result-indistinguishable transfer of decision. In particular, we present a protocol for the language GI (graph isomorphism) which extends to all known random self-reducible languages. This class of languages include many well-known problems, e.g. quadratic residuosity, discrete log, decision Diffie-Hellman problem. Our protocols are quite versatile: when combined with techniques in [3], they give protocols for the language of all true (poly-size) monotone formulae over membership statements in such random self-reducible languages. We also investigate *constant-round* transfers of decision while keeping the properties of perfect zero-knowledge and perfect result-indistinguishability. We modify our protocol into a 5-round protocol by designing a careful parallelization.

2 Notations and definitions

Basic definitions. If A and B are two interactive probabilistic Turing machine, by pair (A,B) we denote an interactive protocol. Let x be an input common to A and B. By $\text{tr}_{(A,B)}(x)$ we denote the transcript of the interaction between A and B on input x, that is, the messages written on B's communication tape during such interaction. By $\text{OUT}_B(\text{tr}_{(A,B)}(x))$ we denote B's output given the transcript $\text{tr}_{(A,B)}(x)$. We define $View_B(x)$, B's view of the interaction with A on input x, as the probability space that assigns to pairs $(R; \text{tr}_{(A,B(R))}(x))$ the probability that R is the content of B's random tape and that $\text{tr}_{(A,B(R))}(x)$ is the transcript of an execution of protocol (A,B) on input x given that R is B's random tape. Similarly, define (A,B)-$View(x)$, an observer's view of the interaction between A and B on input x, as the probability space that assigns to pairs $(\text{tr}_{(A,B)}(x))$ the probability that $\text{tr}_{(A,B)}(x)$ is the transcript of an execution of protocol (A,B) on input x. If L is a language, by $\chi_L : \{0,1\}^* \to \{0,1\}$ we denote the indicator function for the language L (i.e., $\chi_L(x) = 1$ if and only if $x \in L$).

The definition of Result-Indistinguishable Zero-Knowledge Transfers of Decision. A result-indistinguishable zero-knowledge transfer of decision is an interactive protocol in which a prover convinces a poly-bounded verifier whether a string x belongs to a language L or not, and enjoying the following two security properties: 1) no additional information is revealed to any verifier, and 2) any observer of the communication between prover and verifier cannot derive any information at all (not even whether $x \in L$ or not). The formal definition for result-indistinguishable zero-knowledge transfers of decision has four requirements: regularity, soundness, zero-knowledge, and result-indistinguishability. Formally:

Definition 1. *Let P be a probabilistic Turing machine and V a probabilistic polynomial-time Turing machine that share the same input and can communicate with each other. Also, let C be a probabilistic Turing machine having access to the communication between P and V. Let L be a language. We say that a pair (P,V) is a* PERFECTLY RESULT-INDISTINGUISHABLE *and* PERFECT ZERO-KNOWLEDGE TRANSFER OF DECISION *for L if*

1. *(Regularity) For all x, and all constants c,*
$\mathrm{Prob}(\mathrm{OUT}_V(\mathrm{tr}_{(P,V)}(x)) = (\mathrm{ACCEPT}, \chi_L(x))) \geq 1 - |x|^{-c}$.

2. *(Soundness) For all x, for all constants c and all P',*
$\mathrm{Prob}(\mathrm{OUT}_V(\mathrm{tr}_{(P',V)}(x)) = (\mathrm{ACCEPT}, 1 - \chi_L(x))) \leq |x|^{-c}$.

3. *(Perfect Zero-Knowledge) For any Turing machine V', there exists a probabilistic Turing machine $S_{V'}$ (called the V-simulator) running in expected polynomial-time who is given $\chi_L(x)$ as additional input, such that for all x, the probability spaces $\mathrm{View}_{V'}(x)$ and $S_{V'}(x, \chi_L(x))$ are equal.*

4. *(Perfect Result-Indistinguishability) There exists a probabilistic Turing machine M (called the C-simulator) running in probabilistic polynomial-time such that for all x, the probability spaces (P,V)-$\mathrm{View}_C(x)$ and $M(x)$ are equal.*

Some basic propositions. We give two propositions that follow directly from the above definition. Note that *public-coin* protocols are such where the verifier's messages consist only of his random coins.

Proposition 2. *Let L be a language having a transfer of decision (A,B) satisfying the requirements of regularity and soundness in Definition 1. If protocol (A,B) is public-coin then (A,B) does not satisfy the requirement of result-indistinguishability in Definition 1.*

Proposition 3. *Let L be a language having a zero-knowledge transfer of decision (A,B). Then there exists a zero-knowledge proof system of membership (P,V) for L such that (P,V) has the same number of rounds as (A,B).*

Recall that in [9] it was proved that zero-knowledge proofs of membership require at least 4 rounds unless the corresponding language is in BPP. Then the latter proposition suggests that in order to construct a *constant-round* zero-knowledge transfer of decision, we have to consider only protocols having at least 4 rounds.

3 Result-indistinguishable zk transfers of decision

Our main result is a protocol for all known random self-reducible languages and their monotone composition, we start with the graph isomorphism language.

3.1 A transfer of decision for GI

An informal description. The common input to prover P and verifier V is a pair of graphs (G_0, G_1). We can view (P,V) as made of a sequential composition of $3n$ iterations of an atomic protocol (A,B), which in turn can be divided into three phases. In the first phase B randomly chooses a bit b and a graph G isomorphic to G_b, and sends it to A. In the second phase, B proves to A that graph G has been correctly constructed, without revealing any information about bit b and the permutation chosen. In the third phase, A checks that B's proof is accepting; now, if $G_0 \approx G_1$ then A randomly chooses a bit g; otherwise, if $G_0 \not\approx G_1$ then A computes bit b such that $G \approx G_b$ and sets $g = b$. In both cases A proves in zero-knowledge to B that $G \approx G_g$, and if this proof is not convincing then B rejects. At the end of the $3n$ iterations of protocol (A,B), V accepts if B has never rejected. Furthermore, if in at least n iterations it holds that $b \neq g$, V outputs 1 (meaning that he is convinced that $G_0 \approx G_1$); otherwise V outputs 0 (meaning that he is convinced that $G_0 \not\approx G_1$).

Implementation of (A,B). The implementation of the first phase of protocol (A,B) is simple. We observe that the second phase can be implemented by using a 'witness-indistinguishable' subprotocol, as done in [12] in their zero-knowledge proof system of membership for the language of quadratic non-residuosity (see [5, 6] for the notion of witness-indistinguishability). In particular, we will use the protocol of [10] used in the middle of a zero-knowledge proof of membership for graph non-isomorphism. Then we observe that the subprotocol in the third phase which allows P to convince V that $G \approx G_g$, for some bit g, can be implemented by using a three-steps protocol, as done in [12] for the language of quadratic residuosity or in [10] for the language of graph isomorphism.

A formal description of (P,V): Let n be an integer and $m = n \log n$. The protocol (P,V) is made of $3n$ sequential repetitions of subprotocol (A,B), which is described in Figure 1. Now, if any verification by B is not satisfied, V outputs: (REJECT), and halts. Otherwise, denote by b_i the bit chosen by V in step V1 of the i-th execution of subprotocol (A,B), and by g_i the bit computed by P in step P2 of the i-th execution of subprotocol (A,B). Then V computes the number s of indices $i \in \{1, \ldots, 3n\}$ such that $b_i = g_i$; if it holds that $s \geq 2n$ then V outputs: (ACCEPT,1); otherwise V outputs: (ACCEPT,0). We obtain the following

Theorem 4. *The protocol (P,V) is a perfectly result-indistinguishable and perfect zero-knowledge transfer of decision for GI.*

The rest of the subsection proves the Theorem. Clearly, V's program can be performed in polynomial time. Now we prove the requirements in Definition 1.

Regularity. We show that for all pairs (G_0, G_1) of graphs, if P and V follow their protocol, then V accepts and outputs $\chi_{GI}(G_0, G_1)$ with probability greater than $1 - n^{-c}$, for any constant c. We analyze two cases. First assume $G_0 \approx G_1$; now, since V follows his protocol, P will be convinced by V's witness-indistinguishable proof that graph H has been correctly computed. Then H is isomorphic to one of

The Protocol (A,B)

Input to A and B: (G_0, G_1), where G_0, G_1 are n-node graphs.

B1: Uniformly choose bit b and a permutation β and compute $H = \beta(G_b)$;
  for $j = 1, \ldots, m$,
   uniformly choose bit a_j and two permutations α_{j0}, α_{j1};
   compute graphs $A_{j0} = \alpha_{j0}(G_{a_j})$ and $A_{j1} = \alpha_{j1}(G_{1-a_j})$;

$A \leftarrow B$: $(H, (A_{10}, A_{11}), \ldots, (A_{m0}, A_{m1}))$.

A1: For $j = 1, \ldots, m$,
   uniformly choose bit c_j;

$A \rightarrow B$: $(c_1, \ldots, c_m)$.

B2: For $j = 1, \ldots, m$,
   if $c_j = 0$ then set $\sigma_j = (\alpha_{j0}, \alpha_{j1})$;
   if $c_j = 1$ then set $\sigma_j = \beta \circ \alpha_{j,b\oplus a_j}^{-1}$;

$A \leftarrow B$: $(\sigma_1, \ldots, \sigma_m)$.

A2: For $j = 1, \ldots, m$,
   if $c_j = 0$ then
    let $\sigma_j = (\eta_{j0}, \eta_{j1})$;
    check that $A_{j0} = \eta_{j0}(G_{a_j})$ and $A_{j1} = \eta_{j1}(G_{1-a_j})$, for some bit a_j;
   if $c_j = 1$ then check that $H = \sigma_j(A_{j0})$ or $H = \sigma_j(A_{j1})$;
   if the above verifications are not satisfied then halt;
   if $G_0 \approx G_1$ then randomly choose a bit g;
   if $G_0 \not\approx G_1$ then
    compute bit b and permutation β such that $H = \beta(G_b)$ and set $g = b$;
   set $L_0 = H$ and $L_1 = G_g$;
   uniformly choose a bit t and a permutation τ and set $T = \tau(L_t)$;

$A \rightarrow B$: $((L_0, L_1), T)$.

B3: Uniformly choose a bit l;

$A \leftarrow B$: (l).

A3: If $l = t$ then set $\rho = \tau$;
   if $l = 1 - t$ then compute ρ such that $T = \rho(L_l)$;

$A \rightarrow B$: (ρ).

B4: Check that $T = \rho(L_l)$.

Figure 1: A result-indistinguishable transfer of decision for GI

G_0, G_1, and the statement $L_0 \approx L_1$ is true, since L_0 is isomorphic to H and L_1 is isomorphic to a randomly chosen graph between G_0, G_1. Moreover, it holds that $b_i = g_i$ with probability $1/2$, and therefore the number s of indices $i \in \{1, \ldots, 3n\}$ such that $b_i = g_i$ will be at least $2n$ with exponentially small probability (using Chernoff bounds). This guarantees that V outputs (ACCEPT,1) with probability greater than $1 - n^{-c}$, for any constant c. Now, assume $G_0 \not\approx G_1$; then P can compute bit b and permutation β such that $H = \beta(G_b)$. This implies that the statement $L_0 \approx L_1$ is true, since L_0 is isomorphic to H and L_1 is chosen isomorphic to G_b. We observe that $b_i = g_i$ for all $i = 1, \ldots, 3n$, and thus V outputs (ACCEPT,0) with probability 1.

Soundness. We show that for any P' and any input pair (G_0, G_1), the probability that V's output is (ACCEPT,$1 - \chi_{GI}(G_0, G_1)$) is negligible. First, consider case $G_0 \approx G_1$ and assume that V accepts. Then notice that V outputs (ACCEPT,0) only when it holds that $g_i = b_i$, for at least $2n$ values of $i \in \{1, \ldots, 3n\}$; however, since $G_0 \approx G_1$, and the subprotocol in the second phase of (P,V) is witness-indistinguishable, bit b_i cannot be computed by any P' better than by random guessing. Therefore, for any P', the probability that $g_i = b_i$, for at least $2n$ values of index i, is smaller than n^{-c}, for any constant c (using Chernoff bounds). Now, consider case $G_0 \not\approx G_1$ and assume that V accepts. Then notice that V outputs (ACCEPT,1) only when it holds that $g_i \neq b_i$, for at least n values of $i \in \{1, \ldots, 3n\}$; however, since $G_0 \not\approx G_1$, the statements $L_{i0} \approx L_{i1}$, for all i such that $g_i \neq b_i$, are all false, and thus the probability that V accepts in this case is smaller than n^{-c}, for all constants c (using Chernoff bounds).

Perfect zero-knowledge. Now we informally describe a simulator $S_{V'}$ such that, for all pairs (G_0, G_1), the probability spaces $S_{V'}(G_0, G_1)$ and $\text{View}_{V'}(G_0, G_1)$ are equal. Since protocol (P,V) is constructed as a sequential repetition of an atomic protocol, it will be enough to describe the program of $S_{V'}$ simulating only such atomic protocol (in this description we will also omit the index of messages denoting the number of iteration).

The algorithm $S_{V'}$. First of all $S_{V'}$ feeds V' with a uniformly chosen random tape R; then he receives from V' graph H and the witness-indistinguishable proof of knowledge certifying that this graph has been correctly constructed (during this proof, $S_{V'}$ acts as a verifier of such proof and can run P's program, since it can be performed in polynomial time). Now, if the proof is not convincing then $S_{V'}$ outputs the conversation obtained so far, and halts. If the proof is convincing and $\chi_{GI}(G_0, G_1) = 1$ then $S_{V'}$ runs the extractor for the proof of knowledge in order to compute bit b and permutation β such that $H = \beta(G_b)$. Now, $S_{V'}$ can compute pair (L_0, L_1) as follows: graph L_0 is computed as done by P in the protocol (i.e., $L_0 = H$), and graph L_1 is computed as uniformly chosen among graphs isomorphic to G_b if $\chi_{GI}(G_0, G_1) = 1$ or isomorphic to G_g for some random bit g, otherwise. Now, the remaining steps of $S_{V'}$ consist of simulating the atomic proof by P that $L_0 \approx L_1$, and can simulated by using the rewinding technique, as follows. First $S_{V'}$ computes a graph T uniformly among those isomorphic to L_t, for some random bit t; then he receives bit l from V'; now, if $t = l$ then $S_{V'}$ sends the permutation between T and L_t, otherwise he rewinds V' until after he has computed graphs L_0, L_1 and tries again until $t = l$. Finally $S_{V'}$ outputs the conversation obtained.

To prove that the perfect zero-knowledge requirement is satisfied, we need to show that algorithm $S_{V'}$ is expected polynomial time, and his output $S_{V'}(G_0, G_1)$ is identically distributed to $\text{View}_{V'}(G_0, G_1)$, for all input pairs (G_0, G_1).

Lemma 5. For all V', algorithm $S_{V'}$ runs in expected polynomial time.

Proof. We observe that $S_{V'}$ only runs polynomial-time instructions and the extractor for the proof of knowledge by V', which runs in expected polynomial time. When simulating the third phase of (P,V), the simulation is iterated with some probability. Now, we can show the probability to be at most $1/2$, from which the Lemma follows.

Lemma 6. For all pairs (G_0, G_1) and any V', the probability spaces $S_{V'}(G_0, G_1)$ and $\text{View}_{V'}(G_0, G_1)$ are equal.

Proof. Clearly the verifier's random tape is uniformly distributed in both spaces and the messages sent by the verifier are computed equally in both spaces. Now, we consider the messages sent by the prover. It is simple to check that the random bits sent by the prover during the executions of the witness-indistinguishable subprotocol executed in the second phase of (P,V) are also computed in the same way in both spaces. This is true also for graphs L_0, L_1 and for the other messages of P (argument omitted).

Perfect result-indistinguishability. To prove this we exhibit an efficient simulator M such that, on input (G_0, G_1), outputs a probability space $M(G_0, G_1)$ which is equal to the view of an observer C of the conversation during the execution of the protocol (P,V) on input G_0, G_1. In this case a description of the atomic protocol (A,B) suffices. Informally, first M simulates the first two phases of (P,V) by executing the same instructions by P and V. That is, he will compute a graph H as $H = \beta(G_b)$ for random b and β, and simulate the witness-indistinguishable proof that H has been correctly constructed, using b, β. Now, the simulator M computes graphs L_0, L_1 as follows: L_0 is set equal to H, and L_1 is uniformly chosen among the graphs isomorphic to G_b. Now, M simulates the proof by P that L_0 is isomorphic to L_1 as follows: he chooses T uniformly among graphs isomorphic to L_t, for some random bit t, sets the message by the verifier equal to bit t, and sets the final message by P equal to the permutation between T and L_t. We get (proof omitted):

Lemma 7. For any (G_0, G_1), the probability spaces $M(G_0, G_1)$ and (P,V)-$\text{View}(G_0, G_1)$ are equal.

3.2 Extensions and remarks

Random self-reducible languages (RSR) [1] are widely used in cryptography [12, 10, 15, 3]. The protocol for GI easily extends, and we have:

Theorem 8. *Let L be any of the mentioned random self-reducible languages. There exists a perfectly result-indistinguishable and perfect zero-knowledge transfer of decision for L.*

Theorem 9. *Let L be any of the mentioned random self-reducible languages. There exists a perfectly result-indistinguishable and perfect zero-knowledge transfer of decision for the language $\Phi(L)$ of true (poly-size) monotone formulae over L-membership statements.*

The above is obtained by combining the technique of Section 3.1 with the constructions in [3].

In [4] it has been shown that the proof of membership for the graph non-isomorphism language is a proof of computational ability (a notion first defined in [17]) of deciding whether two graphs are isomorphic or not. This is shown by using a new proof for a direct product theorem for proofs of computational ability. Using the same techniques, we can show that all result-indistinguishable transfers of decision given so far are proofs of computational ability.

4 Constant-round transfers of decision

Let (P,V) be the protocol presented in this section, we show the following

Theorem 10. *Let L be any of the mentioned random self-reducible languages. The protocol (P,V) is a 5-round perfect result-indistinguishable and perfect zero-knowledge transfer of decision for language L.*

An informal description of a 6-round protocol. We start with an informal 6-round transfer of decision for GI and then show how to further reduce a round. The usual first attempt to construct a constant-round zero-knowledge protocol is to directly parallelize an atomic protocol achieving soundness with a constant error probability (this does not succeed in general [9]). Directly parallelizing our constant-error protocol (A,B) of Section 3.1, it is not clear how to simulate a parallel third phase which goes as following: The prover computes graphs L_0, L_1 which depend on the first two phases of the protocol; then the prover shows that $L_0 \approx L_1$, as follows: he sends to the verifier a graph T uniformly chosen among the graphs isomorphic to L_t, for some random bit t; the verifier sends a random bit l and the prover answers with a permutation ρ such that $T = \rho(L_l)$. Now, notice that the rewinding technique used to simulate the atomic protocol is not appropriate for its parallelization, as it may require a superpolynomial number of iterations. This problem is the same as in the zero-knowledge proof of membership for GI, pointed out in [2]. In their paper they give a technique for the verifier to commit to his random bits. We solve the problem for our protocol by applying a variation of their technique (a variation is needed since it is possible to see that a direct application of the same technique won't work). Specifically, the following steps are added to a direct parallel repetition of our protocol (A,B): the prover will send in his step A1 a pair of graphs M_0, M_1, each uniformly chosen among graphs isomorphic to H, the graph sent by the verifier in his step B1; now, the verifier will commit in step B2 to each bit l by sending a graph E uniformly chosen among the graphs isomorphic to M_l; also, in step B3 the verifier will reveal the commitments to bits l by sending the permutation between graph M_l and graph E; finally, in step A3 the prover will send a permutation between graphs M_0, M_1.

The variations with respect to the technique in [2] are two: first, the prover sends one pair of graphs M_0, M_1 to the verifier for each parallel repetition of the protocol (instead than just a single pair); second, the graphs M_0, M_1 are computed as isomorphic to the graph H sent by the verifier (instead than isomorphic to the input graphs). The resulting protocol is a 6-round perfectly result-indistinguishable and perfect zero-knowledge transfer of decision for GI.

A 5-round protocol. In order to obtain a 5-round protocol we would like to properly interleave the steps of the 6-round protocol. First we observe that a way to obtain a 5-round protocol is to somehow delay the current first step V1 of the verifier, where he sends graph H and start proving that it is correctly constructed. This may be a problem because in his first step P1, the prover sends graphs M_0, M_1 which are both computed as isomorphic to the graph H, and it does not seem possible to further delay this step from the prover. Instead, we observe that graph H will be proved by the verifier as isomorphic to either G_0 or G_1, and therefore the prover can compute graphs M_0, M_1 as follows. In step P1 he will compute two pairs of graphs (M_{00}, M_{01}) and (M_{10}, M_{11}), where the first pair is made of graphs isomorphic to G_0 and the second pair is made of graphs isomorphic to G_1. Now, the prover and the verifier will continue the execution of the original protocol twice (in parallel): first using pair (M_{00}, M_{01}), and then using pair (M_{10}, M_{11}). This is done until the prover verifies that graphs H sent by the verifier are indeed isomorphic to one of G_0, G_1; after this verification, he can choose one of the two pairs to run the protocol as before. Another problem occurs when the prover has to compute graphs L_0, L_1 in step P2. Because of the previous modifications, the pair L_0, L_1 is computed before the proof that the verifier has honestly computed graphs H is terminated; on the other hand in the simulation this pair needs to be computed only after such proof has terminated (since the simulator needs to run the extractor for the proof to compute the bit b chosen by the verifier). We solve this problem as follows. The prover will compute two pairs of graphs (L_{00}, L_{01}) and (L_{10}, L_{11}), where graphs L_{00}, L_{10} are isomorphic to G, graph L_{01} is isomorphic to G_0 and graph L_{11} is isomorphic to G_1; clearly, exactly one of the two pairs is computed as the prover would do in the original protocol. Then, as done before, the prover and the verifier will continue the execution of the original protocol twice (in parallel): first using pair (L_{00}, L_{01}), and then using pair (L_{10}, L_{11}). Again, this is done until the prover verifies that graphs H sent by the verifier are indeed isomorphic to one of G_0, G_1; after this verification, the prover (and thus the simulator that can run the extractor) can choose one of the two pairs to run the protocol as before.

A formal description of (P,V): Let n be an integer and $m = n \log n$. (P,V) is made of $3n$ independent parallel repetitions of subprotocol (A,B), described in

Figure 2. Now, if any verification by B is not satisfied, V outputs: (REJECT), and halts. Otherwise, denote by b_i the bit chosen by V in step V1 of the i-th parallel execution of subprotocol (A,B), and by g_i the bit computed by P in step P3 of the i-th parallel execution of subprotocol (A,B). Then V computes the number s of indices $i \in \{1, \ldots, 3n\}$ such that $b_i = g_i$; if it holds that $s \geq 2n$ then V outputs: (ACCEPT,1); otherwise V outputs: (ACCEPT,0). A formal proof of Theorem 10 is omitted due to lack of space.

References

1. M. Abadi, J. Feigenbaum, and J. Kilian, *On Hiding Information from an Oracle*, STOC 87.
2. M. Bellare, S. Micali, and R. Ostrovsky, *Perfect Zero Knowledge in Constant Rounds*, STOC 90.
3. A. De Santis, G. Di Crescenzo, G. Persiano and M. Yung, *On Monotone Formula Closure of SZK*, FOCS 94.
4. G. Di Crescenzo and R. Impagliazzo, *Proof of Membership vs. Proofs of Knowledge*, manuscript, March 1996.
5. U. Feige, A. Fiat, and A. Shamir, *Zero-Knowledge Proofs of Identity*, Journal of Cryptology, vol. 1, 1988, pp. 77–94.
6. U. Feige and A. Shamir, *Witness-Indistinguishable and Witness-Hiding Protocols*, STOC 90.
7. L. Fortnow, *The Complexity of Perfect Zero Knowledge*, STOC 87.
8. Z. Galil, S. Haber, and M. Yung, *Minimum-Knowledge Interactive Proofs for Decision Problems*, SIAM Journal on Computing, vol. 18, n.4, pp. 711–739 (previous version in FOCS 85).
9. O. Goldreich and H. Krawczyk, *On the Composition of Zero-Knowledge Proof Systems*, ICALP 1990.
10. O. Goldreich, S. Micali, and A. Wigderson, *Proofs that Yield Nothing but their Validity or All Languages in NP Have Zero-Knowledge Proof Systems*, Journal of the ACM, vol. 38, n. 1, 1991, pp. 691–729.
11. O. Goldreich, S. Micali, and A. Wigderson, *How to play any mental game*, STOC 88.
12. S. Goldwasser, S. Micali, and C. Rackoff, *The Knowledge Complexity of Interactive Proof-Systems*, SIAM Journal on Computing, vol. 18, n. 1, February 1989.
13. R. Impagliazzo and M. Yung, *Direct Minimum Knowledge Computations*, CRYPTO 87.
14. M. Naor and M. Yung, *Public-Key Cryptosystems Provably Secure against Chosen Ciphertext Attack*, STOC 90.
15. M. Tompa and H. Woll, *Random Self-Reducibility and Zero-Knowledge Interactive Proofs of Possession of Information*, FOCS 87.
16. A. Yao, *Theory and Applications of Trapdoor Functions*, FOCS 85.
17. M. Yung, *Zero-Knowledge Proofs of Computational Power*, Eurocrypt 89.

The Protocol (A,B)

Input to A and B: (G_0, G_1), where G_0, G_1 are n-node graphs.

A1: Uniformly choose bit d and permutations $\alpha_{00}, \alpha_{01}, \alpha_{10}, \alpha_{11}$ and compute $M_{00} = \alpha_{00}(G_d)$, $M_{01} = \alpha_{01}(G_d)$, $M_{10} = \alpha_{10}(G_{1-d})$, $M_{11} = \alpha_{11}(G_{1-d})$;

$A \to B$: $((M_{00}, M_{01}), (M_{10}, M_{11}))$.

B1: Uniformly choose bit b and a permutation π and compute $H = \pi(G_b)$;
 for $j = 1, \ldots, m$,
  uniformly choose bit a_j and permutations α_{j0}, α_{j1};
  compute graphs $A_{j0} = \alpha_{j0}(G_{a_j})$ and $A_{j1} = \alpha_{j1}(G_{1-a_j})$;
 uniformly choose bits e_0, e_1 and permutations η_0, η_1;
 compute graphs $E_0 = \eta_0(M_{0,e_0})$ and $E_1 = \eta_1(M_{1,1-e_1})$;

$A \leftarrow B$: $(H, (A_{10}, A_{11}), \ldots, (A_{m0}, A_{m1}), (E_0, E_1))$.

A2: For $j = 1, \ldots, m$,
  uniformly choose bit c_j;
 set $L_{00} = H, L_{01} = G_0$ and $L_{10} = H, L_{11} = G_1$;
 uniformly choose bits t_0, t_1 and permutations τ_0, τ_1 and set $T_0 = \tau_0(L_{0t_0})$, $T_1 = \tau_1(L_{1t_1})$;

$A \to B$: $((c_1, \ldots, c_m), (L_{00}, L_{01}, T_0), (L_{10}, L_{11}, T_1))$.

B2: For $j = 1, \ldots, m$,
  if $c_j = 0$ then set $\sigma_j = (\alpha_{j0}, \alpha_{j1})$;
  if $c_j = 1$ then set $\sigma_j = \beta \circ \alpha_{j,b\oplus a_j}^{-1}$;

$A \leftarrow B$: $((\sigma_1, \ldots, \sigma_m), (e_0, \eta_0), (e_1, \eta_1))$.

A3: For $j = 1, \ldots, m$,
  if $c_j = 0$ then
   let $\sigma_j = (\alpha_{j0}, \alpha_{j1})$;
   check that $A_{j0} = \alpha_{j0}(G_{a_j})$ and $A_{j1} = \alpha_{j1}(G_{1-a_j})$, for some bit a_j;
  if $c_j = 1$ then check that $H = \sigma_j(A_{j0})$ or $H = \sigma_j(A_{j1})$;
  if the above verifications are not satisfied then halt;
 if $G_0 \approx G_1$ then randomly choose a bit g;
 if $G_0 \not\approx G_1$ then compute bit b and permutation π such that $H = \pi(G_b)$ and set $g = b$;
 compute a permutation ρ such that $T_g = \rho(L_{g,g\oplus e_g})$;
 compute a permutation γ such that $M_{g0} = \gamma(M_{g1})$;

$A \to B$: (g, ρ, γ).

B3: Check that $T_g = \rho(L_{g,g\oplus e_g})$ and $M_{g0} = \gamma(M_{g1})$.
 If some verification is not successful then output: REJECT and halt.

Figure 2: A 5-round result-indistinguishable transfer of decision for GI

Bounded Size Dictionary Compression: SC^k-Completeness and NC Algorithms

Sergio De Agostino and Riccardo Silvestri

Dipartimento di Scienze dell'Informazione, Università di Roma "La Sapienza"
Via Salaria 113, 00198 Roma, Italy.
e-mail:{agos, silver}@dsi.uniroma1.it

Abstract. We study the parallel complexity of a bounded size dictionary version (LRU deletion heuristic) of the LZ2 compression algorithm. The unbounded version was shown to be P-complete. When the size of the dictionary is $O(\log^k n)$, the algorithm is shown to be hard for the class of problems solvable simultaneously in polynomial time and $O(\log^k n)$ space (that is, SC^k). We also introduce a variation of this heuristic that turns out to be the first natural SC^k-complete problem (the original heuristic belongs to SC^{k+1}). In virtue of these results, we argue that there are no practical parallel algorithms for LZ2 compression with LRU deletion heuristic or any other heuristic deleting dictionary elements in a continuous way. For simpler heuristics (SWAP, RESTART, FREEZE), practical parallel algorithms are given.

1 Introduction

Textual substitution methods are among the most practical and effective for lossless text compression. *Textual substitution* replaces substrings in the text with *pointers* to copies, called *targets*, that are stored in a *dictionary*. The encoded string will be a sequence of pointers and uncompressed characters. The *static* method is when the dictionary is known in advance (see the book of Storer for references [17]). By contrast, with the *dynamic* method (often called "LZ2" method due to the work of Ziv and Lempel [19]) the dictionary may be constantly changing as the data is processed. A special way to change dynamically is the *sliding* dictionary method (often called "LZ1" method due to the work of Lempel and Ziv [12]), where the dictionary is a window that passes continuously from left to right over the input.

In this paper, we consider the LZ2 method. The LZ2 algorithm learns substrings by reading the input string from left to right with an *incremental parsing* procedure. The dictionary is empty, initially. The procedure adds a new substring to the dictionary as soon as a prefix of the still unparsed part of the string does not match a dictionary element and replaces the prefix with a pair comprising a pointer to the dictionary and the last uncompressed character. For example, the parsing of the string $abababaaaaaa$ is a, b, ab, aba, aa, aaa and the coding is $0a, 0b, 1b, 3a, 1a, 5a$ (the pointer value for the first target in the dictionary is 1 and 0 represents the empty string). While parallel algorithms have been designed for compression with sliding dictionaries (Crochermore and Ryter [6]), the LZ2 method seems hardly parallelizable, since the LZ2 compression algorithm is P-complete (De Agostino [7]).

Several variations of the LZ2 algorithm have been designed (see the books of Storer [17][18] and Bell, Cleary and Witten [1]). The main issue for implementation purposes is to bound the dictionary size. A strategy that can achieve good compression ratio with small memory is the LRU *deletion heuristic* that discards the least recently used dictionary element to make space for the new substring.

We study the parallel complexity of the LRU deletion heuristic applied to the LZ2 compression algorithm. We argue that there is no practical parallel algorithm for this strategy, even if the dictionary size is constant. Such claim exploits a hardness result given in section 4. When the size of the dictionary is $O(\log^k n)$, the strategy is shown to be log-space hard for the class of problems solvable simultaneously in polynomial time and $O(\log^k n)$ space (that is, SC^k). Since its sequential complexity is polynomial in time and $O(\log^k n \log \log n)$ in space, the problem belongs to SC^{k+1}. As far as we know this is the second natural problem in SC^{k+1} that is shown to be hard for SC^k. In (Dessmark, Lingas and Maheswari [8]) such kind of result was attained for the multilist layering problem. Moreover, we introduce a variation of this heuristic that turns out to be the first natural SC^k-complete problem. As it is observed in (Cook [5]), the word "natural" should preclude having "$\log^k n$" in the statement of a problem. However, we remark that in our problem the presence of a bound in the instance is intrinsic to its definition (the LRU deletion heuristic does not make sense without a bound). We also give parallel algorithms working with a dictionary of size κ either in $O(\log n)$ time with $2^{O(\kappa \log \kappa)} n$ processors or in $2^{O(\kappa \log \kappa)} \log n$ time with $O(n)$ processors. Due to the need of a large bound for the size of the dictionary and in virtue of the hardness result we infer that there are no practical parallel algorithms for LZ2 compression with LRU deletion heuristic or any other heuristic deleting dictionary elements in a continuous way, unless $SC \subseteq NC$. For simpler heuristics (SWAP, RESTART, FREEZE), practical parallel algorithms are given.

2 Relating SC^k-Hardness to NC Algorithms

First, we introduce the basic definitions that we need to make our argues. We define a circuit C as a labeled, acyclic, directed graph with maximum indegree 2. Nodes with indegree 0 are labeled as *inputs*. The other nodes are labeled by one of the boolean functions OR, AND or NOT. The nodes of outdegree 0 are the *outputs*. The *depth* of a circuit C is the length of the longest path in C. A circuit is *synchronous* if its gates can be divided into levels such that all inputs to the gates at level ℓ either are input gates or are from gates at level $\ell - 1$. The *width* of a synchronous circuit is the maximum of the number of gates at any level.

SC is the class of languages that can be recognized in sequential polynomial time and polylogarithmic space (Cook [3], Cook [4]). More precisely, $SC^k =$ DTIME-SPACE$(n^{O(1)}, \log^k n)$ and $SC = \bigcup_k SC^k$. From (Pippenger [14]), we have a characterization of SC^k as the class of languages recognizable by synchronous circuits of polynomial size and width $O(\log^k n)$. NC, the set of languages that can be recognized in polylogarithmic time with a polynomial number of processors (Pippenger [14]), is a more popular class that can be viewed in some sense symmetric to SC (Dymond and Cook [9]). Pippenger also gives a characterization of NC as the set of languages recognizable by circuits of polynomial size and polylogarithmic depth.

Turing machine time and circuit size are polynomially related (Pippenger and Fischer [15]), as are Turing machine space and circuit depth (Borodin [2]). One might suspect that these two relations should hold simultaneously ($SC = NC$), but it is observed in (Cook [4], Dymond and Cook [9], Johnson [11]) that when resource bounds must be satisfied simultaneously, this can substantially change their effects and one might conjecture that SC and NC are incomparable. Candidates in NC−SC and SC−NC are given in Cook [3] and Ruzzo [16]. Recently, Dessmark, Lingas and Maheshwari [8] show that the multilist layering problem is P-complete and the bounded version with $O(\log^k n)$ lists belongs to SC^{k+1} and is hard for SC^k. It follows that this problem can be seen as a candidate in SC−NC. When the number of lists is constant, say κ, they give an NC algorithm with $O(n^\kappa)$ processors. We want to point out that κ figures as an exponent in the parallel complexity of the problem. This is not by accident, if we believe that SC is not included in NC. In fact, the SC^k-hardness would imply the exponentiation of κ. Hence, it is unlikely that practical NC algorithms exist for large values of κ. Observe that the P-completeness of the problem, that holds when κ is superpolylogarithmic, does not suffice to infer this exponentiation since κ can figure as multiplicative factor of the time function.

As in P-completeness theory [10], we can define an algorithm A to be hard (complete) for SC^k if the problem of computing the output of A is hard (complete) for SC^k. In section 4, we show that the LZ2 algorithm with LRU deletion heuristic on a dictionary of size $O(\log^k n)$ is hard for SC^k and belongs to SC^{k+1}. Moreover, we give an NC algorithm for dictionaries of constant size κ requiring $O(\log n)$ time and $2^{O(\kappa \log \kappa)} n$ processors. Since a dictionary has to be filled up with at least thousands of elements to achieve good compression, it follows from the SC^k-hardness that practical parallel algorithms are unlikely for LZ2 compression with LRU deletion heuristic. In the same section, the SC^k-completeness of a relaxed version of LRU is presented. Practical parallel algorithms for simpler heuristics are described in the next section.

3 Practical Parallel Algorithms

Simple choices for the deletion heuristic include:

- FREEZE: once the dictionary is full, freeze it and do not allow any further entries to be added.

- RESTART: stop adding further entries when the dictionary is full; when the compression ratio starts deteriorating clear the dictionary and learn new strings.
- SWAP: when the *primary* dictionary first becomes full, start an *auxiliary* dictionary, but continue compression based on the primary dictionary; when the auxiliary dictionary becomes full, clear the primary dictionary and reverse their roles.

A dictionary can be represented with a *trie* data structure. Given an alphabet Σ, a trie is a tree where edges are labeled by elements of Σ in such a way that children are connected to their parent via edges that have distinct labels. In this way the set of strings of a dictionary can be represented by the nodes of the trie. The root represents the empty string. At each step we find the longest match in the dictionary as a path from the root to a leaf. The dictionary can be updated by adding a new leaf to the trie.

Theorem 1. *The LZ2 algorithm with FREEZE deletion heuristic on a dictionary of size k can be computed in $O(km + \log n)$ time with $O(n)$ processors on a EREW PRAM, where n is the length of the input string and m is the maximum match length.*

Proof. One processor fills up the dictionary in $O(km)$ time, by storing the dictionary in a trie. Then, the longest match for each position of the remaining suffix is computed in parallel with $O(n)$ processors in $O(m)$ time. Link position i to the position next to the end of the match computed in i. If the match ends the string, link i to a dummy final element. We obtain a tree rooted in this dummy final element and we need to find the path from the leaf representing the first position of the remaining suffix to the root in order to have the parsing. A procedure to compute a path from a leaf to the root of a tree is given in (Olariu, Schwing and Zhang [13]), requiring $O(\log n)$ time and $O(n)$ processors on a PRAM EREW. □

Theorem 2. *The LZ2 algorithm with RESTART deletion heuristic can be computed in $O(km + \log n)$ time with $O(n^2)$ processors on a EREW PRAM.*

Proof. Starting from each position, one processor fills up a dictionary in $O(km)$ time. Then, for each dictionary $O(n)$ processors compute the parsing on the corresponding remaining suffix in $O(m + \log n)$ time as in the proof of Theorem 1 and check where the compression ratio starts deteriorating. Finally, we build a tree by linking position i to the one where the dictionary computed from i is cleared out and compute the path giving the blocks providing the parsing of the string. □

For the RESTART heuristic the number of processor is quadratic. It becomes linear if the dictionary is cleared as soon as it becomes full.

Theorem 3. *The LZ2 algorithm with SWAP deletion heuristic can be computed in $O(km + \log n)$ time with $O(n)$ processors on a EREW PRAM.*

Proof. Starting from each position, one processor fills up a dictionary in $O(km)$ time. Then, compute the blocks as in the version of RESTART using a linear number of processors. Then, recompute the parsing on each block with the dictionary computed on the previous block. □

The SWAP and RESTART heuristics can be viewed as discrete versions of LRU. In fact, the dictionaries depend only on small segments of the input string and this is what makes possible a practical parallel algorithm.

4 The Parallel Complexity of the LRU Deletion Heuristic

We showed in the introduction of the paper how the LZ2 algorithm parses the example string *abababaaaaa*. If we bound the dictionary size with 3, at the fourth algorithmic step a, b, ab is the partial parsing, $0a, 0b, 1b$ is the partial coding and the dictionary is filled up with the three elements a, b, ab. The LRU deletion heuristic defines a string as "used" if it is a match or a prefix of a match and replaces the least recently used leaf of the trie representing the dictionary with the new element. Hence at the fourth step, first b is discarded and then aba is added. Finally, aba is replaced with aa and ab with aaa.

Theorem 4. *The LZ2 algorithm with LRU deletion heuristic on a dictionary of size $O(\log^k n)$ can be computed in polynomial time and $O(\log^k n \log \log n)$ space.*

Proof. The trie may take $O(\log^k n)$ space since the degree of each node is bounded by the alphabet cardinality. The $\log \log n$ factor is essentially due to keep the information needed for the LRU deletion heuristic. □

In the next theorem, we give the SC^k-hardness result. The proof is very involved. For the sake of the reader, we advice to look first at the P-completeness proof for the unbounded case (De Agostino [7]).

Theorem 5. *Given an input string drawn over a ternary alphabet, the LZ2 algorithm with LRU deletion heuristic on a dictionary of size $O(\log^k n)$ is hard for SC^k.*

Proof (sketch). We show the SC^k-hardness of the following problem: given two strings Z and T, does T belong to the final dictionary produced parsing Z by the LZ2 algorithm with LRU deletion heuristic on a dictionary of size $O(\log^k n)$? The reduction will be from the following circuit value problem: Given a synchronous circuit, whose gates $c_1 \cdots c_n$ are ordered by level, with width $O(\log^k n)$ and values for its inputs, what is the value of its output? Without loss of generality, c_i is either an input gate with value 0 or 1, a NOT gate or an OR gate. Moreover, by replicating the inputs of the circuit we can assume that the fan-out of an input gate is 1 and has level $\ell - 1$ if the gate receiving the input has level ℓ. We can also assume that there are exactly w gates at each level by adding dummy gates (disconnected input gates). Let the ternary alphabet be $\{a, b, c\}$. We reduce the circuit to a string $Z = SY_1X_1 \cdots Y_nX_n$, where $S = abcabba$, Y_i depends only on i and X_i is associated to gate c_i. We will prove that $T = aca^{2w}b$ belongs to the final dictionary produced parsing Z by the LZ2 algorithm with LRU deletion heuristic on a dictionary of size $27w + 5$ iff the output of the circuit is 1. For each gate, exactly nine strings will be parsed. We consider the circuit as divided into consecutive triple of levels numbered starting from zero. By adding dummy gates, we can assume that the last gate of each triple is not an OR. When the parsing associated with the first triple is computed, the dictionary becomes full. A gate c_j outputs a 1 if and only if a given substring U of X_j is parsed. If c_i receives an input from c_j, X_i contains U as a substring and to simulate the circuit it is necessary to have the substrings parsed in X_j in the dictionary when X_i is parsed. This requires that the size of the dictionary be large enough to include the parsed substrings associated to a triple of levels in the circuit.

We need some notations. Let $i(j)$ be the number of gates c_h, $h \leq i$, having the output of c_j as input. Let $f(i)$ be the number of input or NOT gates c_j that belong to the same triple of c_i and $j \leq i$. Let

$$\ell(i) = \begin{cases} i - 3w \cdot t(c_i) & \text{if } i \leq 3w \cdot t(c_i) + 2w \\ i - 3w \cdot t(c_i) - 2w & \text{otherwise} \end{cases}$$

where $t(c_i)$ is the number of the triple c_i belongs to (clearly, $3w \cdot t(c_i) + 1 \leq i \leq 3w \cdot t(c_i) + 3w$). We distinguish six cases in defining the strings Y_i and X_i as in Fig. 1.

By parsing S, we learn a, b, c, ab, ba. By parsing Y_1 we add to the dictionary $ca\ abc, bc, bb$, and abb. Then the substring X_1, associated to the input gate c_1, is parsed into the substrings $bab, baba, babaa, bba$ ($abba$) iff its input is 1 (0). Let c_m be the first OR gate in the topological order of the circuit to receive input values both equal to 1 or 0. We can verify that, for $2 \leq i \leq \min\{m, 2w\}$, when Y_i is parsed the substrings $ab^{2\ell(i)-1}$, $b^{2\ell(i)-1}$, $ab^{2\ell(i)}$, $ab^{2\ell(i)-1}a$ and $b^{2\ell(i)}$ are added to the dictionary and the first character of X_i starts a new dictionary element. In fact, if c_i is an INPUT gate, with $1 \leq i \leq 2w$, then the substring X_i is parsed into the substrings $bab^{f(i)}, bab^{f(i)}a, \ bab^{f(i)}aa, \ b^{2i}a$ ($ab^{2i}a$) iff the input value is 1 (0). Therefore, if c_i is a NOT gate with input from the INPUT gate c_j and $w + 1 \leq i \leq 2w$, then the parsing X_i is $bab^{f(i)}, bab^{f(i)}a, b^{2\ell(j)}aa^{i(j)}$ ($b^{2\ell(j)}aa^{i(j)-1}$), $b^{2\ell(i)}a$ ($ab^{2\ell(i)}a$) iff the output value of c_i is 0 (1). Since an OR gate needs just one input equal to 1 to have 1 as its own output value, if c_i is an OR gate with inputs from INPUT gates or NOT gates and $w + 1 \leq i \leq 2w$, then by parsing X_i the substring $b^{2\ell(i)}a$ is added iff its output value is 1. It follows that these conditions verify also for the gates receiving inputs from an OR gate. The inputs of c_m are both equal to 1 (0) and if $m < 2w$ then the substring $b^{2\ell(m)}aa$ ($ab^{2\ell(m)}aa$) is added to the dictionary, where the last a is prefix of Y_{m+1}. The substrings $b^{2\ell(m)+1}$, $b^{2\ell(m)+1}a$, $b^{2\ell(m+1)}$, $ab^{2\ell(m+1)}$ and $ab^{2\ell(m+1)}$ are added by parsing the suffix of Y_{m+1}. Thus the substrings $b^{2\ell(m+1)}$ and $ab^{2\ell(m+1)}$ are in the dictionary and the first character of X_{m+1} starts a new dictionary element. Hence, the parsing of X_i observes the rules above, for $m \leq i \leq 2w$.

For $2w + 1 \leq i \leq \min\{\max\{2w, m\}, 3w\}$, when Y_i is parsed the substrings $abcb^{2\ell(i)-1}$, $bcb^{2\ell(i)-1}$, $abcb^{2\ell(i)}$, $abcb^{2\ell(i)-1}a$ and $bcb^{2\ell(i)}$ are added to the dictio-

First level of an even triple

$$Y_i = \begin{cases} caabcbcbbabb & \text{if } c_i \text{ is the first gate of the level} \\ ab^{2\ell(i)-1}b^{2\ell(i)-1}ab^{2\ell(i)}ab^{2\ell(i)-1}ab^{2\ell(i)} & \text{otherwise} \end{cases}$$

- c_i INPUT gate with value 1: $X_i = bab^{f(i)}bab^{f(i)}abab^{f(i)}aab^{2\ell(i)}a$
- c_i INPUT gate with value 0: $X_i = bab^{f(i)}bab^{f(i)}abab^{f(i)}aaab^{2\ell(i)}a$
- c_i NOT gate with input from c_j: $X_i = bab^{f(i)}bab^{f(i)}aaca^{2\ell(j)}ba^{i}(j)b^{2\ell(i)}a$
- c_i OR gate with inputs from c_j and c_k, then:

i) c_j and c_k both either OR gate or INPUT gate: $X_i = aca^{2\ell(j)}ba^{i}(j)b^{2\ell(i)}aaca^{2\ell(k)}ba^{i}(k)b^{2\ell(i)}a$

ii) c_j either OR gate or INPUT gate and c_k NOT gate: $X_i = aca^{2\ell(j)}ba^{i}(j)b^{2\ell(i)}abaca^{2\ell(k)}ba^{i}(k)b^{2\ell(i)}a$

iii) c_j and c_k both NOT gate: $X_i = baca^{2\ell(j)}ba^{i}(j)b^{2\ell(i)}abaca^{2\ell(k)}ba^{i}(k)b^{2\ell(i)}a$

Second level of an even triple

$$Y_i = ab^{2\ell(i)-1}b^{2\ell(i)-1}ab^{2\ell(i)}ab^{2\ell(i)-1}ab^{2\ell(i)} \text{ and:}$$
- c_i INPUT gate with value 1: $X_i = bab^{f(i)}bab^{f(i)}abab^{f(i)}aab^{2\ell(i)}a$
- c_i INPUT gate with value 0: $X_i = bab^{f(i)}bab^{f(i)}abab^{f(i)}aaab^{2\ell(i)}a$
- c_i NOT gate with input from c_j: $X_i = bab^{f(i)}bab^{f(i)}ab^{2\ell(j)}aa^{i}(j)b^{2\ell(i)}a$
- c_i OR gate with inputs from c_j and c_k, then:

i) c_j and c_k both either OR gate or INPUT gate: $X_i = b^{2\ell(j)}aa^{i}(j)b^{2\ell(i)}ab^{2\ell(k)}aa^{i}(k)b^{2\ell(i)}a$

ii) c_j either OR gate or INPUT gate and c_k NOT gate: $X_i = b^{2\ell(j)}aa^{i}(j)b^{2\ell(i)}aab^{2\ell(k)}aa^{i}(k)b^{2\ell(i)}a$

iii) c_j and c_k both NOT gate: $X_i = ab^{2\ell(j)}aa^{i}(j)b^{2\ell(i)}aab^{2\ell(k)}aa^{i}(k)b^{2\ell(i)}a$

Third level of an even triple

$$Y_i = abcb^{2\ell(i)-1}bcb^{2\ell(i)}abcb^{2\ell(i)-1}abcb^{2\ell(i)} \text{ and:}$$
- c_i INPUT gate with value 1: $X_i = bab^{f(i)}bab^{f(i)}abab^{f(i)}aabcb^{2\ell(i)}a$
- c_i INPUT gate with value 0: $X_i = bab^{f(i)}bab^{f(i)}abab^{f(i)}aaabcb^{2\ell(i)}a$
- c_i NOT gate with input from c_j: $X_i = bab^{f(i)}bab^{f(i)}ab^{2\ell(j)}aa^{i}(j)bcb^{2\ell(i)}a$
- c_i OR gate with inputs from c_j and c_k, then:

i) c_j and c_k both either OR gate or INPUT gate: $X_i = b^{2\ell(j)}aa^{i}(j)bcb^{2\ell(i)}ab^{2\ell(k)}aa^{i}(k)bcb^{2\ell(i)}a$

ii) c_j either OR gate or INPUT gate and c_k NOT gate: $X_i = b^{2\ell(j)}aa^{i}(j)bcb^{2\ell(i)}aab^{2\ell(k)}aa^{i}(k)bcb^{2\ell(i)}a$

iii) c_j and c_k both NOT gate: $X_i = ab^{2\ell(j)}aa^{i}(j)bcb^{2\ell(i)}aab^{2\ell(k)}aa^{i}(k)bcb^{2\ell(i)}a$

First level of an odd triple

$$Y_i = \begin{cases} cbbacacaabaa & \text{if } c_i \text{ is the first gate of the level} \\ ba^{2\ell(i)-1}a^{2\ell(i)-1}ba^{2\ell(i)}ba^{2\ell(i)-1}ba^{2\ell(i)} & \text{otherwise} \end{cases}$$

- c_i INPUT gate with value 1: $X_i = aba^{f(i)}aba^{f(i)}baba^{f(i)}bba^{2\ell(i)}b$
- c_i INPUT gate with value 0: $X_i = aba^{f(i)}aba^{f(i)}baba^{f(i)}bbba^{2\ell(i)}b$
- c_i NOT gate with input from c_j: $X_i = aba^{f(i)}aba^{f(i)}bbcb^{2\ell(i)}ab^{i}(j)a^{2\ell(i)}b$
- c_i OR gate with inputs from c_j and c_k, then:

i) c_j and c_k both either OR gate or INPUT gate: $X_i = bcb^{2\ell(i)}ab^{i}(j)a^{2\ell(i)}bbcb^{2k}ab^{i}(k)a^{2\ell(i)}b$

ii) c_j either OR gate or INPUT gate and c_k NOT gate: $X_i = a^{2\ell(i)}bb^{i}(j)a^{2\ell(i)}babcb^{2k}ab^{i}(k)a^{2\ell(i)}b$

iii) c_j and c_k both NOT gate: $X_i = abcb^{2\ell(i)}ab^{i}(j)a^{2\ell(i)}babcb^{2k}ab^{i}(k))a^{2\ell(i)}b$

Second level of an odd triple

$$Y_i = ba^{2\ell(i)-1}a^{2\ell(i)-1}ba^{2\ell(i)}ba^{2\ell(i)-1}ba^{2\ell(i)} \text{ and:}$$
- c_i INPUT gate with value 1: $X_i = aba^{f(i)}aba^{f(i)}baba^{f(i)}bba^{2\ell(i)}b$
- c_i INPUT gate with value 0: $X_i = aba^{f(i)}aba^{f(i)}baba^{f(i)}bbba^{2\ell(i)}b$
- c_i NOT gate with input from c_j: $X_i = aba^{f(i)}aba^{f(i)}ba^{2\ell(j)}bb^{i}(j)a^{2\ell(i)}b$
- c_i OR gate with input from c_j and c_k, then:

i) c_j and c_k both either OR gate or INPUT gate: $X_i = a^{2\ell(j)}bb^{i}(j)a^{2\ell(i)}ba^{2k}bb^{i}(\ell(k))a^{2\ell(i)}b$

ii) c_j either OR gate or INPUT gate and c_k NOT gate: $X_i = a^{2\ell(j)}bb^{i}(j)a^{2\ell(i)}bba^{2k}bb^{i}(\ell(k))a^{2\ell(i)}b$

iii) c_j and c_k both NOT gate: $X_i = ba^{2\ell(j)}bb^{i}(j)a^{2\ell(i)}bba^{2k}bb^{i}(\ell(k))a^{2\ell(i)}b$

Third level of an odd triple

$$Y_i = baca^{2\ell(i)-1}aca^{2\ell(i)-1}baca^{2\ell(i)}baca^{2\ell(i)-1}baca^{2\ell(i)} \text{ and:}$$
- c_i INPUT gate with value 1: $X_i = aba^{f(i)}aba^{f(i)}baba^{f(i)}bbaca^{2\ell(i)}b$
- c_i INPUT gate with value 0: $X_i = aba^{f(i)}aba^{f(i)}baba^{f(i)}bbbaca^{2\ell(i)}b$
- c_i NOT gate with input from c_j: $X_i = aba^{f(i)}aba^{f(i)}ba^{2\ell(j)}bb^{i}(j)aca^{2\ell(i)}b$
- c_i OR gate with inputs from c_j and c_k, then:

i) c_j and c_k both either OR gate or INPUT gate: $X_i = a^{2\ell(j)}bb^{i}(j)aca^{2\ell(i)}ba^{2\ell(k)}bb^{i}(k)aca^{2\ell(i)}b$

ii) c_j either OR gate or INPUT gate and c_k NOT gate: $X_i = a^{2\ell(j)}bb^{i}(j)aca^{2\ell(i)}bba^{2\ell(k)}bb^{i}(k)aca^{2\ell(i)}b$

iii) c_j and c_k both NOT gate: $X_i = ba^{2\ell(j)}bb^{i}(j)aca^{2\ell(i)}bba^{2\ell(k)}bb^{i}(k)aca^{2\ell(i)}b$

Fig. 1.

nary and the first character of X_i starts a new dictionary element. In fact, if c_i is an INPUT gate, with $2w + 1 \leq i \leq 3w$, then the substring X_i is parsed into the substrings $bab^{f(i)}, bab^{f(i)}a$, $bab^{f(i)}aa$, $bcb^{2i}a$ $(abcb^{2i}a)$ iff the input value is 1 (0). Therefore, if c_i is a NOT gate with input from the INPUT gate c_j, then the parsing X_i is $bab^{f(i)}, bab^{f(i)}a$, $b^{2\ell(j)}aa^{i(j)}$ $(b^{2\ell(j)}aa^{i(j)-1})$, $bc^{2\ell(i)}a$ $(abcb^{2\ell(i)}a)$ iff the output value of c_i is 0 (1). If c_i is an OR gate with inputs from c_j and c_k, then by parsing X_i the substring $bcb^{2\ell(i)}a$ is added iff its output value is 1.

If $2w \leq m \leq 3w$ then the substring $bcb^{2\ell(m)}aa$ $(abcb^{2\ell(m)}aa)$ is added to the dictionary $(b^{2\ell(m)}aa$ $(ab^{2\ell(m)}aa)$ if $m = 2w)$, where the last a is prefix of Y_{m+1}. The substrings $bcb^{2\ell(m)+1}$, $bcb^{2\ell(m)+1}a$, $bcb^{2\ell(m+1)}$, $abcb^{2\ell(m+1)}$ and $abcb^{2\ell(m+1)}$ are added by parsing the suffix of Y_{m+1}. Thus the substrings $bcb^{2\ell(m+1)}$ and $abcb^{2\ell(m+1)}$ are in the dictionary and the first character of X_{m+1} starts a new dictionary element. Hence, the parsing of X_i observes the previous rules for $m \leq i \leq 3w$.

Now, the dictionary is full. We give the parsing corresponding to the next triple of levels in the circuit as if enough space for the new substrings is still available in the dictionary and show later that is the same as if we applied the LRU deletion heuristic.

By parsing Y_{3w+1} we add to the dictionary five substrings that are cb, bac, ac, aa, baa (observe that this follows also from the assumption that the last gate of a triple is not an OR).

If c_{3w+1} is an INPUT gate, the substring X_{3w+1} is parsed into the substrings aba, $abab$, $abab$ and aab $(baab)$ iff its input is 1 (0).

For $3w + 2 \leq i \leq \min\{\max\{3w + 1, m\}, 5w\}$, when Y_i is parsed the substrings $ba^{2\ell(i)-1}$, $a^{2\ell(i)-1}$, $ba^{2\ell(i)}$, $ba^{2\ell(i)-1}b$ and $a^{2\ell(i)}$ are added to the dictionary and the first character of X_i starts a new dictionary element. In fact, if c_i is an INPUT gate, with $3w + 1 \leq i \leq \min\{m, 5w\}$, then the substring X_i is parsed into the substrings $aba^{f(i)}, aba^{f(i)}b$, $aba^{f(i)}bb$, $a^{2i}b$ $(ba^{2i}b)$ iff the input value is 1 (0). Therefore, if c_i is a NOT gate with input from c_j and $3w + 1 \leq i \leq 4w$, then the parsing of X_i is $aba^{f(i)}, aba^{f(i)}b$, $bcb^{2\ell(j)}ab^{i(j)}$ $(bcb^{2\ell(j)}ab^{i(j)-1})$, $a^{2\ell(i)}b$ $(ba^{2\ell(i)}b)$ iff the output value of c_i is 0 (1). If c_i is an OR gate, instead, the substring $a^{2\ell(i)}b$ is added iff its output value is 1.

if c_i is a NOT gate with $4w + 1 \leq i \leq \min\{m, 5w\}$, then the parsing of X_i is $aba^{f(i)}, aba^{f(i)}b$, $a^{2\ell(j)}bb^{i(j)}$ $(a^{2\ell(j)}bb^{i(j)-1})$, $a^{2\ell(i)}b$ $(ba^{2\ell(i)}a)$ iff the output value of c_i is 0 (1). If c_i is an OR gate the substring $a^{2\ell(i)}b$ is added iff its output value is 1.

If $3w \leq m \leq 5w$ then the substring $a^{2\ell(m)}bb$ $(ba^{2\ell(m)}bb)$ is added to the dictionary, where the last b is prefix of Y_{i+1}. The substrings $a^{2\ell(i)+1}$, $a^{2\ell(i)+1}b$, $a^{2\ell(i+1)}$, $ba^{2\ell(i+1)}$ and $ba^{2\ell(i+1)}$ are added by parsing the suffix of Y_{i+1}. Thus the substrings $a^{2\ell(i+1)}$ and $ba^{2\ell(i+1)}$ are in the dictionary and the first character of X_{i+1} starts a new dictionary element. Hence, the parsing of X_i observes the rules above, for $m \leq i \leq 5w$.

For $5w + 1 \leq i \leq \min\{\max\{5w, m\}, 6w\}$, when Y_i is parsed the substrings $baca^{2\ell(i)-1}$, $aca^{2\ell(i)-1}$, $baca^{2\ell(i)}$, $baca^{2\ell(i)-1}b$ and $aca^{2\ell(i)}$ are added to the dictionary and the first character of X_i starts a new dictionary element. In fact, if c_i is an INPUT gate then the substring X_i is parsed into the substrings $aba^{f(i)}, aba^{f(i)}b, aba^{f(i)}bb, aca^{2^i}b$ $(baca^{2^i}b)$ iff the input value is 1 (0). Therefore, if c_i is a NOT gate with input from c_j, then the parsing of X_i is $aba^{f(i)}, aba^{f(i)}b$, $a^{2\ell(j)}bb^{i(j)}$ $(a^{2\ell(j)}aa^{i(j)-1})$, $aca^{2\ell(i)}b$ $(baca^{2\ell(i)}b)$ iff the output value of c_i is 0 (1). If c_i is an OR gate with inputs from c_j and c_k, then by parsing X_i the substring $aca^{2\ell(i)}b$ is added iff its output value is 1.

If $5w \leq m \leq 6w$ then the substring $aca^{2\ell(m)}bb$ $(baca^{2\ell(m)}bb)$ is added to the dictionary $(a^{2\ell(m)}bb$ $(ba^{2\ell(m)}bb)$ if $m = 2w)$, where the last a is prefix of Y_{m+1}. The substrings $aca^{2\ell(m)+1}$, $aca^{2\ell(m)+1}b$, $aca^{2\ell(m+1)}$, $baca^{2\ell(m+1)}$ and $baca^{2\ell(m+1)}$ are added by parsing the suffix of Y_{m+1}. Thus the substrings $aca^{2\ell(m+1)}$ and $baca^{2\ell(m+1)}$ are in the dictionary and the first character of X_{m+1} starts a new dictionary element. Hence, the parsing of X_i observes the previous rules for $m \leq i \leq 6w$.

Each substring Y_iX_i is parsed into nine phrases. Let us consider the two sets of dictionary elements (with cardinality $27w$) learned for $1 \leq i \leq 3w$ and $3w + 1 \leq i \leq 6w$. Observe that $a, b, ab, ba, abc, c, cc, bcb^\ell$ for $0 \leq \ell \leq 2w$ and $bcb^{2\ell(i)}a$ for $i \in A = \{2w + 1 \leq i \leq 3w : c_i \text{ outputs } 1\}$ are the prefixes shared by elements of the two sets and that $18w$ phrases learned when parsing $Y_1X_1 \cdots Y_{3w}X_{3w}$ are removed from the dictionary when $Y_{3w+1}X_{3w+1} \cdots Y_{5w}X_{5w}$ is parsed. None of these phrases is one of these shared prefixes and after they are removed there are $7w-1-|A|$ of such phrases left that are removed when parsing $Y_{5w+1}X_{5w+1} \cdots Y_{6w}X_{6w}$. The other phrases removed, that are $2w + |A| + 1$, include the phrases learned when parsing $Y_{3w+1}X_{3w+1} \cdots Y_{4w}X_{4w}$ that are extensions of the shared prefixes and are at most $2w$. Since the circuit is synchronous, the removing of the phrases does not alter the parsing described above. Furthermore, the number of shared prefixes left equals the number of removed phrases learned when parsing $Y_{3w+1}X_{3w+1} \cdots Y_{4w}X_{4w}$. Therefore, it can be seen that the parsing and removal of phrases follow the same rules for the other triples of levels of the circuit. In conclusion, the output of c_n is 1 iff the phrase $aca^{2w}b$ is in the final dictionary. □

We present, now, a relaxed version of LRU that is more sophisticated than SWAP since it removes elements in a continuous way as the original LRU but relaxes on the choice of the element. This relaxation makes the problem complete for SC^k when the dictionary size is $O(\log^k n)$. The relaxed version (RLRU) of the LRU heuristic is:

RLRU: Label the ith element added to the dictionary with the integer $\lceil i \cdot \ell/k \rceil$, where k is the dictionary size and $\ell < k$ is the maximum number of distinct labels allowed. At the generic i^{th} step, when the dictionary is full, remove one of the leaves with the smallest label in the trie storing the dictionary and add the new leaf. Let λ be the greatest label among all the dictionary elements. Label the new leaf with λ if $\lceil i \cdot \ell/k \rceil = \lceil (i-1)\ell/k \rceil$. If $\lceil i \cdot \ell/k \rceil > \lceil (i-1)\ell/k \rceil$, label the leaf with $\lambda + 1$ and if $\lambda + 1 > \ell$ decrease by 1 all the labels greater or equal to 2.

Theorem 6. *Given an input string drawn over a ternary alphabet, the LZ2 algorithm with RLRU deletion heuristic (with $\ell = 3$) on a dictionary of size $O(\log^k n)$ is complete for SC^k.*

Proof (sketch). The problem belongs to SC^k since the number of distinct labels used in the heuristic is constant. The reduction shown in Theorem 5 works also for the relaxed version if we define $S = abcabbac^2c^3c^4c^5c^6c^7c^8$, the dictionary size is $27w$ and the number of gates at the first level of the circuit is $w - 1$. $\square$

In the next theorems, we give parallel algorithms for LRU and RLRU.

Theorem 7. *The LZ2 algorithm with LRU deletion heuristic on a dictionary of constant size κ can be parallelized either in $O(\log n)$ time with $2^{O(\kappa \log \kappa)}n$ processors or in $2^{O(\kappa \log \kappa)} \log n$ time with $O(n)$ processors.*

Proof. The number of all the possible dictionaries with the associated information needed for the LRU deletion heuristic is $2^{O(\kappa \log \kappa)}$. For each of these dictionaries the greedy match is computed on each position of the input string. Link the pair (p, D), where p is a position of the string and D is one of the dictionaries, to the pair $(p+\ell+1, D')$ where ℓ is the length of the match relative to (p, D) and D' is the updating of D. If the match is at the end of the string the pair is linking to a special node $\bar{v}$. The parsing of the string is given by the path from $(1, \emptyset)$ to $\bar{v}$ which can be computed by pointer jumping. $\square$

Theorem 8. *The LZ2 algorithm with RLRU deletion heuristic on a dictionary of constant size κ can be parallelized either in $O(\log n)$ time with $2^{O(\kappa)}n$ processors or in $2^{O(\kappa)} \log n$ time with $O(n)$ processors.*

Proof. As in the proof of Theorem 7. $\square$

5 Conclusions

The SC^k-completeness result of RLRU suggests that relaxations of LRU that keep removing elements from the dictionary in a continuous way do not have practical parallel algorithms. Even if $\ell < 3$, we strongly believe that practical parallel algorithms do not exist although the problem does not seem to be complete for SC^k when the dictionary size is $O(\log^k n)$.

As we mentioned earlier, the multilist layering problem was the first problem in SC^{k+1} to be shown hard for SC^k. We could not find any reasonable variation on this problem that would make it SC^k-complete.

In conclusion we want to point out the two relevant aspects of proving the SC^k-hardness of a problem (if we believe $SC \not\subseteq NC$). First, it gives evidence for problems not proved to be P-complete that they are not in NC. Moreover, for possible variations in NC of these problems it might imply that there are not practical parallel algorithms. This last implication cannot follow from a more traditional P-completeness result.

Acknowledgements

We wish to thank Raymond Greenlaw and Pierre McKenzie for helpful comments and discussions.

References

1. Bell, T.C., J.G. Cleary and I.H. Witten [1990]. *Text Compression*, Prentice Hall.
2. Borodin, A. [1977]. "On Relating Time and Space to Size and Depth", *SIAM Journal on Computing* 6, 733–744.
3. Cook, S.A. [1979]. "Deterministic CFL's are accepted simultaneously in Polynomial Time and Log Squared Space", *11th Ann. ACM Symposium on Theory of Computing*, 338–345.
4. Cook, S.A. [1981]. "Towards a Complexity Theory of Synchronous Parallel Computation", *Enseignement Mathematique* 27, 99-124.
5. Cook, S.A. [1985]. "A Taxonomy of Problems with Fast Parallel Algorithms", *Information and Control* 64, 2-22.
6. Crochemore, M. and W. Rytter [1991]. "Efficient Parallel Algorithms to Test Square-freeness and Factorize Strings", *Information Processing Letters* 38, 57–60.
7. De Agostino, S. [1994]. "P-complete Problems in Data Compression", *Theoretical Computer Science* 127, 181–186.
8. Dessmark, A., A. Lingas and A. Maheshwari [1995]. "Multi-list Layering: Complexity and Applications", *Theoretical Computer Science* 141, 337–350.
9. Dymond, P.W. and S.A. Cook [1989]. "Complexity Theory of Parallel Time and Hardware", *Information and Computation* 80, 205–226.
10. Greenlaw, R., H.J. Hoover and W.L. Ruzzo [1995]. *Limits to Parallel Computation*, Oxford University Press.
11. Johnson, D.S. [1990]. "A Catalog of Complexity Classes" *Handbook of Theoretical Computer Science: Algorithms and Complexity*, MIT Press/Elsevier (van Leeuwen J., editor), 67–162.
12. Lempel, A. and J. Ziv [1977]. "A Universal Algorithm for Sequential Data Compression", *IEEE Transactions on Information Theory* 23, 337–343.
13. Olariu, S., J.L. Schwing and J. Zhang [1992]. "Optimal Parallel Algorithms for Problems Modeled by a Family of Intervals", *IEEE Transactions on Parallel and Distributed Systems* 3, 364–374.
14. Pippenger, N. [1979]. "On Simultaneous Resource Bounds", *20th Ann. Symposium on Foundations of Computer Science*, 307–311.
15. Pippenger, N. and M.J. Fischer [1979]. "Relations Among Complexity Measures", *Journal of the ACM* 26, 361–381.
16. Ruzzo, W.L. [1981]. "On Uniform Circuit Complexity", *Journal of Computer and System Sciences* 22, 365–383.
17. Storer, J.A. [1988]. *Data Compression: Methods and Theory* (Computer Science Press).
18. Storer, J.A. [1992]. "Massively Parallel Systolic Algorithms for Real-Time Dictionary-Based Text Compression" *Image and Text Compression*, Kluwer Accademic Publishers (Storer J.A., editor), 159–178.
19. Ziv, J. and A. Lempel [1978]. "Compression of Individual Sequences via Variable Rate Coding", *IEEE Transactions on Information Theory* 24, 530–536.

Expressive Completeness of LTrL on Finite Traces: An Algebraic Proof

Raphaël Meyer and Antoine Petit

LSV, URA 2236 CNRS, ENS de Cachan,
61, av. du Prés. Wilson, F-94235 Cachan Cedex
{rmeyer,petit}@lsv.ens-cachan.fr

Abstract. Very recently a new temporal logic, for Mazurkiewicz traces, denoted *LTrL*, has been defined by Thiagarajan and Walukiewicz [15]. They have shown that this logic is equal in expressive power to the first order theory of finite and infinite traces thus filling a prominent gap in the theory.

We propose in this paper a entirely new, algebraic, proof of this result in the case of finite traces only. Our proof generalizes Cohen, Perrin and Pin's work on finite sequences [2], using as a basic tool a new extension of the wreath product principle on traces [7].

As a major consequence of our proof we show that, when dealing with finite traces only, no past modality is necessary to obtain a expressively complete logic. Precisely, we prove that the logic $LTrL_{red}$, obtained from *LTrL* by not using the past modularity, has the same expressive power as the first order theory on finite traces.

Topics: logic in computer science, automata and formal languages, theory of parallel and distributed computation

1 Introduction

A run of a distributed system can be viewed, in many settings, as a partial order between the events of the system. Two events are ordered if and only if their executions depend causally one of the other. The partial orders that arise in this fashion, are frequently Mazurkiewicz traces [9,3]. A major interest of this model lies on the fact that a trace can be seen either as a labelled ordered graph expressing directly the partial order or as an equivalence class of sequences, each of them representing a linearization of the partial order.

Through this second approach, properties of distributed runs can be described with LTL, the propositional temporal logic of linear time [11]. If the required proprety is insensitive to the choice of linearizations (i.e. either all the members of an equivalence class satisfy the formula or none), it suffices to check the property for just one element of each equivalence class. From a practical point of view, many of the so-called partial order reductions techniques in the verification process are based on this principle [6,10,17].

In a natural way, and in order to exploit directly the partial order underlying to a trace, a good amount of research have focused on developing temporal logics that can be directly interpreted over traces seen as labelled partially ordered graphs (rather than as sets of sequences). After the first attempt of Thiagarajan [14], several such logics have been proposed in the literature [4,1,12]. Until recently, none of them was completely satisfactory since none was expressively complete, i.e had the same expressive power as the first order theory of finite and infinite traces. This was a gap in the generalisation of the theory of sequences to the theory of traces.

Very recently Thiagarajan and Walukiewicz [15] have defined a new temporal logic patterned on LTL, and denoted by $LTrL$, with precisely this expressive power. This logic uses classical $Next$ and $Until$ operations but also a restricted past modality making thus an annoying difference with LTL. Past modalities were also used extensively in the logic $TPLO$, due to Ebinger [4], which is expressively complete when interpreted on finite traces, only. Therefore past modalities can seem necessary to get expressively complete logics on traces.

We propose in this paper an entirely new proof of the completeness of $LTrL$ when interpreted on finite traces only. As a major consequence, we show that, when dealing with finite traces only, no past modality is necessary to obtain a logic expressively complete. Precisely, we prove that the logic $LTrL_{red}$, obtained from $LTrL$ by not using the past modality, has the same expressive power as the first order theory $FO(<)$ of finite traces.

The main difficulty is to show that any trace langage $FO(<)$-definable is also $LTrL$-definable. To this purpose, our proof generalizes Cohen, Perrin and Pin's results on finite sequences [2]. More precisely, it is known that $FO(<)$-definable trace languages coincide with aperiodic trace languages [16,8,4], that is trace languages recognized by aperiodic monoids. Note that these two results form an nice extension to traces to the famous Schützenberger's theorem. Thus, we first study the aperiodic trace languages. Using as a basic tool a recent extension of the wreath product principle to traces [7], we propose a decomposition theorem of aperiodic trace languages. Thus extending the PTL logic of [2], we introduce as as intermediary step a completely past oriented logic on traces, denoted by $PTrL$ and we prove that any aperiodic trace languages is $PTrL$-definable. We conclude easily our proof using the fact that a trace language is $PTrL$-definable if and only if its mirror is $LTrL_{red}$-definable and that aperiodic trace languages are closed under the mirror operator.

2 Finite traces and temporal logics

2.1 Basic definitions

A *dependence alphabet* is a pair (Σ, I) where Σ is a finite set and $I \subseteq \Sigma \times \Sigma$ is a symmetric and irreflexive relation called the *independence relation*. Elements of Σ are called *actions*; two actions $a, b \in \Sigma$ are said *independent* if $(a, b) \in I$. The complementary relation $D = \Sigma \times \Sigma \setminus I$ is called the *dependence relation*.

A *trace* on (Σ, I) is a Σ-labelled partially ordered set (poset) that respects the dependence relation. More formally, let $(E, \leq, \lambda)$ be a Σ-labelled poset, that is: $\leq$ is a partial order on E and λ is a labelling function from E into Σ. For every subset Y of E, we define $\downarrow Y = \{x \in E \mid \exists y \in Y, x \leq y\}$. If Y is a singleton $\{y\}$ we shall write $\downarrow y$ instead of $\downarrow \{y\}$. We also define the *covering relation* $\lessdot$ on $E \times E$: $x \lessdot y$ if $x < y$ and $\forall z \in E, x \leq z \leq y$ implies $z = x$ or $z = y$.

Then a trace over (Σ, I) is a Σ-labelled poset $u = (E, \leq, \lambda)$ satisfying:

$(T1)\quad \forall e \in E,\ \downarrow e$ is a finite set

$(T2)\ \forall e, e' \in E, e \lessdot e' \Rightarrow (\lambda(e), \lambda(e')) \in D$

$(T3)\ \forall e, e' \in E, (\lambda(e), \lambda(e')) \in D \Rightarrow e \leq e'$ ou $e' \leq e$

In the sequel, we shall only deal with *finite* traces, i.e. traces $u = (E, \leq, \lambda)$ such that the set E is finite. Elements of E are called *events*. A *configuration* of a trace $u = (E, \leq, \lambda)$ is a finite subset $c \subseteq E$ satisfying $\downarrow c = c$. We denote by C_u the set of configurations of a trace u. The *transition relation* $\rightarrow_u \subseteq C_u \times \Sigma \times C_u$ is given by: $c \xrightarrow{a}_u c'$ iff there exists $e \in E$ such that $\lambda(e) = a$, $e \notin c$ and $c' = c \cup \{e\}$.

If $u = (E, \leq_E, \lambda_E)$ and $v = (F, \leq_F, \lambda_F)$ are two finite traces, the *concatenation* of u and v is the finite trace $uv = (G, \leq, \lambda)$ where $G = E \cup F$, $\lambda = \lambda_E \cup \lambda_F$ and $\leq$ is the partial order induced by the cover relation $\lessdot_E \cup \lessdot_F \cup \{(x, y) \in E \times F \mid (\lambda_E(x), \lambda_F(y)) \in D\}$. The set of all finite traces over (Σ, I) forms a monoid with respect to concatenation, denoted by $\mathbf{M}(\Sigma, I)$. It is well known that the monoid $\mathbf{M}(\Sigma, I)$ can equivalently be defined as the quotient of the free monoid Σ^* by the congruence $\sim_I$ induced by the relation $\{(ab, ba) \mid (a, b) \in I\}$. In the sequel, we denote by χ the canonical surjection from Σ^* onto $\mathbf{M}(\Sigma, I)$.

2.2 Expressiveness of trace logics

In a recent paper, Thiagarajan and Waluckiewicz [15] have proposed a new temporal logic on (finite and infinite) traces, called *LTrL*, that is expressively complete, i.e. equivalent to the first-order logic $FO(<)$ on (finite and infinite) traces. The set of formulas belonging to *LTrL* is defined inductively by:

$$LTrL(\Sigma, I) ::= \underline{tt} \mid \neg \alpha \mid \alpha \vee \beta \mid @\alpha \mid \alpha \mathcal{U} \beta \mid @^{-1}\underline{tt}$$

The semantics of these formulas are defined inductively: let $u = (E, \leq, \lambda)$ be a trace and c a configuration of u, then

- $u, c \models \underline{tt}$.
  Furthermore, the boolean connectives $\neg$ and $\vee$ have the usual interpretations.
- $u, c \models @\alpha$ iff $\exists c' \in C_u$, $c \xrightarrow{a}_u c'$ and $u, c' \models \alpha$.
- $u, c \models \alpha \mathcal{U} \beta$ iff $\exists c' \in C_u$, $c \subseteq c'$, such that $u, c' \models \beta$ and $\forall c'' \in C_u$, $c \subseteq c'' \subset c'$ implies $u, c'' \models \alpha$.
- $u, c \models @^{-1}\underline{tt}$ iff $\exists c' \in C_u$ such that $c' \xrightarrow{a}_u c$.

In $LTrL$, a formula ϕ is *satisfied* by a trace u if $u, \emptyset \models \phi$ holds. The language defined by ϕ is $L(\phi) = \{u \in \mathbf{M}(\Sigma, I) \mid u, \emptyset \models \phi\}$. A trace language $L \subseteq \mathbf{M}(\Sigma, I)$ is said $LTrL$-definable if there exists a formula ϕ of $LTrL$ such that $L = L(\phi)$.

Remark that the operators $@$ and $\mathcal{U}$ are *future-oriented* operators, while the $@^{-1}\underline{tt}$ formulas refer to the near past. Our proof of the expressive completeness of $LTrL$ will show that, when dealing with finite traces, only the future-oriented operators are needed. We denote by $LTrL_{red}$ the subset of formulas of $LTrL$ which do no contain any $@^{-1}\underline{tt}$ subformula. A trace language $L \subseteq \mathbf{M}(\Sigma, I)$ is said $LTrL_{red}$-definable if there exists a formula ϕ of $LTrL_{red}$ such that $L = L(\phi)$.

Recall that

$$FO(<) ::= R_a(x) \mid x \leq y \mid \neg\phi \mid \phi \vee \phi' \mid (\exists x)\phi$$

If ϕ is a sentence of $FO(<)$, it defines the trace language $L(\phi) = \{u \in \mathbf{M}(\Sigma, I) \mid u \models_I^{FO} \phi\}$. A language $L \subseteq \mathbf{M}(\Sigma, I)$ is said $FO(<)$-definable if there exists a sentence ϕ of $FO(<)$ such that $L = L(\phi)$.

The main result of [15] claims:

Theorem 1. *Let $L \subseteq \mathbf{M}(\Sigma, I)$ be a trace language. Then L is $LTrL$-definable if and only if L is $FO(<)$-definable.*

Note that Thiagarajan and Walukiewicz prove in fact a deeper result since they show that the above theorem holds also for languages of finite and infinite traces. It is not very difficult, although technical, to show that an $LTrL$-definable language is $FO(<)$-definable. On the contrary, it is difficult to prove the reciprocal assertion for which we propose in the present paper an new algebraic proof. Our proof is a generalization of Cohen, Perrin and Pin's results [2] showing the equivalence between the expressive powers of PTL and $FO(<)$ on finite sequences. We use in a crucial way the notion of aperiodic trace languages that we will now recall.

3 Aperiodic trace languages

A *transformation monoid* (tm for short) is a pair (P, S), where P is a finite set and S is a subset of applications from P into itself that has a monoid structure (i.e. that is closed by composition and contains the identity on P). This tm is *aperiodic* if there exists some $n > 0$ such that for all $s \in S$, $s^{n+1} = s^n$.

Remark 2. Any monoid S can be considered as a transformation monoid (S, S), the application from S to S induced by some element $s \in S$ being simply the right multiplication by s.

Example 3. Let $U_2 = \{(1, \alpha, \beta), (1, \alpha, \beta)\}$ with the following composition: $\alpha.\alpha = \beta.\alpha = \alpha$, $\beta.\beta = \alpha.\beta = \beta$, and $x.1 = 1.x = x$ for all $x \in U_2$, and the action $x[y] = y.x$. The tm U_2 is aperiodic since all its elements are idempotent. This particular monoid plays a central role in the sequel.

Let M be a monoid, and let $L \subseteq M$, then L is *recognized* by a tm (P, S) if there exists a morphism $\eta : M \to S$, an element $p_0 \in P$ and a subset $F \subseteq P$ such that $L = \{u \in M \mid \eta(u)[p_o] \in F\}$.

Using this notion of languages recognized by a transformation monoid, we can now recall the following characterization of $FO(<)$-definable trace languages, that extends Schützenberger's theorem on finite words. Note that this deep result has been obtained in two parts. First, it has been shown that $FO(<)$-definable trace languages coincide with star-free trace languages [4,16]. Besides, star-free trace languages are exactly the languages of traces which are recognized by aperiodic transformation monoid [8].

Theorem 4. *Let $L \subseteq \mathbf{M}(\Sigma, I)$ be a trace language. Then L is $FO(<)$-definable if and only if L is recognized by an aperiodic transformation monoid.*

A tm (P, S) *divides* a tm (Q, T) if there exists a surjective partial function $\Phi : Q \to P$ such that for any $s \in S$, there exists some $t \in T$ such that $s[\Phi(q)] = \Phi(t[q])$ for every $q \in Q$. We will use in the sequel the following classical property of this *division*:

Proposition 5. *Let (P, S) and (Q, T) be two transformation monoids; if L is recognized by (P, S) and (P, S) divides (Q, T) then L is also recognized by (Q, T).*

Given two tms $X = (P, S)$ and $Y = (Q, T)$, the *wreath product* of X by Y is the tm $(P \times Q, S^Q \times T)$ where for any element $f \in S^Q$ (that is any application from Q into S) and any element $t \in T$, the application (f, t) is defined on $P \times Q$ as follows. For any $(p, q) \in P \times Q$:

$$(f, t)[(p, q)] = (f(q)[p], t[q]).$$

The composition of two such applications (f_1, t_1) and (f_2, t_2) is given by:

$$(f_1, t_1)(f_2, t_2) = (f, t_1 * t_2) \text{ with } \forall q \in Q, \ f(q) = f_1(q) + f_2(t_1[q])$$

where the $+$ sign denotes the composition in S, while the $*$ sign denotes the composition in T. The wreath product of $X = (P, S)$ by $Y = (Q, T)$ is denoted by $Z = X \circ Y = (P \times Q, S^Q \times T)$. The wreath product is an associative operation on transformation monoids.

With this product, aperiodic transformation monoids are characterized as follows (this result is a consequence of the famous Krohn-Rhodes theorem [5]):

Theorem 6. *A transformation monoid is aperiodic if and only if it divides a wreath product of the form $U_2 \circ U_2 \circ \ldots \circ U_2$.*

4 Wreath product principle on traces

Let $X = (P, S)$ and $Y = (Q, T)$ be two transformation monoids and let $Z = X \circ Y = (P \times Q, S^Q \times T)$ be their wreath product. Let $L \subseteq \mathbf{M}(\Sigma, I)$ be a trace language recognized by Z. Then, by definition, there exists a morphism η from $\mathbf{M}(\Sigma, I)$ into $S^Q \times T$, a state (p_0, q_0) of $P \times Q$ and a subset $F \subseteq P \times Q$ such that

$$L = \{u \in \mathbf{M}(\Sigma, I) \mid \eta(u)[(p_0, q_0)] \in F\}$$

We would like to express L from languages recognized by either the tm X or the tm Y. Thus, let η_1 and η_2 be the component functions of η defined by, for any u in $\mathbf{M}(\Sigma, I)$, $\eta(u) = (\eta_1(u), \eta_2(u))$ with $\eta_1(u) \in S^Q$ and $\eta_2(u) \in T$.

From the definition of the product in $S^Q \times T$, it is immediate to verify that η_2 is a morphism from $\mathbf{M}(\Sigma, I)$ into T. On the contrary, η_1 is not a morphism since, for any traces u, v, $\eta_1(uv)$ is the function which associates with any element $q \in Q$, the element $\eta_1(u)[q] + \eta_1(v)[\eta_2(u)[q]]$.

Let $B = Q \times \Sigma$. Recall that χ denotes the canonical surjection from Σ^* onto $\mathbf{M}(\Sigma, I)$. We define the function $\sigma' : \Sigma^* \to B^*$ by $\sigma'(\epsilon) = (q_0, \epsilon)$ and for any letters $a_1, \ldots, a_k$ in σ:

$$\sigma'(a_1 \ldots a_k) = (q_0, a_1)(\eta_2(\chi(a_1))[q_0], a_2) \ldots (\eta_2(\chi(a_1 \ldots a_{k-1}))[q_0], a_k)$$

For any letters $a, b \in \Sigma$, it holds $\sigma'(ab) = (q_0, a)(\eta_2(\chi(a))[q_0], b)$ and $\sigma'(ba) = (q_0, b)(\eta_2(\chi(b))[q_0], a)$. Therefore, in order to factorize the function σ' on $\mathbf{M}(\Sigma, I)$ we define on B^* the congruence $\sim$ as the smallest congruence such that, for any $q \in Q$ and any $(a, b) \in I$:

$$(q, a)(\eta_2(\chi(a))[q], b) \sim (q, b)(\eta_2(\chi(b))[q], a)$$

Let ρ denote the canonical surjection from B^* onto $B^*/\sim$. Thus, we factorize the function σ' to a function σ from $\mathbf{M}(\Sigma, I)$ into $B^*/\sim$ by defining for any trace $\chi(w)$ in $\mathbf{M}(\Sigma, I)$:

$$\sigma(\chi(w)) = \rho(\sigma'(w))$$

Now, we can state the wreath product principle on traces [7] which generalizes a similar principle on words [13].

Theorem 7. *Let L be a trace language of* $\mathbf{M}(\Sigma, I)$ *recognized by the wreath product* $Z = X \circ Y$, *and let* σ *be the function defined above. Then L is a finite union of languages of the form* $U \cap \sigma^{-1}(V)$ *where* $U \subseteq \mathbf{M}(\Sigma, I)$ *is a trace language recognized by Y and* $V \subseteq B^* / \sim$ *is a language recognized by X.*

In the next section, we will apply the previous theorem in the special case where X is the tm U_2 as defined in Example 3 and use the following particular form of languages of $B^* / \sim$ recognized by U_2.

Lemma 8. *If a language* $L \subseteq \rho(B^*)$ *is recognized by* U_2 *then it is a boolean combination of languages of the form* $\rho(B^* b C^*)$, *where* $b \in B$ *and* $C \subseteq B$.

Proof. Since L is recognized by U_2, there exists by definition a morphism $\phi : B^*/\rho \to \{1, \alpha, \beta\}$, an element $x \in \{1, \alpha, \beta\}$, and a subset P of $\{1, \alpha, \beta\}$ such that $L = \{w \in B^*/\rho \mid \phi(w)[x] \in P\}$. From Remark 2, $\phi(w)[x]$ equals to the product $x.\phi(w)$. Thus L is the union of the languages $L_{x,y} = \{w \mid x.\phi(w) = y\}$ for $y \in P$.

For any $z \in \{1, \alpha, \beta\}$, let B_z be the subset of B defined as $B_z = \{b \in B \mid \phi(\rho(b)) = z\}$. From the multiplication law in $\{1, \alpha, \beta\}$ (see example 3), it is without difficulty to verify that: $L_{1,1} = \rho(B_1^*)$, $L_{\alpha,1} = L_{\beta,1} = \emptyset$, $L_{1,\alpha} = L_{\beta,\alpha} = \rho(B^* B_\alpha B_1^*)$, $L_{1,\beta} = L_{\alpha,\beta} = \rho(B^* B_\beta B_1^*)$, $L_{\alpha,\alpha} = \rho(B^* B_\alpha B_1^*) \cup \rho(B_1^*)$, and $L_{\beta,\beta} = \rho(B^* B_\beta B_1^*) \cup \rho(B_1^*)$.

Now, since $\rho(B_1^*)$ is clearly the complement of the union of the sets $\rho(B^* B_\alpha B_1^*)$ and $\rho(B^* B_\beta B_1^*)$, then $L_{x,y}$ is always a boolean combination of those two sets. Moreover, each of them can be seen as an union of languages of the form $\rho(B^* b B_1^*)$, with $b \in B$, which concludes the proof. $\square$

Our next step is to prove that these languages $\rho(B^* b C^*)$ are $LTrL_{red}$-definables. Unfortunately, this can not be done directly and we have to define first a new logic $PTrL$. We will first show that the languages $\rho(B^* b C^*)$ are $PTrL$-definables and then that $PTrL$ and $LTrL_{red}$ have the same expressive power.

5 A past-oriented trace logic

We define a new logic on traces, $PTrL$, that uses only past temporal operators. This logic is patterned on the PTL logic on finite sequences defined by Cohen, Perrin and Pin [2]. The formulas of $PTrL$ are defined by:

$$PTrL ::= \underline{tt} \mid \neg \alpha \mid \alpha \vee \beta \mid @^{-1}\alpha \mid \alpha \mathcal{S} \beta.$$

Given a trace $u = (E, \leq, \lambda)$, the semantics of these formulas is defined inductively as follow:

- $\underline{tt}$, $\neg$ and $\vee$ have the same interpretations as in $LTrL$.
- $u, c \models @^{-1}\alpha$ iff $\exists c' \in C_u$ such that $c' \overset{a}{\to} c$ and $u, c' \models \alpha$.
- $u, c \models \alpha S \beta$ iff $\exists c' \in C_u$, such that $c' \subseteq c$ and $u, c' \models \beta$, and $\forall c'' \in C_u$, $c' \subset c'' \subseteq c$ implies $u, c'' \models \alpha$.

Since $PTrL$ is a past-oriented logic, a $PTrL$-formula ψ is *satisfied* by a trace $u = (E, \leq, \lambda)$ if $u, E \models \psi$ holds. The language defined by ψ is $L(\psi) = \{u = (E, \leq, \lambda) \in \mathbf{M}(\Sigma, I) \mid u, E \models \psi\}$.

We will prove in the next section that $LTrL_{red}$ and $PTrL$ have the same expressive power. We first characterize the expressive power of $PTrL$. As a first step, we show that some particular trace languages are $PTrL$-definables. We keep the notations of the previous section.

Lemma 9. *Any language of the form* $\sigma^{-1}(\rho(B^* b C^*))$ *is* $PTrL$-*definable.*

Proof. As first step, we consider the case where C is the empty set that is the case of languages of the form $\sigma^{-1}(\rho(B^* b))$. Since $b \in Q \times \Sigma$, let $b = (q, a)$. Let us denote by L the language of traces u such that $\eta_2(u)[q_0] = q$. Since η_2 is a morphism from $M(\Sigma, D)$ into Y, and since $Y \in C$, the language L is $PTrL$-definable. Let ϕ be a $PTrL$-formula such that $L = L(\phi)$.
Let $u = \chi(a_1 \ldots a_k)$ be a trace. If $u \in \sigma^{-1}(\rho(B^* b))$ then $\sigma'(a_1 \ldots a_k) \sim vb$ for some v in B^*. Thus $\exists b_1 \ldots b_k \sim_I a_1 \ldots a_k$ such that $\sigma'(b_1 \ldots b_k) = vb$. From the definition of σ', it follows immediately that $b_k = b$ and $\eta_2(\chi(b_1 \ldots b_k))[q_0] = q$. Thus $u \in La$.
In the opposite way, it is easy to see that if $u \in La$, then $u \in \sigma^{-1}(\rho(B^* b))$. We have thus shown that $\sigma^{-1}(\rho(B^* b)) = La$. Then it is straightforward to verify that La is defined by the formula $@^{-1}\phi$.

As second step, we deal with the languages of the form $\sigma^{-1}(\rho(B^* C))$ that will be used in third step. Since σ^{-1} commutes with union, we have the following equality:

$$\sigma^{-1}(\rho(B^* C)) = \bigcup_{c \in C} \sigma^{-1}(\rho(B^* c))$$

and therefore, using the first step, $\sigma^{-1}(\rho(B^* C))$ is $PTrL$-definable.

As third step, we consider the general case. From first step, there exists some $PTrL$-formula ϕ such that $L(\phi) = \sigma^{-1}(\rho(B^* b))$ and from second step, there exists some $PTrL$-formula ψ such that $L(\psi) = \sigma^{-1}(\rho(B^* C))$. We thus claim that:

$$\sigma^{-1}(\rho(B^* b C^*)) = L(\psi \, S \, \phi).$$

Indeed, $\sigma(u)$ belongs to $\rho(B^* b C^*)$ iff there exists a prefix c of u such that $\sigma(c) \in \rho(B^* b)$ and such that for every prefix c' of u, if c is a strict prefix of c' then $\sigma(c') \in \rho(B^* C)$. This is equivalent to saying that $\exists c \in C_u$ such that $u, c \models \phi$ and $\forall c' \in C_u$, $c \subset c' \subseteq E$ implies $u, c' \models \psi$. This leads directly to the above equality. $\square$

Using the previous lemma, we can now claim:

Theorem 10. *Let $L \subseteq M(\Sigma, I)$ be a trace language. If L is recognized by an aperiodic transformation monoid, then L is PTrL-definable.*

Proof. Let $\mathcal{A}$ be the class of aperiodic transformation monoids, and let $\mathcal{C}$ be the class of transformation monoids X such that every trace language recognized by X is PTrL-definable. We will show that $\mathcal{A} \subseteq \mathcal{C}$.

Using Theorem 6, it is sufficient to prove the following steps::

1. The trivial tm $(1,1)$ is in $\mathcal{C}$.
2. $\mathcal{C}$ is closed under division.
3. If $X \in \mathcal{C}$ then $U_2 \circ X \in \mathcal{C}$.

Proof of (1): This tm recognizes only the empty set and $M(\Sigma, I)$, which can be defined respectively by the PTrL formulas $\neg \underline{tt}$ and $\underline{tt}$.

Proof of (2): It X divides $Y \in \mathcal{C}$ and if L is recognized by X then L is also recognized by Y (Proposition 5). Since $Y \in \mathcal{C}$, L is PTrL-definable, and so X belongs to $\mathcal{C}$.

Proof of (3): Let $Y \in \mathcal{C}$ and $Z = U_2 \circ Y$. We have to show that Z belongs to $\mathcal{C}$. Let $L \in M(\Sigma, I)$ be recognized by Z. By the wreath product principle (Theorem 7), L is a finite union of languages of the form $U \cap \sigma^{-1}(V)$, where U is a language of $M(\Sigma, D)$ recognized by Y and V a language of $B^* / \sim$ recognized by U_2.

Since $Y \in \mathcal{C}$, U is PTrL-definable. Moreover, PTrL-definable languages are closed under boolean operations. Therefore, it remains to prove that, when V is recognized by U_2, the languages of the form $\sigma^{-1}(V)$ are PTrL-definable. From Lemma 8, V is a boolean combination of languages of the form $\rho(B^* b C^*)$, with $b \in B$ and $C \subseteq B$. Since the function σ^{-1} commutes with the boolean operators, we only need to show that the languages of the form $\sigma^{-1}(\rho(B^* b C^*))$ are PTrL-definable which is precisely the result of Lemma 9. □

6 $LTrL_{red}$ is expressively complete

In order to prove that $LTrL_{red}$ is expressively complete, we first study the relation between the expressive powers of PTrL and $LTrL_{red}$.

Let $u = (E, \leq, \lambda) \in M(\Sigma, I)$ be a finite trace, and let $\tilde{u}$ be the finite trace $\tilde{u} = (E, \widetilde{\leq}, \lambda)$ where $\forall x, y \in E$, $x \widetilde{\leq} y$ iff $y \leq x$. The trace $\tilde{u}$ is said to be the *mirror* of u.

If $L \subseteq M(\Sigma, I)$, let $\widetilde{L} = \{\tilde{u} \mid u \in L\}$ is the *mirror language* of L.

Remark 11. Assume that L is trace language which is $FO(<)$–definable. From Theorem 4, L is recognized by some aperiodic transformation monoid (P, S). Let $\widetilde{S}$ be the set of applications of S equipped with the inverse composition law: $\widetilde{s} \,\widetilde{\circ}\, \widetilde{s'} = s' \circ s$. Then it is easy to verify that $(P, \widetilde{S})$ is still aperiodic and that it recognizes $\widetilde{L}$. Therefore, $\widetilde{L}$ is still $FO(<)$–definable.

This mirror operator provides a useful link between $PTrL-$ and $LTrL_{red}-$definabilities.

Proposition 12. *Let $L \subseteq \mathbf{M}(\Sigma, I)$ be a trace language and let $\widetilde{L}$ be its mirror language. Then L is $LTrL_{red}$-definable if and only if $\widetilde{L}$ is $PTrL$-definable.*

Proof. (Sketch) Let $u \in \mathbf{M}(\Sigma, I)$ and ϕ be a $LTrL_{red}$ formula. Let $\widetilde{u}$ be the mirror trace of u, and let $\widetilde{\phi}$ be the $PTrL$ formula obtained from ϕ by replacing each occurrence of @ by @$^{-1}$ and each occurrence of $\mathcal{U}$ by $\mathcal{S}$. Note that if $c \in C_u$, then $\overline{c} = E \setminus c \in C_{\widetilde{u}}$. Then it can be shown by induction on the length of ϕ that $u, c \models \phi$ iff $\widetilde{u}, \overline{c} \models \widetilde{\phi}$. The proposition follows easily. $\qquad\square$

We are now ready to prove our main result.

Theorem 13. *Let $L \in \mathbf{M}(\Sigma, I)$ be a trace language. Then L is $LTrL_{red}$- definable if and only if L is $FO(<)$-definable.*

Proof. First, let L be defined by an $LTrL_{red}$ formula ϕ. Then ϕ is also an $LTrL$ formula, so L is $LTrL$-definable and thus $FO(<)$-definable (it is the "easy" way as shown in [15]).

Now, assume that L is $FO(<)$-definable. By Theorem 4, L is recognized by a finite aperiodic transformation monoid. Since the reverse of an aperiodic tm is also aperiodic, $\widetilde{L}$ is recognized by a finite aperiodic tranformation monoid. Using Theorem 10, we deduce that $\widetilde{L}$ is thus $PTrL$-definable. Then, Proposition 12 shows that L is $LTrL_{red}$-definable. $\qquad\square$

7 Conclusion

We give in this paper a new algebraic proof of the expressively completeness of $LTrL$ when dealing on finite traces only. A major consequence of this proof is to show that no past modality is needed to obtain a logic expressively complete. Precisely, we have proved that the logic $LTrL_{red}$, obtained from $LTrL$ by not using the past modality, has the same expressive power than the first order theory $FO(<)$ of finite traces. Our work is a generalization of Cohen, Perrin and Pin's results on finite sequences [2]. Since these results have been extended by the authors to infinite sequences, a natural and exciting challenge will be to prove the complete result of Thiagarajan and Walukiewicz for finite and infinite traces using this algebraic approach.

References

1. R. Alur, D. Peled, and W. Penczek. Model-checking of causality properties. In *Proceedings of LICS'95*, pages 90–100, 1995.
2. J. Cohen, D. Perrin, and J. E. Pin. On the expressive power of temporal logic. *Journal of Computer and System Sciences*, 46:271–295, 1993.
3. V. Diekert and G. Rozenberg, editors. *The Book of Traces*. World Scientific, Singapore, 1995.
4. W. Ebinger. *Charakterisierung von Sprachklassen unendlicher Spuren durch Logiken*. Dissertation, Institut für Informatik, Universität Stuttgart, 1994.
5. S. Eilenberg. *Automata, Languages, and Machines*, volume B. Academic Press, New York and London, 1976.
6. P. Godefroid. Partial-order methods for the verification of concurrent systems. In *LNCS 1032*, pages 176–185. Springer, 1996.
7. G. Guaiana, R. Meyer, A. Petit, and P. Weil. An extension of the wreath product principle to traces. *to appear*, 1997.
8. G. Guaiana, A. Restivo, and S. Salemi. On aperiodic trace languages. In *Proceedings of STACS'91*, number 480 in Lecture Notes in Computer Science, pages 76–88, 1991.
9. A. Mazurkiewicz. Concurrent program schemes and their interpretations. DAIMI Rep. PB 78, Aarhus University, Aarhus, 1977.
10. D. Peled. Partial-order reductions: model checking using representatives. In *Proceedings of MFCS'96*, number 1113 in Lecture Notes in Computer Science, pages 93–112, 1996.
11. A. Pnueli. The temporal logics of programs. In *Proceedings of the 18th IEEE FOCS, 1977*, pages 46–57, 1977.
12. R. Ramanujam. Locally linear time temporal logic. In *Proceedings of LICS'96*, pages 118–128, 1996.
13. H. Straubing. The wreath product and its applications. In *Proceedings of the LITP Spring School on Theoretical Computer Science (in LNCS vol. 386)*, pages 15–24, 1988.
14. P. S. Thiagarajan. A trace based extension of linear time temporal logic. In *Proceedings of the 9th Annual IEEE Symposium on Logic in Computer Science (LICS'94)*, pages 438–447, 1994.
15. P. S. Thiagarajan and I. Walukiewicz. An expressively complete linear time temporal logic for mazurkiewicz traces. In *Proceedings of the 12th Annual IEEE Symposium on Logic in Computer Science (LICS'97)*, 1997.
16. W. Thomas. Automata on infinite objects. In J. v. Leeuwen, editor, *Handbook of Theoretical Computer Science*, pages 133–191. Elsevier Science Publishers, 1990.
17. A. Valmari. A stubborn attack on state explosion. In *Formal Methods in System Design*, pages 1:297–322, 1992.

On Uniform DOL Words*

Anna E. Frid[1]

Novosibirsk State University
2, Pirogova st., Novosibirsk, 630090, Russia
e-mail: frid@math.nsc.ru

Abstract. We introduce the wide class of *marked* uniform DOL words and study their structure. The criterium of circularity of a marked uniform DOL word is given, and the subword complexity function is found for the uncircular case as well as for the circular one.
The same technique is valid for a wider class of uniform DOL sequences which includes $(p, 1)$-Toeplitz words (see [4]).

1 Introduction

DOL word w is an infinite word on a finite alphabet Σ which is a fixed point of a morphism $\varphi : \Sigma^* \to \Sigma^*$; i. e.,

$$w = \lim_{i \to \infty} \varphi^i(a)$$

for $a \in \Sigma$. The class of DOL words has been extensively studied and contains famous examples concerning pattern avoidance, such as the cube-free Thue-Morse word on the two-letter alphabet and a square-free word on the three-letter alphabet.

A DOL word w is called *uniform* if all the words $\varphi(a_i)$, $a_i \in \Sigma$, are of the same length. In this paper, we deal with *marked* DOL words; i. e., fixed points of uniform morphisms with all the images of letters beginning with distinct symbols and ending with distinct symbols. The class of marked DOL words contains the Thue-Morse word.

To this point mostly *circular* DOL words have been studied (the definition of circularity is given in Section 2). In Section 3, we formulate the criterium of circularity of a marked uniform DOL word and introduce a parameter called *uncircularity number* which is equal to 0 if and only if the DOL word is circular. We study a DOL word with an arbitrary uncircularity number.

The *subword complexity* $f(n)$ of a word w is the number of distinct words of length n which occur in w. We find the subword complexity of a marked uniform DOL word. The circular case has been studied for instance by Tapsoba [12] and Mossé [11], and an explicit formula for the subword complexity of a circular marked uniform DOL word was given in [5]. In [6] the case of the binary alphabet was studied.

* Supported in part by the Russian Foundation for Basic Research (Grant 96-01-01800) and Federal Aim Program "Integration" (Grant 473)

In this paper, we examine a DOL word with an arbitrary uncircularity number on an arbitrary alphabet and find the set of words which occur in it.

In Section 7, we introduce the class of *buffer* DOL words, which is a generalization of the class of marked words. It includes all the uniform DOL words on the binary alphabet (see [6]) and Toeplitz words containing one hole in the pattern [4]. An explicit formula for the subword complexity of a buffer uniform DOL word is given.

We consider the problem as applied to an infinite DOL word. However, the results can be extended obviously to an arbitrary DOL language generated by a marked or buffer uniform morphism.

2 Basic Definitions

Let $\Sigma = \{a_1, \ldots, a_q\}$ be an alphabet, and $\varphi : \Sigma^* \to \Sigma^*$ be a morphism. We say that φ is *uniform* if all the words $\varphi(a_i)$, $i = 1, \ldots, q$, are of the same length $m(\varphi) = m \geq 2$. We shall refer to images $\varphi(a_i)$ of letters as *blocks*, and to the length m as the *block length*.

An infinite word $w = w(\varphi)$ is called a *fixed point* of the morphism φ if $w = \lim_{i \to \infty} \varphi^i(a)$ for some $a \in \Sigma$. The sequence w is called also a *DOL sequence* or a *DOL word*. Without loss in generality we put $a = a_1$ and assume $\varphi(a_1)$ to begin with a_1. We assume also that all the symbols of the alphabet Σ occur in w.

We shall refer to a fixed point of a uniform morphism as a *uniform* DOL word. In this paper, we consider only uniform DOL words.

We denote the empty word by λ, and the length of a finite word v by $|v|$.

A *factor* (or *subword*) of a finite or infinite word u is such a finite word $v \in \Sigma^*$ that $u = s_1 v s_2$ for some words s_1 and s_2; if $s_1 = \lambda$, then v is called a *left factor* of u. We call a finite word v *allowable* (in a word u) if it occurs in u as a factor. The set of all words allowable in u is denoted by $A(u)$.

A word u is called a *power* of a word v if $u = v \ldots v = v^k$ for some integer k. If $u = v^k v'$, where v' is a left factor of v, then u is called a *fractional power* of v.

The *subword complexity* (or simply *complexity*) $f_u(n) = f(n)$ of a word u is the number of its distinct factors of length n; i. e., of words of length n which are allowable in u.

We say that morphism φ is *marked* if its blocks are of the form

$$\varphi(a_i) = a_{\pi(i)} b_i a_{\sigma(i)} \text{ for all } i \in \{1, \ldots, q\}.$$

Here π and σ are permutations ($\pi, \sigma \in S_q$), and b_i are arbitrary words on Σ. We shall refer to a fixed point of a marked morphism as a *marked* DOL word.

For $u \in \Sigma^+$ and such $i, j \geq 0$ that $i + j < |\varphi(u)|$, we define the word $\psi_{ij}(u)$ as obtained from $\varphi(u)$ by erasing i letters from the left and j letters from the right. We say that a non-empty word v admits an *interpretation*

$$s = (b_0 b_1 \ldots b_{n+1}, i, j)$$

in $w(\varphi)$ if $v = \psi_{ij}(b_0 b_1 \ldots b_{n+1})$, where $0 \le i < |\varphi(b_0)|$, $0 \le j < |\varphi(b_{n+1})|$, $b_k \in \Sigma$, and the word $b_0 b_1 \ldots b_{n+1}$ is allowable in $w(\varphi)$: $b_0 \ldots b_{n+1} \in A(w)$. The word $b_0 \ldots b_{n+1}$ is called the *anchestor* of the interpretation s.

Let $u \in A(w(\varphi))$. We say that (u_1, u_2) is a *synchronization point* of u (for φ) on w if $u = u_1 u_2$ and

$$\forall v_1, v_2 \in \Sigma^*, \forall s \in A(w) \; \exists s_1, s_2 \in A(w)$$

$$[v_1 u v_2 = \varphi(s) \implies (s = s_1 s_2, v_1 u_1 = \varphi(s_1), u_2 v_2 = \varphi(s_2))]$$

(see [3]). We call an allowable word $u \in A(w)$ *circular* if it has at least one synchronization point on w. Note, that if φ is marked, then circularity on w means that u admits a unique interpretation in w. If φ is uniform with the block length m, and $w = w_1 w_2 \ldots$, then circularity means that for every occurence $u = w_l \ldots w_{l+|u|-1}$ of the word u in w, l modulo m is the same number.

We call a DOL word w *circular* with *synchronization delay* L if every word of length at least L which is allowable in w is circular. This notion was introduced in the papers of Mignosi and Séébold [9] and Cassaigne [3]; see also Mossé [10], [11].

In [5] a formula for subword complexity of a circular marked uniform DOL word was found. This brings up the question: Is a given marked uniform DOL word circular or not? Actually, it may be difficult to recognize circularity directly.

3 Criterium of Circularity of a Marked Uniform DOL Word

To formulate the criterium, we introduce cyclic shift mappings C_b on Σ^*: if $u = u_1 u_2 \ldots u_n$, $u_i \in \Sigma$, then $C_b(u) = u_{b+1} \ldots u_n u_1 \ldots u_b$.

Let φ be a uniform morphism of the block length m. We shall refer to the set $M = \{A_1, \ldots, A_k\}$ of words $A_1, \ldots, A_k \in \Sigma^*$ as a *φ-chained set* if the following conditions are fulfilled:

1. $M \ne \emptyset$;
2. All the words $A_1, \ldots, A_k$ are of the same length l, and numbers l and m are coprime: $(l, m) = 1$;
3. All the symbols which occur in words $A_1, \ldots, A_k$ are distinct;
4. $\exists b \ge 0 : \forall i \in \{1, \ldots, k\} \; \varphi(A_i) = C_b(A_{i+1}^m)$;
   here we suppose that $A_{k+1} = A_1$.

Let a symbol s occur in the word A_i' of a φ-chained set M' and in the word A_j'' of a φ-chained set M''. It follows from the definition of φ-chained set that all the symbols which occur in words of M' occur in words of M'', and vice versa. We do not distinguish the sets M' and M''.

Theorem 1 (criterium of circularity of a marked uniform DOL word). *A marked DOL word $w(\varphi)$ is circular if and only if no φ-chained sets do exist.*

Proof. We give just an idea of the proof.

Let u be an uncircular finite word; i. e., let u admit two distinct interpretations: $u = s_0\varphi(b_1)\ldots\varphi(b_n)s_{n+1} = s_0'\varphi(b_1')\ldots\varphi(b_n')s_{n+1}'$, where $|s_0| < |s_0'|$. If we know b_1, then we can reconstruct the symbol b_2. Actually, if b_1 is known, then the symbol of $\varphi(b_1)$ which is the first symbol of $\varphi(b_1')$ is known. Because φ is marked, we can reconstruct b_1' from this symbol. Analogously, we obtain b_2 from b_1'.

Thus, every next block in an interpretation of u is uniquely determined by the previous one. Because of this, the anchestor of any interpretation of $\varphi(b_1)\ldots\varphi(b_n)$ is a fractional power of a word $v = v_1\ldots v_l$, where all the symbols $v_1,\ldots,v_l$ are distinct.

Let an infinite word $w(\varphi)$ be uncircular; i. e., let w contain uncircular factors which are as long as wished. It follows from the arguments above that w contains all the powers of some word $v_1 = v_{11}\ldots v_{1l_1}$, where $l_1 \leq q$, and all the symbols v_{1i} are distinct.

So, all the powers v_1^k are allowable and periodic with minimal period equal to l_1. The anchestor of any interpretation of the word with minimal period l_1 is a fractional power of some word v_2 of length $l_2 = \dfrac{l_1}{(l_1,m)}$, where m is the block length of φ. To provide occurence of every power of v_1, we must provide occurence of every power of v_2. Considering the anchestor of an interpretation of a power of v_2 and repeating this procedure, we obtain that the lengths must be constant: $l_1 = l_2 = \ldots = l$. So, $(l,m) = 1$, and the word v_1 is an element of some φ-chained set. Thus, if $w(\varphi)$ is uncircular, a φ-chained set exists.

Conversely, if a φ-chained set M exists, then $w(\varphi)$ contains every power of every element of M. These powers are uncircular, and their lengths grow infinitely. So, $w(\varphi)$ is uncircular.

Let φ be a marked morphism, $M_1,\ldots,M_r$ be all the distinct φ-chained sets; i. e., let all the symbols of the words of M_i be different from the symbols of the words of M_j, $i \neq j$. Let M_i consist of k_i words of length l_i. There exist exactly

$$d(\varphi) = d = \sum_{i=1}^{r} k_i l_i$$ distinct symbols which occur in the elements of φ-chained

sets. We shall refer to them as φ-*chained symbols*, and to their number $d(\varphi)$ as the *uncircularity number* of φ. Actually, it follows from Theorem 1 that $d = 0$ if and only if the fixed point of φ is circular.

We shall call a word φ-*chained* if it is a factor of a power of an element of a φ-chained set. Otherwise we call a word φ-*unchained*. It can be easily seen that for every n there exist exactly d φ-chained words of length n; they begin with d distinct φ-chained symbols, and so they end. φ-chained words are uncircular.

It follows from the proof of the Criterium that

Corollary 2. *Let φ be a marked morphism. Then there exist such a number $L(\varphi) = L$ that for every $n \geq L$ the only uncircular words of length n which are allowable in $w(\varphi)$ are φ-chained words.*

If the sequence $w(\varphi)$ is circular, then the length $L(\varphi)$ is the synchronization delay of φ. That is why we can refer to $L(\varphi)$ as the *synchronization delay* of the (possibly uncircular) DOL word w.

4 Some Examples of Application of the Criterium

In this section, we consider some examples of fixed points of marked morphisms and examine their circularity.

Example 1 (Thue-Morse sequence).

$$\Sigma_1 = \{a, b\}, \begin{cases} \varphi_1(a) = ab \\ \varphi_1(b) = ba, \end{cases}$$

$$w_1 = abbabaabbaababba\ldots.$$

There can be one-letter or two-letter words in a φ-chained set on the two-letter alphabet. As $m = 2$ and $(2, 2) \neq 1$, then only the one-letter variant remains. It can be easily seen that there are no φ_1-chained sets, and w_1 is circular. This well-known fact can be proved also by enumeration of all the allowable words of length 4: all of them are circular. Furthermore, the circularity of the Thue-Morse word follows immediately from the main result of Mignosi and Séébold [9].

The subword complexity of the Thue-Morse sequence was found in [2], [8], and later in [1].

Example 2.

$$\Sigma_2 = \{a, b, c\}, \begin{cases} \varphi_2(a) = aba \\ \varphi_2(b) = cbc \\ \varphi_2(c) = bcb, \end{cases}$$

$$w_2 = abacbcababcbcbcbabacbcaba\ldots.$$

It can be easily seen that $M = \{bc\}$ is a φ_2-chained set, and so the sequence is uncircular with the uncircularity number $d = 2$. Actually, for all n the word $(bc)^n$ is allowable in w_2 and uncircular. It follows from Corollary 2 that the length of each uncircular subword of w_2 except $(bc)^n$, $(bc)^n b$, $(cb)^n$, and $(cb)^n c$ is bounded; it can be shown that it is at most 3. So, the synchronization delay $L_2 = 4$.

Example 3. The following morphism g on the four-letter alphabet $\Sigma = \{a, b, c, d\}$ was obtained by Keränen [7] and avoids abelian squares. Its blocks are of the form $g(\sigma(x)) = \sigma(g(x))$ for all $x \in \Sigma$, where σ is a cyclic permutation of letters in Σ, and

$$g(a) = abcacdcbcdcadcdbdabacabadbabcbdbcbacbcdcacb -$$
$$abdabacadcbcdcacdbcbacbcdcacdcbdcdadbdcbca.$$

The morphism is marked, and its fixed point is circular, which follows immediately from the main result of [9]. Actually, a g-chained set cannot occur, because of the structure of the blocks.

5 Subword Complexity of Marked Uniform DOL Words

Recall that the function of subword complexity $f_w(n) = f(n)$ counts the number of distinct words of length n which occur in a word w. The subword complexity of a circular uniform DOL word has been studied, for instance, by Tapsoba [12] and Mossé [11]. In [5] the complexity function of all circular marked DOL words was found. In [6] the subword complexity of all uniform DOL words on the 2-letter alphabet was calculated. Now we extend these results to all marked uniform DOL words.

From this point on we assume that w is an unperiodic fixed point of a marked morphism φ with the block length m. Let d be an uncircularity number of w. In particular, the formula below (and its proof) is valid for $d = 0$; i. e., for circular DOL words considered in [5].

Let L be the synchronization delay of w introduced in Section 3. We define the *structure ordering number* K of the DOL word w as the least integer satisfying the inequality $m(K - 1) + 1 \geq L$. We assume here that $K \geq m - 1$ (otherwise, we put $K = m - 1$).

Note, that for every $n \geq K$ there exists a unique triplet of *decomposition parameters* $(p(n), k(n), \Delta(n))$, where $p(n)$ is a nonnegative integer, $k(n) \in \{K, \ldots, m(K-1)\}$, and $\Delta(n) \in \{1, \ldots, m^{p(n)}\}$, such that n can be decomposed as

$$n = m^{p(n)}(k(n) - 1) + \Delta(n).$$

Theorem 3. *For all $n \geq K$*

$$f(n + 1) = \Delta(n)f(k(n) + 1) + (m^{p(n)} - \Delta(n))f(k(n)) - d(m^{p(n)} - 1).$$

Note, that we can find the entire function $f(n)$ from the block length m, the structure ordering number K, and the values of $f(k)$ for $k \leq m(K-1)+1$.

Example 4. Let us calculate the subword complexity $f_2(n)$ of the word w_2 (see Ex. 2).

Since $L_2 = 4$, we put $K = K_2 = 2$. To write the formula for $f_2(n)$, we must know $f_2(k)$ for $k = 2, 3, 4$; it can be seen that $f_2(2) = 6$, $f_2(3) = 10$, and $f_2(4) = 14$. Now we consider successively two cases.

1. Let $k(n) = 2$; i. e., let $n = 3^{p(n)} + \Delta(n)$ for some $\Delta(n) \in \{1, \ldots, 3^{p(n)}\}$. Then

$$f_2(n + 1) = 10\Delta(n) + 6(3^{p(n)} - \Delta(n)) - 2(3^{p(n)} - 1) =$$
$$4(3^{p(n)} + \Delta(n)) + 2 = 4n + 2.$$

2. Let $k(n) = 3$; i. e., let $n = 2 \cdot 3^{p(n)} + \Delta(n)$ for some $\Delta(n) \in \{1, \ldots, 3^{p(n)}\}$. Then

$$f_2(n + 1) = 14\Delta(n) + 10(3^{p(n)} - \Delta(n)) - 2(3^{p(n)} - 1) =$$
$$4(2 \cdot 3^{p(n)} + \Delta(n)) + 2 = 4n + 2.$$

We have obtained that

$$\begin{cases} f_2(1) = 3, \\ f_2(n) = 4n - 2 \quad \forall n \geq 2. \end{cases}$$

6 Proof of the Formula for Subword Complexity

First of all, we introduce some auxiliary notions. An allowable word u of length n is called *special* if it can be extended to an allowable word of length $n + 1$ by appending a letter to the right in at least two different ways; i. e., if both words ua and ub are allowable for some distinct letters $a, b \in \Sigma$. The number $c(u)$ of these letters is called the *speciality degree* of the word u:

$$c(u) = |\{a|a \in \Sigma, ua \in A(w)\}|.$$

The number of allowable words of length n with speciality degree j is denoted by $B_j(n)$.

We introduce also the function $s(n)$ of first differences of the complexity:

$$s(n) = f(n + 1) - f(n).$$

This function is an important tool for studying subword complexity.

Proposition 4. *For every $l \geq 0$*

$$s(l) = B_2(l) + 2B_3(l) + \ldots + (q - 1)B_q(l).$$

Proof. The set of allowable words of length $l + 1$ can be derived from the set of allowable words of length l by appending letters to the right. A word of length l with speciality degree j can be extended by j different ways; i.e., gives $j - 1$ additional words of length $l + 1$. From this the formula above follows.

Proposition 5. *Let $l \geq K$. Then*

$$\forall j \in \{2, \ldots, q\} \, \forall i \in \{0, \ldots, m - 1\} \, B_j(l) = B_j(ml - i).$$

Proof. Let v be a word of length l; then the length of a word $\psi_{i0}(v)$ is at least L. It follows from Corollary 2 that if $\psi_{i0}(v)$ is not φ-chained, then it is circular. On the other hand, $\psi_i(v)$ is φ-chained if and only if v is φ-chained.

Since $K \geq m - 1$, a φ-chained word v of length $l \geq K$ may occur in the sequence only as a factor of a positive power of an element of a φ-chained set. Thus, if there exist exactly d_j φ-chained words of length l with the speciality degree j, then there exist exactly d_j φ-chained words of every bigger length of the speciality degree j.

Let $\psi_{i0}(v)$ be φ-unchained; then it is circular, and its unique interpretation is $(v, i, 0)$. Every occurrence of $\psi_{i0}(v)$ in w corresponds to an occurrence of v. Because φ is marked, the speciality degrees of v and $\psi_{i0}(v)$ are equal.

On the other hand, every φ-unchained word u of length $ml - i$ is circular. Since the morphism is marked, it follows that if u is special with speciality degree $j \geq 2$, then it has the synchronization point (u, λ). So, u contains symbols of exactly l blocks and ends synchronously with a block end. It means that $u = \psi_{i0}(v)$ for some φ-unchained word v of length l with the speciality degree j.

Thus, for every $i \in \{0, \ldots, m-1\}$ ψ_{i0} is a one-to-one mapping of the set of circular words of length l with the speciality degree j onto the set of circular words of length $ml - i$ with this speciality degree. The cardinal numbers of these sets are equal:

$$B_i(l) - d_j = B_j(ml - i) - d_j.$$

The next corollaries follow immediately from Propositions 4 and 5:

Corollary 6. *For all $l \geq K$ and for every $i \in \{0, \ldots, m-1\}$*

$$s(ml - i) = s(l).$$

Corollary 7. *For all $l \geq K$*

$$s(l) = s(k(l)).$$

Proposition 8. *For every $l \geq K - 1$*

$$f(ml + 1) - d = m(f(l + 1) - d).$$

Proof. The function $f(n) - d$ counts the number of φ-unchained words of length n. Since $l \geq K - 1$, we have $ml + 1 \geq L$, and every φ-unchained word v of length $ml + 1$ is circular. It contains at least one symbol of exactly $l + 1$ blocks. Since φ is marked, v admits a unique interpretation $(u, m - i + 1, i)$ in w, where $|u| = l + 1$, and $i \in \{0, \ldots, m - 1\}$:

$$v = \psi_{m-i,i-1}(u).$$

The word u is allowable, and every allowable φ-unchained word of length $l + 1$ will be obtained in such a way exactly m times (as i takes m different values). So,

$$f(ml + 1) - d = m(f(l + 1) - d).$$

The following derivation is similar to Subsection 7.3 in [6].

From the definitions of first differences and decomposition parameters, we obtain:

$$f(n + 1) = f(K) + \sum_{i=K}^{n} s(i)$$

$$= f(K) + \sum_{p(i)=0}^{p(n)-1} \sum_{k(i)=K}^{m(K-1)} \sum_{\Delta(i)=1}^{m^{p(i)}} s(i) +$$

$$+ \sum_{k(i)=K}^{k(n)-1} \sum_{\Delta(i)=1}^{m^{p(n)}} s\left(m^{p(n)}(k(i) - 1) + \Delta(i)\right) +$$

$$+ \sum_{\Delta(i)=1}^{\Delta(n)} s\left(m^{p(n)}(k(n) - 1) + \Delta(i)\right)$$

Using Corollary 7, we extend this continued equality to

$$f(n+1) = f(K) + \sum_{p=0}^{p(n)-1} \sum_{\Delta=1}^{m^p} \left(\sum_{k=K}^{m(K-1)} s(k) \right) +$$

$$+ \sum_{\Delta=1}^{m^{p(n)}} \left(\sum_{k=K}^{k(n)-1} s(k) \right) + \sum_{\Delta=1}^{\Delta(n)} s(k(n))$$

$$= f(K) + (1 + \ldots + m^{p(n)-1}) \sum_{k=K}^{m(K-1)} s(k) + m^{p(n)} \sum_{k=K}^{k(n)-1} s(k) +$$

$$+ \Delta(n)s(k(n))$$

$$= f(K) + \frac{m^{p(n)} - 1}{m - 1}(f(m(K-1)+1) - f(K)) +$$

$$+ m^{p(n)}(f(k(n)) - f(K)) + \Delta(n)(f(k(n)+1) - f(k(n)))$$

Using Proposition 8, we obtain $f(m(K-1)+1) - f(K) = (m-1)(f(K)-d)$, and

$$f(n+1) = \Delta(n)f(k(n)+1) + (m^{p(n)} - \Delta(n))f(k(n)) - d(m^{p(n)} - 1).$$

7 Buffer Morphisms

In general, the criterium of circularity can be extended to a somewhat wider class of DOL words. We refer to a uniform morphism φ as a *buffer* one if its blocks are of the form

$$\varphi(a_i) = d_l a_{\pi(i)} s_i a_{\sigma(i)} d_r \text{ for all } i \in \{1, \ldots, q\}.$$

Here π and σ are permutations ($\pi, \sigma \in S_q$), d_l and d_r are arbitrary words of nonnegative lengths b_l and b_r respectively, and s_i are arbitrary words of length $m - b_l - b_r - 2$.

We call the number $b = b_l + b_r$ the *buffer length* of the morphism φ. So, a marked morphism is a buffer morphism with zero buffer length.

Theorem 9. *Let φ be a buffer morphism. Then its fixed point $w(\varphi)$ is circular if and only if no φ-chained sets do exist.*

Proof. The idea of the proof is the same as in the case of marked DOL words.

The uncircularity number of a buffer morphism can be defined just as in the marked case. However, it follows from the definition of φ-chained set that all the images of φ-chained symbols should begin with different letters and end with different letters. So, if the buffer length is not equal to zero, then the uncircularity number d is at most 1.

Analogously to Corollary 2, we obtain

Corollary 10. *Let a buffer DOL word $w(\varphi)$ of buffer length $b > 0$ be uncircular. Then there exist such a number $L(\varphi) = L$ and a letter $a \in \Sigma$ that for every $n \geq L$ the only uncircular word of length n is a^n.*

The number L is called the *synchronization delay* of w.

Note, that every injective on letters uniform morphism on the 2-letter alphabet is a buffer one. So, Theorem 9 enables all the fixed points of binary uniform morphisms to be classified (see [6]).

Another example of buffer DOL words is given by Toeplitz words of type $(p, 1)$. It was shown by Cassaigne and Karhumäki in [4] that every such a word is a DOL word or, more specifically, a fixed point of a morphism φ defined by the condition

$$\varphi(a_i) = d_l a_i d_r, \ a_i \in \Sigma.$$

In other terms, φ is a buffer morphism with the buffer length $b = m - 1$. As it was shown in [4], all the Toeplitz words are circular.

The subword complexity of a buffer DOL word can be found in quite similar way as of a marked DOL word, and looks similarly to the binary case.

Actually, we define the *structure ordering number K* of a buffer DOL word w as the least integer satisfying the inequality $m(K-1)+b+1 \geq L$, where L is the synchronization delay of w. For every $n \geq K$, the triplet $(p(n), k(n), \Delta(n))$ can be defined here as follows: $p(n)$ is a nonnegative integer, $k(n) \in \{K, \ldots, m(K-1) + b\}$, $\Delta(n) \in \{1, \ldots, m^{p(n)}\}$, and

$$n = m^{p(n)}(k(n) - 1) + b\frac{m^{p(n)} - 1}{m - 1} + \Delta(n).$$

The only distinction from the marked case is in the decomposition of n. After this correction, the Theorem 3 is valid for buffer DOL words. As applied to Toeplitz words, the formula for the subword complexity agrees with the recurrence relation found in [4].

8 Concluding Remarks

In this paper, we have found more than the subword complexity function of a marked uniform DOL word. We have described the set of words allowable in it. Actually, all the allowable words of length greater than the synchronization delay L are divided into two groups. The main group consists of circular φ-unchained words, increasing in number as the length grows. Here the law of this increase has been found. The other group contains d uncircular φ-chained words of each length.

This classification can be useful for further studying of DOL words.

Generally speaking, instead of fixed point $w(\varphi)$, we may consider an arbitrary DOL language $w(\varphi, a)$, $a \in \Sigma^*$, generated by a marked (or buffer) morphism φ:

$$w(\varphi, a) = \{\varphi^i(a) | i \geq 0\}.$$

All the results above extend immediately to a DOL language.

9 Acknowledgements

I am grateful to S. V. Avgustinovich for helpful and stimulating discussions, and to the anonimous referees for useful comments. I would like also to thank D. Chubarov for computer support.

References

1. Avgustinovich S. V.: The number of distinct subwords of fixed length in the Morse-Hedlund sequence. Sibirsk. zhurnal issledovaniya operatsii. 1 no. 2 (1994) 3-7
2. Brlec, S.: Enumeration of factors in the Thue-Morse word. Diskr. Appl. Math. 24 (1989) 83-96
3. Cassaigne, J.: An algorithm to test if a given circular HDOL-language avoids a pattern. IFIP World Computer Congress'94. Elsevier (North-Holland) 1 (1994) 459-464
4. Cassaigne, J., Karhumäki, J.: Toeplitz Words, Generalized Periodicity and Periodically Iterated Morphisms. COCOON'95, Lect. Notes Comp. Sci., no. 959 (1995)
5. Frid, A.: On the subword complexity of symbolic sequences generated by morphisms. Diskretnyi analiz i issledovaniye operatsii 4 no.1 (1997) 53-59 (in Russian)
6. Frid, A.: The Subword Complexity of Fixed Points of Binary Uniform Morphisms. FCT'97, Lect. Notes Comp. Sci., no. 1279 (1997) 178-187
7. Keränen, V.: Abelian Squares are Avoidable on 4 Letters. ICALP'92, Lect. Notes Comp. Sci., no. 623 (1992)
8. de Luca, A., Varriccio, S.: Some combinatorial properties of the Thue-Morse sequence and a problem of semi groups. Theoret. Comp. Sci. 63 (1989) 333-348
9. Mignosi, F., Séébold, P.: If a DOL language is k-power-free then it is circular. ICALP'93, Lect. Notes Comp. Sci., no. 700 (1993)
10. Mossé, B.: Puissances de mots et reconnaissabilité des points fixes d'une substitution. Theoret. Comp. Sci. 99 (1992) 327-334
11. Mossé, B.: Reconnaissabilité des substitutions et complexité des suites automatiques. Bulletin de la Société Mathématique de France 124 (1996) 329-346
12. Tapsoba, T.: Automates calculant la complexité des suites automatiques. Journal de Théorie des nombres de Bordeaux 6 (1994) 127-134

Series-Parallel Posets: Algebra, Automata and Languages[1]

K. Lodaya
Institute of Mathematical Sciences
CIT Campus
Chennai 600 113 - India
kamal@imsc.ernet.in

P. Weil
LIAFA, Univ. Paris-7 & CNRS
2 place Jussieu
75251 Paris Cedex 05 - France
weil@liafa.jussieu.fr

Abstract. In order to model concurrency, we extend automata theory from the usual word languages (sets of labelled linear orders) to sets of labelled series-parallel posets — or, equivalently, to sets of terms in an algebra with two product operations: sequential and parallel. We first consider languages of posets having *bounded width*, and characterize them using *depth-nilpotent* algebras. Next we introduce *series-rational expressions*, a natural generalization of the usual rational expressions, as well as a notion of *branching* automaton. We show both a Myhill-Nerode theorem and a Kleene theorem. We also look at generalizations.

Introduction

In this paper, we seek to extend automata theory from the usual languages over words (labelled linearly ordered sets) to (labelled) posets which include more information regarding concurrency. This has been done earlier; we explain how our approach differs below.

Let A be a finite nonempty alphabet. A language over A accepted by a finite automaton is said to be *regular*. Languages can also be described by expressions using concatenation, union and iteration; such languages are called *rational*. Yet another way of describing languages is by morphisms into a finite algebra; such languages are *recognizable*. The fundamental theorems of automata theory (Kleene, Myhill, Nerode) show that the regular, rational and recognizable languages coincide. (These terms are used in an interchangeable way. Since we need to distinguish them, we have fixed on this terminology. Please bear with us if you use the terms in a different way.)

Our generalization is from words to posets. We will adopt an intermediate framework, and consider the so-called series-parallel posets (*sp*-posets for short) [11, 7] as the structures accepted by our restricted automata. The advantage this yields is that we can work with an algebra in which two different operations coexist: sequential product (concatenation) and parallel product [3]. The book by Büchi [4] advocates such a fully universal algebra approach (although devoid of any attempted application to the modelling of parallelism or concurrency). Absence of communication is a lacuna in our work.

Our starting point is to introduce an expression $e_1 \| e_2$ for the parallel product. When rational expressions are extended to allow the use of the $\|$ operator,

---

[1] Part of this work was done while the second author was visiting the Institute of Mathematical Sciences, in Chennai.

the corresponding (poset) languages are called *series-rational*. Using morphisms into a finite algebra yields a definition of *recognizable sp*-languages. But recognizability turns out to be a broader notion than series-rational, allowing for instance the language $a^{\oplus} := \{a, a\|a, a\|a\|a, \ldots\}$.

Series-rational *sp*-languages are easily seen to have *bounded width*, that is, the "parallel-width" of the elements of these languages is uniformly bounded. The algebraic counterpart to bounded width turns out to be interesting in its own right. We have to work with *depth-nilpotent sp*-algebras, where the parallel operation is nilpotent and the nesting within parallel products forms a partial order (see Section 3 for the precise definitions). We first characterize the bounded-width languages. Putting recognizability and bounded width together leads to a Myhill-Nerode theorem.

Theorem 1 *An sp-language L has bounded width iff it is recognized by a morphism φ into a depth-nilpotent sp-algebra, such that $0 \notin \varphi(L)$.*

Theorem 2 *A bounded width sp-language is series-rational iff recognizable.*

Next we introduce a kind of branching automaton with fork and join transitions in addition to the usual transitions on a letter of the alphabet. The *sp*-languages accepted by such automata are called *regular*. Again, our machine model has to rule out unbounded parallelism to disallow languages like $a^{\oplus}$ above. By restricting the use of the fork transitions, we define the *fork-regular sp*-languages. We now have the analogue of the Kleene theorem.

Theorem 3 *An sp-language is series-rational iff it is fork-regular.*

What happens when we drop the different restrictions which we have adopted for dealing with bounded width languages? Thatcher and Wright [13] gave a way of constructing generalized rational expressions for algebras with many operations. We are able to show that:

Theorem 4 *The series-rational sp-languages are properly included in the recognizable and the regular ones, which are in turn properly included in the generalized rational ones.*

Related work. It should be clear to the reader familiar with concurrency theory that we are looking at Petri nets and related objects [12]. Our branching automata (Section 4) are a restricted form of nets, and fork-regularity corresponds to 1-safe behaviour. We do not have a characterization of the class of nets we deal with. From the process algebra community, too, attempts have been made to work with series-parallel posets [1, 2].

The algebraically informed reader may see our languages as a special kind of *trace* languages, on which there is a major body of work (see [6] for a recent summary). In traces, the alphabet is extended with an independence relation which allows commutativity of actions to represent concurrency. However, while the independence relation allows coding to represent communication between parallel processes — which is completely missing from our approach — it also prevents two actions from being serially as well as parallelly related. Our approach makes this possible, even for languages such as the solution of the equation $X = a(b\|aX)$.

A very general algebraic approach using operations on hypergraphs has been developed by Courcelle [5]. It is easy to express series-parallel languages in this approach, but automata models operating on graphs have proved hard to find. Our notion of branching automata (Section 4) provides a small step forward on the tightrope, without losing balance with algebra (Section 1).

1 *sp*-posets, *sp*-algebras and *sp*-languages

Let A be an alphabet. An *A-labelled poset* is a poset $(P, \leq)$ equipped with a labelling function $\lambda : P \to A$. This is a generalization of words, which are finite linearly ordered A-labelled sets. We are in fact considering isomorphism classes of labelled posets (*pomsets* [11]). Our posets are always finite and the underlying sets of distinct posets are assumed to be pairwise disjoint.

Let $\mathcal{P} = (P, \leq_P, \lambda_P)$ and $\mathcal{Q} = (Q, \leq_Q, \lambda_Q)$ be A-labelled posets. The *parallel product* of these posets $\mathcal{P}\|\mathcal{Q}$ is $(P \cup Q, \leq_P \cup \leq_Q, \lambda_P \cup \lambda_Q)$. The *sequential product* of the same posets $\mathcal{P} \circ \mathcal{Q}$ is $(P \cup Q, \leq_P \cup \leq_Q \cup (P \times Q), \lambda_P \cup \lambda_Q)$. It is easily verified that both products are associative, that they have the same unit (the empty set), and that the parallel product is commutative.

We say that an A-labelled poset is *series-parallel*, an *sp-poset* for short, if either it is empty, or it can be obtained from singleton A-labelled posets by using a finite number of times the operations of sequential and parallel product.

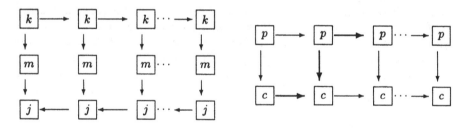

Fig. 1. Posets: series-parallel and not

Example 1.1 Figure 1 shows an example of an *sp*-poset. Indeed, let P_n be the poset in Figure 1 having n occurrences of k. Then $P_1 = \{k\} \circ \{m\} \circ \{j\}$ (usually written kmj) and $P_{n+1} = \{k\} \circ (\{m\}\|P_n) \circ \{j\}$ (written $k(m\|P_n)j$).

The right hand picture in Figure 1 shows a producer-consumer system, a simple example of message-passing. This poset is not series-parallel. The trouble can be traced to the N-shaped subposet shown using thick lines. A poset is N-*free* if it does not contain N as a subposet. That is, more formally, if it does not contain 4 distinct elements a, b, c and d among which the only order relations are $a < c$, $a < d$ and $b < d$. Observe that this property is independent of any labelling of the poset. It is well known that an A-labelled poset $(P, \leq, \lambda)$ is an sp-poset if and only if $(P, \leq)$ is N-free [8, 14].

sp-**algebras.** The notion of sp-poset is generalized in the following fashion. An sp-$algebra$ is a set S equipped with two binary operations, denoted $\circ$ and $\|$ and called respectively $sequential$ and $parallel\ product$, such that $(S, \circ)$ is a semigroup (that is, the sequential product is associative), and such that $(S, \|)$ is a commutative semigroup. Note that no distributivity is assumed, thus $ab\|ab$ is distinct from $(a\|a)(b\|b)$. The discussion in the previous section shows that the class of sp-posets labelled by the alphabet A, forms an sp-algebra.

It is clear from this definition that sp-algebras form a variety of algebras (of signature $(2, 2)$), so there exists a free sp-algebra over each set A, denoted $SP(A)$. This free algebra was characterized by Bloom and Ésik [3].

Proposition 1.2 $SP(A)$ *is (isomorphic to) the sp-algebra of non-empty A-labelled sp-posets.*

sp-**terms.** The elements of the free sp-algebra over A, namely $SP(A)$, can be identified with A-labelled sp-posets according to Proposition 1.2. We mildly abuse notation and call them sp-$terms$ as well.

Let $x \in SP(A)$ be an sp-term. We say that x is a $sequential\ term$ if it cannot be written as a parallel product $x = y\|z$. We say that x is a $parallel\ term$ if it cannot be written as a sequential product $x = yz$ $(y, z \in SP(A))$.

The semigroup $(SP(A), \circ)$ is freely generated by the parallel terms, and the commutative semigroup $(SP(A), \|)$ is freely generated by the sequential terms. So, if $x \in SP(A)$ is an sp-term, then x admits a factorization of the form $x = x_1\|\ldots\|x_n$, where $n \geq 1$ and each x_i is a sequential term, and this factorization is unique up to the order of the factors. The sp-term x also admits a unique factorization of the form $x = x_1 \circ \ldots \circ x_n$, where $n \geq 1$ and each x_i is a parallel term. These factorizations are called, respectively, the $parallel$ and the $sequential$ factorizations of x.

Example 1.3 Let us follow up on the sp-posets of Example 1.1 above. For $n = 3$, the sp-poset P_n can be written $P_3 = k\left(m\|\left(k\left(m\|(kmj)\right)j\right)\right)j$. For each $n \geq 1$, P_{n+1} is a sequential term, whose factorization has length 3: $P_{n+1} = k\circ(m\|P_n)\circ j$. The first and last factor are letters, and the second one is a parallel term whose factorization has length 2.

The width $\mathrm{wd}(x)$ of an sp-term $x \in SP(A)$ is inductively defined as follows: $\mathrm{wd}(a) = 1$ for $a \in A$; $\mathrm{wd}(x_1 x_2) = \max\{\mathrm{wd}(x_1), \mathrm{wd}(x_2)\}$; $\mathrm{wd}(x_1\|x_2) = \mathrm{wd}(x_1) + \mathrm{wd}(x_2)$. The width of the sp-term P_3 in the example above is 3. Intuitively, the width of an sp-term x is the number of processors needed to realize x.

sp-**languages.** An sp-$language$ is a subset of some free sp-algebra $SP(A)^1$ (with an identity element for both products, denoted 1). The family of $series$-$rational$

sp-languages over the alphabet A is defined to be the least class of sp-languages containing the finite sp-languages and closed under union, the sequential and parallel products and sequential iteration (Kleene star). Although our results pertain only to the series-rational languages, we define the *series-parallel-rational* languages as those also closed under parallel iteration $L^\oplus$, defined to be the set $\{t_1\|t_2\|\dots\|t_n \mid n \geq 1,\ t_i \in L, 1 \leq i \leq n\}$.

We say that an sp-language $L \subseteq SP(A)^1$ has *bounded width* if there exists an integer k such that, for each element $x \in L$ we have $\mathrm{wd}(x) \leq k$.

Fact 1.4 *Every series-rational sp-language has bounded width.*

There are bounded-width sp-languages which are not series-rational, e.g. non-rational languages contained in the free monoid A^*, since such languages have width 1. From Fact 1.4 we also know that the complement of a series-rational sp-language is not series-rational.

2 Recognizable sp-languages

As is usual with algebras, we define a *morphism* from an sp-algebra S to an sp-algebra T to be a mapping which preserves the two products. We say that an sp-language $L \subseteq SP(A)$ is *recognized* by a morphism φ from $SP(A)$ into an sp-algebra S if $L = \varphi^{-1}\varphi(L)$. In that case, we also say that L is *recognized* by the sp-algebra S. An sp-language L is *recognizable* if it is recognized by a *finite* sp-algebra. If our sp-languages may contain 1, then we must use morphisms from $SP(A)^1$ into (finite) sp-algebras with identity.

Example 2.1 Let $a \in A$. Then the sp-language $\{a\}$ is recognized by the 2-element sp-algebra $S = \{x, 0\}$ given by $x \circ x = x\|x = 0$. (Consider the morphism from $SP(A)$ into S which maps a to x and the other letters of A to 0.)

Let $a^\oplus$ be the sp-language consisting of all the parallel powers of a. It is recognized by the 2-element sp-algebra $\{x, 0\}$ given by $x\|x = x$ and $xx = 0$. One needs to consider the morphism from $SP(A)$ into S mapping a to x and the other letters to 0.

This example can easily be generalized. Note that the free semigroup A^+ and the free commutative semigroup $A^\oplus$ are naturally embedded in $SP(A)$ (the latter as the semigroup generated by A under the parallel product). Then an sp-language L contained in A^+ (resp. $A^\oplus$) is recognizable if and only if L is a recognizable subset of A^+ (resp. $A^\oplus$).

The sp-language $\{a^n b^n \mid n \geq 1\}$ is not recognizable, since it is not recognizable in A^+. Similarly, the sp-language $(a\|b)^\oplus$, which consists of all the parallel powers of $a\|b$, is not recognizable. Indeed it is not recognizable in $A^\oplus$ (its inverse image in A^+, the set of all words having as many a as b, is not recognizable).

Let $L \subseteq SP(A)$ be an sp-language. If $\xi \notin A$, we define a ξ-term to be a term in $SP(A \cup \{\xi\})$ which has exactly one occurrence of ξ. Such terms represent "contexts". If c is a ξ-term and t is a term, then $c[\xi, t]$ denotes the result of the substitution of t for the occurrence of ξ in c.

The *syntactic congruence* of L is defined as follows. Let $u, v \in SP(A)$. Intuitively, we say that u is $\sim_L$-equivalent to v if, for each sp-term $t \in L$ such that u (resp. v) is a subterm of t, replacing u by v (resp. v by u) in t yields another element of L. More precisely, we let $u \sim_L v$ if, for each ξ-term c (where $\xi \notin A$), we have $c[\xi, u] \in L \iff c[\xi, v] \in L$.

The quotient sp-algebra $\mathrm{Synt}(L) = SP(A)/\sim_L$ is called the *syntactic sp-algebra* of L and the projection morphism $\mu_L: SP(A) \to \mathrm{Synt}(L)$ is called the *syntactic morphism* of L. This morphism recognizes L.

As usual, an absorbing element for both products is called a *zero*. Such an element is unique if it exists, and is written 0.

Lemma 2.2 *Let L be an sp-language and let s be an sp-term which is not a subterm of any element of L. Then $\mathrm{Synt}(L)$ has a zero and $\mu_L(s) = 0$.*

3   Bounded-width sp-languages

There are recognizable sp-languages which are not series-rational. For instance, the series-parallel-rational sp-language $a^\oplus$ is recognizable by Example 2.1, but not series-rational by Fact 1.4. To rule out languages like $a^\oplus$, we give an algebraic characterization of the sp-languages having bounded width.

Consider an sp-algebra S. If $s, t \in S$, we let $s \prec t$ if and only if $s = u(p\|(qtr))v$ for some $u, p, q, r, v \in S^1$ such that $p\|(qtr) \neq qtr$. The relation $<_n$ ($\leq_n$) is defined to be the (reflexive) transitive closure of $\prec$, and it is called the *parallel nesting relation*. $\leq_n$ is a quasi-order on S, but need not be a partial order.

Example 3.1 Let $x, y \in SP(A)$ be sp-terms. Then $s <_n t$ if and only if t is a subterm of s and at least one context of t in s (a ξ-term) is not a sequential term having ξ as a factor in its sequential decomposition. In particular, $\leq_n$ is a partial order on $SP(A)$, and $s <_n t$ implies $\mathrm{wd}(s) > \mathrm{wd}(t)$.

We will be concerned with sp-algebras S in which, for each $s \in S$, there exists an integer n such that each $<_n$-chain with least element s has length at most n. Such algebras are called *depth-graded*. Example 3.1 shows that this is the case if $S = SP(A)$, the free sp-algebra.

In addition, if S is a depth-graded sp-algebra, there is a notion of *parallel nesting depth*, or simply *depth*, for the elements of S:

$$\mathrm{dp}(s) = \begin{cases} 1 & \text{if } s \text{ is } \leq_n\text{-maximal,} \\ 1 + \max\{\mathrm{dp}(t) \mid s <_n t\} & \text{otherwise.} \end{cases}$$

In particular, the depth of s is the maximum cardinality of a $<_n$-chain with least element s. Note that the maximal $<_n$-chains are $\prec$-chains.

If L is a subset of a depth-graded sp-algebra S, we let $\mathrm{dp}(L) = \sup_{x \in L} \mathrm{dp}(x)$. If $\mathrm{dp}(L) < \infty$, we say that L has *bounded depth*.

It is easily verified (see Example 3.1) that if $x \in SP(A)$ is an sp-term, then $dp(x) \leq wd(x)$. However, we do not necessarily have equality between the width and the depth of an sp-term: if $u = a(a\|a)$ and if x is the parallel product of n copies of u, then one verifies that $wd(x) = 2n$ and $dp(x) = n+1$. Yet very wide sp-terms must be very deep, as is shown by:

Proposition 3.2 *An sp-language has bounded width iff it has bounded depth.*

Depth-nilpotent sp-algebras

We say that an sp-algebra S has *bounded depth* if there exists an integer d such that each $\prec$-chain is of length at most d. If S is a bounded-depth sp-algebra, then S is trivially depth-graded, $dp(S) < \infty$ and any element of maximal depth is $\leq_n$-minimal. The converse also holds:

Lemma 3.3 *An sp-algebra S has bounded depth if and only if it is depth-graded and it has a $\leq_n$-minimal element.*

Recall that a semigroup is said to be *nilpotent* if it has a zero and there exists an integer $c \geq 1$ such that any product of c or more elements is equal to 0. In that case, it has a zero which is its only idempotent [10]. Now we say that an sp-algebra S is *depth-nilpotent* if it has bounded depth and 0 is the only idempotent element of S for the parallel product. We can show that a bounded depth sp-algebra S is depth-nilpotent if and only if the semigroup $(S, \|)$ is nilpotent. This leads to the following property, and our first main theorem.

Lemma 3.4 *Let $\varphi: T \to S$ be a morphism between sp-algebras, and assume S is depth-nilpotent. If $t, t' \in T$ and $t \prec t'$, then either $\varphi(t) = 0$ or $\varphi(t) \prec \varphi(t')$.*

Theorem 1 *Let L be a sp-language. The following properties are equivalent.*

(1) *L has bounded width;*
(2) $\mathrm{Synt}(L)$ *is depth-nilpotent and $0 \notin \mu_L(L)$;*
(3) *L is recognized by a morphism $\varphi: SP(A) \to S$ into a depth-nilpotent sp-algebra S, such that $0 \notin \varphi(L)$.*

Since recognizable sp-languages have finite syntactic sp-algebras, Theorem 1 immediately implies the following result.

Corollary 3.5 *Let L be a sp-language. The following properties are equivalent.*

(1) *L is recognizable and it has bounded width;*
(2) $\mathrm{Synt}(L)$ *is finite depth-nilpotent and $0 \notin \mu_L(L)$;*
(3) *L is recognized by a morphism $\varphi: SP(A) \to S$ into a finite depth-nilpotent sp-algebra S, such that $0 \notin \varphi(L)$.*

We complete this section by connecting bounded-width, recognizability and series-rationality. As an aside, note that finite depth-nilpotent algebras can recognize (using the zero) complements of recognizable bounded-width languages as well (e.g. $SP(A)$). It can be shown that finite depth-nilpotent sp-algebras form a pseudovariety, and using the next theorem, that series-rational languages form a positive variety (in the sense of [10]).

Theorem 2 *An sp-language is series-rational if and only if it is recognizable and bounded-width.*

Proof sketch. By adapting the classical proof for languages in the free semigroup, we prove that recognizable languages are closed under the two products, boolean operations and sequential iteration. The forward direction then follows from Fact 1.4. To prove the converse, we use the algebraic characterization of bounded-width recognizable sp-languages established in Corollary 3.5 and a McNaughton-Yamada-like construction. □

4 Branching automata and regular sp-languages

In order to provide an automaton characterization of series-rational sp-languages, we introduce a new kind of automata.

A *branching automaton* over A is a tuple $\mathcal{A} = (Q, T_{seq}, T_{fork}, T_{join}, I, F)$ where Q is the (finite) set of states, I and F are subsets of Q, respectively the set of initial and of final states, $T_{seq} \subseteq Q \times A \times Q$ is the set of sequential transitions, $T_{fork} \subseteq Q \times \mathcal{P}_{ns}(Q)$ and $T_{join} \subseteq \mathcal{P}_{ns}(Q) \times Q$ are respectively the sets of fork and join transitions. Here $\mathcal{P}_{ns}(Q)$ ("nonempty and nonsingleton") stands for subsets of Q of cardinality at least 2.

An element $(q, a, q') \in T_{seq}$ is said to be an a-labelled transition from q to q'. As usual, we write $q \xrightarrow{a} q'$. An element $(q, \{q_1, \ldots, q_n\})$ of T_{fork} is said to be a *fork transition of arity* n. We denote it by $q \to \{q_1, \ldots, q_n\}$. Similarly, an element $(\{q_1, \ldots, q_n\}, q)$ of T_{join}, written $\{q_1, \ldots, q_n\} \to q$, is said to be a *join transition of arity* n.

The existence of a *run* of $\mathcal{A}$ on the sp-poset t from state p to state q is inductively defined as follows: there is a run on 1 from p to q only if $p = q$, and there is a run on letter a from p to q if $p \xrightarrow{a} q$. If a poset t has sequential factorization $t_1 \ldots t_n$ ($n \geq 2$), there is a run on t from p to q if there exist states $p = p_0, p_1, \ldots p_n = q$ and there exist runs on t_m from p_{m-1} to p_m, for each $1 \leq m \leq n$.

Suppose now that a poset t has parallel factorization $t_1 \| \ldots \| t_n$ ($n \geq 2$). There is a run on t from p to q if there is a fork at p, sub-runs for the factors and then a matching join ending at q. However, the automaton can do the forking in levels. We formalize this as follows:

Let $\pi = \{r_1, \ldots, r_s\}$ be a subset of $\{1, \ldots, n\}$. By t_π we mean the parallel term $t_{r_1} \| \ldots \| t_{r_s}$. Now we say there is a run on t from p to q if there exists a partition $\{\pi_1, \ldots, \pi_m\}$ of $\{1, \ldots, n\}$ and there are states p_ℓ, q_ℓ, for $1 \leq \ell \leq m$ such that there is a fork $k = p \to \{p_1, \ldots, p_m\}$, there are runs on t_{π_ℓ} from p_ℓ to q_ℓ, $1 \leq \ell \leq m$, and there is a matching join $j = \{q_1, \ldots, q_m\} \to q$. Figure 2

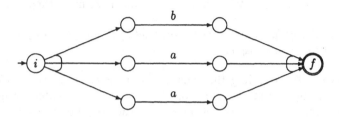

Fig. 2. Automaton accepting $a\|a\|b$

shows an automaton with a run on $a\|a\|b$ from state i to state f. The partition of $\{1, 2, 3\}$ here is simple: it just has three singletons $\{\{1\}, \{2\}, \{3\}\}$.

Getting back to the previous definition, we say that the states and the transitions forming the sub-runs on the t_{π_ℓ} are *nested within* the fork k in the run on t. In the example in Figure 2, all states except i and f are within the fork in any run. There are no nested forks.

Finally, a branching automaton $\mathcal{A}$ *accepts* the sp-poset t if there exists a run of $\mathcal{A}$ on t from an initial state to a final state. An sp-language L is *regular* if it is the set of sp-posets accepted by a finite branching automaton.

As in the case of recognizable languages (Section 2), the notion of regularity is too general. It is easy to find a branching automaton accepting the non-bounded-width language $a^\oplus$. Further, Figure 3 shows that even the non-

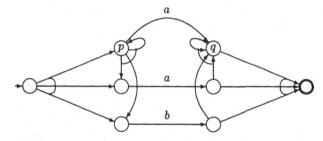

Fig. 3. An automaton which is not fork-acyclic

recognizable language $a\|(a\|b)^\oplus$ (parallel products of a and b having one more a than b) is accepted by an automaton. Note the looping fork at state p and the looping join at state q in Figure 3. It is the presence of a loop from inside one branch of a fork to another (or to outside the fork) that allows unbounded parallel-width terms to be accepted.

A run ρ of $\mathcal{A}$ on t is *fork-acyclic* if in the induced sub-run of ρ on any parallel sub-term of t with outermost fork k, k does not occur nested within itself. An sp-language is *fork-regular* if it is the set of sp-posets of the fork-acyclic runs from an initial state to a final state of a finite branching automaton. If all runs of an automaton are fork-acyclic, we call the automaton *fork-acyclic*. The automaton in Figure 2 is fork-acyclic. We can now prove our next main theorem.

Theorem 3 *An sp-language is series-rational iff it is accepted by a fork-acyclic automaton, iff it is fork-regular.*

Proof sketch. By extending the usual inductive construction for rational languages, we can prove that series-rational languages are accepted by finite fork-acyclic automata. To prove the converse, we inductively build up expressions for languages using a set of forks. The base case, when no forks are used, is provided by the usual McNaughton-Yamada construction for regular languages. $\square$

5 Generalizations

All three notions which we have used and related so far in the context of bounded-width languages, namely recognizability, series-rationality and fork-regularity, admit natural generalizations.

For recognizability, dropping the depth-nilpotency condition, we get the recognizable sp-languages, recognized by a finite sp-algebra. As we saw, $a^\oplus$ is a recognizable language which is not bounded-width. Dropping the fork-acyclicity condition, our automata accept the regular sp-languages, which include the language $a^\oplus$ as well as non-recognizable languages like $a\|(a\|b)^\oplus$ (Figure 3).

We borrow our generalization of series-rationality from [4] and [13]. Let B be an alphabet containing A. The generalized sp-rational expressions over B are defined by closing the letters of B using this rule:
If e_1 and e_2 are generalized rational expressions, and $\xi \in B$, then $e_1 \cup e_2$, $e_1 \circ e_2$, $e_1\|e_2$, $e_1 \circ_\xi e_2$ and $e_1^{*\xi}$ are generalized rational expressions.
With each generalized rational expression, we can associate an sp-language in $SP(B)$. We indicate how this is done for the expressions using the ξ variables.

If e_1 and e_2 are generalized rational expressions, then $L(e_1 \circ_\xi e_2)$ is obtained from the elements of $L(e_2)$ by substituting an element of $L(e_1)$ for each occurrence of ξ (different occurrences may be replaced by different elements). $L(e_1^{*\xi}) = L(e_1)^{*\xi}$, where for any sp-language U, we let $U^{*\xi} = \bigcup_{i \geq 0} U_i^\xi$, with $U_0^\xi = \{\xi\}$ and $U_{i+1}^\xi = (U_0^\xi \cup \cdots \cup U_i^\xi) \circ_\xi U$.

The regular as well as the generalized rational sp-languages are closed under both sequential and parallel iteration. ($L^* = 1 \circ_\xi (\xi L)^{*\xi}$ and $L^\oplus = L \circ_\xi (\xi\|L)^{*\xi}$.) The regular languages properly include the series-parallel-rational languages, e.g. $1 \circ_\xi (a\|(b\xi))^{*\xi}$ is a separating language. The generalized rational languages include the context-free languages in A^* (e.g. $1 \circ_\xi (a\xi b)^{*\xi} = \{a^n b^n \mid n \geq 0\}$). Using a result of Thatcher and Wright [13], and pushing the McNaughton-Yamada construction of Theorem 3 one step further, we show:

Theorem 4 *The series-rational sp-languages are properly included in the recognizable and the regular ones, which are in turn properly included in the generalized rational ones.*

series-rational	$\subset$ depth-nilpotent recognizable	$\subset$	recognizable
$(a\|b)^*$	$SP(A)$		$a^{\oplus}$
$\cap$			$\cap$
series-parallel-rational $\subset$	regular	$\subset$	generalized rational
$(a\|b)^{\oplus}$	$1_{\circ_\xi}(a\|b\xi)^{*\xi}$		$1_{\circ_\xi}(a\xi b)^{*\xi}$

Theorem 2 and Theorem 3 showed that the three formalisms of algebraic recognizability, rational expressions and automaton acceptance can be made to coincide in the context of bounded-width sp-languages. When generalized, these formalisms open up a hierarchy. We are left with the question of characterizing the smaller subclasses within the generalized rational sp-languages. We conjecture that the recognizable sp-languages are included in the regular ones.

Our work can be generalized to algebras with another operation $+$ representing choice. Boudol and Castellani [2] have shown that the free algebra (modulo associativity of $\circ$ and associativity and commutativity of $\|$ and $+$) is characterized by a class of labelled event structures satisfying an "X-property" (going beyond N-freeness). From an example of Milner [9, Example 3], it appears that a syntax of right-linear ξ-expressions has to be used.

References

1. L. Aceto. Full abstraction for series-parallel pomsets, in *Proc. TAPSOFT 91, LNCS* **493**, Springer (1991) 1–40.
2. G. Boudol and I. Castellani. Concurrency and atomicity, *TCS* **59** (1988) 25–84.
3. S. Bloom and Z. Ésik. Free shuffle algebras in language varieties, *TCS* **163** (1996) 55–98.
4. J.R. Büchi. *Finite automata, their algebras and grammars: Towards a theory of formal expressions* (D. Siefkes, ed.), Springer (1989).
5. B. Courcelle. Graph rewriting: an algebraic and logical approach, in *Handbook of Theoretical Computer Science* B (J. van Leeuwen, ed.), Elsevier (1990) 193–242.
6. V. Diekert and G. Rozenberg. *The book of traces*, World Scientific (1995).
7. J.L. Gischer. The equational theory of pomsets, *TCS* **61** (1988) 199–224.
8. J. Grabowski. On partial languages, *Fund. Inform.* **IV** (1981) 427–498.
9. R. Milner. A complete inference system for a class of regular behaviours, *JCSS* **28** (1984) 439–466.
10. J.-E. Pin. Syntactic semigroups, in *Handbook of Formal Language Theory* 1 (G. Rozenberg and A. Salomaa, eds.), Springer (1997) 679–746.
11. V. Pratt. Modelling concurrency with partial orders, *IJPP* **15**(1) (1986) 33–71.
12. W. Reisig. *Petri nets, an introduction*, Springer (1985).
13. J.W. Thatcher and J.B. Wright. Generalized finite automata with an application to a decision problem of second order logic, *Math. Syst. Theory* **2** (1968) 57–82.
14. J. Valdes, R.E. Tarjan and E.L. Lawler. The recognition of series-parallel digraphs, *SIAM J. Comput.* **11**(2) (1981) 298–313.

On the Expected Number of Nodes at Level k in 0-balanced Trees

Rainer Kemp

Johann Wolfgang Goethe-Universität, Fachbereich Informatik
D - 60054 Frankfurt am Main, Germany

Abstract. An ordered tree with height n and m leaves is called 0-balanced if all leaves have the same level. We compute the average number of nodes (with specified degree) appearing at a given level in a 0-balanced ordered tree as well as in a 0-balanced t-ary ordered tree.

With respect to the former class we shall show that the average rate of increase of nodes amounts to $\rho := (m-1)n^{-1}$ passing from one level to the next one. The same fact holds for nodes with a degree one at a large level. The average number of nodes with a degree two and that one with a degree greater than two tends to ρ and zero for large levels, respectively.

The class of 0-balanced t-ary trees corresponds to the set of all code trees associated with all n-block codes with m code words over a given alphabet with cardinality t. In that case, we shall show that all nodes with maximal degree t are concentrated at levels smaller than $\log_t(m)$, and all nodes with degree one appear at levels greater than $\log_t(m)$, on the average. This result implies that the *first* $\log_t(m)$ positions in the m code words appearing in an n-block code over a given alphabet with cardinality t are sufficient to separate these words, on the average, provided that all those codes are equally likely.

1 Introduction and Preliminaries

We recall that a family $\mathcal{F}$ of rooted trees is said to be *simply generated* if the generating function $E(z) := \sum_{s>0} t(s)z^s$ of the number $t(s)$ of all trees $\tau \in \mathcal{F}$ with s nodes satisfies a functional equation of the form $E(z) = z\Theta(E(z))$, where $\Theta(y) := \sum_{\lambda \geq 0} c_\lambda y^\lambda$ is a regular function when $|y| < R < \infty$ with $c_0 = 1$, $c_\lambda \geq 0$ for $\lambda \in \mathbb{N}$, and $c_j > 0$ for some $j \geq 2$ ([8]). This definition obviously includes the most common classes of trees such as *extended t-ary trees* ($\Theta(y) := 1 + y^t$, $t \in \mathbb{N}\backslash\{1\}$), *$t$-ary trees* ($\Theta(y) := (1+y)^t$, $t \in \mathbb{N}\backslash\{1\}$), *ordered trees* or *plane trees* ($\Theta(y) := (1-y)^{-1}$), *unary–binary trees* ($\Theta(y) := 1+y+y^2$) and *unbalanced 2–3–trees* ($\Theta(y) := 1+y^2+y^3$). Given the regular function Θ, the corresponding simply generated family of trees $\mathcal{F}(\Theta)$ is completely characterized: The elements of the set $\mathcal{DEG}(\Theta) := \{\lambda \in \mathbb{N}_0 \mid \langle y^\lambda; \Theta(y)\rangle \neq 0\}$[1] are the allowed node degrees in the trees appearing in $\mathcal{F}(\Theta)$ and $c_\lambda = \langle y^\lambda; \Theta(y)\rangle$, $\lambda \in \mathbb{N}_0$, reflects whether

---

[1] The abbreviation $\langle z_1^{n_1} \ldots z_m^{n_m}; f(z_1, \ldots, z_m)\rangle$ denotes the coefficient of $z_1^{n_1} \ldots z_m^{n_m}$ in the expansion of $f(z_1, \ldots, z_m)$ at $(z_1, \ldots z_m) = (0, \ldots, 0)$.

different orderings of the λ edges of a node with degree λ are taken into account in distinguishing between the trees in $\mathcal{F}(\Theta)$. Here, the *degree deg(x)* of a node x is the number of its sons. Using the suggestive terminology introduced in [1], the close connection between the function Θ and the structure of the trees in $\mathcal{F}(\Theta)$ is illustrated in Fig. 1.

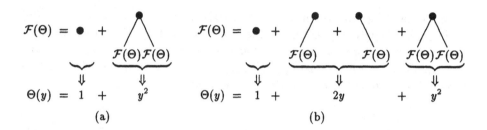

Fig. 1. The recursive definition of extended binary trees [(a)] and of binary trees [(b)] together with the function Θ characterizing the corresponding classes of trees.

Given a tree $\tau \in \mathcal{F}(\Theta)$, the *level lev(x)* of a node x appearing in τ is equal to the number of nodes on the path from the root to the node x excluding the root; the root has the level zero. The tree $\tau \in \mathcal{F}(\Theta)$ is said to be *0-balanced* if all nodes with degree zero ($=$ *leaves*) have the same level. It has the *height $h(\tau) := n$* if all leaves have the same level $n \in \mathbb{N}_0$.

In the following, we shall consider the set $\mathcal{F}_{n,m}(\Theta)$ consisting of all 0-balanced simply generated trees in $\mathcal{F}(\Theta)$ with m leaves and height n. For example, Fig. 2 presents the set $\mathcal{F}_{3,3}(\Theta)$ with $\Theta(y) := 1 + y + y^2$ and $\Theta(y) := (1 - y)^{-1}$. Note that classes of 0-balanced ordered trees have been extensively discussed in [3] and [4, 3.8].

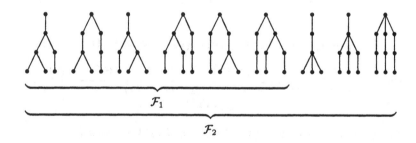

Fig. 2. All 0-balanced unary-binary trees ($\mathcal{F}_1$) and 0-balanced ordered trees ($\mathcal{F}_2$) with $m = 3$ leaves and height $n = 3$.

It is not hard to show that the average behaviour of interesting parameters defined on trees in $\mathcal{F}_{n,m}(\Theta)$ can be expressed by means of the iterate of some order p of the function $\Theta(y) := \Theta(y) - 1$ defined by $\Theta^{<0>} := id$ and $\Theta^{<p>} := \Theta \circ \Theta^{<p-1>}$ for $p \geq 1$; here, id denotes the identity mapping. For this purpose, let

$$F_n(\vec{x}_0, \ldots, \vec{x}_{n-1}, z) := \sum_{\substack{\vec{r}_k \in \mathbb{N}_0^*, \, 0 \leq k < n \\ m \geq 0}} t(\vec{r}_0, \ldots, \vec{r}_{n-1}, m) \, \vec{x}_0^{\vec{r}_0} \cdots \vec{x}_{n-1}^{\vec{r}_{n-1}} \, z^m$$

be the generating function of the number $t(\vec{r}_0, \ldots, \vec{r}_{n-1}, m)$ of all $\tau \in \mathcal{F}_{n,m}(\Theta)$ with $r_{k,\lambda}$ nodes with degree $\lambda \in \mathbb{N}$ at level $k \in [0 : n[$. Here, the λ-th component of a vector $\vec{\rho}_k$, $k \in \mathbb{N}_0$, is denoted by $\rho_{k,\lambda}$, i. e., $\vec{\rho}_k := (\rho_{k,1}, \rho_{k,2}, \rho_{k,3}, \ldots)$, and $\vec{\rho}_k^{\vec{c}_k}$ stands for the monomial $\rho_{k,1}^{c_{k,1}} \rho_{k,2}^{c_{k,2}} \rho_{k,3}^{c_{k,3}} \ldots$. Now, a moment's reflection shows that the function F_n satiesfies the recurrence

$$F_0(z) = z$$
$$F_n(\vec{x}_0, \ldots, \vec{x}_{n-1}, z) = \sum_{\ell \geq 1} \langle y^\ell; \Theta(y) \rangle \, x_{0,\ell} \, (F_{n-1}(\vec{x}_1, \ldots, \vec{x}_{n-1}, z))^\ell. \quad (1)$$

For example, if we want to enumerate the number $a_{n,m}(q)$ of all 0-balanced trees $\tau \in \mathcal{F}_{n,m}(\Theta)$ with q nodes, we simply have to set $\vec{x}_k := u\vec{1}$, $0 \leq k < n$, and $z := zu$; we immediately find $a_{n,m}(q) = \langle z^m u^q; A_n(z, u) \rangle$, where $A_0(z, u) := F_0(zu) = zu$ and $A_n(z, u) := F_n(u\vec{1}, u\vec{1}, \ldots u\vec{1}, zu) = u[\Theta(F_{n-1}(u\vec{1}, \ldots u\vec{1}, zu)) - 1]$, i.e., $A_n(z, u) = f_u^{<n>}(zu)$ with $f_u(x) := u\Theta(x)$. Choosing $u = 1$, we obtain

$$|\mathcal{F}_{n,m}(\Theta)| = \langle z^m; A_n(z, 1) \rangle = \langle z^m; \Theta^{<n>}(z) \rangle. \quad (2)$$

In the following, we shall focus on two parameters defined on 0-balanced trees $\tau \in \mathcal{F}_{n,m}(\Theta)$ describing their average shape. These parameters have been analyzed in [8] for simply generated trees and in [2] for ordered trees.

Lemma 1. *Assume that all 0-balanced simply generated trees appearing in $\mathcal{F}_{n,m}(\Theta)$ are equally likely.*
(a) The average number $N_{n,m}(\lambda, k)$ of nodes with degree λ at level k is given by [2]

$$N_{n,m}(\lambda, k) = \begin{cases} m \, \delta_{k,n} \, \delta_{\lambda,0} & \text{if } (\lambda, k) \in (\{0\} \times [0 : n]) \cup (\mathbb{N}_0 \times \{n\}) \\ \langle y^\lambda; \Theta(y) \rangle \dfrac{\langle z^m; [\Theta^{<n-k-1>}(z)]^\lambda \, \Theta^{<k>\,\prime}(\Theta^{<n-k>}(z)) \rangle}{\langle z^m; \Theta^{<n>}(z) \rangle} & \\ & \text{if } (\lambda, k) \in \mathbb{N} \times [0 : n[\end{cases}.$$

(b) The average number $N_{n,m}(k)$ of nodes at level k is given by

$$N_{n,m}(k) = \frac{\langle z^m; \Theta^{<n-k>}(z) \, \Theta^{<k>\,\prime}(\Theta^{<n-k>}(z)) \rangle}{\langle z^m; \Theta^{<n>}(z) \rangle}, \quad k \in [0 : n].$$

Proof. (a) The result for $(\lambda, k) \in (\{0\} \times [0 : n]) \cup (\mathbb{N}_0 \times \{n\})$ is clear because all nodes with degree zero appear at level n in a 0-balanced tree $\mathcal{F}_{n,m}(\Theta)$.

---

[2] Here, $\delta_{i,j}$ denotes Kronecker's delta, i.e., $\delta_{i,j} = 1$ if $i = j$, and $\delta_{i,j} = 0$ if $i \neq j$.

Next, let $(\lambda, k) \in \mathbb{N} \times [0 : n[$. Starting with (1), we have to set $\vec{x}_s := \vec{1}$ for $s \in [0 : n[\setminus \{k\}$, and $\vec{x}_k := (1, \ldots, 1, x_{k,\lambda}, 1, \ldots)$. The corresponding generating function $F_n(\vec{1}, \ldots, \vec{1}, \vec{x}_k, \vec{1}, \ldots, \vec{1}, z)$ enumerates the number of all 0-balanced trees in $\mathcal{F}_{n,m}(\Theta)$ with respect to the number of nodes with degree λ at level k (variable $x_{k,\lambda}$) and the number of leaves (variable z). Now, the recurrence (1) immediately yields $F_0(z) = z$ and

$$F_{n-i}(\vec{1}, \ldots, \vec{1}, \vec{x}_k, \vec{1}, \ldots, \vec{1}, z) = \Theta(F_{n-i-1}(\vec{1}, \ldots, \vec{1}, \vec{x}_k, \vec{1}, \ldots, \vec{1}, z))$$
$$\text{for } 0 \le i < k,$$

$$F_{n-k}(\vec{1}, \ldots, \vec{1}, \vec{x}_k, \vec{1}, \ldots, \vec{1}, z) = \Theta(F_{n-k-1}(\vec{1}, \ldots, \vec{1}, z)) - \langle y^\lambda; \Theta(y) \rangle$$
$$\times (1 - x_{k,\lambda}) (F_{n-k-1}(\vec{1}, \ldots, \vec{1}, z))^\lambda,$$

$$F_{n-i}(\vec{1}, \ldots, \vec{1}, z) = \Theta(F_{n-i-1}(\vec{1}, \ldots, \vec{1}, z)) \text{ for } k < i < n.$$

Solving that recurrence, we find

$$F_n(\vec{1}, \ldots, \vec{1}, \vec{x}_k, \vec{1}, \ldots, \vec{1}, z)$$
$$= \Theta^{<k>}\left(\Theta^{<n-k>}(z) - \langle y^\lambda; \Theta(y) \rangle (1 - x_{k,\lambda}) [\Theta^{<n-k-1>}(z)]^\lambda\right).$$

Since

$$|\mathcal{F}_{n,m}(\Theta)| \, N_{n,m}(\lambda, k) = \left\langle z^m; \left. \frac{\partial}{\partial x_{k,\lambda}} F_n(\vec{1}, \ldots, \vec{1}, \vec{x}_k, \vec{1}, \ldots, \vec{1}, z) \right|_{x_{k,\lambda}=1} \right\rangle,$$
$$\underbrace{= \langle y^\lambda; \Theta(y) \rangle [\Theta^{<n-k-1>}(z)]^\lambda \, \Theta^{<k> \, '}(\Theta^{<n-k>}(z))}$$

the proof of part (a) of our lemma is completed.

(b) Obviously, the average number of nodes at level k in a 0-balanced tree $\tau \in \mathcal{F}_{n,m}(\Theta)$ is given by $N_{n,m}(k) = \sum_{\lambda \ge 0} N_{n,m}(\lambda, k)$. Inserting the explicit expression for $N_{n,m}(\lambda, k)$ into that relation, an elementary computation yields the result stated in part (b) of our lemma. □

It should be obvious that expected values of other interesting parameters defined on 0-balanced simply generated trees with height n and m leaves can be expressed by the iterates of the function Θ. For example, we find by Lemma 1(b) for the average number $N_{n,m}$ of all nodes

$$N_{n,m} = \frac{\langle z^m; \sum_{0 \le k \le n} \Theta^{<n-k>}(z) \, \Theta^{<k> \, '}(\Theta^{<n-k>}(z)) \rangle}{\langle z^m; \Theta^{<n>}(z) \rangle}.$$

In the same way, the average internal and external (degree) path length of a 0-balanced tree $\tau \in \mathcal{F}(\Theta)$ can be expressed by means of the functions $\Theta^{<p>}$, $p \in \mathbb{N}_0$. But in all cases, we need more detailed information about the iterates $\Theta^{<p>}$ in order to obtain information about the behaviour of the parameter in discussion. In the next section, we shall consider two non-trivial cases.

2 Applications

2.1 0-Balanced Ordered Trees with Height n and m Leaves

The class $\mathcal{F}_{n,m}(\Theta)$ of 0-balanced ordered trees with height n and m leaves is characterized by the function $\Theta(y) = \Theta(y) - 1 = y(1-y)^{-1}$. It can easily be verified that $\Theta^{<p>}(y) = y(1-py)^{-1}$, $p \geq 0$. Thus, the relation $|\mathcal{F}_{n,m}(\Theta)| = \langle z^m; \Theta^{<n>}(z) \rangle = \langle z^m; z(1-nz)^{-1} \rangle = n^{m-1}$ holds by (2). Note that this result already follows by applying d'Hospital's rule (for $z \to 1$) to the generating function $A_{n+1}(z,y) := \sum_{\nu \geq 0} \sum_{m \geq 0} a_{n+1}(\nu,m) z^\nu y^m$ of all 0-balanced ordered trees with ν nodes, m leaves and height $n+1$ given in [3; formula (2)].

The following theorem presents explicit expressions for the expected values $N_{n,m}(\lambda, k)$ and $N_{n,m}(k)$ introduced in Lemma 1.

Theorem 1. *Assume that all 0-balanced ordered trees with height n and m leaves are equally likely.*
(a) The average number of nodes with degree λ at level k is given by[3]

$$N_{n,m}(\lambda, k) = \frac{k}{(k+1)^\lambda}\left[1 + k\frac{m-1}{n}\right] + \frac{k(k+1-k\lambda)}{(k+1)^{\lambda+1}}$$
$$+ \frac{1}{(k+1)^2 n^{\lambda-1}}\left(1 - \frac{k+1}{n}\right)^{m-\lambda}$$
$$\times \left[\binom{m-1}{\lambda-1} + k S_{m,\lambda,k}\left(\frac{k+1}{n}\right)\right]$$

for $(\lambda, k) \in [1:m] \times [0:n[$. Here, $S_{m,\lambda,k}(x)$ denotes the sum $S_{m,\lambda,k}(x) := \sum_{1 \leq i < \lambda} \binom{m-1}{\lambda-1-i}[k(i-1)-2]x^{-i}(1-x)^i$.
(b) The average number of nodes at level k is given by

$$N_{n,m}(k) = 1 + k\frac{m-1}{n}, \quad k \in [0:n].$$

Proof. (a) Using $\langle y^\lambda; \Theta(y) \rangle = 1$, $\lambda \in \mathbb{N}_0$, and inserting the above given closed-form expression $\Theta(y) = y(1-py)^{-1}$, $p \geq 0$, into the relation presented in part (a) of Lemma 1, we immediately find $N_{n,m}(\lambda, k) = m\,\delta_{k,n}\delta_{\lambda,0}$ for $(\lambda, k) \in (\{0\} \times [0:n]) \cup (\mathbb{N}_0 \times \{n\})$, and

$$N_{n,m}(\lambda, k) = n^{-(m-1)}\left\langle z^m; \frac{z^\lambda[1-(n-k)z]^2}{[1-(n-k-1)z]^\lambda[1-nz]^2}\right\rangle$$

for $(\lambda, k) \in \mathbb{N} \times [0:n[$. Introducing the function

$$F_{n,k}(z,y) := \sum_{\lambda \geq 1}\frac{z^\lambda[1-(n-k)z]^2}{[1-(n-k-1)z]^\lambda[1-nz]^2}y^\lambda$$
$$= \frac{[1-(n-k)z]^2}{[1-nz]^2}\frac{zy}{1-(n-k-1+y)z}, \quad k \in [0:n[, \quad (3)$$

---
[3]In this paper, we constantly use the convention $0^0 := 1$.

the above expression for $N_{n,m}(\lambda, k)$, $(\lambda, k) \in \mathbb{N} \times [0:n[$, can be rewritten as

$$N_{n,m}(\lambda, k) = n^{-(m-1)} \langle z^m y^\lambda; F_{n,k}(z,y) \rangle. \tag{4}$$

Now, computing the partial fraction decomposition of $F_{n,k}(z,y)$ with respect to the variable z, we find by a lengthy computation

$$F_{n,k}(z,y) = F_{n,k}^{(1)}(z,y) + F_{n,k}^{(2)}(z,y) + F_{n,k}^{(3)}(z,y) + F_{n,k}^{(4)}(z,y), \tag{5}$$

where

$$F_{n,k}^{(1)}(z,y) = \frac{y(n-k)^2}{n^2(1+k-n-y)}, \qquad F_{n,k}^{(2)}(z,y) = \frac{yk^2}{n^2(1+k-y)} \frac{1}{(1-nz)^2},$$

$$F_{n,k}^{(3)}(z,y) = -\frac{k(2k^2 + 2k - 2n - kn)y + 2ky^2(n-k)}{n^2(1+k-y)^2} \frac{1}{1-nz},$$

$$F_{n,k}^{(4)}(z,y) = -\frac{y(1-y)^2}{(1+k-y)^2(1+k-n-y)} \frac{1}{1-(n-k-1+y)z}.$$

Note that $F_{n,k}(0,y) = 0$. Using the evaluations $(1-\xi x)^{-(p+1)} = \sum_{s \geq 0} \binom{p+s}{s} \xi^s x^s$ ([7; p.90]) for $(\xi, x) \in \{(\frac{1}{k+1}, y), (n, z), (n - k - 1 + y, z)\}$, $p \in \{0, 1\}$, we immediately obtain explicit expressions for $\langle z^m y^\lambda; F_{n,k}^{(s)}(z,y) \rangle$, $(m, \lambda, s) \in \mathbb{N} \times \mathbb{N} \times \{2, 3, 4\}$. Now, inserting these expressions into the right-hand side of equation (5), we find the explicit result established in part (a) of our theorem. (b) Using again the closed-form expression $\Theta(y) = y(1-py)^{-1}$, $p \geq 0$, part (b) of Lemma 1 yields $N_{n,m}(k) = n^{-(m-1)} \langle z^m; \frac{z[1-(n-k)z]}{[1-nz]^2} \rangle$, $m \geq 1$, $k \in [0:n[$. Thus, $N_{n,m}(k) = n^{-(m-1)} \langle z^m; F_{n,k}(z,1) \rangle$, where $F_{n,k}(z,y)$ is the function given in (3). Now, using (5) and the explicit expressions for $\langle z^m; F_{n,k}^{(s)}(z,1) \rangle$, $(m, s) \in \mathbb{N} \times \{2, 3, 4\}$, derived in part (a) of this proof, a simple computation leads to the result stated in part (b) of the theorem. This completes the proof. $\qquad\square$

Starting with the expression for $N_{n,m}(\lambda, k)$ presented in the preceding theorem, the following result can easily be derived ([6]).

Corollary 1. *Assume that all 0-balanced ordered trees with height n and m leaves are equally likely and let $\rho := \frac{m-1}{n} \in]0, \infty[$ fixed. The average number of nodes with degree λ at level k is given by*

$$N_{n,m}(\lambda, k) = \frac{k}{(k+1)^\lambda}[1+k\rho] + k\frac{k+1-k\lambda}{(k+1)^{\lambda+1}} + \mathcal{O}(e^{-k\rho})$$

for $\lambda \in \mathbb{N}$ fixed and $0 \leq k < n$. $\qquad\square$

In the light of the results of the preceding theorem and corollary, the above considerations about the average number of nodes (with a given degree λ) can be made precisely as follows:

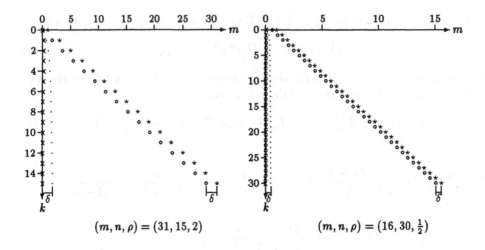

$(m, n, \rho) = (31, 15, 2)$ $(m, n, \rho) = (16, 30, \frac{1}{2})$

Fig. 3. The average number of nodes at level k [$\star$ $\star$ $\star$] and the average number of nodes with degree one [o o o], two [$\cdot$ $\cdot$ $\cdot$] and three [$\times$ $\times$ $\times$] at level k in a 0-balanced ordered tree with m leaves and height n for various values of m and n. Here, δ tends to $\rho := \frac{m-1}{n}$ for large k.

- The average number $N_{n,m}(k)$ of nodes at level k grows *linearly* in k, i.e., passing from one level to the next one, the average rate of increase of nodes amounts to $\rho = \frac{m-1}{n}$;

- The quota of nodes with degree one at level k is equal to $\frac{N_{n,m}(1,k)}{N_{n,m}(k)} = \frac{k}{k+1} + \mathcal{O}(\frac{1}{k^2})$, on the average, i.e., almost all nodes at a large level k have the degree one;

- The average number of nodes at level k with a degree greater than one is equal to $N_{n,m}(k) - N_{n,m}(1, k) = \rho + \mathcal{O}(\frac{1}{k})$, i.e., that number tends to the constant ρ for large k;

- The average number of nodes at level k with degree two is also given by $N_{n,m}(2, k) = \rho + \mathcal{O}(\frac{1}{k})$, i.e., all nodes at level k with a degree greater than one have the degree equal to two for large k, on the average;

- The average number of nodes at level k with a degree λ greater than two is equal to $\frac{k^2}{(k+1)^\lambda} + \mathcal{O}(\frac{1}{k^{\lambda-1}})$, i.e., that number tends to zero for large k;

- The expected number of nodes with a degree λ at a large level k only depends on the proportion $\rho = \frac{m-1}{n}$, i.e., the expected values $N_{n,m}(k)$ and $N_{n,m}(\lambda, k)$ with the same ρ have the same behaviour, on the average.

The behaviour of the quantities $N_{n,m}(k)$ and $N_{n,m}(\lambda, k)$ is illustrated in Fig. 3 for some values of m and n.

2.2 0-Balanced t-ary Ordered Trees with Height n and m Leaves

The class $\mathcal{F}_{n,m}(\Theta)$ of 0-balanced t-ary ordered trees with height n and m leaves is characterized by the function $\Theta(y) = \Theta(y) - 1 = (1+y)^t - 1$, $t \geq 2$ fixed. A simple calculation shows that $\Theta^{<p>}(y) = (1+y)^{t^p} - 1$, $p \geq 0$. Thus, by (2) $|\mathcal{F}_{n,m}(\Theta)| = \langle z^m; \Theta^{<n>}(z) \rangle = \langle z^m; (1+z)^{t^n} - 1 \rangle = \binom{t^n}{m}$. This class of trees is more interesting than that one discussed in the previous subsection because it corresponds to the set of all different n-block codes with m code words over a given alphabet with cardinality t. The set $\mathcal{F}_{n,m}(\Theta)$ consists of all code trees.

Using Lemma 1 and the explicit expression for $\Theta^{<p>}(y)$ given above, a straightforward compution yields the following explicit result ([6]).

Theorem 2. *Assume that all 0-balanced t-ary ordered trees with height n and m leaves are equally likely.*
(a) The average number of nodes with degree λ at level k is given by

$$N_{n,m}(\lambda, k) = t^k \frac{\binom{t}{\lambda}}{\binom{t^n}{m}} \sum_{0 \leq i \leq \lambda} (-1)^{\lambda-i} \binom{\lambda}{i} \binom{t^n [1 - t^{-k}(1 - \frac{i}{t})]}{m}$$

for $(\lambda, k) \in [1:t] \times [0:n[$.
(b) The average number of nodes at level k is given by

$$N_{n,m}(k) = t^k \left[1 - \frac{\binom{t^n(1-t^{-k})}{m}}{\binom{t^n}{m}} \right], \quad k \in [0:n]. \qquad \square$$

In order to discuss the behaviour of the expected values $N_{n,m}(\lambda, k)$ and $N_{n,m}(k)$, we have to study the quantity

$$A_{n,m}(k, t, \alpha) := \frac{\binom{t^n(1-\alpha t^{-k})}{m}}{\binom{t^n}{m}}, \quad \alpha \in [0,1], \ 1 \leq k \leq n, \tag{6}$$

because by Theorem 2

$$N_{n,m}(\lambda, k) = t^k \binom{t}{\lambda} \sum_{0 \leq i \leq \lambda} (-1)^{\lambda-i} \binom{\lambda}{i} A_{n,m}\left(k, t, 1 - \frac{i}{t}\right) \tag{7}$$

and

$$N_{n,m}(k) = t^k [1 - A_{n,m}(k, t, 1)]. \tag{8}$$

The following technical lemma has been proved in [6].

Lemma 2. *Let $\alpha \in]0,1]$ fixed, $t \geq 2$ fixed, $m \leq t^{(1-\delta)n}$ for $\delta \in]0,1]$, and $k \in [1:n]$.*
(a) As $n \to \infty$, the following approximation holds for $m \leq [t^n(\alpha^{-1}t^k - 1)]^{\frac{1}{2}+\epsilon}$ with $0 < \epsilon < \frac{1}{4}\delta$:

$$A_{n,m}(k, t, \alpha) = (1 - \alpha t^{-k})^m \, e^{-\frac{m^2}{2t^n(\alpha^{-1}t^k - 1)}} \left[1 + \mathcal{O}\left(\frac{m^3}{t^{2n+k}}\right) + \mathcal{O}\left(\frac{m}{t^{n+k}}\right) \right].$$

(b) As $n \to \infty$, the following approximation holds for all m:

$$A_{n,m}(k,t,\alpha) = (1 - \alpha t^{-k})^m \left[e^{-\frac{m^2}{2t^n(\alpha^{-1}t^k-1)}} + \mathcal{O}\left(\frac{1}{t^{\frac{1}{2}(n-k)}}\right) \right]. \qquad \square$$

Using (7), (8) and the approximations presented in the preceding lemma, the asymptotical behaviour of the expected values $N_{n,m}(\lambda, k)$ and $N_{n,m}(k)$ can be computed. A lengthy computation ([6]) yields the following result.

Theorem 3. *Assume that all 0-balanced t-ary ordered trees with height n and m leaves are equally likely. Moreover, let $m \leq t^{(1-\delta)n}$ with $\delta \in]0,1]$, and $t \geq 2$ fixed. We have for large n:*
(a) If $t^k = o(m)$, $m \to \infty$, then the average number $N_{n,m}(\lambda, k)$ of nodes with degree λ at level k and the average number $N_{n,m}(k)$ of nodes at level k are given by

$$N_{n,m}(\lambda, k) \sim \begin{cases} o(t^k) & \text{if } 1 \leq \lambda < t \\ t^k & \text{if } \lambda = t \end{cases} \qquad \text{and} \quad N_{n,m}(k) \sim t^k.$$

(b) If $m = o(t^k)$, $m \to \infty$, then the average number $N_{n,m}(\lambda, k)$ of nodes with degree λ at level k and the average number $N_{n,m}(k)$ of nodes at level k are given by

$$N_{n,m}(\lambda, k) \sim \begin{cases} o(m) & \text{if } 1 < \lambda \leq t \\ m & \text{if } \lambda = 1 \end{cases} \qquad \text{and} \quad N_{n,m}(k) \sim m.$$

In addition, the asymptotical relation $N_{n,m}(\lambda, k) \sim o(1)$, $1 < \lambda \leq t$, holds for $m^2 = o(t^k)$, $m \to \infty$. $\qquad \square$

The preceding theorem tells us that the average number of nodes appearing at a level of a 0-balanced t-ary ordered tree with height n and m leaves has the following asymptotical behaviour for large $m \leq t^{\beta n}$, $\beta \in [0, 1[$:

- If k grows weaker than $\log_t(m)$ then the average number $N_{n,m}(k)$ of nodes at level k is asymptotically equal to t^k, i.e., the subtree of height $\approx \log_t(m)$ at the root is a *complete extended t-ary tree*, on the average. Indeed, the average number $N_{n,m}(\lambda, k)$ of nodes with degree $\lambda \in [1 : t[$ is $o(t^k)$, and only the nodes with the maximal degree t appear at the levels just discussed.

- If k grows stronger than $\log_t(m)$ then the average number $N_{n,m}(k)$ of nodes at level k is asymptotically equal to m, i.e., only nodes with degree one appear at these levels. Here, $N_{n,m}(\lambda, k) = o(m)$ for $2 \leq \lambda \leq t$.

The behaviour of the expected values $N_{n,m}(\lambda, k)$ and $N_{n,m}(k)$ is depicted in Fig. 4 for $(n, m, t) = (30, 10000, 3)$.

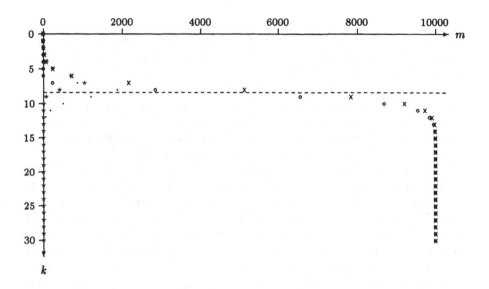

Fig. 4. The average number of nodes at level k [$\times$ $\times$ $\times$] and the average number of nodes with degree one [o o o], two [$\cdot$ $\cdot$ $\cdot$] and three [$\star$ $\star$ $\star$] at level k in a 0-balanced ternary ordered tree with $m = 10000$ leaves and height $n = 30$. The dashed line [- - -] corresponds to the level $k = \log_3(m) \approx 8.38361$.

Thus, almost all nodes with the maximal degree t are concentrated at the levels less than k, where $k = o(\log_t(m))$, and almost all nodes with degree one appear at the levels greater than k with $\log_t(m) = o(k)$, on the average. These observations imply the following illustrative interpretation: Our result says that the *first* k positions with $\log_t(m) = o(k)$ in the m code words appearing in an n-block code over a given alphabet with cardinality t are sufficient to separate these words, on the average, provided that all those codes are equally likely.

The discussed distribution of the nodes in a 0-balanced t-ary ordered tree with height n and m leaves gives rise to the conjecture that the average number of nodes $N_{n,m}$ of such a tree shows a behaviour like $N_{n,m} \approx m[n - \log_t(m)] + \sum_{0 \le k < \log_t(m)} t^k = n\,m - m\log_t(m) + \mathcal{O}(m)$. The proof of that conjecture is very technical; it consists of a series of estimations of non-trivial sums ([6]).

Theorem 4. *Assume that all 0-balanced t-ary ordered trees with height n and m leaves are equally likely. Moreover, let $m \le t^{(1-\delta)n}$ with $\delta \in\,]0,1]$, and $t \ge 2$ fixed. The average number of nodes in such a tree is given by $N_{n,m} = -m\log_t\left(\frac{m}{t^n}\right) + \mathcal{O}(m)$.* $\qquad\square$

The previous result has the following illustrative interpretation: Given a set $\mathcal{L}$ consisting of $m := |\mathcal{L}|$ words of length n over an ordered alphabet T with cardinality $|T| := t$, then the quantity $\mathbb{E}[X_{\text{sufl}}(\mathcal{L})] := \frac{1}{m}(|\text{INIT}(\mathcal{L})| - 1)$ with $\text{INIT}(\mathcal{L}) := \{u \in T^* | (\exists v \in T^*)(uv \in \mathcal{L})\}$ is a measure for the average running time per word to generate the words in $\mathcal{L}$ lexicographically ([4]). If we construct the code tree associated with $\mathcal{L}$, we obtain a 0-balanced ordered t-ary tree with height n and m leaves and vice versa. The number of nodes in that code tree is just equal to $|\mathsf{INIT}(\mathcal{L})|$. So, if all sets $\{\mathcal{L} \subseteq T^n | \mathcal{L} = |m|\}$, $m \leq t^{\beta n}$, $\beta \in [0, 1[$, and all words in these sets are equally likely then the average running time per word to generate the strings in such a set lexicographically is equal to $-\log_t(\frac{m}{t^n}) + \mathcal{O}(1)$, i.e., assuming that all n-block codes with m code words over a given alphabet with cardinality t are equally likely, then the average running time per word to generate these code words lexicographically is given by $-\log_t(\frac{m}{t^n}) + \mathcal{O}(1)$.

In [5], the author has introduced the so-called 'average minimal prefix-length' of a formal language $L_n \subseteq T^n$. It corresponds to the average length of the shortest prefix which must be read in order to decide whether or not a word $w \in T^n$ belongs to L_n. Assuming that the set of all n-block codes with m code words over a given alphabet with cardinality t and all these code words are equally likely, a similar computation as needed in the proof of Theorem 4 yields that the average minimal prefix-length of the formal language consisting of the m code words is equal to $\log_t(m) + \mathcal{O}(1)$, where $m \leq t^{\beta n}$, $\beta \in [0, 1[$. This result shows that the prefixes of length $\log_t(m)$ are sufficient to separate the code words, on the average.

Finally note that the investigations presented in this paper can be made for other families $\mathcal{F}(\Theta)$ of simply generated trees provided that some knowledge about the behaviour of the iterates $\Theta^{<p>}$ is available. But, in general, this seems to be a difficult task.

References

[1] Flajolet, Ph.: Analyse d'algorithmes de manipulation d'arbres et de fichiers, *Cahiers BURO* **34-35** (1981) 1-209.

[2] Kemp, R.: The Expected Number of Nodes and Leaves at Level k in Ordered Trees, *LNCS* **145** (1983) 153-163.

[3] Kemp, R.: Balanced Ordered Trees, *Random Structures and Algorithms* **5** (1) (1994) 99-121.

[4] Kemp, R.: Generating Words Lexicographically: An Average-Case Analysis, Preprint, JWG-Universität Frankfurt am Main, FB 20, 1995 (to appear in *Acta Informatica*).

[5] Kemp, R.: On Prefixes of Formal Languages and Their Relation to the Average-Case Complexity of the Membership Problem, *Journal of Automata, Languages and Combinatorics* **1** (1996), 259-303.

[6] Kemp, R.: On the Expected Number of Nodes at Level k in 0-balanced Ordered Trees, Tech. Rep. 1/97, JWG-Universität Frankfurt am Main, FB 20, 1997.

[7] Knuth, D. E.: *The Art of Computer Programming.* Vol.1, 2nd ed. (Addison-Wesley, 1973).

[8] Meir, M. and Moon, J. W.: On the Altitude of Nodes in Random Trees. *Can. Math.* **30** (1973), 997-1015.

Cell Flipping in Permutation Diagrams

Martin Charles Golumbic[1] and Haim Kaplan[2]

[1] Dept. of Math. and Computer Science, Bar-Ilan University, Ramat Gan, Israel.
email: golumbic@cs.biu.ac.il
[2] AT&T Labs, 180 Park Avenue, Florham Park, NJ 07932, USA. email:
hkl@research.att.com

Abstract. Permutation diagrams have been used in circuit design to model a set of single point nets crossing a channel, where the minimum number of layers needed to realize the diagram equals the clique number $\omega(G)$ of its permutation graph, the value of which can be calculated in $O(n \log n)$ time. We consider a generalization of this model motivated by "standard cell" technology in which the numbers on each side of the channel are partitioned into consecutive subsequences, or *cells*, each of which can be left unchanged or flipped (i.e., reversed). We ask, for what choice of flippings will the resulting clique number be minimum or maximum. We show that when one side of the channel is fixed (no flipping), an optimal flipping for the other side can be found in $O(n \log n)$ time for the maximum clique number. We prove that the general problem is NP-complete for the minimum clique number and $O(n^2)$ for the maximum clique number. Moreover, since the complement of a permutation graph is also a permutation graph, the same complexity results hold for the independence number.

1 Introduction

1.1 Background

A *permutation diagram* D consists of two horizontal lines L_1 and L_2 each having n distinguished points labelled by a permutation of the numbers $1, 2, \ldots, n$ and n straight line segments connecting i on L_1 with i on L_2. We call each of these points a *position* and its label is called the *contents* of the position. The labels of L_1 are denoted by i, j, and z, etc. The labels of L_2 are denoted by i', j', and z', etc.

The *permutation graph* $G(D) = (V, E)$ of diagram D is an undirected graph with vertices $V = \{1, 2, \ldots, n\}$ and edges $(i, j) \in E$ if and only if segment $i - i'$ intersects segment $j - j'$ in D. See Figure 1.

We may also observe that the number of edges $e = |E|$ equals the number of intersections of segments in D. Further background on permutation graphs can be found in Golumbic[3].

Permutation diagrams have been used in a number of application areas including in circuit design to model single point nets crossing a channel. The clique number $\omega(G)$ of the permutation graph $G = G(D)$ is the size of the largest complete subgraph. We also define the clique number of a diagram D to be equal

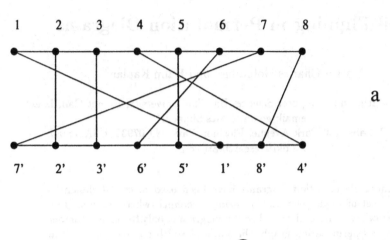

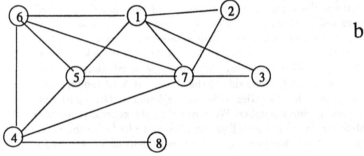

Fig. 1. (a) Permutation Diagram D. (b) The Permutation Graph $G(D)$.

to $\omega(G(D))$. The clique number of a permutation diagram equals the minimum number of layers on which the diagram D can be realized (the segments are partitioned into layers where segments on the same layer may not intersect). For a given diagram, $\omega(G)$ can be calculated in $O(n \log n)$ time [2] (see also [3]). We review this algorithm in Section 4.

1.2 Problem definition and results

In this paper, we consider a generalization of the permutation diagram model in which the numbers on each line L_k are partitioned into consecutive subsequences $c_1, \ldots, c_r, c'_1, \ldots, c'_{r'}$, called *cells*, each of which can be left unchanged or flipped (i.e., reversed). Each choice σ of which cells to flip will yield a different permutation diagram D_σ with its corresponding permutation graph $G_\sigma = G(D_\sigma)$ and clique number $\omega(G_\sigma)$. For a position s, denote by $c(s)$ and $c'(s)$ the cells containing position s on L_1 and L_2, respectively. Note that the contents of position s on L_1 in D_σ may vary between two possible values according to whether σ flips

$c(s)$ or not; similarly for the contents of each position s on L_2. If $c(s) = c_l$ and $c(t) = c_k$ we denote $c(s) < c(t)$ if $l < k$, similarly for c'. We raise the question of finding the following two quantities over all possible choices σ of flippings for a given partitioned permutation diagram $\mathcal{D} = < D, \{c_i\}, \{c'_j\} >$:

$$MAXFLIP(\mathcal{D}) = \max_\sigma \omega(G_\sigma)$$
$$MINFLIP(\mathcal{D}) = \min_\sigma \omega(G_\sigma)$$

The $MAXFLIP$ and $MINFLIP$ problems are to find $MAXFLIP(\mathcal{D})$ and $MINFLIP(\mathcal{D})$, respectively, given a partitioned permutation diagram $\mathcal{D}$.

Example: Let the labels on L_1 be $[3, 4, 7], [2, 6], [1, 5, 8]$ where the brackets indicate cells, and let the labels on L_2 be $[6', 2', 5'], [4', 1', 7', 8', 3']$. Position 6 on L_1 has contents 1 if $c(6)$ is not flipped, and contents 8 if $c(6)$ is flipped. The clique number is 4 with no flipping but is reduced to 3 if we flip $[2, 6]$, or increased to 5 if we flip $[2, 6], [6', 2', 5']$ and $[4', 1', 7', 8', 3']$.

If there are $k = r + r'$ cells, then both $MAXFLIP$ and $MINFLIP$ can be solved in an exhaustive manner by solving all 2^k possible choices. For small values of k (say $k < 5$) this may be reasonable, but for larger values (say $k > 10$) this may be unreasonable. We study the computational complexity of these problems independent of the number k. We give an $O(n^2)$ algorithm for $MAXFLIP$ and prove that $MINFLIP$ is NP-complete. We also show that for the special case of flipping the cells only on one line, $MAXFLIP$ can be solved in $O(n \log n)$.

Note that the same complexity results immediately hold for the analogous independence (or stability) number problems on flipping a partitioned permutation diagram. This follows since the complement of a permutation graph is also a permutation graph, (see [3]).

We mention a very simple case which can be solved in an efficient greedy manner, namely, when each cell is of size at most two. In this case, each cell $[i, j]$ on L_1 is checked once: if segments $i - i'$ and $j - j'$ do not cross, then leave $[i, j]$ unchanged for $MINFLIP$ and flip it for $MAXFLIP$. Do the same for each cell $[i', j']$ on L_2. The effect of each such flip decision is local, it either adds the edge (i, j) to the permutation graph or deletes it without changing any other edges. Moreover, this represents all the degrees of freedom in the problem. Since adding edges can only increase the clique size and deleting edges can only decrease the clique size, the following result holds.

Proposition 1. *If each cell is of size at most two, then $MAXFLIP$ and $MINFLIP$ can be solved in $O(n)$ time.*

1.3 Motivation and Applications

In the computer aided design of VLSI circuits using "standard cell" technology, a stage is reached where cell placement on horizontal rows has already been performed, and the only remaining degree of freedom is replacing some of the cells with their "mirror image" with respect to a vertical axis, i.e., cell flipping. Every problem considered for a fixed channel can also be studied in its cell flipping versions. For example, minimum width channel routing in the jog-free Manhattan model [4] is NP-hard even without cell flipping, so its cell flipping versions would also be NP-hard. But the minimum channel density problem is $O(p)$ for a fixed channel with p pins, and its flipping version can be solved in $O(p \log n)$ time [1].

2 An $O(n^2)$ Dynamic Programming Algorithm for $MAXFLIP$

We start out describing a simple quadratic algorithm that computes the clique number of a (unpartitioned) permutation diagram. Later on we show how to generalize this algorithm to the partitioned case.

Let s be a position on L_1, t a position on L_2, and $tr(s,t)$ the trapezoid defined by positions s and n on L_1 and 1 and t on L_2 (see Figure 2). Define $C(s,t)$ to be the clique number of the permutation diagram induced by lines in $tr(s,t)$. The algorithm computes $C(s,t)$ for every s and t in increasing order of t and decreasing order of s. Initially, $C(s,1) = 1$ if $tr(s,1)$ contains a line and $C(s,1) = 0$ otherwise. Assume $C(k,l)$ have already been computed for every $l < t$ and $k \geq s$, let g be the line incident with position t in L_2, and let k be the position in which g meets L_1. Then $C(s,t) = C(s,t-1)$ if $k < s$; otherwise, $C(s,t)$ is the maximum among $C(s,t-1)$ and $C(k,t-1)+1$.

Now, keeping the algorithm above in mind, we turn back to the partitioned case. Observe that when we fix the directions of the cells $c(s)$ and $c'(t)$ the set of line segments with both endpoints within $tr(s,t)$ is determined. We denote by a 0/1 pair $\delta = (\delta_1, \delta_2)$ a direction assignment to $c(s)$ and $c'(t)$. The assignment δ flips $c(s)$ if and only if $\delta_1 = 1$, and flips $c'(t)$ if and only if $\delta_2 = 1$. Given a direction assignment δ we denote by $L(s,t,\delta)$ the set of segments contained in $tr(s,t)$. Let σ' be a flipping of the cells $c(s) < c \leq c(n)$ and $c'(1) \leq c' < c'(t)$. For each σ', $L(s,t,\delta)$ induces a permutation diagram obtained by flipping $c(s)$ and $c'(t)$ according to δ and flipping cells $c(s) < c \leq c(n)$, and $c'(1) \leq c' < c'(t)$ according to σ'. We denote by $C(s,t,\delta)$ the maximum over all possible σ''s of the clique number of the permutation diagram induced by $L(s,t,\delta)$.

Our dynamic programming algorithm computes $C(s,t,\delta)$ for every possible s, t, and δ. The computation proceeds in increasing order of t, the lower right endpoint of the trapezoid, and decreasing order of s, the upper left endpoint of the trapezoid. The details are as follows.

We initialize $C(s,1,\delta)$ to one if $L(s,1,\delta)$ contains one line segment and to zero otherwise. Fix s, $\delta = (\delta_1, \delta_2)$, and assume that we have already computed $C(k,l,\delta)$ for every $l < t$ and $k \geq s$.

If $c'(t-1) = c'(t)$ let $A = C(s,t-1,\delta)$. Otherwise, $c'(t-1)$ precedes $c'(t)$, and we let A be the maximum among $C(s,t-1,\delta')$ and $C(s,t-1,\delta'')$ where $\delta' = (\delta_1, 1)$ and $\delta'' = (\delta_1, 0)$; i.e., δ' flips $c'(t-1)$ and δ'' does not flip $c'(t-1)$, and both δ' and δ'' give the same direction to $c(s)$ as δ.

Let g be the line segment incident with position t on L_2 when $c'(t)$ is directed as specified by δ. If g is not in $L(s,t,\delta)$ then we can finish setting $C(s,t,\delta) = A$.

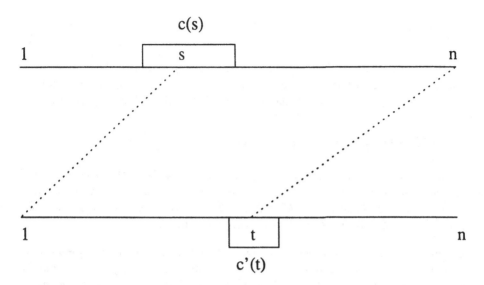

Fig. 2. The trapezoid $tr(s,t)$ defined by a position s on L_1 and position t on L_2.

Otherwise, the line segment g is in $L(s,t,\delta)$, we denote by $c(g)$ the cell that contains the L_1-endpoint of g. Note that since g is in $L(s,t,\delta)$, $c(s) \leq c(g)$. We continue according to the appropriate of the following two cases.

Case 1: The cell $c(g)$ is different from $c(s)$, i.e. $c(s) < c(g)$.
Let k be the position of the L_1-endpoint of g when $c(g)$ is flipped, and let l be the position of the L_1-endpoint of g when $c(g)$ is not flipped. If $c'(t-1) = c'(t)$ then assign B to one plus the maximum among $C(k, t-1, \delta')$ and $C(l, t-1, \delta'')$ where $\delta' = (1, \delta_2)$ and $\delta'' = (0, \delta_2)$. If $c'(t-1) < c'(t)$ then assign B to be one plus the maximum among $C(k, t-1, (1,0))$, $C(k, t-1, (1,1))$, $C(l, t-1, (0,0))$, $C(l, t-1, (0,1))$.

Case 2: Cell $c(g)$ is the same as $c(s)$.
We perform a computation similar to the above, the difference stems from the fact that the direction of $c(s)$ is now determined by δ. Let k be the position of the L_1-endpoint of g when the direction of $c(g)$ is fixed as defined by δ. If $c'(t-1) = c'(t)$ then assign B to be one plus $C(k, t-1, \delta)$, and if $c'(t-1) \neq c'(t)$ then assign B to be one plus the maximum among $C(k, t-1, (\delta_1, 0))$, $C(k, t-1, (\delta_1, 1))$.

In either one of the two cases above we finish by setting $C(s,t,\delta)$ to be the maximum among A and B.

It is straightforward to prove by induction on t that $C(s,t,\delta)$ are computed correctly. Computation of every entry in the table C takes $O(1)$ time and since the size of C is $O(n^2)$ our algorithm runs in $O(n^2)$ time. The space complexity is $O(n)$ since we need to store the values in C only for two consecutive values of t at any one time.

The maximum of the four entries $C(1, n, (0,0))$, $C(1, n, (0,1))$, $C(1, n, (1,0))$, and $C(1, n, (1,1))$ is the answer to MAXFLIP. Therefore, we have proved the following theorem.

Theorem 2. *There exists an algorithm for MAXFLIP that runs in $O(n^2)$ time using $O(n)$ space.* $\square$

By maintaining with each entry $C(s, t, \delta)$ a flipping for which the permutation diagram of $L(s, t, \delta)$ has clique number $C(s, t, \delta)$ our algorithm can be adapted to produce a flipping for which the optimum MAXFLIP value is obtained.

3 $MINFLIP$ is NP-complete

In this section, we give a reduction from 3-SAT to MINFLIP. Given a 3-SAT instance I with variable $X_1, \ldots, X_n$ and clauses $C_1, \ldots, C_m$ we construct a MINFLIP instance $\mathcal{D}(I)$ such that $MINFLIP(\mathcal{D}(I)) \leq k$ if and only if I is satisfiable. Our construction proves the following theorem.

Theorem 3. *Given a MINFLIP instance $\mathcal{D}$ and an integer k, the problem of deciding whether $MINFLIP(\mathcal{D}) \leq k$ is NP-complete. This problem remains NP-complete even if k is fixed, $k \geq 7$, and the cells in $\mathcal{D}$ are of length at most $2k + 4$.*

Given a 3-SAT formula I with n variables $X_1, \ldots, X_n$ and m clauses $C_1, \ldots, C_m$ we construct $\mathcal{D}(I)$ as follows. On L_1, $\mathcal{D}(I)$ has m *clause cells* $c_1, \ldots, c_m$ each of length 14, $2m$ *special cells* $s_1, \ldots, s_m$ and $t_1, \ldots, t_m$ each of length 1, and nm *variable cells* x_{ij}, $1 \leq i \leq n$, $1 \leq j \leq m$ whose lengths will be described below. Similarly, on L_2, $\mathcal{D}(I)$ has cells c_i', s_i', t_i', and x_{ij}' with the same lengths. The length of x_{ij}, $1 < j \leq m$, and x_{ij}', $1 \leq j < m$ is 18 if $X_i \in C_j$ or $\overline{X_i} \in C_j$ and 16 otherwise. The length of x_{i1} is 10 if $X_i \in C_1$ or $\overline{X_i} \in C_1$ and 8 otherwise, and the length of x_{im}' is 10 if $X_i \in C_m$ or $\overline{X_i} \in C_m$ and 8 otherwise.

The L_1 cells are ordered

$$s_1\, x_{11}\, x_{21} \ldots x_{n1}\, t_1\, c_1,\ s_2\, x_{12} \ldots x_{n2}\, t_2\, c_2,\ \ldots,\ s_m\, x_{1m} \ldots x_{nm}\, t_m\, c_m,$$

and the L_2 cells are ordered

$$c_1'\, s_1' x_{11}'\, x_{21}' \ldots x_{n1}'\, t_1', c_2'\, s_2'\, x_{12}' \ldots x_{n2}'\, t_2',\ \ldots,\ c_m'\, s_m'\, x_{1m}' \ldots x_{nm}'\, t_m'.$$

In order to complete the description of $\mathcal{D}(I)$ we have to specify the labelings of L_1 and L_2. On L_1 we label the positions consecutively with $1, \ldots, n$ in the natural order. We label L_2 as follows. Partition each cell x_{ij}, $1 < j \leq m$, and x_{ij}', $1 \leq j < m$, into three regions. The second and third regions are of length 8 each, and the first region is of length two if x_{ij} or x_{ij}', respectively, is of length 18 and empty otherwise. Similarly, we partition x_{i1} and x_{im}' into three regions where the third region is of length 8, the second region is empty and the first region is of length two if x_{ij} or x_{ij}', respectively, is of length 10 and empty otherwise.

Fill-in the 8 positions in the last region of x_{ij}' with the same labels as of the positions in the last region of x_{ij}, and the positions of the second region of x_{ij}', $1 \leq j < m$, with the labels of the second region of x_{ij+1}. The positions in the first region of x_{ij}' are filled with labels from c_j as we will specify shortly.

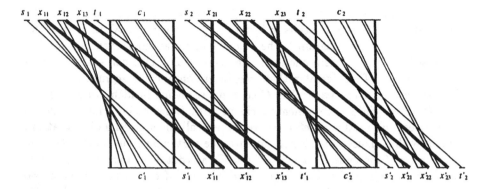

Fig. 3. The partitioned permutation diagram $\mathcal{D}(I)$ that corresponds to $I = (X_1 \vee X_2 \vee X_3) \wedge (\overline{X_1} \vee \overline{X_2} \vee X_3)$. Thick segments denote bundles of either four or eight parallel segments.

We also partition each clause cell into three regions, the first is of length 4, the second of length 6, and the last of length 4. We fill-in the first and the last regions in c'_j as the first and the last regions of c_j. The middle region in c'_j is further partitioned into three pairs that we fill-in as follows.

Assume C_j contains literals of the variables X_{j_1}, X_{j_2}, and X_{j_3} where $j_1 < j_2 < j_3$. We fill-in the first pair in the middle region of c'_j with the labels from the first region of $x_{j_3 j}$, the second pair of c'_j with the labels from the first region of $x_{j_2 j}$, and the third pair of c'_j with labels from the first region of $x_{j_1 j}$. The order of the labels within each pair in the middle region of c'_j is reversed if the corresponding variable is not negated in C_j. Otherwise, the order is the same as in the adjacent variable cell.

We now return and specify how to fill-in the labels in the first region of x'_{ij} when it is not empty. We still assume that C_j contains literals of the variables X_{j_1}, X_{j_2}, and X_{j_3} where $j_1 < j_2 < j_3$. We fill-in the first region of $x'_{j_1 j}$ with the labels of the first pair of the middle region of c_j, the first region of $x'_{j_2 j}$ with the labels of the second pair of c_j, and the first region of $x'_{j_3 j}$ with the labels of the third pair of c_j. The order of the labels within each pair is the same as the order of the adjacent pair on c_j if the corresponding variable is not negated in C_j and the order is in reverse otherwise. This completes the definition of $\mathcal{D}(I)$. Figure 3 shows the partitioned diagram that corresponds to the 3-SAT formula $(X_1 \vee X_2 \vee X_3) \wedge (\overline{X_1} \vee \overline{X_2} \vee X_3)$.

It is straightforward to verify that the construction is polynomial. The following lemma proves that we can flip cells in $\mathcal{D}(I)$ to obtain a diagram whose corresponding permutation graph has a small clique number if and only if I is satisfiable. We denote by V_i, $1 \leq i \leq n$, the set of cells associated with the variable X_i, i.e. $V_i = \{x_{ij}, x'_{ij} \mid 1 \leq j \leq m\}$.

Lemma 4. *There is a subset of cells to flip in $\mathcal{D}(I)$ such that the corresponding permutation graph has clique number at most seven if and only if I is satisfiable.*

Proof. Assume I is satisfiable, and let f be a satisfying truth assignment. We show that if one flips all the cells in V_j for every true variable X_j in f, then the resulting permutation diagram D_f has clique number at most seven. Assume for a contradiction that C is a clique of size eight in D_f. One can easily observe that any clique of size greater than two must contain line segments incident with clause cells, so in particular, C contains segments incident with a clause cell. Moreover, since the segments incident with two different clause cells do not intersect, no matter which cells are flipped, all the segments of C that are incident with a clause cell are incident with one particular clause cell c. Assume $c = c_l$ or $c = c_l'$ for some $1 \leq l \leq m$. Clique C can contain at most two segments that do not connect the cell c with a variable cell. If $c = c_l'$ one of these two segments is the segment between s_l and s_l' or a segment between x_{il} and x_{il}' for some $1 \leq i \leq n$, and if $c = c_l$ then the segment is either the one connecting t_l and t_l' or a segment between x_{il} and x_{il}' for some $1 \leq i \leq n$. The second segment in C that does not connect C with a variable cell is a segment connecting c_l and c_l'. Thus, since C has size eight it must contain all the six segments connecting c to variable cells. These segments are pairwise crossing if and only if I does not satisfy clause C and that is a contradiction.

For the converse, assume that there exists a subset S of the cells such that when flipping S the resulting diagram has clique number at most 7. We claim that for every variable X_i in our 3-SAT formula the cells in V_i are either all in S or all not in S. To prove the claim assume that this is not the case, then there exists $1 \leq i \leq n$ and a pair of adjacent cells in V_i one in S and the other not in S. The segments connecting these two cells form a clique of size 8 in contradiction with the assumption that the clique number of our permutation diagram is at most 7. A similar argument shows that for each clause C_j the two clause cells c_j and c_j' are either both flipped or both not flipped.

We define a truth assignment f for our 3-SAT formula by letting $X_i = 1$ if $V_i \subseteq S$ and $X_i = 0$ if $V_i \cap S = \emptyset$. We claim that this assignment satisfies all clauses. To prove the claim, assume for a contradiction that f does not satisfy clause C_l. If c_l and c_l' are left unchanged then the six segments connecting c_l' with variable cells combined with one of the 4 segments connecting the left side of c_l' with c_l, and the segment between s_l and s_l' would form a clique of size eight. Symmetrically, if c_l and c_l' are flipped then the six segments connecting c_l with variable cells together with one of the segments between c_l and c_l', and the segment between t_l and t_l' would form a clique of size eight. Thus we obtain a contradiction to our assumption that when flipping the cells is S the corresponding permutation diagram has clique number at most seven. $\square$

The proof of Theorem 2 for the case $k = 7$ follows directly from our construction and Lemma 4. It is straightforward to adapt our construction such that it would work for any fixed $k > 7$.

4 $O(n \log n)$ Algorithm for One Side MAXFLIP

When only L_1 is partitioned into cells that can be flipped, one may observe that without loss of generality we may rename the positions on L_2 to be labeled in increasing order $1', \ldots, n'$. The problem MAXFLIP is then equivalent to finding a flipping σ such that the length of the maximum decreasing subsequence of the integer sequence induced by σ on L_1 is maximized. In this section we show how to solve MAXFLIP by extending the algorithm for finding a maximum decreasing subsequence (or equivalently finding the clique number of a permutation diagram) described by Fredman [2]. The running time of the flipping algorithm that we obtain is of the same order as the ordinary running time of the algorithm for maximum decreasing subsequence, that is $O(n \log n)$. We start by reviewing the algorithm in [2], and subsequently show how to use it for MAXFLIP.

We find the maximum length of a decreasing subsequence of $a_1, \ldots, a_n$ by processing the a_i's in order while maintaining an array T of length n. Initially $T(j) = -\infty$ for every $1 \leq j \leq n$, and after processing a_i, $T(j)$ is the largest number that ends a decreasing subsequence of length j in $a_1, \ldots, a_i$. While processing a_i, $1 \leq i \leq n$, we search for the largest length $0 < l < n$ such that $T(l) > a_i$ (assume $l = 0$ if $T(j) < a_i$ for every $1 \leq j \leq n$). If $T(l+1) < a_i$ then we assign $T(l+1) = a_i$. Since we can search for l using binary search in $O(\log n)$ time, the running time of this algorithm is $O(n \log n)$. It is straightforward to modify this algorithm such that it also produces a decreasing sequence of maximum length.

Our algorithm for MAXFLIP will use the algorithm for maximum decreasing subsequence as a procedure that we denote by MDSS. We run MDSS on the sequences defined by the individual cells, and each time we use it we supply it with an initialized array to work on. In contrast with the regular MDSS algorithm, the array is not initialized to $-\infty$ but rather with values based on computations performed on previous cells. Let the cells on L_1 be $c_1, \ldots, c_k$ from left to right; assume c_i also denotes the integer sequence defined by cell c_i when it is left unchanged and $\overline{c_i}$ denotes the sequence defined by c_i when it is flipped. An application of MDSS with an array T on integer sequence A is denoted by $MDSS(T, A)$.

Define $F(i, j)$ to be the maximum integer that terminates a decreasing subsequence of length j of an integer sequence defined by some flipping of $c_1, \ldots, c_i$. Let T and T_f be two arrays that we use to communicate input and output with MDSS, initialized with all entries equal to $-\infty$. We start out computing $F(1, j)$ by running $MDSS(T, c_1)$ then $MDSS(T_f, \overline{c_1})$, and setting $F(1, j) = \max\{T(j), T_f(j)\}$ for every $1 \leq j \leq |c_1|$. Assume we have computed $F(i-1, j)$ for some $i < k$ and every j; we compute $F(i, j)$ as follows. We initialize $T(j) = T_f(j) = F(i-1, j)$, and then run $MDSS(T, c_i)$ and $MDSS(T_f, \overline{c_i})$. Finally, we set $F(i, j) = \max\{T(j), T_f(j)\}$.

Example: Let the input sequence be $[8, 5], [3, 11, 7, 1, 4], [6, 10, 9, 2]$ where the cells are indicated by brackets. While describing our algorithm on this input, we assume that all tables are initialized with $-\infty$ and that each entry not explicitly referred to remains $-\infty$. After executing $MDSS(T, c_1)$ and $MDSS(T_f, \overline{c_1})$, $T(1) = 8$, $T(2) = 5$, and $T_f(1) = 8$. Taking the maximum, we obtain $F(1, 1) = 8$

and $F(1,2) = 5$. The second iteration starts by setting $T(1) = T_f(1) = F(1,1)$, and $T(2) = T_f(2) = F(1,2)$. After executing $MDSS(T, c_2)$ and $MDSS(T_f, \overline{c_2})$, $T(1) = 11$, $T(2) = 7$, $T(3) = 4$, $T(4) = 1$, and $T_f(1) = 11$, $T_f(2) = 7$, $T_f(3) = 4$, $T_f(4) = 3$. Taking maxima, we obtain $F(2,1) = 11$ and $F(2,2) = 7$, $F(2,3) = 4$, and $F(2,4) = 3$. Similarly, the reader can check that after the last iteration $F(3,1) = 11$, $F(3,2) = 10$, $F(3,3) = 9$, $F(3,4) = 3$, and $F(3,5) = 2$. We conclude that the length of a maximum decreasing subsequence of any sequence that can be obtained by flipping some of the cells c_1, c_2, c_3 is 5. Indeed if we do not flip c_1, flip c_2, and either flip or do not flip c_3 we obtain a sequence that contains $8, 5, 4, 3, 2$ as a subsequence. $\square$

It is straightforward to prove by induction on i that the table $F(i,j)$ is computed correctly. To achieve $O(n \log n)$ running time we implement MDSS such that instead of destructively modifying its input array, it returns a list of the array entries changed together with the final value of each. Note that the length of the output list is bounded by the length of the sequence MDSS is applied to. The maximum computation that follows each pair of calls to MDSS is then carried out only for indexes of changed entries. This implementation ensures that for each element in the original sequence, we perform at most two binary searches and two maximum operations, so the $O(n \log n)$ upper bound on the running time follows.

5 Other Cell Flipping Problems

Boros, Hammer and Minoux [1] investigate the cell flipping problem to minimize channel density for arbitrary nets. Their solution, based on pseudo-Boolean optimization methods, has complexity $O(p \log n)$ for a single channel with p pins and n nets and $O(p(n/c)^c)$ for a stack of c channels. This problem includes the flipping problem on partitioned interval graph representations to minimize the clique number.

References

1. E. Boros, P.L. Hammer, and M. Minoux. Optimal cell flipping to minimize channel density in VLSI design and a class of pseudo-boolean optimization problems. Technical Report 2, RUTCOR, Rutgers University, 1995.
2. M.L. Fredman. On computing the length of longest increasing subsequences. *Discrete Mathematics*, 11:29–35, 1975.
3. M.C. Golumbic. *Algorithmic Graph Theory and Perfect Graphs*. Academic Press, New York, 1980.
4. A.S. LaPaugh. *Algorithms for integrated circuit layout: an analytic approach*. PhD thesis, Dept. of Electrical Engineering and Computer Science, M.I.T., Cambridge, Mass, USA, 1980.

Construction of Non-intersecting Colored Flows Through a Planar Cellular Figure

Marius Dorkenoo[1], Marie-Christine Eglin-Leclerc[1], Eric Rémila[1,2]

[1] Grima, IUT Roanne, Université J. Monnet
20 avenue de Paris, 42334 Roanne cedex, France
[2] LIP, ENS-Lyon, CNRS ura 1398
46 allée d'Italie, 69364 Lyon cedex 07, France
email: {dorkenoo, leclermc, remila}@univ-st-etienne.fr

Abstract. we give a linear time algorithm which, given a simply connected figure of the plane divided into cells, whose boundary is crossed by some colored inputs and outputs, produces non-intersecting directed flow lines which match inputs and outputs according to the colors, in such a way that each edge of any cell is crossed by at most one line. The main tool is the notion of height function, previously introduced for the tilings. This notion appears as the extension of the notion of potential of a flow in a planar graph.

1 Introduction

In 1990, W. P. Thurston [8] produced a very interesting algorithm to tile a polyomino (i. e. a simply connected figure, formed from the union of a finite set of unit cells of the square lattice) with dominoes (formed from two cells which share an edge). Using the notion of tiling group, introduced in [2], W. P. Thurston first proves that the set of the tilings of a fixed polyomino P can be partially ordered, using a height function associated to a tiling, defined on the vertices of the cells of P. Afterwards, special properties of extremal tilings are exhibited and these properties permit to construct such a tiling.

W. P. Thurston's ideas have been taken again and generalized, which permitted to obtain other results about tilings on regular lattices ([5],[7]), but W. P. Thurston's ideas did not seem easy to apply in a irregular lattice : the only result on a partially irregular lattice was the result of Th. Chaboud [1], which produced an algorithm of tiling with (generalized) dominoes formed from two cells of the lattice. This author only assumes that the cells of the lattice are 2-colorable and all the cells of the lattice have the same number of edges.

In this paper, we use W. P. Thurston's method in a problem of colored flows : given a figure F divided in cells with colored inputs and outputs at the boundary of F, we want to match inputs and outputs by non-intersecting directed lines such that each edge of any cell is crossed by at most one line. Notice that the regularity of the structure of F is not a one of our hypothesises.

This paper is divided as follows : in section 2, we present our problem and give an algebraic interpretation of it. This algebraic model permits to apply Thurston method : creation of an order on the solution (section 3), study of minimal solutions according to this order (section 4), construction of a minimal solution (section 5). In section 6, we present two similar problems and the elements which permits to produce the algorithms which solve them.

Consequently, this paper is a bridge between flow problems and tiling problems and this bridge is done by the theory of groups.

2 The problem

2.1 Definitions

A cell is a polygon of the plane R^2. A cellular figure is a finite union of cells with disjoined interiors, such that two intersecting distinct cells share either a vertex or a complete edge. Two distinct cells of a figure are neighbors if they share an edge.

A path of cells is a sequence of cells such that two successive cells are neighbors. A figure F is connected if for each pair (f_1, f_2) of cells, there exists a path of F starting in f_1 and finishing in f_2. Moreover, F is simply connected if for each pair (P, P') of points of R^2 which are not elements of F, there exists a polygonal line F starting in P and finishing in P' which does not meet F. Thus, if F is simply connected, one can describe its boundary by a sequence of edges (called the boundary cycle) such that two successive edges have a common vertex and each edge of the boundary appears exactly once in the sequence.

A boundary condition of F is a set of colored arrows which cross the boundary of a simply connected figure F (incoming in F or outgoing from F), in such a way that each edge of the boundary is crossed at most once.

The colored inputs-outputs (CIO for short) problem is the following : given a simply connected figure F and a boundary condition of F, can we join all the incoming arrows with all the outgoing arrows by non-intersecting colored polygonal lines (called *flow lines*) of the interior of F, with respect of the colors (i.e. each colored flow line join two arrows of its own color) in such a way that each interior edge is crossed by at most one line, and if it is possible, how can we exhibit a solution ?

The applicative motivations are very clear : imagine that a planar ground has to be crossed by pipes for water, gas, or anything else, and that incoming and outgoing places are fixed. How can the pipes be placed ? One can assume that these pipes have a fixed thickness or, for safety, to distinct pipes cannot be too much close. The above problem is a discretization of this concrete situation, the conditions of thickness or safety being interpreted by the condition about the crossing of each edge. Similar problems naturally arise in the design of electronic processors.

2.2 Algebraic interpretation

The first step to solve the CIO problem is the algebraic translation of it. This needs some preliminaries.

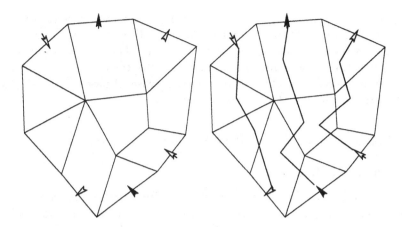

Fig. 1. an input of the problem an a corresponding solution

An extended solution of the CIO problem is obtained adding colored directed cycles to a (classical) solution in such a way that each edge remains crossed at most once. We will see later that, given an extended solution, one can easily find a (classical) solution of the CIO problem.

We first (arbitrarily) fix an orientation of the edges of F. For each edge e, this orientation permits to define the right side of e and the left side of e (which are semi-planes hose common frontier is a straight line which contains e). Let $C = \{c_0, c_1, ..., c_p\}$ be the set of the colors used and G be the free group generated by C. From an extended solution, each edge of F can be labeled by an element of G, as follows: if the edge e is crossed by a line (or cycle) of color c_i from left to right, then e is labeled by c_i, if the edge e is crossed by a line (or cycle) of color c_i from right to left, then e is labeled by c_i^{-1}, if the edge e is not crossed, then e is labeled by 1 (i. e. the unit element of G).

For each cell f, and each edge e of the contour of f, one define $lab_f(e)$ as follows : $lab_f(e) = lab(e)$ if the fixed orientation of e is the same as the orientation of e given by the clockwise contour path of f, and $lab_f(e) = lab(e)^{-1}$ otherwise).

Hence $lab_f(e) = c_i$ (respectively c_i^{-1}) if and only if a line of color c_i incomes in f (outgoes from f) through edge e.

For each cell f, and each vertex v of the contour of f, one define the clockwise contour word of f starting in v (which is denoted by $g_f(v)$) as follows : let $e_1, e_2, ..., e_p$ be the clockwise path of the f starting and finishing at v, we have $g_f(v) = lab_f(e_1)lab_f(e_2)...lab_f(e_p)$.

Remark that, for each face f and each vertex v of f, $g_f(v) = 1$.

Conversely, assume that each edge of F is labeled by an element of G, which is either a color (of type c_i), the inverse of a color (of type c_i^{-1}) or the unit element 1 in such a way that for each edge e of the boundary of F, $lab(e)$ is the label induced by the boundary condition.and for each face f and each vertex v of the contour of f, the element $g_f(v)$ of G, defined as above, is equal to 1. Such a labeling canonically gives a (possibly extended) solution of the CIO problem, as follows :

For each face f, choose a vertex v of its contour. We use a current sequence of labels which, for initialization is empty. From v, follow the clockwise contour of f and successively, for each edge e, read $lab_f(e)$. If $lab_f(e) = lab_f(e')^{-1}$, (where $lab_f(e')$ denotes the last label of our current sequence) the edges e and e' are joined by a colored line which crosses f. Otherwise, if moreover $lab_f(e) \neq 1$, place $lab_f(e)$ at the end of the current sequence.

3 The ordered lattice of the extended solutions

3.1 The group function

Proposition 1. *let S be an extended solution. There exists a function f_S (called a group function) from the set of the vertices of F to G such that, for each pair (v, v') of vertices such that there exists an edge e which is directed from v to v', then $f_S(v') = f_S(v)lab(e)$.*

This proposition is a particular case of the following theorem (with the free group G), which is an extension of a theorem of J. H. Conway and J. C. Lagarias about tilings [2].

Theorem 2. *let Γ be simply connected a planar cellular figure with directed edges and H be a group. Assume that each edge e of Γ is labeled by an element of H, denoted by $lab(e)$. For each boundary cycle $b = (e_1, e_2, ..., e_p)$, which is a contour of a finite face of Γ, one canonically associate the element : $g(b) = lab_f(e_1)lab_f(e_2)...lab_f(e_p)$ If for each cycle b as above, we have : $g(p) = 1$, then there exists a function f from V to H such that for each pair (v, v') of vertices such that there exists a directed edge e from v to v', then $f_S(v') = f_S(v)lab(e)$.*

Proof. by induction on the number of cells. If Γ is reduced to a unique cell, then the result is obvious, from the hypothesis on $g(b)$. Remark that if f is a group function and u is a element of H, then function f_u defined by $f_u(v) = f(v)u$ is also a group function. This yields that, if a group function exists, then for each vertex v, there exists a group function f'_v such that $f'_v(v) = 1$. Now assume that G' has at least two finite faces. Then there exists a path p_0 of Γ joining two vertices of the infinite face of Γ, whose edges are not edges of the infinite face. Such a path canonically defines two connected subgraphs Γ_1 and Γ_2 of Γ such that Γ_1 and Γ_2 only share the path p_0. By induction hypothesis, there exists a group function f_1, defined on Γ_1 and a group function f_2, defined on Γ_2 Let v_0 be any vertex of p_0. From the remark above, function f_1 and f_2 can be chosen in such a way that $f_1(v_0) = f_2(v_0) = 1$. In such a case, we can define the function f on the vertices of Γ by : if v is a vertex of Γ_1 then $f(v) = f_1(v)$, if v is a vertex of Γ_2, then $f(v) = f_2(v)$, There is no conflict since $f_1(v) = f_2(v)$ for each vertex of p_0.

For the following, we fix a vertex w_0 on the boundary of F and, given an extended solution S, f_S denotes the function defined by the above proposition such that $f_S(w_0) = 1$. This convention yields that, for each vertex v of the boundary of F and for each pair (S, S') of solutions satisfying the same boundary condition, $f_S(v) = f_{S'}(v)$.

3.2 Height function

As each finitely presented group, G can be represented by a directed graph (called the Cayley graph of G) whose vertices are the elements of G and two vertices v and v' are joined by a directed edge labeled by c_i, from v to v', if $v' = vc_i$.

Since G is free, its Cayley graph is an infinite tree, which yields that one can construct an integer height function h from G to Z using, for example, with the following conditions:

- for each integer n of Z, $h(c_0^n) = n$,
- each element g of G has exactly one neighbor g' (called the father of g) such that $h(g') = h(g) - 1$,
- the other neighbors $g"$ of g (which are called the sons of g) are such that $h(g") = h(g) + 1$.

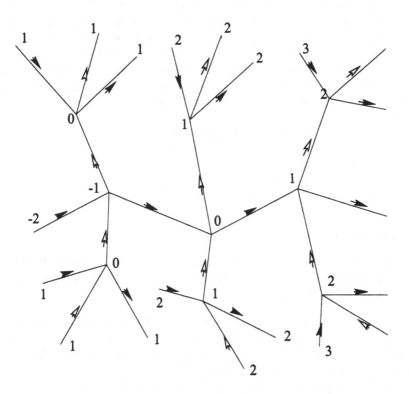

Fig. 2. the free group G with the induced height

We can now define the height function h_S induced by an extended solution S, which is a function from the set of the vertices of F to Z such that, for each vertex v of F, $h_S(v) = h(f_S(v))$. Remark that, for each vertex v of the boundary of F and for each pair (S, S') of solutions, $h_S(v) = h_{S'}(v)$.

3.3 Order on the solutions

We say that a solution S is lower than a solution S' if for each vertex v of F $h_S(v) \leq h_{S'}(v)$. By this way we define a (partial) order on the set of the solutions of the CIO input-output problem for figure F. Since this set is finite, there exists at least one minimal solution S_0. We study the properties of such a solution in the next section.

4 Properties of a minimal solution

4.1 Rotations

Let v be an interior vertex of F and S be a (possibly extended) solution satisfying a boundary condition. For each edge e an extremity of which is v ,we define $lab_v(e)$ by the conditions : if v is the terminal extremity of the edge e (directed by the fixed orientation), then $lab_v(e) = lab(e)$, otherwise $lab_v(e) = lab(e)^{-1}$.

Assume that for a fixed label g, the set of the directed edges a vertex of which is v can be divided into two subsets : the set S_g of the edges e such that $lab_v(e) = g$, and the set S_1 of the edges e' such that $lab_v(e') = 1$.

Another solution S' satisfying the same boundary condition can be deduced from S as follows the edges e of the set S_g are such that $lab_v(e) = 1$ the edges e' of the set S_1 are such that $lab_v(e') = g^{-1}$ the other edges have the same label for S and for S'

The transformation as described above to pass from S to S' and the inverse transformation are called rotations. Notice that for each vertex v such that $w \neq v$ we have : $h_S(v) = h_{S'}(v)$ and $|h_S(v) - h_{S'}(v)| = 1$. This yields that S and S' are consecutive elements of the lattice of the previous section (if, for instance, $h_S(v) = h_{S'}(v) + 1$, then S is a successor of S').

4.2 The key-lemma

Lemma 3. (key-lemma) *Let S_0 be a minimal solution, h_0 denote the height function induced by S_0 and w be an interior vertex of F. There exists a neighbor w' of w such that $h_0(w') = h_0(w) + 1$.*

Proof. Let f_0 denote the group function induced by S_0 and assume that the lemma is false for a vertex w. Let $f'_0(w)$ denote the father of $f_0(w)$ and $g_0(w)$ denote the element of G such that $f'_0(w) = f_0(w)g_0(w)$ (which yields that either $g_0(w) = c_i$ or $g_0(w) = c_i^{-1}$ for some integer i).

Let w' be a neighbor of w. If $h_0(w') = h_0(w)$, then $f_0(w') = f_0(w)$, thus the edge joining w and w' is labeled by 1. If $h_0(w') = h_0(w) - 1$, then, necessarily,

$f_0(w') = f_0'(w)$, which yields that if w is the origin vertex of the edge joining w and w', then this edge is labeled by $g_0(w)$ and if w is the terminal extremity of the edge joining w and w', then this edge is labeled by $(g_0(w))^{-1}$

Thus a rotation can be done in the neighboring of w. A solution S_1, which is lower than S_0, is created by this way, which contradicts the minimality of S_0.

Remark that the key-lemma implies that all the vertices of F of maximal height are in the boundary of F.

5 The algorithm

We now can produce an algorithm which, given a simply connected cellular figure F with a boundary condition, either gives a solution of the CIO problem (if such a solution exists) or indicates that there is no solution. When a solution exists, this algorithm constructs a minimal solution S_0.

5.1 Description

– initialization : Arbitrarily assign a direction to each edge of F, fix a vertex v_0 of the boundary of F. Afterwards, from v_0, compute $f_0(v)$ and $h_0(v)$ for each vertex v of the boundary of F with respect of the boundary conditions. If a contradiction appears (two neighboring vertices, v and v', which are such that ($f_0(v)$ and $f_0(v')$ are neither equal nor neighbors in the Cayley graph of G), there is no solution.
Let M be an integer variable which, for initialization, is equal to the highest value obtained for h_0 on the boundary of F.

– main loop : Each vertex v for which group function and height function are not previously defined such that the height one of its neighbor w is M takes the father of $f_0(w)$ for image by f_0 and, consequently, $M - 1$ or image by h_0.
The variable M is decreased of one unit.

The main loop is repeated until the group function and the height function are defined for each vertex of F.

– control: For each edge e of F, control that if v and v' denote the vertices of e, then there exits a label g such that ($f_0(v) = f_0(v')g$ (i.e. ($f_0(v)$ are $f_0(v')$ are either equal or neighbors in the Cayley graph of G).

If no contradiction has appeared until the end of the control, we can claim that a extended solution exists, from the key-lemma and, using the routine described at the end of section 2, for each cell of F, one constructs an extended solution.

If we want a classical solution (with no cycle), it suffices to destroy the cycles as follows : Take an input of the boundary of F and confirm its label and follow the line starting at this input and confirm the labels of this line. Repeat these instructions for all the inputs. Afterwards, delete the non-confirmed inputs.

5.2 Time complexity

Let n denote the number of cells, m denote the number of edges, and n denote the number of vertices of F.

Let n denote the number of vertices and m denote the number of edges of F.

The intialization costs at most $O(m)$ time units. The execution of the main loop costs $O(m)$ time units : the neighborhood of each vertex of height M has to be explored to define new values of the group function, thus each edge is used twice (once in each direction) The execution of the control costs $O(m)$ time units The execution of the routine of section 2 for each cell of F costs $O(m)$ time units, since each interior edge is used twice and each edge of the boundary is used once, according to the number of cells which share the edge . The deletion of cycles costs $O(m)$ time units since each edge is used at most once during the confirmation process.

Thus , the algorithm has a time complexity in $O(m)$. We recall that , from planarity, $m \leq 3n + 6$., thus the algorithm also has a time complexity in $O(n)$.

6 Similar problems

In this section we give two examples of similar problems which can be solved by the same way as the CIO problem, with an $O(n)$ algorithm.

6.1 Colored inputs problem

The colored inputs problem is defined as follows : given a simply connected cellular figure F with colored inputs which cross the boundary, can we match these inputs joining them by non-intersecting (undirected) colored lines of the interior of F, with respect of the colors, in such a way that each interior edge is crossed by at most 1 line, and if it is possible, how can a solution be exhibited?

This problem can be solved by a very similar way of the CIO problem, with the group $G' = < c_0, c_1, ...c_p | c_0^2 = c_1^2 = ... = c_p^2 = 1 >$ (i.e. the group generated by the colors such that each true equality in G' is can be deduced from $c_0^2 = c_1^2 = ... = c_p^2 = 1$). The Cayley graph of G' is a tree when for each element g of G' and each integer i, the edge incoming in g labeled by c_i and the edge outgoing from g labeled by c_i are identified. (This can be done since $gc_i = gc_i^{-1}$ which is a consequence of the equality $c_i^2 = 1$). This yields that a height function, which induces a partial order on the solutions, can be defined. Notice that, for this problem, the edges have not to be directed.

6.2 The saturated inputs outputs problem

The saturated (uncolored) inputs outputs problem is defined as follows : given a simply connected cellular figure F with an (incoming or outgoing) arrow in each edge of the boundary, can we put an arrow on each interior edge e,which crosses e, in such a way that for each cell f of F, there is as much arrows incoming in f as arrows outgoing from f ?

For this problem, the group used is the additive group Z, which makes that group functions and height functions are confused. This problem is a flow problem (in the classical acceptation) and the height function here appears as the "potential function" introduced by R. Hassim ([3], [4]) to produce an algorithm a maximum flow on a planar network.

Remark that in the particular case when the figure is a piece of the square lattice of the plane, a solution of the saturated input-output problem is a solution of the ice model (or six-vertex model) of physicians [6] : an atom of oxygen is in each cell, an atom of hydrogen is in each edge and the arrows indicates the interatomic connections in a cristal of ice.

7 Concluding remarks

These algorithms prove that the main reason which makes that W. P. Thurston method on tilings holds is not the regurality of the lattice used, since in our problems, one can use an irregular figure. On the other hand, we prove that the notion of potential of a flow can be extended to colored flows. This makes a bridge between flows and some tilings. They both appear as labelings of edges of a cellular figure by elements of a group G such that for each cell f, the contour word of f is equal to the unit element of G.

References

1. T. Chaboud : Pavages et Graphes de Cayley, Ph. D. Thesis, Ecole Normale Supérieure de Lyon (1995).
2. J. H. Conway, J. C. Lagarias : Tiling with Polyominoes and Combinatorial Group Theory, Journal of Combinatorial Theory A 53 (1990) 183-208.
3. R. Hassim : maximum flows in (s, t) planar network, Information Processing Letters 13 (1981) p 107.
4. R. Hassim, D. B. Johnson : An $O(n \log^2 n)$ algorithm for maximum flow in undirected planar networks, SIAM journal of Computing, 14, (1985) 612-624.
5. C. Kenyon, R. Kenyon : Tiling a polygon with rectangles, Proceedings of 33rd FOCS (1992) 610-619.
6. J. Kondev, Ch. L. Henley : Kac-Moody symmetries of critical ground states, Nuclear Physics B 464 [FS] (1996) 540-575.
7. E. Rémila : Tiling a figure using a height in a tree, Proceedings of the 7th annual ACM-SIAM Symposium On Discrete Algorithms (SODA), SIAM eds, Philadelphia (1996) 168-174.
8. W. P. Thurston : Conway's tiling group, American Mathematical Monthly (1990) 757-773.

Dedicated to G. J. Chaitin
for his 50th Birthday

Recursively Enumerable Reals and Chaitin Ω Numbers*

Cristian S. Calude, Peter H. Hertling, Bakhadyr Khoussainov, Yongge Wang

Department of Computer Science, University of Auckland, Private Bag 92019, Auckland, New Zealand
Emails: {cristian,hertling,bmk,wang}@cs.auckland.ac.nz

Abstract. A real α is called recursively enumerable if it is the limit of a recursive, increasing, converging sequence of rationals. Following Solovay [23] and Chaitin [10] we say that an r.e. real α *dominates* an r.e. real β if from a good approximation of α from below one can compute a good approximation of β from below. We shall study this relation and characterize it in terms of relations between r.e. sets. Solovay's [23] Ω-like numbers are the maximal r.e. real numbers with respect to this order. They are random r.e. real numbers. The halting probability of a universal self-delimiting Turing machine (Chaitin's Ω number, [9]) is also a random r.e. real. Solovay showed that any Chaitin Ω number is Ω-like. In this paper we show that the converse implication is true as well: any Ω-like real in the unit interval is the halting probability of a universal self-delimiting Turing machine.

1 Introduction and Notation

Algorithmic information theory, as developed by Chaitin [8, 9, 11], Kolmogorov [16], Solomonoff [22], Martin-Löf [19], and others (see Calude [4]), gives a satisfactory description of the quantity of information of individual finite strings and infinite sequences. The same quantity of information may be organised in various ways; in order to quantify the degree of organisation of the information in a string or a sequence, Bennett [2], Juedes, Lathrop, and Lutz [13], and others, have considered the computational depth. Roughly speaking, the computational depth of an object is the amount of time required for an algorithm to derive the

* The first and third authors were partially supported by AURC A18/XXXXX/ 62090/F3414056, 1996. The second author was supported by the DFG Research Grant No. HE 2489/2-1, and the fourth author was supported by an AURC Post-Doctoral Fellowship.

object from its shortest description. Bennett [2] showed that the characteristic sequence χ_K of the halting problem is strongly deep, while no random sequence is strongly deep. Investigating this matter further, Juedes, Lathrop, and Lutz [13] have considered the notion of "usefulness" of infinite sequences. A sequence x is useful if all recursive sequences can be computed with oracle access to x within a fixed recursive time bound. For example χ_K is useful, while no recursive or random sequence is useful.

It is well known that the halting probability of a universal self-delimiting Turing machine, called Chaitin Ω number (see Chaitin [9, 12], Calude [4]), is random, but χ_K is not; Ω and χ_K contain the same quantity of information but codified in vastly different ways. As we noted before, χ_K is useful but Ω is not useful in the sense of Juedes, Lathrop, and Lutz [13]. However, when one is interested in approximating sequences[1] Ω is more "useful" than χ_K; it is one of the aims of this paper to give a mathematical sense to this statement.

A real number is called r.e. if it is the limit of a recursive, increasing, converging sequence of rationals. R.e. reals are extensively used in computable analysis, see Weihrauch [25] and Ko [15]. We will characterize r.e. reals in various ways. In order to compare the "usefulness" of r.e. reals for approximation purposes, Solovay [23] (see also Chaitin [10]) has introduced a *domination* relation $\leq_{dom}$ between real numbers, which we shall define precisely in Section 3. Informally, a real α dominates a real β (written as $\beta \leq_{dom} \alpha$) if from a good approximation of α from below one can compute a good approximation of β from below. The relation $\leq_{dom}$ is transitive and reflexive, hence it naturally defines a partially ordered set $\langle \mathbf{R}_{r.e.} ; \leq_{dom} \rangle$ whose elements are the $=_{dom}$-equivalence classes. We shall see that this partially ordered set is an upper semilattice. It has a minimum element which is the class containing exactly all recursive reals. We study this relation $\leq_{dom}$ further and characterize it in terms of certain reducibilities between r.e. sets. Solovay [23] (see also Chaitin [10]) called an r.e. real Ω-*like* if it dominates every r.e. real. He showed that every Chaitin Ω number is Ω-like. In this paper we prove the converse implication by showing that any Ω-like real in the unit interval is the halting probability of a universal self-delimiting Turing machine. Thus, the semilattice of $=_{dom}$-classes of r.e. reals under $\leq_{dom}$ has a maximum element, which is the equivalence class containing exactly all Chaitin Ω numbers. This shows the strength of all Ω's for approximation purposes: from a good approximation of Ω one can obtain a good approximation of any r.e. real, and no other reals have this property. Consequently, Ω contains more information than any non-Ω-like r.e. real.

In the following section we review some fundamental notions and facts from algorithmic information theory. In Section 3 we give various characterizations of r.e. reals and introduce the domination relation and prove basic facts about it. We compare it with Turing reducibility and characterize it in terms of another reducibility between sets of strings. In Section 4 we show that every Ω–like real is in fact the halting probability of a universal self–delimiting Turing machine.

---

[1] As in constructive mathematics, see Bridges and Richman [3], Weihrauch [25] and Ko [15], and many other areas.

We close this section by introducing some notation. By $\mathbf{N}, \mathbf{Q}$ and $\mathbf{R}$ we denote the set of nonnegative integers, the set of rational numbers, and the set of real numbers, respectively. If X and Y are sets, then $f : X \stackrel{o}{\rightarrow} Y$ denotes a possibly partial function defined on a subset of X. Let $\Sigma = \{0,1\}$ denote the binary alphabet; Σ^* is the set of (finite) binary strings and Σ^ω is the set of infinite binary sequences. The length of a string x is denoted by $|x|$. Let $string_n$ $(n \geq 0)$ be the nth string under the quasi-lexicographical ordering on Σ^*. For a sequence $\mathbf{x} = x_0 x_1 \cdots x_n \cdots \in \Sigma^\omega$ and an integer number $n \geq 0$, $\mathbf{x}(n)$ denotes the initial segment of length $n + 1$ of $\mathbf{x}$ and x_i denotes the ith digit of $\mathbf{x}$, i.e., $\mathbf{x}(n) = x_0 x_1 \cdots x_n$. Lower case letters k, l, m, n will denote nonnegative integers, and x, y, z strings. By $\mathbf{x}, \mathbf{y}, \cdots$ we denote infinite sequences from Σ^ω; finally, we reserve α, β, γ for reals. Capital letters are used to denote subsets of Σ^*. For a set $A \subseteq \Sigma^*$, we denote by χ_A the infinite characteristic sequence of A, that is, $(\chi_A)_n = 1$ if $string_n \in A$ and $(\chi_A)_n = 0$ otherwise.

We assume that the reader is familiar with Turing machine computations, including oracle computations. We use K to denote the halting problem, that is, $string_n \in K$ if and only if the nth Turing machine halts on the input $string_n$. We say that a language A is Turing reducible to a language B, and we write $A \leq_T B$, if there is an oracle Turing machine M such that $M^B(string_n) = (\chi_A)_n$, for all n. For further notation we refer to Calude [4].

2 Complexity and Randomness

In this section, we review some fundamentals of algorithmic information theory that we will use in this paper. We are especially concerned with self-delimiting (Chaitin/program-size) complexity and algorithmic randomness. The advantage of the self-delimiting version of the descriptive complexity is that it gives a precise characterization of algorithmic probability and random sequences.

A *self-delimiting* Turing machine is a deterministic Turing machine such that the program set $\mathrm{PROG}_M = \{x \in \Sigma^* \mid$ on input x, the machine M halts after finitely many steps$\}$ is prefix-free, i.e., a set of strings with the property that no string in it is a proper prefix of another. It follows by Kraft's inequality that, for every self-delimiting Turing machine M,

$$\Omega_M = \sum_{x \in \mathrm{PROG}_M} 2^{-|x|} \leq 1.$$

The number Ω_M is called the *halting probability* of M. In what follows we will omit the adjective "self-delimiting", since this is the only type of Turing machine considered in this paper.

Definition 1. Let M be a Turing machine. The *program-size complexity* of the string $x \in \Sigma^*$ (relative to M) is $H_M(x) = \min\{|y| \mid y \in \Sigma^*, M(y) = x\}$, where $\min \emptyset = \infty$.

It was shown by Chaitin [9] that there is a self-delimiting Turing machine U that is *universal*, in the sense that, for every self-delimiting Turing machine M, there is a constant c_M (depending upon U and M) with the following property: if $x \in \mathrm{PROG}_M$, then there is an $\tilde{x} \in \mathrm{PROG}_U$ such that $U(\tilde{x}) = M(x)$ and $|\tilde{x}| \leq |x| + c_M$. Clearly, every universal machine produces every string. For two universal machines U and V, we see $H_U(x) = H_V(x) + O(1)$. The halting probability Ω_U of a universal machine U is called *Chaitin Ω number*; for more about Ω_U see Bennett [1], Calude, Salomaa [7], Calude, Meyerstein [6]. In the rest of the paper, unless stated otherwise, we will use a fixed universal machine U and will omit the subscript U in $H_U(x)$ and Ω_U. We will also abuse our notation by identifying the real number Ω with the infinite binary sequence which corresponds to Ω (i.e., the infinite[2] binary expansion of Ω without "0.").

We conclude this section with a brief discussion of (algorithmically) random infinite binary sequences.[3] Random sequences were originally defined by Martin-Löf [19] using constructive measure theory. Complexity-theoretic characterizations of random sequences have been obtained by Chaitin [9] (see also Levin [18], Schnorr [20]). A Martin–Löf test is an r.e. set $A \subseteq \mathbf{N} \times \Sigma^*$ satisfying the following measure-theoretical condition: $\mu(A^{(i)}) \leq 2^{-i}$, for all $i \in \mathbf{N}$ where we define $A^{(i)} = \{\alpha \in \Sigma^\omega \mid$ there is a finite prefix x of α with $(i, x) \in A\}$.[4] Here μ is the usual product measure on Σ^ω given by $\mu(\{w\}\Sigma^\omega) = 2^{-|w|}$, for $w \in \Sigma^*$. An infinite sequence $\mathbf{x}$ is *random* if, for every Martin-Löf test A, $\mathbf{x} \notin \bigcap_{i \geq 0} A^{(i)}$. This can also be expressed via the program-size complexity: $\mathbf{x}$ is random if and only if there exists a constant $c > 0$ such that $H(\mathbf{x}(n)) > n - c$, for every integer $n > 0$, and if and only if $\lim_{n \to \infty} (H(\mathbf{x}(n)) - n) = \infty$.

Theorem 2. (Chaitin [9]) *For every universal machine U, the halting probability Ω_U is random.*

3   R.E. Reals and Domination

It is the aim of this section to compare the information contents of r.e. reals. A real α is called *r.e.* if there is a recursive, increasing sequence of rationals which converges to α.[5] We start with several characterizations of r.e. reals.

For a prefix-free set $A \subseteq \Sigma^*$ we define a real number by $2^{-A} = \sum_{x \in A} 2^{-|x|}$, which, due to Kraft's inequality, lies in the interval $[0, 1]$. For a set $X \subseteq \mathbf{N}$ we define the number $2^{-X-1} = \sum_{n \in X} 2^{-n-1}$. This number also lies in the interval $[0, 1]$. If we disregard all finite sets X, which lead to rational numbers 2^{-X-1}, we get a bijection $X \mapsto 2^{-X-1}$ between the class of infinite subsets of $\mathbf{N}$ and the real numbers in the interval $(0, 1]$. If $0.\mathbf{y}$ is the binary expansion of a real α with infinitely many ones, then $\alpha = 2^{-X_\alpha - 1}$ where $X_\alpha = \{i \mid y_i = 1\}$. Clearly, if X_α is r.e., then the number $2^{-X_\alpha - 1}$ is r.e., but the converse is not true as the

---

[2] This expansion is unique since by Theorem 2, Ω is random and, hence, irrational.

[3] The interested reader is referred to Calude [4] and Wang [24] for more details.

[4] See Calude [4] for a detailed motivation.

[5] Note that the property of being r.e. depends only on the fractional part of the real number.

Chaitin Ω numbers show. We characterize r.e. reals α in terms of prefix-free r.e. sets of strings[6] and in terms of the sets X_α.

Theorem 3. *For a real $\alpha \in (0,1]$ the following conditions are equivalent:*

1. *The number α is r.e.*
2. *There is a recursive, non–decreasing sequence of rationals $(a_n)_{n \geq 0}$ which converges to α.*
3. *The set $\{p \in \mathbf{Q} \mid p < \alpha\}$ of rationals less than α is r.e.*
4. *There is an infinite prefix-free r.e. set $A \subseteq \Sigma^*$ with $\alpha = 2^{-A}$.*
5. *There is a total recursive function $f : \mathbf{N}^2 \to \{0,1\}$ such that*
   (a) *If for some k, n we have $f(k,n) = 1$ and $f(k, n+1) = 0$ then there is an $l < k$ with $f(l,n) = 0$ and $f(l, n+1) = 1$.*
   (b) *We have: $k \in X_\alpha \iff \lim_{n \to \infty} f(k,n) = 1$.*

In order to compare the information contents of r.e. reals, Solovay [23] has introduced the following definition.

Definition 4. (Solovay [23] and Chaitin [10]) The real α is said to *dominate* the real β if there are a partial recursive function $f : \mathbf{Q} \xrightarrow{o} \mathbf{Q}$ and a constant $c > 0$ with the property that if p is a rational number less than α, then $f(p)$ is (defined and) less than β, and it satisfies the inequality $c(\alpha - p) \geq \beta - f(p)$. In this case we write $\alpha \geq_{dom} \beta$ (or $\beta \leq_{dom} \alpha$).

Roughly speaking, a real α dominates a real β if there is an effective way to get a good approximation to β from below from any good approximation to α from below. For r.e. reals this can also be expressed as follows.

Lemma 5. *An r.e. real α dominates an r.e. real β if and only if there are recursive, non-decreasing sequences (a_i) and (b_i) of rationals and a constant c with $\lim_n a_n = \alpha$, $\lim_n b_n = \beta$, and $c(\alpha - a_n) \geq \beta - b_n$, for all n.*

Theorem 6. (Solovay [23]) *Let $\mathbf{x}, \mathbf{y} \in \Sigma^\omega$ be two infinite binary sequences such that both $0.\mathbf{x}$ and $0.\mathbf{y}$ are r.e. reals and $0.\mathbf{x} \geq_{dom} 0.\mathbf{y}$. Then, $H(\mathbf{y}(n)) \leq H(\mathbf{x}(n)) + O(1)$.*

Next, we formulate a few results which will be useful in discussing the lattice structure of r.e. reals.

Lemma 7. *Let α, β and γ be r.e. reals. Then the following conditions hold:*

1. *The relation $\geq_{dom}$ is reflexive and transitive.*
2. *For every α, β one has $\alpha + \beta \geq_{dom} \alpha$.*
3. *If $\gamma \geq_{dom} \alpha$ and $\gamma \geq_{dom} \beta$, then $\gamma \geq_{dom} \alpha + \beta$.*

The second and third statement are true also if addition is replaced by multiplication and only positive r.e. reals are considered. By Theorem 6 we obtain

---

[6] Note that the prefix-free r.e. sets $A \subseteq \Sigma^*$ are exactly the domains of self-delimiting Turing machines.

Corollary 8. *The sum of a random r.e. real and an r.e. real is a random r.e. real.*

Also, the product of a positive random r.e. real with a positive r.e. real is a random r.e. real. Hence, addition (and multiplication on positive numbers) preserves the property of being random and r.e. in contrast with the fact that addition and multiplication do not preserve randomness. For example, if α is a random number, then $1 - \alpha$ is random as well,[7] but $\alpha + (1 - \alpha) = 1$ is not random.

In the following we discuss the lattice structure of r.e. reals under the domination relation. For two reals α and β, $\alpha =_{dom} \beta$ denotes the conjunction $\alpha \geq_{dom} \beta$ and $\beta \geq_{dom} \alpha$. For a real α, let $[\alpha] = \{\beta \in \mathbf{R} \mid \alpha =_{dom} \beta\}$; $\mathbf{R}_{r.e.} = \{[\alpha] \mid \alpha \text{ is an r.e. real}\}$. By Lemma 7 the least upper bound of any two classes containing r.e. reals α and β, respectively, is the class containing the r.e. real $\alpha + \beta$. We conclude

Theorem 9. *The structure $\langle \mathbf{R}_{r.e.} ; \leq_{dom} \rangle$ is an upper semi-lattice.*

Now we compare the domination relation $\leq_{dom}$ on r.e. reals with the Turing reducibility. For every infinite sequence $\mathbf{x} = x_0 x_1 x_2 \ldots \in \Sigma^\omega$ such that $0.\mathbf{x}$ is an r.e. real, let $A_{\mathbf{x}} = \{v \in \Sigma^* \mid 0.v \leq 0.\mathbf{x}\}$ and $A_{\mathbf{x}}^{\#} = \{string_n \mid x_n = 1\}$. Then, obviously, $A_{\mathbf{x}}$ is an r.e set which is Turing equivalent to $A_{\mathbf{x}}^{\#}$.[8] In the following, we establish the relationship between domination and Turing reducibility.

Lemma 10. *Let $\mathbf{x}, \mathbf{y} \in \Sigma^\omega$ be two infinite binary sequences such that both $0.\mathbf{x}$ and $0.\mathbf{y}$ are r.e. reals and $0.\mathbf{x} \geq_{dom} 0.\mathbf{y}$. Then $A_{\mathbf{y}} \leq_T A_{\mathbf{x}}$.*

Does the converse of Lemma 10 hold true? The negative answer will be given in Corollary 27. Let $\langle RE ; \leq_T \rangle$ denote the upper semi-lattice structure of the class of r.e. sets under the Turing reducibility. A *strong homomorphism* from a partially ordered set $(X, \leq)$ to another partially ordered set $(Y, \leq)$ is a mapping $h : X \to Y$ such that

1. For all $x, \tilde{x} \in X$, if $x \leq \tilde{x}$, then $h(x) \leq h(\tilde{x})$.
2. For all $y, \tilde{y} \in Y$, if $y \leq \tilde{y}$, then there exist $x, \tilde{x}$ in X such that $x \leq \tilde{x}$ and $h(x) = y$, $h(\tilde{x}) = \tilde{y}$.

Lemma 10 shows that the mapping $0.\mathbf{x} \mapsto A_{\mathbf{x}}$ is a homomorphism. One can show that it is even a strong homomorphism.

Theorem 11. *There exists a strong homomorphism from $\langle \mathbf{R}_{r.e.} ; \leq_{dom} \rangle$ onto $\langle RE ; \leq_T \rangle$.*

We continue with the characterization of the domination relation between r.e. real numbers in terms of prefix-free r.e. sets of strings. We consider only infinite prefix-free r.e. sets. By $R.E.$ we denote the class of all infinite prefix-free r.e. subsets of Σ^*. First, we consider a relation between r.e. sets which is very close to the domination relation, but will turn out to be not equivalent.

---

[7] The number $1 - \alpha$ does not need to be r.e. if α is r.e.

[8] Note that $A_{\mathbf{x}}^{\#}$ is not necessarily an r.e. set.

Definition 12. Let $A, B \in R.E.$ The set A *strongly simulates* B (shortly, $B \leq_{ss}$ A) if there exist a partial recursive function $f : \Sigma^* \xrightarrow{o} \Sigma^*$ and a constant $c > 0$ such that $A = \mathrm{dom}(f)$, $B = f(A)$, and $|x| \leq |f(x)| + c$, for all $x \in A$. If $A \leq_{ss} B$ and $B \leq_{ss} A$, then we say that A and B are $\sim_{ss}$-*equivalent*.

One observes immediately that the relation $\leq_{ss}$ is reflexive and transitive. Hence, it defines a partially ordered set $\langle R.E._{ss}, \leq_{ss} \rangle$ where $R.E._{ss}$ is the set of $\sim_{ss}$-equivalence classes of $R.E.$ Our next goal is to see how the strong simulation relation $\leq_{ss}$ and $\leq_{dom}$ are related. The following lemma is straightforward.

Lemma 13. *If A, B are infinite prefix-free r.e. sets and $B \leq_{ss} A$, then 2^{-A} dominates 2^{-B}.*

The next result shows that the converse implication in Lemma 13 is in some sense true as well. It will also be important in the following section. Therefore we give its proof.

Theorem 14. *Let α be an r.e. real in the interval $(0, 1]$, and B be an infinite prefix-free r.e. set. If α dominates 2^{-B}, then there is an infinite prefix-free r.e. set A with $\alpha = 2^{-A}$ and $B \leq_{ss} A$.*

Proof. Let (y_i) be a one-to-one recursive enumeration of B and (a_n) be an increasing recursive sequence of positive rationals converging to α. In view of the domination property of α, there are an increasing, total recursive function $f : \mathbf{N} \to \mathbf{N}$ and a constant $c \in \mathbf{N}$ such that, for each $n \in \mathbf{N}$,

$$2^c \cdot (\alpha - a_n) \geq 2^{-B} - \sum_{i=0}^{f(n)} 2^{-|y_i|}. \tag{1}$$

Without loss of generality, we may assume that $a_0 \geq \sum_{i=0}^{f(0)} 2^{-|y_i|-c}$ (otherwise we may take a large enough c).

We construct a recursive sequence $(n_i)_{i \geq 0}$ of numbers and a recursive double sequence $(m_{i,j})_{i,j \geq 0}$ of elements in $\mathbf{N} \cup \{\infty\}$. These numbers will be the lengths of the strings in A. The numbers n_i serve in order to guarantee that $B \leq_{ss} A$. The numbers $m_{i,j}$ are used "to fill" the set A up in order to get exactly $\alpha = 2^{-A}$.

Construction of (n_i): We define $n_i = |y_i| + c$, for all i.

Begin of construction of $(m_{i,j})$.

Stage 0. Let $m_{i,j} = \infty$, for all $i \leq f(0)$ and $j \in \mathbf{N}$.

Stage s ($s \geq 1$). If

$$a_s \leq \sum_{i=0}^{f(s)} 2^{-n_i} + \sum_{i=0}^{f(s-1)} \sum_{j=0}^{\infty} 2^{-m_{i,j}},$$

then let $m_{i,j} = \infty$, for all i with $f(s-1) < i \le f(s)$ and $j \in \mathbf{N}$. Otherwise, let $m_{i,j} = \infty$, for all i with $f(s-1) < i < f(s)$ and $j \in \mathbf{N}$, and let $(m_{f(s),j})_{j \in \mathbf{N}}$ be recursively defined in such a way that

$$\sum_{j=0}^{\infty} 2^{-m_{f(s),j}} = a_s - \left(\sum_{i=0}^{f(s)} 2^{-n_i} + \sum_{i=0}^{f(s-1)} \sum_{j=0}^{\infty} 2^{-m_{i,j}} \right).$$

End of construction of $(m_{i,j})$.

One proves the following equation

$$\alpha = \sum_{i=0}^{\infty} \left(2^{-n_i} + \sum_{j=0}^{\infty} 2^{-m_{i,j}} \right). \tag{2}$$

by distinguishing the following two cases:

Case 1. There are infinitely many s with $a_s = \sum_{i=0}^{f(s)} \left(2^{-n_i} + \sum_{j=0}^{\infty} 2^{-m_{i,j}} \right)$. For this case, it is straightforward that the equation (2) holds.

Case 2. The inequality $a_s < \sum_{i=0}^{f(s)} \left(2^{-n_i} + \sum_{j=0}^{\infty} 2^{-m_{i,j}} \right)$ holds true for almost all $s \in \mathbf{N}$. Then the estimate "$\le$" in (2) is obvious. For "$\ge$" observe that in this case there is a largest $s_0 \ge 0$ with $a_{s_0} = \sum_{i=0}^{f(s_0)} \left(2^{-n_i} + \sum_{j=0}^{\infty} 2^{-m_{i,j}} \right)$. By (1) we have $\alpha - a_{s_0} \ge \sum_{i=f(s_0)+1}^{\infty} 2^{-n_i}$. Hence, by the construction, the estimation "$\ge$" in (2) follows.

Let $h : \mathbf{N} \to \{(i,j) \in \mathbf{N}^2 \mid m_{i,j} \ne \infty\}$ be a recursive bijection. We define a recursive sequence $(\bar{n}_i)$ by $\bar{n}_{2i} = n_i$ and $\bar{n}_{2i+1} = m_{h(i)}$. By the Kraft-Chaitin theorem (see Chaitin [9], Calude, Grozea [5]) and (2), combined with $0 < \alpha < 1$, we can construct a one-to-one recursive sequence (x_i) of strings with $|x_i| = \bar{n}_i$ such that the set $\{x_i \mid i \in \mathbf{N}\}$ is prefix-free. We set $A = \{x_i \mid i \in \mathbf{N}\}$ and, using (2), obtain $2^{-A} = \alpha$. Finally we define a recursive function $g : A \to B$ by $g(x_{2i}) = y_i$ and such that $|g(x_{2i+1})| \ge |x_{2i+1}|$, for all i (this is possible because B is infinite and r.e.). Obviously, $g(A) = B$, and $|x| \le |g(x)| + c$, for all $x \in A$. This shows $B \le_{ss} A$. $\qquad\square$

Lemma 13 and Theorem 14 imply

Theorem 15. *The mapping h defined by $h(A) = 2^{-A}$ is a strong homomorphism from $\langle R.E._{ss}, \le_{ss} \rangle$ onto $\langle \mathbf{R}_{r.e.}, \le_{dom} \rangle$.*

But the next result shows that h cannot be one-to-one.

Theorem 16. *There exist infinite prefix-free r.e. sets A and B with $2^{-A} = 2^{-B} = 1$ but $A \not\le_{ss} B$ and $B \not\le_{ss} A$.*

However, by relaxing the strong simulation relation one can characterize the domination relation by a simulation relation between prefix-free r.e. sets. A sequence $D_0, D_1, D_2, \ldots$ of finite subsets of Σ^* is called a *strong array* (Soare [21]) if there is a total recursive function g such that with respect to a standard bijection ν from $\mathbf{N}$ onto the set of all finite subsets of Σ^* we have $D_i = \nu(g(i))$ for all i. An *effective, finite partition* of an infinite r.e. set A is a strong array $D_0, D_1, D_2, \ldots$ of finite, pairwise disjoint subsets of A with $\bigcup_i D_i = A$.

Definition 17. Let A and B be infinite, prefix-free, r.e. sets. We say that A *simulates* B if there are two effective, finite partitions (D_i) of A and (E_i) of B, respectively, and a constant $c > 0$ such that $c \cdot (2^{-D_i}) \geq 2^{-E_i}$, for all i.

Theorem 18. *Let A, B be infinite prefix-free r.e. sets. Then A simulates B if and only if 2^{-A} dominates 2^{-B}.*

4 Random R.E. Reals and Ω-like Reals

In this section, we study random r.e. reals and Ω-like reals, which were introduced by Solovay [23]. Chaitin [10] has given a slightly different definition. We start with Chaitin's definition.

Definition 19. (Chaitin [10]) An r.e. real α is called Ω-*like* if it dominates all r.e. reals.

Solovay's original manuscript [23] contains the following definition.

Definition 20. (Solovay [23]) A recursive, increasing, and converging sequence (a_i) of rationals is called *universal* if for every recursive, increasing and converging sequence (b_i) of rationals there exists a number $c > 0$ such that $c \cdot (\alpha - a_n) \geq \beta - b_n$ for all n, where $\alpha = \lim_n a_n$ and $\beta = \lim_n b_n$.

Solovay called a real α Ω-like if it is the limit of a universal recursive, increasing sequence of rationals. We shall see that both definitions are equivalent. One implication follows immediately from Lemma 5.

Lemma 21. *If a real α is the limit of a universal recursive, increasing sequence of rationals, then it is Ω-like.*

By modifying slightly a proof of Solovay [23] we obtain the following result.

Theorem 22. *Let U be a universal machine. Every recursive, increasing sequence of rationals converging to Ω_U is universal.*

Thus, every Chaitin Ω number is Ω-like in Solovay's sense. The converse of Theorem 22 holds true even for Ω-like numbers in Chaitin's sense.

Theorem 23. *Let $0 < \alpha < 1$ be an Ω-like real. Then there exists a universal machine U such that $\Omega_U = \alpha$.*

Proof. Let V be a universal machine. Since α is Ω-like it dominates 2^{-PROG_V}. By Theorem 14 there exist a prefix-free r.e. set A with $2^{-A} = \alpha$, a recursive function $f : A \to \text{PROG}_V$ with $A = \text{dom}(f)$ and $f(A) = \text{PROG}_V$, and a constant $c > 0$ with $|x| \leq |f(x)| + c$, for all $x \in A$. We define a machine U by $U(x) = V(f(x))$. The universality of V implies that also U is universal. We have $\alpha = 2^{-A} = 2^{-\text{PROG}_U} = \Omega_U$. $\qquad\qquad\square$

Theorem 24. *Let $0 < \alpha < 1$ be an r.e. real. The following conditions are equivalent:*

1. *For some universal Turing machine U, $\alpha = \Omega_U$.*
2. *The real α is Ω-like.*
3. *There exists a universal recursive, increasing sequence of rationals converging to α.*
4. *Every recursive, increasing sequence of rationals with limit α is universal.*

Proof. This follows from Lemma 21, Theorem 22, and Theorem 23.   □

The following result was proved by Solovay [23] for Ω-like numbers. If follows immediately from Theorem 6 and Theorem 24.

Corollary 25. *Let U and V be two universal machines. Then $H(\Omega_U(n)) = H(\Omega_V(n)) + O(1)$.*

In analogy with Corollary 8 we obtain from Lemma 7 and Theorem 24 the following corollary.

Corollary 26. *The fractional part of the sum of an Ω number and an r.e. real is an Ω number.*

Also, the fractional part of the product of an Ω number with a positive r.e. real is an Ω number. Since Ω is random (Theorem 2) every Ω-like number is random (Solovay [23]). Now we can answer the question raised after Lemma 10. The sets A_Ω and A_{χ_K} are defined as before Lemma 10.

Corollary 27. *The following statements hold:*

1. $0.\chi_K \not\geq_{dom} \Omega$,
2. $A_\Omega =_T A_{\chi_K} =_T K$.

Proof. It is well-known that χ_K is not random, and hence, by Theorem 2, not an Ω number. Theorem 24 and the transitivity of $\geq_{dom}$ show the first claim. $A_\Omega \leq_T K =_T A_{\chi_K}$ is clear and $A_{\chi_K} \leq_T A_\Omega$ follows from Lemma 10.   □

Does there exist a random r.e. real which is not Ω-like? We conjecture that this is true. Kucera [17] (see also Kautz [14]) has observed that $0'$ is the only r.e. degree which contains a random real.[9] But Corollary 27 shows that it splits into different $=_{dom}$-classes.

References

1. C. H. Bennett, M. Gardner. The random number omega bids fair to hold the mysteries of the universe, *Scientific American* 241(1979), 20–34.

---

[9] Here we identify a real $0.x$ with the set $A_x^\#$.

2. C. H. Bennett. Logical depth and physical complexity, in R. Herken (ed.). *The Universal Turing Machine. A Half-Century Survey*, Oxford University Press, Oxford, 1988, 227–258.

3. D. S. Bridges, F. Richman. *Varieties of Constructive Mathematics*, Cambridge University Press, Cambridge, 1987.

4. C. Calude. *Information and Randomness. An Algorithmic Perspective*, Springer-Verlag, Berlin, 1994.

5. C. Calude and C. Grozea. Kraft-Chaitin inequality revisited, *J. UCS* 5(1996), 306–310.

6. C. Calude, F. W. Meyerstein. Is the universe lawful?, *Chaos, Solitons & Fractals*, (to appear).

7. C. Calude, A. Salomaa. Algorithmically coding the universe, in G. Rozenberg, A. Salomaa (eds.), *Developments in Language Theory*, World Scientific, Singapore, 1994, 472–492.

8. G. J. Chaitin. On the length of programs for computing finite binary sequences, *J. Assoc. Comput. Mach.* 13(1966), 547-569. (Reprinted in: [11], 219–244)

9. G. J. Chaitin. A theory of program size formally identical to information theory, *J. Assoc. Comput. Mach.* 22(1975), 329-340. (Reprinted in: [11], 113–128)

10. G. J. Chaitin. Algorithmic information theory, *IBM J. Res. Develop.* 21(1977), 350-359, 496. (Reprinted in: [11], 44–58)

11. G. J. Chaitin. *Information, Randomness and Incompleteness, Papers on Algorithmic Information Theory*, World Scientific, Singapore, 1987. (2nd ed., 1990)

12. G. J. Chaitin. *The Limits of Mathematics*, Springer-Verlag, Singapore, 1997.

13. D. Juedes, J. Lathrop, and J. Lutz. Computational depth and reducibility, *Theoret. Comput. Sci.* 132 (1994), 37–70.

14. S. Kautz. *Degrees of Random Sets*, PhD Thesis, Cornell University, Ithaca, 1991.

15. Ker-I. Ko. *Complexity Theory of Real Functions*, Birkhäuser, Boston, 1991.

16. A. N. Kolmogorov. Three approaches for defining the concept of "information quantity", *Problems Inform. Transmission* 1(1965), 3–11.

17. A. Kucera. Measure, Π_1^0–classes and complete extensions of PA, H.-D. Ebbinghaus, G. H. Müller, G. E. Sacks (eds.), *Recursion Theory Week* (Oberwolfach, 1984), Lecture Notes in Math. 1141, Springer–Verlag, Berlin, 1985, 245–259.

18. L. A. Levin. On the notion of a random sequence, *Soviet Math. Dokl.* 14(1973), 1413–1416.

19. P. Martin-Löf. The definition of random sequences, *Inform. and Control* 9(1966), 602–619.

20. C. P. Schnorr. Process complexity and effective random tests, *J. Comput. System Sci.* 7(1973), 376–388.

21. R. I. Soare. *Recursively Enumerable Sets and Degrees*, Springer-Verlag, Berlin, 1987.

22. R. J. Solomonoff. A formal theory of inductive inference, Part 1 and Part 2, *Inform. and Control* 7(1964), 1–22 and 224–254.

23. R. Solovay. *Draft of a paper (or series of papers) on Chaitin's work ... done for the most part during the period of Sept. - Dec. 1974*, unpublished manuscript, IBM T. J. Watson Research Center, Yorktown Heights, New York, May 1975, 215 pp.

24. Y. Wang. *Randomness and Complexity*, PhD Thesis, Universität Heidelberg, Germany, 1996.

25. K. Weihrauch. *Computability*, Springer-Verlag, Berlin, 1987.

Uniformly Defining Complexity Classes of Functions

Sven Kosub, Heinz Schmitz*, and Heribert Vollmer

Theoretische Informatik, Universität Würzburg,
Am Exerzierplatz 3, 97072 Würzburg, Germany

Abstract. We introduce a general framework for the definition of function classes. Our model, which is based on polynomial time nondeterministic Turing transducers, allows uniform characterizations of FP, $\mathrm{FP}^{\mathrm{NP}}$, counting classes ($\#{\cdot}\mathrm{P}$, $\#{\cdot}\mathrm{NP}$, $\#{\cdot}\mathrm{coNP}$, GapP, $\mathrm{GapP}^{\mathrm{NP}}$), optimization classes ($\max{\cdot}\mathrm{P}$, $\min{\cdot}\mathrm{P}$, $\max{\cdot}\mathrm{NP}$, $\min{\cdot}\mathrm{NP}$), promise classes (NPSV, $\#_{\mathrm{few}}{\cdot}\mathrm{P}$, $\mathrm{c}\#{\cdot}\mathrm{P}$), multivalued classes (FewFP, NPMV) and many more. Each such class is defined in our model by a certain family of functions. We study a reducibility notion between such families, which leads to a necessary and sufficient criterion for relativizable inclusion between function classes. As it turns out, this criterion is easily applicable and we get as a consequence e.g. that there are oracles A, B, such that $\min{\cdot}\mathrm{P}^A \not\subseteq \#{\cdot}\mathrm{NP}^A$, and $\max{\cdot}\mathrm{NP}^B \not\subseteq \mathrm{c}\#{\cdot}\mathrm{coNP}^B$ (note that no structural consequences are known to follow from the corresponding positive inclusions).

1 Introduction

Starting with the work of Valiant [24], Krentel [18, 19], and Toda [23], complexity classes of functions have gained considerable interest. Krentel showed that by considering functional problems in their original form (e.g. "compute the length of the shortest traveling salesperson tour") instead of artificially coding them into yes-no-problems ("is the shortest traveling salesperson tour shorter than k?") allows one to make finer distinctions between different NP-complete problems than known previously. Toda showed that the class $\#\mathrm{P}$ defined by Valiant has an unexpected power: every set from the polynomial hierarchy is Turing reducible to some $\#\mathrm{P}$ function. Thus to consider function classes turns out to be a worthwhile study.

But how to compare function classes? And how do they relate to set classes? For this end—also going back to Toda's seminal paper [23]—operators have been used widely. Vollmer and Wagner [27, 28] introduced a pair of operators establishing a one-one correspondence between Krentel's maximization classes [19] and the classes that form the polynomial hierarchy. These results have been pushed a bit further in [26, 21, 9]. The general framework for what is now known as the *operator method to separate function classes* can be described as follows: Given two function classes $\mathcal{F}_1$ and $\mathcal{F}_2$ and an operator $\mathcal{O}$ transforming them into

---

* Supported by the Deutsche Forschungsgemeinschaft (DFG), grant Wa 847/1-2.

set classes $\mathcal{K}_1$ and $\mathcal{K}_2$, the following is clear: If $\mathcal{O}$ is monotonic, then $\mathcal{F}_1 \subseteq \mathcal{F}_2$ implies $\mathcal{K}_1 \subseteq \mathcal{K}_2$. Thus if we knew $\mathcal{K}_1 \neq \mathcal{K}_2$ then we would have proven $\mathcal{F}_1 \neq \mathcal{F}_2$. Unfortunately results of the form $\mathcal{K}_1 \neq \mathcal{K}_2$ are very rare in computational complexity theory. But of course under the above circumstances, if we have evidence against $\mathcal{K}_1 \subseteq \mathcal{K}_2$, then this also serves as evidence against $\mathcal{F}_1 \subseteq \mathcal{F}_2$. An oracle separating $\mathcal{K}_1$ from $\mathcal{K}_2$ will also separate $\mathcal{F}_1$ from $\mathcal{F}_2$. Along these lines a lot of structural results were given in the above mentioned papers.

However this approach sometimes fails. A prominent example is the case of the classes $\#\cdot\mathrm{NP}$ and $\min\cdot\mathrm{P}$, left as an open problem in [9]. It is known that $\#\cdot\mathrm{NP} \not\subseteq \min\cdot\mathrm{P}$ unless $\mathrm{P} = \mathrm{PP}$ (this also yields an oracle B such that $\#\cdot\mathrm{NP}^B \not\subseteq \min\cdot\mathrm{P}^B$), but the inclusion in the other direction is open. All known operators transform $\min\cdot\mathrm{P}$ and $\#\cdot\mathrm{NP}$ into set classes $\mathcal{K}_1$ and $\mathcal{K}_2$ resp., such that $\mathcal{K}_1 \subseteq \mathcal{K}_2$. Thus here it turns out that it can be very difficult to find a suitable operator $\mathcal{O}$. One might argue that the failure to find such an operator shows that the inclusion $\min\cdot\mathrm{P} \subseteq \#\cdot\mathrm{NP}$ is likely to hold, but also this could not be proved. It seems that a simple transformation to the set side does not help in this case.

In the present paper we try to find a new way to compare function classes without reducing them to set classes. We introduce a uniform framework for the definition of function classes. We do this by using one single computation model (nondeterministic polynomial time transducers) and considering different evaluation modes for this model. Thus, a function class is given by specifying how the computation of such a machine is evaluated to compute the function value (more formally below: by giving a so called *generator*). Almost all up to now considered classes of functions out of the polynomial time context (in particular all of the above-mentioned classes) are definable in our way. We think that using one single computation model for the definition of different classes may make reasoning about them and comparing them to one another easier. It may make clear similarities and differences which otherwise are hidden in the peculiarities of particular models. We give a necessary and sufficient criterion for the separability of function classes by oracles. Classes in our model are given by generators. We reduce the question of comparing function classes to the question of comparing their generators by a certain reducibility. As we will see, our criterion is easily applicable, and we use it to attack (among others) the above mentioned open question: We construct an oracle A such that $\min\cdot\mathrm{P}^A \not\subseteq \#\cdot\mathrm{NP}^A$.

It should be remarked that our framework is very much inspired by the leaf language approach to define complexity classes, introduced independently in [5] and in [25] and later applied successfully in a number of different contexts, see [11, 13, 10, 12] and many more. These papers are only concerned with classes of sets, and our paper can be regarded as a generalization of their approach to the case of function classes. The above mentioned criterion for oracle separations is a generalization of corresponding theorems in [5, 25]. In fact, our Theorem 12 gives their result as a corollary. In a similar way as done for set classes in [5, 25], Theorem 12 below proves once and for all the diagonalization part of oracle separations of function classes. Thus when applying our theorem to prove a separation one no longer has to worry about the cumbersome details of a stage construction but can concentrate solely on the combinatorial problems.

The status of oracle separations of course is not clear in the general case, since we know of separations also in cases where the unrelativized classes collapse. However in the case of a uniform computation model as we use it here we believe that oracle separations still may contribute to our knowledge about the complexity of the problems under consideration. On the other hand of course, a uniform framework for definability is interesting in its own and may lead to other results. We come back to this point in Sect. 7.

A full version of this paper, including proofs of all claims, is available as Technical Report No. 183 of the Institut für Informatik, Universität Würzburg.

2 Preliminaries

In this paper we use the alphabet $\Sigma = \{0, 1\}$. As usual we identify Σ^* with the set of natural numbers (e.g. by binary encoding). For $\ell, r \in \mathbb{N}$, we denote the corresponding integer interval by $[\ell, r] =_{\text{def}} \{\ell, \ell+1, \ldots, r\}$.

Let $\Omega =_{\text{def}} \{ \vec{x} = (x_1, \ldots, x_k) \mid k \geq 1 \wedge x_j \in \Sigma^* \text{ for } 1 \leq j \leq k \}$ be the set of all finite vectors (sequences) of elements of Σ^*. We have to consider certain subsets of Ω given as follows: For a function $s \colon \mathbb{N} \to \mathbb{N}$, let $\Omega_s =_{\text{def}} \{ \vec{x} = (x_1, \ldots, x_k) \mid \vec{x} \in \Omega \wedge \max_{1 \leq j \leq k} x_j \leq s(k) \}$. For $\vec{x} = (x_1, \ldots, x_k)$, the dimension of $\vec{x}$ is defined as dim $\vec{x} =_{\text{def}} k$.

We assume the reader to be familiar with basic complexity classes and reducibilities. Refer to the standard literature, e.g. [4, 20, 1]. Let FP be the class of total functions computed in polynomial time by deterministic Turing transducers. With FP^α we denote the class of functions computed in polynomial time by oracle transducers with oracle function α. An oracle machine is equipped with an oracle query tape, on which it writes the intended query x, and an oracle answer tape, on which it receives $\alpha(x)$ in one step.

To investigate sub-linear time reductions we use transducers with index access to their input. These machines have besides their work-tapes some special tapes: an index tape and an input tape. With i written on the index tape, the i-th argument of the machine's input vector appears in one step on the input tape. Finding a blank symbol on the input tape tells the machine that there was no i-th argument. On a third special tape the machine gets as input the length of its input vector, i.e. the sum of the lengths of the arguments. With FPL we denote the class of functions computable by such index transducers in polylogarithmic time (in the length of the input vector).

For functions ϕ and ψ, let $\phi \simeq \psi$ denote that for all inputs x, $\phi(x)$ is defined if and only if $\psi(x)$ is defined, and if $\phi(x)$ is defined then $\phi(x) = \psi(x)$. For a given input x we also use the notation $\phi(x) \simeq \psi(x)$ with the obvious meaning. For functions f and g, let $f \geq_{a.e.} g$ denote that f is greater than or equal to f almost everywhere, i.e. there is some x_0 such that for all $x \geq x_0$ we have $f(x) \geq g(x)$. We define the operator $[\cdot, \cdot]$ for pairs of total functions as follows: For all x, let $[f, g](x) =_{\text{def}} (f(x, 0), f(x, 1), \ldots, f(x, g(x)))$.

3 How to Define Function Classes

In order to emphasize the essential characteristics of different function classes we introduce the general notion of a *generator*.

Definition 1. A pair (F, Φ) is said to be a *generator* if and only if F is a sequence of functions $\{f_k\}_{k=1}^{\infty}$ where $f_k : (\Sigma^*)^k \longrightarrow B$, and $\Phi \subseteq \Omega$.

Note that we will fix B later to capture the range of F when describing concrete function classes and that we use sequences of functions to deal with different arities. For notational reasons we set $F(x_1, \ldots, x_k) =_{\text{def}} f_k(x_1, \ldots, x_k)$ for any sequence of functions F. Moreover, for a generator (F, Φ) we will call F the *generator function* and Φ the *generator set*. Next, we define the function class induced by a given generator in the context of nondeterministic polynomial time computations.

Definition 2. Let (F, Φ) be a generator. A function $f : \Sigma^* \longrightarrow B$ belongs to the function class (F, Φ)-FP if and only if there are functions $g_1, g_2 \in$ FP such that for every $x \in \Sigma^*$ the following two conditions hold:

1. $[g_1, g_2](x) \in \Phi$,
2. $f(x) \simeq F([g_1, g_2](x))$.

Note that the first condition allows us to capture promises about the computation model. A similar approach for the definability of total, non-promise function classes was suggested by Borchert in [3].

Remark 3. *We will study relativized versions (F, Φ)-FP$^{\alpha}$ for some oracle function α defined similar in a natural way by choosing g_1 and g_2 from FP$^{\alpha}$.*

So far, generators are a very general notion since only little is required for F and Φ. As it turns out, the generators we use all have a special property which we call *constant invariance* (for short *c-invariance*).

Definition 4. A generator (F, Φ) is said to be *c-invariant* if and only if there is a constant $\gamma \in \Sigma^*$ such that for every $k \geq 1$ and for every $\vec{x} = (x_1, \ldots, x_k) \in \Omega$ it holds that

1. $F(x_1, \ldots, x_k) = F(x_1, \ldots, x_k, \gamma)$,
2. $\vec{x} \in \Phi$ if and only if $(\vec{x}, \gamma) \in \Phi$.

In Table 1 we give a list of function classes with their defining generators. Neither can we restate the original definitions here, nor can we give proofs for their characterizations, both due to lack of space. Note that most of the proofs are easy to see from the original definitions, so we mention the appropriate papers within the list. An exception might be the characterization of FP^{NP} which follows immediately from Corollary 6.2 in [28].

For the notations in Table 1 let $\vec{x}$ be a tuple of k components and let index variables range between 1 and k. We define a function span$_+$ as span$_+(\vec{x}) =_{\text{def}}$

(F,Φ)-FP	$F(\vec{x})$		Φ

Let $B =_{\text{def}} \mathbb{N}$, and refer to [24, 15, 16, 9, 21, 8, 17].

FP	x_1		Ω
FP^{NP}	x_{2i-1} if $x_{2i} = \max x_{2j}$ and i maximal with this property		Ω
#·P	$\sum x_j$		Ω
#·NP	$\text{span}_+(\vec{x})$		Ω
#·coNP	$x_1 - \text{span}_+(\vec{x})$	$\{\, \vec{x} \mid x_1 = \max x_j \,\}$	Ω
#·P^{NP}	$\#\{\, z \mid z > 0 \wedge z \equiv 1(2) \wedge (\exists j)(x_j = z) \wedge (\forall j)(x_j \neq z-1) \,\}$		Ω
max·P	$\begin{cases} \max\{\, j \mid x_j > 0 \,\}, & (\exists j)(x_j > 0) \\ 0, & \text{otherwise} \end{cases}$		Ω
max·NP	$\max x_j$		Ω
min·P	$\begin{cases} \min\{\, j \mid x_j > 0 \,\}, & (\exists j)(x_j > 0) \\ k+1, & \text{otherwise} \end{cases}$		Ω
min·NP	$\max\{(\min x_j) - 1, 0\}$		Ω
NPUV	$\begin{cases} \max\{\, x_j - 1 \mid x_j > 0 \,\}, & \text{if } \text{span}_+(\vec{x}) = 1 \\ \bot, & \text{otherwise} \end{cases}$		Ω
#$_{\text{few}}$·P	$\sum x_j$	$\{\, \vec{x} \mid \#\{x_1, \ldots, x_k\} \leq \log k \,\}$	
c#·P	$\sum x_j$	$\{\, \vec{x} \in \Omega_1 \mid \{\, j \mid x_j > 0 \,\} = [\ell, r] \,\}$	
c#·NP	$\text{span}_+(\vec{x})$	$\{\, \vec{x} \mid \{x_1, \ldots, x_k\} \setminus \{0\} = [\ell, r] \,\}$	
c#·coNP	$x_1 - \text{span}_+(\vec{x})$	$\{\, \vec{x} \mid x_1 = \max x_j \,\} \cap \{\, \vec{x} \mid [1, \max x_j] \setminus \{x_1, \ldots, x_k\} = [\ell, r] \,\}$	

Let $B =_{\text{def}} \mathbb{Z}$, and refer to [6, 8].

GapP	$\#\{\, j \mid x_j > 1 \,\} - \#\{\, j \mid x_j = 1 \,\}$	Ω
GapP^{NP}	$\#\{\, z \mid z > 0 \wedge z \equiv 1(2) \wedge (\exists j)(x_j = z) \,\} - \#\{\, z \mid z > 0 \wedge z \equiv 0(2) \wedge (\exists j)(x_j = z) \,\}$	Ω

Let $B =_{\text{def}} 2^{\mathbb{N}}$, and refer to [2, 22].

NPMV	$\begin{cases} \{\, x_j - 1 \mid x_j > 0 \,\}, & (\exists j)(x_j > 0) \\ \bot, & \text{otherwise} \end{cases}$		Ω
FewFP	$\begin{cases} \{\, x_j - 1 \mid x_j > 0 \,\}, & (\exists j)(x_j > 0) \\ \bot, & \text{otherwise} \end{cases}$	$\{\, \vec{x} \mid \text{span}_+(\vec{x}) \leq \log k \,\}$	
NPSV	$\begin{cases} \{\, x_j - 1 \mid x_j > 0 \,\}, & (\exists j)(x_j > 0) \\ \bot, & \text{otherwise} \end{cases}$	$\{\, \vec{x} \mid \text{span}_+(\vec{x}) \leq 1 \,\}$	

Table 1. Characterizations of some function classes

$\#(\{x_1, \ldots, x_k\} \setminus \{0\})$. The constant witnessing the c-invariance of min$\cdot$P may be any integer $\gamma \geq 1$, in any other case take $\gamma = 0$. Note that $\#\cdot$coNP is characterized by a c-invariant promise generator. This characterization was chosen in order to reflect the intuitive understanding of lacking outputs. Actually, $\#\cdot$coNP is not a promise class, and so there is a non-promise generator for this class. For instance, $\#\cdot P^{NP}$ is known to be equal to $\#\cdot$coNP [15, 16], and so we have two characterizations for the same class within our framework.

We conclude this section with a technical lemma. It shows that we can restrict ourself in the polynomial time setting to generator sets with elements that satisfy a certain relationship between the number of components and their values. Recall the definition of Ω_s where the value of the largest component is less or equal to a function s applied to the number of components.

Lemma 5. *Let (F, Φ) be a c-invariant generator, and let s be a function satisfying $s \geq_{a.e.}$ id. Then (F, Φ)-FP $= (F, \Phi \cap \Omega_s)$-FP.*

4 Reducibilities

It will be our aim in the upcoming sections to compare function classes w.r.t. their complexity. To do the comparison we establish certain relations between the defining generators. This is made precise formally by defining suitable *reducibilities* between generators.

Definition 6. *Let (F, Φ) and (G, Γ) be generators. Then we say that (F, Φ) is polylogarithmic time component-wise many-one reducible to (G, Γ), in symbols $(F, \Phi) \leq_m^{FPL} (G, \Gamma)$, if and only if there exist functions $d_1, d_2 \in$ FPL such that $[d_1, d_2](\vec{x}) \in \Gamma$ and $F(\vec{x}) \simeq G([d_1, d_2](\vec{x}))$ for every $\vec{x} \in \Phi$.*

It is not hard to see that *if $(F, \Phi) \leq_m^{FPL} (G, \Gamma)$, then* the function class given by (F, Φ) is contained in the one given by (G, Γ), in fact the inclusion is even relativizable. (This is made precise in Corollary 13 in the next section.) However, it does not seem possible to give a statement in terms of this reducibility, which is *equivalent* to the inclusion of the corresponding function classes. For this we need a second notion of reducibility as follows.

Definition 7. *Let (F, Φ) and (G, Γ) be generators and let s be any function. We say that (F, Φ) is s-partially polylogarithmic time component-wise many-one reducible to (G, Γ), in symbols $(F, \Phi) \leq_{s\text{-}m}^{FPL} (G, \Gamma)$, if and only if $(F, \Phi \cap \Omega_s) \leq_m^{FPL} (G, \Gamma)$.*

Before we address the question of comparing function classes, let us first mention some properties of the reducibilities. The following relations between total and partial reducibility and between different forms of partial reducibilities are obvious:

Proposition 8. *Let (F, Φ) and (G, Γ) be generators.*

1. *If $(F, \Phi) \leq_{m}^{\mathrm{FPL}} (G, \Gamma)$ then $(F, \Phi) \leq_{s\text{-}m}^{\mathrm{FPL}} (G, \Gamma)$ for every function s.*
2. *Let $s_1 \leq s_2$. If $(F, \Phi) \leq_{s_2\text{-}m}^{\mathrm{FPL}} (G, \Gamma)$ then $(F, \Phi) \leq_{s_1\text{-}m}^{\mathrm{FPL}} (G, \Gamma)$.*

Next we notice that both reducibility notions fulfill some form of transitivity:

Theorem 9. *1. $\leq_{m}^{\mathrm{FPL}}$ is transitive.*
2. *Let $(F, \Phi), (G, \Gamma)$ and (H, Π) be generators where (G, Γ) is c-invariant, and let $s_2 \geq_{a.e.}$ id. If $(F, \Phi) \leq_{s_1\text{-}m}^{\mathrm{FPL}} (G, \Gamma)$ for an arbitrary function s_1, and $(G, \Gamma) \leq_{s_2\text{-}m}^{\mathrm{FPL}} (H, \Pi)$, then $(F, \Phi) \leq_{s_1\text{-}m}^{\mathrm{FPL}} (H, \Pi)$.*

Corollary 10. *Let $(F, \Phi), (G, \Gamma)$ and (H, Π) be generators. If $(F, \Phi) \leq_{s\text{-}m}^{\mathrm{FPL}} (G, \Gamma)$ for an arbitrary function s and $(G, \Gamma) \leq_{m}^{\mathrm{FPL}} (H, \Pi)$, then $(F, \Phi) \leq_{s\text{-}m}^{\mathrm{FPL}} (H, \Pi)$.*

Corollary 11. *Restricted to c-invariant generators, $\leq_{s\text{-}m}^{\mathrm{FPL}}$ is transitive for every function $s \geq_{a.e.}$ id (thus in particular, $\leq_{id\text{-}m}^{\mathrm{FPL}}$ is transitive).*

5   Comparing Function Classes

In this section, we state our main technical result.

Theorem 12. *Let (F, Φ) be a c-invariant generator, and let (G, Γ) be a generator. Then the following statements are equivalent:*

1. *$(F, \Phi) \leq_{s\text{-}m}^{\mathrm{FPL}} (G, \Gamma)$ for every function s.*
2. *$(F, \Phi) \leq_{id\text{-}m}^{\mathrm{FPL}} (G, \Gamma)$.*
3. *$(F, \Phi)\text{-}\mathrm{FP}^{\alpha} \subseteq (G, \Gamma)\text{-}\mathrm{FP}^{\alpha}$ for every polynomially length-bounded function α.*

With this theorem we state a result similar to [5, 25] for function classes. The proof however is more involved since we deal with more complicated objects. We omit the proof due to lack of space and mention two immediate corollaries. First as already pointed out in the previous section, to show that one generator (totally) reduces to another proves a relativizable inclusion between the corresponding function classes.

Corollary 13. *Let (F, Φ) be a c-invariant generator, and let (G, Γ) be a generator. If $(F, \Phi) \leq_{m}^{\mathrm{FPL}} (G, \Gamma)$ then $(F, \Phi)\text{-}\mathrm{FP}^{\alpha} \subseteq (G, \Gamma)\text{-}\mathrm{FP}^{\alpha}$ for every function α.*

Of more importance for the applications however is the following corollary.

Corollary 14. *Let (F, Φ) be a c-invariant generator, and let (G, Γ) be a generator. If there exists s such that $(F, \Phi) \nleq_{s\text{-}m}^{\mathrm{FPL}} (G, \Gamma)$, then there exists an α such that $(F, \Phi)\text{-}\mathrm{FP}^{\alpha} \nsubseteq (G, \Gamma)\text{-}\mathrm{FP}^{\alpha}$.*

Thus to separate two function classes all we have to do is find a function s for which we can show that their generators do not reduce to one another s-partially.

Finally let us mention that in the above we talked about separability via oracle functions. However in our model there is no difference between oracle separations by functions and oracle separations by sets. To be more precise:

Proposition 15. *Let (F, Φ) and (G, Γ) be generators. Then the following statements are equivalent:*

1. *(F, Φ)-FP$^\alpha \subseteq (G, \Gamma)$-FP$^\alpha$ for every polynomially length-bounded function α.*
2. *(F, Φ)-FP$^\alpha \subseteq (G, \Gamma)$-FP$^\alpha$ for every function α.*
3. *(F, Φ)-FP$^A \subseteq (G, \Gamma)$-FP^A for every $A \subseteq \Sigma^*$.*

When we consider zero-one valued generator functions, i.e. classes of characteristic functions which are classes of sets, then our main theorem yields as a corollary the separability criterion given in [5, 25].

6 Applications

We give two applications of our criterion and show the existence of separating oracles. In both cases no structural consequences, e. g. via the *operator method*, are known to follow from the positive inclusions. First, we turn to the open question $\min \cdot \mathrm{P} \subseteq \# \cdot \mathrm{NP}$ from [9]. Our criterion easily yields a relativized world where the answer is negative. (We want to remark that independently the existence of such an oracle has been proved in [7] by a suitable diagonalization.)

Theorem 16. *There is an oracle A such that $\min \cdot \mathrm{P}^A \not\subseteq \# \cdot \mathrm{NP}^A$.*

Proof. Suppose that such an oracle A does not exist, i.e. for $F_1(\vec{x}) =_{\mathrm{def}} \min\{ j \mid x_j > 0 \}$, $F_2(\vec{x}) =_{\mathrm{def}} \mathrm{span}_+(\vec{x})$ (see Table 1) and for all s we have $(F_1, \Omega) \leq_{s\text{-}m}^{\mathrm{FPL}} (F_2, \Omega)$ by Theorem 12 and Proposition 15. Let $s(x) =_{\mathrm{def}} 1$ for all x and $d_1, d_2 \in$ FPL be the functions witnessing the reduction. We start with a vector $\vec{x}_0 =_{\mathrm{def}} (0, 0, \ldots, 0, 1) \in \Omega_s$ having k components, i.e. $F_1(\vec{x}_0) = k$. Let I_{x_0} be the set of the k smallest indices j in $[d_1, d_2](\vec{x}_0)$ such that $d_1(\vec{x}_0, j)$ increases the span in the sequence of computations $d_1(\vec{x}_0, 0), \ldots, d_1(\vec{x}_0, d_2(\vec{x}_0))$.

W.l.o.g. assume that during the computation of $d_2(\vec{x}_0)$ only the k-th and the $p(\log k)$-leftmost components of the input are asked (for some polynomial p). Consider vectors $\vec{x}_i$ like $\vec{x}_0$ with an additional 1 in component $i + p(\log k)$ for $i = 1, \ldots, (k - p(\log k) - 1)$. Then we have $|\vec{x}_0| = |\vec{x}_i|$, $d_2(\vec{x}_0) = d_2(\vec{x}_i)$ and there are $k - p(\log k) - i$ indices $j \in I_{x_0}$ such that during the computation of $d_1(\vec{x}_0, j)$ component $i + p(\log k)$ has to be asked in order to distinguish between $\vec{x}_0$ and $\vec{x}_i$. So for the k computations $d_1(\vec{x}_0, j)$ for $j \in I_{x_0}$ the number of accesses to input components has to be quadratic in $(k - p(\log k))$. This is a contradiction, since each of these computations can only access a polylogarithmic number of components. $\square$

For our second separation example, we turn to the context of the so called "cluster sets." Cluster sets are sets A of pairs $\langle x, y \rangle$ such that for every fixed x the set $A_x = \{\, y \mid \langle x, y \rangle \in A \,\}$ forms an integer interval. $c\#{\cdot}C$ is defined to consist of those functions f such that there is a cluster set $A \in C$ and a polynomial p such that for all x, $f(x) = \#\{\, y \mid |y| = p(|x|) \wedge \langle x, y \rangle \in A \,\}$. (Note that the difference to the definition of $\#{\cdot}C$ is that here we require A to be a *cluster set*.) For the case of $c\#{\cdot}P$ this means that functions from this class are defined by Turing machines where accepting paths are neighbored. The classes $c\#{\cdot}P$, $c\#{\cdot}NP$ and $c\#{\cdot}coNP$ are characterized in Table 1 in Sect. 3. These classes are related to the other well-known function classes in [17]. However, the question if $\max{\cdot}NP \subseteq c\#{\cdot}coNP$ remained open (observe that $\max{\cdot}NP \subseteq \#{\cdot}coNP$ [15, 16]).

Theorem 17. *There is an oracle A such that $\max{\cdot}NP^A \not\subseteq c\#{\cdot}coNP^A$.*

Proof. Again, suppose that such an oracle A does not exist, so for the generators (F_1, Ω) and (F_2, Φ_2) from Table 1 of $\max{\cdot}NP$ and $c\#{\cdot}coNP$, respectively, and for all s we have $(F_1, \Omega) \leq_{s\text{-}m}^{FPL} (F_2, \Phi_2)$. Let $s(x) =_{\text{def}} 2^x$ and $d_1, d_2 \in FPL$ witness the reduction for all $\vec{x} \in \Omega_s$. We start with a vector $\vec{x}_0 =_{\text{def}} (10, 10, \ldots, 10, 0, \ldots, 0, 0)$ having l components '10', which is the value of 2 in binary, and a total of k (we will fix l soon). Then, the set of values in $[d_1, d_2](\vec{x}_0)$ is a set $\{1, 2, \ldots, max\}$ with $max = d_1(\vec{x}_0, 0)$ and having a gap of two consecutive values. W.l.o.g. we can assume that max is odd. So there exists an index j such that $d_1(\vec{x}_0, j) \in \{\frac{max-1}{2}, \frac{max-1}{2} + 1, \frac{max-1}{2} + 2\}$.

Next, we will turn $\vec{x}_0$ into a vector $\vec{x}_c$ for some c by placing c in one of the components where we had a '0' before while switching a certain number of '10' entries to '0' to preserve the length of the vector. This increases the gap from size two to c. Now, the contradiction for the computation $d_1(\vec{x}_0, j)$ follows if we can ensure by appropriate choices for l, c and the components we switch that

1. $|\vec{x}_0| = |\vec{x}_c|$, $d_1(\vec{x}_0, 0) = max = d_1(\vec{x}_c, 0)$ and $d_2(\vec{x}_0) = d_2(\vec{x}_c)$,
2. there are super-polylogarithmic many positions to place c in $\vec{x}_c$, and
3. $d_1(\vec{x}_0, j)$ is always in the gap of size c—no matter where the gap is in $[1, 2, \ldots, max]$—so it must be that $d_1(\vec{x}_0, j) \neq d_1(\vec{x}_c, j)$.

This can be achieved by choosing $c =_{\text{def}} \frac{max-1}{2} + 2$, $l = |c| + t$ where t is the number of components accessed during the computations of $d_1(\vec{x}_0, 0)$, $d_2(\vec{x}_0)$, and $d_1(\vec{x}_0, j)$, and switching only components not accessed during these computations. We omit further technical details. Note that $\vec{x}_c \in \Omega_s$. □

7   Conclusion

We presented a framework for the uniform definability of function classes. With a number of examples we made clear that virtually all (polynomial time) function classes which are of current topical interest in complexity theory can be defined using our machinery. This allows us to contribute to the study of the relations between such function classes, either from an oracle separation point of view or from a more structural point of view. Let us look at an example for this second point.

Consider the two classes $\mathrm{FP}^{\mathrm{NP}}[O(\log n)] \subseteq \mathrm{FP}^{\mathrm{NP}}_{\mathrm{tt}}$. Relatively few is known about the inverse inclusion. One feels that it should imply $P = NP$ [14]. Comparing the two generators in our model might lead to new insight. Unfortunately, right now we have no characterization for these classes. But if we define Φ to be the set of all tuples $(x_1, x_2, \ldots, x_{2k})$ for which $\max_{0 \leq j < k} x_{2j+1} \leq \log k$ and $x_{2j+1} = x_{2i+1} \implies x_{2j+2} = x_{2i+2}$ for all i, j, and let F be the generator function from the definition of $\mathrm{FP}^{\mathrm{NP}}$ (see Table 1), then the following holds:

$$\mathrm{FP}^{\mathrm{NP}}[O(\log n)] \subseteq (F, \Phi)\text{-}\mathrm{FP} \subseteq \mathrm{FP}^{\mathrm{NP}}_{\mathrm{tt}}$$

Thus we have a third class in between, not known to be equal to one of the other two. Such a class was not known so far, and any further progress with respect to the relations between these three classes would be very interesting. One could hope that by either proving a collapse of two of the three classes, or giving generators that characterize $\mathrm{FP}^{\mathrm{NP}}[O(\log n)]$ or $\mathrm{FP}^{\mathrm{NP}}_{\mathrm{tt}}$ will shed some light on the question if $\mathrm{FP}^{\mathrm{NP}}[O(\log n)] \overset{?}{=} \mathrm{FP}^{\mathrm{NP}}_{\mathrm{tt}}$.

Besides finding more characterizations it will of course be interesting to see if our criterion will turn out to be as useful for the examination of function classes as the leaf language criterion from [5, 25] was in the case of classes of sets.

Acknowledgment. Thanks are due to Klaus W. Wagner, Würzburg, for helpful hints.

References

1. J. L. Balcázar, J. Díaz, and J. Gabarró. *Structural Complexity I*. Texts in Theoretical Computer Science. Springer-Verlag, Berlin Heidelberg, 2nd edition, 1995.
2. R. V. Book, T. Long, and A. Selman. Quantitative relativizations of complexity classes. *SIAM Journal on Computing*, 13:461–487, 1984.
3. B. Borchert. *Predicate classes, promise classes, and the acceptance power of regular languages*. PhD thesis, Naturwissenschaftlich-Mathematische Fakultät, Universität Heidelberg, 1994.
4. D. P. Bovet and P. Crescenzi. *Introduction to the Theory of Complexity*. International Series in Computer Science. Prentice Hall, London, 1994.
5. D. P. Bovet, P. Crescenzi, and R. Silvestri. A uniform approach to define complexity classes. *Theoretical Computer Science*, 104:263–283, 1992.
6. S. Fenner, L. Fortnow, and S. Kurtz. Gap-definable counting classes. *Journal of Computer and System Sciences*, 48:116–148, 1994.
7. C. Glaßer and G. Wechsung. Relativizing function classes. Manuscript, 1997.
8. L. Hemaspaandra and H. Vollmer. The satanic notations: counting classes beyond #P and other definitional adventures. *Complexity Theory Column 8, ACM SIGACT-Newsletter*, 26(1):2–13, 1995.

9. H. Hempel and G. Wechsung. The operators min and max on the polynomial hierarchy. In *Proceedings 14th Symposium on Theoretical Aspects of Computer Science*, volume 1200 of *Lecture Notes in Computer Science*, pages 93–104. Springer-Verlag, 1997.

10. U. Hertrampf. Classes of bounded counting type and their inclusion relations. In *Proceedings 12th Symposium on Theoretical Aspects of Computer Science*, volume 900 of *Lecture Notes in Computer Science*, pages 60–70. Springer-Verlag, 1995.

11. U. Hertrampf, C. Lautemann, T. Schwentick, H. Vollmer, and K. W. Wagner. On the power of polynomial time bit-reductions. In *Proceedings 8th Structure in Complexity Theory*, pages 200–207, 1993.

12. U. Hertrampf, H. Vollmer, and K. W. Wagner. On the power of number-theoretic operations with respect to counting. In *Proceedings 10th Structure in Complexity Theory*, pages 299–314, 1995.

13. B. Jenner, P. McKenzie, and D. Thérien. Logspace and logtime leaf languages. In *9th Annual Conference Structure in Complexity Theory*, pages 242–254, 1994.

14. B. Jenner and J. Torán. Computing functions with parallel queries to NP. *Theoretical Computer Science*, 141:175–193, 1995.

15. J. Köbler. *Strukturelle Komplexität von Anzahlproblemen*. PhD thesis, Universität Stuttgart, Fakultät für Informatik, 1989.

16. J. Köbler, U. Schöning, and J. Torán. On counting and approximation. *Acta Informatica*, 26:363–379, 1989.

17. S. Kosub. On cluster machines and function classes. Technical Report 172, Institut für Informatik, Universität Würzburg, 1997.

18. M. W. Krentel. The complexity of optimization functions. *Journal of Computer and System Sciences*, 36:490–509, 1988.

19. M. W. Krentel. Generalizations of OptP to the polynomial hierarchy. *Theoretical Computer Science*, 97:183–198, 1992.

20. C. H. Papadimitriou. *Computational Complexity*. Addison-Wesley, Reading, MA, 1994.

21. H. Schmitz. Nichtdeterministische Polynomialzeit-Berechnung von Funktionen. Master's thesis, Institut für Informatik, Universität Würzburg, 1996.

22. A. Selman. A taxonomy on complexity classes of functions. *Journal of Computer and System Sciences*, 48:357–381, 1994.

23. S. Toda. PP is as hard as the polynomial time hierarchy. *SIAM Journal on Computing*, 20:865–877, 1991.

24. L. G. Valiant. The complexity of enumeration and reliabilty problems. *SIAM Journal of Computing*, 8(3):411–421, 1979.

25. N. K. Vereshchagin. Relativizable and non-relativizable theorems in the polynomial theory of algorithms. *Izvestija Rossijskoj Akademii Nauk*, 57:51–90, 1993. In Russian.

26. H. Vollmer. On different reducibility notions for function classes. In *Proceedings 11th Symposium on Theoretical Aspects of Computer Science*, volume 775 of *Lecture Notes in Computer Science*, pages 449–460. Springer-Verlag, 1994.

27. H. Vollmer and K. W. Wagner. The complexity of finding middle elements. *International Journal of Foundations of Computer Science*, 4:293–307, 1993.

28. H. Vollmer and K. W. Wagner. Complexity classes of optimization functions. *Information and Computation*, 120:198–219, 1995.

Recognizability Equals Monadic Second-Order Definability for Sets of Graphs of Bounded Tree-Width*

Denis Lapoire **

Bordeaux-I University, LaBRI (CNRS Laboratory no 1304),
351 cours de la Libération, 33405 Talence, France.

Abstract. We prove that for each k, there exists a MSO-transduction that associates with every graph of tree-width at most k one of its tree-decompositions of width at most k. Courcelle proves in (*The Monadic second-order logic of graphs, I: Recognizable sets of finite graphs*) that every set of graphs is recognizable if it is definable in Counting Monadic Second-Order logic. It follows that every set of graphs of bounded tree-width is CMSO-definable if and only if it is recognizable.

A fundamental theorem by Büchi [2] states that a language of words is recognizable iff it is definable by some formula in a *monadic second-order logic* (MSOL). This result is extended to finite ranked ordered trees by Doner [7], and to sets of finite unranked unordered trees by Courcelle [3]. This last result uses an extension of MSOL, called *counting* monadic second-order logic (CMSOL), that allows counting of cardinality of sets modulo fixed integers. These results relate an algebraic aspect, namely *recognizability*, to a logical one.

For graphs (by graph, we mean a finite graph), similar relationships have been investigated. On the one hand, a graph can be viewed as a logical structure, hence we have a notion of a CMSO-definable set of graphs. On the other hand, Bauderon and Courcelle [1] propose an algebraic structure over sets of graphs. The notion of a recognizable set of graphs follows, as an instance of the general notion of recognizability introduced by Mezei and Wright [11]. A fundamental theorem by Courcelle [5] states that every CMSO-definable set of graphs is recognizable.

In the same paper, he proves that the converse is false and conjectures that this one holds for an interesting class of graphs. More precisely, Robertson and Seymour [12], in their study of minors, introduce the notions of *tree-width* and *tree-decomposition*. Courcelle[3] conjectures that:

Conjecture 1. *If a set of graphs of bounded tree-width is recognizable, then it is CMSO-definable.*

Such a conjecture has already been proved in several restricted cases. Courcelle [5] shows it holds for graphs of tree-width at most 2. Kaller [10] shows it holds for graphs of tree-width at most 3. Kabanets [9] shows it holds for graphs of path-width at most k. To address this conjecture, Courcelle [5] introduces the notion of a *MSO-definable*

---

* Research partly supported by the EC TMR Network GETGRATS (General Theory of Graph Transformation Systems) through the Universities of Bremen and Leiden.
** e-mail: lapoire@LaBRI.U-Bordeaux.fr

transduction, *MSO-transduction* for short. A MSO-transduction transforms a logical structure S into a logical structure S' by defining S' inside S by means of MSO-formulas. In [5], Courcelle calls *strongly context-free* each set of graphs L admitting a MSO-transduction which transforms L into a set of algebraic terms of values L and proves that Conjecture 1 holds if the following one holds:

Conjecture 2. *For each k, the set of graphs of tree-width at most k is strongly context-free.*

The fact that Conjecture 2 implies Conjecture 1 can be quickly explained as below. Let $k \geq 0$ and T_k be the set of algebraic terms presented in [5] that denote graphs of tree-width at most k. Let f_k be the homomorphism that associates with every term of T_k the graph it denotes. Let g_k the MSO-transduction induced by Conjecture 2. The converse of any algebraic homomorphism preserves obviously recognizability. Every recognizable subset of T_k is CMSO-definable (see [3]). The converse of any MSO-transduction preserves CMSO-definability (see [5]). Then, every recognizable set L of graphs of bounded tree-width, equal to $g_k^{-1}(f_k^{-1}(L))$, is CMSO-definable.

Our main result is to prove Conjecture 1 by proving Conjecture 2. For this purpose, we will not consider algebraic terms but very similar objects, that are *e-tree-decompositions* which denote *e-hypergraphs*. An *e*-hypergraph is an unoriented unlabelled hypergraph with a distinguished edge, called its *source-edge*. An *e-tree-decomposition* X is a tree-decomposition of some *e*-hypergraph, which is denoted by **val**(X). The set of all *e*-tree-decompositions (resp. of width at most some $k \geq 0$) is denoted by $\mathcal{T}$ (resp. $\mathcal{T}_k$), two *e*-tree-decompositions of same value are said *equivalent*. A set $L \subseteq \mathcal{T}$ is *MSO-parsable* if the converse of the restriction of **val** on L is a MSO-transduction. In order to prove Conjecture 2, Theorem 34 establishes:

1. *For each k, $\mathcal{T}_k$ contains an equivalent MSO-parsable subset.*

To prove Theorem 34, we introduce an algebra Π that produces sets of tree-decompositions. Π verifies two properties which involve Theorem 34. These properties, expressed by Theorems 33 and 28, are:

2. *Every operation of Π preserves MSO-parsability.*
3. *For each k, Π produces an equivalent subset of $\mathcal{T}_k$.*

The algebra Π is defined by six classes of n-ary operations on $\mathcal{P}(\mathcal{T})$. The first one is the nullary operation that associates the set of all *atomic* *e*-tree-decompositions, that have a unique node. The second kind of operation is a MSO-transduction that intersects every subset of $\mathcal{T}$ with some given MSO-definable set. The third one is the MSO-transduction +, that "adds" a unique vertex to every *e*-tree-decomposition. To define the fourth class of operations, we introduce an higher-order substitution $\otimes$ that associates with every couple $(u, v) \in \mathcal{P}(\mathcal{T})^2$ the set of all *e*-tree-decompositions obtained by refining some $X \in u$ thanks the set v, indeed by replacing simultaneously all the nodes of X by *e*-tree-decompositions of v. An important property of $\otimes$ is the fact that it preserves MSO-parsability, under an additional condition. Hence, we enrich Π with each operation of the form $(u, v) \mapsto u \otimes (v \cap Type_k)$ for some k, where $Type_k$ contains every $X \in \mathcal{T}$ having a source-edge of degree at most k. Theorem 32 proves:

4. *For each k, $(u, v) \mapsto u \otimes (v \cap Type_k)$ preserves MSO-parsability.*

The last operations of Π require the notion, due to Courcelle [4], of an *internally connected* e-hypergraph G, that is such that every pair of elements of its domain is not separated by the extremities of its source-edge. Concerning the fifth class of Π, let us just say that each of its operations is a restriction of the hyperedge-substitution introduced in [1] (see also Habel and Kreowski [8]) and can be viewed as a derived operation of $(u, v) \mapsto u \otimes (v \cap Type_k)$ for some k. In a natural way, we extend internal connectivity on $\mathcal{T}$ (its so defined subset is denoted $\mathcal{T}_{i.c}$) and introduce the notion of a *critical* edge in an e-hypergraph G: an edge e is critical if it is needed to internally connect the extremities of the source-edge of G. That permits to consider a subset of $\mathcal{T}_{i.c}$: the set $\mathcal{T}_{i.c}^{nc}$ of all *nowhere-critical* e-tree-decompositions. For each $k \geq 0$, let $Rank_k$ be the set of all $X \in \mathcal{T}$ whose every arc and every edge has a degree bounded by k. The last class of Π contains every nullary operation defined by each set of the form $\mathcal{T}_{i.c}^{nc} \cap Rank_k$ for some $k \geq 0$. Theorem 31 states:

5. *For each k, $\mathcal{T}_{i.c}^{nc} \cap Rank_k$ is MSO-parsable.*

Π produces an equivalent subset of $\mathcal{T}_k$ in four steps. First step concerns the internally connected e-tree-decompositions, which are *linear*, indeed having a path-structure. Note that this result is similar with (but different from) the result of Kabanets [9]. Second step concerns the internally connected e-tree-decompositions, which are *quasi linear*, indeed obtained by substituting a finite number of linear sets. Trivially, Π produce such sets from linear ones. Third step concerns the internally connected e-tree-decompositions. In this case, we consider $\mathcal{T}_{i.c}^c$ the set of all *critical* e-tree-decompositions, that are internally connected and critical "everywhere". This set verifies two properties (see Theorem 21, 24 and Lemma 23), which permit to produce $\mathcal{T}_{i.c} \cap \mathcal{T}_k$ from the linear subset of $\mathcal{T}_{i.c} \cap \mathcal{T}_k$:

6. $\mathcal{T}_{i.c} = \mathcal{T}_{i.c}^{nc} \otimes \mathcal{T}_{i.c}^c$.
7. *For every k, $\mathcal{T}_{i.c}^c \cap \mathcal{T}_k$ is equivalent with a quasi linear subset of $\mathcal{T}_{i.c} \cap \mathcal{T}_k$.*

The fourth step is jumped by using the fact that every e-tree-decomposition of $\mathcal{T}_k$ can be rewritten into an equivalent e-tree-decomposition of $\mathcal{T}_k$ that is internally connected, except, possibly, "at the root".

The paper is organized as follows. In Sections 1 and 2, we recall the definitions of (hyperedge)-substitution on graphs, of internal connectivity and of a tree-decomposition. In Section 3, we define criticality, quasi linearity, the higher-substitution $\otimes$, we show: $\mathcal{T}_{i.c} = \mathcal{T}_{i.c}^{nc} \otimes \mathcal{T}_{i.c}^c$ and we compare criticality and quasi linearity. In Section 4, we define Π and study its power. In Section 5, we shows that each operation of Π preserves MSO-parsability and establish our main result. All complete and detailed proofs are available at URL http://www.labri.u-bordeaux.fr/Annuaire/lapoire

Notation

We denote by $[i, j]$ the set of integers $\{i, i+1, \ldots, j\}$ and by $[n]$ the interval $[1, n]$. Let A be a set. The cardinality of A is denoted by $\mathbf{card}(A)$, its powerset by $\mathcal{P}(A)$. The set of nonempty sequences of elements of A is denoted by A^+, and sequences are denoted by $(a_1, \ldots, a_n)$ with commas and parentheses. We use $:=$ for "equal by definition" i.e, for introducing new notations, and $:\Leftrightarrow$, similarly, for introducing logical conditions. A binary relation $R \subseteq A \times B$ is also called a *transduction*. The *domain of* R is $\mathbf{Dom}(R) := \{a \in A \mid (a, b) \in R\}$, and the *image of* R is $\mathbf{Im}(R) :=$ $\{b \in B \mid (a, b) \in R\}$. The composition of two relations $R \subseteq A \times B$, and $S \subseteq$ $B \times C$ is denoted by $S \circ R \subseteq A \times C$. R is *functional* if $\mathbf{card}(\{b \mid (a, b) \in R\}) \leq 1$ for each $a \in \mathbf{Dom}(R)$. We identify functional relations $R \subseteq A \times B$ with partial functions $R : A \to B$. If two partial functions $f : A \to B$ and $g : A' \to B'$ coincide on $\mathbf{Dom}(f) \cap \mathbf{Dom}(g)$, we denote by $f \cup g$ their common extension into a partial function $A \cup A' \to B \cup B'$. By a mapping, we mean a total function.

1 Graphs

We deal with a certain class of unoriented hypergraphs, called simply *e-hypergraphs*, in which we distinguish an edge, its "source-edge".

Definition 1. A *hypergraph* G is a tuple $(\mathbf{V}_G, \mathbf{E}_G, \mathrm{vert}_G)$, where $\mathbf{V}_G$ is the finite set of *vertices*, $\mathbf{E}_G$ is the finite set of *edges*, supposed disjoint with $\mathbf{V}_G$, and vert_G is a mapping $\mathbf{E}_G \to \mathcal{P}(\mathbf{V}_G)$, defining the set of *extremities of e*. A vertex x and an edge e are *incident* if x is an extremity of e. Two distinct vertices are *adjacent* if they are incident to the same edge. The *degree* of an edge is the number of its extremities. A *graph* is a hypergraph, all edges edges of which are of degree 2. The *empty graph* is the tuple $(\emptyset, \emptyset, \emptyset)$ denoted by $\emptyset$.

Definition 2. Let G and H be two hypergraphs. G is a *subhypergraph* of H, or is *contained* in H, if $\mathbf{V}_G \subseteq \mathbf{V}_H$, $\mathbf{E}_G \subseteq \mathbf{E}_H$ and if $\mathrm{vert}_G(d) = \mathrm{vert}_H(d)$, for each $d \in \mathbf{E}_G$. If G and H be two subhypergraphs of a common hypergraph, the *union of G and H*, denoted by $G \cup H$, is the minimal hypergraph that contains G and H, the *intersection of G and H*, denoted by $G \cap H$, is the maximal subhypergraph of both G and H.

For each set $D \subseteq \mathbf{V}_G \cup \mathbf{E}_G$, we denote by $G \backslash D$ (resp. $(G : D)$) the maximal (resp. minimal) subhypergraph of G that does not contain (resp. contains) any element of D (as edge or as vertex). If D is a singleton $\{x\}$, $G \backslash D$ (resp. $(G : \{x\})$) is denoted by $G \backslash x$ (resp. $(G : x)$).

A *path* of G is a sequence $p = (o_1, \ldots, o_m) \in (\mathbf{V}_G \cup \mathbf{E}_G)^+$ for some $m \geq 1$, with o_i and o_{i+1} incident for every $i \in [m-1]$. The *initial extremity* (resp. *terminal extremity*) of p is o_1 (resp. o_m). p is *elementary* if two edges o_i and o_j with $1 \leq i < j \leq m$ are distinct.

Definition 3. An *e-hypergraph* H is a tuple $(\mathbf{e}_H, \mathbf{V}_H, \mathbf{E}_H, \mathrm{vert}_H)$ consisting of a hypergraph $(\mathbf{V}_H, \mathbf{E}_H, \mathrm{vert}_H)$, denoted by $\mathbf{G}_H$, and an edge $\mathbf{e}_H \in \mathbf{E}_H$, the *source-edge* of H. H shall be identified with the pair $(\mathbf{e}_H, \mathbf{G}_H)$. The set of all e-hypergraphs is denoted by $\mathcal{G}$. A *source* (resp. *internal vertex*) of H is any vertex of $\mathrm{vert}_H(\mathbf{e}_H)$ (resp. $\mathbf{V}_H \backslash \mathrm{vert}_H(\mathbf{e}_H)$). A *subhypergraph* of H is a subhypergraph of $\mathbf{G}_H \backslash \mathbf{e}_H$ (that is not an e-hypergraph!). For each $d \in \mathbf{E}_H \backslash \mathbf{e}_H$, we define $H \backslash d := (\mathbf{e}_H, \mathbf{G}_H \backslash d)$.

Let us recall the (hyperedge)-substitution introduced by Bauderon and Courcelle [1] (see also Habel and Kreowski [8]). The above definition is more simple than for abstract sourced-hypergraphs: the edge of H which is replaced is the source-edge of K, the substitution does not fuse vertices of H. Lemma 5 expresses a well-known property of this operation: it is "commutative and associative".

Definition 4 (Substitution). Let $H, K \in \mathcal{G}$ with $(\mathbf{G}_H : \mathbf{e}_K) = \mathbf{G}_H \cap \mathbf{G}_K = (\mathbf{G}_K : \mathbf{e}_K)$ and $\mathbf{e}_H \neq \mathbf{e}_K$. We denote by $H[K]$ the e-hypergraph $(\mathbf{e}_H, (\mathbf{G}_H \cup \mathbf{G}_K) \backslash \mathbf{e}_K)$. A property φ is *substitution-closed in* $\mathcal{G}$ if for all $H, K \in \mathcal{G}$ that verify φ, $H[K]$ verifies φ too, if $H[K]$ is defined.

Let $H, K_1, \ldots, K_m \in \mathcal{G}$ for some $m \geq 1$ such that $(H[K_1] \ldots)[K_m] = (H[K_{\pi(1)}] \ldots)[K_{\pi(m)}]$ for each permutation π on $[m]$, we denote by $H[K_1, \ldots, K_m]$ the e-hypergraph $(H[K_1] \ldots)[K_m]$.

Lemma 5. *Let G, H, K be three e-hypergraphs. If $(G[H])[K]$ is defined, it is equal either to $G[H[K]]$ or to $(G[K])[H]$.*

Now, let us recall the notion of internal connectivity due to Courcelle [4].

Definition 6 (Internal connectivity). Let $H \in \mathcal{G}$. A *path of H* is a path of G_H. It is *internal in H* if it belongs to I^+, $I^+ \times S$, $S \times I^+$ or $S \times I^+ \times S$ with $S := \text{vert}_H(e_H)$ and $I := (E_H \cup V_H) - (\{e_H\} \cup \text{vert}_H(e_H))$. A subhypergraph K of H is *internally connected in H* if $K \neq \emptyset$ and if every pair of elements of $V_K \cup E_K$ is the pair of extremities of some path of K internal in H. H is *internally connected* if $G_H \backslash e_H$ is internally connected in H. We denote by $\mathcal{G}_{i.c}$ the set of all internally connected e-hypergraphs.

Lemma 7. *Internal connectivity is substitution-closed in $\mathcal{G}$.*

2 Tree-decomposition

In this section, we recall the notion of a tree-decomposition, introduced by Robertson and Seymour in [12] in their study of graph minors. We deal with e-tree-decompositions, which are tree-decompositions of e-hypergraphs. These objects are close to algebraic terms used in [5]: if we consider the unique node that contains the source-edge, we obtain a rooted tree. We use the symbol **val** to denote the valuation mapping $\mathcal{T} \to \mathcal{G}$. In a natural way, we extend the substitution on $\mathcal{T}$ into a new one, which is "commutative and associative" and which commutes with **val**.

Definition 8. A *forest* is a nonempty graph T with no elementary cycle. A *tree* is a connected forest. A vertex of a tree is called a *node*. An edge of a tree is called an *arc*. A *rooted tree* is a pair $R = (T, r)$ consisting of a tree T and a distinguished node r called the *root*. Let s and t be two nodes of a rooted-tree (T, r). The node s is a *child* (resp. the *parent*) of a node t, if s and t are adjacent and if every path of T from r to s (resp. t) contains t (resp. s). A *leaf of* (T, r) is a node with no children.

Definition 9 (Tree-decomposition). A *tree-decomposition* is a pair (T, g) where T is a tree and where g associates to every node $t \in V_T$ a hypergraph $g(t)$ such that:

- $E_{g(s)} \cap E_{g(t)} = \emptyset$, for all distinct nodes s, t of T.
- for all nodes s, u of T and every node t of the elementary path of T from s to u, we have: $V_{g(s)} \cap V_{g(u)} \subseteq V_{g(t)}$.

The *width* of (T, g) is denoted by $\underline{wd}(T, g)$ and is the maximum of $\text{card}(V_{g(t)}) - 1$ taken over all $t \in V_T$. The *tree-width* of a hypergraph G, denoted by $\underline{twd}(G)$, is the minimum width of all tree-decompositions (T, g) such that $\bigcup_{t \in V_T} g(t) = G$. For every tree-decomposition (T, g) and every subset $U \subseteq V_T$ (resp. subhypergraph $U \subseteq T$), we denote by $g(U)$ the hypergraph $\bigcup_{t \in U} g(t)$ (resp. $\bigcup_{t \in V_U} g(t)$).

Definition 10. An *e-tree-decomposition* X is a tuple $(e_X, \mathbf{T}_X, \mathbf{g}_X)$, where $(\mathbf{T}_X, \mathbf{g}_X)$ is a tree-decomposition and e_X an edge of $\mathbf{g}_X(\mathbf{T}_X)$, the *source-edge of* X. The tree $\mathbf{T}_X$ is supposed to be disjoint with the graph $\mathbf{g}_X(\mathbf{T}_X)$, denoted by $\mathbf{G}_X$. We denote by $\mathcal{T}$ the set of all *e*-tree-decompositions and, for every $k \geq -1$, by $\mathcal{T}_k$ the set $\{X \in \mathcal{T} \mid \underline{wd}(X) \leq k\}$.

Let $X \in \mathcal{T}$. The *value* of X is the pair $(e_X, \mathbf{G}_X)$, denoted by $\mathbf{val}(X)$. An *edge* (resp. *vertex, source, internal vertex*) of X is an edge (resp. vertex, source, internal vertex) of $\mathbf{val}(X)$. The *root of* X, denoted by r_X, is the unique node t of X such that $e_X \in \mathbf{E}_{\mathbf{g}_X(t)}$. An *arc* (resp. *node, leaf*) of X is an arc (resp. node, leaf) of $(\mathbf{T}_X, r_X)$. Two *e*-tree-decompositions X and Y are *equivalent* if $\mathbf{val}(X) = \mathbf{val}(Y)$. Two subsets $u, v \subseteq \mathcal{T}$ are *equivalent* if $\mathbf{val}(u) = \mathbf{val}(v)$.

Definition 11 presents the operation of substitution: $\mathcal{T} \times \mathcal{T} \to \mathcal{T}$ and its natural extension on $\mathcal{P}(\mathcal{T})$. Lemmas 12 and 13 establish that $[\,]$ is "commutative and associative" and commutes with **val**.

Definition 11. Let $Y, Z \in \mathcal{T}$ such that: $e_Y \neq e_Z$, $(\mathbf{G}_Y : e_Z) = \mathbf{G}_Y \cap \mathbf{G}_Z = (\mathbf{G}_Z : e_Z)$ and $\mathbf{T}_Y \cap (\mathbf{G}_Z \cup \mathbf{T}_Z) = \emptyset = (\mathbf{G}_Y \cup \mathbf{T}_Y) \cap \mathbf{T}_Z$. We denote by $Y[Z]$ the *e*-tree-decomposition (e_Y, T, g) where T is obtained from $\mathbf{T}_Y \cup \mathbf{T}_Z$ by adding the edge e_Z with extremities r_Z and the unique node s of Y such that $e_Z \in \mathbf{g}_Y(s)$ and where g associates with every node s of T the hypergraph $\mathbf{g}_Y(s) \backslash e_Z$ if s is a node of Y and $\mathbf{g}_Z(s) \backslash e_Z$, otherwise.

A property φ is *substitution-closed in* $\mathcal{T}$ if for all $Y, Z \in \mathcal{T}$ that verify φ, $Y[Z]$ verifies φ too, if $Y[Z]$ is defined.

For all $Y, Z_1, \ldots, Z_m \in \mathcal{T}$, $(Y[Z_1] \ldots)[Z_m]$ is denoted by $Y[Z_1, \cdots, Z_m]$, if it is defined and if it is equal to $(Y[Z_{\pi(1)}] \ldots)[Z_{\pi(m)}]$, for each permutation π on $[m]$. For all subsets u, v of $\mathcal{T}$, we denote by $u[v]$ the set $u \cup v \cup \{Y[Z_1, \ldots, Z_m] \mid Y \in u, Z_1, \ldots, Z_m \in v, m \geq 1\}$.

Lemma 12. *Let* X, Y, Z *be three e-tree-decompositions. If* $(X[Y])[Z]$ *is defined, it is equal either to* $X[Y[Z]]$ *or to* $(X[Z])[Y]$.

Lemma 13. *Let be an e-tree-decomposition of the form* $Y[Z_1, \ldots, Z_m]$ *for some* $m \geq 1$. *Then,* $\mathbf{val}(Y[Z_1, \ldots, Z_m]) = \mathbf{val}(Y)[\mathbf{val}(Z_1), \ldots, \mathbf{val}(Z_m)]$.

Let us define the converse operation of substitution.

Definition 14. Let $X \in \mathcal{T}$ and T be a subtree of $\mathbf{T}_X$. The *e*-tree-decomposition *generated by* X *and* T is the tuple (e, T, g), denoted by $X|T$, where:

- e is e_X if $r_X \in \mathbf{V}_T$ and, otherwise, the unique arc incident in $\mathbf{T}_X$ with some node of T and some node of the connected component of $\mathbf{T}_X \backslash \mathbf{V}_T$ that contains r_X.
- g associates to every node t of T the union $\mathbf{g}_X(t) \cup \bigcup_{d \in D_t} G_d$ where D_t is the set of all arcs incident in $\mathbf{T}_X$ with t that do not belong to T and where G_d is for each arc d of X the unique connected hypergraph such that $\mathbf{E}_{G_d} = \{d\}$ and $\mathbf{V}_{G_d} = \mathbf{V}_{\mathbf{g}_X(u)} \cap \mathbf{V}_{\mathbf{g}_X(v)}$, with u and v the extremities of d in $\mathbf{T}_X$.

An *e*-tree-decomposition Y is *contained in* X, denoted by $Y \subseteq X$, if $Y = X|T$ for some subtree T of $\mathbf{T}_X$. X is *atomic*, if it contains a unique node. For each $u \subseteq \mathcal{T}$, we define $\mathbf{atom}(u) := \{X|t : X \in u, t \in \mathbf{V}_{\mathbf{T}_X}\}$. The *type* of X is its number of sources. The *rank* of X is $\max\{\mathbf{card}(\mathbf{vert}_H(d)) \mid H \in \mathbf{val}(\mathbf{atom}(\{X\})), d \in \mathbf{E}_H\}$.

For each k, the set of all $X \in \mathcal{T}$ of type (resp. rank) at most k is denoted by $Type_k$ (resp. $Rank_k$).

A property φ is *hereditary in* $\mathcal{T}$ if for every e-tree-decomposition X that verifies φ, every e-tree-decomposition contained in X verifies φ.

3 Quasi linearity and criticality

In this section, we introduce the higher-order substitution $\otimes$ and the notions of a (everywhere) critical e-tree-decomposition and of a nowhere critical one. These two notions induce two sets that suffice to produce the set of all internal-connected e-tree-decompositions. Finally, we define the notion of a quasi linear set and compares such sets with "critical" ones.

Before to define criticality, let us extend, in a very simple way, the internal connectivity to e-tree-decompositions.

Definition 15. An e-tree-decomposition Y is *internally connected* if $\mathbf{val}(X)$ is internally connected, for every $X \subseteq Y$. Their set is denoted by $\mathcal{T}_{i.c}$.

The next result is the consequence of the previous definition and of Lemma 7.

Lemma 16. *Internal connectivity is hereditary and substitution-closed in* $\mathcal{T}$.

An edge d is critical in an e-hypergraph if it is needed to internally connect its sources. That enables us to define two kinds of internally connected e-tree-decompositions: the subset of $\mathcal{T}_{i.c}$ denoted by $\mathcal{T}_{i.c}^{\mathbf{nc}}$ (resp. $\mathcal{T}_{i.c}^{\mathbf{c}}$) of all e-tree-decompositions that contains nowhere (resp. everywhere) critical arcs.

Definition 17 (Criticality). An edge d is *critical* in an e-hypergraph H if $d \in \mathbf{E}_H \backslash e_H$ and every internally connected subhypergraph of $H \backslash d$ does not contain every source of H. An e-tree-decomposition X is *critical* if $X \in \mathcal{T}_{i.c}$ and if every arc d of X is critical in $\mathbf{val}(X|Q)$, where Q is the maximal subtree of $\mathbf{T}_X \backslash d$ that contains r_X. Their set is denoted by $\mathcal{T}_{i.c}^{\mathbf{c}}$. An e-tree-decomposition X is *nowhere critical* if $X \in \mathcal{T}_{i.c}$ and if for each arc d of X and each $Y \subseteq X$, d is not critical in $\mathbf{val}(Y)$. Their set is denoted by $\mathcal{T}_{i.c}^{\mathbf{nc}}$.

Note that $\{\mathcal{T}_{i.c}^{\mathbf{nc}}, \mathcal{T}_{i.c}^{\mathbf{c}}\}$ is not a partition of $\mathcal{T}_{i.c}$: the union is strictly contained in $\mathcal{T}_{i.c}$, the intersection is $\mathbf{atom}(\mathcal{T}_{i.c})$. The next fact can be viewed as the "dual" of Lemma 7 (that states that internal connectivity is substitution-closed in $\mathcal{G}$).

Lemma 18. *Criticality is hereditary in* $\mathcal{T}$.

In order to define $\otimes$, we define how to "contract" an arc in any $X \in \mathcal{T}$.

Definition 19. Let $X \in \mathcal{T}$. The e-tree-decomposition *obtained from X by contracting an arc d of X* is the tuple (e_X, T, g) where T is obtained from $\mathbf{T}_X \backslash d$ by identifying s with t, renamed t. where g associates with every node u of T the hypergraph $\mathbf{g}_X(u)$ if $u \neq t$ and $\mathbf{g}_X(\{s, t\})$, otherwise, and where s and t are the extremities of d in $\mathbf{T}_X$ with t the parent of s.

Definition 20 (Higher-order substitution). For all subsets u, v of $\mathcal{T}$, we denote by $u \otimes v$ the set of all $X \in \mathcal{T}$ that contains a set of arcs D such that u contains the e-tree-decomposition obtained from X by contracting each edge of $\mathbf{E}_{\mathbf{T}_X} - D$, one by one, and v contains every e-tree-decomposition generated by X and some connected component of $\mathbf{T}_X \backslash D$.

The following equality relates all the notions that have been presented in this section. Its proof requires Lemma 16 and 18 and does not present any difficulty.

Theorem 21. $\mathcal{T}_{i.c} = \mathcal{T}_{i.c}^{nc} \otimes \mathcal{T}_{i.c}^{c}$.

Now, present the notion of a quasi linear set. A quasi linear set can be viewed as a set of e-tree-decompositions having tree-structures of bounded complexity, by considering as measure of complexity, for example, the path-width.

Definition 22 (Quasi linearity). An e-tree-decomposition is *linear* if it contains a unique leaf. Their set is denoted by $\mathcal{L}$. We define $\mathcal{L}_0 := \emptyset$ and, for each $l \geq 0$, we define $\mathcal{L}_{l+1} := \mathcal{L}[\mathcal{L}_l]$. Every subset of $\mathcal{L}_l$ for some $l \geq 0$ is said *quasi linear*.

Lemma 23. *For all $k, l \geq 0$, the set $\{X \in \mathcal{L}_l \cap \mathcal{T}_k \mid \mathbf{val}(X) \in \mathcal{G}_{i.c}\}$ is equivalent with a subset of $\mathcal{L}_{l \cdot (k+1)} \cap \mathcal{T}_{i.c} \cap \mathcal{T}_k$.*

The next result is the most difficult one(see Example 25). It proves that the "critical" subset of $\mathcal{T}_{i.c} \cap \mathcal{T}_k$ can be rewritten into a quasi linear subset of $\mathcal{T}_k$, and, by using Lemma 23, into a quasi linear subset of $\mathcal{T}_{i.c} \cap \mathcal{T}_k$.

Theorem 24. *For every k, $\mathcal{T}_{i.c}^{c} \cap \mathcal{T}_k$ is equivalent with a subset of $\mathcal{L}_{2 \cdot (1+k)^2} \cap \mathcal{T}_k$.*

Example 25. Let T be a tree with n leaves $L = \{l_1, \ldots, l_n\}$. Rather to rename the leaves, we can suppose that each arc of T partition L in $\{\{l_i \mid i \in I\}, \{l_i \mid i \in [n] - I\}\}$ for some subset $I \subset [n]$. Let $H := (e_1, G) \in \mathcal{G}$ with G a circuit of the form $(s_1, e_1, s_2, \ldots, e_n, s_1)$ for some edges $e_1, \ldots, e_n$ and some vertices $s_1, \ldots, s_n$. Let X be the unique e-tree-decomposition of the form (e_1, T, g) that is minimal w.r.t $\sum_{t \in V_T} \mathbf{card}(\mathbf{V}_{g(t)})$ to verify $H = \mathbf{val}(X)$ and $g(l_i) = (G : e_i)$ for each $i \in [n]$. Clearly, X belongs to $\mathcal{T}_{i.c}^{c} \cap \mathcal{T}_2$ and admits an equivalent $Y \in \mathcal{L} \cap \mathcal{T}_{i.c} \cap \mathcal{T}_2$.

4 An algebra of sets of e-tree-decompositions

In this section, we define an algebra Π and we study its power. One of the operations of Π intersects each subset of $\mathcal{T}$ with some given MSO-definable set. For this purpose, we recall monadic second-order logic. A survey can be found in [6].

Definition 26. Let R be a ranked alphabet such that each $r \in R$ has a rank $\rho(r)$ in $\mathbb{N}_+$. A symbol $r \in R$ is considered as a $\rho(r)$-ary relation symbol. A R-*structure* is a tuple $S = (\mathbf{D}_S, (r_s)_{r \in R})$ where $\mathbf{D}_S$ is a finite set, called the *domain of S*, and r_S is a subset of $\mathbf{D}_S^{\rho(r)}$ for each r in R. We denote by $\mathcal{S}(R)$ the set of R-structures.

Let S be an R-structure for some alphabet R. The formulas of *monadic second-order logic* (called *MSO-formulas* for short) are written with variables of two types, namely lower case letters $x, y, \ldots$ called *object variables*, denoting elements of $\mathbf{D}_S$, and upper case letters $X, Y, \ldots$ called *set variables*, denoting subsets of $\mathbf{D}_S$. The atomic formulas are of the form $x = y$, $x \in X$, $r(x_1, \ldots, x_n)$ (where $r \in R$ and $n = \rho(r)$), and formulas are formed with propositional connectives and quantifications over the variables. For every finite set W of variables, we denote by $L(R, W)$ the set of all formulas that are written with relational symbol from R and have their free variables in W. We also denote by $L(R)$ the set of closed formulas $L(R, \emptyset)$.

Let $\varphi \in L(R, W)$ and let γ be a W-assignment in S (i.e., $\gamma(X)$ is a subset of $\mathbf{D}_S$ for every set variable X in W, and $\gamma(x) \in \mathbf{D}_S$ for every object variable x in W). We write $(S, \gamma) \models \varphi$ if and only if φ holds for S for γ. We write $S \models \varphi$ if φ is closed. A set of R-structures L is *MSO-definable* if there is a formula φ in $L(R)$ such that L is the set of all R-structures S such that $S \models \varphi$.

Any $H \in \mathcal{G}$ is represented by the R-structure $|H| = (\mathbf{V}_H \cup \mathbf{E}_H, (r_{|H|})_{r \in R})$ with $R = \{\mathbf{vr}, \mathbf{ed}, \mathbf{sr}, \mathbf{ic}\}$, where $\mathbf{vr}_{|H|}$, $\mathbf{ed}_{|H|}$ and $\mathbf{sr}_{|H|}$ are the unary predicates that define the vertex-set, the edge-set and the source-edge, respectively, and where $\mathbf{ic}_{|H|}(x, y) :\Leftrightarrow y \in \mathbf{vert}_H(x)$ is the binary incidence relation. Clearly, for all $G, H \in \mathcal{G}$, $|G| = |H|$ if and only if $G = H$.

An e-tree-decomposition X is represented by the R'-structure $|X| = (\mathbf{V}_{G_X} \cup \mathbf{E}_{G_X} \cup \mathbf{V}_{T_X} \cup \mathbf{E}_{T_X}, (r_{|X|})_{r \in R'})$ with $R' = \{\mathbf{vr}, \mathbf{ed}, \mathbf{sr}, \mathbf{nd}, \mathbf{ar}, \mathbf{ic}, \mathbf{mp}\}$, where $\mathbf{vr}_{|X|}$, $\mathbf{ed}_{|X|}$, $\mathbf{nd}_{|X|}$, $\mathbf{ar}_{|X|}$ and $\mathbf{sr}_{|X|}$ are the unary predicates that define the vertex-set, the edge-set, the node-set, the arc-set and the source-edge, respectively, where $\mathbf{ic}_{|X|}(x, y) :\Leftrightarrow y \in \mathbf{vert}_{G_X}(x) \vee y \in \mathbf{vert}_{T_X}(x)$ is the binary incidence relation, and where $\mathbf{mp}_{|X|}(x, y) \Leftrightarrow y \in \mathbf{V}_{g_X(x)} \cup \mathbf{E}_{g_X(x)}$ is the binary mapping relation. Clearly, for all $X, Y \in \mathcal{T}$, $|X| = |Y|$ if and only if $X = Y$. Hence, $|X|$ "contains" $|H|$, the value of X.

Definition 27. We denote by Π the algebra $(\mathcal{P}(\mathcal{T}), \mathcal{F})$ with $\mathcal{F}$ the set of operations $\{\mathbf{1}, +\} \cup \{\mathbf{m}_\varphi \mid \varphi \in L(R')\} \cup \{\mathbf{p}_k, \mathbf{n}_k, \mathbf{o}_k \mid k \geq 0\}$ where:

- $\mathbf{1}$ is the nullary operation $\mapsto \mathbf{atom}(\mathcal{T})$.
- $\mathbf{m}_\varphi$ is the unary operation $u \mapsto \{X \in u : |X| \models \varphi\}$, for each $\varphi \in L(R')$.
- $+$ is the unary operation $u \mapsto u \cup \{X \in \mathcal{T} \mid \exists s \in \mathbf{V}_{G_X} : (e_X, \mathbf{T}_X, \mathbf{g}_X \backslash s) \in u\}$, where for each $X \in \mathcal{T}$, each vertex s of X and each node t of X, $(\mathbf{g}_X \backslash s)(t)$ is the hypergraph $(\mathbf{V}_G \backslash s, \mathbf{E}_G, \{(d, \mathbf{vert}_G(d) \backslash s) \mid d \in \mathbf{E}_G\})$ with $G := \mathbf{g}_X(t)$.

and, where for each $k \geq 0$:

- $\mathbf{n}_k$ is the binary operation $(u, v) \mapsto u \otimes (v \cap Type_k)$.
- $\mathbf{o}_k$ is the binary operation $(u, v) \mapsto u[v \cap Type_k \cap \{X \in \mathcal{T} \mid \mathbf{val}(X) \in \mathcal{G}_{i.c}\}]$.
- $\mathbf{p}_k$ is the nullary operation $\mapsto \mathcal{T}_{i.c}^{nc} \cap Rank_k$.

A subset of $\mathcal{T}$ is *produced by* Π if it is denoted by some finite and well-formed term built with symbols of $\mathcal{F}$.

Theorem 28. *For every k, Π produces an equivalent subset of $\mathcal{T}_k$.*

Sketch of proof. The proof is made by induction by using the MSO-definability of $\mathcal{L}$, $\mathcal{T}_{i.c}$ and $\mathcal{T}_k$ for any k and the fact that for each $k \geq 0$, we have:

1. $\mathcal{T}_k$ contains and is equivalent with $\mathbf{o}_{k+1}(\mathbf{1} \cap \mathcal{T}_k, u)$ with $u = \mathcal{T}_{i.c} \cap \mathcal{T}_k$.
   (This is a consequence of Lemma 23.)
2. $\mathcal{T}_{i.c} \cap \mathcal{T}_k = \mathbf{n}_{k+1}(\mathbf{p}_{k+1}, u)$ with $u = \mathcal{T}_{i.c}^c \cap \mathcal{T}_k$.
   (This is an obvious consequence of Theorem 21.)
3. $\mathcal{T}_{i.c}^c \cap \mathcal{T}_k$ is equivalent with a subset of $\mathbf{o}_{k+1}[\ldots \mathbf{o}_{k+1}[u, u] \ldots, u]$ ($2 \cdot (1+k)^3$ times) with $u = \mathcal{L} \cap \mathcal{T}_{i.c} \cap \mathcal{T}_k$.
   Direct consequence of Theorem 24 and Lemma 23.
4. $\mathcal{L} \cap \mathcal{T}_{i.c} \cap \mathcal{T}_k = \mathbf{n}_{k+1}(\mathbf{p}_{k+1} \cap \mathcal{L}, u) \cap \mathcal{L}$ with $u = \mathcal{L} \cap \mathcal{T}_{i.c}^c \cap \mathcal{T}_k$.
   Obvious consequence of Theorem 21 applied in the linear case.
5. $\mathcal{L} \cap \mathcal{T}_{i.c}^c \cap \mathcal{T}_k$ is equivalent with a subset of $+^{2+6 \cdot k}(u) \cap \mathcal{L} \cap \mathcal{T}_{i.c} \cap \mathcal{T}_k$ with $u = \mathcal{L} \cap \mathcal{T}_{i.c} \cap \mathcal{T}_{k-1}$.
   Obtained in a similar and more simple way, than for Theorem 24. $\qquad\square$

5 Every operation of Π preserves MSO-parsability

In this last section, we establish our main result. For this purpose, we will review, briefly, the notion of a MSO-transduction of relational structures introduced in [5]. It transforms a structure S into a structure S' by defining S' "inside" S by means of MSO-formulas. More precisely, S' is defined inside an intermediate structure made of k disjoint copies of S, for some fixed k. This makes it possible to construct S' with a domain larger than that of S (larger within the factor k).

Definition 29 (MSO-transduction). Let R and R' be two ranked alphabets of relation symbols. Let W be a finite set of set variables. A (R', R)-*definition scheme* is a tuple of formulas of the form $\Delta = (\varphi, \psi_1, \ldots, \psi_k, (\theta_{r,j})_{r \in R', j \in [k]^{\rho(r)}})$ where:

- $k > 0$.
- $\varphi \in L(R, W)$.
- $\psi_i \in L(R, W \cup \{x_1\})$ for $i \in [k]$.
- $\theta_{r,j} \in L(R, W \cup \{x_1, \ldots, x_{\rho(r)}\})$ for $r \in R'$, $j \in [k]^{\rho(r)}$.

Let $S \in \mathcal{S}(R)$ and γ be a W-assignment in S. An R'-structure S' is *defined by* Δ *in* (S, γ), denoted by $S' = \mathbf{def}_\Delta(S, \gamma)$, if:

- $(S, \gamma) \models \varphi$.
- $\mathbf{D}_{S'} = \{(d, i) \mid d \in \mathbf{D}_S, i \in [k], (S, \gamma, d) \models \psi_i\}$.
- for each $r \in R'$: $r_{S'} = \{((d_1, i_1), \ldots, (d_t, i_t)) \mid (S, \gamma, d_1, \ldots, d_t) \models \theta_{r,j}\}$, where $j = (i_1, \ldots, i_t)$ and $t = \rho(r)$.

The *transduction defined by* Δ is the relation denoted by $\mathbf{def}_\Delta$ that contains every pair of the form $(S, S') \in \mathcal{S}(R) \times \mathcal{S}(R')$ with $S' = \mathbf{def}_\Delta(S, \gamma)$ for some assignment γ in S. A transduction f is *MSO-definable*, a *MSO-transduction* for short, if there is a definition scheme Δ such that $f = \mathbf{def}_\Delta$ or such that for every $(a, b) \in f$, there is $(a, c) \in \mathbf{def}_\Delta$ with b and c isomorphic (with the usual notion of isomorphism).

Definition 30. A set $u \subseteq \mathcal{T}$ is *MSO-parsable* if $\{(|\mathbf{val}(X)|, |X|) \mid X \in u\}$ is a MSO-transduction.

Our proof requires a few properties of MSO-transductions. These ones, due to Courcelle [5,6], establish the MSO-definability of the composition of two MSO-transductions and the MSO-definability of the inverse image of a MSO-definable set of structures under a MSO-transduction.

Now, let us consider the constants of Π. Obviously, $\mathbf{atom}(\mathcal{T})$ is MSO-parsable. The other constants are the object of the following result:

Theorem 31. *For every k, the set $\mathcal{T}_{i.c}^{nc} \cap Rank_k$ is MSO-parsable.*

The previous theorem generalizes a well-known result due to Courcelle: the MSO-parsability of $\mathcal{T}_{i.c}^{nc}$ enables us to define a class of context-free MSO-parsable graph-grammars that contains the "regular" one defined in [5].

Now, let us show a remarkable property of $\otimes$: it preserves MSO-parsability, under an additional condition.

Theorem 32. *For each k, $(u, v) \mapsto u \otimes (v \cap Type_k)$ preserves MSO-parsability.*

The operation $+$ and the MSO restriction, namely $u \mapsto u \cap \{X \models \varphi\}$, where φ is given, are two MSO-transductions and, thus, preserve MSO-parsability. As a consequence of Theorem 32, we prove that $(u, v) \mapsto u[v \cap \mathcal{T}_{i.c} \cap Type_k]$ preserves MSO-parsability. Then, Theorem 31 and 32, involve:

Theorem 33. *Every operation of Π preserves MSO-parsability.*

By Definition 29, a subset of $\mathcal{T}$ is MSO-parsable if it contains an equivalent MSO-parsable subset. Hence, Theorems 28 and 33 involve the following result:

Theorem 34. *For every k, $\mathcal{T}_k$ contains an equivalent MSO-parsable set.*

Clearly, Theorem 28 can be extended to each set of the form $\mathcal{T}_{i.c} \cap \mathcal{T}_k$ and $\mathcal{L}_l \cap \mathcal{T}_{i.c} \cap \mathcal{T}_k$ for some k, l. Thus, by Theorem 33, all these sets are MSO-parsable. In particular: $\mathcal{L} \cap \mathcal{T}_{i.c} \cap \mathcal{T}_k$. Note that this result, which concerns "linear and internally connected k-trees", is similar with the result of Kabanets [9] that concerns "k-paths".

As a consequence of Theorem 34 and of the fact that e-tree-decompositions are very near (in a logical point of view) with algebraic terms used in [5], it comes:

Corollary 35. *For every k, the set of graphs of tree-width at most k is strongly context-free.*

Before concluding, let us recall a fundamental result due to Courcelle [3]:

Theorem 36 (Courcelle). *Each CMSO-definable set of graphs is recognizable.*

Thanks the previous result and another result mentioned in the introduction, we obtain our main result:

Theorem 37. *Every set of graphs of bounded tree-width is CMSO-definable if and only if it is recognizable.*

Acknowledgments

I am very thankful to J. Engelfriet and H.-J. Kreowski for their constant and efficient supports. I thank B. Courcelle, F. Drewes, G. Senizergues and the referees, for helpful comments.

References

1. M. Bauderon and B. Courcelle. Graph expressions and graph rewritings. *Math. Systems Theory*, 20:83–127, 1987.
2. J. Büchi. Weak second order logic and finite automata. *S. Math. Logik Grundlagen Math.*, 5:66–92, 1960.
3. B. Courcelle. The monadic second order logic of graphs, I : Recognizable sets of finite graphs. *Inf. and Comp.*, 85:12–75, 1990.
4. B. Courcelle. The monadic second-order logic of graphs, IV : Definability properties of equational graphs. *Annals of Pure and Applied Logic*, 49:193–255, 1990.
5. B. Courcelle. The monadic second-order logic of graphs, V : On closing the gap between definability and recognizability. *Theoret. Comput. Sci.*, 80:153–202, 1991.
6. B. Courcelle. Monadic second order definable graph transductions : A survey. *Theoret. Comput. Sci.*, 126:53–75, 1994.
7. J. Doner. Tree acceptors and some of their applications. *J. Comput. System Sci.*, 4:406–451, 1970.
8. A. Habel and H.-J. Kreowski. May we introduce to you, hyperedge replacement. *LNCS 291*, pages 15–26, 1987.
9. V. Kabanets. Recognizability equals definability for partial k-trees. In *ICALP '97*, pages 805–815, 1997.
10. D. Kaller. Definability equals recognizability of partial 3-trees. In (WG '96), editor, *Workshop on Graph-Theoretic Concepts in Computer Science*, pages 239–253, 1996.
11. J. Mezei and J. Wright. Algebraic automata and context-free sets. *Inform. and Control*, 11:3–29, 1967.
12. N. Robertson and P. D. Seymour. Graph minors. III. planar tree-width. *J. Combin. Theory Ser. B*, 36:49–64, 1984.

Index of Authors

Lecture Notes in Computer Science

For information about Vols. 1–1289

please contact your bookseller or Springer-Verlag